Audel™ Millwrights and Mechanics Guide

5th Edition

Thomas Bieber Davis

Carl A. Nelson

Wiley Publishing, Inc.

Vice President and Executive Group Publisher: Richard Swadley
Vice President and Executive Publisher: Robert Ipsen
Vice President and Publisher: Joe Wikert
Executive Editor: Carol Long
Development Editor: Emilie Herman
Editorial Manager: Kathryn A. Malm
Production Editor: Angela M. Smith
Text Design & Composition: TechBooks

Published simultaneously in Canada

For general information on our other products and services please contact our Customer Care Department within the United States at (800) 762-2974, outside the United States at (317) 572-3993 or fax (317) 572-4002.

Wiley also publishes its books in a variety of electronic formats. Some content that appears in print may not be available in electronic books.

Library of Congress Control Number: 2003105647

ISBN: 978-0-470-63801-9

SKY10064482_011024

Introduction

Industry today depends on complex mechanical and electrical machinery to produce products. These machines require professional attention during the installation, startup, and scheduled maintenance phases. In the past, the job of servicing the mechanical components of this machinery fell to the professional millwright/mechanics. Today industry demands the talents of a new breed of craftsperson—the *multicraft mechanic*. This person must be able to handle mechanical, electrical, welding, carpentry, pipefitting, preventive maintenance, and even complex troubleshooting tasks. This book is intended for that individual.

This comprehensive book also includes chapters not found in any other source; such as blacksmithing; sheet metal layout; air compressors; saw sharpening; hydraulics and pneumatic repair; valves and pumps; bearings and bushings installation and lubrication; packing and seals; structural steel; belt, gear and chain drives; as well as extensive troubleshooting charts for various types of equipment.

Millwrights and Mechanics Guide, Fifth Edition has been updated to include chapters on vibration measurement, preventive maintenance, electric motor troubleshooting, fan maintenance, field sketching, dial indication alignment, precision measurement, lubrication, and more. It also includes a chapter on safety—a vital component in the workplace of today. The constant aim of the book has to present each subject as clearly, concisely, and simply as possible. To accomplish this, numerous sketches, photographs, and examples are used to aid the millwright/mechanic in the performance of day-to-day tasks.

The Audel name has long been associated with trades books that make sense and give the craftsperson the right information quickly and in the most practical manner. The prior edition of *Millwrights and Mechanics Guide* is often called the Bible of the Mechanical Trades. This new edition follows the Audel tradition of providing the most information in one source possible.

It is suggested that readers should also investigate the purchase of the *Audel Mechanical Trades Pocket Manual*—a vest pocket guide for the trades, written by the same author.

—Thomas Bieber Davis

About the Author

Thomas Bieber Davis is a Mechanical Engineer and is a principal in the firm of Maintenance Troubleshooting located in Newark, Delaware. He has been active in field troubleshooting of industrial electrical and mechanical machinery throughout the United States for over 30 years. Mr. Davis has done previous writing for various industrial publications in the areas of predictive maintenance, rotating equipment diagnostics, and bearing failure analysis. His practical insight into maintenance repair is based on consulting assignments at the request of professional maintenance departments for over 400 major companies from the chemical, factory, petrochemical, mining, and steel industries. This wide-ranging background gives a unique style to his writing and is very understandable to the millwright/mechanic who requires in-depth explanations of "what to do," "how to do it," and "when to do it" for all types of industrial machinery.

Contents

Chapter 1

Safety

It's hard to walk past the plant gate or the security office without seeing a sign indicating how many days since the last lost-time accident at the facility. Most employers take great pride in their safety record and preach safety as a way of life. This section includes some tips and information for keeping your operations safe.

OSHA

The Occupational Safety and Health Act (OSHA) in the United States covers workplace conditions for employees. If you are engaged in maintenance, repair, or installation activities, your employer probably has details on the particular parts of the OSHA standards that apply to you.

As a mechanical tradesperson, you might find yourself involved in many activities that require you to think out particular safety issues for your own special type of work. For example, you might be servicing a pump and have all the parts and the manufacturer's manuals, but it would make a great deal of difference in your approach to the job if the pump were handling acid instead of water. You have to consider the safety details for each job, using your trade skills combined with your knowledge of the process, location, or conditions.

Employers are required by OSHA to provide a safe and healthy workplace for their employees. In addition, they are also required to provide safety training if employees are expected to handle any hazardous materials. The employee has the right to request information about any specific health or safety hazard in the workplace. In addition, the employee has the right to information on what procedures will be followed if the employee is involved in an accident or might be exposed to a toxic material. For the most part, the Act promotes communication between the employer and employee so they can work together to reduce workplace hazards and minimize risks.

Lock Out and Tag

When you shut down a machine, process, pipeline, or electrical apparatus to inspect it or perform a repair, you need to *lock out and tag* a piece of equipment so that it cannot be accidentally started or energized. Usually, the start/stop of the switchgear controlling the piece of equipment is physically disabled with a lock (Figure 1-1).

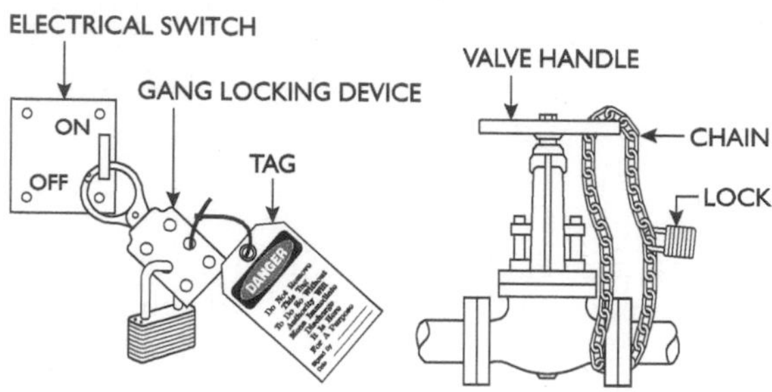

Figure 1-1 Lock out and tag.

The employee working on the equipment usually holds the key to the lock. The lock itself has a tag identifying whose lock is being used. In the case of a pipeline, the valve controlling the flow into the line is closed and a lock placed on the handle or bonnet with an appropriate tag attached. If two or more people are working at the same machine, then each additional person also places a lock on the equipment. A gang-locking device can be used to hold many locks on the job. In addition, it's a smart idea to try the local start/stop switch at the machine site or to open the valve downstream from the locked main valve as a double check to make sure that the machine or process is "safed-out."

It is generally true that a lock out and tag procedure should be followed if service or maintenance is required on any machinery or equipment where injury could take place from either a sudden release of stored energy or an unexpected startup of the machine. Some sample jobs that will usually have a lock out and tag procedure include:

- Inspection or testing of machine parts.
- Change-out of oil in a large gearbox or other reservoir.
- Repairing or testing electrical equipment.
- Working on a steam line, pressurized lines, or lines handling hazardous fluid.

- Freeing up jams in production line equipment.
- Teardown or major repair of mechanical equipment.
- Clearing product from process equipment.

While the particular lock out and tag procedure used at a specific plant or worksite might change depending on the location, the following ideas are good ones:

1. Make sure that each and every employee involved in the job has his or her own lock installed. Two people using one lock is never a good idea.
2. Always test the equipment by pressing the local start/stop switch or by slightly cracking open a downstream valve to make sure equipment doesn't run or material will not flow. Just because a motor starter's nameplate, located in a motor control room, matches the nameplate on the remote machine does not mean that misidentification could not happen. Always do a double check to make sure the device that is locked out is really "dead."
3. When you leave the job at the end of the shift and the next shift continues the work, make sure to remove your lock. It is never a good idea to "cut the lock off" for an employee who has left the job. This should be done only as a last resort and only when you are absolutely sure the employee has left the jobsite for the day.
4. If many people are on the same job, the use of a gang-locking device makes sense. That way there are no excuses for each individual not having a lock in place.
5. Use a lockout lock or lockout tag only for its intended purpose. The locks should have unique characteristics that identify them as safety locks. Safety locks should never be used for security of a toolbox or a locker. Obtain different locks for that purpose.
6. Open *your* lock with *your* key. Don't ever send another person to do it. Safety is a personal possession.

MSDA

Another safety issue for maintenance or installation personnel is the use of chemicals or hazardous materials. In the United States, Material Safety Data Sheets (MSDA) are used to identify the details of each particular chemical and to list safe handling techniques. These sheets, usually available in the workplace, indicate if rubber boots, face shield, respirator, dust mask, or goggles might be required to work safely around a substance. Keep in mind that oil or grease can be classified as chemicals, too. Removed parts of a machine often need to be cleaned after disassembly and require chemical degreasers or solvents.

A little-known fact is that MSDA sheets must also include first aid measures. These procedures are written for people who may not have received first aid training. With many first aid procedures, the first few seconds or minutes count. It's a good idea to review any MSDA sheet for each new material that you intend to use or contact to become familiar with how to handle it and what to do if an accident occurs—before an incident occurs.

Use of Protective Equipment

Often the maintenance worker needs gloves, a hard hat, steel-toe shoes, earplugs, or other articles of apparel to guard against injury to the eyes, feet, head, or ears. Jobs such as grinding, drilling, nailing, painting, or welding mandate protective gear (Figure 1-2).

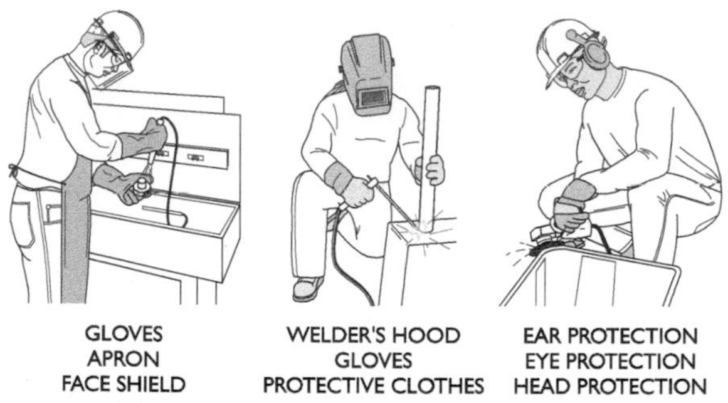

Figure 1-2 Use of protective equipment.

Some special jobs require more expensive and extensive protective equipment. Examples might be nonsparking boots where explosive mixtures are present, special glasses to be used around laser light, or cut-resistant gloves for use with sharp knives or saws.

Confined-Space Entry

While the actual OSHA definition of a confined space is rather lengthy, the Act generally defines a confined space as an area that is large enough for an employee to enter totally and perform work and that also has limited means for entry and exit. Entering a tank, a pipeline, a manhole, or a silo would be excellent examples of confined-space entry (Figure 1-3). Usually a permitted type of entry system is required for work in a confined space. Although the particular permit may not be the same at different workplace locations, it usually includes the use of a buddy system (two people assigned with one standing watch), checking the confined space for toxic materials and oxygen levels, and informing others when entry is necessary. Typically, work permits are dated and valid for a particular time period—usually only one shift. In many cases, artificial ventilation or the use of a safety harness or safety line may be necessary. Many industrial accidents in the past have been related to confined-space entry. Because of this fact, a separate section of the OSHA requirements was instituted to make sure that employers and employees pay particular attention to the unique hazards presented by confined spaces.

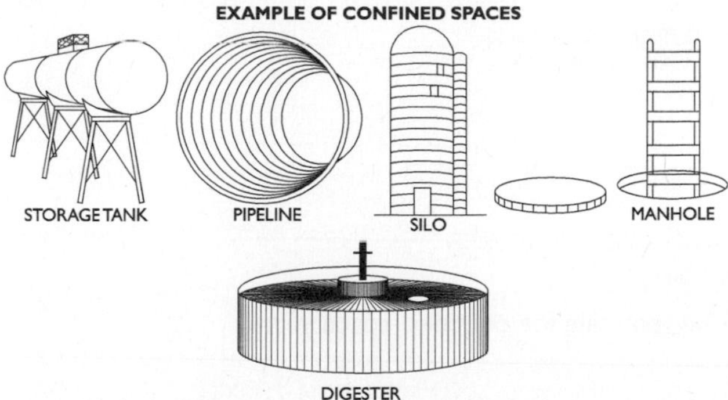

Figure 1-3 Examples of confined spaces.

The atmosphere in a confined space is of particular concern. Most people immediately think of the potential for flammable or toxic fumes within a confined space, but they forget about the most important factor: the lack of oxygen. The oxygen level of a confined space can decrease because of work being done—such as welding, cutting, or brazing—or it can be decreased by certain chemical reactions (rusting) or through bacterial action (fermentation). Figure 1-4 shows an oxygen scale and indicates safe and unsafe levels for occupants of a confined space. Any atmosphere with less than 19.5 percent oxygen should not be entered without an approved self-contained breathing apparatus.

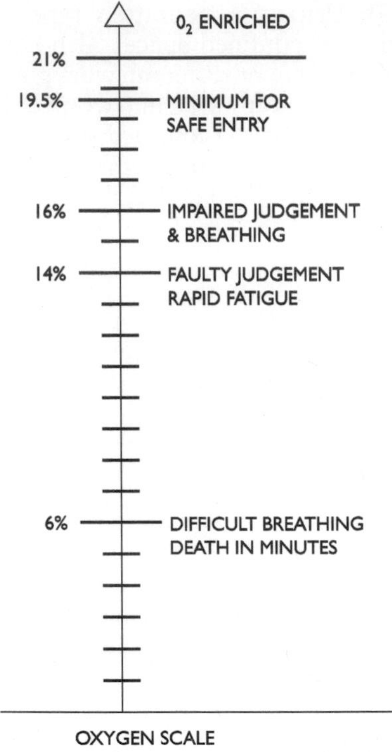

Figure 1-4 Oxygen scale for confined-space work.

Chapter 2

Drawing and Sketching

Drawings and sketches are universally used to communicate technical information. They are the graphic language used by everyone concerned with mechanical operations. They convey information, size, location, accuracy, etc., and by their use, mechanisms ranging from the simple to the most complex can be graphically described. They constitute a universal language because of the general standardization of techniques throughout the world. For this reason, the language of drawings is understood in spite of native tongue and measurement differences. Drawings or sketches are the foundation of production interchangeability, enabling identical parts to be produced at widely separated locations. They are the guide for construction, assembly, installation, etc., of all forms, types, and kinds of objects.

Because drawings and sketches are such a vital and basic means of communication, it is essential that they be understood by the mechanic. He or she must be able to interpret or "read" drawings and make simple sketches. This is very important because thoughts and ideas are almost impossible to express with words alone—but frequently can be illustrated by a simple sketch.

The term *drawing* usually refers to a graphic illustration made by a draftsman with the aid of scales, instruments, etc. When made freehand with pencil or pen, without mechanical devices, they are usually referred to as *sketches*. In either case, basic principles and concepts as well as generally accepted conventions are followed.

The word *blueprint* is loosely used to describe many different types of drawing reproductions. It originally described a specific type that had white lines on a blue background. The blueprint process of drawing reproduction is still widely used and is one of the oldest methods of duplicating drawings. In reproducing drawings by this method, the blueprint is made on paper that is coated with a light-sensitive material called an emulsion. A drawing made on special paper that will allow passage of light, or a transparent tracing of a drawing, is placed on the paper and exposed to strong light. The lines on the drawing or tracing hold back the light and leave an impression on the blueprint paper. After exposure to light, the paper is passed through a developer and washed with water. Except for shrinkage, the blueprint is an exact duplicate of the original drawing. This process is commonly referred to as a "wet process." Also derived from the original blueprint process is the term "blueprint reading." This term is now applied to the general interpretation or reading of all forms of drawing, sketches, and prints or drawing duplicates.

Another process that has become widely used in recent years, probably because it is a so-called dry process, is ammonia development of prints. This is also a contact method of reproduction and results in a print with a white background. The lines are usually blue or black, although other color lines may also be produced. In this process, the coated paper, after exposure to light, is developed by exposure to ammonia fumes.

One of the first engineering uses of small computers was to produce drawings. This is commonly called computer-assisted design (CAD). CAD is an extremely powerful tool. Creating and modifying drawings using computers offers a phenomenal timesaving advantage over "hand" preparation. CAD uses a set of *entities* for constructing a drawing. An entity is a drawing element such as a line, circle, or text string. CAD allows the operator to enter commands to tell the computer which entity to draw. In many cases, symbol libraries are available for making specialized drawings that pertain to the piping, electrical, or electronic trades. After the drawing is completed on the computer, it is sent to a plotter to print out for use in the field.

Lines

Drawings are made up of a combination of lines of different lengths and intensity. These define the shape, size, and details of an object. It is possible, through the correct use of lines, to graphically describe an object so that it can be accurately visualized by any person with a basic understanding of drawing principles. Since lines are the basis for all drawings or sketches, a listing of lines used for various purposes may be called an alphabet of lines. Figure 2-1 and Figure 2-2 illustrate the use of these lines.

Projection

Mechanical drawings and sketches are *not* a true representation of what the human eye actually sees. What is seen by the eye is most nearly represented by what are called "perspective drawings," or sketches. The portion closest to the observer is the largest, while the parts farthest away are the smallest. Lines and surfaces become smaller and closer together as distance from the eye increases, seeming to disappear at a point on an imaginary horizon called a

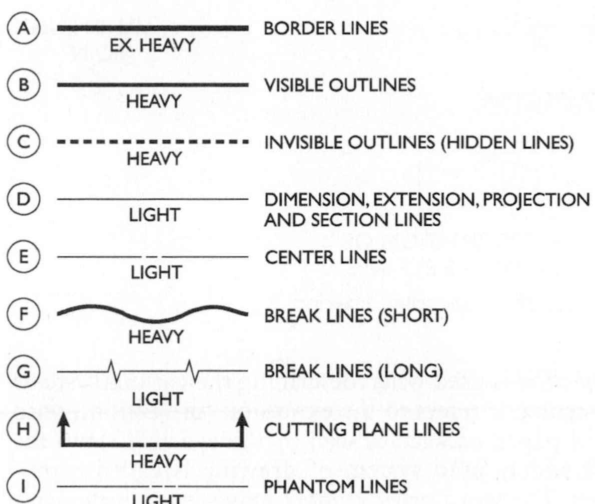

Figure 2-1 Alphabet of lines.

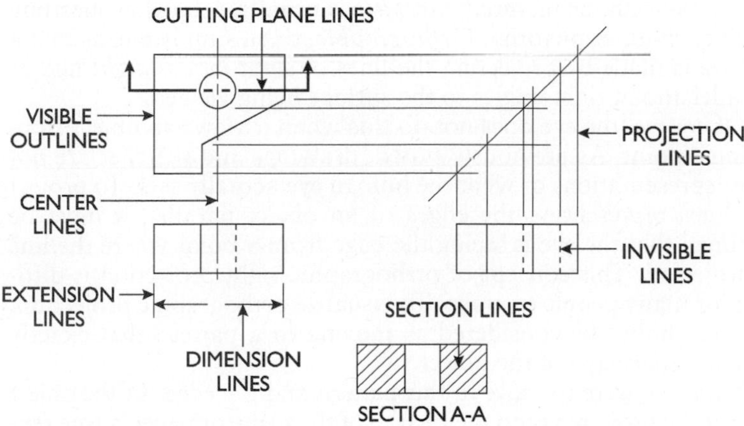

Figure 2-2 Use of lines.

"vanishing point." Perspective drawing or sketching is pictorial in nature, often influenced by artistic talent and not readily lending itself to mechanical methods and drawing practices.

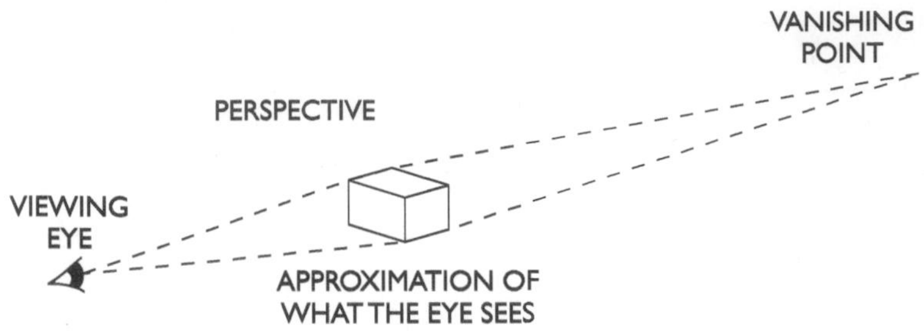

Figure 2-3 Elements of perspective drawing.

The word *projection* is used when describing the various systems of mechanical drawing. It refers to the extension (projection), onto a plane or sheet of paper, of what is seen by the eye as it views the object. The most widely used system of drawing is called *orthographic projection.* The word *orthographic* may seem a little overpowering. For purposes of making it more understandable, the word can be divided into two parts, *ortho* and *graphic.* The first part, *ortho*, is derived from the Greek *orthos* and means "straight." The second part, *graphic*, is from the Greek *graphein*, meaning "to write, diagram, or picture." *Graphic* is usually defined as meaning lifelike, vivid, or pictorial. *Orthographic*, then, simply means that a *picture* is made by projecting the lines of sight in a *straight* line or parallel and at right angles to the surface being viewed.

Of course, the eye does not do this when it views an object from a single point. As previously stated, drawings and sketches are *not* true representations of what the human eye actually sees. To project the lines representing the edges of an object parallel, it must be assumed that the eye is seeing the edge from a point where the line is projected. This concept of orthographic sight projection is difficult for many people to grasp. To visualize orthographic projection, the eye should be considered as moving in a pattern that exactly matches the shape of the object.

If the eye were to move in this pattern and the edges of the object were to be projected onto the surface of the plane or paper, a true representation of the outline of the object could be drawn on the paper. Such a drawing represents only the outline of the surface of the object. Additional drawings from other viewing points would be necessary to show the various features of the object, such as thickness variations and other surface shapes. These additional drawings, as well as the original, are called "views" (each taken from a specific viewing position). Several such drawings, or views, are required to show various

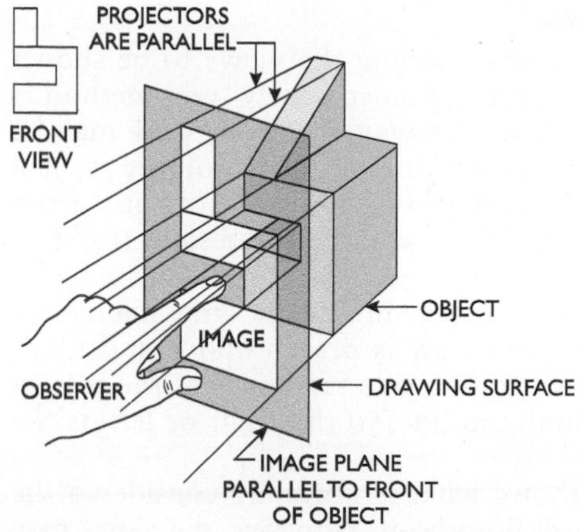

Figure 2-4 Orthographic projection.

features of an object clearly. Therefore, what is commonly called a "drawing" may be made up of several related drawings, usually referred to as "views." Standardized principles and practices govern the positioning of these views in relation to one another on a drawing.

To show size, shape, and location of the various parts of an object, a number of views are necessary. In orthographic projection, each view shows the part of the object as it is seen by looking directly at the surface. Each view, therefore, represents the true shape and size of the surface of the object being viewed. Probably the most widely used combination of views is that of the top, front, and right side. They are arranged so that each view represents the surface adjacent to it, as shown in Figure 2-5.

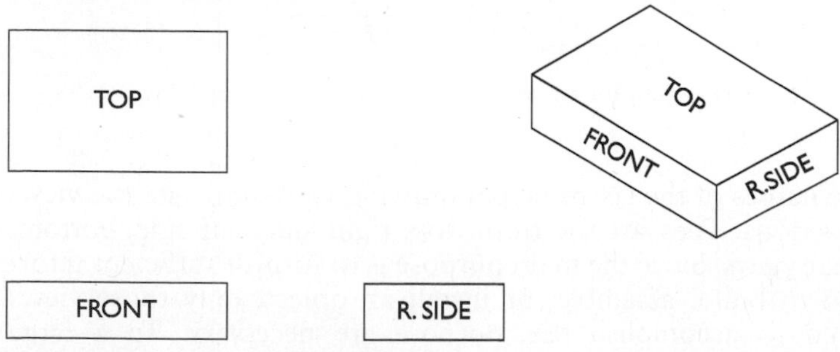

Figure 2-5 Three-view drawing.

Arrangement of Views

Several methods of selecting or obtaining the views to be shown on a drawing are used. Probably the most widely used method is the direct or natural method. In this way, the front view may be thought of as a view obtained by looking at the object from a position directly in front of it, the top view by looking at it from above, and the side views by looking at it from the sides (right or left).

The views are arranged on the drawing in the same manner in which they are viewed. The top view is drawn above and vertically in line with the front view. The side views are at the side of the front view and horizontally in line, to the right or left as the case may be.

Usually the front view shows what is normally considered the front of the object in its natural position. However, the views may be selected to give the best description of the object's shape. In this case, the front view might have no relationship with the actual front position of the part as it fits into an assembled position with other components. See Figure 2-6.

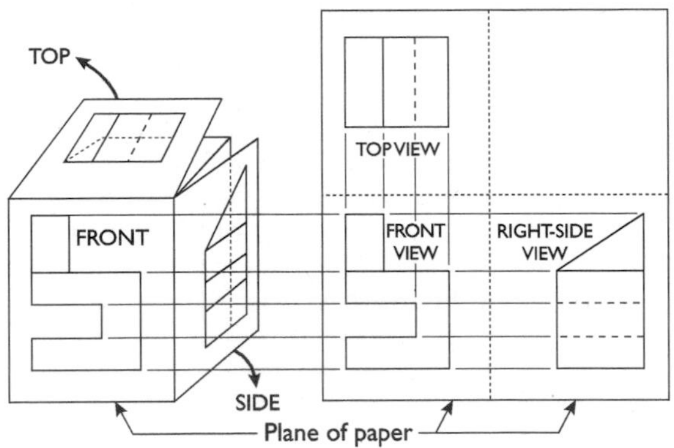

Figure 2-6 Arrangement of views.

The names of the six principal drawing views indicate the viewing positions. They are the front, top, right side, left side, bottom, and rear views. Since the main purpose is to furnish sufficient information to build, assemble, or install an object, only those views required to accomplish this purpose are necessary. To a large degree, this is determined by how simple or complex the part may be. See Figure 2-7.

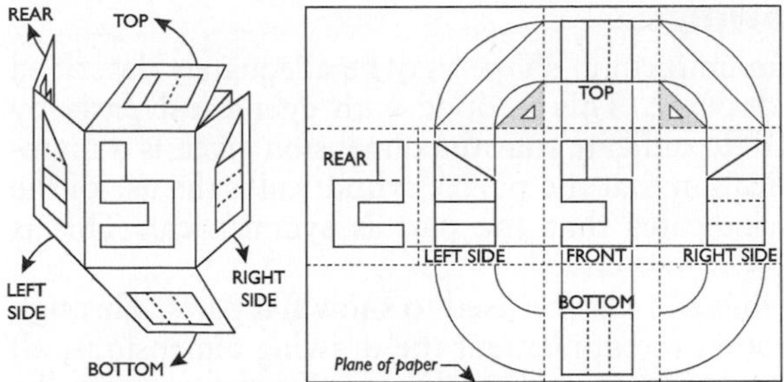

Figure 2-7 The six principal views.

Two-View Drawings

Simple symmetrical flat objects and cylindrical parts, such as shafts and sleeves, can be adequately shown with two views. Usually, the front view can be shown in combination with either a right or left side view or with a top or bottom view. See Figure 2-8.

The selection of views is usually made on the basis of the combination that shows the object to best advantage. When a choice of combinations will do this, the object should be shown in its natural position in the front view.

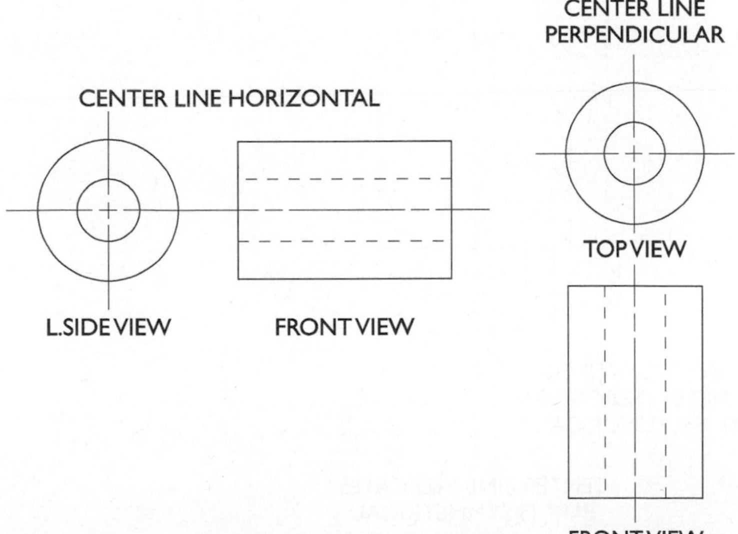

Figure 2-8 Two-view drawing.

One-View Drawings

Some parts that are uniform in shape may be adequately described with a one-view drawing. This is done with cylindrical parts by using the letter "D" to indicate that the dimension given is a diameter. A further indication that the part is cylindrical is the use of the centerline, which indicates that the part is symmetrical. This is shown in Figure 2-9.

One-view drawings can also be used to show flat parts. Through the use of simple notes to supplement the drawing dimensions, all the information necessary to accurately describe the part can be furnished, as illustrated in Figure 2-10.

Hidden Lines

When viewing the surfaces of an object, many of the edges and intersections in the object are not visible. To be complete, a drawing must include lines that represent these edges and intersections. Lines made up of a series of small dashes, called invisible edge or hidden lines, are used for this purpose in Figure 2-11.

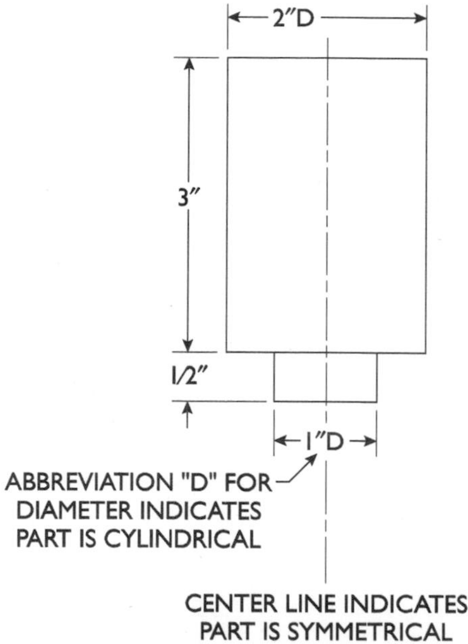

Figure 2-9 One-view cylinder drawing.

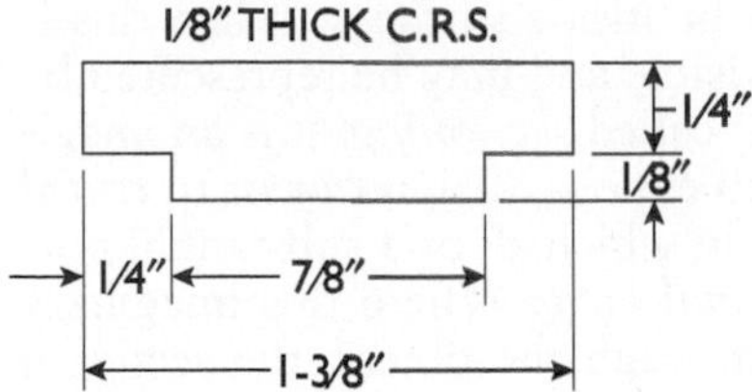

Figure 2-10 One-view flat-object drawing.

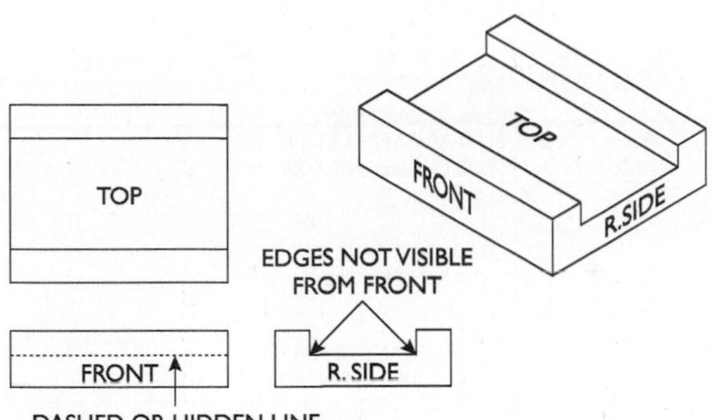

Figure 2-11 Hidden lines.

Auxiliary Views

When the surfaces of an object are at right angles to one another, regular views are adequate for their representation. However, when one or more are inclined and slant away from either the horizontal or vertical plane, regular views will not show the true shape of the inclined surface. To show the true shape, an *auxiliary view* is used. In an auxiliary view, the slanted surface is projected to a plane that is parallel to it. Shapes that would appear in distorted form in the regular view appear in their true shape and size in the auxiliary view. Usually only the inclined surface is shown in the auxiliary view, presented in its true shape and size as shown in Figure 2-12.

Full-Section Views

As the details inside a part become more complex, the invisible lines to show these details become more numerous. This may reach a point where the drawing appears confused and is difficult to interpret. A technique used to simplify such a drawing is to cut away a portion of the object and expose the inside surfaces. When this is done, the invisible edges become visible and may be represented by solid object lines. This technique is called *sectioning*. It is an imaginary cut made through an object to expose the interior or to reveal the shape of some portion. A view in which all or a substantial portion is sectioned is called a *sectional view*. Where this imaginary cutting plane passes completely through the object, the sectional view is called a *full section*. The exposed surfaces through which the imaginary cut is made are identified by slant lines called *section* or *crosshatch* lines. This is completely illustrated in Figure 2-13.

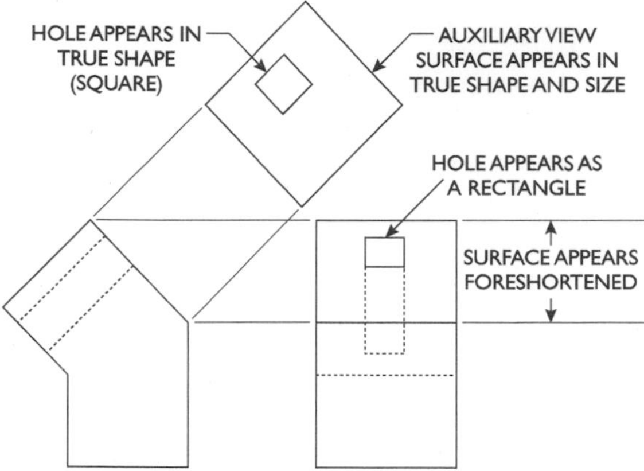

Figure 2-12 Auxiliary view.

Full-Section Cutting Plane

To indicate the edge of the cutting plane and the direction in which the section is viewed, a cutting plane line is used. It is a heavy line with one heavy and two short dashes. The direction in which the section is viewed is shown by arrowheads. This is shown in Figure 2-14. Letters are placed near each arrowhead, in the drawing, to identify the section.

Half-Section Views

Greater clarity may sometimes be accomplished by using a half-section view. This is one in which the imaginary cutting plane extends only halfway across the object. It has the advantage of showing both the interior and exterior of the object on one view without the use of dashed lines, as shown in Figure 2-15.

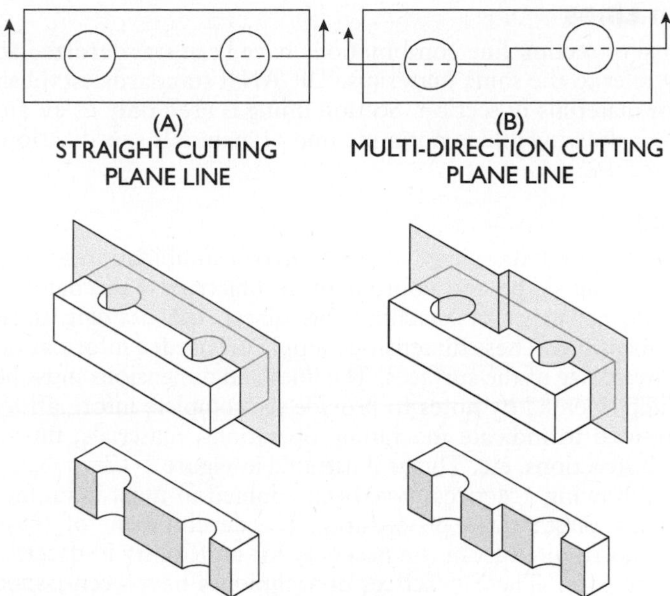

Figure 2-13 Full-section cutting plane.

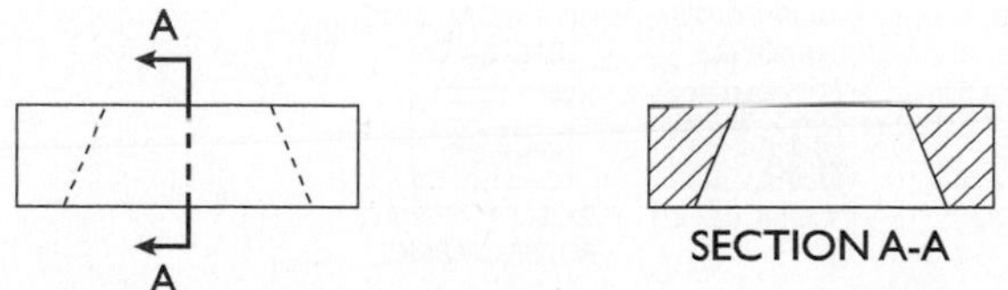

Figure 2-14 Full-section view.

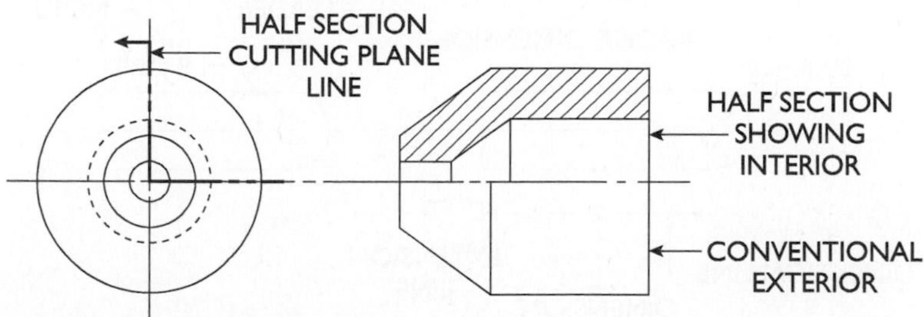

Figure 2-15 Half-section view.

Section Lines

Crosshatch or section line combinations have been standardized so that they refer to the same materials. The ANSI standards establish a code for materials in section. Section lining is used only as an aid in reading a drawing and is not a method of materials specification. See Figure 2-16.

Drafting

As previously stated, drawings are made up of a combination of lines that describe the shape and contour of an object or a mechanism. However, to construct or machine this object, the drawing must include dimensions. These dimensions supply the needed information as to the exact size of the surfaces. The lines and dimensions must be further supplemented by notes to provide the complete information. Notes are used to indicate machining operations, materials, finish, assembly instructions, etc. This is illustrated in Figure 2-17.

Certain drawing practices have been adopted in most manufacturing plants as standard procedures. The development of these standards has resulted from the necessity for uniformity in describing machine parts. These practices or techniques have been issued

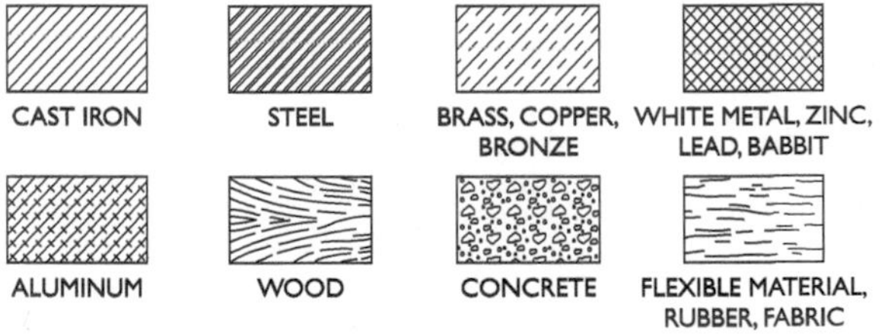

Figure 2-16 Section lines.

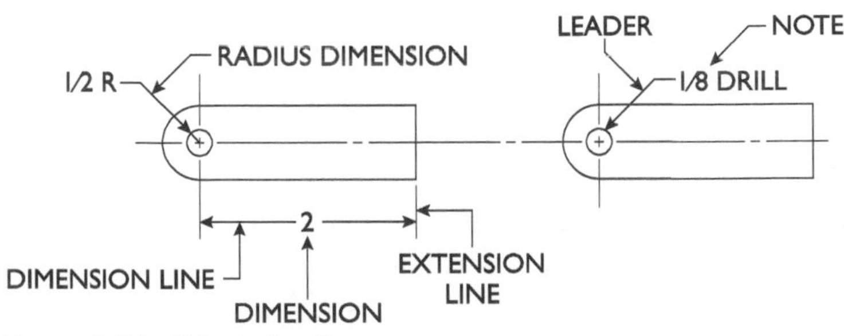

Figure 2-17 Dimension lines.

as standards by the American National Standards Institute (ANSI) and are referred to as the language of drafting.

Two basic dimensioning methods are used to give a distance or size. In the first, a dimension is placed between two points, lines, or planes or between some combination of points, lines, and planes. The dimension lines indicate the direction in which the dimension applies, and the extension lines refer the dimension to the view.

In the second dimensioning method, a note provides the information or dimension. A fine line, called a leader, extends from the note to the feature on the drawing to which the note applies. When notes apply to the object as a whole, they are placed in a convenient location on the drawing without using leaders.

Dimension lines, extension lines, and leaders are made with fine or lightweight lines—similar in width to centerlines. The fine lines provide a contrast with the heavier object lines. See the illustration shown in Figure 2-17.

Arrowheads are usually drawn freehand. The solid head is usually preferred to the open head style. The solid head is drawn narrower and longer than the open head and usually with straight sides. See the illustration shown in Figure 2-18.

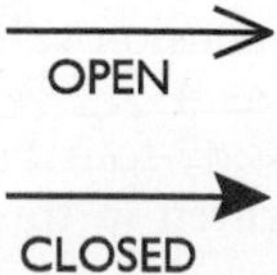

Figure 2-18 Arrowheads.

Extension lines are fine lines that extend from the view to a dimension placed outside the view. They should not touch the view but start about 1⁄16 in. away from it. They should extend about 1⁄8 in. beyond the last dimension line. When dimensioning between centers, the centerlines may be extended outside the view and used as extension lines. This is illustrated in Figure 2-19.

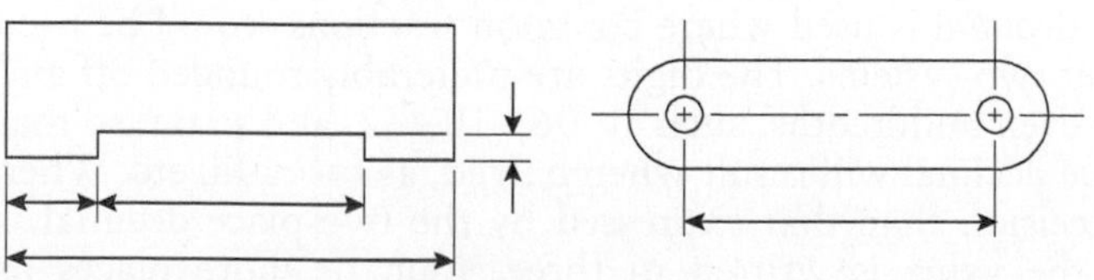

Figure 2-19 Dimensioning a drawing.

Leaders are fine lines (preferably straight rather than curved) leading from a dimension, or a note, to the point on the drawing to which it applies. An arrowhead is used on the terminating or pointing end of the line but not on the note end. The leader should start at the beginning or end of the note with a short horizontal bar. Leaders should not start in the middle of a note. See Figure 2-20.

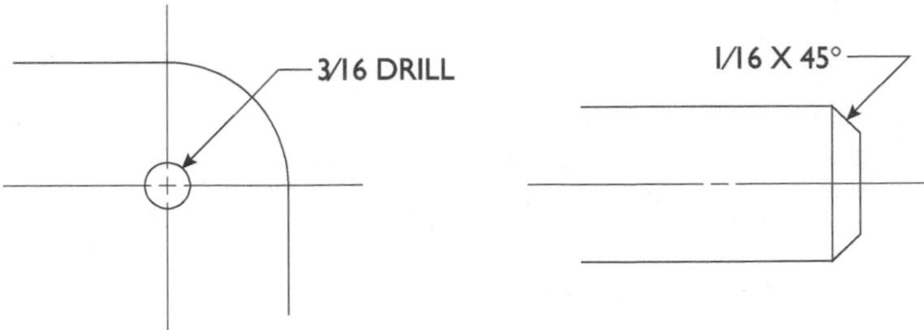

Figure 2-20 Leaders are used to show dimensions.

Three systems for writing dimension values are in general use. Most widely used for general drawing, including architectural and structural work, is the *common fraction system.* All dimension values are written as units and common fractions, such as $4\frac{1}{2}$, $2\frac{3}{8}$, $\frac{5}{16}$, $\frac{9}{32}$, and $\frac{1}{64}$. A tape, rule, or scale graduated in sixty-fourths of an inch can be used for layout when values are written in this manner.

When the degree of precision calls for fractional parts of an inch smaller than those found on the ordinary steel scale, the "common-fraction" or "decimal-fraction" system is used. In this system, values are given in units and fractions for distances not requiring an accuracy closer than $\frac{1}{64}$ in. When distances require greater precision, the value is given in units and decimal fractions, such as 2.250, 1.375, 0.062, etc.

The third system is the "complete decimal" system and uses decimals for all dimensional values. It has the advantages of the metric systems while using the inch as its base. When using this system, a two-place decimal is used where common fractions would be used in the other two systems. The digits are preferably rounded off and written to even hundredths, such as .06, .10, .52, and so on, so that a two-place decimal will result when halved, as for radii, etc. When greater precision than that expressed by the two-place decimal is required, the value is written in three, four, or more places as required. The use of this system is increasing, particularly for machine-type drawing.

When large values must be shown, the feet and inch system may be used. The measuring units are identified by using the foot and the inch marks: 10'6". When the dimension is in even feet, it should be indicated that there are no inches, as in 12' 0". If there are no full inches but rather fractions of inches, it should be indicated as 9' 0¼". The general practice in machine drawing, and recommended by the ANSI, is that dimensions up to and including 72 in. be given in inches and greater lengths in feet and inches. When dimensions are all in inches, the inch mark is preferably omitted.

The dimensioning of a drawing should be complete enough so that very little computation will be necessary. Duplication of dimensions should be avoided unless such duplication will make the drawing clearer or easier to read.

Drawing Scales

Objects are not always drawn to actual size. They may be too large to fit on a standard drawing sheet or too small for all details to show clearly. In either case, it is still possible to represent the object by simply reducing or enlarging the size to which the drawing is made. This practice does not affect the dimensions since the actual part dimensions are shown on the drawing. When the practice is followed, a note is placed on the drawing (or in the title box) to indicate the "scale" used. The scale is the ratio of the drawing size to the actual size of the object. For instance, if a drawing 6" long represents a part 12" long, the scale 6" = 12" is referred to as "half scale" and is commonly noted on drawings as ½" = 1". Other common scales used in machine drawing are the one-quarter scale (¼" = 1"), one-eighth scale (⅛" = 1"), and the double scale (2" = 1").

In architectural drawing, the scale ratios used are much higher to enable the drawing of large plans on a comparatively small sheet of paper. The most commonly used scale on building drawings is the ¼" = 1'0", referred to as ¼ in. equals one foot. A drawing made to this scale is 1/48 actual size. If the construction work is very large, a scale of ⅛" = 1'0" (1/96 of the actual size) may be used. Unlike machine drawings, which should never be measured, it is sometimes necessary to measure building drawings. When this is done, using a regular carpenter's rule on a drawing made to the scale of ¼" = 1'0", every ¼" on the rule represents a foot on the job. Thus, ⅛" would equal 6", and 1/16" would equal 3", etc.

Standard Reduced Scales

Machine Drawing

1"=1" (Full size)
¾"=1" (¾ size)
½"=1" (½ size)
¼"=1" (¼ size)
⅛"=1" (⅛ size)

Architectural Drawing

12"=1'0" (Full size)
6"=1'0" (½ size)
3"=1'0" (¼ size)
1"=1'0" (1/12 size)
¾"=1'0" (1/16 size)
½"=1'0" (1/24 size)
⅜"=1'0" (1/32 size)
¼"=1'0" (1/48 size)
⅛"=1'0" (1/96 size)

Screw Threads

Screw threads, in one form or another, are used on practically all engineering products. Consequently, when making working drawings and sketches, there is the repeated necessity to show and specify screw threads. They are widely used on fasteners (bolts, machine screws, etc.), on devices for making adjustment, and for the transmission of power. For these various purposes, a number of thread forms are in use. See Figure 2-21.

Because a true representation of a screw thread is an extremely laborious operation, it is almost never done when making working drawings. Instead, threads are given a symbolic representation providing illustrations suitable for general understanding and use.

There are two basic types of screw threads: external threads produced on the outside of a part and internal threads produced on the inside of a part. Both types are represented on drawings by three methods (see Figure 2-22). The first, and least used, method is the "detailed presentation," which approximates the true appearance. The second, called the "schematic representation," is nearly as effective as the pictorial and is much easier to draw. The third

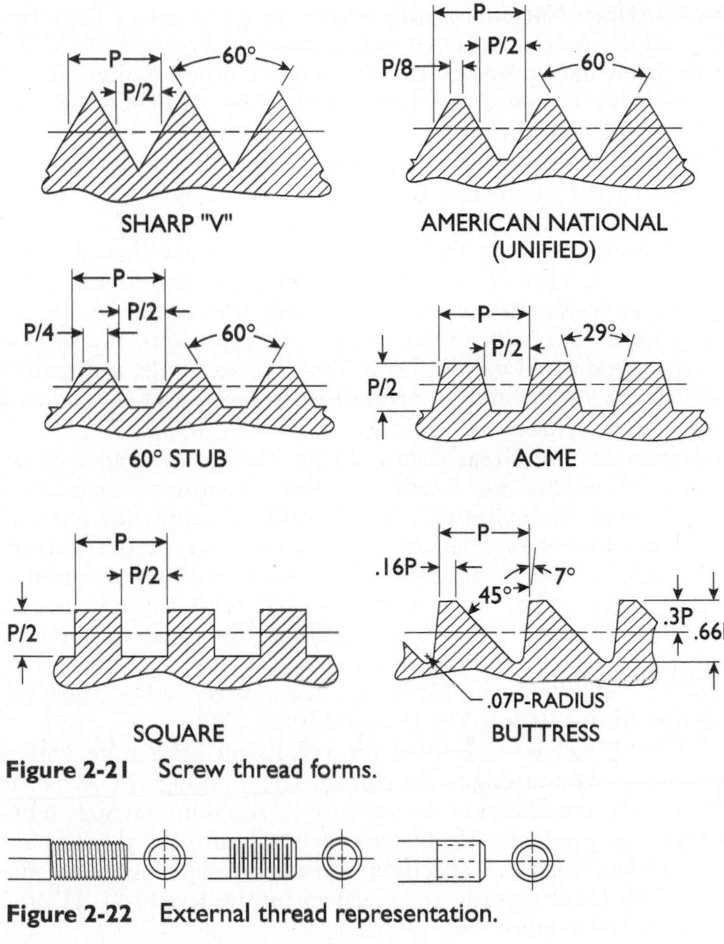

Figure 2-21 Screw thread forms.

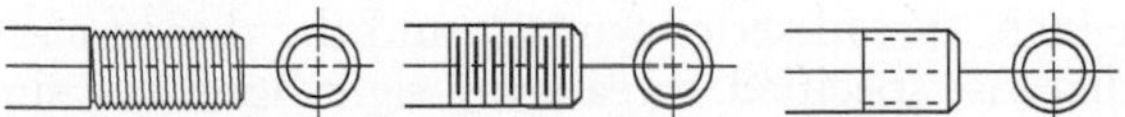

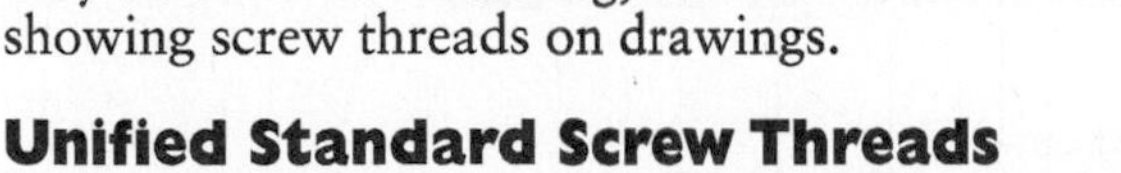

Figure 2-22 External thread representation.

method, called "simplified representation" on the grounds of economy and ease of rendering, is the most commonly used method for showing screw threads on drawings.

Unified Standard Screw Threads

The most widely used thread form is one that resembles a "V." The angle between the sides of the thread is 60°, and the bottom and top of each thread is flat. This style thread, which originated as a sharp "V" thread, has been subjected to a series of standardizations. The earlier standards, such as the U.S. Standard, National

Form, American National Form, Society of Automotive Engineers (SAE), and the American Society of Mechanical Engineers (ASME), have all been incorporated into the latest Unified Screw Thread Standard. This standard has been adopted by the United States, Canada, and Great Britain.

The standard establishes several series, which are groupings distinguished by the number of threads per inch applied to a specific diameter. See Figure 2-23.

The symbols used are (UNC), denoting a coarse-thread series; (UNF), denoting a fine-thread series; (UNEF), denoting an extra-fine series; and various constant-pitch series with 4, 6, 8, 12, 16, 20, 28, and 32 threads per inch. In the constant-thread series, the threads per inch are indicated within the symbol. For example, a thread of the 8-thread series would be symbolized (8UN), and a 12-thread series would be symbolized (12UN).

The standard also establishes thread classes that specify the amount of allowance and tolerance. These are commonly referred to as "class of fit." Classes 1A, 2A, and 3A apply to external threads only, and classes 1B, 2B, and 3B apply to internal threads only. Classes 1A and 1B are used where quick and easy assembly is necessary. They permit easy assembly even with slightly bruised and/or dirty threads. Classes 2A and 2B are the most widely used thread standards for general applications and all types of fasteners. Classes 3A and 3B are suitable for applications requiring closer tolerance than those afforded by classes 2A and 2B.

A screw thread is designated on a drawing by a note with a leader and arrow pointing to the thread. The minimum information required is the specification in sequence of the nominal size, number of threads per inch, thread series symbol, and the thread class number. Unless otherwise specified, threads are right-hand and single lead. Left-hand threads are specified by the letters "L H" following the class symbol.

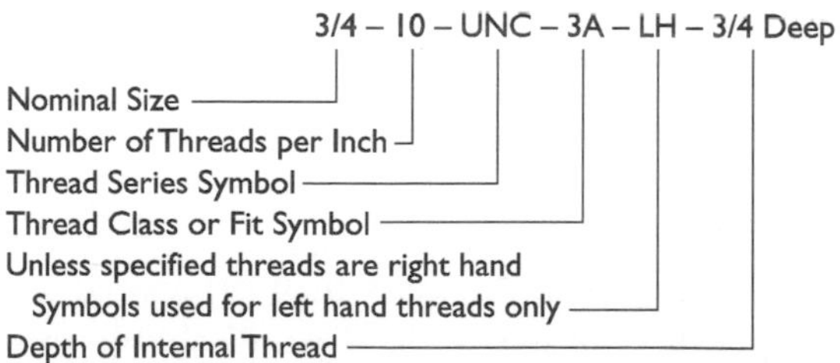

Working Drawings

The purpose of a working drawing is to give the information necessary to construct the object or mechanism represented in the drawing. The drawings of the different parts of the machine are called "detailed drawings." Each part is in detail and is represented in an unmistakable manner, with all of the dimensions and needed information of the parts written in.

The drawings of the complete machine are called "assembly drawings." They show the whole arrangement of the machine, indicating the relative position of its parts; they may also be made to show the motions of the movable parts. In preparing a detail drawing, the first point is to decide on the number of views required to illustrate the shape of the object and its parts in a complete, and at the same time simple and easily understood, manner.

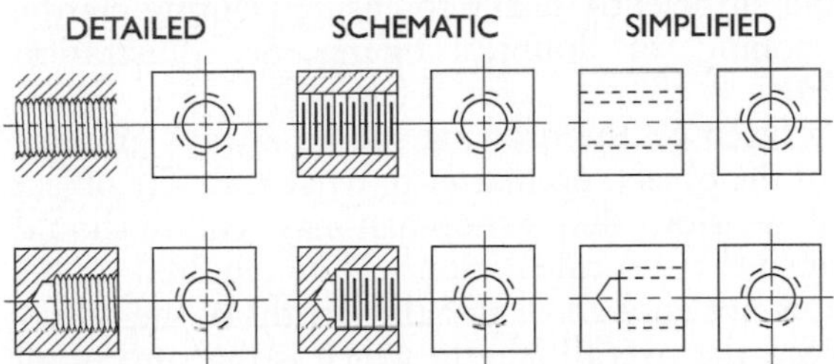

Figure 2-23 Internal thread representation.

After deciding on the manner of views, the selection of scale must be decided to enable the placing of all the required views of the object within the space of the paper. As to the number of views required for an object, no definite rule can be laid down. It depends on the form and character of the figure and must be decided by the best judgment of the draftsman.

After ascertaining the most important dimensions of the mechanism, a general drawing of the main body should be executed, omitting the smaller parts until after this particular drawing is made. The larger and more important parts are produced first, followed by the smaller parts, which are to be attached to the larger parts. The materials of which the parts are to be made should be indicated by special remarks, notes, etc. The methods of work that are to be followed by the workman should be indicated, sometimes to the extent of pointing out what machine tool is to be used for the work.

Dimensioning Drawings

Putting the dimensions on a drawing correctly is not only one of the most important but also one of the most difficult parts of the work of a draftsman. The draftsman will put in only those dimensions required by the shopman. The manner in which this is done must depend on the method to be used by the workman in constructing the part. For this reason, an acquaintance with the methods adopted in shop practice as well as the tools to be used is essential. Every dimension necessary to the execution of the work should be clearly stated by figures on the drawing so that no measurements need be taken in the shop by scale. All measurements should be given with reference to the base, or starting point, from which the work is laid out and also with reference to centerlines.

All figured dimensions on the drawings must be plain, round vertical figures, not less than one-eighth-inch high, formed by a line of uniform width and sufficiently heavy to ensure printing clarity, omitting all thin, sloping, or doubtful figures. See illustration shown in Figure 2-24.

The dimensions written on the drawing should always give the actual finished size of the object, no matter to what scale the object may be drawn. All dimensions that a shopman may require should be put on a drawing so that no calculation will be required on his part. For instance, it is not enough to give the lengths of the different parts of the object; the overall length, which is the sum of all these lengths, should also be marked as shown in Figure 2-25. The figures giving the dimensions should be placed on the dimension lines and not on the outline of the object. The dimension lines should have arrowheads at each end, and the points of these arrowheads should always touch the lines, the distance between which is indicated by the dimension as illustrated in Figure 2-26. The figure should be placed in the middle of the dimension at right angles to that line, so as to read from either the bottom or from the right-hand side of the drawing. The arrowheads should be put inside the lines, from which the distance as given in the dimension is figured.

When the space between these lines is too small for the figures, as shown in Figure 2-26, the arrowheads may be placed outside and the figures put outside as illustrated. The dimension lines should also be put in the drawing, very near to the spaces or lines to which they refer. When the view is complicated, dimension lines drawn within it might tend to make it still more obscure and difficult to understand. In such a case, the dimension lines should be carried outside the view and extension lines drawn from the arrowheads to the points between which the dimension is given.

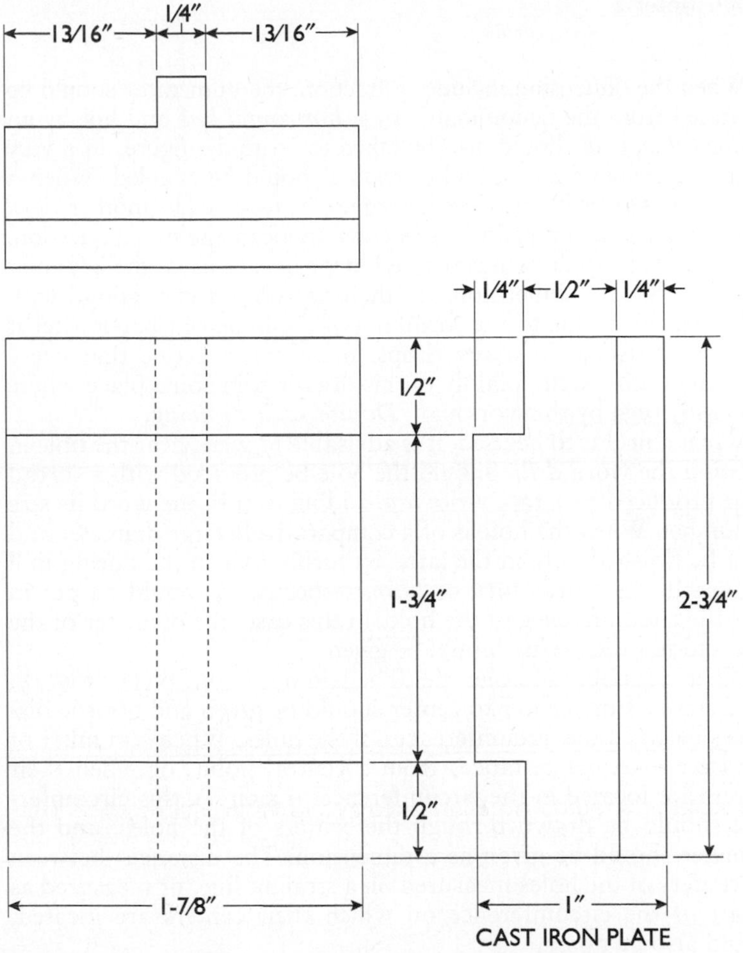

1/4″
13/16″ 13/16″
1/4″ 1/2″ 1/4″
1/2″
1-3/4″
2-3/4″
1/2″
1-7/8″
1″
CAST IRON PLATE

Figure 2-24 Example of drawing a cast-iron plate in full size.

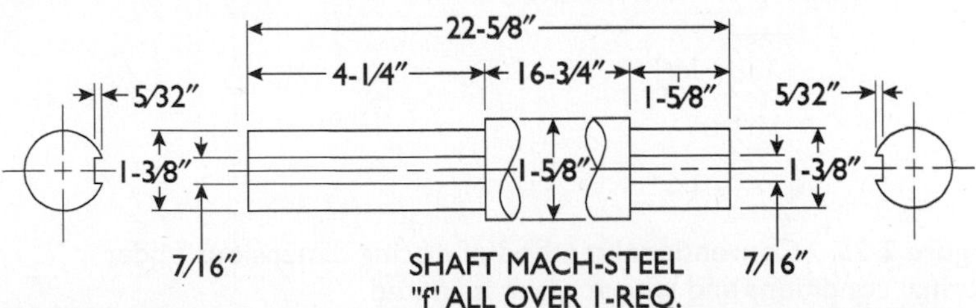

Figure 2-25 Drawing of a typical machine shaft with both end views shown. It is customary in this style of drawing to break the part to conserve space.

When the dimension includes a fraction, the numerator should be separated from the denominator by a *horizontal line* and not by an *inclined line*; care should also be taken to write the figures in a very clear and legible manner, and crowding should be avoided. When a dimension is given in one view, it need not be repeated in another view, except when such a repetition is essential to locate the size in question.

For most shop drawings where blueprints are used, the beginner will find that the dimensions on the print do not correspond with the scale; this is due to the shrinking of the blueprint paper after it has been washed. In many shops, there exists a rule that every draftsman must write plainly on his drawing, in some place where it is easily seen by the workman, *Do not scale drawing.*

When a drill is to be used, it is advisable to write near the hole in question the word *drill.* Should the hole be provided with a thread to be produced by a tap, write *tap,* adding to this one word its size or number. When the hole is of a comparatively large diameter and must be finished only in the lathe by turning, or in the boring mill by boring, the words *turn* or *bore*, respectively, should be put in near the circumference of the hole. In this case, the diameter of the hole and not the radius should be given.

When a number of holes are to be laid out in one piece of work, the distance from center to center should be given and not the distances between the circumferences of the holes. When a number of holes are at equal distances from a central point, or when their centers are located in the circumference of a circle, this circumference should be drawn through the centers of the holes, and the diameter should be given as a dimension. The distances between the centers of the holes measured on a straight line, or measured as a part of the circumference on which their centers are located, should also be noted.

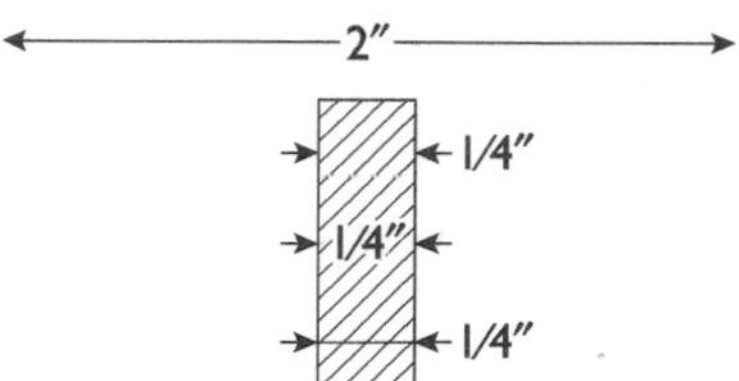

Figure 2-26 Conventional method of placing dimensions under normal conditions and where space is limited.

In practice, reference letters are sometimes used instead of dimensions, as shown here:

D = diameter of shaft, 2½ in.
L = length of bearing, 3¾ in.
T = thickness of collar, ⅞ in.
d = diameter of collar, 3½ in.

It is preferable to give the diameters of turned and bored work on a section, instead of an end drawn separately; confusion is sometimes caused by a number of radial dimensions. See Figure 2-27.

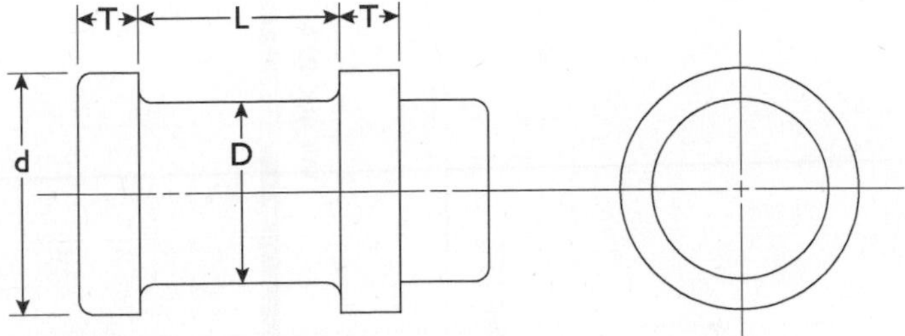

Figure 2-27 Alternate method of showing dimensions.

Architectural Drawings

The term *plans* as applied to a set of drawings, especially architectural drawings, has become firmly established, although it is an erroneous use of the term. A plan is a plan, and a set of drawings or blueprints showing, for instance, the construction of a house will consist of one or more *plans* and one or more *elevations*. The difference should be

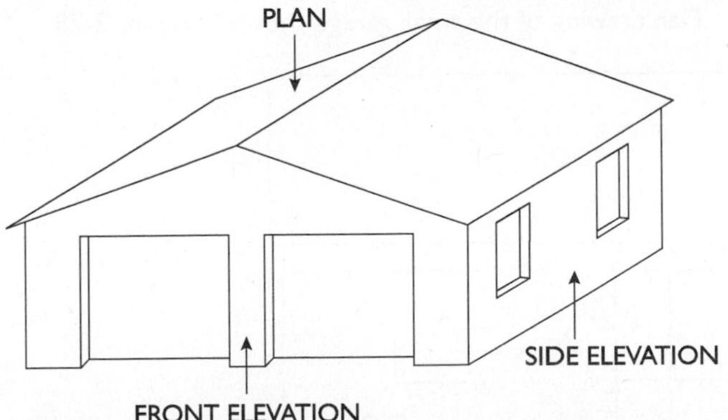

Figure 2-28 Cabinet projection of a small garage. Arrows show direction of sight for plan and elevation.

clearly understood, as has already been explained. A plan is an orthographic projection of the horizontal parts of an object, as distinguished from an elevation. An elevation is the orthographic projection of the vertical parts of an object, as distinguished from a plan.

The accompanying drawings of a small garage, Figures 2-28, 2-29, and 2-30, will illustrate the distinction between plan and elevation. It

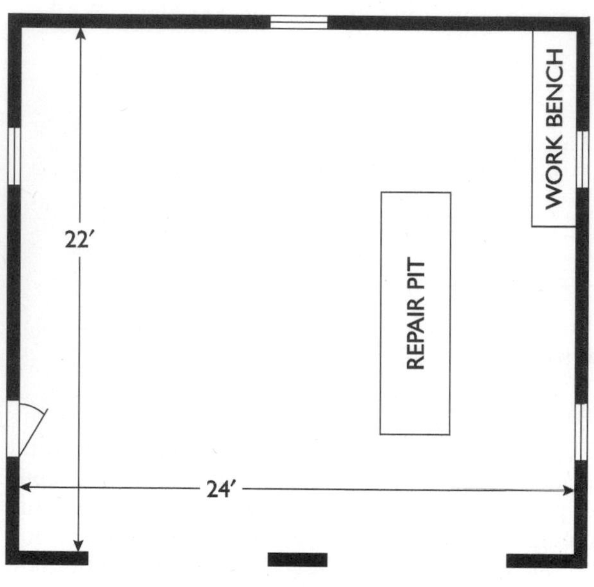

Figure 2-29 Plan drawing of the small garage shown in Figure 2-28.

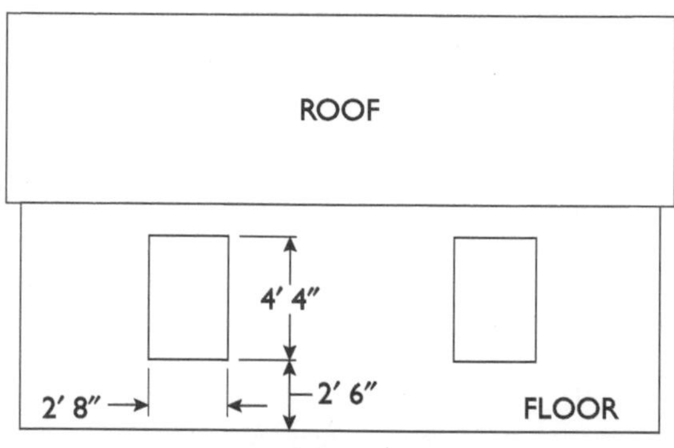

Figure 2-30 Side elevation of the same garage.

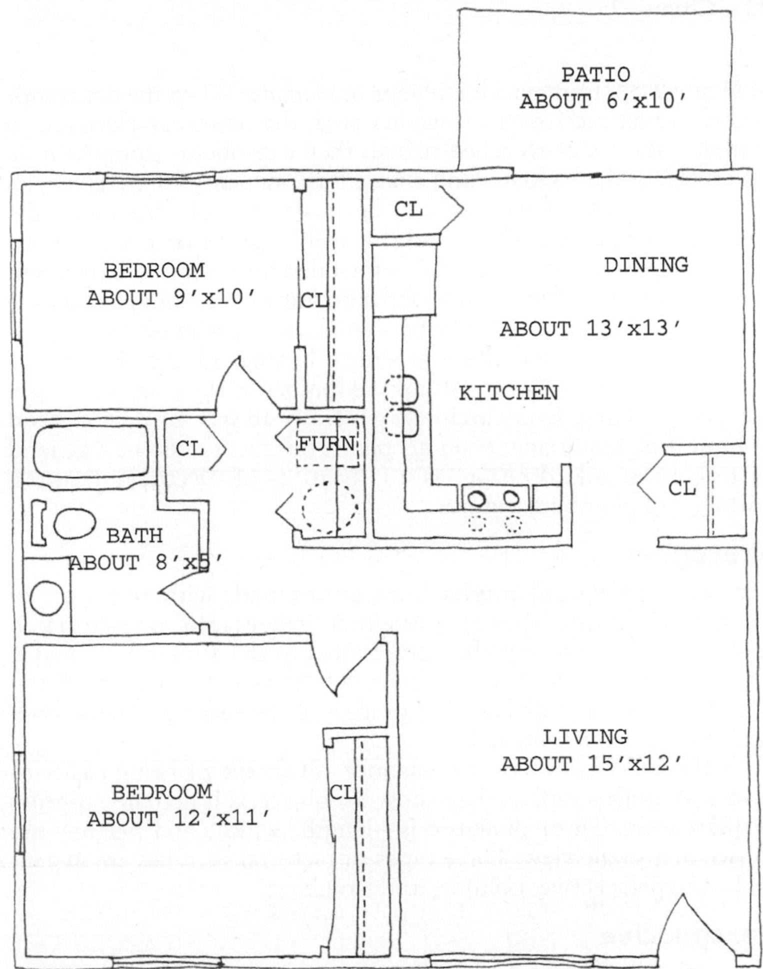

Figure 2-31 Architect's freehand sketch of a house floor plan.

will be noted that several elevations may be necessary to denote which side of the building is shown, these being called front elevation, side elevation, etc. Evidently both the plan and elevations are necessary to fully show the building. Thus, while the plan gives the dimensions of the floor or outside horizontal dimensions and positions of the doors and windows, it is necessary to consult the elevations to determine the height of the building, pitch of roof, etc.

An architect's preliminary freehand sketch of a plan for a house appears in Figure 2-31. Here, just the sizes of the various rooms are stated without any dimension lines to confuse his client. However, if the design is accepted, he will work up a working drawing as shown in Figure 2-32, from which a blueprint is made. When the contractor and carpenter receive this (together with the necessary elevation to complete the set of so-called plans), they can obtain from them an exact idea of the building, and since all the dimensions are given, an accurate estimate of the cost can be made and the structure built. Comparing Figures 2-31 and 2-32, it will be seen that each drawing serves the purpose for which it was intended. In Figure 2-32 the client gets a better view of the outline of the building than would be the case were it obscured by a multiplicity of dimensions, as in Figure 2-31.

On the other hand, the carpenter, having all the dimensions shown in Figure 2-32, can cut up the lumber to the proper size and erect the building. Every architect generally adopts his own method of indicating requirements on his plans, but there must be a general pattern of standardization as to location of electrical fixtures, wiring, and plumbing fixtures.

Sketching

Sketches are freehand graphic illustrations made without the aid of mechanical devices. They may be either orthographic or pictorial in style. Orthographic sketches are probably the simplest to make. The orthographic projection system is followed as in regular drawing, the same types of lines are used, and the same practices, conventions, etc., are followed.

Pictorial sketches have the distinct advantage of being easier to read and understand, as they show an object as it actually appears to the viewer. Three dimensions—length, width, and height—are shown in a single view. Three types of pictorial sketches are in general use—perspective, oblique, and isometric.

Perspective

Perspective sketches most nearly represent what is seen by the eye. The portion of the object closest to the observer is the largest, while the parts farthest away are the smallest. Lines and surfaces become smaller and closer together as the distance from the eye increases, seeming to disappear at a point on an imaginary horizon called a "vanishing point." Perspective sketching is the most difficult type of pictorial sketching, influenced as much by artistic talent as by technical practices or methods. The oblique and isometric sketches are the most practical types for the mechanic to use. See Figure 2-33.

Oblique

The oblique sketch is probably the easiest and most useful style of pictorial sketch for use by the average mechanic. It has the advantage of

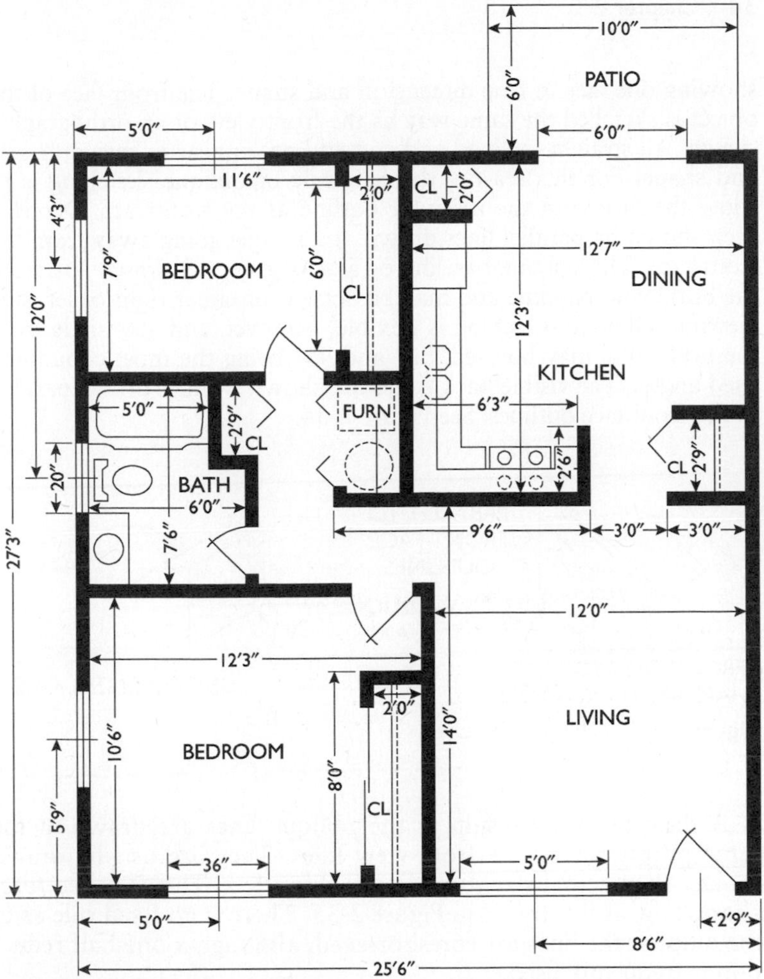

PATIO
10'0"
6'0"
6'0"
5'0"
11'6"
2'0"
CL
2'0"
4'3"
7'9"
BEDROOM
6'0"
CL
12'7"
DINING
12'3"
12'0"
KITCHEN
5'0"
2'9"
CL
FURN
6'3"
2'6"
2'9"
CL
20"
BATH
6'0"
7'6"
9'6"
3'0"
3'0"
27'3"
12'0"
12'3"
2'0"
LIVING
10'6"
BEDROOM
14'0"
8'0"
5'9"
CL
36"
5'0"
5'0"
2'9"
8'6"
25'6"

Figure 2-32 Complete architect's drawing showing all dimensions plus all door and window openings.

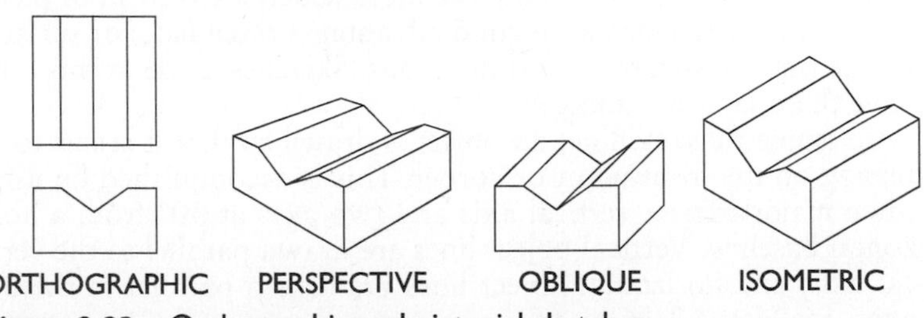

Figure 2-33 Orthographic and pictorial sketches.

showing one face in true dimension and shape. The front face of the object is sketched the same way as the front view of an orthographic sketch. All straight inclined and curved lines appear in their true size and shape. For this reason, the first rule of oblique sketching is to show the face with the irregular outline as the front face. Depth is then shown by parallel lines drawn at an angle going away from the front face. The oblique baseline or axis is generally drawn 45° from the horizontal baseline and may be drawn for either right or left side viewing. Oblique sketching is flexible, however, and any angle from the horizontal may be used, 30° and 45° being the most commonly used angles. The visible back edges are shown by lines drawn parallel to the front face outlines. See Figure 2-34.

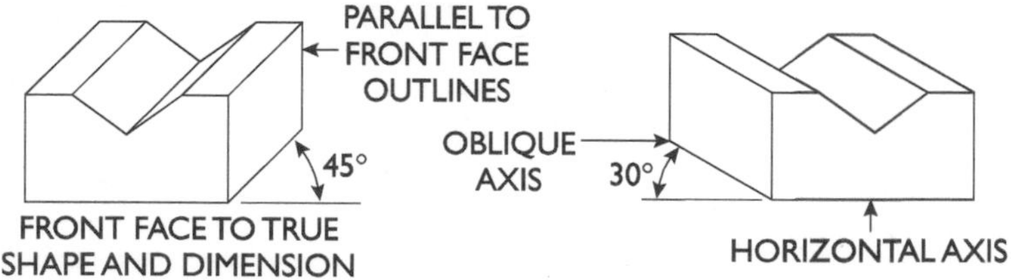

Figure 2-34 Oblique sketches.

A distorted view results if the oblique lines are drawn in the same proportion as the front-view lines. They are usually drawn shorter than actual size to minimize this effect. This is called *foreshortening,* as illustrated in Figure 2-35. There is no fixed rule as to the amount the lines are foreshortened, although a one-half reduction is frequently used.

Isometric

Isometric drawing is probably the most widely used form of pictorial drawing. It presents (to good advantage) three faces or surfaces at one time, drawn to scale dimensions. Sketches made in this form have these same advantages.

In isometric sketching, an object is drawn so that it seems to be resting on the front point or corner. This is accomplished by using three major axes: a vertical axis and two axes at 30° from a horizontal baseline. Vertical object lines are drawn parallel to the vertical axis, and horizontal object lines are drawn parallel to the 30° axes. See Figure 2-36.

The height, width, and depth of the object are measured on three axes. All vertical object lines must be drawn parallel to the vertical axis. All horizontal object lines must be drawn parallel to the 30° axes. The first lines drawn should be to box in the whole object or a major portion. See Figure 2-37.

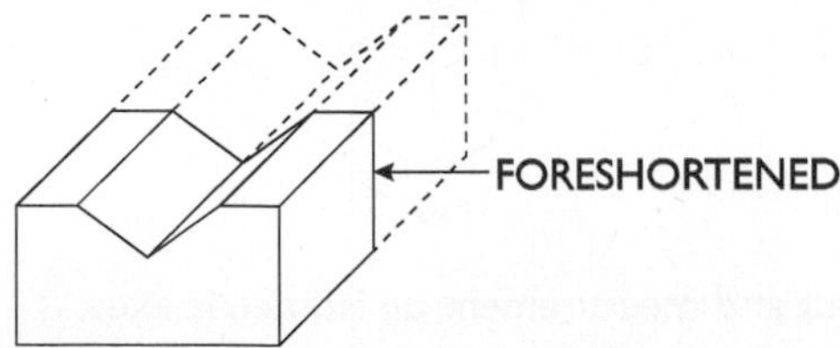

Figure 2-35 Foreshortened oblique sketch.

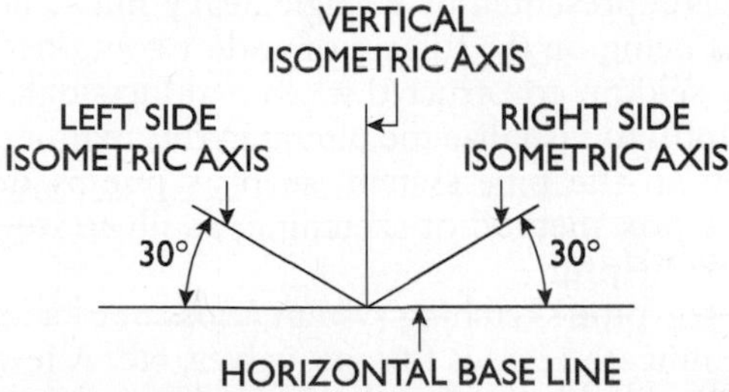

Figure 2-36 Isometric axes.

Measurements can be made only on the three isometric axes. Angular lines, curves, etc., do not appear in true length in an isometric sketch. A simple method of developing these lines is to lay off distances on the isometric axes and join the points to form the line.

Because all vertical and horizontal lines are drawn parallel, the sketch is not in perspective and the object appears slightly distorted.

Isometric Pipe Sketching

One of the most useful forms of isometric sketching is the single-line pipe sketch. With a few simple lines, properly drawn, the location, shape, and direction of pipelines can be clearly illustrated. Valves and fittings, etc., can be clearly shown in their relative positions with symbols. Dimensions and notes, applied to such a sketch, will complete the information, clearly picturing and describing the pipeline or pipe assembly, as shown in Figure 2-38.

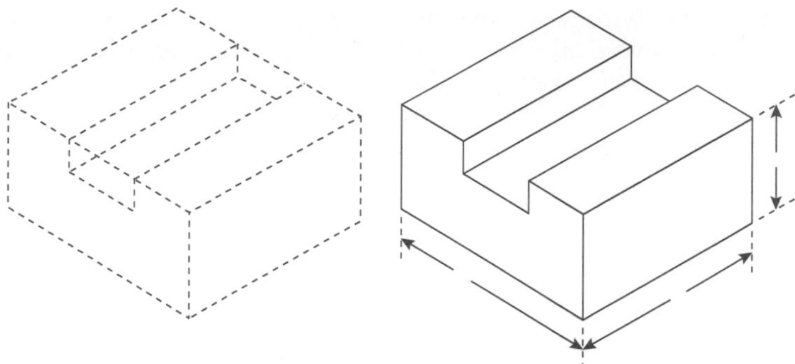

Figure 2-37 Isometric layout and measurement on isometric axes.

Isometric rectangular construction lines, or boxes, can be useful to show piping systems in a manner that is easy to visualize. In this system of sketching, pipe is represented by a single heavy line. The pipelines are thought of as being on the surface of and/or contained in a box. While boxes are seldom constructed when making single-line pipe sketches, it is helpful to visualize the piping in this manner. Also, if other components in the pipe system, such as pumps or tanks, must be shown, the box method of sketching, as illustrated in Figure 2-39, can be most helpful.

In the single-line isometric pipe sketching system, cross marks or symbols are employed to indicate various fittings, valves, etc. A few simple symbols are commonly used, although some organizations have individual practices and tables of symbols. Examples of widely used symbols are given in Figures 2-40 through 2-44.

When sketching the heavy single line, which represents pipe, an organized direction plan must be employed to avoid confusion. As the pipeline or pipelines to be sketched may run in various directions, the plan must first establish the relative directions of the pipe to be sketched and then indicate the directions on the sketch. An *orientation diagram* is the basis for such a plan. The diagram clearly establishes the directions for the sketcher to follow, and the sketcher, by indicating these directions on the sketch, clarifies the sketch for the reader.

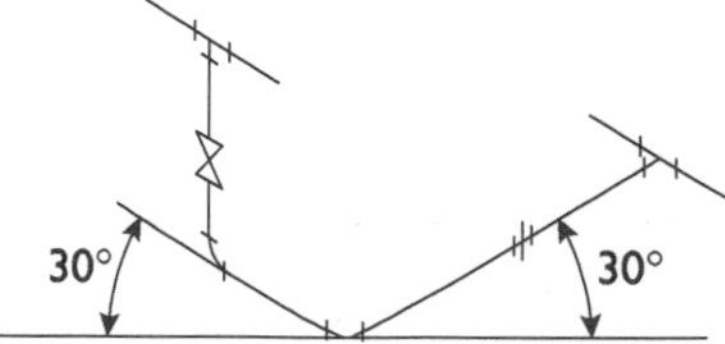

Figure 2-38 Single-line isometric pipe sketch.

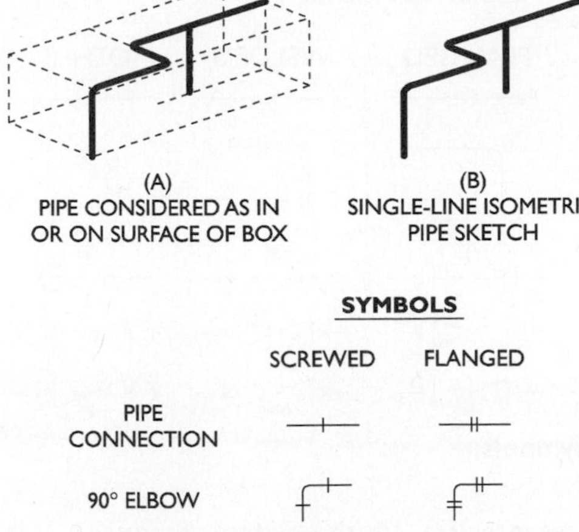

Figure 2-39 (A) Pipe considered as in or on the surface of a box. (B) Single-line isometric pipe sketch.

Figure 2-40 Screwed fittings are illustrated with a single perpendicular line.

Figure 2-41 Flanged fittings are illustrated with double perpendicular cross marks.

Figure 2-42 Welded connections are illustrated with 45-degree cross marks placed on the object lines.

Figure 2-43 Other connections are illustrated with an open circle, usually with a notation as to type.

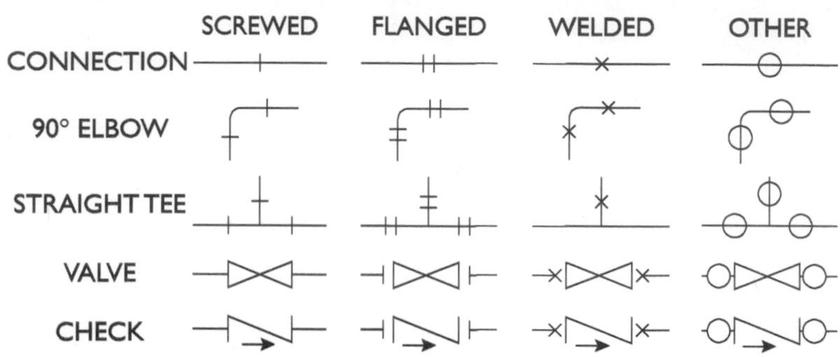

Figure 2-44 Examples of symbols.

The diagram is a representation of the isometric axes, each labeled to indicate relative position and direction. The vertical axis is always labeled UP at the top and DOWN at the bottom. The horizontal axes are labeled to conform to the viewing position. Labels such as FRONT, BACK, LEFT, RIGHT, NORTH, SOUTH, EAST, and WEST may be used, as shown in Figures 2-45 and 2-46.

The point of observation, or viewing point, for an isometric pipe sketch is always

1. Midway between the two horizontal isometric axes.
2. Directly in front of the vertical isometric axis.
3. Slightly above the object being viewed.

Construction and labeling of the orientation diagram to indicate the viewing point and relative directions is an essential preliminary step in the making of a single-line isometric pipe sketch.

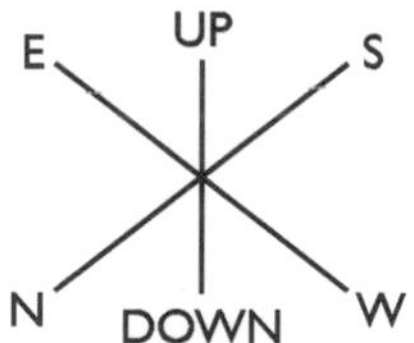

Figure 2-45 The diagram may be viewed as a weather vane with directional points indicated at the ends of the bars. The diagram is drawn in simple form with three axes crossing. Regular (nonisometric) labeling is used.

When selecting a viewing point, it is important to select one that will result in a sketch with maximum clarity. Two sketches of the same system made from different viewing points will result in one being much clearer and easier to understand than the other. The illustrations shown in Figure 2-47 are of the same pipe system sketched from two different viewing points.

When the best viewing point has been selected, the orientation diagram is labeled to indicate directions, and sketching can begin. Following the directions indicated by the diagram, the lines of the sketch are drawn to match the actual directions of the pipelines in the system.

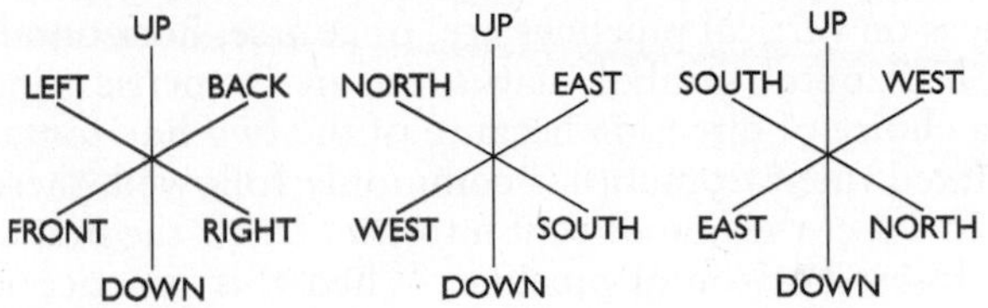

Figure 2-46 Examples of orientation diagrams.

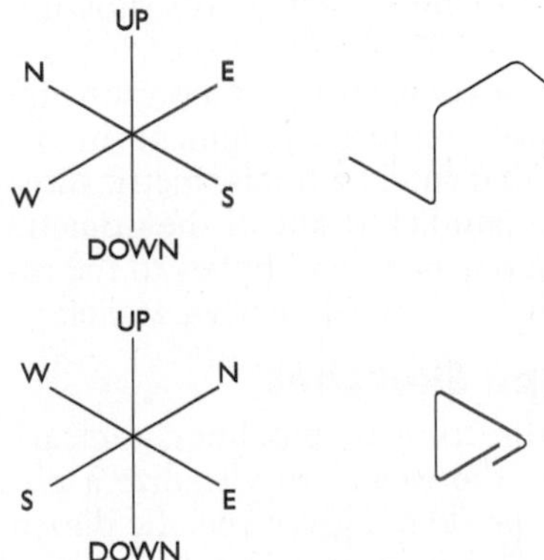

Figure 2-47 Sketching from different viewing points.

Earlier in the chapter it was indicated that the usual practice is to measure actual lengths to scale on the isometric axes when making isometric sketches. This practice need not be followed when making isometric pipe sketches. However, the sketch must be made to look like the pipe system; thus, it cannot be completely out of proportion. It is a good idea for the beginner to make rough measurements to scale until the ability is developed to approximate them with some accuracy.

The representation of fittings and/or connections is done with symbols, the most widely used being the short cross line. The direction in which the cross line should be drawn depends on the direction of the pipeline being sketched. An easy way to determine the cross-line direction is to visualize the pipeline in position and all connections made with flanges. The flange faces on horizontal runs of pipe are, of course, vertical; therefore, when making fitting marks for horizontal pipe runs, they are always in the UP and DOWN (vertical) direction. As there is only one vertical direction in isometric pipe sketching, the marks representing vertical flange faces (flanges on horizontal pipe runs) are always made parallel to the UP and DOWN axis, as shown in Figure 2-48.

The flange faces on vertical pipelines are, of course, horizontal. When drawing horizontal symbol marks on an isometric pipe sketch, there is a choice of direction because of the two horizontal axes. While no fixed rule or practice is commonly followed, there is a simple way to select the line direction to use. Draw the marks parallel to the closest horizontal pipeline. When this practice is followed, the choice of direction confusion is eliminated. Another advantage of this practice is the orderly, uniform appearance of the sketch that results. Illustrations of this practice are shown in Figure 2-49.

The representation of fittings and/or connections on nonisometric lines (neither horizontal nor vertical) requires judgment by the sketcher. In most cases, the pipelines that run in a nonisometric direction on the sketch will appear nearly parallel to one of the isometric axes. The sketcher should follow the practice that applies to the isometric line to which the nonisometric line is most nearly parallel.

Dimensioning Isometric Pipe Sketches

By showing relative position and direction of pipelines, a clearly drawn isometric pipe sketch enables the viewer to visualize a pipe system. In addition, the isometric pipe sketch gives specific dimensional information. To avoid confusion in presenting dimensional information, a basic rule that commonly applies to pipe drawing is followed in isometric pipe sketching. This rule is simply that all dimensions are shown to either "centerlines" or to "points." This means that actual pipe lengths are not shown. The viewer must calculate actual pipe lengths by making allowances for connections, fittings, etc (Figure 2-28).

The distance between a centerline and points is shown with extension and dimension lines. The dimension lines indicate the direction in which the dimension applies, and the extension lines

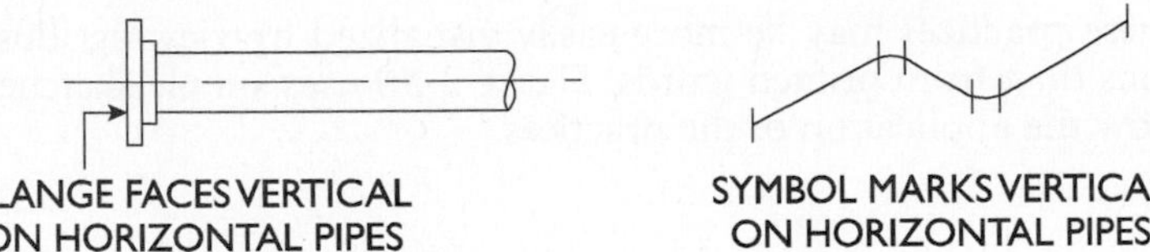

Figure 2-48 Flange faces on horizontal pipe.

refer the dimension to points or lines. Because there are usually several ways measurements between two points may be shown, judgment must be used when dimensioning a pipe sketch. While there are few hard-and-fast rules, some recommended practices should be followed (Figure 2-49):

- Extension and dimension lines should be parallel to isometric axes, with the exception of lines that refer to pipes that do not run horizontally or vertically (nonisometric).
- Extension lines should be drawn in the plane of the surface to which they apply and be parallel to each other.
- Dimension lines should be parallel to the direction of measurement.

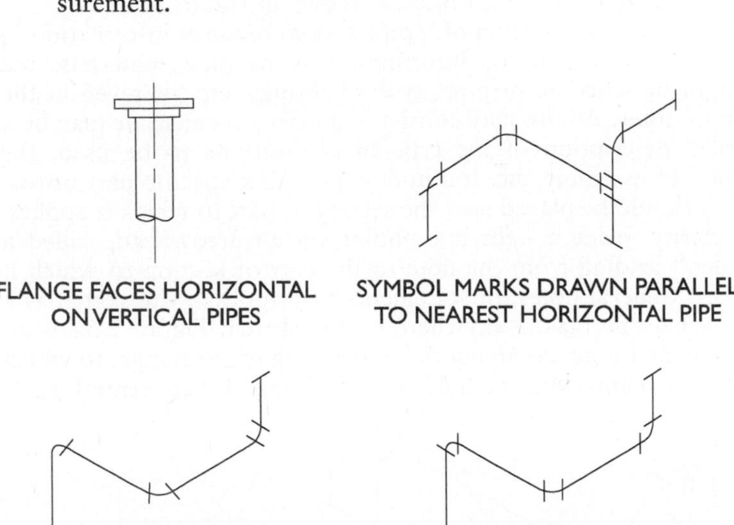

Figure 2-49 Flange faces on vertical pipe.

These practices may be more easily visualized by viewing illustrations than from printed words. Figure 2-50 uses simple sketches to show the application of the practices.

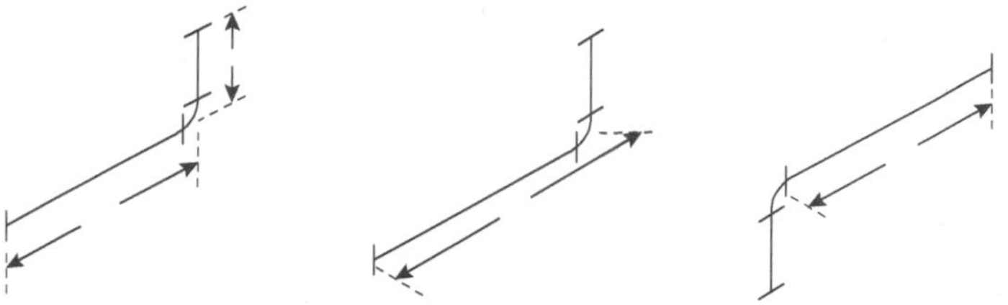

GOOD PRACTICE POOR PRACTICE POOR PRACTICE

Figure 2-50 Single-line pipe sketch dimensioning.

While dimension lines may in some cases extend to objective lines, with arrowheads ending on the object, preferred practice is to have the arrowheads end on extension or centerlines. Crossing extension lines is permissible if it cannot be avoided, but dimension lines should not cross extension lines or dimension lines. A good practice to follow for clarity of dimensioning is that all extension and dimension lines be placed outside the object, as shown in Figure 2-51.

A complete description of a pipe system requires information in addition to dimensions. Information as to pipe materials, the strength or schedule of pipe, type of fittings, etc., is given in the form of notes. A note may consist of a brief statement or may be a detailed description of the type of connections to be used, the method of assembly, etc. If a note applies to a specific part or section, it should be placed near the section or part to which it applies. For clarity, place a light line ending on an arrowhead, called a "leader," leading from the note to the part or section to which it applies. General notes, or statements that apply to the sketch as a whole, may be placed anywhere on the sketch. Figure 2-52 is an example of a note specifying the positioning of the flange, to which the leader points with the bolt holes "on" the vertical centerline.

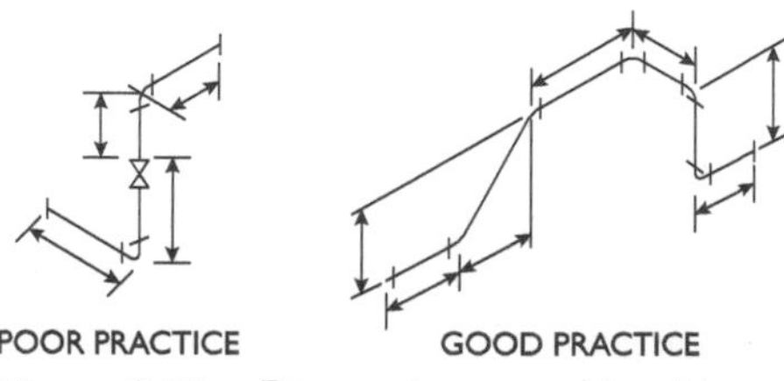

Figure 2-51 Dimensions outside object.

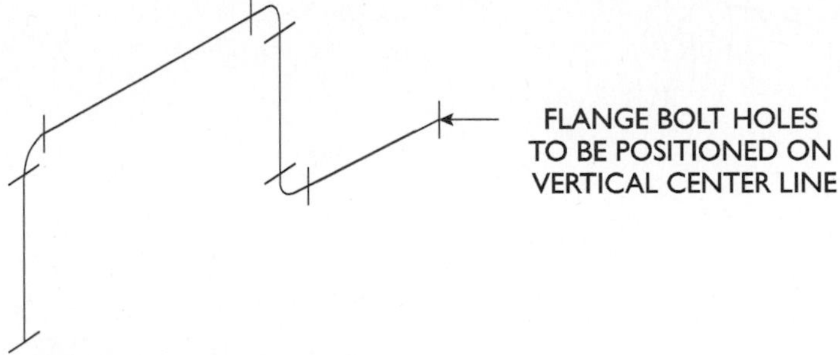

Figure 2-52 Sketch notes.

Field Sketching

Regardless of the skill and proficiency that a craftsperson has in building, repairing, or maintaining equipment, the ability to make a sketch to convey ideas to others is a necessity. Field sketches are critical to convey the installation instructions and show how something was disassembled.

A popular misconception about sketching is that a person must be artistic or able to "draw." This is not so at all. Mechanical, electrical, and piping sketches use simple lines and symbols to depict objects. No artistic shading or other embellishments are needed to show a truthful representation of the object.

Sketching an Object Freehand

All that is needed to produce a freehand sketch are a few good soft-lead pencils, an eraser, and some coordinate paper with four or eight spaces to the inch (often called a quadrille pad). Sketching is just making lines and curves on a piece of paper. It is difficult to draw long, straight lines on paper. Short, light strokes are better. One trick is to place the pencil in a normal writing position and use the third and fourth fingers to ride along the edge of the paper pad to keep the line parallel and straight. Rotating the pad 90° allows the person to use the same technique to make lines perpendicular to the first line that was drawn.

Figure 2-53 shows a picture of a coupling half that was purchased at minimum bore and needs to be bored out for installation on a motor shaft.

Using the sketching techniques just described, use the various edges of the paper to draw a series of straight lines to define some of the borders of the object in the front view. Use very light lines to "box" in the circular shape of the coupling in the side view. An example of this technique is shown in Figure 2-54.

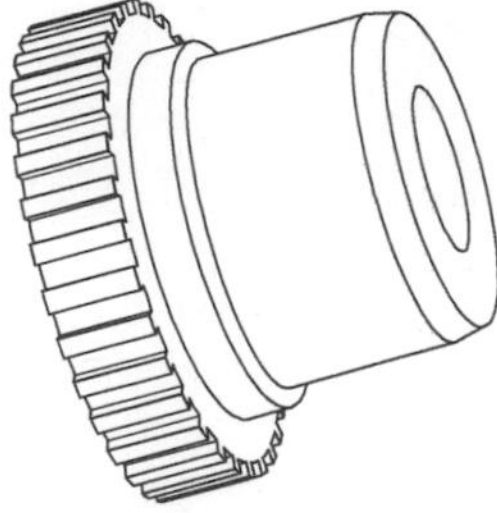

Figure 2-53 Coupling half.

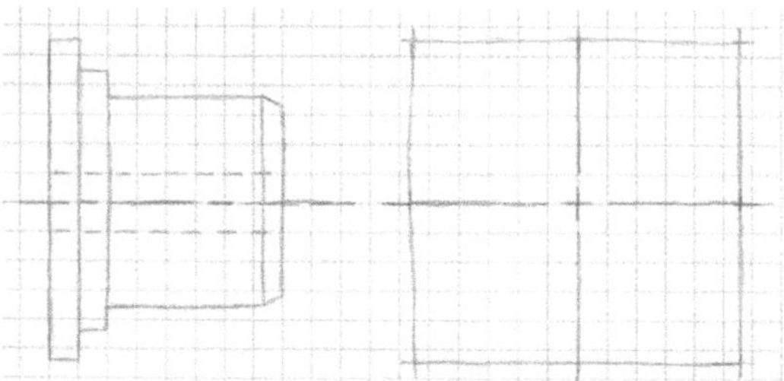

Figure 2-54 Boxing in the shape.

Next, block the circular parts of the coupling by sketching four circular arcs at four points where the circles are tangent to the light lines of the square (Figure 2-55).

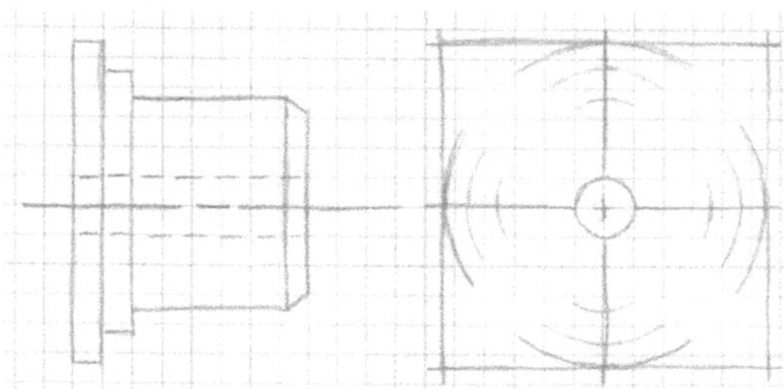

Figure 2-55 Sketching the arcs of the circle.

Finally, using heavier strokes and a good eraser, complete the sketch without instructions or dimensions, as shown in Figure 2-56.

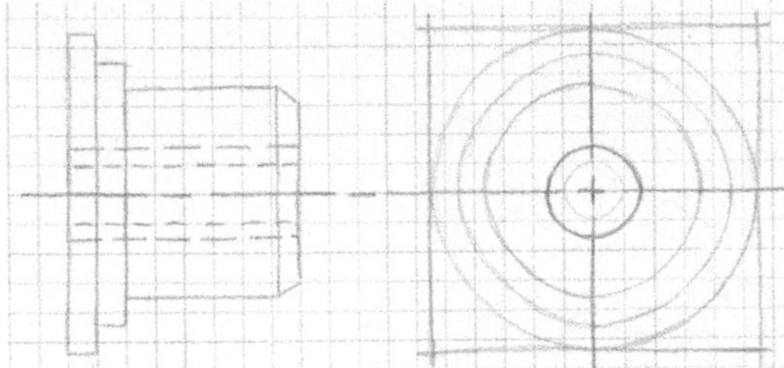

Figure 2-56 Completed sketch without dimensions.

The sketch is now dimensioned. Specific information is required that would allow the machine shop to bore out and place a key seat in the coupling and allow it to be used on the motor (Figure 2-57).

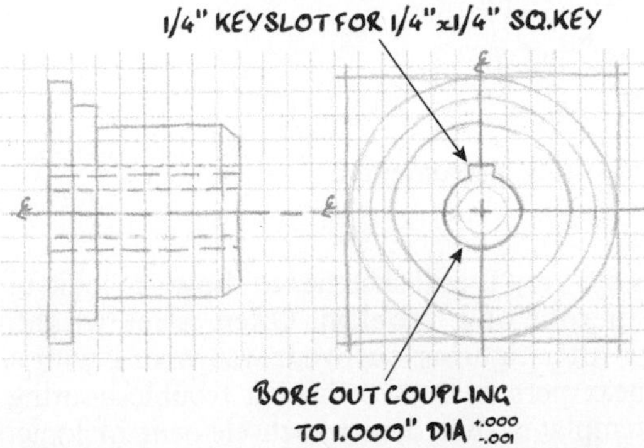

Figure 2-57 Showing dimensions and instructions.

Sketching an Object Using Simple Instruments

Adding a few items, such as a straightedge and compass or even a circle template, will greatly improve a sketch. Many people have trouble lettering a sketch and find the letter template helpful. If the straightedge that is used has measured increments, such as a six-inch rule, then these can be used to proportion the lines more adequately to make the sketch true to life. The compass and circle template make these elements look much more professional. Figure 2-58 shows the same coupling half drawn with the use of a straightedge and compass. The sketch conveys the same information as Figure 2-57, but it has a more professional look and probably took less time to draw.

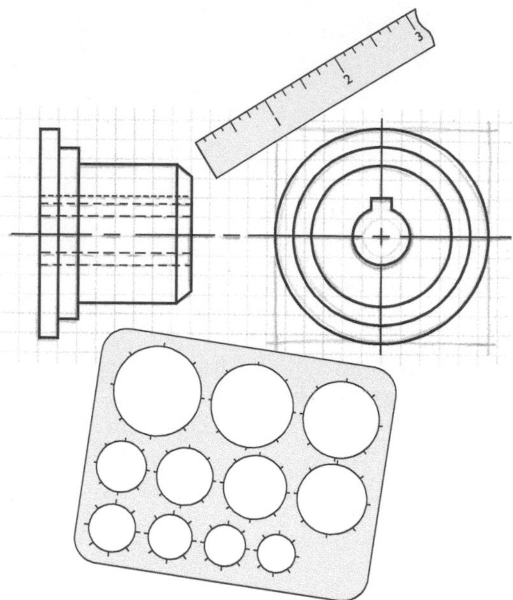

Figure 2-58 Better sketch using instruments.

Troubleshooting trees are very useful sketches to show the logic of determining the extent or root of a problem. Often, a millwright, electrician, or mechanic—after figuring out what was wrong with a machine—can aid the next person by sketching a troubleshooting tree. There are plastic templates that depict each element of logic. Figure 2-59 shows a typical tree used to troubleshoot a simple "light is out" work order. While a craftsperson would probably not develop a troubleshooting tree for this problem, the diagram does depict the proper use of this type of template in the sketching process.

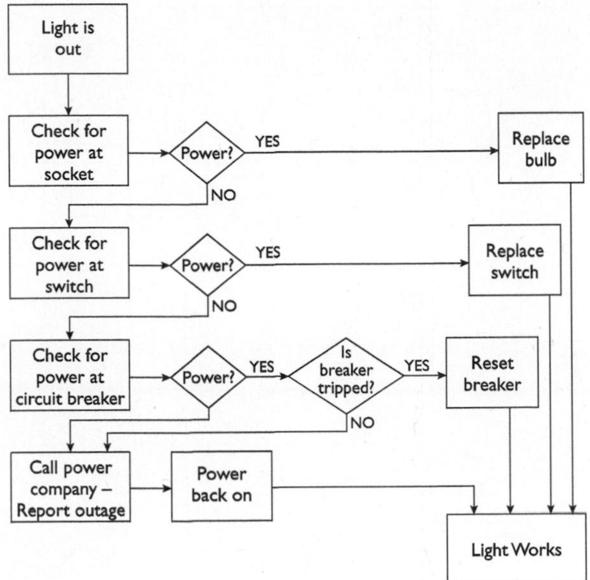

Figure 2-59 Typical troubleshooting tree diagram.

Chapter 3

The Basic Toolbox

The proper use of hand tools is the key to accomplishing many successful jobs. Although each mechanical trade might require specialized tools, a group of basic tools will be the core of any craftsperson's toolbox. These basic tools include the following:

- Wrenches (open end, box, combination)
- Socket wrenches
- Nut drivers
- Adjustable wrenches
- Screwdrivers
- Vises
- Clamps

This set will allow the mechanic to perform most basic repair and installation jobs. At the end of this chapter is a complete list of all the tools needed to complete more complex maintenance jobs that the mechanic or millwright might encounter in industry.

Wrenches—Open End, Box, Combination

Wrenches are one of the most widely used hand tools. They are used for holding and turning bolts, cap screws, nuts, and various other threaded components of a machine. It is important to make sure the wrench holds the nut or bolt with an exact fit. Whenever possible, it is important to *pull* on a wrench handle and keep a good footing to prevent a fall if parts let go in a hurry. Typical types of wrenches needed in the basic toolbox include box wrenches, open-end wrenches, and combination box/open-end wrenches from 1/4 in. to 1 1/4 in. in increments of 1/16 in. to accomplish most standard jobs (Figure 3-1). The same types of wrenches in metric openings would span from 7 to 32 mm. Metric wrenches are stepped in 1-mm increments, but the basic toolbox would usually omit the 20-mm, 29-mm, and 31-mm openings because they are used so infrequently.

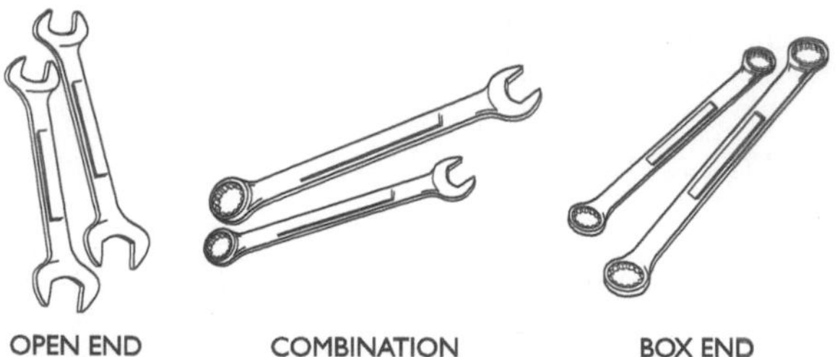

Figure 3-1 Open-end, box, and combination wrenches.

Socket Wrenches

Socket wrenches greatly speed up many jobs. The basic toolbox should contain sockets ranging from 5⁄16 to 1 1⁄4 in. in 1⁄16-in. increments. Socket wrenches are usually sold in sets where the square drive of the ratchet is used to identify the set. Ratchets are typically made with 1⁄4-, 3⁄8-, 1⁄2-, 3⁄4-, and 1-in. square drives. The larger the drive, the greater the capacity of the wrench set. For most work, the basic toolbox needs to have a 1⁄4-in. drive set, a 3⁄8-in. drive set, and a 1⁄2-in. drive set. Larger drives, such as the 3⁄4 in. and 1 in., can be stocked in a central tool crib or kept with a supervisor for use by any craftsperson when the need for these larger units is merited. Sockets are made in regular depth and extra deep (Figure 3-2). Openings may be 12-, 8-, 6-point, or square for the type of work that the wrench set must handle. In addition, jobs that involve the use of an electric or pneumatic impact wrench need extra-heavy wall sockets that will hold up under heavy pounding from the tool.

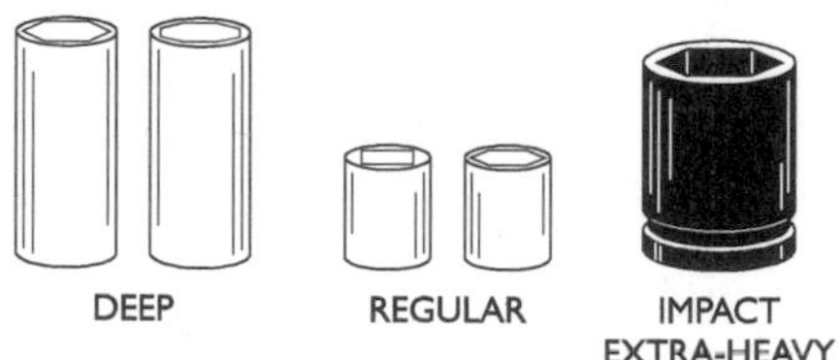

Figure 3-2 Sockets.

Nut Drivers

Nut drivers are a must for electricians, but they are also handy for other jobs. A nut driver looks like a screwdriver but has a socket at the end of the shank (Figure 3-3). The basic toolbox should have a nut driver set, which includes the 3/16-, 1/4-, 5/16-, 11/32-, 3/8-, 7/16-, 1/2-, 17/32-, 9/16-, 5/8-, 11/16-, and 3/4-in. sizes. Often overlooked, these tools speed up the repair or adjustment of equipment and surpass the ability of a ratchet set for loosening or tightening light- to medium-duty assemblies.

Figure 3-3 Using a nut driver.

Adjustable Wrenches

Adjustable wrenches are sometimes called "fits all" wrenches because they cover such a wide range of jobs. These tools are open-end wrenches with an expandable jaw that allows the wrench to be used on many sizes of bolts or nuts. They are available in lengths from 4 to 24 in. Some adjustable wrenches allow the jaw to be locked. These are available in lengths from 6 through 12 in. When using an adjustable wrench, always make sure you apply the force to the fixed jaw (Figure 3-4).

Practically all manufacturers supply parts and repair kits to refurbish adjustable wrenches. The kits are relatively inexpensive compared to the purchase of a brand-new wrench. The basic toolbox should include adjustable wrenches with lengths of 4, 6, 8, and 12 in. These four sizes can cover a wide range of jobs.

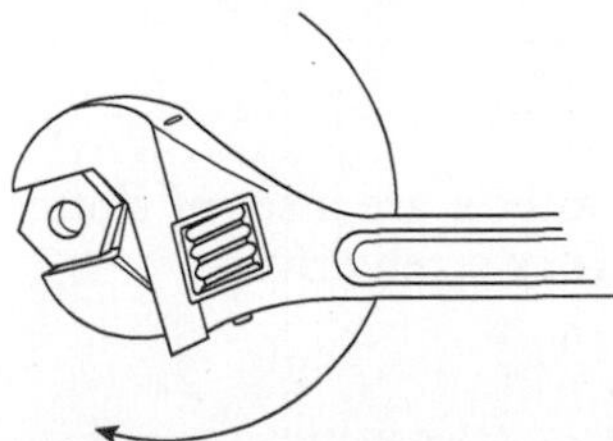

Figure 3-4 Apply the force to the correct jaw.

Screwdrivers

Screwdrivers are the most used tools in a toolbox. They are intended for one simple use: driving and withdrawing threaded fasteners such as wood screws, machine screws, and self-tapping screws. Correct use of a screwdriver involves matching the size of the screwdriver to the job and matching the type of screwdriver to the head of the screw. The first screws that were developed had slotted heads. The proper screwdriver for these types of screws is called a conventional screwdriver, and it can be classified by tip width and blade length. Generally, the longer the screwdriver is, the wider the tip—but not always. Cabinet screwdrivers have long shanks but narrow tips because they are used to drive screws into recessed and counterbored openings that are found in furniture and cabinets. Stubby screwdrivers are available that have very wide tips for use in tight spaces. Most conventional screwdriver tips are tapered. The tip thickness determines the size of the screw that the screwdriver will drive without damaging the screw slot. The taper permits the screwdriver to drive more than one size of screw.

The world of screwdrivers has radically changed since the slotted head screw. Probably the first "new" screw that was developed was the Phillips head screw—a recessed slot design. The recessed slot screw allows a more positive nonslipping drive-up while attaching the fastener. The most common style of recessed head is the Phillips screw. Other recessed screws and screwdrivers are shown in Figure 3-5.

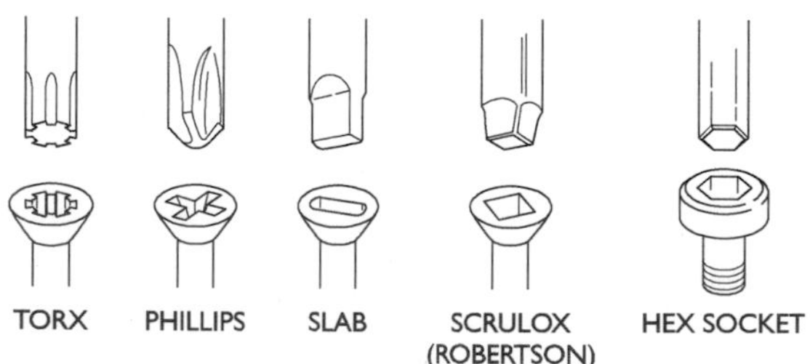

Figure 3-5 Screws and screwdrivers.

Essential screwdrivers for the basic toolbox are a set of slotted screwdrivers, Phillips screwdrivers, and Torx screwdrivers.

Vises

Although a vise would not necessarily be part of the basic toolbox, it certainly would be essential on the tool bench. There are four basic categories of vises: machinist's vise, woodworker's vise, pipe vise, and drill press vise. Each has its purpose and benefits.

The *machinist's vise* is the strongest vise made (Figure 3-6). It is designed to withstand the great strains in industrial work. Models are made with stationary bases and swivel bases and can be equipped with pipe jaws as well as interchangeable jaws. Usually jaw widths start at 3 in. and go to 8 in. for large jobs. Copper (or brass) jaw caps are available to prevent marring of the work.

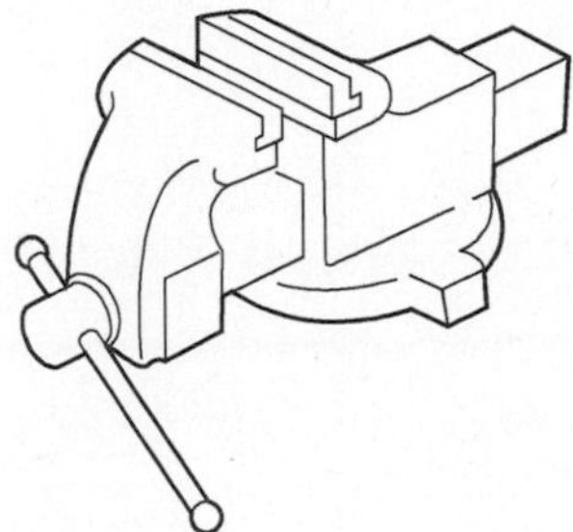

Figure 3-6 Machinist's vise.

The *woodworker's vise* is a quick-acting vise that bolts to the underside of the workbench (Figure 3-7). These vises are equipped with a rapid-action nut that allows the movable jaw to be moved in and out quickly, with the final tightening accomplished by turning the handle a half-turn or so. The jaws on the vise are large, often 7 to 8 in. The jaws are usually metal, but they are intended to be lined with wood (which allows replacement) to protect the work.

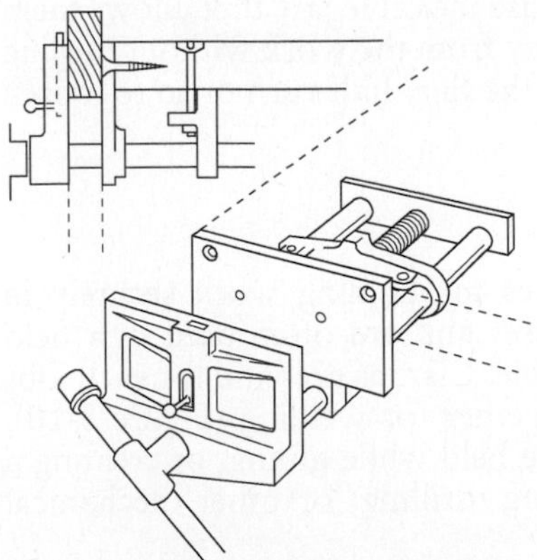

Figure 3-7 Woodworker's vise.

A *pipe vise* is designed to hold pipe or other round material. Most can hold pipe up to 8 in. in diameter. They are usually available with a tripod mount to allow the vise to be portable, but they can be bolted to a workbench as well. The most popular type uses a chain for clamping and can also be used to hold irregular work (Figure 3-8).

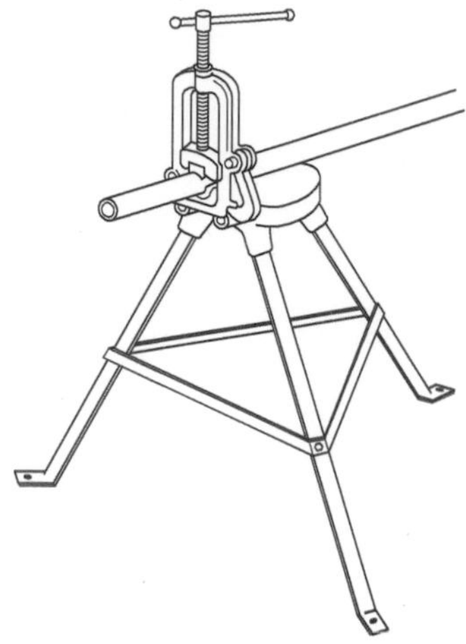

Figure 3-8 Portable pipe vise.

The *drill press vise* is made to accept round, square, or oddly shaped work and hold it firmly in place for drilling (Figure 3-9). The better vises often have a quick-release movable jaw that allows them to be moved up to the work or away from the work without turning the handle. The handle is used for the final half-turn or so to loosen or tighten the jaws.

Clamps

Clamps serve as temporary devices for holding work securely in place. They are first cousins to vises and are often used in a field location where vises are not available. Clamps are vital for such jobs as locking two pieces of metal together for welding (Figure 3-10), securing two pieces that need to be held while gluing, or creating a third hand to allow ease of sawing, drilling, or other mechanical processes.

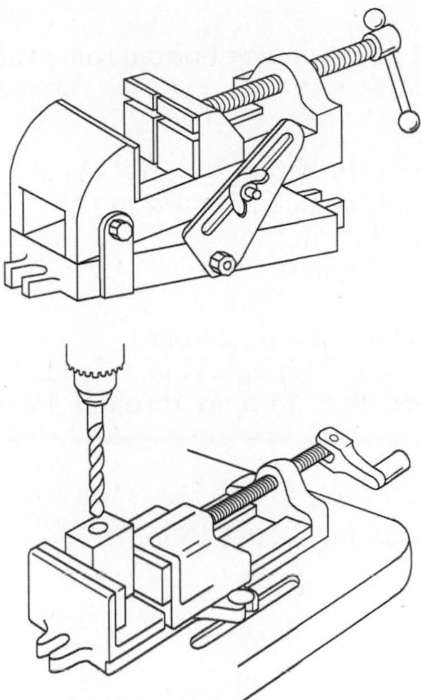

Figure 3-9 Drill press vise.

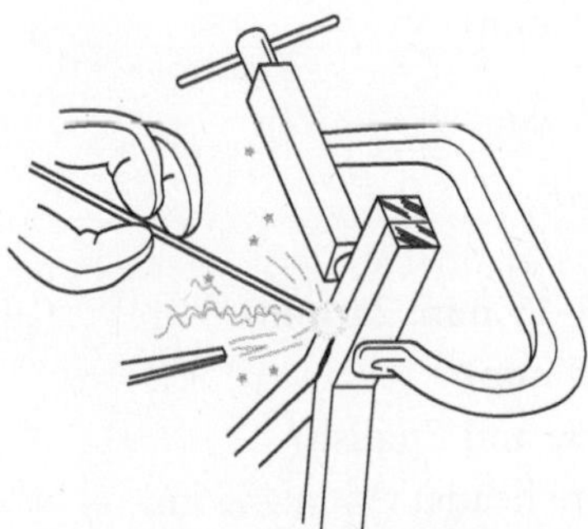

Figure 3-10 Clamping provides a third hand to the job.

Mechanical Trades Tool List

The following is a detailed tool list that can be used for stocking the basic toolbox. Although it is certainly possible to add more tools, this list represents the minimum needed in most industrial or commercial work to perform basic installation, repair, or modification work.

Box and Traveling Pouch

1 Toolbox—10-drawer top cabinet and 8-drawer bottom roll-away
1 Leather Tool Pouch with belt

Wrenches

1 set Open-End Wrenches (1⁄4 in. × 5⁄16 in. through 1 in. × 11⁄4 in.)
1 set Box Wrenches (1⁄4 in. × 5⁄16 in. through 1 in. × 11⁄4 in.)
1 set Combination Wrenches (1⁄4 in.–15⁄16 in.)
1 set Open-End Metric Wrenches (7–25 mm)
1 set Box Metric Wrenches (7–25 mm)
1 set Combination Metric Wrenches (6–22 mm, 24 mm)
1 set Flare Nut Wrenches (3⁄8 in. × 7⁄16 in. through 3⁄4 in. × 7⁄8 in.)
1 set Flare Nut Metric Wrenches (9 × 11 mm through 19 × 21 mm)
1 Folding Allen Wrench Set (small)
1 Folding Allen Wrench Set (large)
5 Adjustable Wrenches (4 in., 8 in., 10 in., 12 in., 18 in.)

Sockets

1 set 1⁄4 in. Drive Socket Set

- With shallow sockets (5⁄32 in. through 3⁄4 in.)
- With deep sockets (3⁄16 in. through 3⁄4 in.)
- With metric shallow sockets (4, 5, 5.5, 6–15 mm)
- With metric deep sockets (4–15 mm)

1 set 3⁄8 in. Drive Socket Set

- With shallow sockets (5⁄16–1 in.)
- With deep sockets (5⁄16–1 in.)
- With metric shallow sockets (9–19 mm, 21 mm)
- With metric deep sockets (4–15 mm)
- With Allen Head sockets (metric and English)
- With 8-point sockets (for square heads) (1⁄4 in.–1⁄2 in.)

1 set 1⁄2 in. Drive Socket Set

- With shallow sockets (3⁄8 in.–11⁄4 in.)
- With deep sockets (3⁄8 in.–11⁄4 in.)
- With metric shallow sockets (9–28 mm + 30 mm + 32 mm)
- With metric deep sockets (13–22 mm + 24 mm)

Pliers
3 pair Slip Joint Pliers ($6\frac{3}{4}$ in., 8 in., 10 in.)
2 pair Arc Joint Pliers (7 in., $9\frac{1}{2}$ in.)
2 pair Locking Pliers (8 in. Straight Jaw)
2 pair Locking Pliers (6 in. Curved Jaw)
1 set Snap Ring Pliers

Screwdrivers and Nut Drivers
1 set Straight Screwdrivers ($\frac{1}{8}$ in. × 2 in., $\frac{3}{16}$ in. × 6 in., $\frac{1}{4}$ in. × 8 in., $\frac{1}{4}$ in. stubby)
1 set Phillips Head Screwdrivers (#1, #2, #3)
1 set Torx Screwdrivers (T-10, T-15, T-20, T-25, T-27, T-30)
1 set Nut Drivers

Alignment and Prying
1 Double-Faced Engineer Hammer (48 oz)
1 Rolling Wedge Bar (16 in.)
1 set Screwdriver Type Pry Bars

Scraping, Filing, Extracting, Punching
1 complete set Thread Files (metric and English)
1 set Screw Extractors
1 set Punches
1 set Cold Chisels
1 set Files (for filing metal)
1 Center Punch
1 Gasket Scraper ($1\frac{1}{2}$ in. face)

Hammers
4 Ball Peen Hammers (8, 12, 16, 30 oz)
1 Claw Hammer, Curved Claw (16 oz)
2 Soft-Faced Mallets (24 oz and 12 oz)

Leveling and Measuring
1 Line Level
1 Torpedo Level (9 in.)
1 Level (24 in.)
1 Machinist's Scale (6 in.)
1 English/Metric Dial Caliper (0–6 in., 0–150 mm)
1 Micrometer (0–1 in.)
1 Plumb Bob ($4\frac{1}{2}$ in. with line)
1 Chalk Line Reel
1 Combination Square
1 Carpenter's Square

1 25 ft Tape (3/4 in. wide)
1 set Feeler Gauges (combination inch and metric)

Cutting and Clamping

1 Utility Knife
1 pair Pipe and Duct Snips (compound action)
1 Hack Saw with blades
2 "C" Clamps (3 in.)
2 "C" Clamps (4 in.)
2 "C" Clamps (6 in.)

Pipe and Tubing

2—10 in. Pipe Wrenches (Aluminum Handle)
2—14 in. Pipe Wrenches (Aluminum Handle)
2—18 in. Pipe Wrenches (Aluminum Handle)
2 Strap Wrenches (6 in. and 12 in.)
1 Tubing Cutter (1/4–1 in.)
1 set Pipe Extractors

Electrical

1 pair Side Cutters (9 1/4 in.)
1 pair Wire Strippers
1 pair Diagonal Cutting Pliers (8 in.)
1 pair Long-Nosed Pliers (8 in.)
1 pair Needle-Nosed Pliers (6 in.)
1 pair Bent Needle-Nosed Pliers (4 1/2 in.)
1 Electrical Multi-Tester (Volts, Ohms, Continuity)
1 Flashlight

Troubleshooting

1 Mechanics Stethoscope
1 Infrared Thermometer

Chapter 4

Portable Power Tools

The field of portable power tools covers everything from small electric hand tools to heavy-duty drilling, grinding, and driving tools. Tool manufacturers have made the largest strides in the field of battery-powered tools.

Battery-Powered Tools

Cordless tools were a novelty when they first arrived on the scene. Some of the earliest entrants to the battery-powered tool market were flimsy and ineffective.

Perhaps the first successful battery-operated portable power tool used in industry was the drill motor, first introduced with two batteries and a charger (Figure 4-1). One battery charged while the other was in use with the tool. Mechanics, recognizing the usefulness of cordless tools, demanded longer battery life between recharging and more power from the tool, but they also wanted lighter weight to promote ease of use. They also sought tools that provided more functions than just drilling.

Figure 4-1 Cordless drill kit.
(Courtesy Milwaukee Electric Tools)

The tool industry responded with drastic and rapid improvements. Currently, battery-powered tools are popular because of their portability, increased power, longer battery life, and overall convenience. Cordless tools are particularly effective where work must be done overhead or in hard-to-reach locations. Their weight and size have been reduced as battery technology has advanced.

The list of effective cordless tools includes screwdriver/drill, saber saw, jigsaw, grinder, sander, soldering iron, and impact wrench. Many tools are now available in kits where the same battery provides power to a range of different tools (Figure 4-2).

Figure 4-2 Multi-purpose kit.
(Courtesy Milwaukee Electric Tools)

Run time from any cordless tool is directly proportional to the amp-hour rating of the battery. The amp-hour rating is like the size of a gas tank in a truck—the larger the tank (amp-hours), the farther the truck can go.

Cordless tools are usually powered by nickel-cadmium batteries (called NI-CADs) or nickel-metal-hydride batteries (abbreviated Ni-Mh). A battery pack has individual cells, each putting out 1.2 volts, soldered together in series.

The working life of a cordless tool battery depends, in large part, on how that battery is charged and especially on how the temperature of the cells in a battery pack rises while charging (Figure 4-3). Heat is the enemy of both Ni-Mh and NI-CAD. If there is too much internal heat, battery lifespan can drop to less than half the potential 1200 to 1500 charge/discharged cycles. Battery chargers for heavy-duty industrial tools are now surprisingly sophisticated, monitoring dozens of parameters during each charge cycle. There are two reasons for this. The first is to minimize damaging heat buildup in cells so the battery cycle life will meet its potential. The other reason is to provide a troubleshooting diagnostic service for that time when the inevitable battery problem does arise.

To extend battery life, avoid anything that boosts tool load and current draw beyond what is necessary, thus reducing excessive cell temperatures. Keeping bits and blades sharp helps achieve this, but it is especially important to avoid prolonged stall conditions with any cordless tool; this is when the motor is loaded but the bit or blade it is driving has become stuck. This battery-frying situation can cause momentary current draw to spike up as high as 70 or 80 amps, with a corresponding drop in battery pack life.

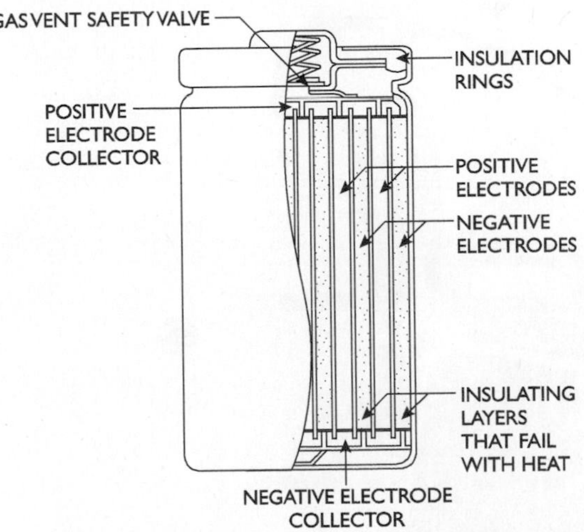

Figure 4-3 Cutaway of a rechargeable battery.

Many "old timers" in the mechanical trades business suggest that tools should be purchased with a total of three rechargeable batteries. One battery can be in use with the tool, the recently discharged battery can sit and cool off, and the third battery (already cooled off) can be in the charger. The cost incurred by the purchase of the third battery is more than offset by the increase in the lifespan of all three batteries because they are never placed in the charger until they cool off.

Electric-Powered Tools

Most mechanics have experience with the more common types of electric-powered tools, and they are knowledgeable in their correct usage and safe operation. There are, however, some more specialized types that mechanics may use only occasionally. Because these specialized tools—as well as all other power tools—are relatively high speed and use sharp-edged cutters, safe and efficient operation requires knowledge and understanding of both the power unit and the auxiliary parts, tools, and so on.

Electromagnetic Drill Press

The *electromagnetic drill press* is the basic equipment used for magnetic drilling. It can be described as a portable drilling machine incorporating an electromagnet with the ability to fasten the machine to ferrous metal work surfaces (Figure 4-4). The

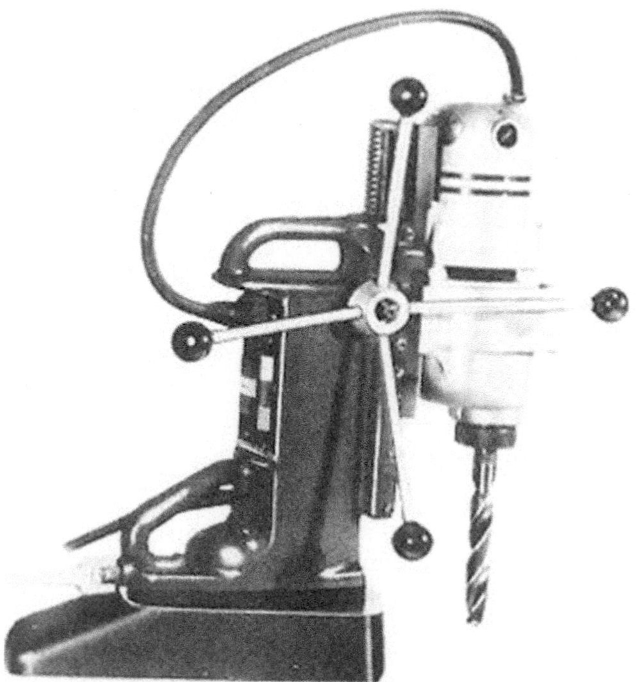

Figure 4-4 Magnetic drill press. *(Courtesy Milwaukee Electric Tools)*

magnetic drill allows you to bring drilling equipment to the work, rather than bring the work to the drilling machine. A major advantage over the common portable drill motor is that it is secured in positive position electromagnetically, rather than depending on the strength and steadiness of the mechanic. This feature allows the mechanic to drill holes with a much greater degree of precision with respect to size, location, and direction, with little operator fatigue.

Magnetic drilling is limited to flat metal surfaces large enough to accommodate the magnetic base in the area where the hole or holes are to be drilled. The work area should be cleaned of chips and dirt to ensure good mating of the magnetic base to the work surface. The unit is placed in the appropriate position, and the drill point is aligned with the center point location. When proper alignment has been established, the magnet is energized to secure the unit. A pilot hold is recommended for drilling holes larger than ½ in. in diameter. The operation proceeds in a manner similar to a conventional drill press. Enough force should be applied to produce a curled chip. Too little force will result in broken chips and increased drilling time; too great a force will cause overheating and shorten the drill life.

You can expand the capacity of magnetic drilling equipment by using carbide-tipped cutters. These tools are related to "hole saws" used for woodworking, but they possess much greater capacity and strength. These are tubular-shaped devices with carbide-tipped multiple cutting edges, which are highly efficient. The alternating inside and outside cutting edges are ground to cut holes rapidly with great precision. Hard carbide cutting tips help them outlast regular high-speed twist drills. They are superior tools for cutting large-diameter holes because their minimal cutting action is fast and the power required is less than when removing all the hole material. Approximately a ⅛ in. wide kerf of material is removed and a cylindrical plug of material is ejected on completion of the cut. When used in conjunction with the magnetic drill, they enable the mechanic to cut large-diameter holes with little effort and great accuracy. Figure 3-5 shows a magnetic drill with a hole-cutting attachment.

The hole cutter makes possible operations in the field that could otherwise be performed only on larger fixed machinery in the shop. The arbor center pin allows accurate alignment for premarked holes. As with most machining operations, a cooling and lubricating fluid should be used when cutting holes with this type of cutter. Ideally, it would be applied with some pressure to furnish

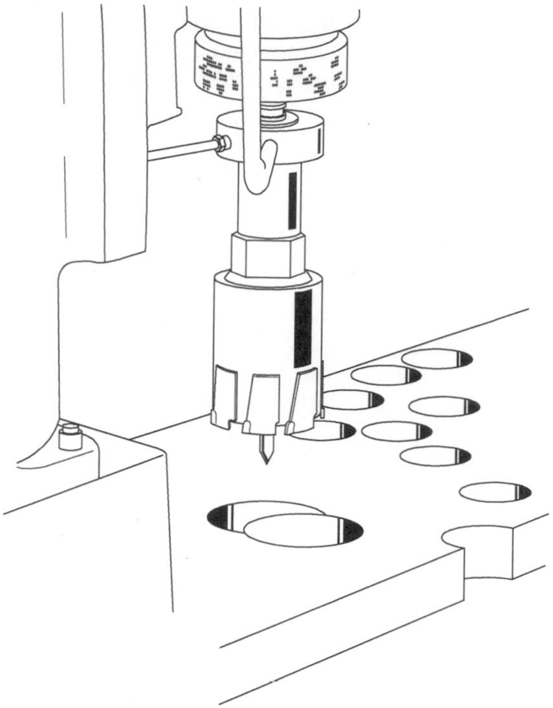

Figure 4-5 Magnetic drill with hole-cutting attachment.

a cutting action as well as cooling and lubricating. This may be accomplished by applying a fluid directly to the cutter and groove or by using an arbor lubricating mechanism. With a special arbor, it is possible to introduce the fluid under pressure. The fluid can be force-fed by a hand pump, which is part of a hand-held fluid container. Introducing the fluid on the inside surfaces of the cutter causes a flushing action across the cutting edges, which tends to carry the chips away from the cutting area and up the outside surface of the cutter.

Diamond Concrete Core Drilling

Diamond concrete core drilling is used to make holes in concrete structures. The tool that is used is the *diamond core bit.* Prior to the development of this tool, holes in concrete structures required careful planning and form preparation or the breaking away of sections of hard concrete and considerable patch-up. The tool is basically a metal tube, on one end of which is a matrix crown embedded with industrial-type diamonds distributed throughout the crown and arranged in a predetermined pattern for maximum cutability and exposure. Bits are made in two styles of construction, the closed back and the open back (Figure 4-6).

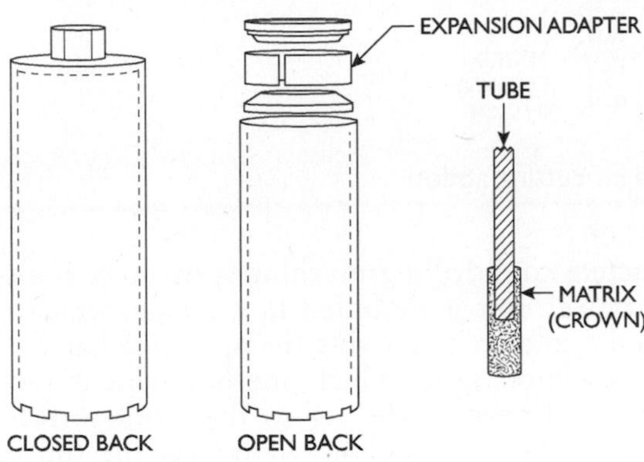

Figure 4-6 Diamond core bits.

There is a decided cost advantage in using the open-back bit, in that the adapters are reusable, offering savings in cost on each bit after the first. Also, if a core is lodged in thc bit, removing the adapter makes the core removal easier. The closed-back bit offers the advantage of simplicity. Installation requires only turning the bit onto the arbor thread, without positioning or alignment issues. As the bit is a single, complete unit, there is no problem with mislaid, lost, or damaged parts.

Successful diamond core drilling requires that several very important conditions be maintained: rigidity of the drilling unit, adequate water flow, and uniform steady pressure. Understanding the action, which takes place when a diamond core bit is in operation, will result in better appreciation of the importance of maintaining these conditions. Figure 4-7 illustrates the action at the crown end of a diamond core bit during drilling. Arrows indicate the water flow inside the bit, down into the kerf slot and around the crown, as particles are cut free and flushed up the outside surface of the bit.

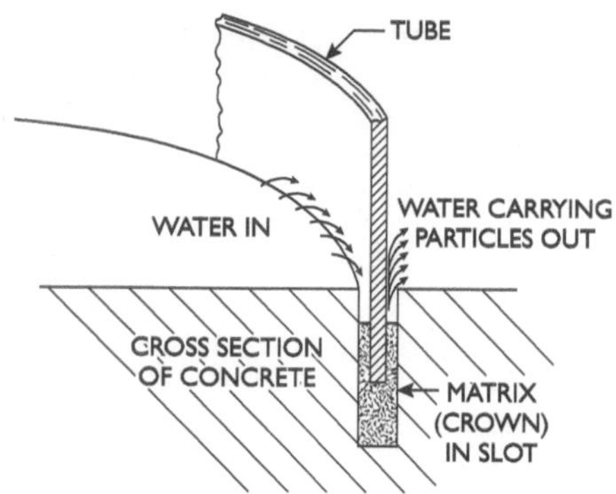

Figure 4-7 Diamond bit cutting action.

The diamond concrete core drilling machine is in effect a special drill press. The power unit is mounted in a cradle, which is moved up and down the column by moving the operating handles. The handles rotate a pinion gear, which meshes with a rack attached to the column. To secure the rig to the work surface, the top of the column includes a jackscrew to lock the top of the column against the overhead with the aid of an extension. Figure 4-8 shows a concrete core drilling rig equipped with a vacuum system, which makes attaching the rig directly to the work surface possible.

The power unit is a heavy-duty electric motor with reduction gears to provide steady rotation at the desired speed. The motor spindle incorporates a water swivel, which allows introduction of water through a hole in the bit adapter to the inside of the core bit. The power unit shown in Figure 4-9 attaches to the cradle of the rig shown in Figure 4-8. The diamond core bits (both open- and closed-back styles) fit the threaded end of the motor spindle.

The rigidity of the drilling rig plays a critical part in successful diamond core drilling. The rig must be securely fastened to the work surface to avoid possible problems. Slight movement may cause chatter of the drill bit against the work surface, fracturing the diamonds. Greater movement will allow the bit to drift from location, resulting in crowding of the bit, binding in the hole, and possible seizure and damage to the bit.

An easy way to anchor the unit is with the jackscrew provided at the top of the column. A telescoping extension, pipe, two-by-four, or other material cut to the appropriate length may be used. This allows the rig to be braced against the opposite wall.

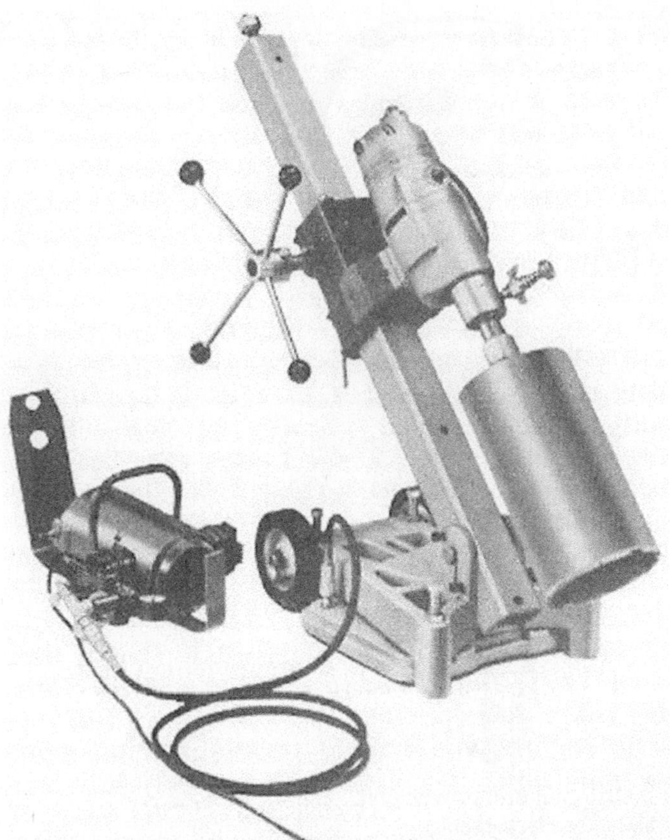

Figure 4-8 Diamond concrete core drilling rig.
(Courtesy Milwaukee Electric Tools)

The versatility of a diamond core drilling rig can be greatly increased with a vacuum system. With this device, it is possible to anchor the unit directly to the work surface, eliminating the need for extensions, braces, or other securing provisions. The rig in Figure 4-8 is equipped with such a system. It consists of a vacuum pump unit and auxiliary hose, gauge, fittings, and more.

To use the vacuum system of attachment, first clean the area where the work is to be performed. Remove any loose dirt or material that might cause leakage of the pad seal to the work surface. Place the rig, with the diamond core bit on the spindle, in the desired location. Loosen the pad nuts to allow the pad to contact

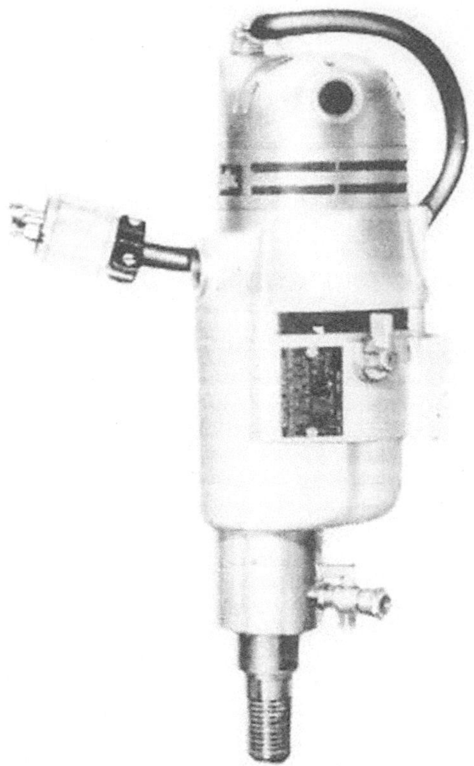

Figure 4-9 Diamond drill power unit.
(Courtesy Milwaukee Electric Tools)

the work surface without restraint. Start the vacuum pump to evacuate the air from inside the pad. sThis produces what is commonly called the "suction," which holds the pad in place. A vacuum gauge indicates the magnitude of the vacuum produced. The graduated gauge face is marked to show the minimum value required for satisfactory operation. A clean, relatively smooth surface should allow building the vacuum value to the maximum. Should the gauge register a value below the minimum required, do not attempt to drill. Check for dirt, porous material, cracks in the surface, or any other condition that might allow air to leak past the pad seal. When the gauge reading indicates that the pad is secure, tighten the pad nuts to fasten the rig base firmly to the vacuum pad. Standing on the base of the rig is not a substitute for good pad fastening. While additional weight will add a little to the downward force, it will not prevent the rig from floating or shifting out of position.

Water—which is vital to the success of diamond core drilling—is introduced through the water swivel, a component of the lower motor housing. The swivel has internal seals, which prevent leakage as the water is directed into the spindle and into the inside of the bit. The preferred water source is a standard water hose (garden hose), which provides dependable flow and pressure. When a standard water hose is not available, a portable pressure tank (used for garden sprayers or gravity feed tanks) can be used. Whatever the arrangement, take care to ensure adequate flow and pressure to handle the job. Depending on the operating conditions, you may need to make provisions to dispose of the used water. On open, new construction, it may be permissible to let the water flow freely with little or no concern for runoff. In other situations, it will be necessary to contain runoff and find a way to dispose of the used water. You can make a water collector ring and pump for this purpose, or you can use the common wet-dry shop vacuum and build a dam with rags or other material.

When the rig is secured and water supply and removal provisions are made, drilling may commence. Starting the hole may present a problem because the bit crown has a tendency to wander, particularly when starting in hard materials and on irregular and inclined surfaces. At the start, the crown may contact only one spot, and thrust tends to cause the bit to walk. Often it is sufficient to guide the bit lightly with a board that is notched at the end. To start, apply light pressure to the bit. After the crown has penetrated the material, pressure may be fully applied. The feed force must be uniformly applied in a steady manner, not jerky or intermittent. Too little pressure will cause the diamonds to polish; too much can cause undue wear. To aid the mechanic in maintaining a steady pressure of the proper amount, most diamond drilling rigs incorporate an ammeter to indicate motor loading. In addition to the regular calibrations, the meter dial has a green area that shows the working range and a red area to indicate when too much pressure is being applied. Enough force should be exerted on the operating handles to keep the ammeter needle in the green area, indicating that the proper bit pressure and drilling speed are being maintained. This prevents overload and ensures optimum bit life.

The importance of constant flowing water during drilling cannot be overstressed. The water pressure and flow must be sufficient to wash cuttings from under the bit crown and up the outside of the bits, as illustrated in Figure 4-7. The water also acts as a coolant, carrying away the heat that might otherwise cause the diamonds to polish or, in an extreme case, cause the bit to burn (to turn blue).

All water connections must be tight and the water flow steady (1–2 gallons per minute). Water flow and bit rotation must *not* stop while the bit is in the hole. The bit should be raised out of the hole *while turning* and then the water and power shut off. Stopping and starting the bit while in the hole could cause binding and damage to the bit. When the bit is cutting freely, the operator can feel movement in to the concrete. The off-flowing water will have a slightly sludgy appearance as it carries away the concrete particles.

Diamond core bits are capable of drilling all masonry materials: concrete, stone, brick, and tile, as well as steel embedded in concrete, such as reinforcing rod and structural steel. They are not, however, capable of continuous steel cutting and so experience considerable wear and/or deterioration of the diamonds if required to do extensive steel cutting. When the bit encounters embedded steel, the mechanic will notice an increased resistance to bit feed. This feed rate should be decreased to accommodate the slower cutting action that occurs when cutting steel. When cutting steel at reduced feed rate, the water becomes nearly clear, and small, gray metal cuttings will be visible in the off-flow water.

Diamond core bits will also cut through electrical conduit buried in concrete. This is a major consideration when cutting holes for changes or revisions in operating areas. Not only is there a possibility of shutting down operations if power lines are cut, but there is a possible electric shock hazard for the mechanic. Common sense dictates that you consider the possibility of buried conduit carefully when doing revision work. If the location of buried lines cannot be determined, drilling should not be attempted unless all power to lines that might be severed is disconnected.

When the concrete being drilled is not too thick, the hole may be completed without withdrawal of the bit. With greater thickness, drilling should proceed to a depth equal to at least two times the diameter of the hole. The bit should then be withdrawn and the core broken out. This may be done with a large screwdriver or pry bar inserted into the edge of the hole and rapped firmly with a hammer. The first section of the core can usually be removed with two screwdrivers, one on each side of the hole, by prying and lifting. If you find that the core is embedded, the bit should be withdrawn and the core and any loose pieces of steel removed. When deeper cores must be removed, a little more thought and effort are required. Larger holes may permit reaching into them and, if necessary, drilling a hole to insert an anchor to aid in removal. If the core cannot be snapped off and removed, it may be necessary to break it out with a demolition hammer.

When drilling through a concrete floor, you should also consider that it is possible for water to escape and for the concrete core to fall from the hole as the bit emerges from the underside of the floor. Make provisions to contain the water and catch the falling core to prevent damage and avoid injury.

Portable Band Saw

The *portable band saw,* while not in common use by mechanics, performs many cutting operations efficiently and with little effort (Figure 4-10). Many field operations that require cutting in place consume excessive time and effort if done by hand. If acetylene torch cutting were used, the resulting rough surfaces might not be acceptable and/or the flame and flying molten metal might not be permissible. Often the object being worked on cannot be taken to another location to perform a sawing operation. In such instances, a portable band saw can do the work in the field. The portable band saw is in effect a lightweight, self-contained version of the standard shop band saw. It uses two rotating rubber-tired wheels to drive a continuous saw blade. Power is supplied by a heavy-duty electric motor, which transmits its power to the wheels through a gear train and worm wheel reduction. This reduces speed and develops powerful torque to drive the saw blade at proper speed. Blade guides, similar in design to the standard shop machine, are built into the unit, as well as a bearing behind the blade to handle thrust loads.

Figure 4-10 Portable band saw.

A new generation of portable band saws with variable speed motors allows the mechanic to adapt the cutting speed to the material. This maximizes cutting performance and blade life. For example, slower speed is better for cutting stainless steel and very hard alloys.

Blades for portable band saws are available in a tooth pitch range from 6 to 24 teeth per inch. The rule of thumb for blade selection is to always have 3 teeth in the material at all times. Using too coarse a blade will cause thin metals to hang up in the gullet between two teeth and to tear out a section of teeth. Too fine a blade will prolong the cutting job, as each tooth will remove only small amounts of metal. Do not use cutting oil. Oil and chips transfer and stick to the rubber tires, causing the blade to slip under load. Chips build up on tires, causing misalignment of the blade.

Blades can be purchased with straight-pitch or variable-pitch tooth arrangements (Figure 4-11). In addition, choices in blade materials range from inexpensive carbon steel to electron beam–welded bimetals used for tough applications. Although the more exotic blades are expensive, they last a longer time and are less prone to breakage. If a very smooth cut is required, keep in mind that variable pitch blades produce much less vibration.

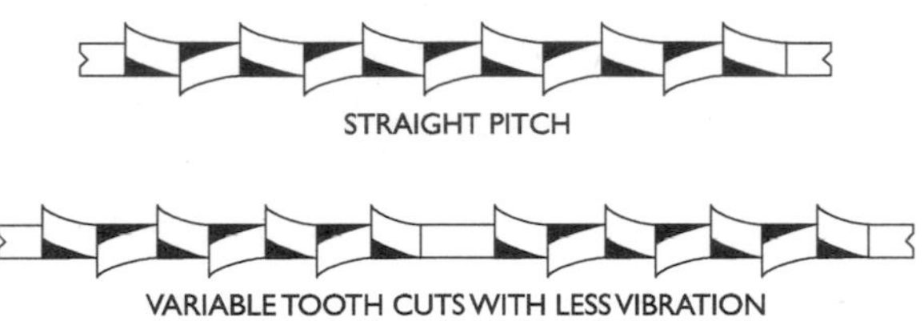

Figure 4-11 Saw teeth.

Hammer Drill

The *hammer drill* combines conventional or rotary drilling with hammer percussion drilling (Figure 4-12). Its primary use is the drilling of holes in concrete or masonry. It incorporates two-way action in that it may be set for hammer drilling with rotation or for rotation only. When drilling in concrete or masonry, special carbide-tipped bits are required. These are made with alloy steel shanks for durability, to which carbide tips are brazed to provide the hard cutting edges necessary to resist dulling. The concrete or masonry is reduced to granules and dust by this combined hammering and rotating action, and shallow spiral flutes remove the material from the hole.

For the most part, the hammer drill is used to drill "blind" holes in concrete that are used as a base for anchors to attach a machine base or structural member to the concrete. Holes are drilled to a

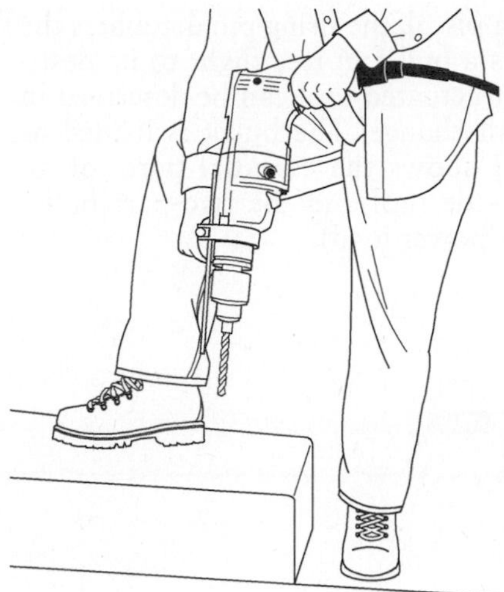

Figure 4-12 Hammer drill.

prescribed depth, and the anchor is inserted and locked in place by a wedging action produced by striking the top or by a screwed arrangement.

A trick for blowing the dust out of the hammer drill hole is to use a turkey baster. Be careful not to get the dust in your eyes, though. The anchor is dropped into the hole and checked to ensure that the hole is not too shallow. Most manufacturers require that the top of the anchor be just level with the concrete before it is set or wedged.

Explosive-Powered Tools

The principle of *explosive-* or *powder-actuated tools* is to fire a fastener into material and to anchor or secure it to another material. Some applications are wood to concrete, steel to concrete, wood to steel, steel to steel, and numerous applications of fastening fixtures and special articles to concrete or steel. Because the tools vary in design details and safe handling techniques, general information and descriptions of the basic tools and accessories are given, rather than specific operation instructions. Because the principle of operation is similar to that of a firearm, safe handling and use must be given the highest priority. An explosively actuated tool, in simple terms, is a "pistol." A pistol fires a round composed of a cartridge with firing cap and powder, which is attached to a bullet.

When you pull the trigger on a pistol, the firing pin detonates the firing cap and powder and sends a bullet in free flight to its destination. The direct-acting powder-actuated tool can be described in the same manner, with one small change: The bullet is loaded as two separate parts. Figure 4-13 shows the essential parts of an explosive-actuated tool system—the tool and the two-part bullet (the combination of fastener and power load).

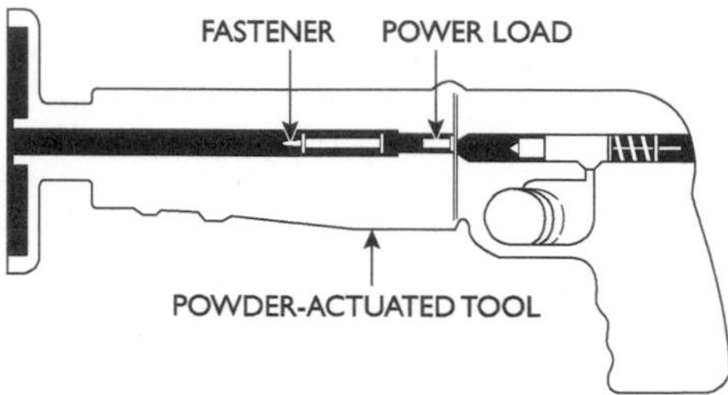

Figure 4-13 Basic parts of a powder-actuated tool.

The fastener is inserted into the tool first. Then, the mechanic inserts the power load. The tool is closed and pressed against the work surface, the trigger is pulled, and the fastener travels in free flight to its destination.

Obviously, a larger fastener will require more power to drive and so would a fastener that must penetrate steel instead of wood. The mechanic uses a chart to match the loads with the fasteners for various penetrations of materials, fastener size, or depth. The end of the explosive charge is often color coded to match the similar color code at the top surface of the fastener.

Safe use of these tools cannot be stressed enough. Some basic rules apply:

- Use the tool at right angles to the work surface.
- Check the chamber to verify that the barrel is clean and free from any obstruction before using the tool.
- Do not use the tool where flammable or explosive vapors, dust, or similar substances are present.
- Do not place your hand over the front (muzzle) end of a loaded tool.
- Wear ear protection and goggles.

Manufacturers' tools may vary in appearance, but they are similar in principle and basic design. They all have a chamber to hold the powder load, a firing pin mechanism, a safety feature, a barrel to confine and direct the fastener, and a shield to confine flying particles. There are two types of tools:

- Direct acting, in which the expanding gas acts directly on the fastener to be driven into the work (Figure 4-14)
- Indirect acting, in which the expanding gas of a powder charge acts on a captive piston, which in turn drives the fastener into the work (Figure 4-15)

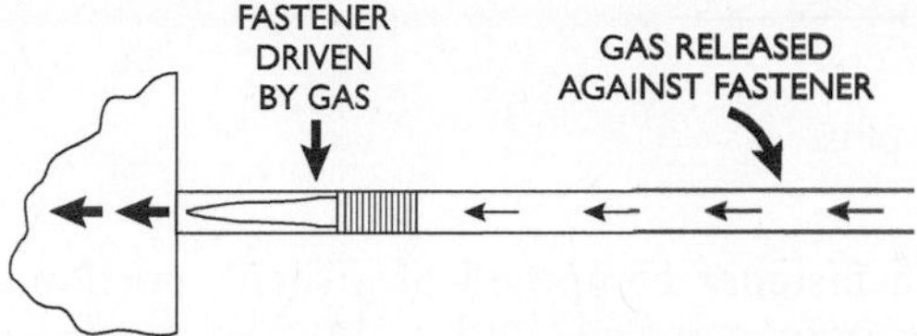

Figure 4-14 Direct-acting operating principle.

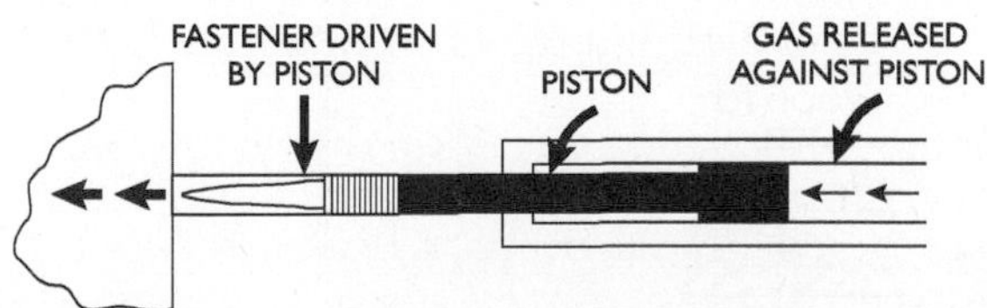

Figure 4-15 Indirect-acting operating principle.

The fasteners used in powder-actuated tools are manufactured from special steel and are heat treated to produce a very hard yet ductile fastener. These properties are necessary to permit the fastener to penetrate concrete or steel without breaking. The fastener is equipped with some type of tip, washer, eyelet, or other guide member. This guide aligns the fastener in the tool, guiding it as it is being driven (Figure 4-16).

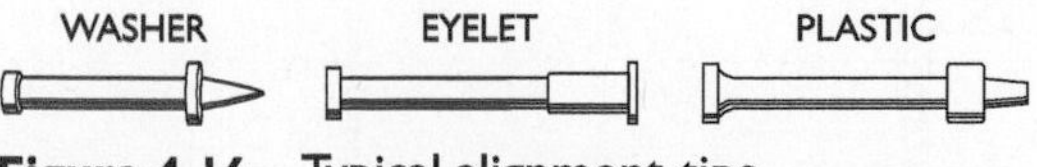

Figure 4-16 Typical alignment tips.

Two types of fasteners are in common use: the *drive pin* and the *threaded stud.* The drive pin is a special nail-like fastener designed to attach one material to another, such as wood to concrete or steel (Figure 4-17). Head diameters are generally $1/4$, $5/16$, or $3/8$ in. For additional head bearings in conjunction with soft materials, washers of various diameters are either fastened through or made a part of the drive-pin assembly.

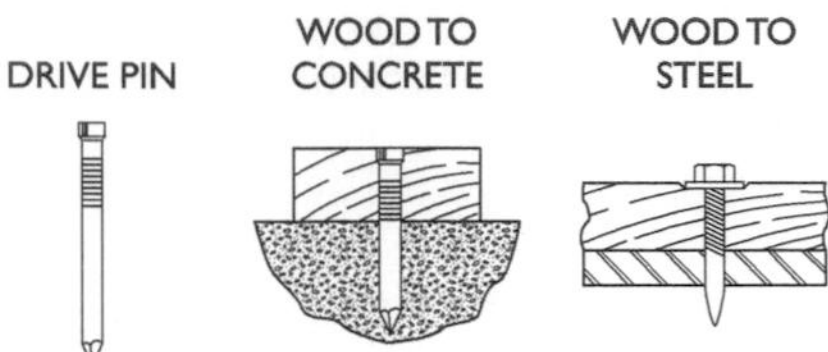

Figure 4-17 Typical drive pins.

The threaded stud is a fastener composed of a shank portion, which is driven into the base material, and a threaded portion to which an object can be attached with a nut (Figure 4-18). Usual thread sizes are #8-32, #10-24, $1/4$"-20, $5/16$"-18, and $3/8$"-16.

Figure 4-18 Typical threaded studs.

A special type of drive pin with a hole through which wires, chains, and so on can be passed for hanging objects from a ceiling is shown in Figure 4-19.

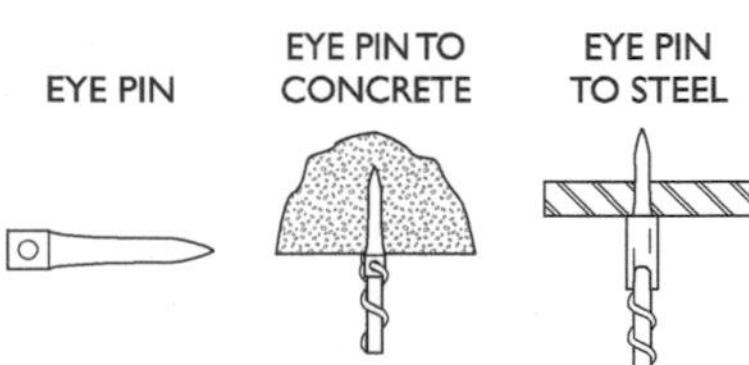

Figure 4-19 Special type eye pins.

Another special type of fastener, in this case a variation of the threaded stud, is the *utility stud*. This is a threaded stud with a threaded collar, which can be tightened or removed after the fastener has been driven into the wood surface. Figure 4-20 shows the stud and its application in fastening wood to concrete.

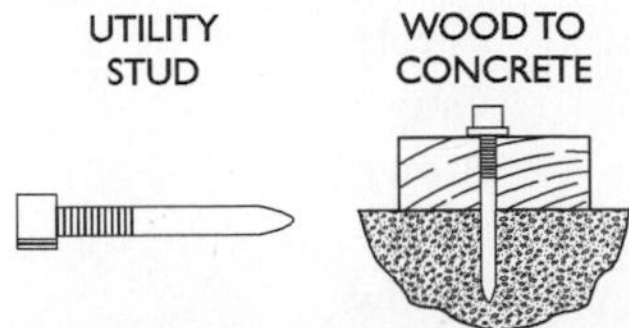

Figure 4-20 Special utility studs.

The *power load* is a unique, portable, self-contained energy source used in powder-actuated tools (Figure 4-21). These power loads are supplied in two common forms: cased or caseless. As the name implies, the propellant in a cased power load is contained in a metallic case. The caseless power load does not have a case, and the propellant is in solid form.

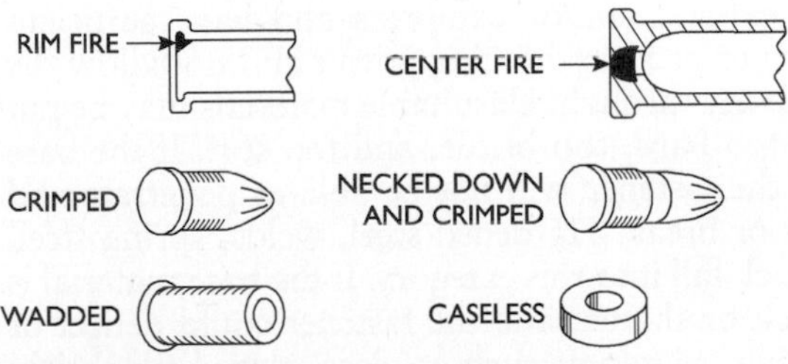

Figure 4-21 Power load construction.

Whatever the type, caliber, size, or shape, a standard number and color code are used to identify the power level or strength of all power loads. The power loads are numbered 1 through 12, with the lightest being #1 and the heaviest being #12. In addition, because there are not 12 readily distinguishable permanent colors, power loads #1 through #6 are in brass-colored cases and #7 through #12 are in nickel-colored cases. A combination of the case color and the load color defines the load level or strength. The number and color identification codes are shown in the following listing:

Power Level	Case Color	Load Color
1	Brass	Gray
2	Brass	Brown
3	Brass	Green
4	Brass	Yellow
5	Brass	Red
6	Brass	Purple
7	Nickel	Gray
8	Nickel	Brown
9	Nickel	Green
10	Nickel	Yellow
11	Nickel	Red
12	Nickel	Purple

When selecting the proper power load to use for an application, start with the lightest power level recommended for the tool. If the first test fastener does not penetrate to the desired depth, the next higher load should be tried until the proper penetration is achieved.

The material into which the fastener shank is driven is known as the base material. In general, base materials are metal and masonry of various types and hardness. Suitable base materials, when pierced by the fastener, will expand and/or compress and have sufficient hardness and thickness to produce holding power and not allow the fastener to pass completely through. Unsuitable materials may be put into three categories: too hard, too brittle, and too soft. If the base material is too hard, the fastener will not be able to penetrate and could possibly deflect or break. Hardened steel, welds, spring steel, marble, and natural rock fall into this category. If the base material is too brittle, it will crack or shatter, and the fastener could deflect or pass completely through. Materials such as glass, glazed tile, brick, and slate fall into this category. If the base material is too soft, it will not have the characteristics to produce holding power, and the fastener will pass completely through. Materials such as wood, plaster, drywall, composition board, and plywood fall into this category.

Shields and special fixtures are important parts of the powder-actuated fastening system and are used for safety and tool adaptation to the job. The shield should be used whenever fastening directly into a base material, such as driving threaded studs or drive pins into steel or concrete. In addition to confining flying particles, the shield also helps hold the tool perpendicular to the work. Medium- and high-velocity class tools are designed so that the tool cannot fire unless a shield or fixture is attached. One of the most important acts that the conscientious user of a powder-actuated fastener system can perform is to see that the tool being used is equipped with the proper safety shield to ensure both safety and good workmanship.

Because masonry is one of the principal base materials suitable for powder-actuated tool fastening, it is important to understand what happens when a fastener is driven into it (Figure 4-22). The holding power of the fastener results primarily from a compression bond of the masonry to the fastener shank. The fastener, on penetration, displaces the masonry, which tries to return to its original form and exerts a squeezing action on the fastener shank. Compression of the masonry around the fastener shank takes place, with the amount of compression increasing in relation to the depth of penetration and the compressive strength of the masonry.

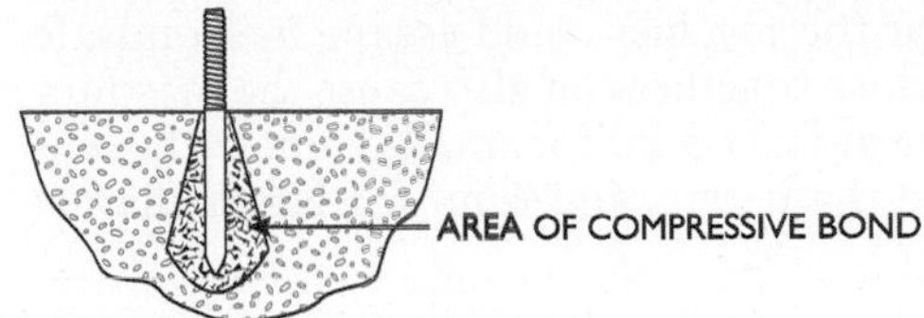

Figure 4-22 Fastener bond in masonry.

When an excessive direct pullout load is applied to a fastener driven into masonry material, failure will occur in either of two ways: The fastener will pull out of the masonry (Figure 4-23a), or failure of the masonry will occur (Figure 4-23b). This illustrates an important relationship between the depth of penetration and the strength of the bond of the fastener shank and the strength of the masonry itself. When the depth of penetration produces a bond on the fastener shank equal to the strength of the masonry, the maximum holding power results.

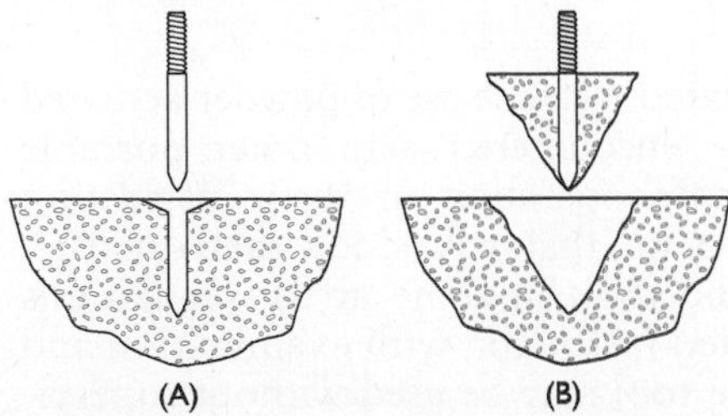

Figure 4-23 Pullout from masonry.

Because the tensile strength of masonry is relatively low, care should be taken not to place it under high tension when driving fasteners. This may occur if fasteners are driven too close to the edge, as illustrated in Figure 4-24.

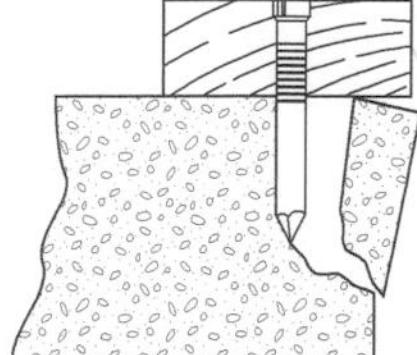

Figure 4-24 Edge failure.

Do not fasten closer than 3 in. from the edge of the masonry. If the masonry cracks, the fastener will not hold, and there is a chance that a chunk of masonry or the fastener could escape in an unsafe way. Setting fasteners too close together can also cause the masonry to crack. Spacing should be at least 3 in. for small-diameter fasteners, 4 in. for medium-sized diameters, and 6 in. for larger diameters, as shown in Figure 4-25.

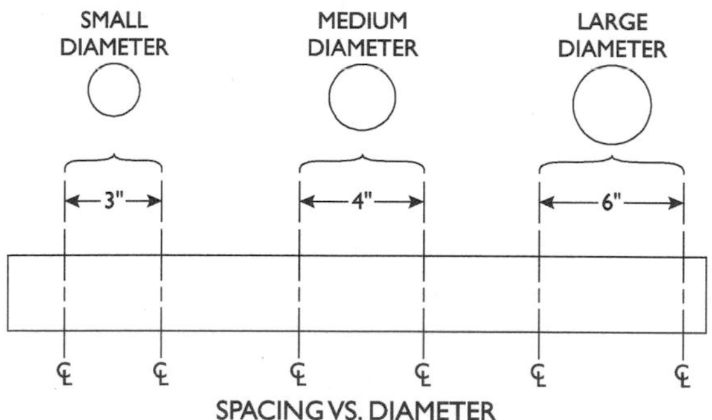

Figure 4-25 Fastener spacing.

The potential hazards associated with the use of powder-actuated tools exceed those commonly encountered with other portable tools. Manufacturers stress safe operation in their instruction material, and it is very important that you study their instructions carefully before using the tool. In some areas, regulations require instruction by a qualified instructor, with examination and issuance of a permit before the tool may be used without supervision. Perhaps the greatest hazard involved is that of release of the fastener. You must exert careful thought and effort to ensure that the fastener is under control at all times and prevented from escaping in free flight.

Fluid Power Portable Tools

The driving force for many portable power tools is called *fluid power*. Portable fluid power tools are usually pneumatic, i.e., using air, or hydraulic, i.e., using oil.

The most popular pneumatic tools are impact types of tools: impact wrenches, jackhammers, and hammer drills. Other types of pneumatic tools deliver rotary motion, which is usually produced by high-speed turbines such as a nut runner or air grinder.

The real workhorses in fluid power tools make use of hydraulics to accomplish the mission. The two basic components of most of these tools are the hydraulic pump, both hand and power operated, and the hydraulic cylinder. The principle that is used in tools involving the hydraulic pump and cylinder is the multiplication of force by hydraulic leverage (Figure 4-26). In practice, a force is exerted on a small surface to generate hydraulic pressure. This pressure is transmitted, usually through a hydraulic hose, to a greater surface to produce a force magnified as many times as the ratio of the power cylinder surface area to the output cylinder area.

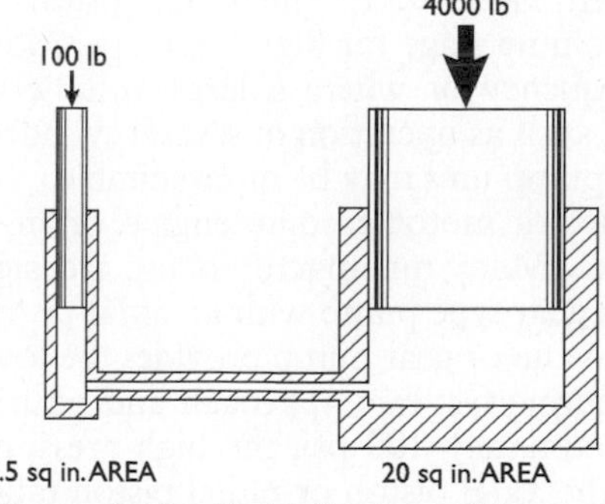

Figure 4-26 Multiplication of force.

This multiplication of force is offset by the corresponding decrease in travel; that is, the small plunger in the diagram must travel 40 in. to displace enough fluid to move the large plunger 1 in.

The primary component in a hydraulic power system is the hydraulic pump. This widely used, hand-operated, single-speed hydraulic pump is a single-piston type designed to develop up to 10,000 pounds per square inch (psi) of pressure. It is commonly constructed with a reservoir body, at one end of which is the power head. The head contains the power cylinder and piston, usually about ⅜ to ½ in. in diameter. Force is developed by use of a lever handle, which enables the effort applied to the handle end to be multiplied many times. Thus, a force of 80 lbs applied to the handle end would result in a force of 1200 lbs acting on the hydraulic piston when under load (Figure 4-27). Under this loaded condition, a pressure of 8000 psi would develop in the hydraulic system.

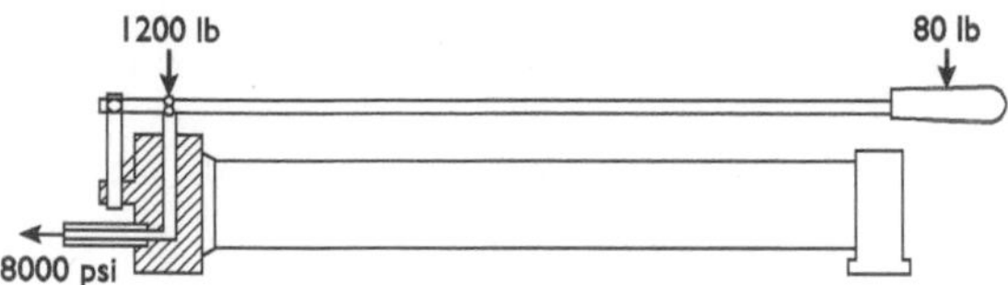

Figure 4-27 Single-acting hand-operated hydraulic pump.

When time or speed is a factor, the two-speed hand pump may be used. The principle of operation is the same as the single-speed pump, except that it has the feature of a second piston of larger diameter. The two-speed operation provides high oil volume at low pressure for rapid ram approach, and when switched to the small piston, it provides a high-pressure, low-volume stage for high-force operation. For applications of greater frequency or where a large volume of hydraulic fluid may be required, such as operation of several cylinders simultaneously, a power-driven pump unit may be more suitable.

Power-driven pump units—electric motor, gasoline engine, or air—are usually of the two-stage type. Many manufacturers use a design that combines a roller-vane or a gear-type pump with an axial-piston or radial-piston type. The roller-vane or gear pump provides the low-pressure, high-volume stage, ensuring fast ram approach and return. When pressure builds to approximately 200 psi, the high-pressure, low-volume stage provided by the axial piston or radial piston takes over to handle high-pressure requirements. Power pumps come equipped with various-sized reservoirs to provide hydraulic fluid for particular applications.

Figure 4-28 shows an electric-driven, two-stage hydraulic pump for use with a single- or double-acting ram. It is equipped with a three-position four-way manual valve for driving double-acting rams or multiple single-acting rams. Compressed air-driven power units as well as gasoline engine-driven units in a great variety of designs are also in wide use.

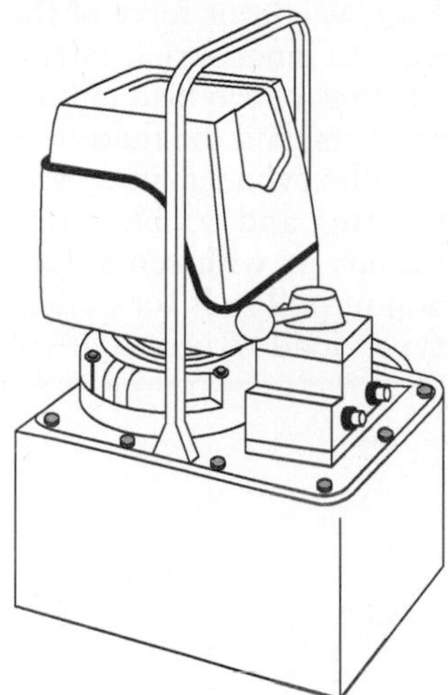

Figure 4-28 Electric hydraulic pump.

The high-pressure hose is vital to the operation of portable hydraulic tools. Most of these hoses are designed for an operating pressure of 10,000 psi, with a bursting strength of 20,000 to 30,000 psi. The lightweight hoses are constructed with a nylon core tube and polyester fiber reinforcement, whereas those designed for heavy-duty, more severe service have one or more layers of braided steel webbing to help withstand the internal hydraulic pressure, as well as the external abuse that may be encountered in some applications.

The pressure gauge is a valuable, although not always necessary, component of the hydraulic power system. Its principal advantage is that it provides a visual indication of the pressure generated by the pump; thus, it can assist in preventing overloading of the hydraulic system as well as the equipment on which work is being done. Gauges, in addition to being graduated in psi, may also include a scale graduated in tonnage. This style of gauge must be matched to the cylinder diameter to obtain correct tonnage values. Correct practice with gauge manufacturers is to show a danger zone on the gauge face. This is done with a red background coloring in the area over 10,000 psi.

The second major component in a hydraulic power system is the hydraulic cylinder. Hydraulic cylinders operate on either a single- or double-acting principle, which determines the type of "return" or piston retraction. Single-acting cylinders have one port and, in their simplest form, retract as a result of the weight or force of the load (load return). They are also made with an inner spring assembly, which enables positive retraction regardless of the load (spring return). Double-acting cylinders have two ports, and the fluid flow is shifted from one to the other, both hydraulic cylinder lifting and retraction (hydraulic return). Both single-acting and double-acting cylinders are manufactured with solid pistons or with center-hole pistons. Center-hole pistons allow insertion of pull rods for pulling applications. The single-acting, push-type, load-return style of hydraulic cylinder is shown in Figure 4-29.

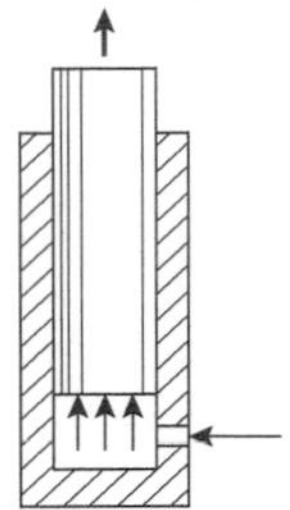

Figure 4-29 Single-acting load-return cylinder.

The single-acting cylinder, either load or spring return, is the most commonly used for hydraulic power tool operation performed by industrial mechanics. In some operations, however, power is required in both the lift and retract directions. The double-acting cylinder provides hydraulic function in both the lifting and lowering modes. Figure 4-30 shows the principle of operation of a double-acting cylinder.

Double-acting cylinders should be equipped with a four-way valve to prevent trapping of the hydraulic fluid in the retract system. Most manufacturers build a safety valve into the retract system to prevent damaging the ram if the top hose is inadvertently left unconnected and the ram is actuated.

All manners of pumps, cylinders, accessories, and special attachments are available to adapt hydraulic power to a multitude of uses. Perhaps the most frequent application the mechanic is concerned with is the disassembly of mechanical components. This usually requires considerable force to remove one closely fitted part from another. One of the great advantages of hydraulic power over

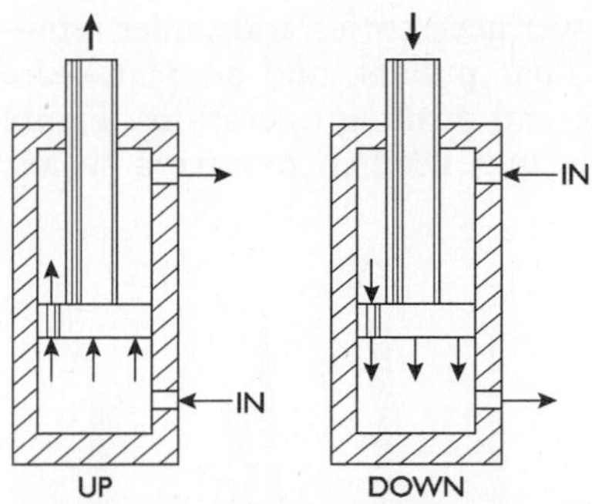

Figure 4-30 Double-acting cylinder in lift and retract modes.

striking or driving to accomplish the operation is the manner in which the force is applied. Instead of heavy blows, which might cause distortion or damage components, the force is applied in a steady, controlled manner. Figure 4-31 shows a portable hydraulic power unit used with a jaw-type mechanical puller. Note that this has the advantage of mechanical positioning and adjustment plus controlled hydraulic power.

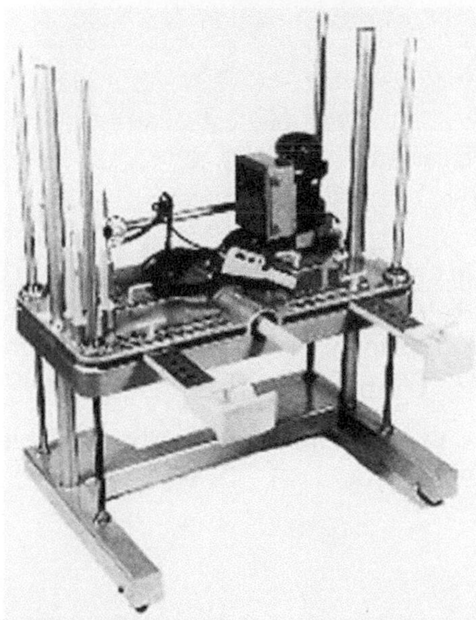

Figure 4-31 Horizontal puller press.
(Copyright Poulan-Puller Press, Inc.)

A great variety of hydraulic power accessories and puller sets—including all types of jaw pullers, bar pullers, and adapters—are available for all manner of pulling and pushing operations. Figure 4-32 shows the use of a hydraulic unit used to remove a sheave from the end of a motor shaft.

Figure 4-32 Portable puller in field use.
(Copyright Poulan-Puller Press, Inc.)

The use of a portable hydraulic puller set greatly enhances the ability of a mechanic to tackle large jobs right at the machine site. Lifting heavy loads, moving machinery, pulling shafting or bearings, and removing pressed-on parts are all jobs where hydraulics plays an important part.

If the portable hydraulic set fails to operate properly, use Table 4-1 to help determine the cause.

Table 4-1 Portable Hydraulic Troubleshooting Chart

Problem	*Possible Cause*
Cylinder will not advance.	Pump release valve open. Coupler not fully tightened. Oil level in pump is low. Pump malfunctioning. Load too heavy for cylinder.
Cylinder advances part way.	Oil level in pump is low. Coupler not fully tightened. Cylinder plunger is binding.
Cylinder advances in spurts.	Air in hydraulic system. Cylinder plunger binding.
Cylinder advances more slowly than normal.	Leaking connection. Coupler not fully tightened. Pump malfunctioning.
Cylinder advances but will not hold.	Cylinder seals leaking. Pump malfunctioning. Leaking connection. Incorrect system setup.
Cylinder leaks oil.	Worn or damaged seals. Internal cylinder damage. Loose connection.
Cylinder will not retract or retracts more slowly than normal.	Pump release valve is closed. Coupler not fully tightened. Pump reservoir overfilled. Narrow hose restricting flow. Broken or weak retraction spring. Cylinder internally damaged.
Oil leaking from external relief valve.	Coupler not fully tightened. Restriction in return line.

Chapter 5

Stationary Power Tools

Three "must have" tools in the shop are the *power hacksaw*, the floor *drill press*, and the hydraulic *shop press*. You must be able to cut material, drill holes for attachment, and press items together. These are the three tools that accomplish these basic mechanical tasks.

Power Hacksaw

No tool gets a workout in the shop like a power hacksaw (Figure 5-1). Cutting steel, pipe, pieces for welding fabrication, and metal conduit are all jobs for this piece of equipment.

The stock is usually held in a vise mounted on the base of the machine. An electric motor supplies the power for the machine. Usually some form of lubricant-coolant is pumped onto the workpiece to cool the cut and flush out the chips.

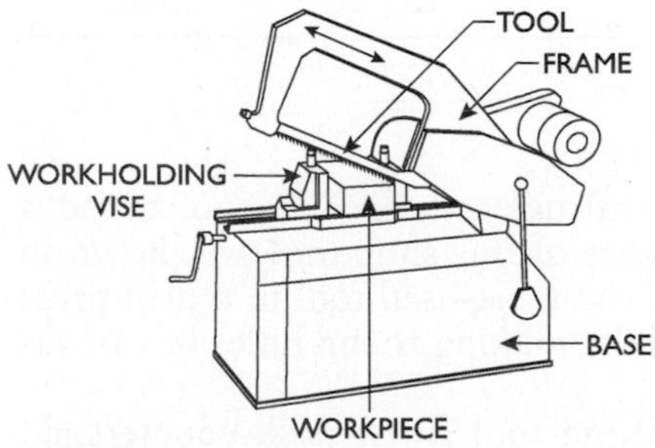

Figure 5-1 Power hacksaw.

The strokes per minute and the feed rate can be controlled and changed depending on the type of material that is being cut. Usually the feed rate is controlled by a weight that is added or moved along a bar to increase the pressure of the saw on the stock to be cut. Power hacksaw blades can be selected with various numbers of teeth per inch. Table 5-1 shows the settings for most saws to obtain quality cuts of various materials.

Table 5-1 Power Hacksaw Setup Recommendations

Material	*Teeth Per Inch*	*Strokes Per Minute*	*Feed Pressure*
Aluminum	4–6	150	Light
Brass, soft	6–10	150	Light
Brass, hard	6–10	135	Light
Cast iron	6–10	135	Medium
Copper	6–10	135	Medium
Carbon tool steel	6–10	90	Medium
Cold-rolled steel	4–6	135	Heavy
Drill rod	10	90	Medium
High-speed steel	6–10	90	Medium
Machinery steel	4–6	135	Heavy
Malleable iron	6–10	90	Medium
Pipe, steel	10–14	135	Medium
Structural steel	6–10	135	Medium
Tubing, brass	14	135	Light
Tubing, steel	14	135	Light

Drill Press

Round holes are drilled in metal using a machine tool called a *drill press*. The main components of this shop tool are shown in Figure 5-2. For the most part, the most-used tool in a drill press is the twist drill, which is used for making round holes in various materials.

The drill press rotates a cutting tool (twist drill, countersink, counterbore) and uses pressure from the operator on the feed lever to advance the drill through the workpiece. The speed of the drill and the pressure on the feed lever vary depending on the material that must be drilled. Usually the workpiece is located on thc table below the drill and is locked into position. Never attempt to start a twist drill without first using a center punch to make an indentation for starting the drill point (Figure 5-3).

A drill press vise can often be used to lock the piece in place to make sure it does not move when the hole is drilled. The use of parallels in the drill press vise (Figure 5-4) keeps the workpiece flat and the hole square to the surface of the workpiece.

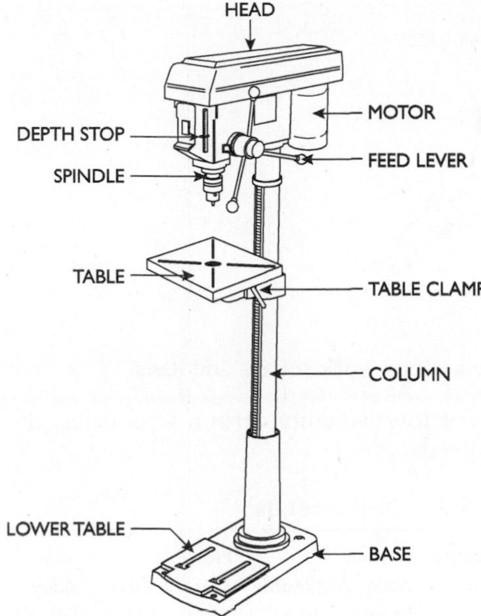

Figure 5-2 Components of a drill press.

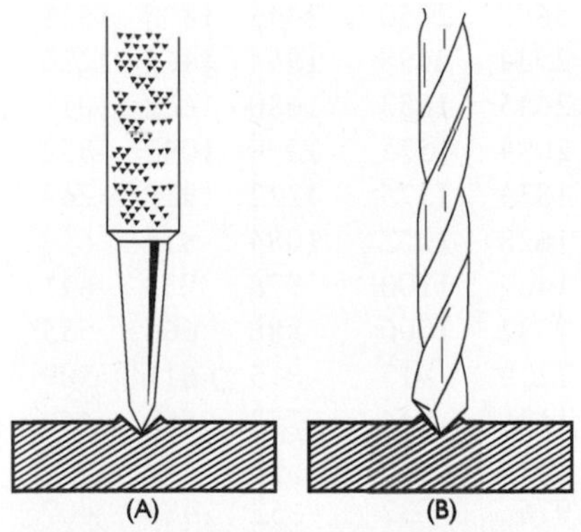

Figure 5-3 Center punch before drilling.

Drill speeds in revolutions per minute (rpm) for various diameter drills and typical workpieces are shown in Table 5-2. It shows figures

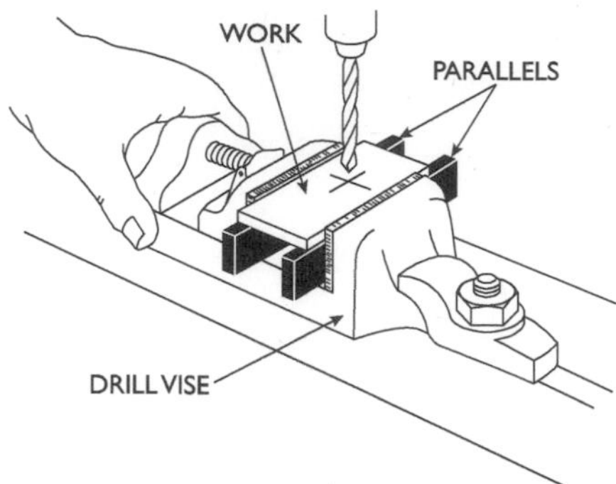

Figure 5-4 Vise with parallels keeps work square and level.

for high-speed twist drills. For lower-quality carbon steel drills, the drill speed rpm should be halved.

Table 5-2 Drill Speeds

Dia. Drill (in.)	Plastics	Soft Metals	Annealed Cast Iron	Mild Steel	Malleable Iron	Hard Cast Iron	Tool Steel	Alloy Steel
1/16	12217	18320	8554	7328	5500	4889	3667	3056
3/32	8142	12212	5702	4884	3666	3245	2442	2038
1/8	6112	9160	4278	3667	2750	2445	1833	1528
5/32	4888	7328	3420	2934	2198	1954	1465	1221
3/16	4075	6106	2852	2445	1833	1630	1222	1019
7/32	3490	5234	2444	2094	1575	1396	1047	872
1/4	3055	4575	2139	1833	1375	1222	917	764
9/32	2712	4071	1900	1628	1222	1084	814	678
5/16	2445	3660	1711	1467	1100	978	733	611
11/32	2220	3330	1554	1332	1000	888	666	555
3/8	2037	3050	1426	1222	917	815	611	509
13/32	1878	2818	1316	1126	846	752	563	469
7/16	1746	2614	1222	1048	786	698	524	437
15/32	1628	2442	1140	976	732	652	488	407
1/2	1528	2287	1070	917	688	611	458	382
9/16	1357	2035	950	814	611	543	407	339
5/8	1222	1830	856	733	550	489	367	306
11/16	1110	1665	777	666	500	444	333	277
3/4	1018	1525	713	611	458	407	306	255

Using a cutting oil or other suitable lubricant will greatly enhance the smoothness of the drilled hole and extend the life of the drill bit.

Drill Terms and Geometry

Twist drills, as we know them today, are the most common and widely used metal-cutting tools.

The parts of a twist drill are identified in Figure 5-5. One particular component is the point angle—the angle included between the cutting lips projected on a plane parallel to the drill axis. The clearance angle, or heel of the drill point, is also shown.

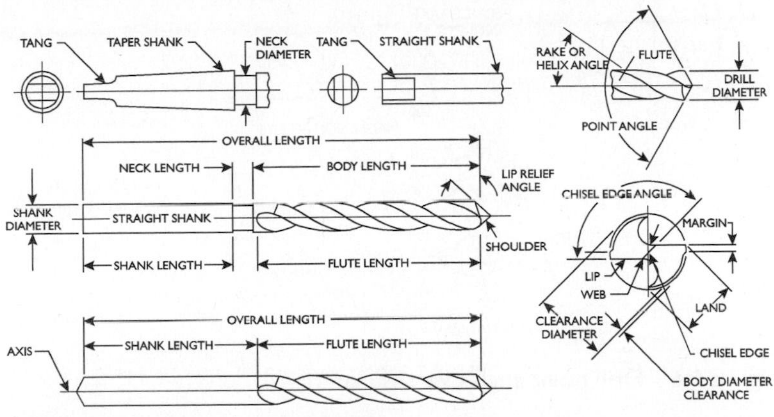

Figure 5-5 Parts of a twist drill.

When drilling steel with a drill press, a twist drill with a point angle of 118° is recommended. Field work on hard or tough materials, especially when using a battery- or electric-powered drill motor, requires a drill point of 135°.

The heel or clearance angle for the 118° point should be about 8–12 degrees. The 135° point should have a clearance angle of about 6–9 degrees. The types of points, clearance, and angles—depending on the material to be drilled—are shown in Figure 5-6. A twist drill cuts by wedging under the material and raising a chip. The steeper the point and the greater the clearance angle, the easier it is for the drill to penetrate the workpiece. The blunter the point and the smaller the lip angle clearance, the greater the support for the cutting edges. From this logic, it makes sense that a greater lip

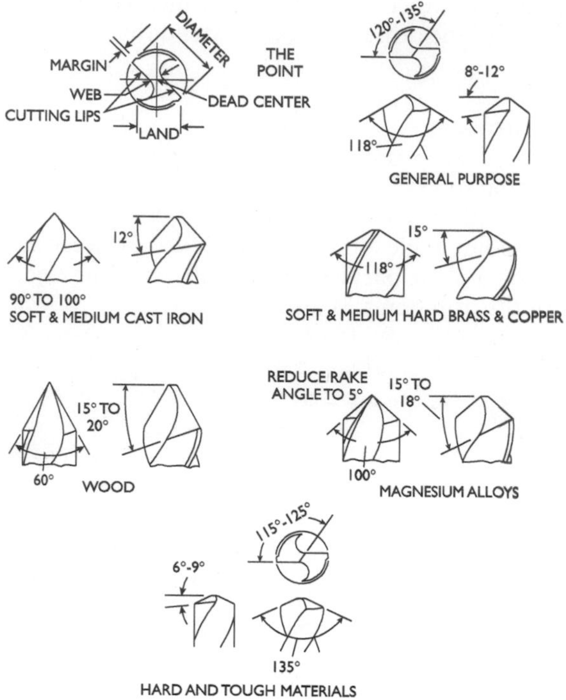

Figure 5-6 Drill point angles versus material.

angle and lesser lip clearance are good for hard and tough materials and that a decreased point angle and increased lip clearance are suitable for softer materials.

When drilling some softer nonferrous metals such as brass or copper, the drill point tends to "bite in," or penetrate too rapidly. This gives tearing and a poor finish to the hole.

Another tricky material to drill is stainless steel. If the craftsperson allows the drill to rotate without adequate pressure on the drill, the stainless will *work harden* and further drilling becomes almost impossible. Always put plenty of pressure on the drill when making a hole in stainless to keep this condition from spoiling the work.

Drill Sharpening

A drill begins to wear as soon as it is placed in operation. The maximum drill wear occurs at the corners of the cutting lips. The web, or chisel-point edge, begins to deform under the heat generated during drilling. The increase in wear at the corners travels back along the lands, resulting in a loss of size and tool life.

Wear occurs at an accelerated rate. When a drill becomes dull, it generates more heat and wears faster. In other words, there is more wear on the twentieth hole than on the tenth, still more on the thirtieth, and this continues. As wear progresses, the torque and thrust required increase. In addition to the accelerated wear, drill breakage due to excessive torque is one of the most common drill failures. In comparison, running a drill beyond its practical cutting life is like driving an automobile with a flat tire.

When regrinding a twist drill, all of the worn sections must be removed. Sharpening the edges or lips only, without removing the worn land area, will not properly recondition a twist drill.

Most twist drills are made with webs that increase in thickness toward the shank (Figure 5-7). After several sharpenings and shortenings of a twist drill, the web thickness at the point increases, resulting in a longer chisel edge (Figure 5-8). When this occurs, it is necessary to reduce the web so that the chisel edge is restored to its normal length. This operation is known as *web thinning*.

There are several common types of web thinning. The method shown in Figure 5-9 is perhaps the most common. The length *A* is usually made about one-half to three-quarters the length of the cutting lip.

The usual method of web thinning involves the use of a pedestal or bench grinder. Start with a clean, sharp grinding wheel. Hold the drill at approximately 30–35 degrees to the axial centerline of the drill.

The corner of the wheel must be lined up with the tip of the web.

The cutting lip must be turned out about 10° to make sure the edge is not ground and the web can be ground away.

After the worn portion of the drill has been removed and the web thinned (if necessary), the surfaces of the point must be reground. These two conical surfaces intersect with the faces of the flutes to form the cutting lips and with each other to form the chisel edge. As in the case of any other cutting tool, the surface back of these cutting lips must not rub on the work and must be relieved in order to permit the cutting edge to penetrate. Without such relief, the twist drill cannot penetrate the work and will only rub around and around.

In addition to grinding the conical surfaces to give the correct point angle and cutting clearance, both surfaces must be ground alike.

Figure 5-7 Web increases toward shank.

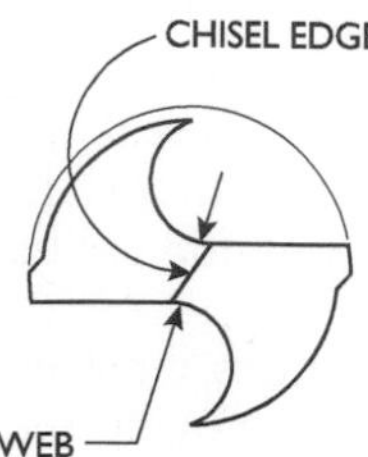

Figure 5-8 Chisel edge.

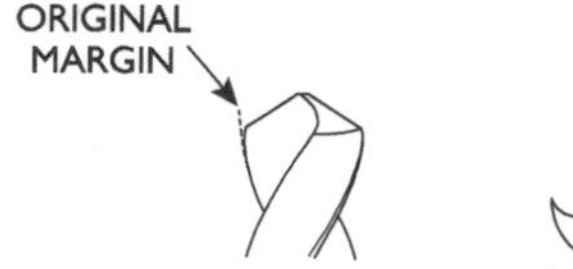

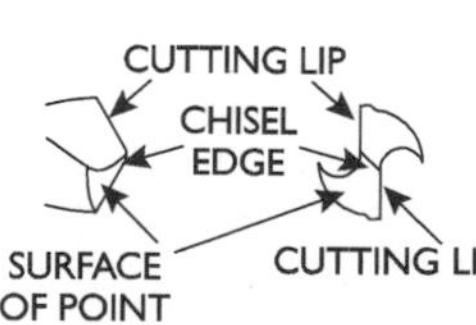

Figure 5-9 Hold these dimensions when web thinning.

Regardless of the point angle, the angles of the two cutting lips (A1 and A2) must be equal. Drill points of unequal angles or lips of unequal lengths will result in one cutting edge doing most of the cutting. This type of point will cause oversized holes, excessive wear, and short drill life.

To maintain the necessary accuracy of drill point angles, lip lengths, lip-clearance angle, and chisel-edge angle, the use of machine point grinding is recommended. The lack of a drill-point grinding machine is not sufficient reason, though, to excuse poor drill points. Drills may be pointed accurately by hand if proper procedure is followed and care is exercised.

Grinding the Drill Point by Hand

The most commonly used drill point is the conventional 118° point. This grind will give satisfactory results for a wide variety of materials and applications.

1. Adjust the grinder tool rest to a convenient height for resting back of forehand on it while grinding drill point.
2. Hold drill between thumb and index finger of left hand. Grasp body of drill near shank with right hand.
3. Place forehand on tool rest with centerline of drill making desired angle with cutting face of grinding wheel (Figure 5-11a) and slightly lower end of drill (Figure 5-11b).
4. Place heel of drill lightly against grinding wheel. Gradually raise shank end of drill while twisting drill in fingers in a counterclockwise rotation and grinding conical surfaces in the direction of the cutting edges. Exert only enough pressure to grind the drill point without overheating. Frequently cool drill in water while grinding.

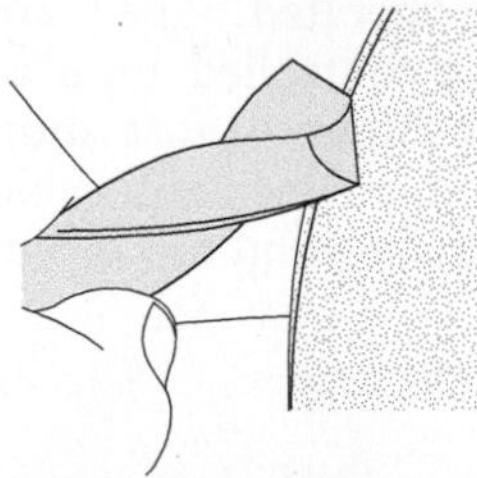

HOLDING THE DRILL AT
A 30-35 DEGREE ANGLE

Figure 5-10 Web thinning on a grinder.

5. Check results of grinding with a gauge to determine whether cutting edges are the same length and at desired angle and that adequate lip clearance has been provided (Figure 5-12).

Hydraulic Press or Arbor Press

No maintenance shop can be without an arbor (mechanical) or hydraulic press. This tool removes and installs interference-fitted

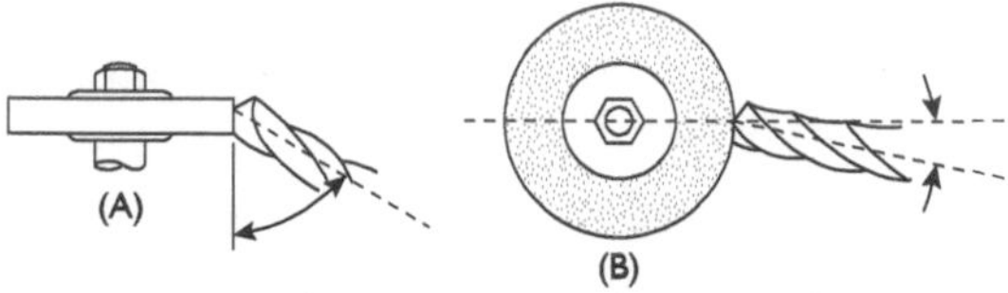

Figure 5-11 Sharpening a drill.

Figure 5-12 Checking the cutting edges and angle.

parts (shaft and bearing, shaft and gear, shaft and sprocket, pin and hub, and others). The shop press can be used for straightening bent machine components to allow their reuse as well as for bending certain components that must be attached by clamped forces.

Shop presses are both hand and power operated. They are almost essential for mechanical work involving controlled application of force to assemble or disassemble machinery. Most shop presses provide a large work area under the ram, a winch and cable mechanism to raise and lower the press bed quickly, and rapid ram advance and return (Figure 5-13).

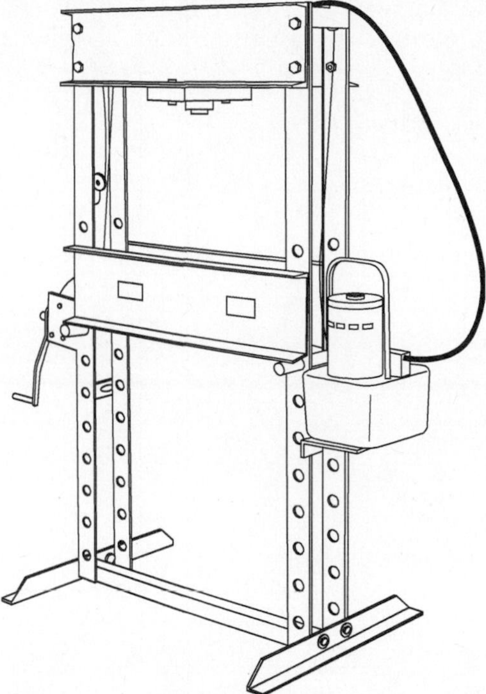

Figure 5-13 Hydraulic shop press.

Chapter 6

Measurement

The most useful tools for machinery measurement are the *steel rule*, *vernier caliper*, *micrometer*, *dial indicator*, *screw-pitch gauge*, and *taper gauge*.

Steel Rule

A plain, flat six-inch *steel rule* is a type commonly used. The inches are graduated into eighths and sixteenths of an inch on one side of the rule and thirty-seconds and sixty-fourths on the other (Figure 6-1).

When taking measurements with the steel rule, place the one-inch index mark—instead of the end of the rule—at one edge of the piece to be measured. Lay the steel rule parallel to the edge of the piece (Figure 6-2). Remember to deduct one inch from the reading.

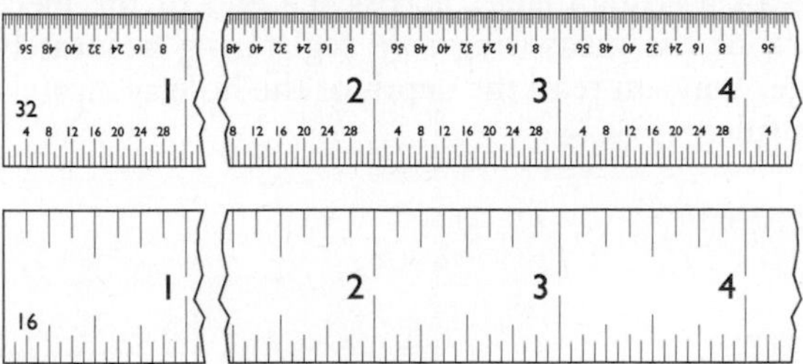

Figure 6-1 Steel rule.

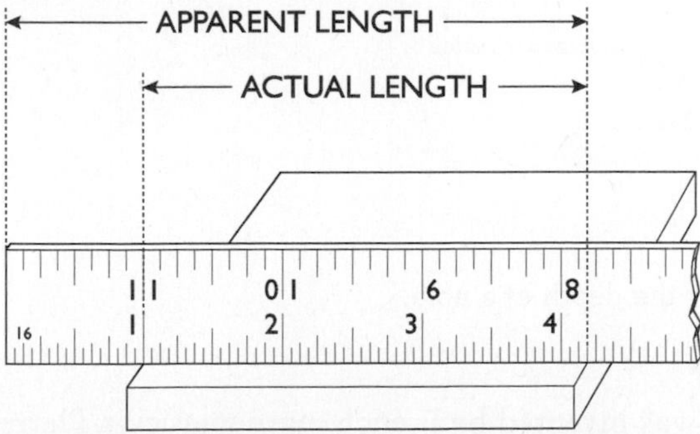

Figure 6-2 Remember to subtract one inch.

Be careful when you set the one-inch mark at the edge of the piece. The edge of the piece should coincide with the center of the index mark on the steel rule (Figure 6-3).

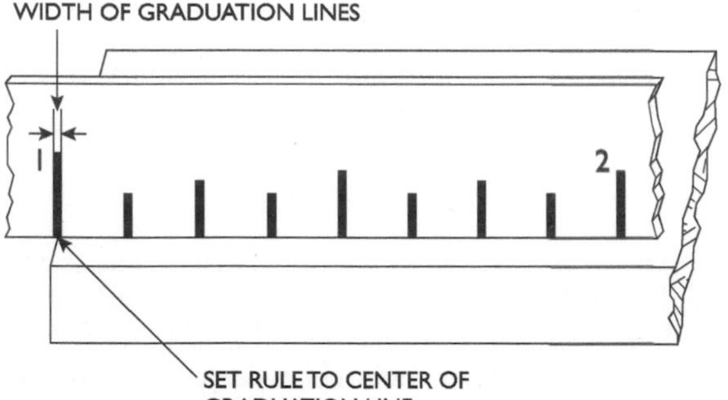

Figure 6-3 Edge of piece coincides with center of the index mark.

Use two steel rules to measure the depth of a hole (Figure 6-4). Place one steel rule, or a straightedge, across the end of the piece with the index mark of the measuring rule registering accurately with the straightedge. You can read the depth of the hole accurately from the measuring rule.

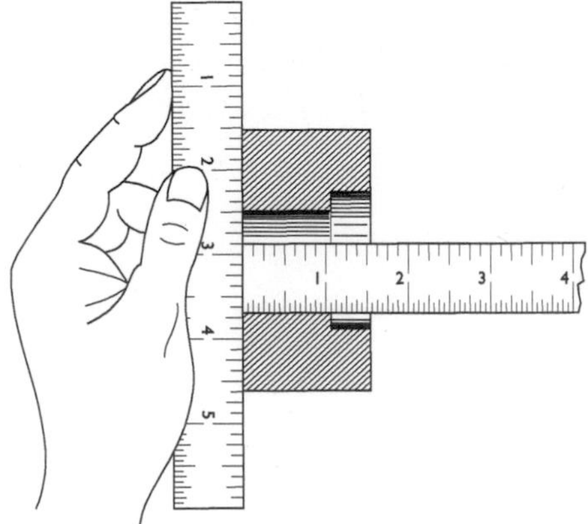

Figure 6-4 Checking the depth of a hole.

Vernier Caliper

The *vernier caliper* was invented by French mathematician Pierre Vernier. It consists of a stationary bar and a movable vernier slide assembly. The stationary rule is a hardened graduated bar with a fixed measuring jaw. The movable vernier slide assembly combines a movable jaw, vernier plate, clamp screws, and adjusting nut.

As shown in Figure 6-5, the bar of the tool is graduated in twentieths of an inch (0.050 in.). Every second division represents a tenth of an inch and is numbered. On the vernier plate is a space divided into 50 parts and numbered 0–50 at intervals of 5. The 50 divisions on the vernier occupy the same space as 49 divisions on the bar.

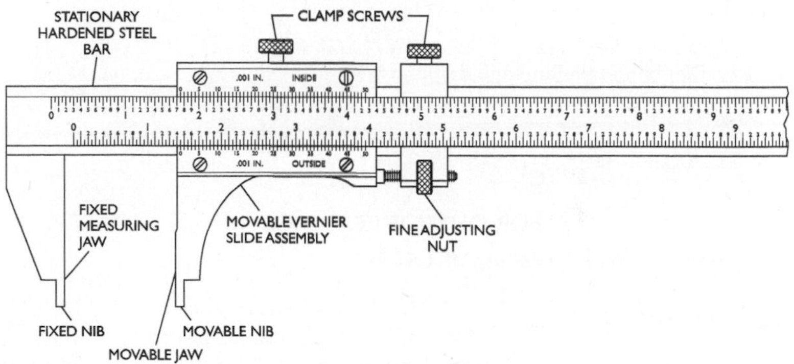

Figure 6-5 Parts of a vernier caliper.

The difference between the width of one of the 50 spaces on the vernier and one of the 49 spaces on the bar is therefore $\frac{1}{50}$ of $\frac{1}{20}$, or $\frac{1}{1000}$ of an inch. If the tool is set so that the 0 line on the vernier coincides with the 0 line on the bar, the line to the right of 0 on the vernier will differ from the line to the right of the 0 on the bar at $\frac{1}{1000}$, the second line will differ by $\frac{2}{1000}$, and so on. The difference will continue to increase $\frac{1}{1000}$ of an inch for each division until line 50 on the vernier coincides with a line on the bar.

To read the tool, note how many inches, tenths (or 0.100 in.), and twentieths (or 0.050 in.) the 0 mark on the vernier is from the 0 mark on the bar. Then, note the number of divisions on the vernier from 0 to a line that exactly coincides with a line on the bar.

Example

In Figure 6-6, the vernier has been moved to the right one and four-tenths and one-twentieth inches (1.450 in.), as shown on the bar, and the fourteenth line on the vernier coincides with a line, as indicated by the stars, on the bar. Therefore, fourteen-thousandths of an inch should be added to the reading on the bar, and the total reading is one and four-hundred-and-sixty-four thousandths inches (1.464 in.).

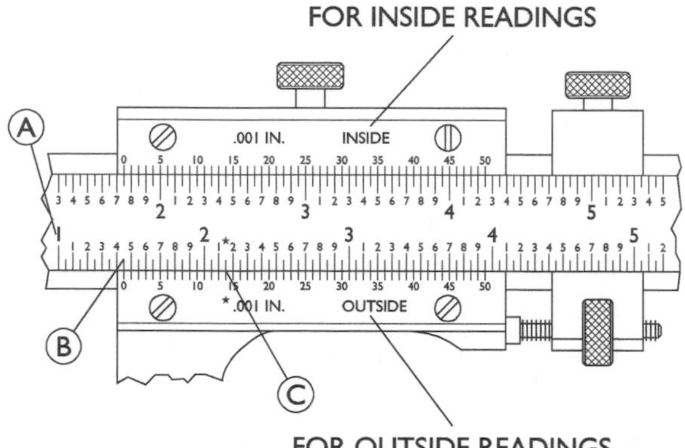

Figure 6-6 Vernier reading of 1.464.

Micrometer

Reading a Micrometer Graduated in Thousandths of an Inch (0.001 in.)

The pitch of the screw thread on the spindle of a micrometer is 40 threads per inch. One revolution of the thimble advances the spindle face toward or away from the anvil face exactly 1⁄40 in., or 0.025 in. (Figure 6-7).

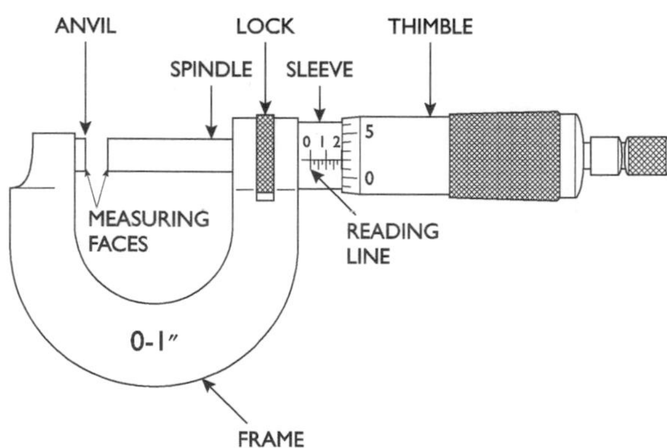

Figure 6-7 Parts of a micrometer.

The reading line on the sleeve is divided into 40 equal parts by vertical lines that correspond to the number of threads on the spindle. Therefore, each vertical line designates 1⁄40 in. or 0.025 in., and every fourth line, which is longer than the others, designates hundreds of thousandths. For example, the line marked "1" represents 0.100 in., the line marked "2" represents 0.200 in., and the line marked "3" represents 0.300 in.

The beveled edge of the thimble is divided into 25 equal parts with each line representing 0.001 in. and every line numbered consecutively. Rotating the thimble from one of these lines to the next moves the spindle longitudinally 1⁄25 of 0.025 in., or 0.001 in.; rotating two divisions represents 0.002 in. Twenty-five divisions indicate a complete revolution, 0.025 in. or 1⁄40 of an inch.

To read a micrometer in thousandths, multiply the number of vertical divisions visible on the sleeve by 0.025 and to this add the number of thousandths indicated by the line on the thimble that coincides with the reading line on the sleeve.

Example

Look at the drawing in the example:

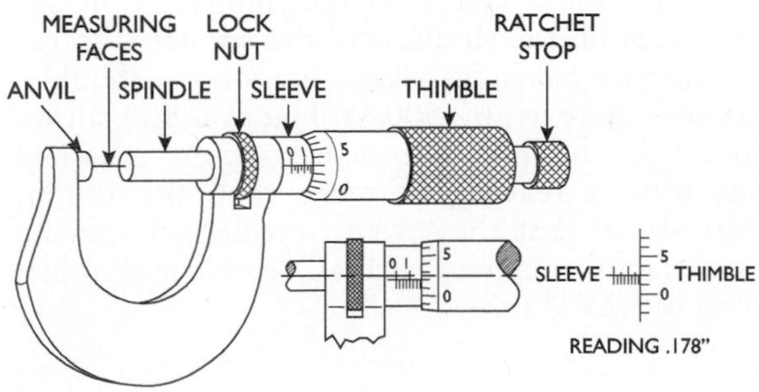

The "1" line on the sleeve is visible, representing	0.100 in.
There are three additional lines visible, each representing 0.025 in.	3×0.025 in. = 0.075 in.
Line "3" on the thimble coincides with the reading line on the sleeve, with each line representing 0.001	3×0.001 in. = 0.003 in.
The micrometer reading is	0.178 in.

Reading a Micrometer Graduated in Ten-Thousandths of an Inch (0.0001 in.)

When measuring shafts for bearing installation, the measurements are usually given to the ten-thousandth of an inch. A micrometer with a vernier scale capable of measuring to the ten-thousandth must be used.

The micrometer is read to the thousandth, but now the vernier micrometer shown has 10 divisions marked on the sleeve occupying the same space as 9 divisions on the beveled edge of the thimble. The difference between the width of one of the 10 spaces on the sleeve and one of the 9 spaces on the thimble is one-tenth of a division on the thimble. Because the thimble is graduated to read in thousandths, one-tenth of a division would be one ten-thousandth. To make the reading, first read to the thousandths as with a regular micrometer, and then see which of the horizontal lines on the sleeve coincides with a line on the thimble. Add to the previous reading the number of ten-thousandths indicated by the line on the sleeve that exactly coincides with the line on the thimble.

Example

In Figure 6.8 (a) and (b), the 0 (zero) on the thimble coincides exactly with the axial line on the sleeve, and the vernier 0 (zero) on the sleeve is the one that coincides with a line on the thimble. The reading is, therefore, an even 0.2500 in. Figure 6.8(c) shows that the 0 (zero) line on the thimble has gone beyond the axial line on the sleeve, indicating a reading of more than 0.2500 in. Checking the vernier shows that the seventh vernier line on the sleeve is the one that exactly coincides with a line on the thimble; therefore, the reading is 0.2507 in.

Dial Indicator

A *dial indicator* is an instrument for indicating size differences rather than for making measurements, as the dial indicator ordinarily is not used to indicate distance. It can be used in combination with a micrometer to measure the exact distance.

The movement of a hand on the dial indicator shows variations in measurements. The dial is graduated in thousandths of an inch; each division on the dial represents the contact point movement of 0.001 of an inch.

The dial indicator is useful for checking shafts for alignment and straightness, checking cylinder bores for roundness and taper, and testing bearing bores.

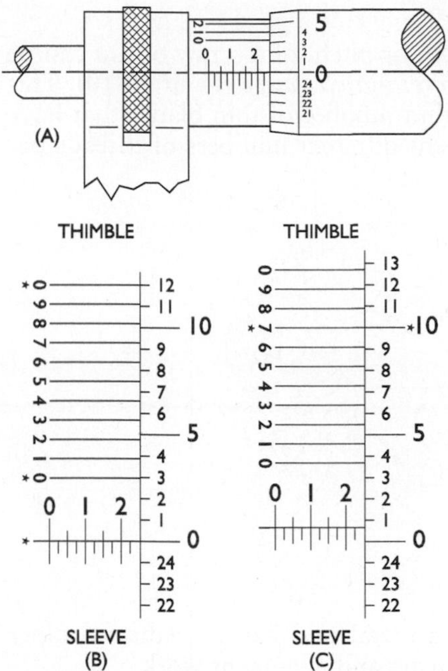

Figure 6.8 The reading on the micrometer is 0.2507 in.

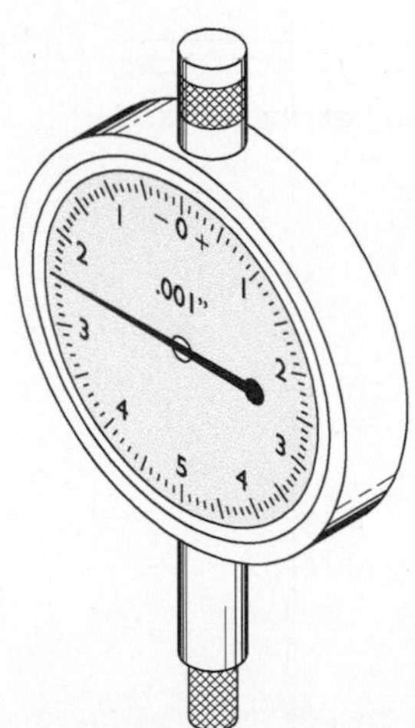

Figure 6-9 Dial indicator.

Screw-Pitch Gauge

The number of threads per inch, or pitch, of a screw or nut can be determined by the use of the *screw-pitch gauge* (Figure 6-10). This device consists of a holder with a number of thin blades that have notches cut on them to represent different numbers of threads per inch.

Figure 6-10 Screw-pitch gauge.

Taper Gauge

This type of gauge is made of metal and has a graduated taper (Figure 6-11). It is used for bearing and alignment work.

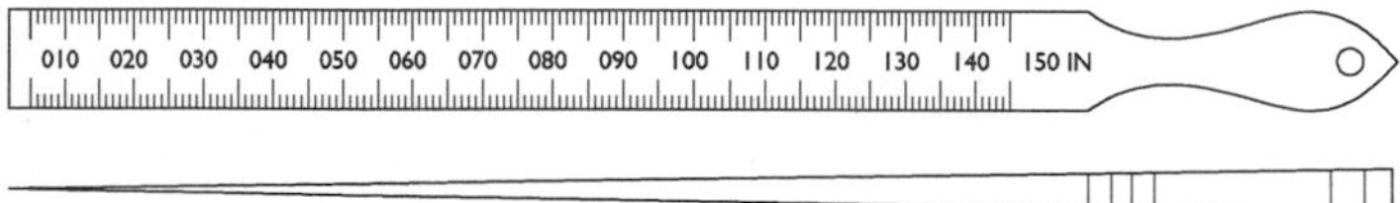

Figure 6-11 Taper gauge used for alignment and bearing work.

Chapter 7

Screw Threads

Screw threads used on fasteners or to mechanically transmit motion or power are made in conformance to established standards. These screw thread standards cover the cross-sectional shape of the thread, a range of diameters, and a specific number of threads per inch for each diameter. The most commonly used standard is the "Unified" standard for screw threads published as an American National Standards Institute (ANSI). The Unified standard threads generally supersede the American National form (formerly known as the United States Standard) that was used for many years. The Unified system was agreed to by the United Kingdom, Canada, and the United States to obtain screw thread interchangeability among these three nations. Unified threads have substantially the same thread form as and are mechanically interchangeable with the former American National threads of the same diameter and pitch. The principal differences between these standards relate to the application of allowances and tolerances. The general form of Unified threads is practically the same as the previous American standard. The external thread may have either a flat or a round root, as shown in Figure 7-1. In practice, the amount of rounding varies as a result of tool wear.

Unified standards are established by various thread series, which are groups of diameter-pitch combinations distinguished by the number of threads per inch applied to a specific diameter.

Coarse-Thread Series

The coarse-thread series is designated by the symbol UNC and is commonly called the *Unified National Coarse* series. It is generally utilized for bolts, screws, nuts, and other general applications. It is applicable to rapid assembly or disassembly or if corrosion or other slight damage might occur.

Fine-Thread Series

The fine-thread series is designated by the symbol UNF and is commonly called the *Unified National Fine* series. It is suitable for the production of bolts, screws, nuts, and other applications where a finer thread than that provided by the coarse series is required. External threads of this series have greater tensile stress area than

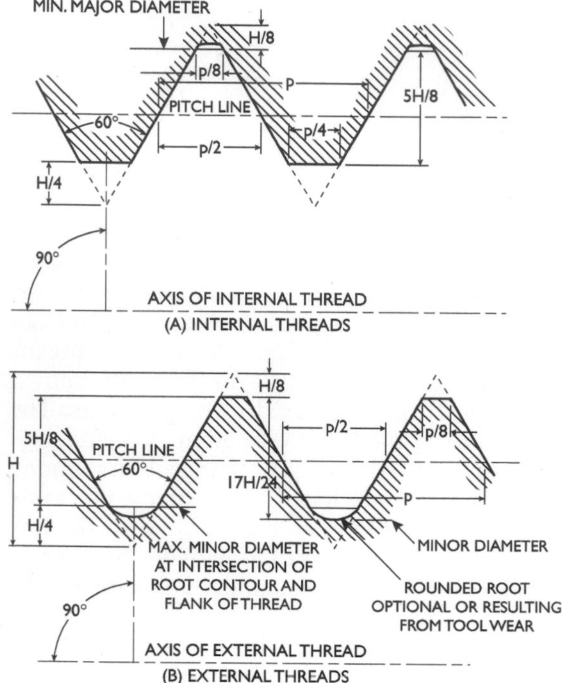

Figure 7-1 Unified standard thread form.

comparable sizes of the coarse series. It is used where the length of engagement is short, where a smaller lead angle is desired, or where the wall thickness demands a fine pitch.

Extra-Fine Thread Series

The extra-fine thread series is designated by the symbol UNEF and is commonly called the *Unified National Extra-Fine* series. It is used where even finer pitches of threads are desirable for a short length of engagement and for thin-walled tubes, nuts, ferrules, or couplings.

Constant-Pitch Series

The constant-pitch series is designated by the symbol UN and is commonly called simply *Unified National.* In this series, various constant pitches are used on a variety of diameters. Preference is given wherever possible to use of the 8-, 12-, and 16-thread series.

Thread Classes

The Unified standard also establishes limits of tolerances called classes. Classes 1A, 2A, and 3A apply to external threads only, and classes 1B, 2B, and 3B apply to internal threads only. Thread classes are distinguished from each other by the amounts of tolerance and allowance they provide. Classes 3A and 3B provide a minimum and classes 1A and 1B a maximum.

Classes 2A and 2B are the most commonly used thread standards for general applications, including production of bolts, screws, nuts, and similar threaded fasteners. Classes 3A and 3B are used when close tolerances are desired. Classes 1A and 1B are used on threaded components where quick and easy assembly is necessary and a liberal allowance is required to permit ready assembly, even with slightly bruised or dirty threads.

Unified Thread Designation

When designating a screw thread, the standard method is to specify in sequence the nominal size, number of threads per inch, thread series symbol, and thread class symbol. For example, a ¾-in. Unified Coarse series thread for a common fastener would be designated as follows:

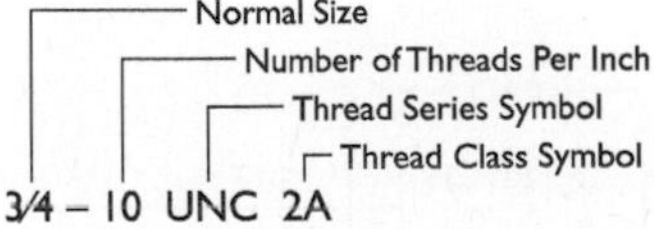

The threads per inch for each UNC, UNF, and UNEF series from #4 diameter to 3-in. diameter are listed in Table 7-1.

Screw Thread Terms

Only the more important screw thread terms are included in the following definitions. Many of these terms are illustrated in Figure 7-1.

> **Major Diameter.** The largest diameter of a screw thread. The term *major diameter* applies to both internal and external threads. The term *major diameter* replaces both the term *outside diameter* as applied to the thread of a screw and the term *full diameter* as applied to the thread of a nut.

Minor Diameter. The smallest diameter of a screw thread. The term *minor diameter* applies to both internal and external threads. It replaces the term *root diameter* as applied to the thread of a screw and *inside diameter* as applied to the thread of a nut.

Pitch Diameter. The diameter of an imaginary cylinder, the surface of which would pass through the threads at such points as to make equal the width of the threads and the width of the spaces cut by the surface of the cylinder.

Pitch. The distance from a point on the screw thread to a corresponding point on the next thread measured parallel to the axis.

Lead. The distance a screw advances axially in one turn. On a single-thread screw, the lead and pitch are the same; on a double-thread screw, the lead is twice the pitch; on a triple-thread screw, the lead is three times the pitch, etc.

Multiple Thread. A screw thread that is formed of two or more single threads, as shown in Figure 7-2. A double thread has two separate or single threads starting diametrically opposite or at points 180° apart. A triple thread has three single threads starting at points 120° apart. A quadruple thread has four single threads starting at points 90° apart. A multiple thread is used to increase the lead of a screw.

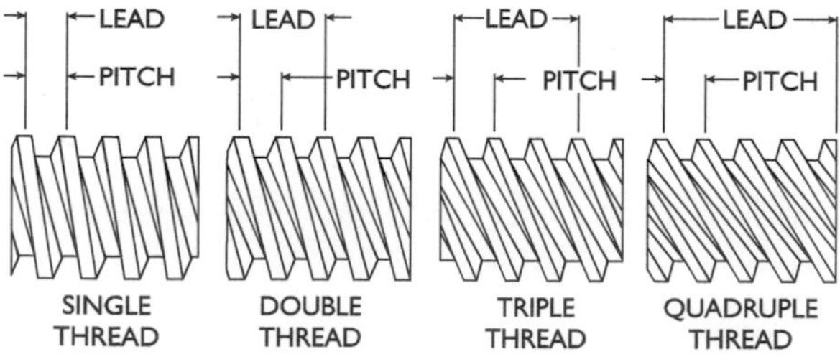

Figure 7-2 Relation of lead and pitch of multiple threads.

Angle of Thread. The angle included between the sides of the thread measured in an axial plane. (Unified thread form is a 60° angle.)

Helix Angle. The angle made by the helix of the thread at the pitch diameter with a plane perpendicular to the axis, as shown in Figure 7-3.

Depth of Thread. The distance between the crest and the root of the thread measured normal to the axis.

Table 7-1 Unified Standard Threads per Inch

Nominal Size	*UNC Threads per Inch*	*UNF Threads per Inch*	*UNEF Threads per Inch*
4	40	48	
5	40	44	
6	32	40	
8	32	36	
10	24	32	
12	24	28	32
1/4	20	28	32
5/16	18	24	32
3/8	16	24	32
7/16	14	20	28
1/2	13	20	28
9/16	12	18	24
5/8	11	18	24
11/16			24
3/4	10	16	20
13/16			20
7/8	9	14	20
15/16			20
1	8	12	20
1 1/16			20
1 1/8	7	12	18
1 3/16			18
1 1/4	7	12	18
15/16			18
1 3/8	6	12	18
1 7/16			18
1 1/2	6	12	18
1 9/16			18
1 5/8			18
1 11/16			18
1 3/4	5		
2	4 1/2		
2 1/4	4 1/2		
2 1/2	4		
2 3/4	4		
3	4		

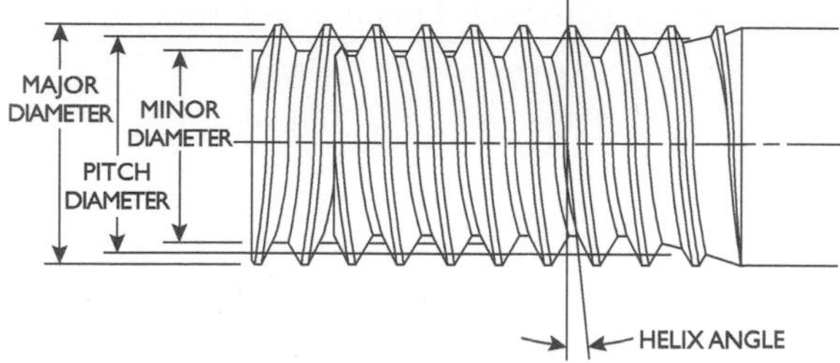

Figure 7-3 Illustrating the helix angle.

Translation Threads

Certain threads are used to move machine parts for adjustment, setting, transmission of power, etc. For these so-called translation threads, a stronger form than the V form is required. The most widely used translation thread forms are the square, the Acme, and the buttress.

Square Thread

Of the three translation thread forms, the square thread is the strongest and most efficient, but it is also the most difficult to manufacture. This is due to its parallel sides, which are difficult to accurately machine. The theoretical proportions of a square external thread are shown in Figure 7-4. The mating nut must have a slightly larger thread space than the screw to allow a sliding fit. Similar clearance must also be provided on the major and minor diameters.

Acme Thread

The Acme form of thread has largely replaced the square thread for most applications. Although it is not quite as strong as the square thread, it is preferred for most applications because it is fairly easy to machine. The angle of an Acme thread, measured in an axial plane, is 29 degrees. The basic proportions of an Acme thread are shown in Figure 7.5. The ANSI for Acme screw threads establishes thread series, fit classes, allowances and tolerances, etc., for this thread form, similar to the standard for the Unified thread form.

Buttress Thread

The buttress thread has one side cut approximately square and the other side slanting. It is used when a thread having great strength along the thread axis in only one direction is required. Because one side is cut nearly perpendicular to the thread axis, there is practically no radial thrust when the thread is tightened. This feature makes this thread form particularly applicable where relatively thin tubular members are screwed together. The basic thread form of a simple design of buttress thread in common use is shown in Figure 7.6. Other buttress thread forms are more complex, with the load side of the thread inclined from perpendicular to facilitate machining. The ANSI buttress thread form angle of inclination is 7°.

Screw Thread Tapping

Cutting internal screw threads with a hand tap, called tapping, is an operation frequently performed by mechanics. This is often a troublesome operation involving broken taps and time-consuming efforts to remove them. Some of these troubles may be avoided by a better understanding of the tapping operation and the tools involved.

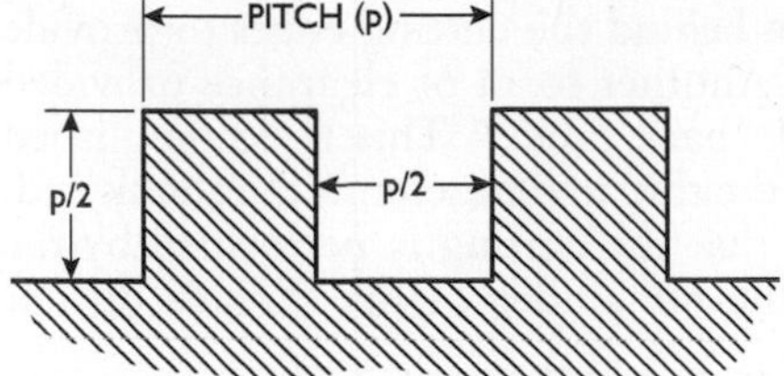

Figure 7-4 Diagram of external square threads.

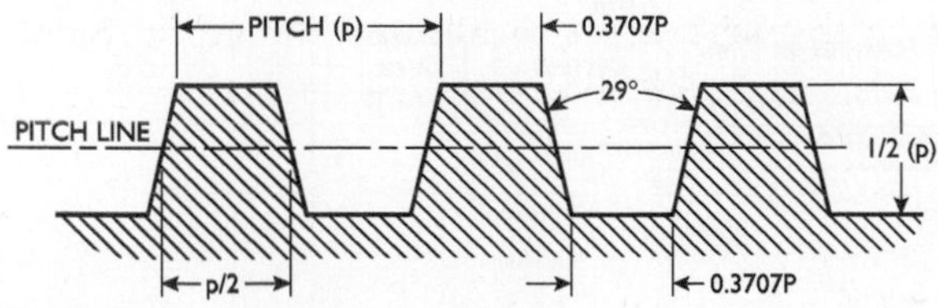

Figure 7-5 The basic proportions and dimensions of the Acme thread.

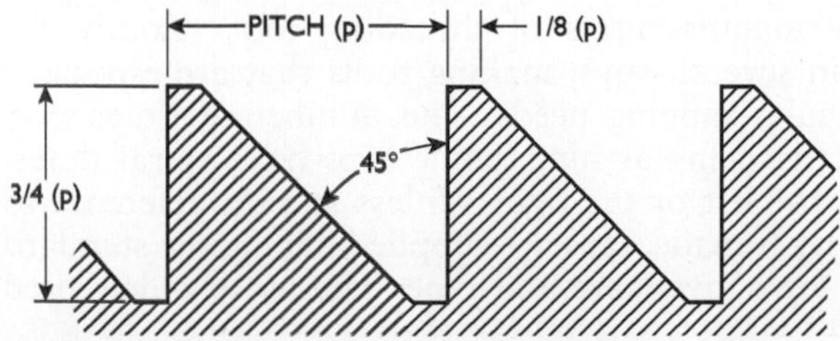

Figure 7-6 The buttress thread design.

Because threads must be made in many types of materials ranging from steel and cast iron to brass, aluminum, and plastics, procedures must be varied. Also, a thread that goes all the way through a part presents a different problem from a thread in a blind hole. The sketch of a typical hand tap in Figure 7-7 indicates the names of the various parts of a tap.

A tap basically consists of a shank, which has flats on the end to hold and drive it, and a threaded body, which does the thread cutting. The threaded body is composed of lands, which are the cutters, and flutes or channels to let the chips out and permit cutting fluid to reach the cutting edges. The threaded body is chamfered or tapered at the point to allow the tap to enter a hole and to spread the heavy cutting operation over several rings of lands or cutting edges. The radial thread relief shown in Figure 7-7 refers to material having been removed from behind the cutting edges to provide clearance and reduce friction. Another form of clearance provided on commercial taps is termed "back taper." This is accomplished with a very slight reduction in the thread diameter at the shank end. This provides thread relief so that the cutting is performed by the forward end of the tap and the remainder of the threads do not drag.

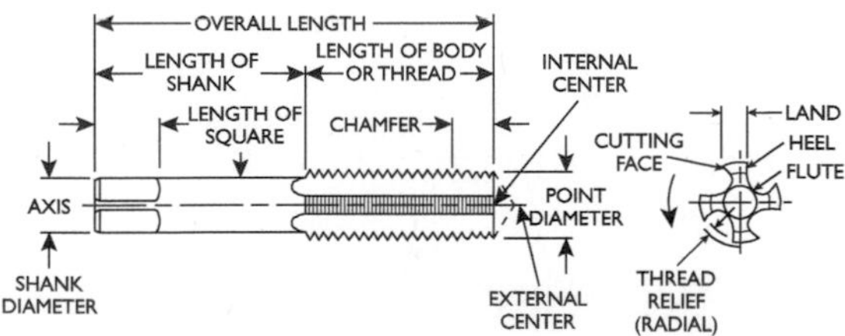

Figure 7-7 Illustrating the typical hand tap.

Since no single tap could possibly meet all the difficult tapping requirements, the manufacturers of threading tools modify the basic tap design in several ways, making tools that are especially suitable for particular tapping needs. The number of flutes may vary from two to as many as nine. Some taps have spiral flutes, winding either to the left or the right. Unless specific reference to the number of flutes is made, taps are supplied having the standard number of flutes for a given size and type. The most widely used sizes for general use are made with four flutes. The flutes deliver lubricant to the tap's cutting edges and provide chip clearance. For tapping especially tough materials, some taps are made with interrupted threads to provide even more chip accommodation and better cutting edge lubrication. These interrupted thread taps have alternate cutting teeth omitted.

The "chamfer" on a tap refers to the angular reduction in diameter of the leading threads at its point. This chamfer allows the tap to enter the hole, and the gradually increasing diameter of the threads serves to lead the threaded body of the tap into the hole as it is turned. The longer the chamfer, the smaller the chip each thread must cut, as the load is distributed over a greater number of cutting edges. Three different chamfer lengths are in common use; these are designated as "taper," "plug," and "bottoming."

Taper taps have the longest chamfer (8 to 10 threads) and are usually used for starting a tapped hole. The taper tap is frequently referred to as a "starting" tap. The plug tap has a 3- or 4-thread chamfer and is used to provide full threads closer to the bottom of a hole than is possible with a taper tap. The bottoming tap has practically no chamfer but when carefully used can tap to the bottom of a hole if it is preceded by a plug tap. Taper taps should always be used for starting a tapped hole. While a tapped hole may be started with a plug tap if care is used, the load of the leading threads is extremely heavy because of the short chamfer. Starting a tapped hole with a bottoming tap should not be attempted.

The first operation in internal threading with a hand tap is to drill the proper diameter hole. The usual method of selecting the drill size is to consult a "tap-drill size chart" as shown in Table 7-2. Note that the chart carries the statement, "Based on approximately 75% full thread."

The notation that the chart is based on approximately 75 percent full thread means that the drill diameter is larger than the minor diameter of the thread to be tapped. The drill size noted in the table will produce a hole sufficiently oversize so that 25 percent of the crest of the internal thread will be missing, as shown in Figure 7-8.

An oversize tap hole is made to provide clearance between the tap hole and the minor diameter of the tap. If this is not done, there will be no clearance; the tap will turn hard and tear the threads, and there is a high risk of breakage. The 25 percent that is missing from the crest of the internal thread does not appreciably reduce its strength.

Table 7-2 Tap-Drill Sizes—Based on Approximately 75% Full Thread

National Coarse and Fine Threads				Taper Pipe		Straight Pipe	
Thread	**Drill**	**Thread**	**Drill**	**Thread**	**Drill**	**Thread**	**Drill**
0–80	3/64	7/16–14	U	1/8–27	R	1/8–27	S
1–64	No.53	7/16–20	25/64	1/4–18	7/16	1/4–18	29/64
1–72	No.53	1/2–12	27/64	3/8–18	37/64	3/8–18	19/32
2–56	No.50	1/2–13	27/64	1/2–14	23/32	1/2–14	47/64
2–64	No.50	1/2–20	29/64	3/4–14	59/64	3/4–14	15/16
3–48	No.47	9/16–12	31/64	1–11 1/2	1 5/32	1–11 1/2	1 3/16
3–56	No.45	9/16–18	33/64	1 1/4–11 1/2	1 1/2	1 33/64	
4–40	No.43	5/8–11	17/32	1 1/2–11 1/2	1 47/64	1 1/2–11 1/2	1 3/4
4–48	No.42	5/8–18	37/64	2–11 1/2	2 7/32	2–11 1/2	2 7/32
5–40	No.38	3/4–10	21/32	2 1/2–8	2 5/8	2 1/2–8	2 21/32
5–44	No.37	3/4–16	11/16	3–8	3 1/4	3–8	3 9/32
6–32	No.36	7/8–9	49/64	3 1/2–8	3 3/4	3 1/2–8	3 25/32
6–40	No.33	7/8–14	13/16	4–8	4 1/4	4–8	4 9/32
8–32	No.29	1–8	7/8				
8–36	No.29	1–12	59/64				
10–24	No.25	1–14	59/64				
10–32	No.21	1 1/8–7	63/64				
12–24	No.16	1 1/8–12	1 3/64				
12–28	No.14	1 1/4–7	1 7/64				
1/4–20	No.7	1 1/4–12	1 11/64				
1/4–28	No.3	1 3/8–6	1 7/32				
5/16–18	F	1 3/8–12	1 19/64				
5/16–24	I	1 1/2–6	1 11/32				
3/8–16	5/16	1 1/2–12	1 27/64				
3/8–24	Q	1 3/4–5	1 9/16				

To tap a hole with a hand tap, begin with the taper tap. Start the tap by placing it in the hole, and then carefully turn it about one half turn until it starts to cut. Check the position of the tap by eye to keep it square with the work surface. Back it up to break the chip, and again turn it in about one half turn. Each time, check the tap to make sure it is square. After the tap is well started, check it with a square. If it is not true, pressure must be exerted in the direction to correct its position on the next several cuts. Use plenty of lubricant and occasionally completely remove the tap from the hole to clean out chips. Continue alternately cutting and reversing the tap to break the chips. When the tap is cutting properly, one can "feel" it as it is turned. If the tap resists turning and there is a springy feeling, it is not cutting properly and should be removed and inspected.

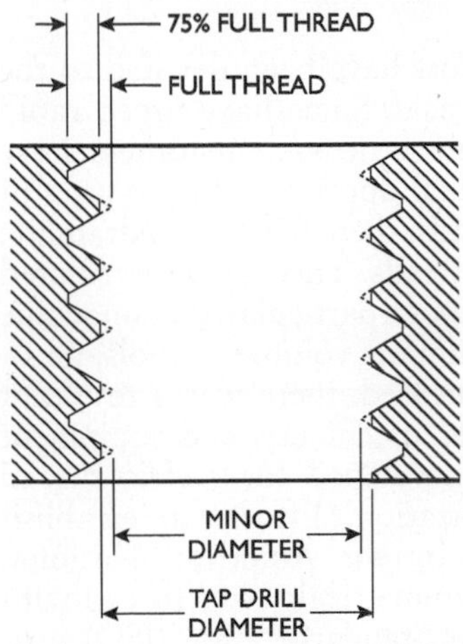

Figure 7-8 Illustrating full thread and 75 percent of full thread.

To tap a blind hole, first use the taper tap, then the plug tap, and finally the bottoming tap. Each tap must be carefully turned in until bottomed. Because the chips cannot fall through, special attention must be paid to chip removal. The tap must be removed more frequently than for a through hole, and the workpiece must be inverted and jarred to remove the chips. If the work cannot be inverted, blow the chips out with air or remove them with a magnet. In special cases, it might even be necessary to use a vacuum for chip removal.

External Threading

Most external screw threads are cut by means of dies because they cut rapidly and are capable of cutting threads to meet most commercial accuracy requirements. There are two general classes of external threading dies: those removed from the thread by unscrewing and those that may be opened so that the cutting edges clear the thread for removal. The nonopening type, usually split with provision for adjustment, commonly called "threading dies," are the type used by millwrights and mechanics. The die is placed in a "diestock," which provides both a means of holding the die and handles for leverage to turn it. External threading with hand dies is similar to tapping, in that care must be taken to start the die square, the same type of cutting "feel" is involved, the die must be occasionally reversed to break the chip, and adequate lubrication is necessary. Split dies are adjustable and may be opened so that the first cut may be more easily made or external thread size varied to suit requirements.

Metric Screw Threads

Much interest and a great deal of effort have been devoted to the subject of metric measure in recent years. Some have urged rapid conversion to worldwide use of the metric system; others have advocated a more cautious and gradual approach. In the area of mechanical fasteners using screw threads, there has been a dramatic increase in the use of metric system threads. This can be attributed to the import and export business, most particularly to machine tool and automobile imports. As the import volume of tools, automobiles, etc., with metric threads increased, there was a resultant increased usage of metric-related tools, machinery, and equipment of all types. Consequently, a need developed for metric thread information and domestic standardization. Efforts to establish standards resulted in the adoption by the American National Standards Institute (ANSI) of a 60° symmetrical screw thread with a basic International Organization for Standardization (ISO) profile. This is termed the ISO 68 profile. In the ANSI standard, this 60° thread profile is designated as the M-Profile. The ANSI M-Profile standard is in basic agreement with ISO screw thread standards and features detailed information for diameter-pitch combinations selected as preferred standard sizes. The basic M-Profile is illustrated in Figure 7-9.

The designations, terms, etc., used in the new metric standard differ in many respects from those used in the familiar Unified standard. Because the Unified standard incorporates terms and practices of long standing with which mechanics have knowledge and experience, an explanation of the M-Profile metric system may best be accomplished by a comparison of metric M-Profile standards with Unified standards.

A major difference between the standards is in the designation used to indicate the thread form. The Unified standard uses a series of capital letters, which not only indicate that the thread is of the Unified form but also indicate whether it is a coarse (UNC), a fine (UNF), an extra-fine (UNEF), or a constant pitch (UN), which is a series of pitches used on a variety of diameters. In the metric M-Profile standard, the capital letter "M" is used to indicate that the thread is metric M-Profile with no reference to pitch classification.

A second major difference is that the Unified standard designates the thread pitch in terms of the number of threads in one inch of thread length. The metric M-Profile standard states the specific pitch measurement, i.e., the dimension in millimeters from the centerline of one thread to the centerline of the next thread.

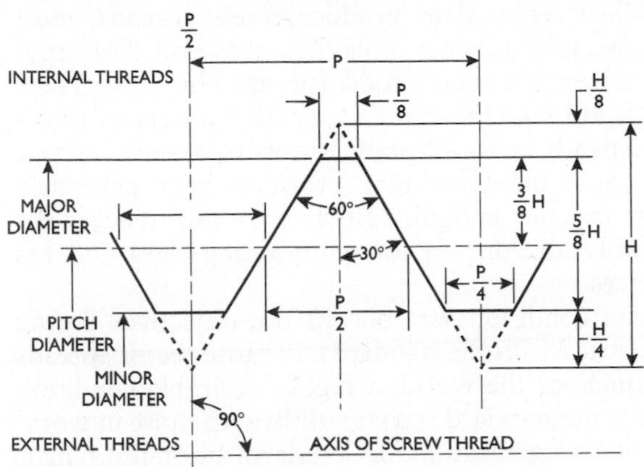

Figure 7-9 Basic M-Profile metric thread.

A third major difference is that the Unified system establishes limits of tolerance called classes. Classes 1A, 2A, and 3A apply to external threads only; classes 1B, 2B, and 3B apply to internal threads only. Classes 3A and 3B provide a minimum, and classes 1A and 1B a maximum. In the metric M-Profile standard, the tolerance designation uses a number and a letter to indicate the pitch diameter tolerance, as well as a number and a letter to designate crest diameter tolerance. This results in four symbols for tolerance designation rather than the two used in the Unified system. In the metric M-Profile standard, the lowercase letter "g" is used for external threads, and the capital letter "H" is used for internal threads. The numbers 4, 5, 6, 7, and 8 are used to indicate the degree of internal thread tolerance; the numbers 3, 4, 5, 6, 7, 8, and 9 indicate the degree of external thread tolerance. Thus, two nominally similar threads are designated as follows:

Unified	***Metric M***
⅜-16-UNC-2A	M10 × 1.5-6g6g

Common shop practice in designating Unified screw threads is to state only the nominal diameter and the threads per inch—for example, ⅜-16 for a coarse thread and ⅜-24 for a fine thread. Most mechanics recognize the thread series by the threads per inch number. As for the thread class symbol, it is usually assumed that if no class number is stated, the thread is a class 2A or 2B. This is a safe assumption because classes 2A and 2B are commonly used tolerances for general applications, including production of bolts, screws, nuts, and similar fasteners. When required to produce screw threads, most mechanics use commercial threading tools (i.e., taps and dies) or, if chasing threads, commercial gauging tools for size checking. These commonly used tools are made to class 2A or 2B tolerances; therefore, the mechanic depends on tool manufacturers to maintain screw thread accuracy. Because manufacturers' products have extremely close tolerances, this practice is highly satisfactory and relieves the mechanic of practically all concern for maintaining sizes and fits within stated tolerances.

To reach this same condition was one of the objectives during efforts to establish metric M-Profile standards. Because metric threads are in wide use throughout the world, a highly desirable condition would be one in which the standard is compatible with those in world use. As the International Organization for Standardization (ISO) metric standard was, for all practical purposes, recognized as the worldwide metric standard, the metric M-Profile standard is in large measure patterned after it. It has a profile in basic agreement with the ISO profile and features detailed information for diameter-pitch combinations selected as preferred standard sizes.

Metric M-Profile screw threads are identified by letter (M) for thread form, followed by the nominal diameter size and pitch expressed in millimeters, separated by the sign × and followed by the tolerance class separated by a dash (-) from the pitch. For example, a coarse-pitch metric M-Profile thread for a common fastener would be designated as follows:

External Thread M-Profile, Right Hand

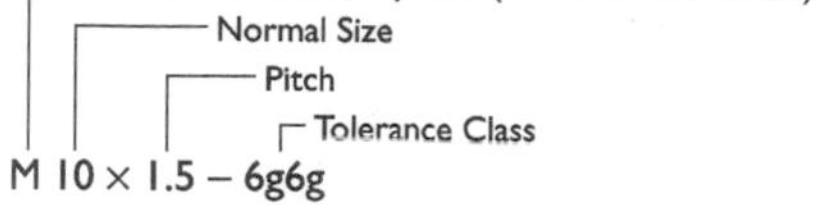

Metric Thread Symbol (ISO 68 Metric Thread)
Normal Size
Pitch
Tolerance Class
M 10 × 1.5 – 6g6g

The simplified international practice for designating coarse-pitch ISO screw threads is to leave off the pitch. Thus, an M14 × 2 thread is simply designated M14. In the ANSI standard, to prevent misunderstanding, it is mandatory to use the value for pitch in all designations. Thus, a 10-mm coarse thread is designated M10 × 1.5. When no tolerance classification is stated, it is assumed to be classification 6g6g (usually stated as simply 6g), which is equivalent to the United classification 2A, the commonly used classification for general applications.

The standard metric screw thread series for general-purpose equipment's threaded components and mechanical fasteners is the coarse-thread series. The diameter-pitch combinations selected as preferred standard sizes in the coarse-pitch series are listed in Table 7-3. These combinations are in basic agreement with ISO standards.

The metric M-Profile designation does not specify series of diameter-pitch combinations as does the Unified system (i.e., coarse, fine, etc.). Although no indication of such grouping is given in the designation of a metric thread, series groupings are recommended. The coarse-pitch series of diameter-pitch combinations shown in Table 7-3 are described as standard metric screw thread series for general-purpose equipment's threaded components and mechanical fasteners.

A second series, called fine-pitch M-Profile metric screw threads and shown in Table 7-4, lists additional diameter-pitch combinations that are standard for general-purpose equipment's threaded components. These combinations, in some instances, list more than one pitch for a nominal diameter. As with the coarse-pitch series, they are in basic agreement with ISO screw standards.

Table 7-3 Standard Coarse-Pitch M-Profile General Purpose and Mechanical Fastener Series

Nominal Size	*Pitch*	*Nominal Size*	*Pitch*
1.6	0.35	20	2.5
2	0.4	24	3.0
2.5	0.45	30	3.5
3	0.5	36	4.0
3.5	0.6	42	4.5
4	0.7	48	5.0
5	0.8	56	5.5
6	1.0	64	6.0
8	1.25	72	6.0
10	1.5	80	6.0
12	1.75	90	6.0
14	2.0	100	6.0
16	2.0		

In preparation for cutting internal threads, a hole must be drilled slightly larger than the thread's minor diameter. The reason for the oversize hole is to provide clearance between the wall of the hole and the roots of the tap threads. This gives chip space and allows free turning of the tap, reducing the tendency for the threads to tear. A practice of long standing when tapping Unified screw threads is to make the hole 25 percent larger than the minor diameter. This results in an internal thread that is 75 percent of standard. The 25 percent that is missing from the crest does not appreciably reduce its strength. Unified thread tap-drill charts, listing tap-drill sizes, are a common shop convenience. Most carry this notation: "Based on approximately 75 percent of full thread."

Table 7-4 Standard Fine-Pitch M-Profile Screw Threads

Nominal Size	*Pitch*		*Nominal Size*	*Pitch*	
8	1		40	1.5	
10	0.75	1.25	42	2.0	
12	1	1.25	45	1.5	
14	1.25	1.5	48	2.0	
15	1		50	1.5	
16	1.5		55	1.5	
17	1		56	2.0	
18	1.5		60	1.5	
20	1		64	2.0	
22	1.5		65	1.5	
24	2		70	1.5	
25	1.5		72	2.0	
27	2		75	1.5	
30	1.5	2.0	80	1.5	2.0
33	2		85	2.0	
35	1.5		90	2.0	
36	2		95	2.0	
39	2		100	2.0	

The metric tap-drill chart (see Table 7-5) lists both coarse and fine diameter-pitch combinations and the recommended drill size in millimeters to produce approximately 75 percent internal metric threads.

Table 7-5 Metric Tap-Drill Chart

Nominal Size	Pitch	Tap-Drill Size	Inch Decimal
1.6	.35	1.25	.050
2	.4	1.6	.063
2.5	.45	2.05	.081
3	.5	2.5	.099
4	.7	3.3	.131
5	.8	4.2	.166
6	1.0	5.0	.198
8	1.0	7.0	.277
8	1.25	6.8	.267
10	.75	9.3	.365
10	1.25	8.8	.346
10	1.5	8.5	.336
12	1.0	11.0	.434
12	1.25	10.8	.425
12	1.75	10.3	.405
14	1.25	12.8	.503
14	1.5	12.5	.494
14	2.0	12.0	.474
16	1.5	14.5	.572
16	2.0	14.0	.553
18	1.5	16.5	.651
20	1.0	19.0	.749
20	1.5	18.5	.730
20	2.5	17.5	.692
22	1.5	20.5	.809
24	2.0	22.0	.868
24	3.0	21.0	.830
25	1.5	23.5	.927

Millimeters × .03937 = inch decimals.

The values in the table result from the following formula:

Millimeter Drill Size for 75% Thread = Major Diameter − (.974 × Pitch)

Thread Repair Using Heli-Coil Inserts

There are many reasons why internal threads on a machine assembly can become stripped or cross-threaded. Repeated use, overtorquing the wrench, or just one mistake by cocking a bolt while reassembling a machine can cause the threads to become useless. An old-timer's fix is to drill the hole out to allow tapping to the next size screw thread and use a larger bolt. Although this certainly will produce a quality repair, this fix is not always possible because of space or assembly requirements. The Heli-Coil insert is a good maintenance fix.

The Heli-Coil insert is a precision-formed screw thread coil of stainless steel wire with a diamond-shaped cross section (Figure 7-10). When they are installed into Heli-Coil tapped holes, they provide permanent conventional 60' internal screw threads that accommodate most standard bolts or machine screws. This procedure consists of choosing an insert that has the same thread size as the damaged thread. The old threads are drilled out and the hole is tapped with a Heli-Coil tap. Special tools are used to insert the proper Heli-Coil insert and to break off the driving tang. The hole can now be reused for assembly using the original-sized bolt. In many cases, the Heli-Coil insert is stronger that the original threaded hole, and at the same time it significantly reduces the possibility of thread wear, seizing, and corrosion.

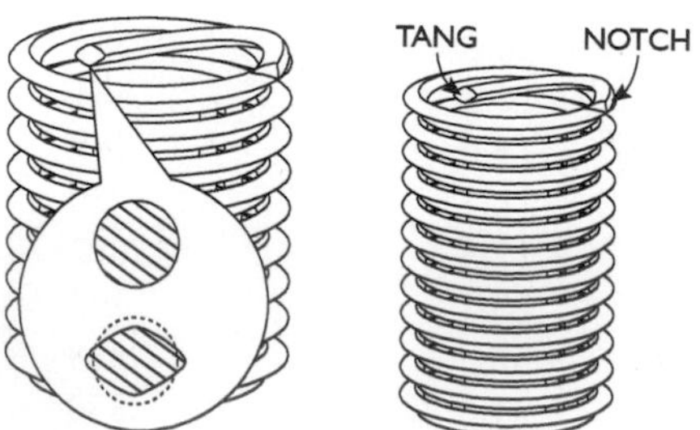

Figure 7-10 The Heli-Coil insert.
Heli-Coil is a registered trademark of Emhart Teknologies, Inc.

Each insert has a tang for installation purposes. It is notched for easy removal so that a through, free-running threaded assembly results. The tang is broken off after installation by simply striking it with a piece of rod. In sizes over ½ in. and in all spark plug applications, use long-nosed pliers, bending the tang up and down until it snaps off at the notch.

Heli-Coil inserts are retained in the hole with spring-like action. In the free state, they are larger in diameter than the tapped hole into which they are installed. In the assembly operation, the force applied to the tang reduces the diameter of the leading coil and permits it to enter the tapped thread. When the torque or rotation is stopped, the coils expand outward with a spring-like action, anchoring the insert permanently in place against the tapped hole. No staking, locking, swaging, key, or interference fits are required. Because the insert is made of wire, it automatically adjusts itself to any expansion or contraction of the parent material.

Repair Procedure

Heli-Coil inserts allow stripped threads to be fixed in three very easy steps, as outlined in Figure 7-11.

1. Drill.
2. Tap.
3. Install.

If there are traces of thread left, they must be cleaned out before the hole can be prepared for the Heli-Coil insert. The hole must be drilled out to the size specified on the repair kit itself.

The hole is then threaded using the tap supplied in the kit. This is a tap specifically designed to prepare a hole for a Heli-Coil insert; it cannot be used for anything else, and no other tap can be used in its place for this purpose.

After tapping, the insert is wound into the hole using the installation tool supplied in the kit. When the insert has been fully installed (that is, when the top coil is ¼ to ½ turn below the surface), the tang is broken off and the result is a stainless-steel thread that is the same size as the original. It is also stronger and more reusable than the original tapped hole.

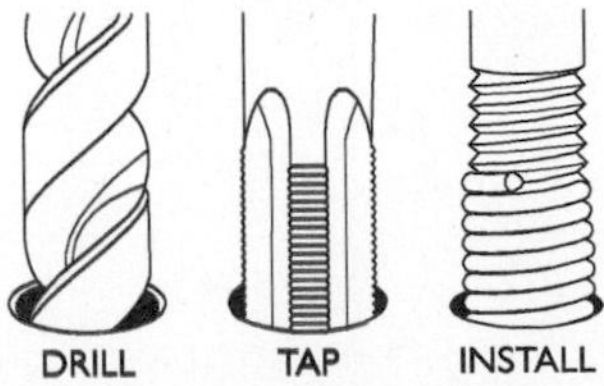

Figure 7-11 Heli-Coil repair procedure.
Heli-Coil is a registered trademark of Emhart Teknologies, Inc.

Chapter 8
Mechanical Fasteners

Machinery and mechanical equipment are assembled and held together by a wide variety of fastening devices. Manufacturers may select from hundreds of thousands of standard fasteners and millions of special sizes, kinds, and shapes. In addition, fasteners made of plastics, ceramics, new alloys, and other materials are constantly being refined and developed. A listing and discussion of mechanical fasteners, therefore, can only cover the common types encountered by mechanics in their day-to-day duties.

Threaded fasteners are by far the most widely used, and the bolt, screw, and stud are the most common threaded fasteners. The bolt is described as an externally threaded fastener designed for insertion through holes in assembled parts. It is normally tightened and released by turning a mating nut. A screw differs from a bolt in that it is supposed to mate with an internal thread into which it is tightened or released by turning its head. These definitions obviously do not always apply, since bolts can be screwed into threaded holes and screws can be used with a nut. The third most common fastener, the stud, is simply a cylindrical rod threaded on either one or both ends or throughout its entire length. Figure 8-1 shows some of the common bolt, screw, and stud types.

Most threaded fasteners are furnished with either coarse threads conforming to Unified National Coarse (UNC) standards or Unified National Fine (UNF) threads. UNF standards specify substantially more threads per inch.

For the majority of applications, fasteners with coarse threads are used because of the following advantages:

- They assemble faster and more easily, providing a better start with less chance of cross-threading.
- Nicks and burrs from handling are less liable to affect assembly.
- They are less liable to seize in temperature applications and in joints where corrosion will form.
- They are less prone to strip when threaded into lower-strength materials.
- Coarse threads can be tapped more easily in brittle materials and/or materials that crumble easily.

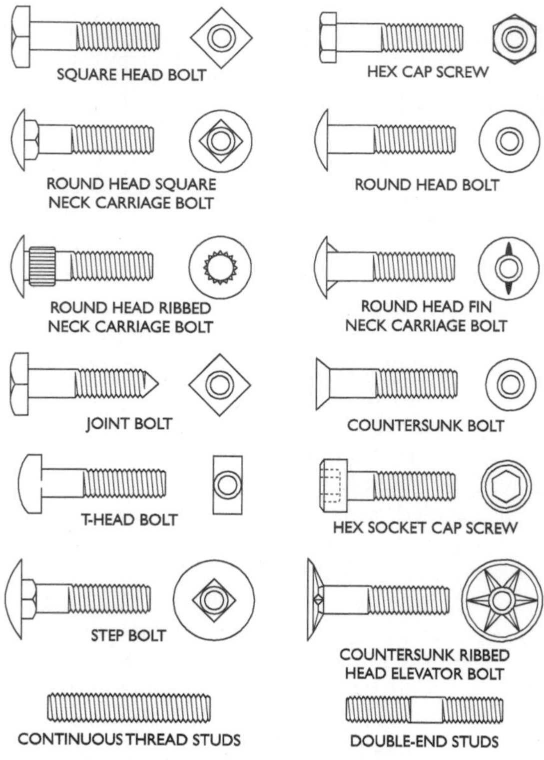

Figure 8-1 Standard bolts, screws, studs.

The use of fine threads may provide superior fasteners for applications where specific strength or other qualities are required. Fine threads provide the following advantages:

- They are about 10 percent stronger than coarse threads because of their greater cross-sectional area.
- In very hard materials, fine threads are easier to tap.
- They can be adjusted more precisely because of their smaller helix angle.
- Where length of engagement is limited, they provide greater strength.
- They may be used with thinner wall thickness because of their smaller thread cross section.

Most threaded fasteners are installed where vibration occurs. This mechanical motion tends to overcome the frictional force between the threads, causing the fastener to back off and loosen. Washers are usually placed beneath the fastener head to help maintain frictional resistance to loosening. Various classes of washers are shown in Figure 8-2.

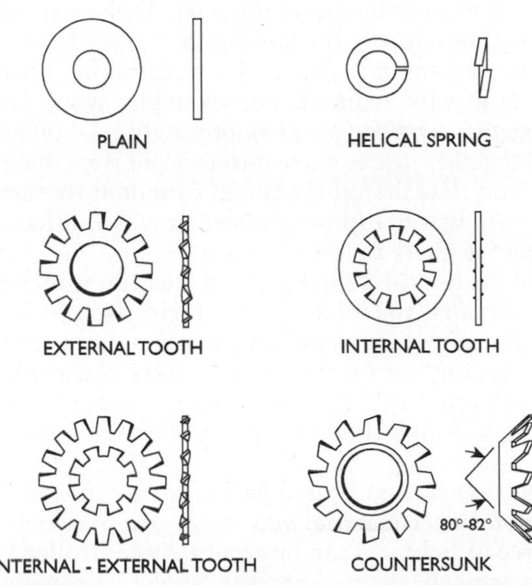

Figure 8-2 Various classes of standard washers.

The primary function of a washer is to provide a surface against which the head of the fastener or the surface of the nut can bear. Flat washers provide this surface and spread the load over an increased holding area. Normally they do not provide much additional locking action to the fastener. Lock washers tend to retard loosening of inadequately tightened fasteners. Theoretically, if fasteners were properly tightened, lock washers would not be necessary. Multiple-tooth locking washers provide greater resistance to loosening because their teeth bite into the surface against which the head or nut bears. Their teeth are twisted to slide against the surface when tightened and hold when there is a tendency to loosen. Lock washers are most effective when the mating surface under the teeth is soft. A harder surface resists the digging action of the teeth.

Tightening Threaded Fasteners

To obtain maximum holding power from a threaded fastener, it must be tightened until its initial tension is greater than the greatest external load that will be imposed on the assembly. If a fastener is over-tightened, it may break off as the tightening force is applied. However, the effects of insufficient tightening may not be evident until a joint separates. Neglecting to apply a high enough preload to the fastener can result in early fatigue and failure. The greater the clamping force (preload) developed, the less will be the tendency of parts to creep or shift, and the possibility of a nut backing off or loosening in service is greatly reduced. For example, when two plates are fastened together by a cap screw tightened to a 10,000-lb preload, they are held together just as if a five-ton weight were placed on top of them. Any force less than this amount cannot move them apart. If, however, the screw is not fully tightened, it will be affected by the parting force and is likely to fail.

Failure results from the imposed external load plus vibration and temperature changes. Because the joint is not sufficiently tightened to resist these separating forces, movement within the joint occurs. An action similar to the simplest form of fatigue takes place—that of metal being bent back and forth. If the fastener does not actually break from this action, it is elongated enough to allow the joint to loosen.

The two most important factors, therefore, in fastener tightening are the strength of the fastener material and the degree to which it is tightened. The degree of tightness can be accurately controlled by the use of a torque wrench to measure the applied tightening torque. The strength of the fastener can be determined by markings on the head of the bolt or screw. These head markings were developed by the SAE and ASTM to denote relative strength and have been incorporated into standards for threaded fastener quality. They are called "grade markings" or "line markings" and are listed in Table 8-1.

As shown in Table 8-1, unmarked bolt heads or cap screws are generally considered to be mild steel. The greater the number of marks on the head, the higher the quality. Thus, bolts of the same diameter vary in strength and require a correspondingly different tightening torque or preload. Torque, the turning effort or force applied to the fastener to preload it or place it in tension, is normally expressed in inch-pounds (in. lbs) or foot-pounds (ft lbs). A one-pound weight or force applied to a lever arm one foot long exerts one foot-pound, or 12 inch-pounds, of torque. See Chapter 29 for information about the torque wrench.

Table 8-1 ASTM and SAE Grade Markings for Steel Bolts and Screws

Grade Marking	*Specification*	*Material*
	SAE—Grade 0	Steel
	SAE—Grade 1	Low-Carbon Steel
	ASTM—A 307	
	SAE—Grade 2	Low-Carbon Steel
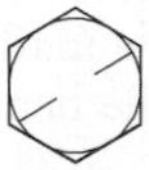	SAE—Grade 3	Medium-Carbon Steel, Cold Worked
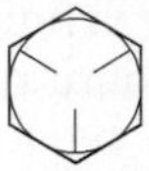	SAE—Grade 5 ASTM—A 449	Medium-Carbon Steel, Quenched and Tempered
	ASTM—A 325	Medium-Carbon Steel, Quenched and Tempered
	ASTM—A 354 Grade BB	Low-Alloy Steel, Quenched and Tempered
	ASTM—A 354 Grade BC	Low-Alloy Steel, Quenched and Tempered
	SAE—Grade 7	Medium-Carbon Alloy Steel, Quenched and Tempered, Roll Threaded After Heat Treatment
	SAE—Grade 8	Medium-Carbon Alloy Steel, Quenched and Tempered
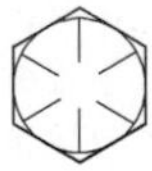	ASTM—A 354 Grade BD	Alloy Steel, Quenched and Tempered
	ASTM—A 490	Alloy Steel, Quenched and Tempered

ASTM Specifications:

A 307—Low-Carbon Steel Externally and Internally Threaded Standard Fasteners.

A 325—High-Strength Steel Bolts for Structural Steel Joints, Including Suitable Nuts and Plain Hardened Washers.

A 449—Quenched and Tempered Steel Bolts and Studs.
A 354—Quenched and Tempered Alloy Steel Bolts and Studs with Suitable Nuts.
A 490—High-Strength Alloy Steel Bolts for Structural Steel Joints, Including Suitable Nuts and Plain Hardened Washers.

SAE Specification

J 429—Mechanical and Quality Requirements for Threaded Fasteners.
Courtesy The American Society of Mechanical Engineers

Several factors that affect the tension developed in a fastener when tightened to a given torque are the type of lubricant, if any, used on the threads; the material from which the fastener is made; the type of plating, if any; the type of washer used under the head; the finish of the thread surfaces; and possibly others.

Because of these variables, it is not possible to establish a definite relationship between torque applied to the fastener and internal tension developed. Manufacturers' specifications should be followed on specific applications. For noncritical applications, general torque tables may be used as a guide for fastener tightening. Such a general table of maximum torque values for fasteners of varying sizes and materials is shown in Table 8-2. This table is based on the fastener retaining a quantity of oil from the manufacturing process. Adding more oil lubrication will make no appreciable difference on the tension developed in the fastener. The values in this table do not apply if special lubricants such as colloidal copper or molybdenum disulphite are used. Use of special lubricants can reduce the amount of friction in the fastener assembly, so the torque applied may produce far greater tension than desired.

Table 8-2 Torque in Foot-Pounds

Fastener Diameter	*Threads per Inch*	*Mild Steel*	*Stainless Steel 18-8*	*Alloy Steel*
1/4	20	4	6	8
5/16	18	8	11	16
3/8	16	12	18	24
7/16	14	20	32	40
1/2	13	30	43	60
5/8	11	60	92	120
3/4	10	100	128	200
7/8	9	160	180	320
1	8	245	285	490

A table of suggested torque values for graded steel fasteners is shown in Table 8-3. The suggested values are based on loading the fastener to approximately 75 percent of its yield strength. Threads should be lightly oiled or have some oil remaining from manufacture.

Bolt-type threaded fasteners require a mating nut to produce tension by rotating and advancing on the bolt threads. Nuts should be of an equal grade of metal with the bolt to provide satisfactory service. Many types of nuts are available; those in common use are illustrated in Figure 8-3. The variety of styles allows different methods of wrenching, strength to suit applications, and provisions for locking.

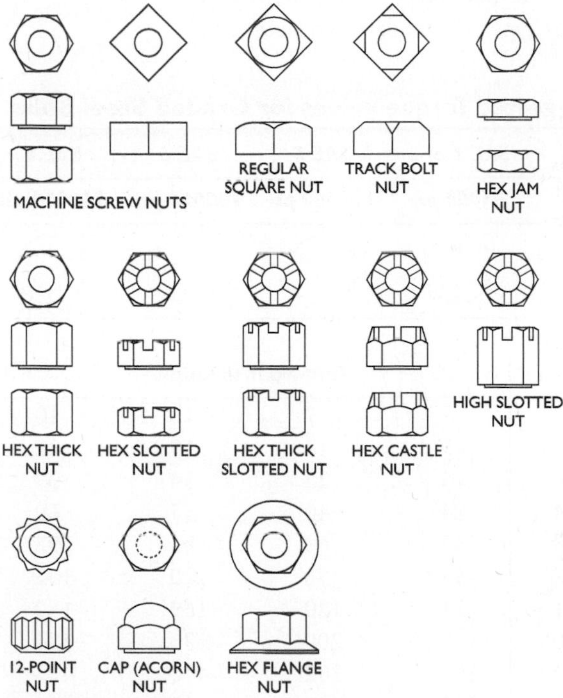

Figure 8-3 Types of standard nuts.

Pins are used in conjunction with threaded fasteners in many applications. After the nut has been properly tightened on the fastener, slots in the nut are lined up with a hole drilled through the fastener body. Some form of holding pin, such as those illustrated in Figure 8-4, is then inserted through the nut slots and fastener hole to prevent the nut from turning in relation to the fastener. The pins shown are not limited to use with fasteners; they are used for many other applications, such as shear pins, for locating and positioning parts, and hinge applications.

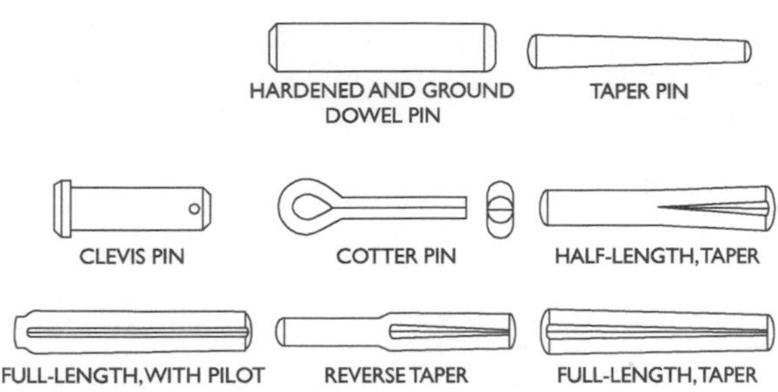

Figure 8-4 Various types of standard pin fasteners.

Table 8-3 Suggested Torque Values for Graded Steel Bolts

Grade		*SAE 1 or 2*	*SAE 5*	*SAE 6*	*SAE 8*
Tensile Strength		*64000 psi*	*105000 psi*	*130000 psi*	*150000 psi*
Grade Mark					
Bolt Dia.	*Thds. per In.*	*Foot-Pounds Torque*			
¼	20	5	7	10	10
5⁄16	18	9	14	19	22
⅜	16	15	25	34	37
7⁄16	14	24	40	55	60
½	13	37	60	85	92
9⁄16	12	53	88	120	132
⅝	11	74	120	169	180
¾	10	120	200	280	296
⅞	9	190	302	440	473
1	8	282	466	660	714

Mechanical Fastener Characteristics

Mechanical fasteners are manufactured in a variety of types and sizes too extensive to enumerate. However, their design features or characteristics are standardized. Fasteners with almost any combination of these characteristics are commercially available. The illustrations will serve to graphically illustrate many of the characteristics that distinguish one fastener from another. Figure 8-5 illustrates the various heads commonly used on threaded fasteners.

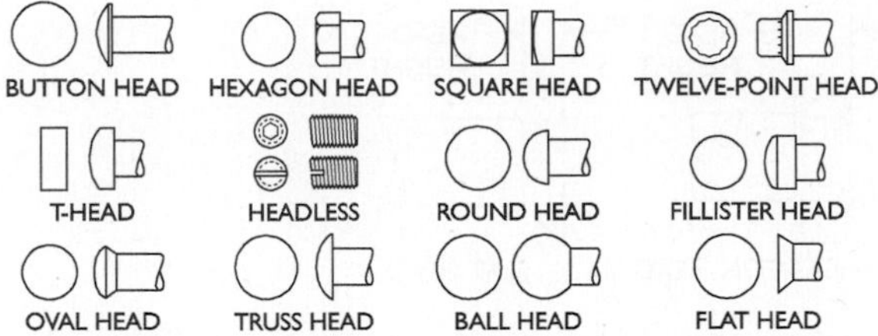

Figure 8-5 Various types of heads used on threaded fasteners.

Points of fasteners fall into the general categories shown in Figure 8-6. The length of fasteners of various styles is measured as shown in Figure 8-7, and measurement of head height is shown in Figure 8-8. Head width across corners and across flats is measured as shown in Figure 8-9. Nut width across flats and nut thickness are measured as shown in Figure 8-10.

The term "neck" is used to define a specialized form of a portion of the body near the head that performs a definite function, such as preventing rotation, or a reduced diameter of a portion of the shank of a fastener. Several common neck styles are shown in Figure 8-11.

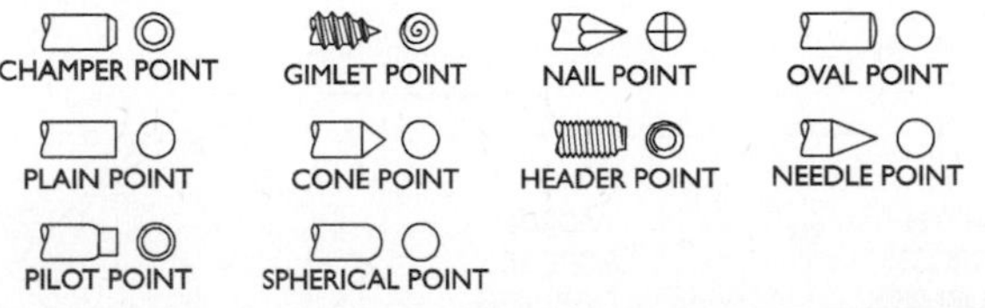

Figure 8-6 A varicty of points on threaded fasteners.

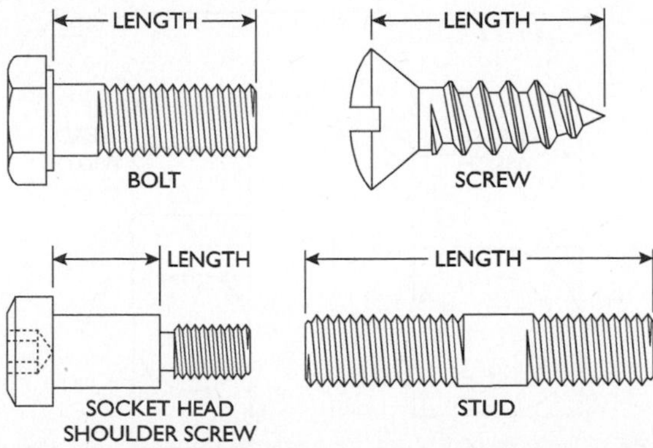

Figure 8-7 Bolt and screw measurement of length.

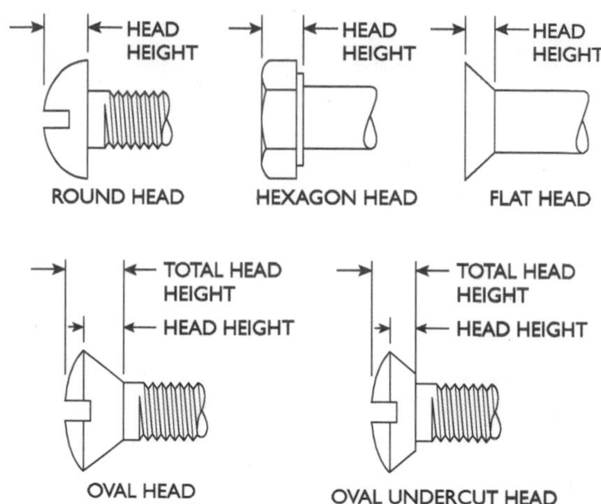

Figure 8-8 Bolt and screw measurement of head thickness.

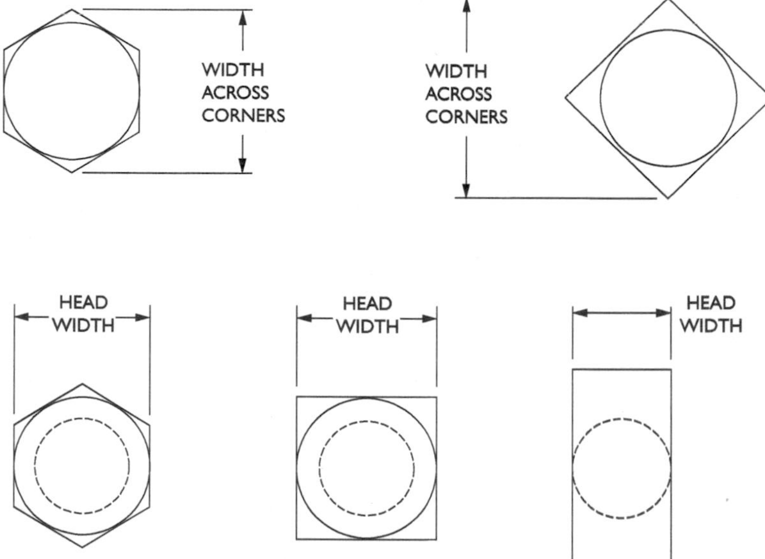

Figure 8-9 Bolt and screw measurement of head width.

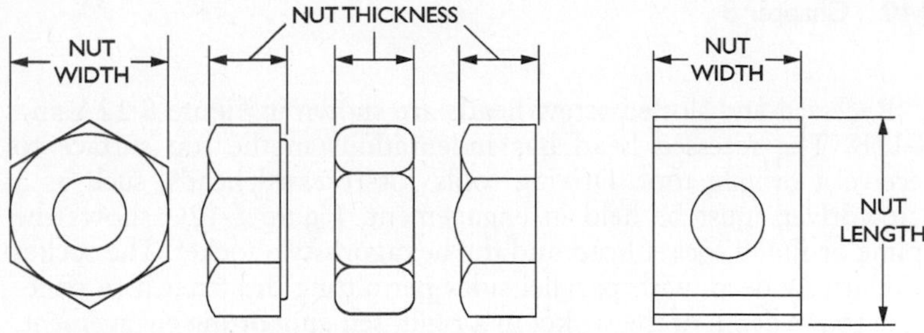

Figure 8-10 Thickness and width measurement of nuts.

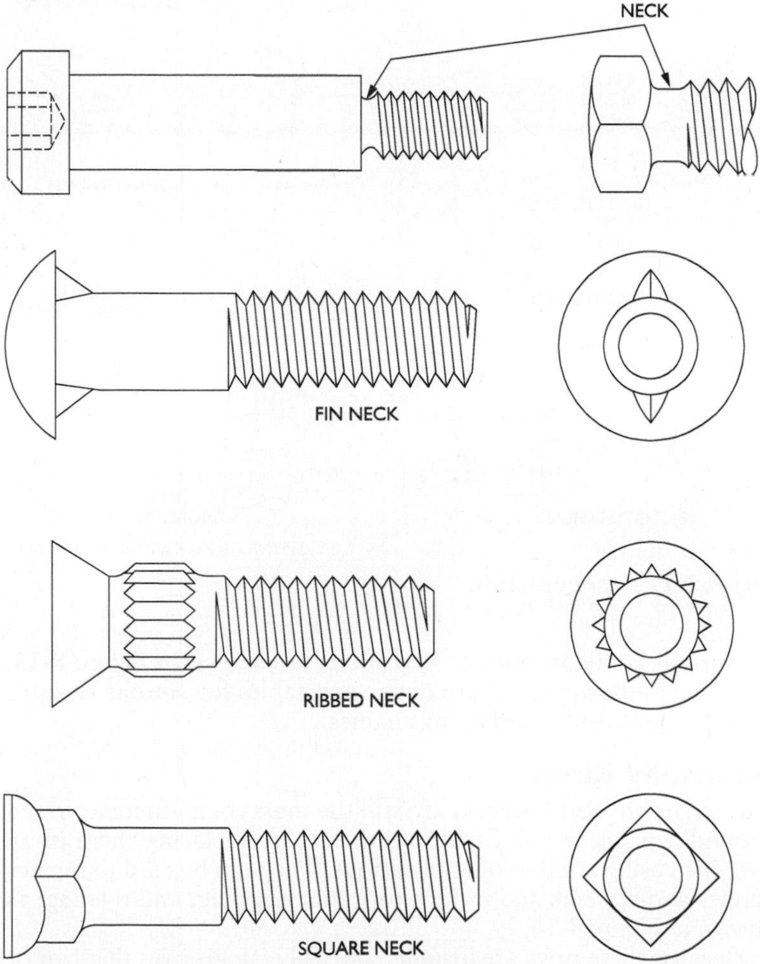

NECK

FIN NECK

RIBBED NECK

SQUARE NECK

Figure 8-11 Various necks used on bolts.

Recessed and slotted screw heads are shown in Figure 8-12A and 8-12B. The recessed head has indentations in the top surface to receive a driving tool. Driving tools for recessed heads, such as a screwdriver, must be held in engagement. Figure 8-12C shows the spline or fluted socket head and the hexagon-style socket. The socket is relatively deep, with parallel sides permitting the wrench to penetrate to the depth of the socket in a rigid, self-supporting engagement.

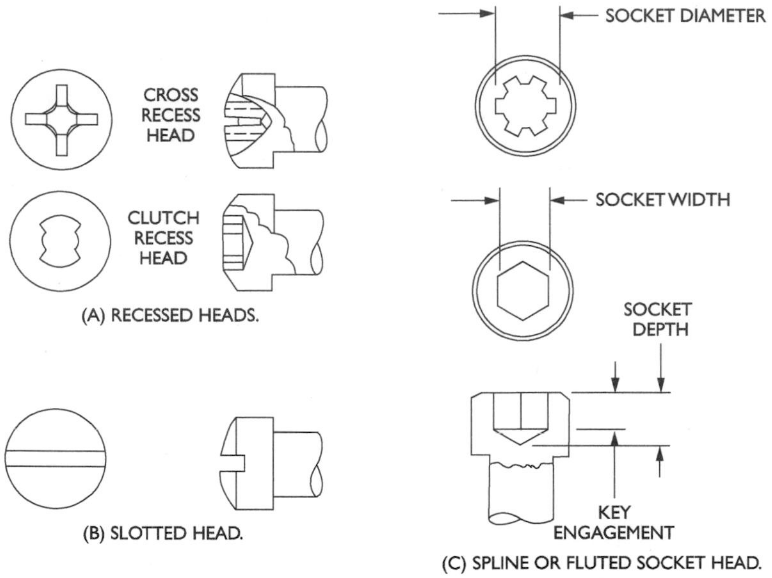

Figure 8-12 Various heads.

Various configurations of bent bolts are shown in Figure 8-13. Included in the appendix are dimension tables for various types of fasteners frequently used by mechanics.

Retaining Rings

Although threaded fasteners are still the most common items found around a plant, metal "retaining rings" are replacing them in an ever-increasing number of applications. They can be used to replace screws and washers and in heat forming, spinning, and riveting, as shown in Figure 8-14.

Because these rings are installed in easily cut grooves that can be machined simultaneously with other production processes, they

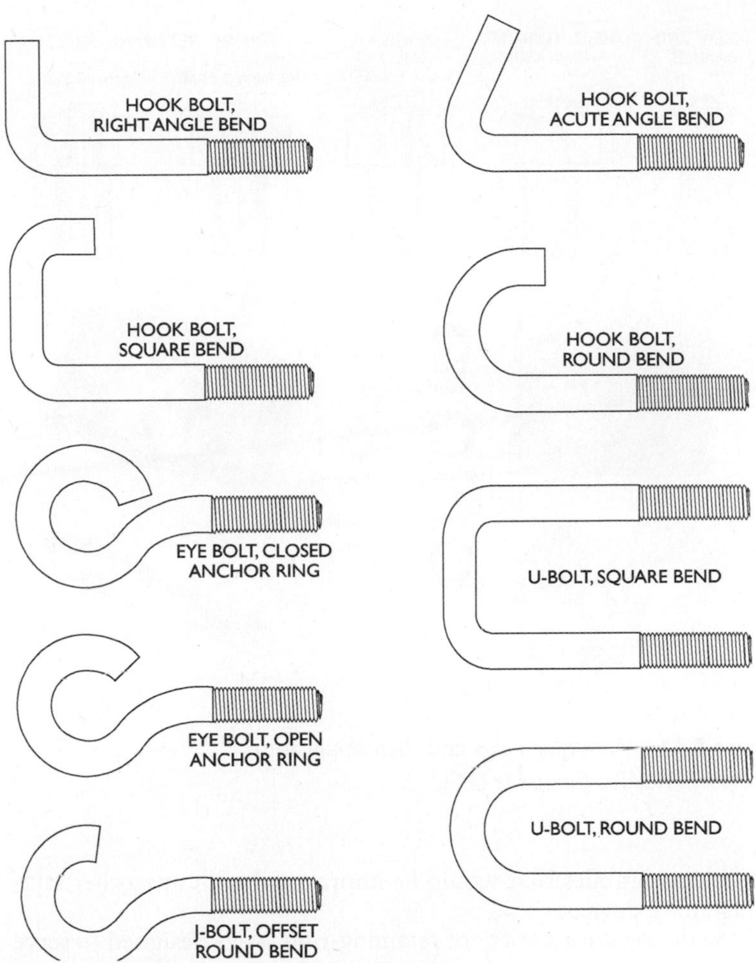

Figure 8-13 Various styles of bent bolts.

eliminate threading, tapping, drilling, and other costly machining operations. An example of this type of design simplification and production cost reduction can be seen in Figure 8-15.

The speed of assembly and disassembly with retaining rings may further reduce manufacturing costs. In addition to the greater economy, the rings often provide a more compact and functional design than other assembly methods. In many cases, they make

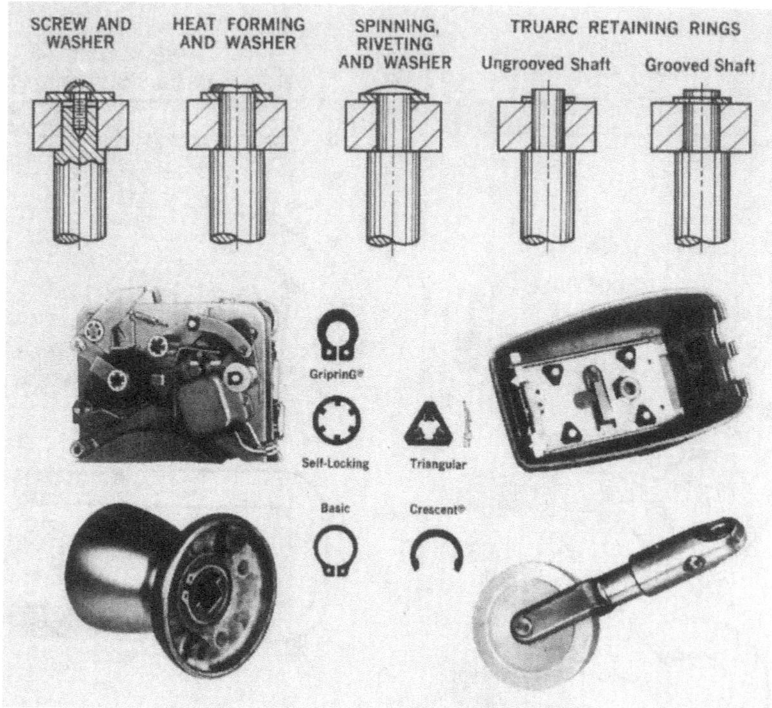

Figure 8-14 Retaining rings and their application. *(Courtesy TRUARC Company LLC)*

possible assemblies that would be impractical with any other style of fastening device.

In wide use are a variety of retaining-ring styles designed to serve as shoulders for accurately locating, retaining, or locking components on shafts and in boxes and housing. They are assembled in an axial direction into precut grooves that ensure precise location of parts. Assembly and disassembly are accomplished by expanding the external rings for assembly over a shaft or by compressing the internal rings for insertion into a bore or housing. Figure 8-16 illustrates a typical external application and the obvious savings in machining. The internal retaining ring and its simplification of design is shown in Figure 8-17.

The great variety of retaining rings in many styles, and their design features, can be seen in the table of standard rings shown in

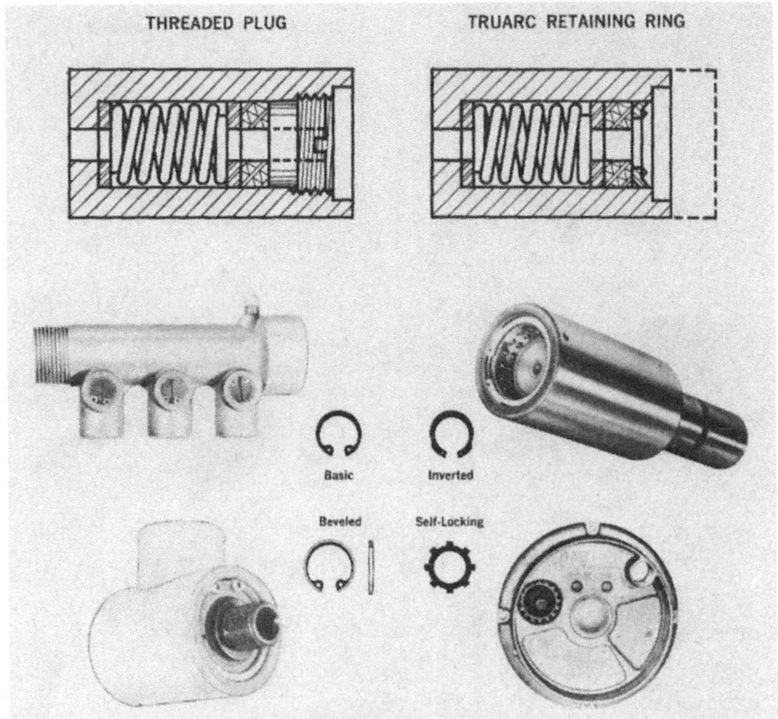

Figure 8-15 Examples of cost-reducing applications of retaining rings. *(Courtesy TRUARC Company LLC)*

Figure 8-18. This guide also gives the names, size ranges, and other pertinent information. Information for specification of ring style, etc., for replacement can also be obtained from this table.

A complete line of assembly and production tools has been developed to ensure rapid, economical installation and removal of retaining rings both on the assembly line and later during field servicing. The tools range from pliers and hand applicators to fully mechanized and automated equipment for high-volume, mass-production ring installations. For fieldwork, such as that performed by millwrights and mechanics, pliers would normally be used for installation or disassembly of retaining rings. Pliers are designed to grasp the ring securely by the lugs and expand or compress it. Three general types of pliers are offered: internal, external, and

Figure 8-16 External application of retaining rings.
(Courtesy TRUARC Company LLC)

convertible. The internal type, shown in use in Figure 8-19, is designed to compress the fasteners for insertion or removal in a bore or housing. The style illustrated has adjustable stops that can be set to limit travel of the tips for automatic alignment with lug holes.

Pliers for external-type retaining rings are designed for assembly or disassembly over a shaft. The pliers shown in use in Figure 8-20 have adjustable stops to limit travel of tips to prevent overspreading the ring.

Convertible pliers are designed to facilitate assembly and disassembly of a number of different types of axially installed

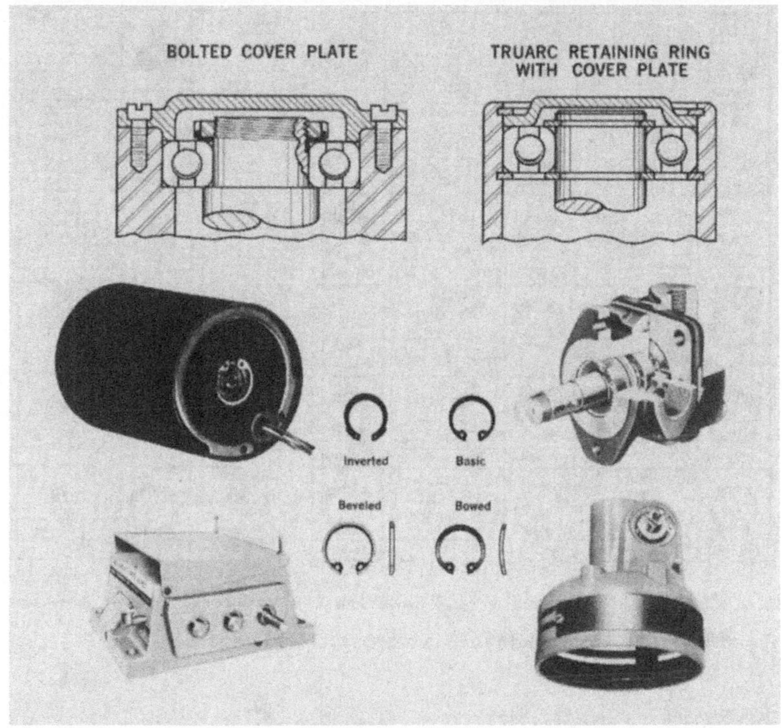

Figure 8-17 Internal application of retaining rings.
(Courtesy TRUARC Company LLC)

retaining rings. They are unique in that they may be used in two ways:

1. To compress internal-type rings for insertion into a bore or housing, as shown in Figure 8-19
2. To expand external-type rings for assembly over a shaft, as shown in Figure 8-20

The procedure for changing settings for pliers is illustrated in Figure 8-21. Convertible pliers are intended primarily for maintenance and repair operations or for production where rings are assembled in small quantities.

Type	Ring	Type	Ring	Type	Ring	Type	Ring
INTERNAL	BASIC **N5000** For housings and bores Size Range: 250 10 0 in / 6 4 254 0 mm	EXTERNAL	BOWED **5101** For shafts and pins Size Range: .188–1.750 in. / 4.8 44.4 mm.	EXTERNAL	REINFORCED **5115** For shafts and pins Size Range: 094- 1 0 in / •	EXTERNAL	TRIANGULAR NUT **5300** For threaded parts Size Range: 6-32 and 8-32 / 10-24 and 10-32 / 1/4-20 and 1/4-28
INTERNAL	BOWED **N5001** For housings and bores Size Range: .250 1.750 in. / 6.4- 44.4 mm.	EXTERNAL	BEVELED **5102** For shafts and pins Size Range: 1 0 10 0 in / 25 4 254 0 mm	EXTERNAL	BOWED E-RING **5131** For shafts and pins Size Range: 110 1 375 in / 2 8 34 9 mm	EXTERNAL	KLIPRING• **5304** **T-5304** For shafts and pins Size Range: .156 1.000 in. / 4.0–25.4 mm.
INTERNAL	BEVELED **•N5002/•N5003** For housings and bores Size Range: ° 1.0—10.0 in. 25.4—254.0 mm / • 1.56—2.81 in. 39.7—71.4 mm	EXTERNAL	CRESCENT® **5103** For shafts and pins Size Range: 125– 2 0 in / 3 2 50 8 mm	EXTERNAL	E-RING **5133** For shafts and pins Size Range: 040- -1.375 in. / 1 0 34 9 mm	EXTERNAL	TRIANGULAR **5305** For shafts and pins Size Range: 062— 438 in. / •
INTERNAL	CIRCULAR **5005** For housings and bores Size Range: 312—2.0 in / •	EXTERNAL	CIRCULAR **5105** For shafts and pins Size Range: 094 1 0 in / •	EXTERNAL	PRONG-LOCK® **5139** For shafts and pins Size Range: .092–.438 in / •	EXTERNAL	GRIPRING® **5555** For shafts and pins Size Range: 079—.750 in. / 2.0—19.0 mm
INTERNAL	INVERTED **5008** For housings and bores Size Range: .750—4.0 in. / 19 0- 101 6 mm	EXTERNAL	INTERLOCKING **5107** For shafts and pins Size Range: .469—3.375 in / 11.9—85.7 mm.	EXTERNAL	REINFORCED E-RING **5144** For shafts and pins Size Range: 094- .562 in / 2.4–14.3 mm.	EXTERNAL	HIGH-STRENGTH **5560** For shafts and pins Size Range: .101— 328 in / •
EXTERNAL	BASIC **5100** For shafts and pins Size Range: 125- 10.0 in / 3 2 254 0 mm.	EXTERNAL	INVERTED **5108** For shafts and pins Size Range: 500 -4 0 in / 12 7 101 6 mm	EXTERNAL	HEAVY-DUTY **5160** For shafts and pins Size Range: 394—2.0 in / 10 0- 50 8 mm.	EXTERNAL	PERMANENT SHOULDER **5590** For shafts and pins Size Range: 250 750 / 6 4 19 0 mm

Figure 8-18 Selector guide for standard retaining rings.
(Courtesy TRUARC Company LLC)

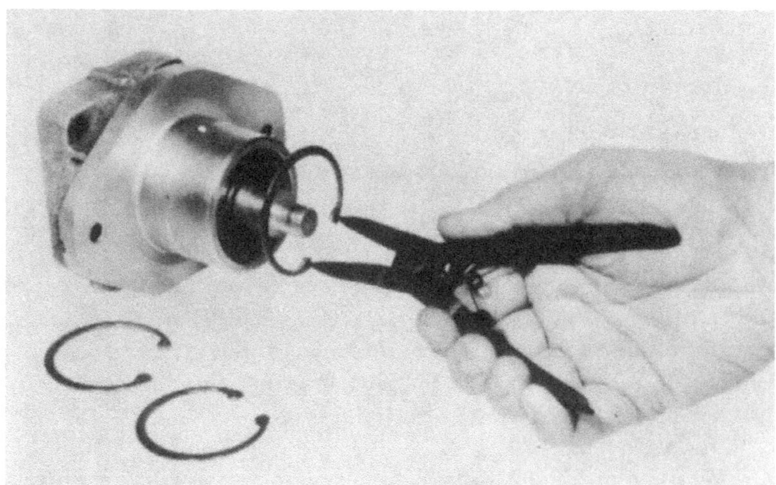

Figure 8-19 Standard internal pliers.
(Courtesy TRUARC Company LLC)

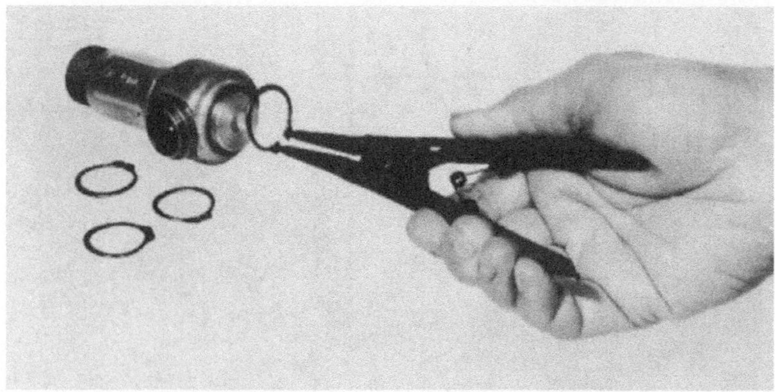

Figure 8-20 Standard external ring pliers.
(Courtesy TRUARC Company LLC)

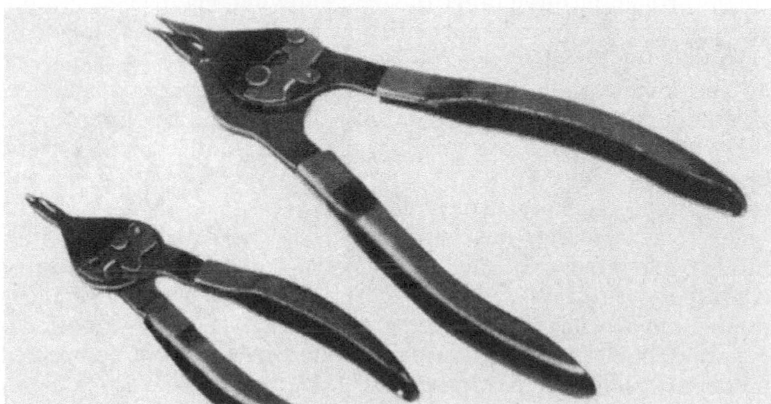

Figure 8-21 Combination internal and external ring pliers.
(Courtesy TRUARC Company LLC)

Retaining rings are made in a wide assortment of types and sizes ranging from the miniature rings used in small precision work to the extremely large rings used in heavy-duty automotive and off-highway equipment. Although they are all assembled in the same manner, the assembly tool must match the ring size. Pliers shown in preceding illustrations are used for what might be classified as the

Figure 8-22 Large-diameter retaining ring pliers.
(Courtesy TRUARC Company LLC)

small-through-medium range of sizes. Shown in Figure 8-22 is a pliers designed for the assembly of rings in a grouping that might be classified as large diameter.

The jaw design in Figure 8-22 is such that adequate mechanical advantage is available to flex large rings, yet the narrow span of the handles provides the necessary comfort and control for the operator.

Retaining rings are also made in extra-large or giant sizes. The pliers for these rings have an added feature, namely the double ratchet that allows locking the pliers in either the expansion or contraction direction. Figure 8-23 shows typical giant rings and a double-ratchet pliers in operating engagement. Double-ratchet pliers are designed to lock the ring in either the spread or contracted mode, depending on the ring type.

Standard pliers for large-size rings are equipped with a double ratchet that reduces the effort necessary to spread or compress the ring and locks the pliers at any given point of expansion or contraction. Thus, it is unnecessary for the operator to maintain constant pressure on the handles. A locking mechanism also serves as a safety device to prevent the ring from springing loose by accident. The safe handling of large rings, such as those shown in Figure 8-24, requires ratchet-type pliers.

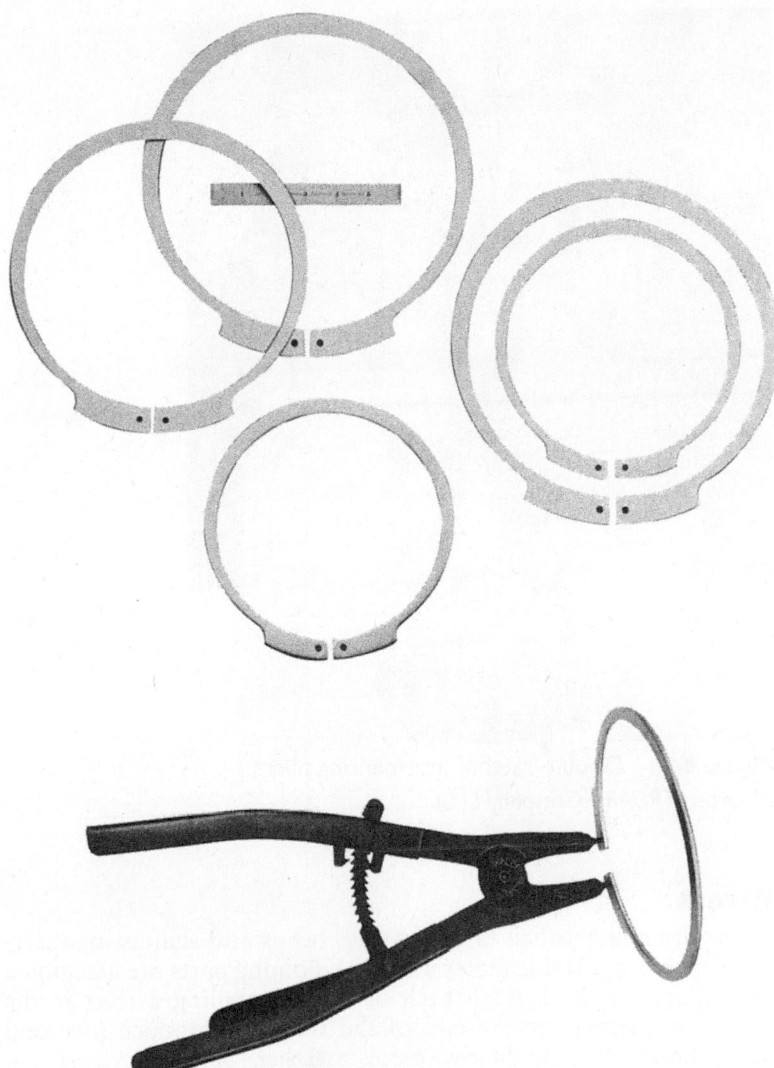

Figure 8-23 Giant-size retaining ring pliers.
(Courtesy TRUARC Company LLC)

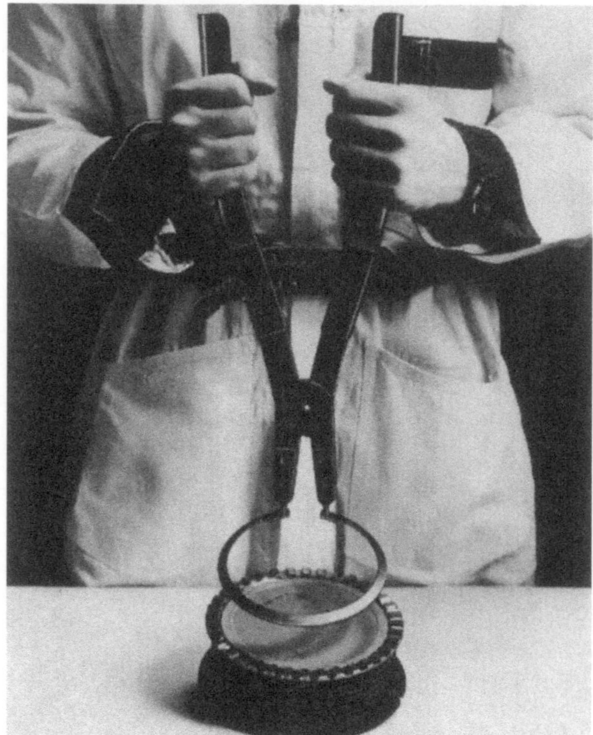

Figure 8-24 Double-ratchet internal ring pliers. *(Courtesy TRUARC Company LLC)*

Rivets

Rivets are nonthreaded fasteners with heads and shanks manufactured from a malleable material. Two adjoining parts are assembled by drilling a hole through each of them, installing a rivet in the holes, and deforming the end of the shank to produce a second head, thereby locking the two pieces together.

Although rivets are still in use by manufacturers, they are less often used in the maintenance of equipment. There is one exception—the use of POP rivets. These handy fasteners are made in three common diameters: 1⁄8, 5⁄32, and 3⁄16 in. Figure 8-25 shows how they work, and Table 8-4 shows selection criteria.

POP rivets are "blind" rivets. That means they are inserted and set from the same side of the work. In Figure 8-25, step 1 shows the POP rivet inserted into a hole in the work. Step 2 shows the jaws of the

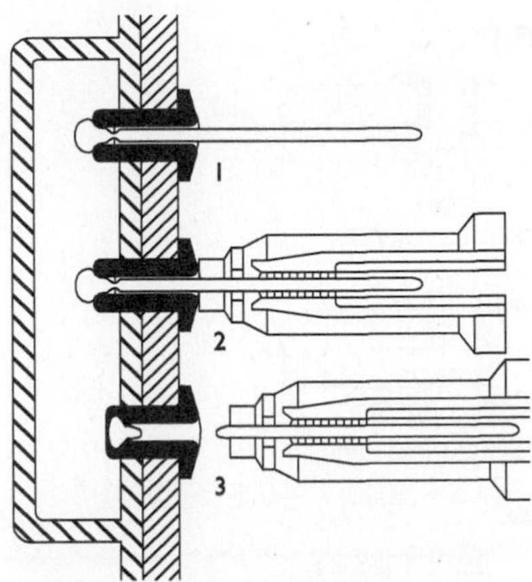

Figure 8-25 How a POP rivet works.
POP is a registered trademark of Emhart Teknologies, Inc.

POP Rivetool gripping the mandrel of the rivet. In step 3, the setting tool is actuated, setting the rivet by pulling the mandrel head into the rivet body, expanding it, and forming a strong, reliable fastening.

Table 8-4 POP Rivet Selection

POP Rivet Type	*Application*
Open-end rivet	Most used for general applications. Covers a wide range of repairs or installations. Available with a dome head, countersunk head, or large flange head.
Closed-end sealing rivet	Prevents the leakage of vapor or liquid. One hundred percent mandrel retention. Provides greater tensile and shear strength than equivalent open-end rivet.
Multigrip rivet	Wide grip range. Optimum clamp-up force. Accommodates oversize and irregularly sized holes.
T-rivet	Designed for structural and similar high-strength applications. High clamp loads. Exceptional pull-up properties.
Load-spreading rivet	Designed for plastics and other brittle or soft materials. Resists pull through or cracking.

POP is a registered trademark of Emhart Teknologies, Inc.

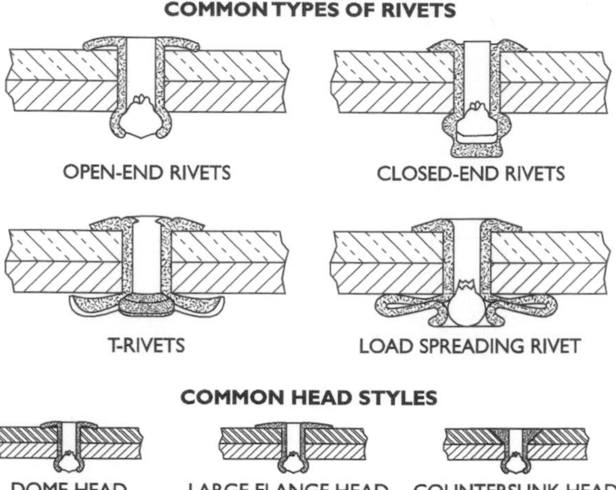

Figure 8-26 Types and styles of POP rivets.
POP is a registered trademark of Emhart Teknologies, Inc.

POP rivets are made in various types and styles to handle a broad range of fastening applications, as shown in Figure 8-26.

Engineering Adhesives

Although adhesives have been around for a long time, engineering strides have been made to allow their use to permit fabrication of assemblies that are mechanically equal to or superior to threaded assemblies.

Although the main use of adhesives is mechanical joining, they also seal and insulate. They may find good use in the assembly of lightweight metals, which can become distorted when using threaded fasteners.

Anaerobic adhesives are used for thread locking and bearing mounting applications. The word "anaerobic" means that the adhesive cures between two metal surfaces only when the air is excluded. The anaerobic adhesives that are available come in different grades, depending on the application. Obviously, a thread-locker must be strong enough to keep threads from loosening under vibration but not so strong that the bolt cannot be removed using a wrench. The anaerobic adhesives used to seal pipe threads or flanges are modified to produce various levels of holding or removal strength.

Other engineered adhesive products that are available are epoxies that have great resistance to chemicals or solvents, cyanoacrylates for joint splicing, and silicone adhesives for use with gaskets.

Chapter 9

Machinery and Equipment Installation

One of the primary work functions of millwrights and mechanics is the installation of machinery and equipment. Drawings stating the location or position of mechanical objects usually make reference to points or surfaces. These may be building columns, walls, or other machinery or equipment. See Figure 9-1.

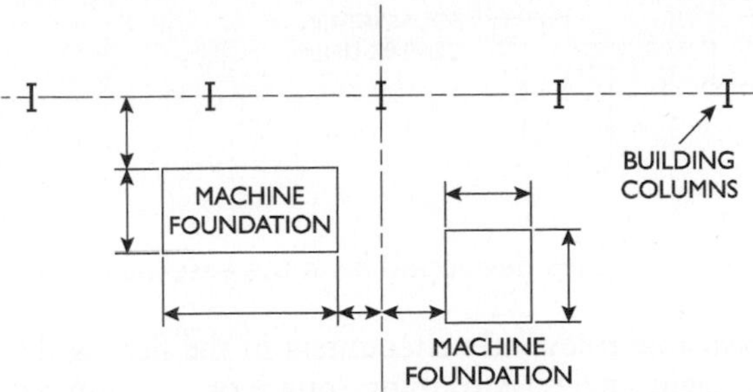

Figure 9-1 Locating machine bases in reference to building columns.

The initial step in preparation for an installation is to locate baselines with respect to the reference points or surfaces given in the drawing. In most cases, two lines at right angles are required, and all dimensions are taken from these baselines. See Figure 9-2.

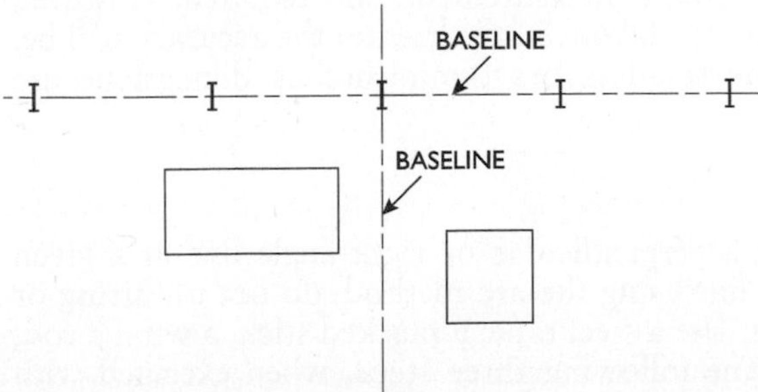

Figure 9-2 Illustrating baselines for machinery layout.

Additional layout lines are developed from the baselines. To ensure accurate locations, a layout should be made whether one is preparing a foundation to support machinery, erecting structural members, or placing equipment on a prepared surface.

For plane-surface layouts, two baselines at right angles are usually sufficient. These baselines may be used as centerlines, or additional centerlines as required may be laid out parallel to the original baselines. See Figure 9-3.

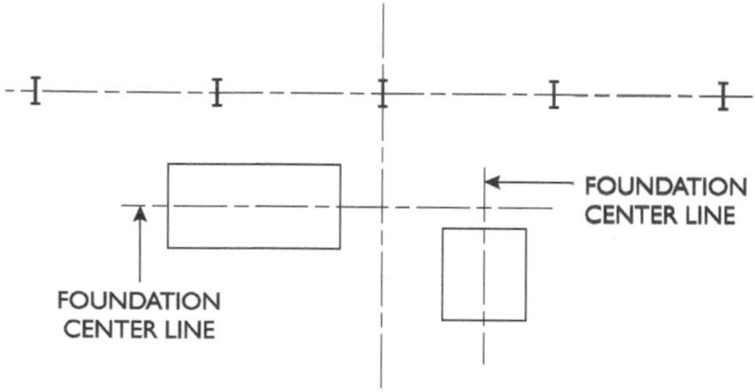

Figure 9-3 Additional layout lines developed from the baseline.

A problem the mechanic frequently encounters in the field is the layout of baselines at right angles. A framing square or steel square, although quite suitable for relatively small work, should not be used for large right-angle layouts. A very slight error at the square may be magnified to unacceptable proportions. A more accurate and dependable method is to develop the layout to suit the size of the job.

There are several methods to laying out right angles that will give very accurate results. The two most suitable for fieldwork are the development by swinging arcs and the 3-4-5 triangle layout. In both methods, only simple measurements are required. However, the larger the scale of the layout is, the greater the accuracy will be, as the effect of the small errors diminishes as dimensions are increased.

Arc Method

When constructing a perpendicular or right-angle line at a given point on a straight line using the arc method, do not use string or line that can stretch. Use a steel tape, a marked stick, a wire, a rod, or similar device. The following three steps, when executed with reasonable care, will give very accurate results.

Step 1

Locate two points, A and B, on the straight line at equal distance from the given point (see Figure 9-4). This may be done by measurement, by using a marked stick, or by swinging a steel tape.

Step 2

Swing arcs from both point A and point B using a radius length about 1½ times the measurement used to locate points A and B. *Exactly* the same radius must be used for both arcs to develop a true right angle. Locate point C where the arcs intersect. See Figure 9-5.

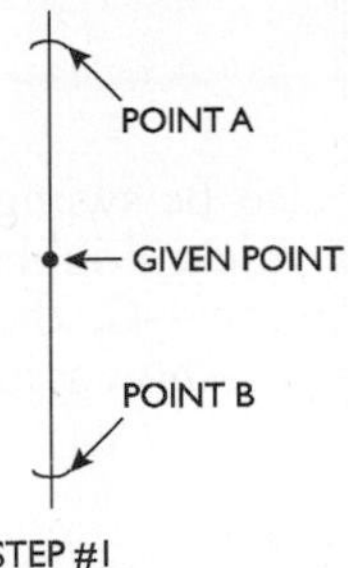

Figure 9-4 A way of locating two points from a given point.

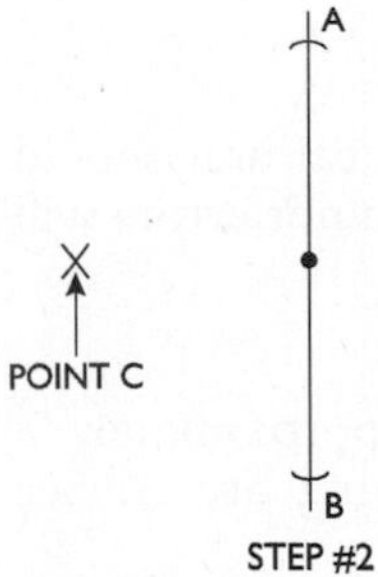

Figure 9-5 Locating the third point.

Step 3

Construct a line from point C through the given point on the line. It will be perpendicular or at right angles to the original straight line. This is illustrated in Figure 9-6.

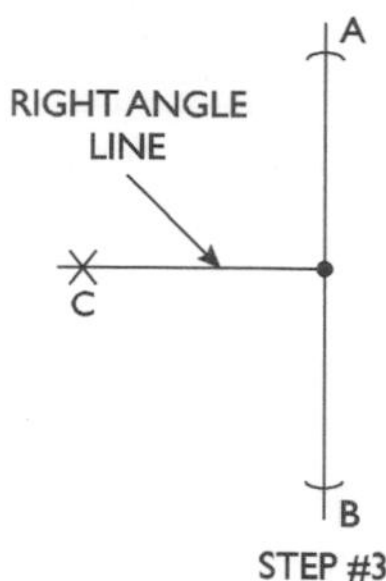

Figure 9-6 A true right-angle line from the starting point or "given point."

If conditions permit, arcs locating point D can also be swung while swinging arcs to locate point C. This will provide a double check on the accuracy of the layout. The three points—C, the given point, and D—should form a straight line, as shown in Figure 9-7.

3-4-5 Method

The 3-4-5 layout method is based on the fact that one angle of a triangle that has sides with a 3-4-5 length ratio is exactly a right angle (90 degrees). The following steps, when executed with reasonable care, will give very accurate results.

Step 1

Select a suitable measuring unit. The largest practical unit should be selected, since the larger the layout is, the less minor errors will affect its accuracy.

Step 2

Measure three units from the given point at approximately a right angle from the straight line, and then swing arc A. See Figure 9-8.

Step 3

Measure four units from the given point along the straight line and locate point B, as shown in Figure 9-9.

Step 4

Measure five units from point B and locate point C on arc line A. See Figure 9-10.

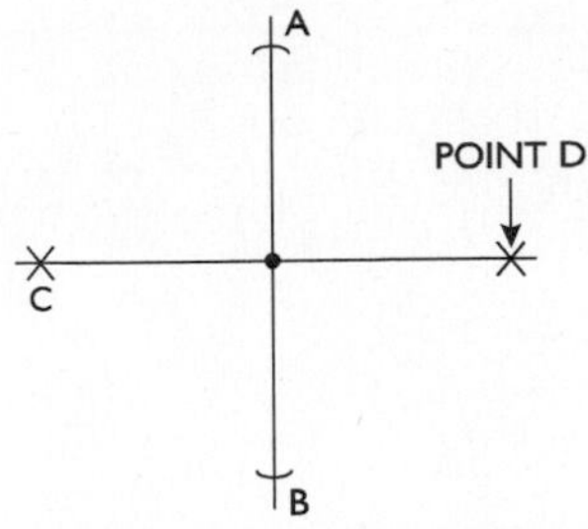

Figure 9-7 Locating point D as a final check for accuracy.

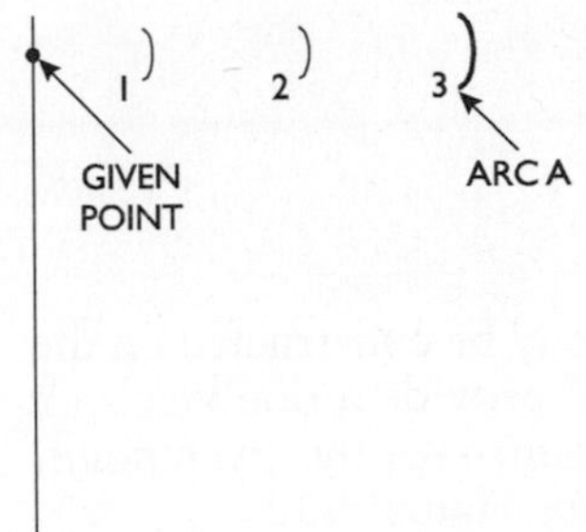

Figure 9-8 Measuring at right angle from a given point.

Step 5

Construct a line from point C through the given point on the line. It will be perpendicular to, or at right angles to, the original straight line. See Figure 9-11.

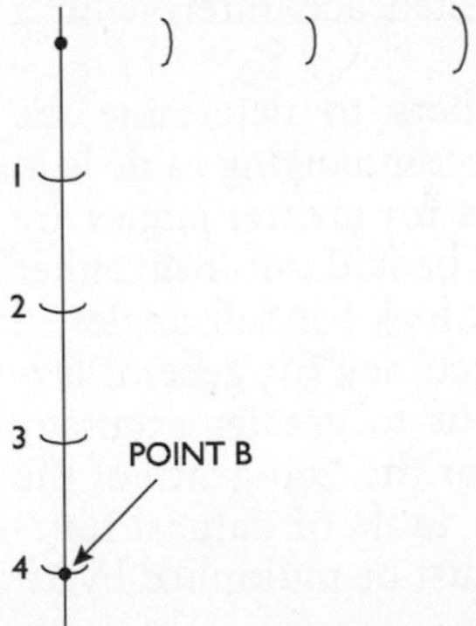

Figure 9-9 Measuring from the given point for point B.

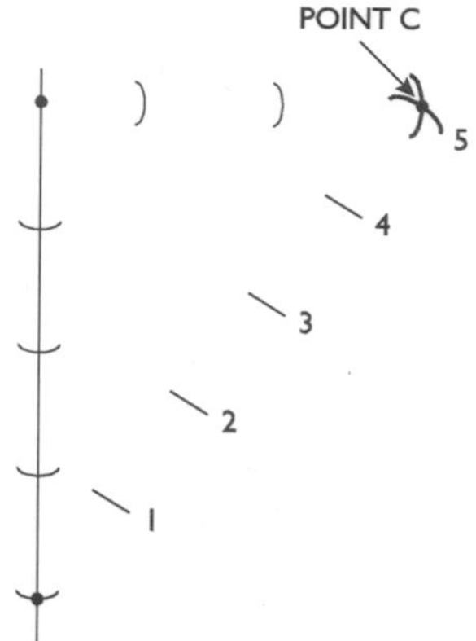

Figure 9-10 Step 4 will locate point C.

If conditions permit, a similar triangle may be constructed on the opposite side of the straight line. This will provide a double check on the accuracy of the layout. The three points—C, the given point, and D—should form a true straight line. See Figure 9-12.

Because it is infrequently encountered, the accurate layout of a line at an angle other than 90 degrees may pose a problem for many mechanics. Although accurate layouts may be accomplished on small work using a standard protractor, this is impractical and inaccurate for large work. A measurement system similar to the 3-4-5 triangle layout system may be applied to this problem. Angle layout by measurement is both practical and easy, as well as being extremely accurate. It may be used in the shop when the dimensions of the work exceed what can be accomplished accurately with a protractor, as well as in the field.

The system employs a table of multipliers to determine the length of one leg of a layout triangle. The accompanying table is in steps of full degrees up to 45 degrees. Values for greater angles are not necessary because the required angle can be laid out from either the horizontal or perpendicular baseline. Values for full angles in two-place decimals should give sufficient accuracy for general layout work. If it is required to perform layout to greater accuracy than that afforded by the table, the values for the "tangent" of the angle may be obtained from a trigonometry table of natural functions. If the tangent value is substituted, it must be multiplied by 10 before being used as a multiplier.

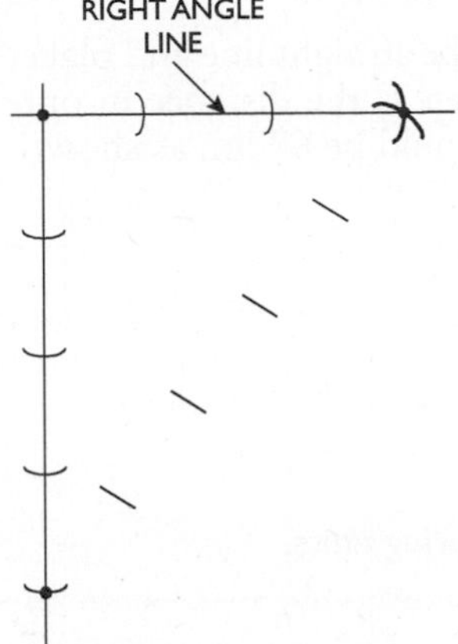

Figure 9-11 Drawing a straight line from point C through given point.

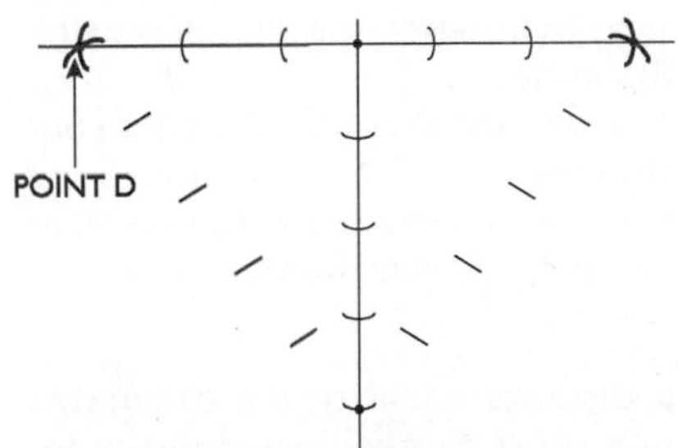

Figure 9-12 Checking accuracy of point C by locating point D.

Perpendicular Distance Method

When constructing a perpendicular or right-angle line at a given point on a straight line using the perpendicular distance method, do not use string or line that can stretch. Use a steel tape, a marked stick, a wire, a rod, or similar device. The following five steps, when executed with reasonable care, will give very accurate results.

Step 1

Select a measuring unit. The largest unit that is practical to use in the layout space should be selected. This will give maximum accuracy, since the effect of minor errors is reduced as overall dimensions are increased. To illustrate the system with an example, let us assume that a measuring unit of 6 in. is selected.

Step 2

Measure 10 units from the given point on the straight line and place point A. Using the 6-in. unit selected in step 1, the distance in our example from the given point to point A would be 60 in., as shown in Figure 9-13.

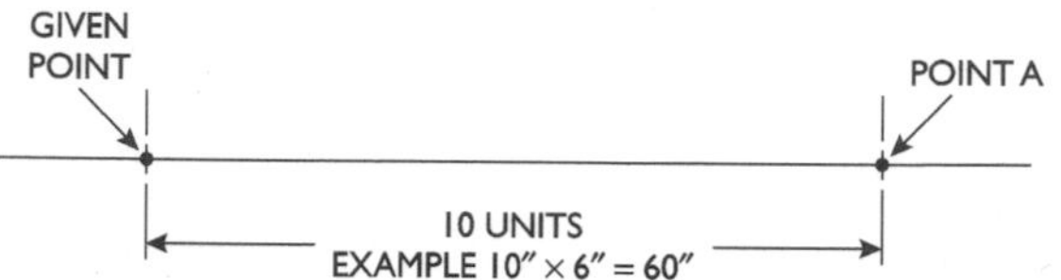

Figure 9-13 Locating point A by using measuring units.

Step 3

From the accompanying table of perpendicular distance multipliers (Table 9-1), select the appropriate value for the angle desired. Calculate the perpendicular distance by multiplying the measuring unit by the value obtained from the table.

For our illustrative example, let us assume that the desired layout angle is 28°. From the table, the multiplier for 28° is 5.32. Multiplying by 6, which is the measuring unit selected in step 1, we determine that 5.32 × 6 equals 31.92 in. This is the perpendicular distance.

Step 4

Measure from the straight line a distance equal to the calculated perpendicular distance and lay out a short parallel line above point A. See Figure 9-14.

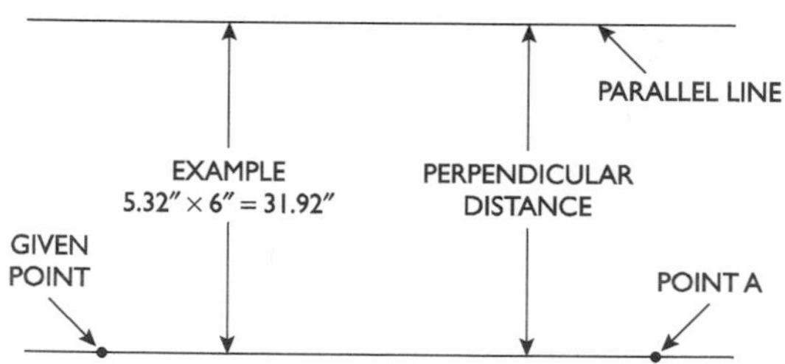

Figure 9-14 Finding point B with the use of Table 9-1.

Step 5

Erect a perpendicular, or right-angle, line from point A on the straight line, and locate point B where the perpendicular line crosses the parallel line. The required angle line may now be constructed through the given point and point B. See Figure 9-15.

Table 9-1 Perpendicular Distance Multipliers

Angle	*Multiplier*	*Angle*	*Multiplier*	*Angle*	*Multiplier*
1	.18	16	2.87	31	6.01
2	.35	17	3.06	32	6.25
3	.52	18	3.25	33	6.49
4	.70	19	3.44	34	6.75
5	.88	20	3.64	35	7.00
6	1.05	21	3.84	36	7.27
7	1.23	22	4.04	37	7.54
8	1.41	23	4.25	38	7.81
9	1.58	24	4.45	39	8.10
10	1.76	25	4.66	40	8.39
11	1.94	26	4.88	41	8.69
12	2.13	27	5.10	42	9.00
13	2.31	28	5.32	43	9.33
14	2.49	29	5.54	44	9.66
15	2.68	30	5.77	45	10.00

Piano-Wire Line

Field layout is commonly performed using line made of twisted or braided natural fibers such as cotton, hemp, or one of the numerous man-made fibers. Although these are adequate for general layout work, there are occasions when piano wire is preferable. Piano wire, which has very high tensile strength, may be drawn extremely taut, providing a straight, true line that does not stretch, loosen, or sag. In cases where layout lines will also be used as reference or alignment lines for machinery, piano wire is far superior.

When properly secured and tightened, a piano-wire line will retain its setting and provide a high degree of accuracy. Because it stays taut and is relatively stationary in space, it is an excellent material for use as a precision reference or baseline. Other lines and/or points may be precisely located from such a line by measurement, use of plumb lines, layout by triangles, etc. A fixed-point location may be easily marked on a wire line by crimping on a small particle of lead or other soft metal.

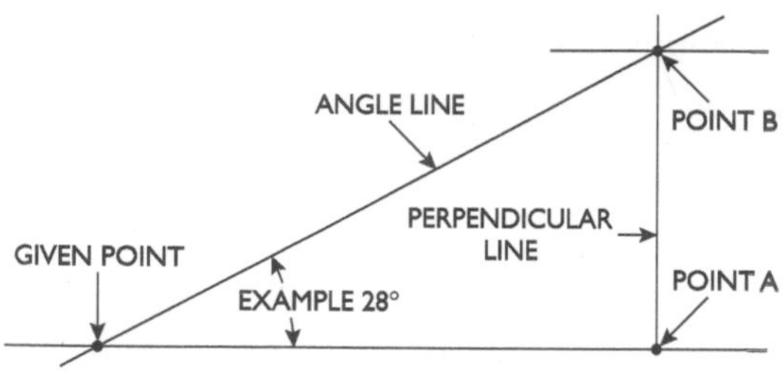

Figure 9-15 Locating point B.

To make a wire line taut, it must be drawn up very tightly and placed under high tensile load. To accomplish this, and to hold the line in this condition, requires rigid fastenings at the line ends and a means of tightening and securing the wire. Provisions must also be made for adjustment if the accuracy of setting, which is the principal advantage of a wire line, is to be accomplished.

Illustrated in Figure 9-16 is a simple device that incorporates provisions for holding, tightening, adjusting, and securing a piano-wire line.

When right-angle piano-wire lines are required, a convenient and practical method to accomplish accurate setting is the 3-4-5 triangle system. A steel square may be used first to obtain an approximate 90°-angle setting. The more accurate 3-4-5 measurement check should then be made. This is done by measuring from the crossing point of the two wires, 3 ft on one wire and 4 ft on the crossing wire. A small particle of soft metal should be crimped on the lines at the 3-ft point and at the 4-ft point. By adjusting the wires so the distance from the point on one wire to the point on the crossing wire is exactly 5 ft, the wires are positioned at a true right angle. Care must be exercised during adjustment so that the crossing point of the two wires is not changed. If larger measurements are possible, use larger multiples of 3-4-5 to reduce the effect of possible measurement errors. See Figure 9-17.

Although the accuracy of a piano-wire line is not required for most foundation layouts, it is almost a necessity for machinery alignment. This is particularly true when various components of a machine are assembled or when independent machine units must be installed in true relative alignment. In such cases, the use of a wire line for the complete job, foundations through final alignment, is recommended.

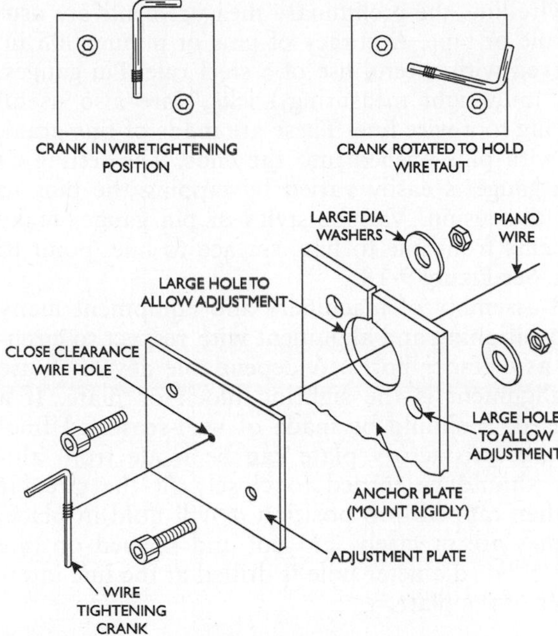

Figure 9-16 Provisions for holding and adjusting piano-wire line.

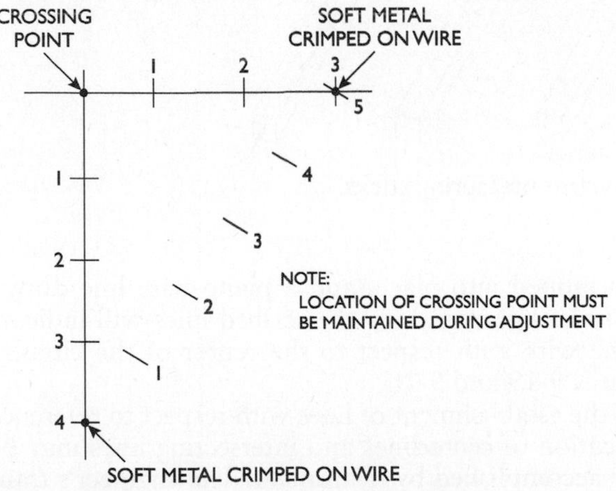

Figure 9-17 Using 3-4-5 triangle system.

When using a wire line, the preliminary measurements are usually made with a rule or tape. Accuracy of plus or minus .005 in. can easily be achieved with careful use of a steel rule. Pin gauges, sometimes called "millwright measuring sticks," are also useful tools when measuring to a wire line. These are made of fine-grain, fairly hard wood with pins pushed into the ends. The setting or adjusting of a pin gauge is easily varied by tapping the pins to obtain the desired dimension. Various styles of pin gauges make possible measurements from line to line, surface to line, point to line, and vice versa. See Figure 9-18.

Installation and assembly of machinery and equipment many times requires its positioning and alignment with respect to circular openings such as cylinder bores. A dependable device to use for this type of alignment is the sighting block or plate. If a wooden plate is used, it should be made of well-seasoned fine-grain wood. A more satisfactory plate can be made from aluminum. The plate should be turned to closely fit the circular opening so that when tapped into position it will hold in place. Two right-angle lines are precisely laid out and scribed on one face of the plate. A ¼ in.-diameter hole is drilled at the line intersection in the center of the plate.

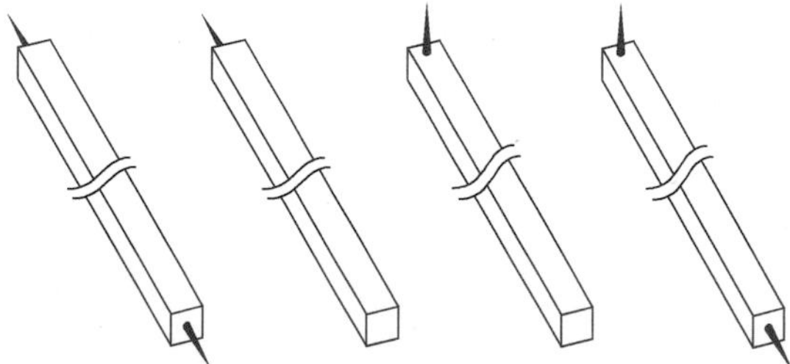

Figure 9-18 Millwright measuring sticks.

With the plate tapped into place, and a piano-wire line drawn taut through the hole, sighting along the scribed lines will indicate the position of the wire with respect to the center of the circular opening. See Figures 9-19 and 9-20.

In many cases, the establishment of lines with respect to reference points and the location of centerlines and intersecting lines may be much more easily accomplished by instrument. The engineer's transit, in common use by surveyors, is an excellent instrument for this

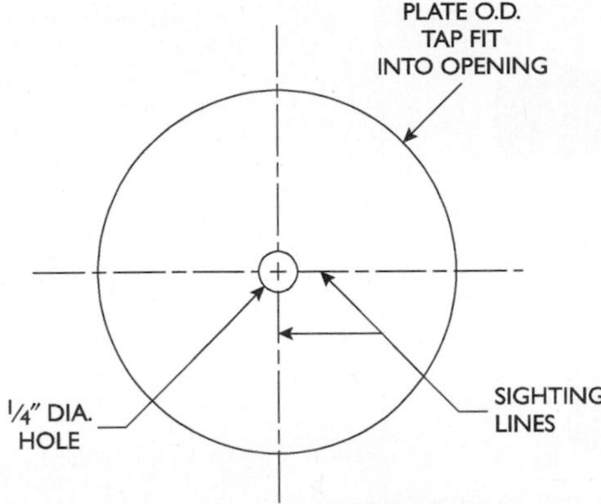

Figure 9-19 Making a sighting plate.

purpose. While the transit shown in Figure 9-21 is designed for surveying work that is much more complex than simple centerline layout, it can accomplish this task with great accuracy.

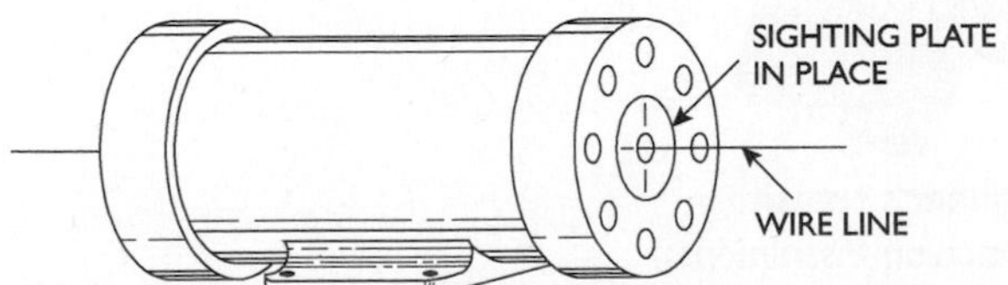

Figure 9-20 A sighting instrument used in conjunction with piano wire.

To use the transit, select a point along the baseline as the reference location from which sightings are to be made. Place the transit approximately over the point and move the tripod legs as required to bring the plumb bob in approximate position. Then, level the instrument approximately and shift it laterally on top of the tripod until the plumb bob is exactly over the reference point. Next, level the instrument by means of the leveling screws and the level tubes, first bringing each level tube approximately to center and then centering each bubble carefully. Now, make a sighting through the telescope at a second point on the baseline to align the instrument with the baseline. Set the horizontal circle graduations to exactly zero with the aid of the vernier. The establishment of a right-angle line

Figure 9-21 Engineer's transit.
(Courtesy Gurley Precision Instruments)

now merely requires the rotation of the instrument 90 degrees and a sighting to locate a point on the desired right-angle line.

Although the transit is normally handled by engineers and surveyors, this elementary operation may be accurately performed by mechanics experienced with precision measuring tools if they have received proper instructions. A much more sophisticated procedure called "industrial surveying" makes use of this instrument and other related instruments for locating, aligning, and leveling in the course of manufacturing operations. The extremely close tolerances of measurement for location, alignment, and elevation required in the manufacture and assembly of large machines, aircraft, space vehicles, etc., has necessitated the development of this skill. Ordinary tools, jigs, and measuring devices are not adequate, and so methods and apparatus similar to those used in surveying have been substituted.

Laser Layout

An important new tool for layout work is the laser layout tool. A transit level similar to a builder's level can now incorporate a laser. These units are fairly expensive, but they can be rented for the few days that might be required for field layout.

Another tool is the laser square or laser protractor, which is usually far less expensive. This device generates precise 90-degree angles with speed and accuracy. The device allows for a one-person operation and projects easy-to-see lines. Useful angles such as 90°, 45°, and 22.5° are easily generated from a built-in protractor. The laser has the ability to generate lines even over wet concrete. The usual working range is about 50 feet. Accuracy is far superior to other methods of layout.

Machinery Foundations

A foundation may generally be defined as the *base* or *support* upon which anything rests. A building, for example, is supported by a base (concrete or other strong material) at its lowest point; it is this base that is termed *foundation.*

All foundations may conveniently be included under one common definition inasmuch as they are independent structures interposed between the ground supporting them and an independent load.

The strength and size of a foundation depend on the size and nature of the supported structure, as well as the nature of the neighboring soil. However, certain fundamental requirements are common to all machinery foundations, and these are as follows:

1. They must carry the applied load without any settlement or crushing.
2. They must maintain true alignment with any power communication mechanism.
3. They must absorb all the vibrations and noise that may be caused by inertia or unbalanced forces.

For a successful analysis of machinery foundation problems, it is necessary to know the type and nature of the machinery to be erected—whether it is, for example, an oil or gas engine, a steam engine, a turbogenerator, or a combination of prime movers and electrical machinery. In each case, it is necessary that the physical size, weight, and other characteristics be fully known before any attempt at designing a foundation can be made.

Foundations for heavy machinery are generally constructed of

1. Solid concrete.

2. Reinforced concrete.
3. Structural steel.

Before determining the most suitable material to use, it is necessary that local conditions such as allowable space, conditions and strength of neighboring supporting structures, and soil conditions be completely known. For heavy machinery, as a rule, the solid-concrete type of structure is the most desirable. In cases where space and economy are the determining factors, the reinforced-concrete or structural-steel type of foundation is most commonly used.

Reinforced-Concrete Foundations

A reinforced-concrete foundation has very favorable characteristics where simplicity of design and constructions, with necessary rigidity, is desired at a reasonable cost. It has found an extensive use in such machinery as turbogenerators where, for example, the rigidity is the most important factor. In machinery of this kind, the *deflection* in the supporting members must be very small, since any appreciable deflection might cause warpage of the turbine casing or a general misalignment of the unit. This would result in serious operating difficulties.

Vibration and Noise Prevention

Vibration and noise have become an acute problem, principally as a result of the increased speed of modern machinery. In nearly all types of machinery, some form of vibration that is perceptible to the human ear, either as noise or vibration, is emitted. This vibration frequently has an adverse effect on the buildings from which it emanates and frequently on neighboring properties as well.

Types of Vibration

In machinery operation, two classes of vibration are in evidence:

1. Primary air vibration communicated directly from the machinery to the air surrounding it
2. Secondary air vibration created indirectly by the foundation

The second class is foundation vibration, which is produced by the machinery and transmitted through building floors and the adjoining walls—sometimes even through the ground for considerable distances. This may be detected either as secondary air vibration or as noise.

Vibration and Noise Control

Care in planning, manufacturing, and testing of machinery is undoubtedly the most fruitful means of avoiding noise problems of the type described. However, in the case of certain machinery, vibration and noise cannot be fully eliminated, although it can be limited to very small proportions by

1. A properly designed foundation.
2. Interposing a suitable slab of resilient material between the foundation block and the machinery or between the foundation block and ground.
3. Isolating the sides of the foundation block from adjacent material by means of air spaces.

A foundation of this type is shown in Figure 9-22. Here, the block is free to move, and the forces remaining after overcoming the inertia of the foundation block are taken up by internal friction. The design of vibration- and noise-control foundations depends on the particular conditions and characteristics of the machinery encountered. Therefore, each case should be carefully analyzed as to the best method for a successful treatment. Among the numerous sound-absorbing materials commonly utilized are cork, rubber, and felt. See Figure 9-23.

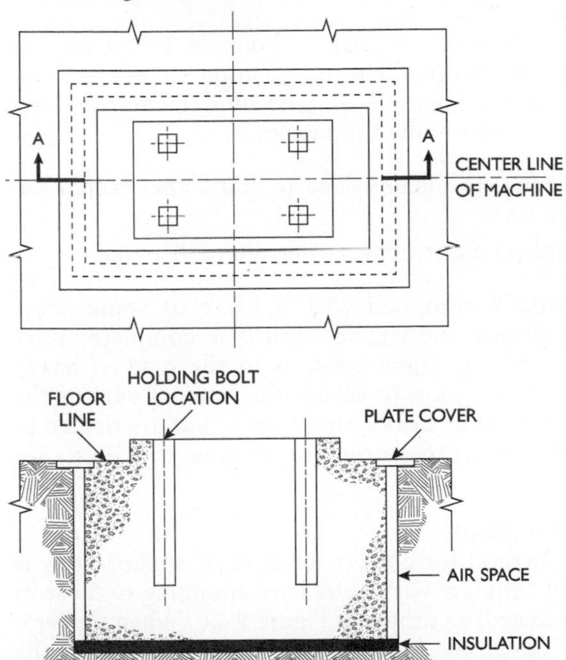

Figure 9-22 Views of a machine foundation with vibration-control features.

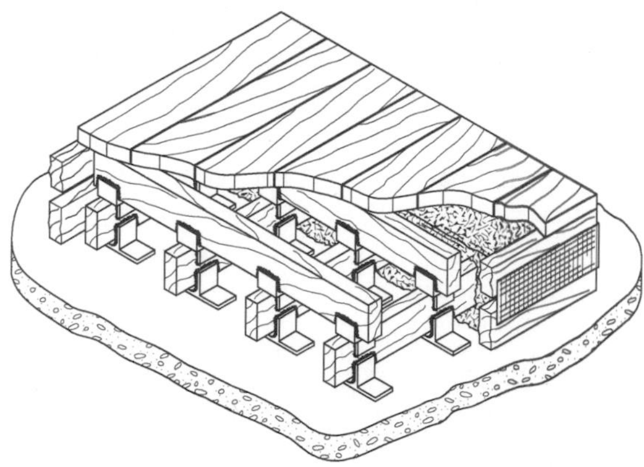

Figure 9-23 Assembly details of a vibration-isolation platform. Various sound-absorbing materials can be used inside the foundation platform.

Foundation Bolts

The purpose of foundation, or anchor, bolts is to secure the machine rigidly to its foundation. The most common types of foundation bolts are shown in Figure 9-24. The fundamental requirements of a foundation bolt are the following:

1. It must be adequately dimensioned to stand the mechanical stresses exerted on it.
2. It must be simple in construction to facilitate its usage.

The bolts are usually equipped with a hook or some other form of fastening device shown, to ensure a complete unity with the foundation block. Sometimes, as in the case of heavy machinery, it is not uncommon to secure the bottom part of the bolt by means of channels or heavy angle irons that are drilled in one or more locations, as required, to receive the bolt. See Figure 9-25.

Methods of Bolt Location

The most common method for correct location of anchor bolts is by making a wooden template with holes corresponding to those in the machine to be fastened, as shown in Figure 9-26. When the template is completed, tin pipes, at least two inches larger than the diameter of the bolts, are suspended centrally from each hole, reaching to the anchor space, so that when the concrete is poured

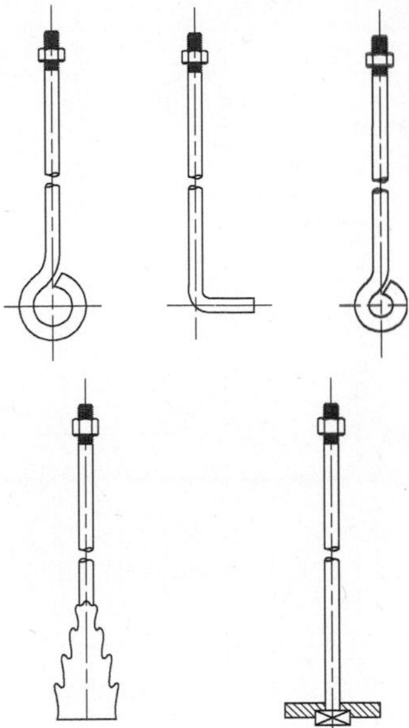

Figure 9-24 Common types of foundation bolts.

there will be a margin of space around each bolt, permitting lateral adjustment to allow for small errors in measurements and to facilitate the removal of bolts in case of breakage.

After this arrangement, together with the necessary forms in place, the concrete is poured and carefully packed to eliminate voids. See Figure 9-27. The foundation should be completed at least fifteen days before the machinery is placed upon it, by which time it should have become sufficiently hard to receive the weight of the machine.

Another method of anchor bolt locating is shown in Figure 9-28. This method is similar to that previously described except that bolt boxes are not used. These are not necessary since the bent end, or hook, on the bolt provides the required anchoring feature. The practice of placing the bolts in sleeves to allow for bolt movement is followed. The inside diameter of the sleeves should be two to three times the diameter of the bolt. As previously described, the bolts and sleeves are suspended in the form from a board frame or template. The holes in the template are located to match the bedplate of the machine to be secured on the foundation. The anchor

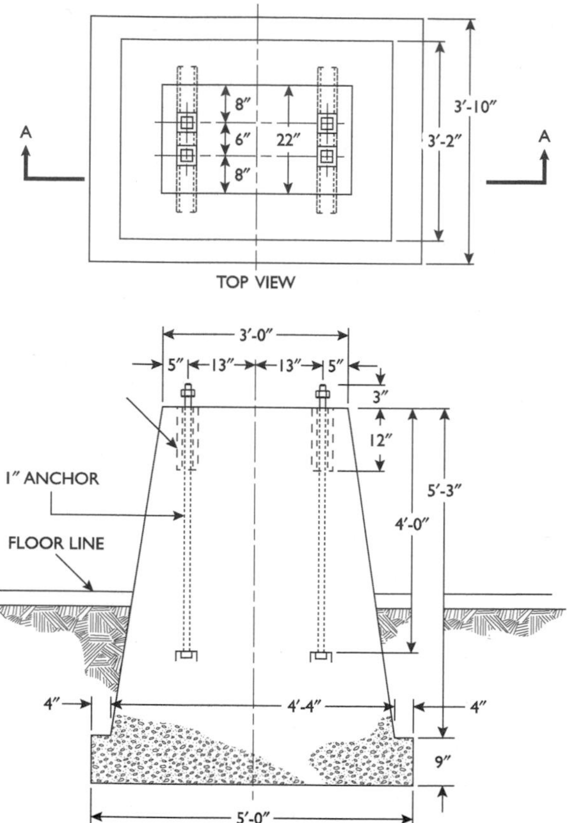

Figure 9-25 Complete design of a typical foundation.

bolts must be of sufficient length and placed so that there will be adequate bolt projection above the foundation. When calculating the projection required, consideration must be given to bedplate thickness, grout, and allowance for washer and nut. Shimming and grouting will be discussed later.

Foundation Materials

Due to the availability of Portland cement, foundations are almost exclusively built of concrete. The materials necessary for the making of concrete are cement, sand, aggregate, water, and in some instances reinforcement. It has, however, become customary to

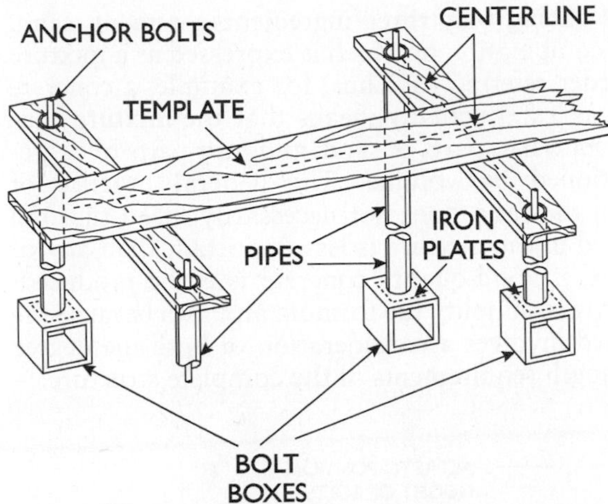

Figure 9-26 View showing a template used to locate foundation bolts.

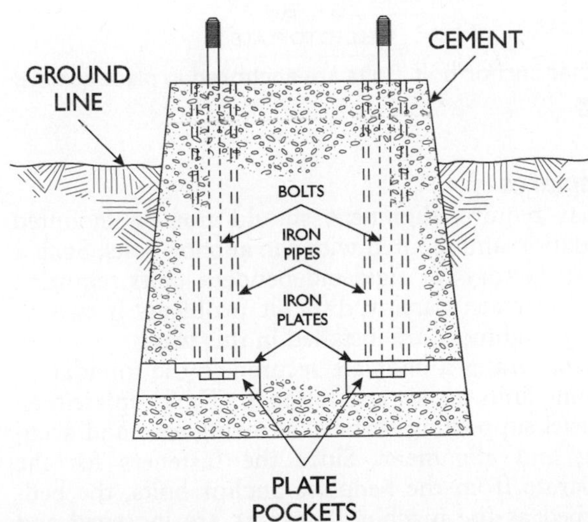

Figure 9-27 Complete foundation showing method of installing anchor bolts.

refer to concrete as having only three ingredients—cement, sand, and aggregate—the combination of which is expressed as a mixture by *volume* in the order referred to. Thus, for example, a concrete mixture referred to as 1:2:4 actually means that the mixture contains *one* part of cement, *two* parts of sand, and *four* parts of aggregate, each proportioned by volume. The general practice of omitting water from the ratio does not necessarily mean that the amount of water used in the mixture is less important than any of the other ingredients, if good-quality concrete is to be produced; water is omitted partly to simplify the formula and also because the amount of water used involves a consideration of both the degree of exposure and strength requirements of the complete structure.

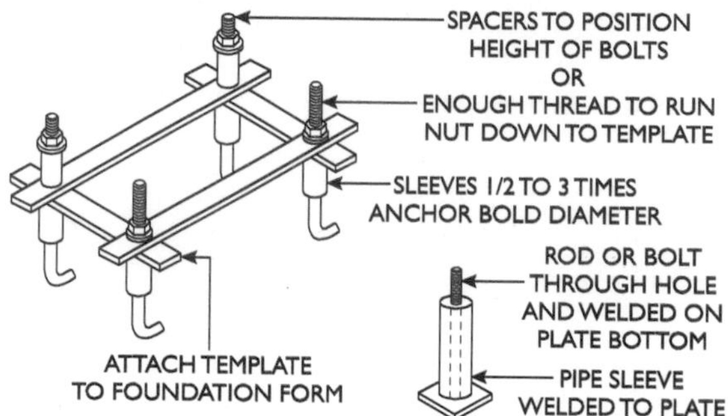

Figure 9-28 Another anchor bolt. Bolts are anchored in place at time of concrete pouring.

Foundation Bedplates

Machine units that require alignment should not be mounted directly on a foundation and secured with the anchor bolts. Such a practice may be satisfactory for rigid independent units requiring no alignment, but it presents many difficult problems if two or more units requiring alignment are installed in this way.

Good design incorporates a bedplate secured to the foundation on which the machine units are set and fastened. The bedplate then provides a solid, level supporting surface, allowing ease and accuracy of shimming and alignment. Since the fasteners for the machinery are separate from the bedplate anchor bolts, the bedplate is not disturbed as the machine fasteners are loosened and tightened during alignment. Figure 9-29 shows the mounting of a bedplate on a foundation with provisions for shims and grout.

Grout

A fluid mixture of mortar-like concrete called "grout" is poured between the foundation and bedplate. This secures and holds the leveling shims and provides an intimate support surface for the bedplate. The foundation should be constructed with an elevation allowance of ¾ to 1½ in. for grout. The bedplate is set on shims, and the grout is poured between the foundation and the bedplates. To have the required body to support the bedplates, the grout should be at least ¾ in. thick. If thinner, it may crack and break up in service. The 1½ in. maximum thickness indicated represents a practical shimming limit.

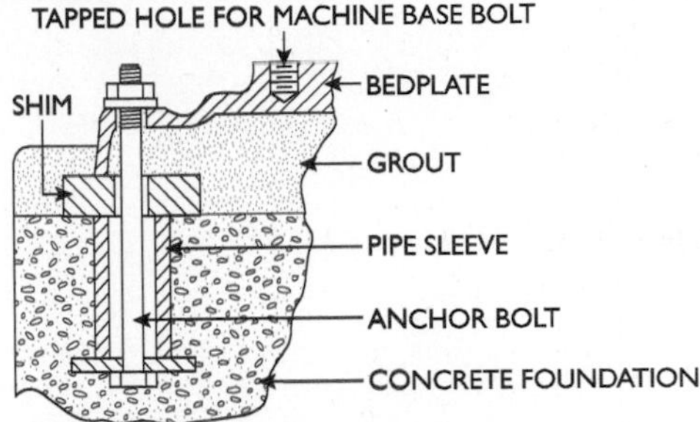

Figure 9-29 The mounting of a bedplate.

Bedplate Shims

The number and location of shims will be determined by the design of the bedplate. Firm support should be provided at points where weight will be concentrated at anchor-bolt locations. Shims should be of generous dimensions and sufficient in number to provide rigid support. A practice sometimes followed when setting a bedplate is to use wedges as shims. When correctly done, this simplifies the leveling of the bedplate as well as providing solid support surfaces. However, if narrow shims with inadequate surface are used, an unsatisfactory installation will result. When shimming with wedges, a double wedge, as shown in Figure 9-30, should be used. They should be placed in line with the bottom surface of the bedplate edge to get full-length contact. Illustrated in Figure 9-31 is the improper practice of using single wedge shims at right angles to the bedplate edge. Only a relatively small area of the shim is supporting the bedplate—unit stress may be too high to support the weight load.

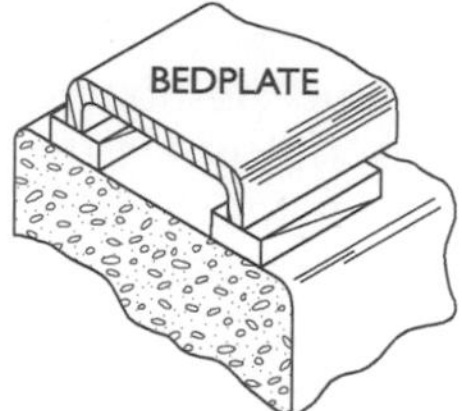

Figure 9-30 A double-wedge shim mount.

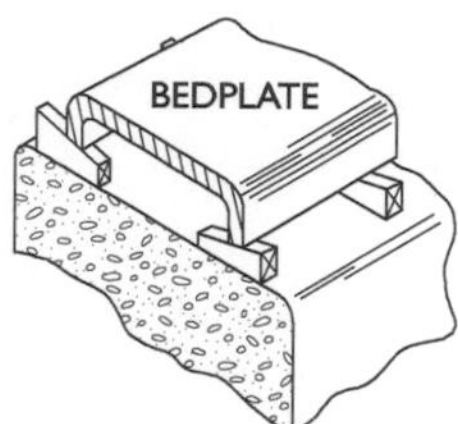

Figure 9-31 Poor shimming practice.

Pouring Grout

Bedplates are usually designed with openings to allow pouring grout. When fabricating bedplates from structural members or plate, openings should be provided. This is especially true if large plates are used, since a means of inserting support shims to give firm bearing is essential.

When the bedplate has been shimmed and leveled, and the anchor bolts have been snugly tightened, a dam is constructed around the foundation to contain the grout. The dam level should be at least ½ in. above the top surface of the shims. Grout should be poured inside the bedplate and thoroughly puddled. The level inside the bedplate will be above the level outside, and unless the bedplate is unusually high, grout will completely fill it without overflowing the dam. The grout should be allowed to set at least 48 hours, after which the anchor bolts should be tightened evenly and securely. See Figure 9-32.

Machinery Mounts

In many manufacturing plants, the machinery is no longer permanently fastened in place with anchor bolts. This is especially true in metal manufacturing industries where frequent model changes or design improvement of the product are made. Flexibility and mobility are a necessity because production lines are constantly being rearranged, with new or improved machines replacing obsolete units.

In some cases where a line of many machines may be shut down if a single unit fails, the mobility capability may be carried even further. Disabled units may be removed from the line and replaced by new or rebuilt replacement units to minimize production downtime. To accomplish this, the machine must be equipped with flexible quick-disconnecting coupling service lines for power, air, water, etc.

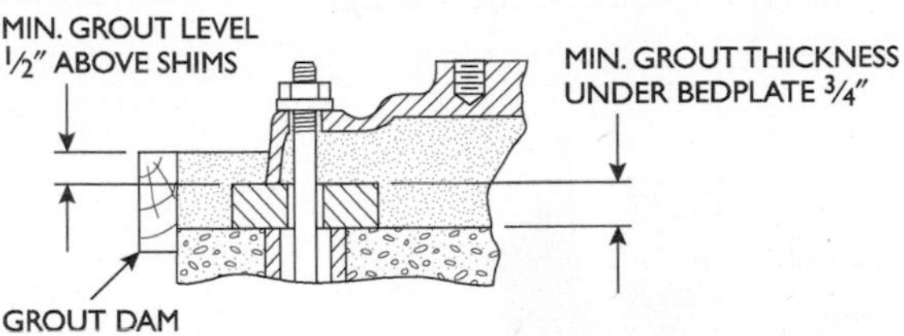

Figure 9-32 Pouring grout into bedplate.

A prime requirement for this kind of mobility is the use of devices that provide support for the machine and have the capability of quick adjustment for leveling and alignment purposes. These devices are called *machinery mounts* because they eliminate the need for anchor bolts and make machines freestanding. This type of machine mounting allows units to be picked up and moved to another location with a minimum of installation work. Using this system of mounting eliminates the need to remove anchor bolts or other fasteners or to break floors. The leveling features built into such mounts allow leveling of the machine to precise limits in a minimum amount of time.

Because the weight of the machine is the primary anchor for this system of mounting, it is very important that the weight be solidly transferred to the floor. Because floor surface is seldom smooth, and flat, mounts with a self-leveling feature should be used. Figure 9-33 shows one type of machine leveler, which incorporates precision vertical leveling features in combination with complete self-alignment to floors sloping in any direction.

Vibration-Control Materials

Because of the ever-increasing variety of industrial machinery, the problem of vibration, shock, and noise has become an important consideration in many machine installations. To overcome these problems, various types of special machine mounts are available. If the problem is not too severe, felt padding may be used as a damping material.

Felt padding is low in cost and easy to install in addition to being an effective vibration- and noise-damping material. All that is necessary is to apply a special cement to both sides of the felt, place it in position, and install the machine. To correct uneven floor surfaces, steel shims should be cemented to the floor and the felt padding cemented to the shim surface before the machine installation. The thickness of felt used should be selected to suit the weight of the machine.

Figure 9-33 Self-leveling machinery mounts.

For ordinary installations, the following is a general guide for thickness selection. Machines weighing up to 800 lbs should use ¼ in. thickness with loading from 5 to 10 psi. Machines weighing 800 to 5000 lbs should use ½ in. felt loaded from 50 to 100 psi. Machines weighing over 5000 lbs should use 1-in. felt loaded at about 150 psi. Unit loading may be controlled by the use of plates under the machine support surfaces to increase contact area, or if the surfaces are larger than necessary, the felt may be cut to the area required to give the desired loading.

Manufacturers can supply felt in different grades for various loading pressures as well as in a variety of thicknesses. Unbalanced weight in a machine may require different grades of felt to be used. Careful selection of felt for such characteristics as grade and thickness may be necessary for machines having a wide variation in weight distribution.

Also available in a variety of densities and thicknesses are elastomeric pads for vibration and noise control. No general guide for their use can be given, as different manufacturers' products vary in construction characteristics and materials. Selection should be made from the manufacturer's specification and recommendations.

Vibration-Control Machinery Mounts

A wide range of machine mounts, designed to control shock, vibration, and noise, are commercially available for situations where appreciable deflection is needed to absorb vibrations and shock. When off-balance conditions are present, spring-type mounts may be most suitable. Shown in Figure 9-34 is a spring unit designed for light- to medium-weight machinery that operates at low frequency and produces large displacement vibrations (compressors, reciprocating pumps, etc). Another style of spring mount with additional features is shown in Figure 9-35. Incorporated in this design are damping features that provide stability by preventing sudden recoil and residual vibration from heavy shock. In addition, it has built-in leveling adjustment and padding to reduce noise transmission.

Another widely used style of machine mount, with vibration and shock-damping capability as well as adjustment and self-leveling features, is shown in Figure 9-36. This style of mount may be used on machinery and equipment ranging from office and laboratory installations to metalworking machines and similar equipment subjected to high-impact shock loading and severe vibrations. This type of mount eliminates anchor bolts or floor lag bolts, yet keeps machinery firmly in place.

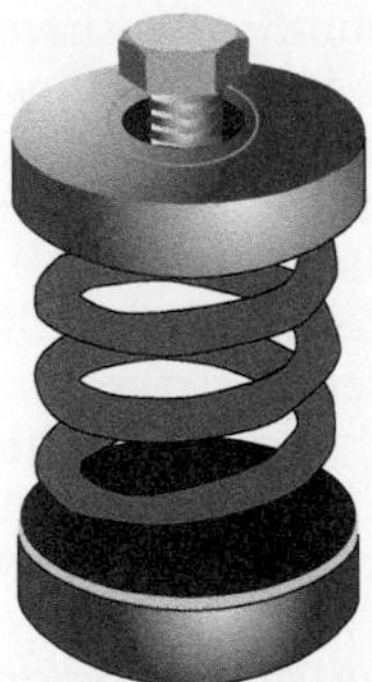

Figure 9-34 Spring-mount support for light- to medium-weight machinery.

Figure 9-35 Spring mount with built-in damping features.

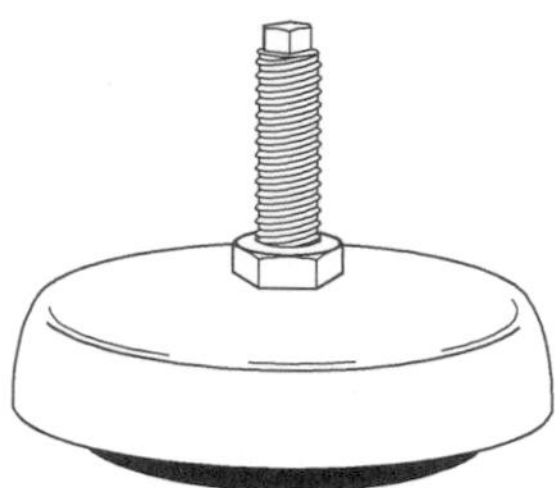

Figure 9-36 A special leveling mount used for heavy and high-impact or severe vibration machinery.

Use of this style of mount makes possible quick and easy relocation of machinery. These mounts also eliminate the problem of shock and vibration from one machine disturbing another. Sensitive high-precision machinery can be located for the best workflow without danger of impact or vibrations from machinery nearby. Noise, as well as vibrations, transmitted from the base of the machine through the floor and building is also reduced.

Another very important benefit resulting from the use of vibration-damping mounts is the reduction of internal stresses in the machine. When vibrating equipment is rigidly bolted to the floor, an amplification of internal stress occurs. The results often are misalignment of machine frames and undue wear on bearings and related parts.

Another mount in use to isolate low-frequency shock and/or vibration is the pneumatic elastomer mount. This is an air-spring style of construction commonly referred to as an air bag.

These mounts are generally used on machines subject to low-frequency shock and/or vibration forces, such as high-speed punch presses, drop hammers, clickers, or forging hammers. By varying the air input to the mount, which is pressurized through a standard tire valve, the support and damping capacity of the unit may be varied.

Vibration-damping mounts and materials include pneumatic elastomer, leveling, compression elastomer, and neoprene pads. Neoprene pads are a simple, low-cost method of vibration-control support; they may be used singly or stacked to provide optimum shock and vibration isolation.

Leveling

One of the basic, and in many cases most important, operations in the installation of machinery and equipment is leveling. The term "level" in this case refers to a line or surface that is exactly horizontal. All buildings are constructed with reasonably level floors and foundations. When installing machinery and equipment, the leveling or placing of the machine on a true horizontal plane is essential to its satisfactory operation.

The tool used to measure a surface or line to determine if it is horizontal is called a "level." It is an instrument incorporating a glass tube containing spirits. A bubble is formed in the tube when it is slightly less than completely filled with the spirits. The tube is then mounted in the body of the level and calibrated to indicate true horizontal position when the bubble is centered. See Figure 9-37.

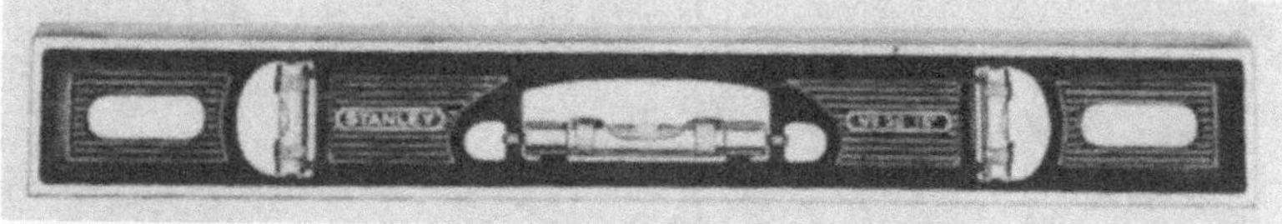

Figure 9-37 Illustrating a horizontal and vertical level. *(Courtesy The Stanley Works)*

When leveling machinery, recommended practice is to raise the low point to match the high point by shimming. When a specific elevation must be attained, the first step must be to establish the high point at the correct height or elevation. Raising the low point by shimming will then bring all points to correct elevation when they reach a level plane.

Too frequently, the leveling operation is a trial-and-error procedure. A shim, estimated to be the required thickness, is installed under the low point of the object being leveled. A check is then made, and level is further adjusted by increasing or decreasing the shim thickness. This procedure, like most trial-and-error methods, requires many adjustments before the desired level plane is obtained.

The correct and more efficient method is to use the level and a thickness or feeler gauge to measure the amount of error or off-level. The leveling shim thickness is then calculated from this measurement. Illustrated in Figure 9-38 is this system of measuring a sloping surface. The feeler under the low end of the level is selected to obtain a zero indication on the level.

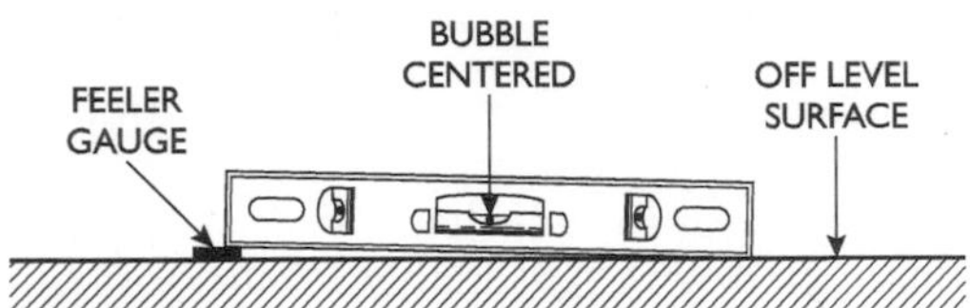

Figure 9-38 Using a feeler gauge to measure amount of error or off-level.

Having determined the amount of off-level, or slope, by measurement, the thickness of the shim necessary to level the object may be easily calculated. In most cases, a simple mental calculation using approximate measurements will give surprisingly accurate results. The following three steps will allow leveling objects the first time to an accuracy acceptable for all general work:

1. Determine the amount off-level by placing a feeler gauge under the low end of the level that will bring the bubble to center.
2. Determine the approximate ratio of the level length to the distance between bearing or shimming points.
3. The thickness of the shims will be the ratio times the feeler gauge measurement. See Figure 9-39.

Using an 18-in. level and a .016-in. feeler gauge under the low end centers the bubble. The distance between bearing points is 65 in., or approximately 3½ times the 18-in. level length. The shim required is 3½ times .016, or .056 in. thick.

There may be occasions when extremely accurate level settings are desired and require greater accuracy in calculating shim thickness. In such situations, a basic requirement is that a precision level be used when measuring. The procedure is similar except that an accurate determination is made of the correction required per inch. Using the example in Figure 9-39 again, the steps are as follows:

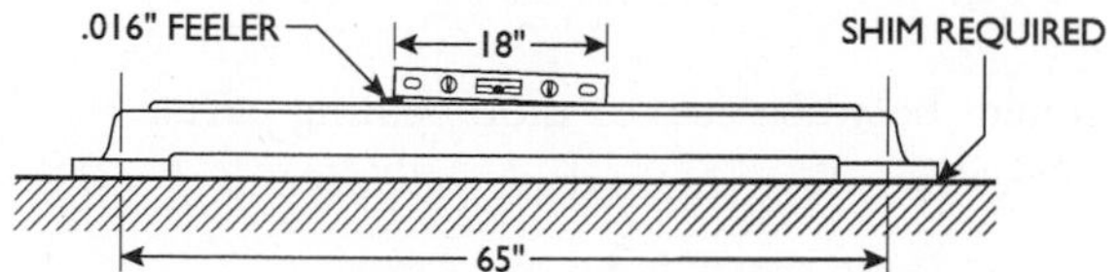

Figure 9-39 Finding the feeler thickness to bring to proper level.

1. Using an 18-in. precision level, a .016-in. feeler gauge under the low end centers the bubble.
2. Dividing 18, which is the length of the level, into the .016-in. measurement indicates that the off-level per inch is .0009 in.
3. Multiplying .0009 in. by 65 determines that a shim of .058 in. will be required at the low bearing point to level the object exactly.

Note that the difference between the shim thickness calculated by the two methods is only .002 in. The accuracy of the first method, which requires only mental calculation using approximate figures, is surprisingly accurate. This situation is true as long as the distance between bearing points where the shim is to be inserted is several times greater than the length of the level. If the situation were reversed and the distance between bearing points were less than the length of the level, the correction per inch method should be used.

Extremely accurate adjustments may also be made by using a precision level, as illustrated in Figure 9-40. The calibrated scale on this measuring instrument may be used to calculate the thickness of shim required to obtain precise settings.

Figure 9-40 A precision level with calibrated scale.
(Courtesy L.S. Starrett Company)

Doweling

After setting the machine on the bedplate and making all the necessary alignment moves, it is often useful to dowel the machine to do the following:

1. Preserve alignment between two or more mating parts.
2. Resist lateral forces tending to separate the parts.

There are three basic types of dowel pin: straight, tapered, and spring-type.

To select a dowel pin:

1. Selection of the type of pin to use—straight, tapered, or spring-type—depends on the application.
2. Where it is used to assemble and locate two mating parts, tapered or spring-type dowels are preferred.
3. Spring-type dowels have the advantage of offering the lowest installation cost because they require only a drilling operation in the two mating pieces, without any subsequent reaming.
4. Taper dowels are recommended where higher stresses are involved or more precise positioning is required. They are not intended for use where parts are to be interchangeably assembled.
5. Straight dowels should be used where parts are frequently disassembled or where more precise positioning is required than is obtainable with spring-type dowels. The original cost of machining the two mating parts for straight dowels is higher because very close tolerances are required on hole diameters, hole locations, and squareness of the holes with the mating surfaces.
6. Taper pins, large end threaded, are for use instead of plain taper pins in applications where the small end of the pin is not accessible for removal by striking with a drift pin.

To install a dowel pin:

1. When the dowel pinhole extends completely through the two mating parts, the pin installation is referred to as a through pin. This type of installation is preferred and shown in Figure 9-41.

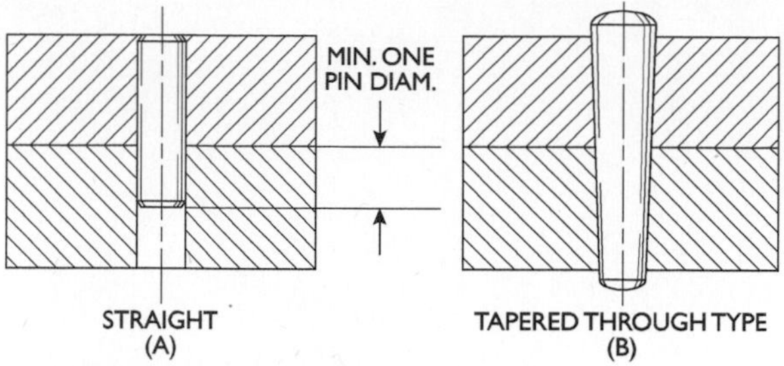

Figure 9-41 Through pin type.

2. Figure 9-42 shows a semi-blind type. The smaller hole serves as a vent and also as a knockout hole.

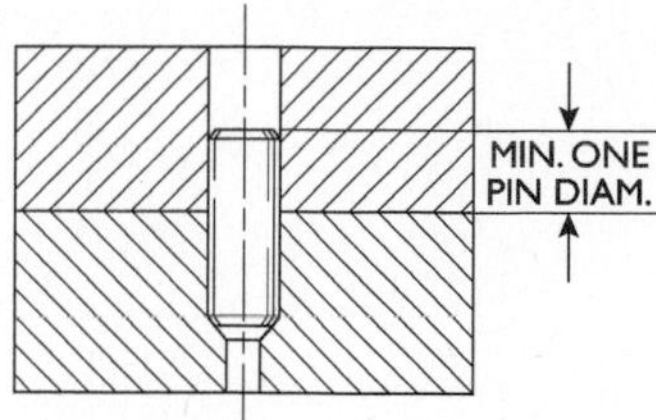

Figure 9-42 Semi-blind type.

3. Figure 9-43 shows a blind type of hole; with this arrangement, a dowel pin should always be a sliding fit in the blind hole and a force fit in the mating part. A blind-type dowel should never be used when it is feasible to employ either the through type or the semi-blind type.

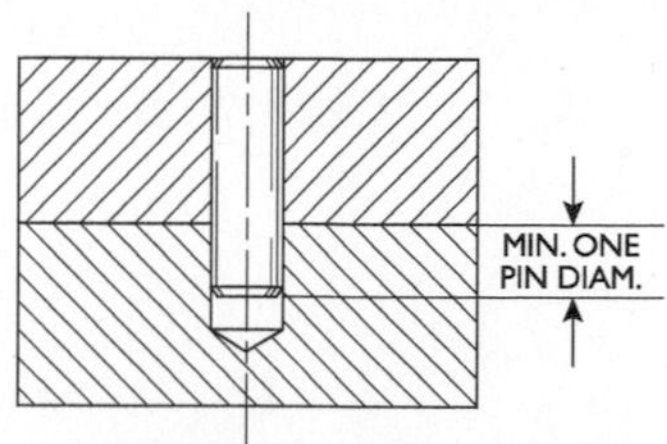

Figure 9-43 Blind type.

Chapter 10

Bearings

The many types and styles of bearings used to support shafts may be divided into two very broad classifications: plain and antifriction bearings. Very simply stated, the plain bearing operates on the principle of sliding motion, and the antifriction bearing operates on the principle of rolling motion.

In the plain bearing, there is a relative sliding movement between shaft and bearing surfaces. A lubricant is used to keep the surfaces separated and overcome friction. If the film of lubricant, shown in Figure 10-1, can be maintained, the surfaces will be prevented from making contact and long bearing life will result. This is extremely difficult to accomplish, and there is usually some surface contact taking place in operating plain bearings. Dissimilar metals with low frictional characteristics have proven most suitable for plain-bearing applications because they are less susceptible to seizure on contact. The common practice is to use a steel shaft and to make plain bearings of bronze, babbitt, or some other material that is softer than steel. A bearing metal that is softer than steel will wear before the shaft and usually can be replaced more easily, quickly, and cheaply.

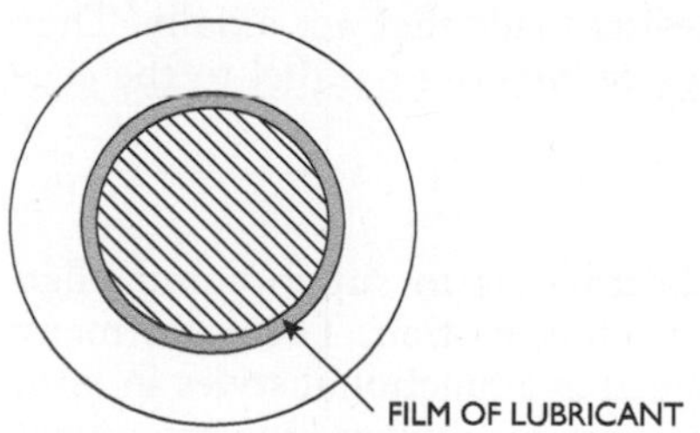

Figure 10-1 Illustrating the film of lubricant on plain bearing.

The antifriction bearing operates on the principle of rolling motion. A series of rollers or balls are interposed between the moving and stationary members as shown in Figure 10-2. These rollers are usually mounted in a cage or separator and enclosed between rings or races. In theory, the rolling elements in antifriction bearings perform the function that the lubricant film performs in the plain bearing. In practice, however, antifriction bearings require lubrication, as some sliding action occurs between rolling elements and races. A sliding action also occurs between contacting surfaces of rolling elements and the cage or separator.

Bearing styles are usually designated by the function they perform. The three principal style categories are radial bearings, thrust bearings, and guide bearings.

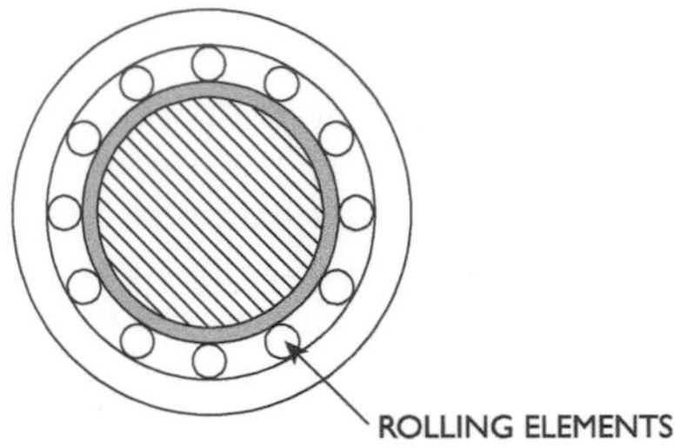

Figure 10-2 The antifriction bearing.

Radial Bearings

The function of radial bearings is to support loads that act radially. These are loads acting at right angles to the shaft centerline. They may be visualized as radiating into or away from a center point.

Thrust Bearings

The thrust bearing supports or resists loads that act axially. They may be termed *endwise loads* because they act parallel to the centerline, toward the shaft ends.

Guide Bearings

The primary function of guide bearings is to support and align members having sliding or reciprocating motion. There are many variations and combinations of the above functional styles in both plain and antifriction bearings. In most cases where the functional style name is used, it indicates a principal function, as many bearings perform more than one function.

Plain Bearings

A bearing, in the mechanical terms used by millwrights and mechanics, is a support for a revolving shaft. The word *bearing* is also used in some cases to describe a complete assembly rather than just the supporting member. Such an assembly of a plain bearing is usually made up of the following parts: the shaft or the surface of the shaft, which is called the journal; the bearing liner, box, sleeve, bushing, or insert (depending on the construction and design); and the box block or pedestal, which contains and supports the liner. If the box or block is of the split type, there would also be a cap fastened by bolts or studs. To clarify the meaning of some of the names and terms commonly used in connection with bearings and associated parts, the following definitions are given.

Journal

That part of a shaft, axle, spindle, etc., that turns in a bearing.

Journal Box

The box or bearing that supports the journal.

Axis

The straight line (imaginary) passing though a shaft on which the shaft revolves or may be supposed to revolve.

Radial

Extending from a point or center in the manner of rays (as the spokes of a wheel are radial).

Thrust

Pressure of one part against another part, or force exerted endwise or axially through a shaft.

Friction

Resistance to motion between two surfaces in contact.

Sliding Motion

Two parallel surfaces moving in relation to each other.

Rolling Motion

Round objects rolling on mating surfaces with theoretically no sliding motion.

Plain bearings are generally very simple in construction and operation, although they vary widely in design and materials. They are amazingly efficient and can support extremely heavy rotating loads. The secret of this tremendous load-carrying ability lies in the tapered oil film developed between the journal and the bearing surface. To understand the formation of this film, a detailed discussion of the factors involved is required.

Friction

Friction has been defined as *the resistance to motion between two surfaces in contact*. It is present at all times where any form of relative motion occurs. It is the governing factor in machine design and operation and in most cases is the major limiting factor in machine speed, capabilities, etc. It is overcome in machinery by separating the moving surfaces with a fluid film or lubricant, usually an oil. Figure 10-3 shows enlarged views of mating surfaces with and without an oil film. Microscopic examination of bearing surfaces shows minute peaks and valleys similar to those shown. The smoother or more highly finished the surfaces, the thinner the oil film required to keep them separated. The *wearing-in* or *running-in* of a plain bearing is a process of wearing off the peaks and filling in the valleys so that extreme high points do not penetrate the film and make contact.

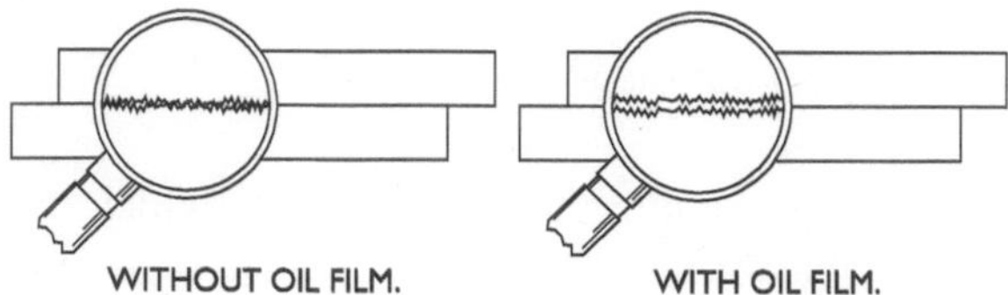

Figure 10-3 Mating surfaces with and without an oil film.

When the surfaces are separated with an oil film, the only friction then is within the fluid, and generally this is of small consequence. If an oil film can be formed and complete separation of the surfaces maintained, the condition is described as thick-film lubrication. In plain bearings, the shaft's rotation generates this essential film between the journal and bearing surfaces. While stationary, the weight of the shaft squeezes out the oil, causing the journal to rest on the bearing surface as shown in Figure 10-4A. When rotation starts, a wedge-shaped film develops under the journal. In a properly designed bearing, the pressure that develops in this wedge lifts the journal away from contact with the bearing, as shown in Figure 10-4B.

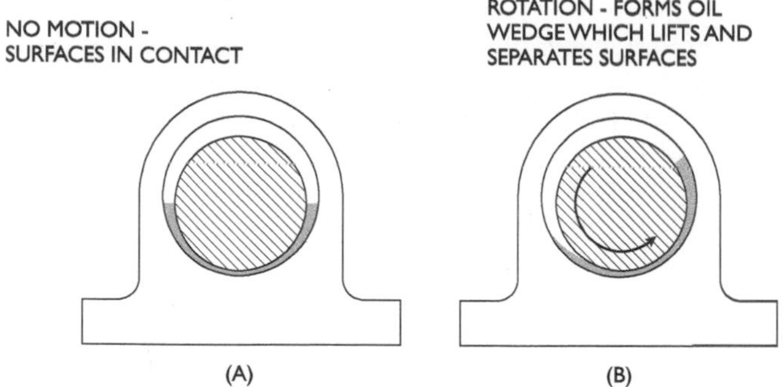

Figure 10-4 Action that takes place on lubricated parts when in motion and at rest.

Although plain bearings are operating at normally rated load and speed, practically no wear occurs because the oil film keeps the bearing surfaces separated.

Hydromatic Action

The hydromatic pressure that is built up in most plain bearings and forms the oil film that makes possible their operation under heavy loads is called hydrodynamic action or hydrodynamic lubrication. This occurs because the lubricant has a tendency to adhere to the rotating journal and is drawn into the space between the journal and the bearing. Because of the viscosity, or the internal resistance to motion of the lubricant, hydraulic pressure is built up in the lubricant film. Pressures as high as 600 psi are not unusual in well-designed bearings. Hydrodynamic action occurs in a plain bearing as the shaft rotates. When at rest, the journal slowly squeezes out most of the lubricant. A condition is reached where there is metal-to-metal contact between the journal surface and the bearing liner. When the journal begins to rotate, it has a tendency to climb up the side of the bearing. A further rotation of the journal causes oil to be drawn into the clearance space between the journal and the liner. The journal is lifted by the oil film that forms and is separated from the liner by the film. As the speed is increased, higher pressure is developed in the oil film, and the journal takes an eccentric position near the opposite side of the bearing. The oil-film pressure in the bearing is highest in the center and falls off to zero at both ends because of end leakage.

Boundary Lubrication

Under ideal conditions of thick-film lubrication, the material of which a plain bearing is made is not too important, provided it possesses sufficient mechanical strength to support the loads imposed without deforming or cracking. There are times, however, when thick-film lubrication does not exist—for example, during starting and stopping or when the bearing is subjected to shock loads, overloads, or misalignment. Under these conditions, what is called *boundary* lubrication exists. The fluid film is broken, and varying degrees of metal-to-metal contact occur. The factors governing the performance of a bearing under conditions of boundary lubrication are the very thin film of oil that may not have been squeezed out and the properties of the bearing materials.

Bearing Materials

When there is relative movement between two contacting metal surfaces, the amount of sliding friction present depends on the properties of the materials. Usually, with two similar metals, friction is high and seizing of the surfaces occurs at relatively low pressures or surface loads. Certain combinations of materials have low frictional qualities and are capable of supporting substantial loads without seizing or welding together. In most machinery, shafts are made of steel and the bearing is made of some other, softer material which, when in contact with steel, has low friction qualities. The most widely used bearing materials are cast iron, bronze, and babbitt. To a lesser degree, wood, plastics, and other synthetic materials are used. These materials make possible plain-bearing operation during boundary lubrication periods because of their ability to withstand sliding contact without excessive friction. Other qualities are also necessary in bearing materials and usually are the reason for selection of a particular material. The following are some of the more important qualities: strength to withstand loads without deforming; an ability to permit embedding of grit or dirt particles that may contaminate the lubricant; an ability to deform to permit distribution of the load over the full bearing surface; an ability to conduct heat so that frictional heat will be dissipated and hot spots, which might cause seizure, will not develop; and a resistance to corrosion.

Surface Finish

Another factor that has a major influence on plain-bearing operation is the surface finish of both the journal and the bearing. Simply stated, the rougher the surfaces, the thicker the film required to separate them. Oil-film thickness, even in very large bearings, does not exceed a couple of thousandths and in small bearings may be only a few ten-thousandths. This gives some indication of the importance of fine surface finishes to satisfactory bearing performance. To obtain suitable surfaces in new bearings, a run-in period is often required even with the high finishes that are possible with present machining operations. This is necessary because a surface that appears to be perfectly smooth to the touch actually is not smooth at all. If the surface cross section were to be highly magnified, it would show a sawtooth irregularity of peaks and valleys. When two such surfaces are required to move on one another, a wearing down or smoothing out of the sawtooth projections occurs. This is called *wearing-in*, *running-in*, or *breaking-in*. When fully run-in, the surface will have achieved a high degree of smoothness. The peaks will have been reduced and flattened, all sharp edges will have been rounded, and the maximum surface variation from peak to valley will have been greatly reduced. When such surfaces have been achieved, a very thin oil film is adequate to keep them separated and the bearing capable of supporting heavy loads.

Bearing Design

Plain bearings are made in a wide variety of types or styles. They may be complete, self-contained units or built into, and part of, a machine assembly. Common styles and types are shown in Figure 10-5.

Journal Bearings

The plain bearing, used for support of radial loads, is commonly referred to as a journal bearing. It takes its name from the portion of the shaft or axle that operates within the bearing. See Figure 10-5A.

Solid Bearing

The most common plain bearing is the sleeve or bushing. See Figure 10-5B.

Split Bearing

Divided in two pieces, the split bearing allows easy removal of the shaft. See Figure 10-5C.

Part Bearings

The part and/or half bearing encircling only part of the journal is used when the principal load presses in the direction of the bearing. Its advantages are low material cost and ease of replacement. See Figure 10-5D.

Thrust Bearings

The thrust bearing supports axial loads and/or restrains endwise movement. A widely used style of thrust bearing is the simple annular ring or washer. Two or more such thrust rings may be combined to mate a hard steel surface with a softer low-friction material. Another common practice is to support thrust loads on the end surface of journal bearings. When the area is too small for the applied load, a flange may be provided. A shoulder is the usual method of supporting the thrust load on the shaft. See Figure 10-5E.

Plain radial bearings, as the name implies, are capable of supporting only radial loads, that is, loads perpendicular to the shaft. If thrust loads are to be supported, they may be handled by simple collars and flanges, or more elaborate arrangements may be used. The simplest style of radial bearing is the plain bushing. They are generally very simple in appearance, being essentially cylindrical sleeves. Bronze is one of the most commonly used materials for plain bearings. It is one of the harder bearing materials generally used for low speeds and heavy loads. One of its principal advantages is that, when used as a solid bearing or bushing material, it can be easily and economically replaced. Plain bushings may also be lined with a softer bearing material, such as babbitt. The principal reason for this is that many relatively soft bearing materials are weak, and their use is limited by their load-carrying ability. The use of bronze, steel, or iron as a backing material provides the strength needed for the complete bearing.

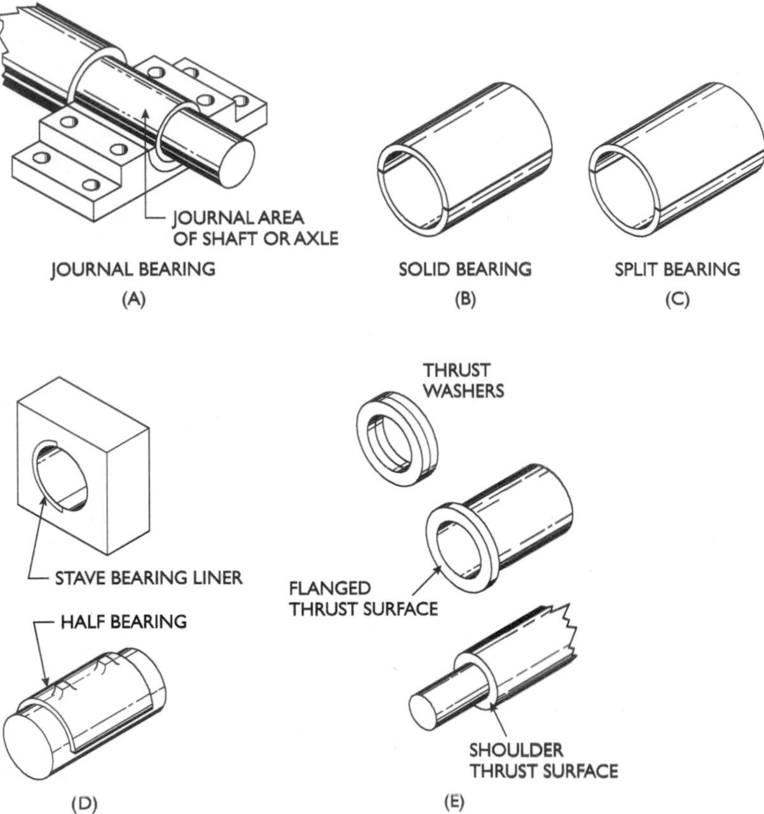

Figure 10-5 Various types and styles of plain bearings.

Lubrication Holes and Grooves

The simplest plain bearings have a hole in the top through which the lubricant travels to the journal. For longer bearings, a groove or combination of grooves extending in either direction from the oil hole, as shown in Figure 10-6, will distribute the oil. Recommended practice is to locate the inlet hole in the center of the bearing in the low-pressure region. On long bearings, two or more inlets may be required. Proper grooving is the most dependable means of distributing the oil so the moving surfaces, as they pass the grooves, will take up a film of oil. In bearings not fed under pressure, this is necessary if a taper oil film is to be developed to provide thick-film lubrication. Frequently, a pocket or reservoir is provided in the low-pressure region to place a volume of oil at the oil-film formation point. Regardless of the system of grooving, it is important that such grooving be confined to the unloaded portion of the bearing surface. If the grooving extends into the loaded or pressure region, the oil film will be disrupted, and boundary lubrication conditions will exist.

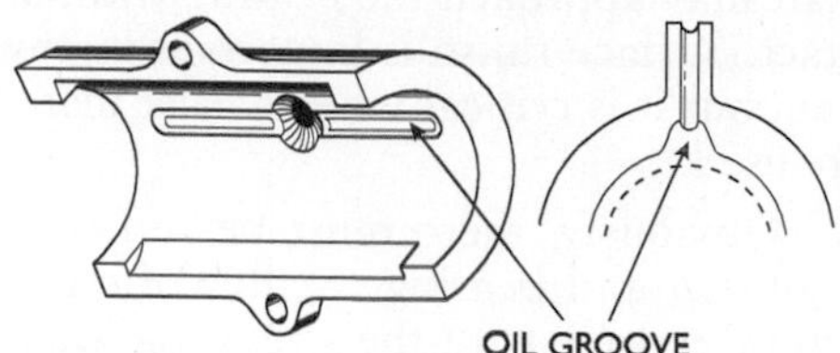

Figure 10-6 Oil hole and groove used in plain bearings.

In cases where plain bearings are grease lubricated, a different grooving system is used. Grooves are inserted at the high-pressure areas in an attempt to supply the entire surface of the journal with a continuous coating of grease, ensuring at least partial lubrication of the bearing.

Bearing Failures

Most industrial machine designers provide good bearings for their equipment. However, there are some cases where bearings are improperly designed, manufactured, or installed. While determination of the exact cause of a bearing failure is often a difficult matter because of the many factors involved, usually the trouble lies in one or more of the following areas:

Unsuitable Materials for either Shaft or Bearing. From a lubrication standpoint, the only requirement for the shaft is sufficient stiffness and hardness plus proper surface finish. The hardness requirements of the shaft depend on the bearing material used. Under normal operating conditions, tin and lead babbitts may be used successfully with soft steel journals. With harder bearing materials, it is frequently necessary to have harder shaft surfaces for satisfactory operation. This is normally a problem for the engineer rather than the mechanic.

Incorrect Grooving. As discussed previously, oil grooving is an important factor in plain-bearing operations. If the grooves are incorrectly located, bearing failures can result. If the oil grooves are located in the high-pressure areas, they act as pressure-relief passages. Thus, the formation of the hydrodynamic film is interfered with, and the bearing's load-carrying capacity is reduced.

Unsuitable Surface Finish. It has been pointed out that if the surface variations of the bearing or shaft are excessive, they will penetrate the oil film. Rough surfaces frequently cause scoring, overheating, and bearing failure. The smoother the finish, the closer the shaft may approach the bearing without danger of metallic contact. Surface finish is important in any plain-bearing application, but it is critical when harder materials such as bronzes are used.

Insufficient Clearance. Obviously, there must be sufficient clearance between the journal and bearing to allow the formation of oil film. Clearances vary with the size of the shaft and type of bearing material, the load carried, and the accuracy of the shaft position desired. In industrial machine design, a diametral clearance of .001" per inch of shaft diameter is often used. This is an average figure and applies to general applications. Where high speeds are encountered or where heavy loading is present, this figure will require adjustment.

Faulty Relining Practices. Many bearing failures can be attributed to faulty practices in relining. This applies chiefly to babbitted bearings rather than precision machine-made inserts. The pouring of babbitted bearings is a job that should be performed under carefully controlled conditions. Some of the reasons for faulty relining are

- Improper preparation of bonding surface
- Poor pouring technique
- Contamination of babbitt
- Pouring bearing to size with journal in place

Operating Conditions. The great majority of bearing failures are the result of abnormal operating conditions or the neglect of some phase of necessary maintenance precautions. When machines are speeded up, heavily overloaded, or used for a purpose other than that for which they were designed, bearings take a beating; careless maintenance, such as failure to properly lubricate or to use improper lubricants, frequently results in bearing failure. A few typical causes of premature failure from abnormal operating conditions or faulty maintenance practices are

- Excessive operating temperatures.
- Foreign material in the oil supply.
- Corrosion of the bearing metal.
- Bearing fatigue.
- Use of unsuitable lubricants.

Excessive Operating Temperature. The strength, hardness, and life of every plain bearing material are affected by excessive temperatures. Thick babbitt liners cannot be operated at as high a temperature as some of the precision thin babbitt inserts. In addition to the effect on the bearing material, high temperature also reduces the viscosity of the oil. As this affects the thickness of the oil film, it does in turn affect the load-carrying capacity of the bearing. Most lubricating oils will oxidize more rapidly at high temperatures, with resultant unsatisfactory performance.

Contamination of the Oil Supply. In cases where it is known that a bearing will probably be exposed to abrasive materials at some time during its life, soft materials such as babbitt are usually used. These materials have the ability to completely embed hard particles and protect the shaft against abrasion. The harder bearing materials are scored and galled by abrasives caught between the journal and bearing. Filters and breathers should be examined, the oil should be checked for contamination, and foreign material collected at the bottom of the bearing sump should be removed to avoid contamination of the oil supply.

Antifriction Bearings

Antifriction bearings, so called because they are designed to overcome friction, are of two types: ball bearings and roller bearings. In the plain bearing, the frictional resistance to sliding motion is overcome by separating the surfaces with a fluid film. The antifriction bearing substitutes rolling motion for sliding motion by the use of rolling elements between the rotating and stationary surfaces and thereby reduces friction to a fraction of that in plain bearings.

Life of Antifriction Bearings

Bearing life is a term used to describe the life expectancy of 90 percent of a group of the same bearings operating under the same load and at the same speed. Because most mechanics are interested in how long a bearing will last, bearing life is usually given in hours of operation. Bearing engineers call the minimum expected life of a bearing the L_{10} life. The term "average life" equals the minimum life, L_{10}, × 5. An average life of 100,000 hours is the same as 20,000 hours L_{10} minimum life.

Any bearing manufacturer's catalog will show the speed limits for each size and style of bearing. These limits are based on specific maximum loads, maximum ambient air temperature, good installation, and proper lubrication. As speed increases for a given size, so does the temperature within a bearing. At some point, either the grease breaks down or the steel begins to destruct.

The life of a bearing depends on many items, but the most important ones are these:

- Speed
- Load
- Proper lubrication
- Proper installation

Life Equation

Each antifriction bearing has a published basic load rating. This is a standard rating, given by the manufacturer, and can be found in its catalog.

Formulas exist for the life of both a ball bearing and a roller bearing:

$$L_{10} = \frac{1{,}00{,}000}{\text{RPM} \times 60} \times \left(\frac{C}{P}\right)^{3} \qquad \text{Ball bearing}$$

$$L_{10} = \frac{1{,}00{,}000}{\text{RPM} \times 60} \times \left(\frac{C}{P}\right)^{10/3} \qquad \text{Roller bearing}$$

Where:

L_{10} is minimum life in hours
C is the published basic load rating in pounds
P is the actual load on the bearing in the machine

For example, consider a ball bearing with a basic load rating of 57,423 lbs, an actual load of 502 lbs, running at 1180 rpm. The L_{10} life is shown in the following calculations:

$$L_{10} = \frac{1{,}00{,}000}{1180 \times 60} \times \left(\frac{5743}{502}\right)^3$$

$$L_{10} = 21{,}148 \text{ Hours}$$

If the same bearing is run under the same load but at half the speed, the life doubles:

$$L_{10} = \frac{1{,}00{,}000}{590 \times 60} \times \left(\frac{5743}{502}\right)^3$$

$$L_{10} = 42{,}296 \text{ Hours}$$

Mathematically, it can be shown that bearing life is inversely proportional to the rpm. If the speed is doubled, the bearing life is halved. If the speed is halved, the bearing life is doubled.

If the actual load on the bearing is changed—for instance, doubled—then the following calculation shows the result. The load used will be 1004 lbs—double the 502 that was used originally:

$$L_{10} = \frac{1{,}00{,}000}{1180 \times 60} \times \left(\frac{5743}{1004}\right)^3$$

$$L_{10} = 2{,}644 \text{ Hours}$$

Now the hours are reduced to one-eighth of 21,148, or only 2644 hours. Load is a very big factor on the life of a bearing. On a ball bearing, if the load is doubled, the life decreases by a factor of 8. On a roller bearing, if the load is doubled, the life decreases by a factor of 10. Controlling the load and the speed is important to extend the life of any machine running on antifriction bearings, but the load is far more influential than the speed.

Proper Lubrication

To keep wear to a minimum, bearings rely on a thin film of lubricant between balls or rollers, races, and retainers to prevent actual surface-to-surface contact. This lubricant also dissipates heat, prevents corrosion, restricts the entry of contaminants, and flushes out any particles that result from wear. The quickest and surest way to determine the proper lubricant for a bearing is to check the manufacturer's manual that was supplied with the equipment.

Grease lubrication is suitable for most low to medium speeds but not normally recommended for the highest speeds. Oil is used at higher speeds. Oil, especially when set up in a system—which includes a pump, cooler, and a reservoir—can carry more heat away from a bearing than grease.

Remember that grease is a combination of oil and thickening agent, sometimes called soap. The viscosity and the amount of thickener that is added determine the thickness of the grease. Two greases with the same thickness rating are not necessarily the exact same combination of elements and may not blend together well. Usually it is preferable to choose one brand of grease and stick with it.

Proper Installation

It is absolutely critical to follow the bearing manufacturer's recommendations about properly installing a bearing. Everything must be clean and dry when you install bearings. Take care to ensure that the bearings are not cocked or misaligned in any way. Accidentally dropping a bearing or hitting it with a hammer where the force is transmitted to the rolling elements is a sure way to doom a bearing to an early death.

The seven mistakes to watch for when installing a bearing are these:

1. Allowing dirt or water to enter the bearing.
2. Overheating the bearing to shrink-fit it to the shaft (250°F is usually the maximum).
3. Hitting or pressing the bearing to cause force to be transmitted to the rolling elements.
4. Cocking the bearing on the shaft so that it is not square to the shaft shoulder.
5. Failing to match mark pillow block housings to make a correct reassembly.
6. Failing to use the spanner wrench to tighten the bearing nuts (hammer and punch not acceptable).
7. Forgetting to install oil to the bearing reservoir after a machine is rebuilt.

What Makes an Antifriction Bearing Work?

A simple experiment that demonstrates how effectively rolling motion reduces friction can be performed as follows:

1. Pile several books on a smooth tabletop and push them across the surface with a thin ruler. Notice that it takes considerable effort to keep them in motion. This is sliding friction.

2. Now place several round pencils under the books. Push them again and notice how much more easily they can be moved backward and forward. This is line-contact rolling motion.
3. Replace the pencils with four marbles of the same size and push the books once more. You will find it requires even less effort to move them. Also, they can be moved in any direction. This is point-contact rolling motion.

In this experiment, the book, the table, and the rolling elements form the basic components of a ball or roller bearing. The books probably ran off the pencils and balls, and the rolling elements were not inclined to track or follow one another or keep the same distance apart. In order to correct this condition, a bearing incorporates a retainer or separator to help keep the rolling elements in their proper position.

Background on Antifriction Bearings

Antifriction bearings, as we know them, were developed in the 19th century. Long before that time, however, the value of round stones and logs for moving heavy objects was known. Primitive man was able to launch heavy boats by rolling them on a series of logs. The Egyptians, in building the pyramids, transported huge blocks of stone by means of rollers. For many centuries, no attempt was made to use balls or rollers in the bearings that supported axles. This would have been practically impossible because of the lack of machinery and know-how to attain the accuracy required in balls, rollers, and races for proper functioning of this type of bearing. The development of antifriction ball and roller bearings was slow until stimulated by the invention of the electric motor and the automobile. The ball bearing found its first widespread use in the bicycle just before the turn of the century, at about the same time the Bessemer steel process was developed. The steel from this process could be highly finished because it was free of impurities. It also had the strength to support the concentrated loads present in ball and roller bearings. In the development of the automobile, ball and roller bearings were found to be ideal for many applications, and today they are widely used in almost every kind of machinery. Both ball and roller bearings have been highly developed to meet the demands of various types of services. Classified by function, ball bearings may be divided into three main groups: radial, thrust, and angular-contact bearings. Radial bearings are designed primarily to carry a load in a direction perpendicular to the axis of rotation. See

Figure 10-7A. Thrust bearings can carry only thrust loads, that is, a force parallel to the axis of rotation tending to cause endwise motion of the shaft. See Figure 10-7B. Angular-contact bearings can support combined radial and thrust loads. See Figure 10-7C.

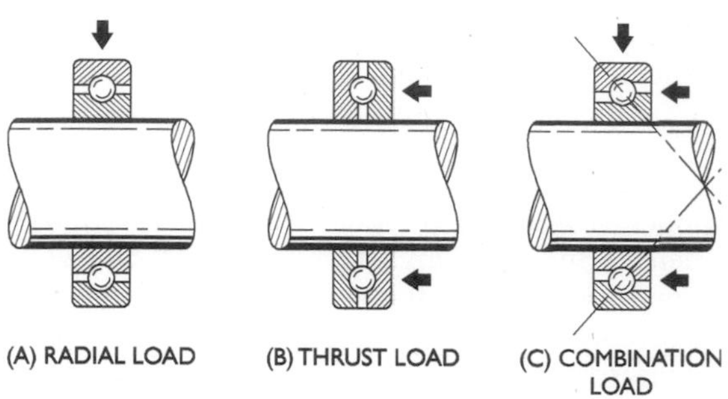

Figure 10-7 Three principal types of ball bearings.

Roller bearings may also be classified by their ability to support radial, thrust, and combination loads. Combination load-supporting roller bearings are not called angular-contact bearings, as they are quite different in design. The taper-roller bearing, for example, is a combination load-carrying bearing by virtue of the shape of its rollers, as shown in Figure 10-8.

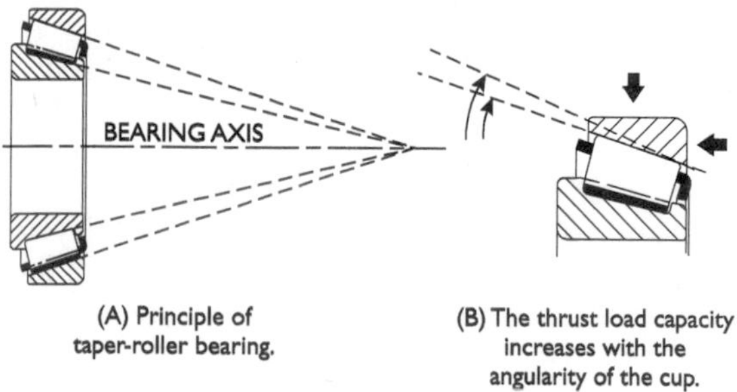

Figure 10-8 The taper-roller bearing.

Nomenclature

Basically, all antifriction bearings consist of two hardened rings called the inner and outer rings, the hardened rolling elements that may be either balls or rollers, and a separator. Bearing size is usually given in terms of what are called boundary dimensions. These are the outside diameter, the bore, and the width. The inner and outer rings provide continuous tracks or races for the rollers or balls to roll in. The separator or retainer properly spaces the rolling elements around the track and guides them through the load zone. Other words and terms used in describing antifriction ball bearings are the face, shoulders, corners, etc. All are illustrated in Figure 10-9. The terms used to describe taper-roller bearings are a little different in that what is normally the outer ring is called the cup, and the inner ring, the cone. The word *cage* is standard for taper-roller bearings rather than separator or retainer. These terms are illustrated in Figure 10-10. Nomenclature for the straight roller bearing is illustrated in Figure 10-11.

Bearing manufacture is governed by two factors: the form of the product and the material of which it is made. The exact form of antifriction bearings depends mainly on the loads and speeds to be met in operation and, to some extent, space limitations of the product into which it is assembled. For unusual applications, bearings have been made of glass, plastic, and other substances, but for 98 percent of ball bearing uses, specially developed steel is best adapted. Bearing steel is a high-carbon chrome alloy with high hardenability and good toughness characteristics in the hardened and drawn state.

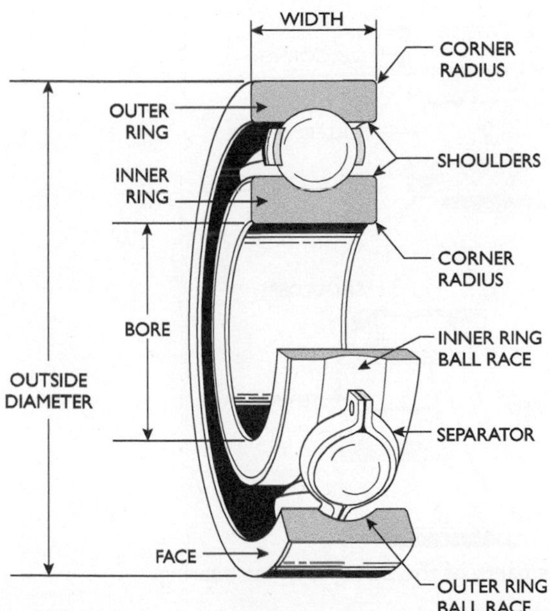

Figure 10-9 Various parts of the antifriction ball bearing.

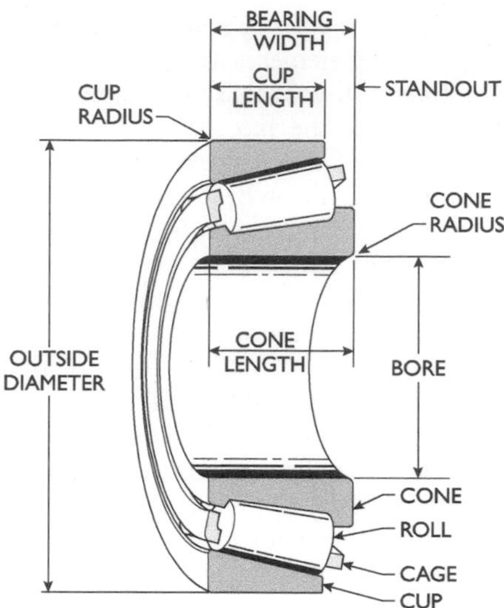

Figure 10-10 Various parts of the taper-roller bearing.

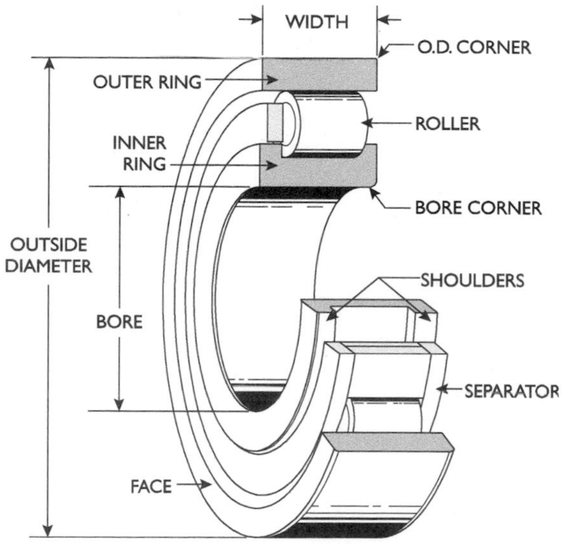

Figure 10-11 Various parts of the straight-roller bearing.

The manufacture of bearings resolves itself into two phases of stock removal. In the first, the general outline of the ball or rings is formed. Relatively large amounts of stock are taken off while removal is easy in the soft state. Balls and rollers are brought nearly to soft size by cold forging in the smaller size and by hot forging in the larger sizes. A soft grinding is performed before heat treatment. In the first phase of stock removal, material must be left on all parts to allow for distortion, growth, peeling, etc., incurred in heat treatment.

After heat treatment, the second stage of stock removal begins. On almost all surfaces, there is a stock allowance to be removed. This layer varies from .005 to .030 in. thick. On removal of the layer, size must be reached simultaneously with attaining the proper surface finish. The tolerance on dimensions to be met is in the magnitude of .002 to .0006 in. for rings and in the vicinity of .00002 in. for balls and rollers. The finishes for general surfaces need to be relatively fine, whereas running surfaces must have an extremely fine, smooth finish. Such requirements are generally met by grinding, honing, or polishing operations or by combinations of these.

The hard stock removal from balls is performed in the sequence of grinding and lapping. The grinding operation, which is rather unusual, can be compared to the formation of a sphere from a ball of clay rolled in the palms of the hands. The top hand represents a circular grinding wheel and the bottom one a cast-iron plate, both of which are circularly grooved to accept several rings or balls. All portions of the surface of all balls are exposed time and again to the cutting grooves of the grinding wheel. Over a period of 20 to 30 hours and with proper adjustment, the machine produces a load of balls that are within .00001 in. of each other for size and within .000005 in. for roundness.

The method of final stock removal for rings varies with the surface. The faces of the rings are finished with a disc grinding wheel while they are held on a circular revolving table by magnetic force. Outside diameters are normally ground in the centerless type of machine where rings are fed through between the cylindrical grinding wheel and rubber driver. Inner rings are ground, two or three at a time, in a machine that automatically stops when the surface comes to specification. Outer race grinding is a unique operation not commonly met in the metalworking industry. Special grinding machines have been designed and built for this purpose that can grind the curvature in two different planes as required. Formation of the inner race is fairly simple. The grinding wheel is diamond dressed to the shape of the race. With wheel and work both rotating, cutting contact is made. Both inner and outer races are polished after grinding.

Radial and lateral internal clearance is necessary in a completed bearing to compensate for press fits, misalignment of machine parts, etc. This means that selective assembly is necessary to provide close control over bearing internal looseness when assembled. Races are sized in groups of one ten-thousandth's tolerance and balls to one twenty-thousandth's tolerance as standard procedure at matching. Highly sensitive special gauges check parts so that suitable size combinations for proper clearance at assembly are matched. The bearing is then assembled, and the retainers or separators are inserted to space the rolling elements. These may be made of pressed steel, pressed bronze, machined bronze, synthetic materials, etc. Seals and shields are inserted at this stage in various combinations as required.

Final inspection verifies principally the running characteristics of the finished product. The bearing is spun on a test motor where any unusual noise or looseness is a cause for rejection. Run-out also must be within specific limits. After final inspection, bearings are protected by a light film of rust-preventive oil, wrapped in oil-proof paper, and packaged for shipment.

Separators

Bearing separators are also called retainers or cages. While the words *separator* and *retainer* are most frequently used in reference to ball and straight-roller bearings, the word *cage* is used in reference to taper-roller bearings, and all are used interchangeably.

The function of the separator is to properly space the rolling elements around the track and guide them through the load zone. There are two operational types of separators—the ball-riding type and the land-riding type. As the name implies, the ball-riding type is supported and driven by the rolling elements themselves. It is designed to conform to the shape of the rolling element, and when assembled, each ball is enveloped in a separate pocket. As the bearing rotates and the balls roll in the race or tracks, the separator rides with them and maintains equal ball spacing. The land-riding type is machined to a close running fit with either the inner or outer ring shoulder and is supported and guided by it. Usually the ball pockets are cylindrical holes in the separator, and line contact occurs between the separator and the balls. The separator is the weakest point in an antifriction bearing, as sliding friction is always present between the separator pockets and the rolling elements.

The simplest and most commonly used separator is a two-piece pressed-steel construction. Two metal ribbons, formed with spherically shaped ball pockets, are joined together between the pockets by riveting, clinching, or staking. These separators are ball-riding or controlled. Although light in weight, they are strong enough to serve the purpose of separating and guiding the balls and are economical to manufacture. Many variations of pressed-steel separators are made. One-piece construction is used where design allows placing of the rolling elements in the separator before assembly of the bearing.

Two-piece cast-bronze ball-riding separators similar to those made of pressed metal are used where heavier load-carrying ability is needed. They are usually used on larger-size bearings and are more expensive than the pressed-steel separators.

Another widely used separator is the solid bronze type, made in one or two pieces depending on bearing design and assembly practice. It is machined all over to close tolerances and is land-riding or ring-controlled. It rides or is piloted on the outside shoulder of the inner ring. It is designed to support heavy-shock loads and for precision bearing applications.

The composition solid-type retainer, usually of phenolic resin material, is generally used for high-speed applications. It can be accurately machined, is strong, and is light in weight. It has the ability to absorb lubricant, thereby decreasing friction and contributing to quiet operation. For ultrahigh speeds, composition retainers are piloted by the outer ring band.

Ball Bearing Dimensions

Basic ball bearings, for general use in all industries, are manufactured to standardized dimensions of bore, outside diameter, and width (boundary dimensions). Tolerances for these critical dimensions and for the limited dimensions for corner radii have been established by industry standards. Therefore, all types and sizes of ball bearings made to industry-standardized specifications are satisfactorily interchangeable with other makes of like size and type.

Most basic ball bearings are available in four different series known as the *extra light*, *light*, *medium*, and *heavy.* The names applied to each series are descriptive of the relative proportions and load-carrying capacities of the bearings. This means that there are as many as four bearings (one in each series) with the same bore size but with different widths, outside diameter, and load-carrying capacities. Relative proportions of bearings in each series with the same bore are illustrated in Figure 10-12.

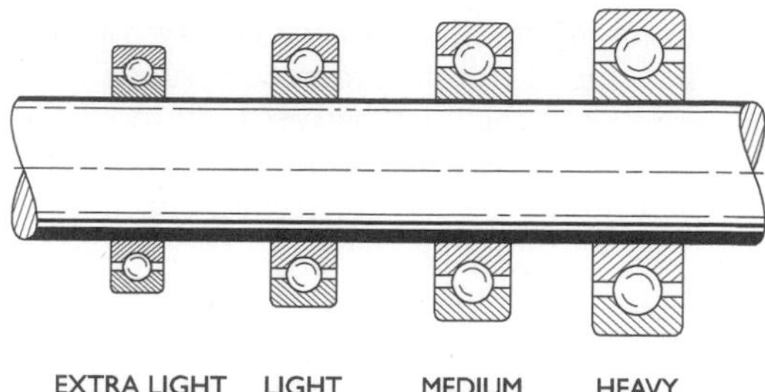

Figure 10-12 Four basic ball bearings with same bore size but different sizes in width and outside diameter.

It is also possible to select as many as four bearings with the same outside diameter (one in each series) with four different bore sizes, widths, and load-carrying capacities. Thus, there is a choice of four different shaft sizes without changing the diameter of the housing. This relationship is illustrated in Figure 10-13.

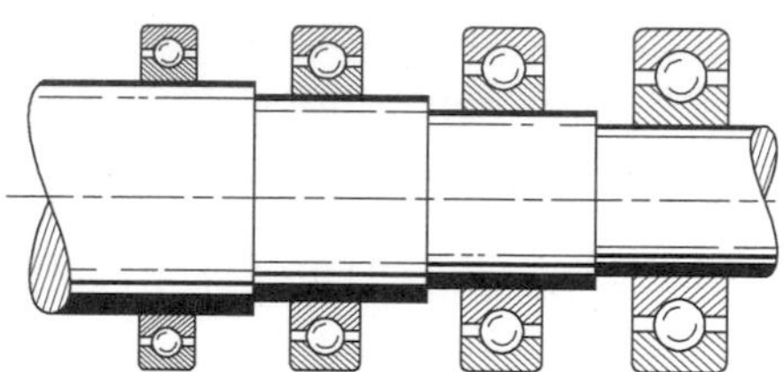

Figure 10-13 Four basic ball bearings with same outside diameter but different sizes in bore and width.

Bearing manufacturers designate the various series by using numbers that they incorporate into their basic bearing numbering systems. The *extra light* series is designated as the 100 series, the *light* as 200, the *medium* as 300, and the *heavy* as the 400 series.

Extra Light Series (100)

As the name implies, this series is designed to give the smallest widths and outside diameters in proportion to the standard bore sizes. The smaller cross-sectional areas of this series necessitate the use of smaller balls than the heavier series, and more of them. Although the load capacity is less than the heavier series, it is adequate for many applications.

Light Series (200) and Medium Series (300)

The light-and medium-series bearings are the most widely used of the four series. They are commonly used in a great variety of machinery and equipment. The great variety of proportions and capacities available in the full line of these two series provides a range of dimensions and sizes to meet most general requirements.

Heavy Series (400)

The heavy-series bearings are used where exceptionally heavy load conditions are encountered. Because their use is rather limited, they are not produced in nearly as wide a selection of types and sizes as the light and medium series.

Antifriction ball bearings were first developed and manufactured in Europe, where the metric system of measurement is used. Many of the European manufacturing plants had standardized dimensions to some extent before wide-scale manufacture of ball bearings started in America. Metric dimensions were widely accepted as standard for ball bearings; therefore, the practice of using metric sizes was continued in American manufacturing plants. Because of this, metric ball bearing sizes are interchangeable throughout the Western world. However, the fact that they are made to even metric sizes results in dimensions that have no relation to inch fractions when converted to the inch system of measurement. For example, a ball bearing with a 60-mm bore measurement measures 2.3633 in. when converted to the inch system. Conversion of millimeter dimensions to inch dimensions is accomplished by multiplying the millimeter dimension by .03937, which is the equivalent of one millimeter in thousandths of an inch. For approximate conversion purposes, one millimeter is roughly ¹⁄₂₅ in.

The basic ball bearing number is made up of three digits. The first digit indicates the bearing series (i.e., 100, 200, 300, or 400). The second and third digits, from 04 and up, when multiplied by 5, indicate the bore in millimeters. An example in each of the four duty series would be the following:

108—Extra Light Series—40 mm bore—1.5748 in.

205—Light Series—25 mm bore—0.9843 in.

316—Medium Series—80 mm bore—3.1496 in.

420—Heavy Series—100 mm bore—3.9370 in.

The bore of bearings with a basic number under 04 are as follows:

00—10 mm—0.3937 in.
01—12 mm—0.4724 in.
02—15 mm—0.5906 in.
03—17 mm—0.6693 in.

Tables of basic bearing numbers and boundary dimensions (bore, OD, width) in millimeters and inches are included in the Appendix.

Bearing Types

Single-Row Deep-Groove Ball Bearings

Single-row deep-groove ball bearings are the most widely used of all ball bearings and probably of all antifriction bearings. They can sustain combined radial and thrust loads, or thrust loads alone, in either direction even at extremely high speeds. They are also termed *Conrad*-type bearings, after the man who first successfully designed and produced this style of bearing. A cross section of the single-row deep-groove bearing is illustrated in Figure 10-14.

Figure 10-14 Single-row deep-groove ball bearing assembly.

Double-Row Deep-Groove Ball Bearings

The double-row deep-groove bearing shown in Figure 10-15 embodies the same principle of design as the single-row bearing.

However, the grooves for the two rows of balls are positioned so that the load through the balls tends to push outward on the outer ring races. This bearing has substantial thrust capacity in either direction and high radial capacity due to two rows of balls.

Figure 10-15 Double-row deep-groove ball bearing.

Angular-Contact Radial-Thrust Ball Bearings

Angular-contact bearings, illustrated in Figure 10-16, can support radial loads when combined with thrust loads in one direction. The inner and outer rings are made with an extra high shoulder on one side only (thrust side). It is designed for combination loads where the thrust component is greater than the capacity of single-row deep-groove bearings. They may be mounted either face-to-face or back-to-back and in tandem for constant thrust in one direction. When mounting in pairs, care must be taken that the bearings have been ground for the style of assembly used.

Self-Aligning Ball Bearings

The internal self-aligning style of ball bearing shown in Figure 10-17 compensates for angular misalignment. This may be due to errors in mounting, shaft deflection, misalignment, etc. This angular movement is possible because the two rows of balls are rolling on the spherical surface of the outer ring rather than in fixed races. Self-aligning bearings can support moderate radial loads and limited thrust loads, since the balls have only point-contact support on the spherical outer ring surface.

Figure 10-16 A radial-thrust ball bearing.

Figure 10-17 A self-aligning ball bearing.

Ball-Thrust Bearings

The ball-thrust bearing can support thrust loads in only one direction, which precludes —radial loading. The thrust load is transmitted through the balls parallel to the axes of the shaft, resulting in very high thrust capacity. The rings of these bearings are commonly known as washers (see illustration in Figure 10-18). To operate successfully, this type of bearing must be at least moderately thrust-loaded at all times. This style of bearing should not be operated at high speeds, as centrifugal force will cause excessive loading of the outer edges of the races.

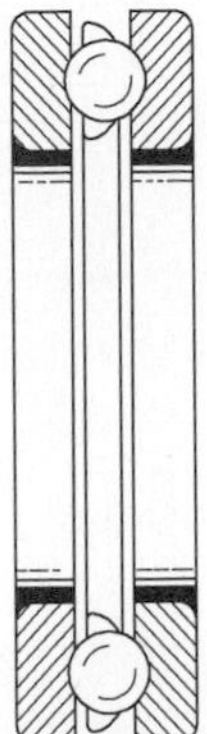

Figure 10-18 A ball-thrust bearing.

Cylindrical-Roller Bearings

The cylindrical-roller bearing has straight cylindrical-shaped rolling elements. These rolling elements are approximately equal in length and diameter, as shown in Figure 10-19. These equal-dimensioned rollers distinguish cylindrical-roller bearings from other roller bearings that have rollers with a much greater length-to-diameter ratio. They are used primarily for applications where heavy radial loads must be supported—loads beyond the capacities of radial ball bearings of comparable sizes. They are ideally suited for heavy loads in the moderate speed ranges because there is a line contact between rolling elements and races. Therefore, the rolling elements are deformed less under heavy load conditions than the point contact of ball bearings of comparable size. There are three basic types of straight cylindrical-roller bearings: the separable outer ring type shown in Figure 10-19A, the separable inner ring type shown in Figure 10-19B, and the nonseparable type shown in Figure 10-19C. Types with separable rings, Figure 10-19A and 10-19B, allow free axial movement of the shaft in relation to the housing.

Spherical-Roller Bearings

The double-row spherical-roller bearing is a self-aligning bearing utilizing rolling elements shaped like barrels, as shown in Figure 10-20. The outer ring has a single spherical raceway. The double-shoulder inner ring has two spherical races separated by a center flange. The rollers are retained and separated by an accurately constructed cage.

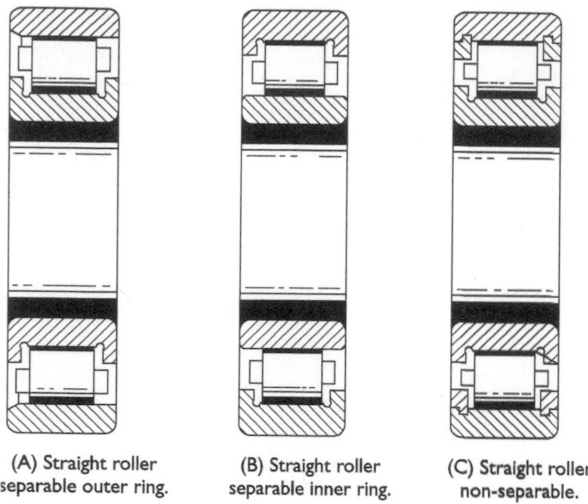

Figure 10-19 Three types of cylindrical bearings.

Figure 10-20 The spherical-roller double-row bearing assembly.

This type of bearing is inherently self-aligning because the assembly of the inner unit, i.e., ring, cage, and rollers, is free to swivel within the outer ring. Thus, there is automatic adjustment that allows successful operation under severe misalignment conditions. It will support a heavy radial load and heavy thrust loads from both directions.

Journal-Roller Bearings

Journal-roller bearings, shown in Figure 10-21, are straight-roller bearings; however, the rollers have a greater length-to-diameter ratio than the rollers of straight cylindrical-roller bearings. They are used in a wide variety of low-speed applications where loads are light to moderate and where space is limited. Because of their length and long line contact, journal bearings develop a considerable amount of internal friction. Therefore, they are made with large internal clearances to compensate for normal heat expansion. The journal-roller bearing takes its name from the fact that a common practice is to use the surface of the shaft as the inner race for the rollers. Journal bearings are also made with an inner race, an outer race, or both.

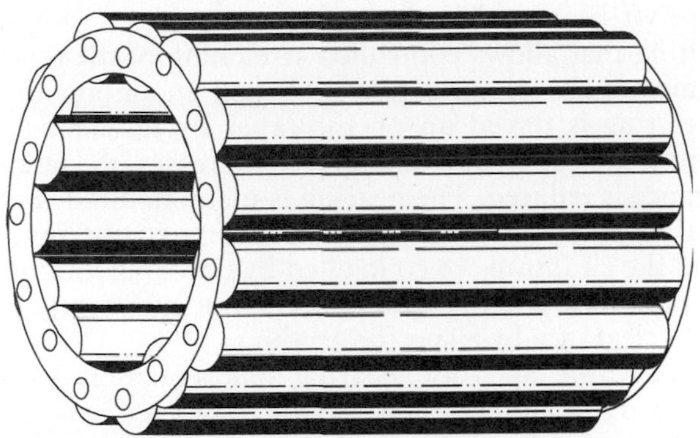

Figure 10-21 A typical journal-roller bearing.

Taper-Bore Spherical-Roller Bearings

The construction and applications for spherical-roller bearings have already been discussed. A specialized style of this bearing, designed specifically to ensure maximum inner-ring grip at assembly, warrants further discussion. Generally speaking, bearing applications have a rotating inner ring and a stationary outer ring. When correctly assembled, the inner ring is sufficiently tight on the shaft to ensure that both inner ring and shaft turn as a unit and that "creeping" of the ring on the shaft does not occur. Should creeping occur, there will be overheating, excessive wear, and erosion between the shaft and the inner ring. For normal applications, the inner ring is press-fitted to the shaft and/or clamped against a shoulder with a locknut. However, on applications subjected to severe shock or unbalanced loading, the usual press-fit or locknut clamping does not grip tightly enough to prevent creeping. For such applications, a design providing maximum grip of the inner ring on the shaft is required. In providing for the tremendous grip that is possible with the method to be discussed, some very important conditions must be controlled:

1. The stress in the inner ring must remain below the elastic limit of the steel.
2. The internal bearing clearance must not be eliminated.
3. There must be a practical mounting method.

The taper-bore self-aligning spherical-roller bearing incorporates the features required. Size for size, its capacity is greater than that of any other type of bearing. The taper bore provides a simple method of mounting that allows controlled stretching of the inner ring to obtain maximum gripping power. When mounting this bearing, the inner ring is forced up on the taper by tightening a locknut. As the inner ring stretches, and its grip increases, the internal bearing clearance is reduced. The bearing is manufactured with sufficient internal clearance to allow this stretching of the inner ring. The grip and the clearance are controlled by checking internal clearance before mounting and by tightening the nut sufficiently to reduce the internal clearance by a specific amount.

Another feature that the taper-bore bearing makes possible is the adapter-sleeve style of mounting. The use of a tapered adapter sleeve allows one to mount the taper-bore bearing on straight cylindrical surfaces. It also provides an easy means of locating the bearing. In many cases, this feature is the reason for using taper-bore bearings in applications where loading is relatively light and the tremendous inner-ring grip feature is not required. Figure 10-22 illustrates the taper-bore spherical-roller bearing with mounting adapter for assembly to straight cylindrical surfaces.

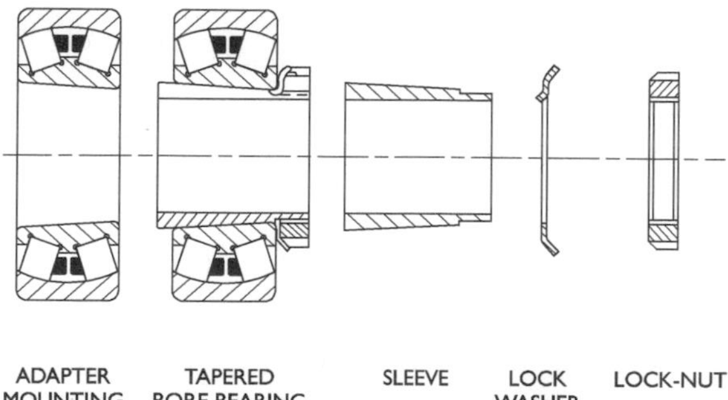

Figure 10-22 Taper-bore bearing with adapter.

Installation of Taper-Bore Spherical-Roller Bearings

Before mounting taper-bore spherical-roller bearings on taper shaft fits, or adapters, the internal clearance must be checked and recorded. The measurement should be made on only one side of the bearing. Recommended practice is to rest the bearing upright on a table and to insert the thickness gauge between the top roller and the inside surface of the outer ring.

Mounting Procedure

1. Check internal clearance between rollers and the outer race on the open side (as noted above). Thickness gauge must be inserted far enough to contact the entire roller surface.
2. Lightly oil surface of bearing bore and install on taper shaft or adapter sleeve *without* lock washer.
3. Check internal clearance as nut is tightened. Nut must be tightened until clearance is reduced by the amount recommended for the bearing size.
4. Remove nut, install lock washer, and retighten nut. Secure lock washer.

To determine the correct amount of clearance reduction for a given size bearing, the manufacturer's specifications should be consulted. If this information is not obtainable, and assembly cannot be delayed, use the table of values for general applications given in Table 10-1. These values should not be used if there are unusual conditions present, such as extremely high temperatures. For these special cases, such as high temperatures or bearings on hollow shafts with steam passing through, manufacturers' specifications and instructions for special applications must be obtained before proceeding.

Table 10-1 Clearance Reduction Table for General Applications

Shaft Diameter	*Clearance Reduction*
1½–3½	.001–.002
3½–6½	.002–.0035
6½–10¼	.0035–.0055

Taper-Roller Bearings

Taper-roller bearings can be classified as a separate group of roller bearings because of their design and construction. They consist fundamentally of tapered (cone-shaped) rollers operating between tapered raceways. They are so constructed, and the angle of all rolling elements so proportioned, that if straight lines were drawn from the tapered surfaces of each roller and raceway, they would meet at a common point on the centerline of the axis of the bearing, as illustrated in Figure 10-23.

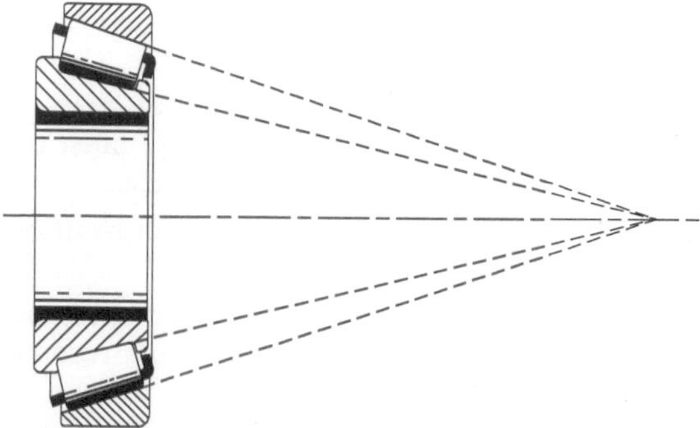

Figure 10-23 Taper-roller bearing.

One major difference between taper-roller bearings (separable types) and most antifriction bearings is that taper rollers are adjustable. In many cases, this is a distinct advantage since it permits accurate control of bearing running clearance. The proper amount of running clearance may be assembled into the bearing at installation to suit the specific application. This feature also permits preloading the bearing for applications where extreme rigidity is required. This adjustable feature also requires that proper procedure be followed by the millwright or mechanic at assembly to ensure correct setting. Another major difference between taper-roller bearings and most other antifriction bearings is that standard taper-roller bearings are made to inch dimensions rather than metric dimensions.

The basic parts of a taper-roller bearing are the *cone,* or inner race; the *taper rollers;* the *cage* (this is usually called the "retainer" or "separator" in other antifriction bearings); and the *cup*, or outer race.

The most widely used taper-roller bearing is a single-row type with cone, rollers, and cage factory-assembled into one unit with the cup independent and separable. Shown in Figure 10-24 is what is commonly called the standard single-row taper-roller bearing. Many additional styles of single- and multiple-row taper-roller bearings are made, including unit assemblies that are factory preadjusted.

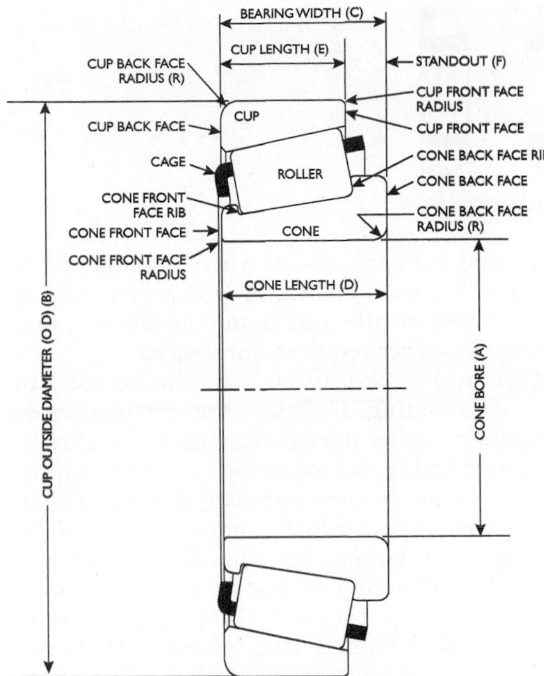

Figure 10-24 Single-row taper-roller bearing.

Taper-Roller Bearing Adjustment

Although each single-row taper-roller bearing is an individual unit, their construction is such that they must be mounted in pairs so that thrust may be carried in either direction. Two systems of mounting are employed, direct and indirect, as shown in Figures 10-25 and 10-26.

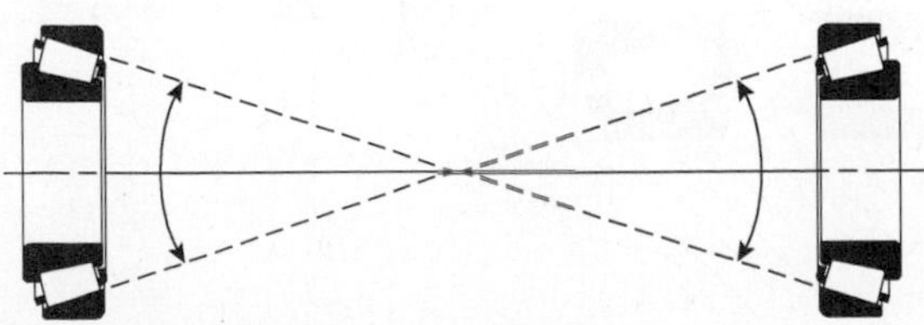

Figure 10-25 Indirect mounting.

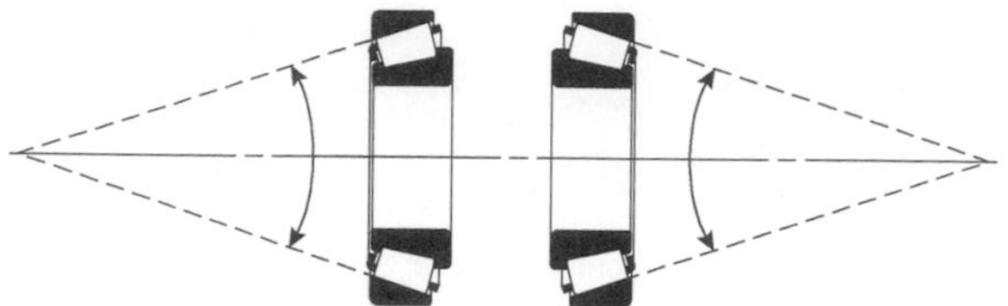

Figure 10-26 Direct mounting.

Indirect mounting is used for applications where maximum stability must be provided in a minimum width. Direct mounting is used where this assembly system offers mounting advantages and where maximum stability is neither required nor desired.

The terms *face-to-face* and *back-to-back* may also be used to describe these systems of mounting. If these terms are used, they must be qualified by stating whether the terms are used in regard to the cup or the cone. Indirect and direct mounting may more appropriately be referred to as *cone-clamped* and *cup-clamped*. Cone-clamped describes indirect mounting, which is normally secured by clamping against the cone. Cup-clamped describes direct mounting, which is normally secured by clamping against the cup.

Many devices are used for adjusting and clamping taper-roller bearing assemblies. Illustrated in Figures 10-27 through 10-32 are

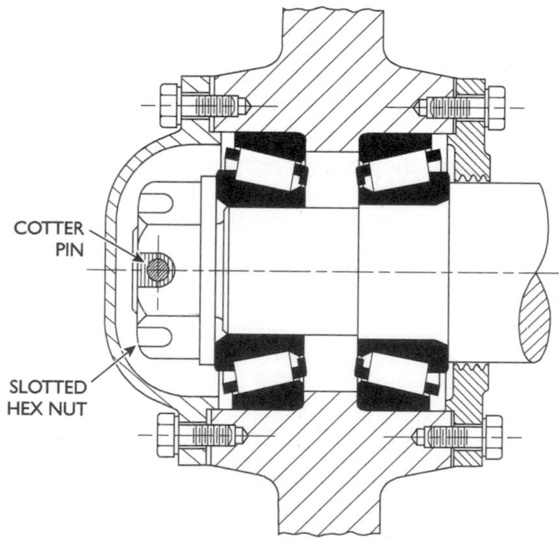

Figure 10-27 Adjustable mounting.

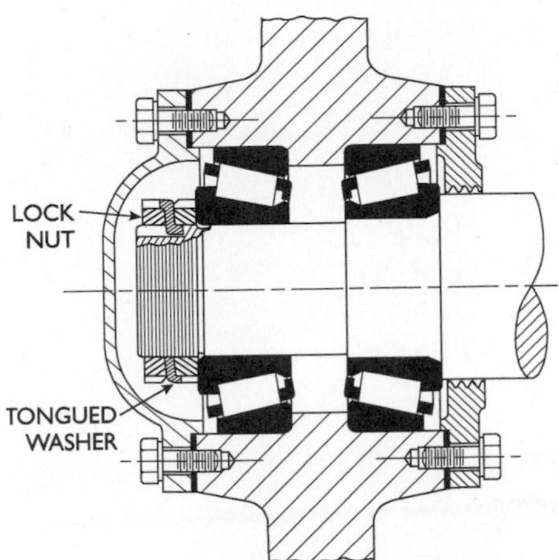

Figure 10-28 Adjustable mounting.

six of the basic devices used for this purpose. There are many others in general use; however, they are variations of these basic designs.

In Figure 10-27, the slotted hex nut and the cotter pin are used to adjust bearings. Here, the adjustment need not be extremely precise, and there is room for the nut. Simply tighten the nut while rotating the outer assembly until a slight bind is obtained in the bearing. This ensures proper seating of all parts. Then, back off the nut one slot and lock with a cotter pin. The result is free-running clearance in the bearing. The bearing is rotated in order to seat rollers against the cone. This should be done when making any bearing adjustment.

In Figure 10-28, two standard locknuts and a tongued washer are used to adjust the bearing. They provide a much finer adjustment than a slotted hex nut. Pull up the inner nut until there is a slight bind on the bearings, and then back off just enough to allow running clearance after the outer, or "jab," nut is tightened. This type of adjustment is used in full-floating rear axles and industrial applications where slotted hex nuts are not practical or desirable.

In Figure 10-29, shims are used between the end of the shaft and the end plate. The shim pack is selected to give proper setting of the bearing. (This will vary with the unit in which the bearings are used.) The end plate is held in place by cap screws. The cap screws are wired together for locking. A slot may be provided in the end plate to measure the shim gap.

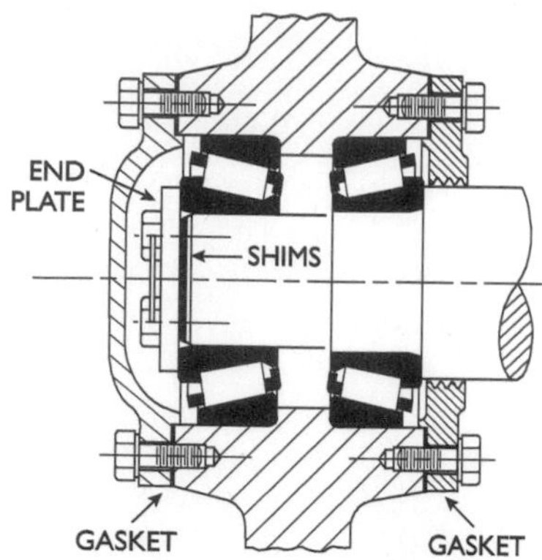

Figure 10-29 Adjustable mounting.

Figure 10-30 shows another adjustment using shims. They are located between the end-cap flange and the housing. Select the shim pack that will give the proper bearing running clearance recommended for the particular application. The end cap is held in place by cap screws, which can be locked with a lock washer (as shown) or wired. This easy method of adjustment is used where press-fitted cones are used with loose-fitted cups.

Shims are also used in the adjustment in Figure 10-31. The difference here is that one cup is mounted in a carrier. This adjusting device is commonly used in gearboxes and drives. It is also used in industrial applications for ease of assembly or disassembly.

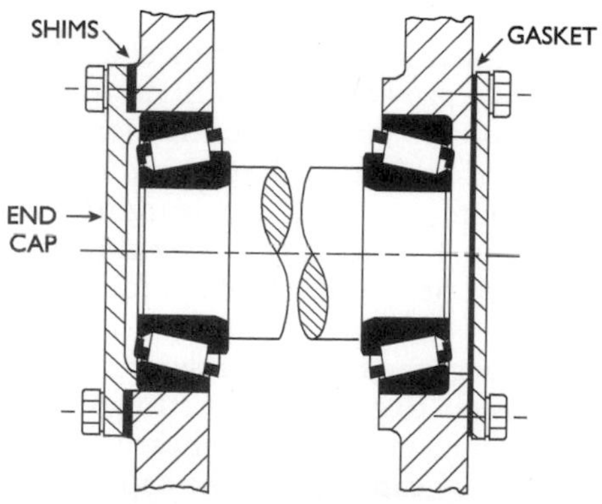

Figure 10-30 Adjustable mounting.

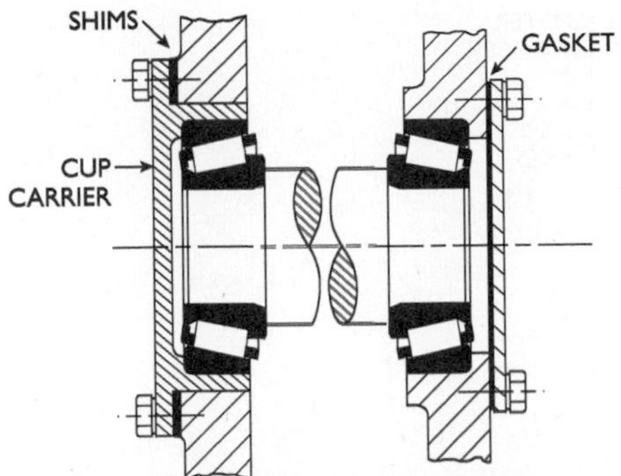

Figure 10-31 Adjustable mounting.

Special Installation Instructions

Three types of bearings require some special mounting installation instructions. The individual bearing, the split pillow block bearing, and the one-piece pillow block bearing are used on many different machines. Specific steps to mount these bearings properly are well worth learning.

Installation of the Individual Bearing

An antifriction bearing requires an interference fit on one of the bearing races, usually the inner race. Take care when mounting a bearing onto a shaft; it is critical that the internal clearance (required for the bearing to spin) is not totally removed when the inner race is stretched to fit over the shaft. Dimensional checks are the key to proper mounting. The bearing manufacturer or machine builder provides checking measurements for the shaft. When replacing a bearing, be sure to check that proper fitting between bearing and shaft is obtained. Figure 10-32 shows an example of the checks to be made on a shaft to ensure proper fitting without loss of internal clearance.

There are two common ways of mounting an individual bearing to a shaft: *arbor* (or *hydraulic*) *press mounting* and *thermal mounting*.

In press mounting, a force capable of moving the bearing onto the shaft is applied to the face of the ring that has the interference fit—usually the face of the inner race. Always exert force on the ring to be mounted to prevent damaging the bearing. Figure 10-33 shows the correct and incorrect methods of applying force.

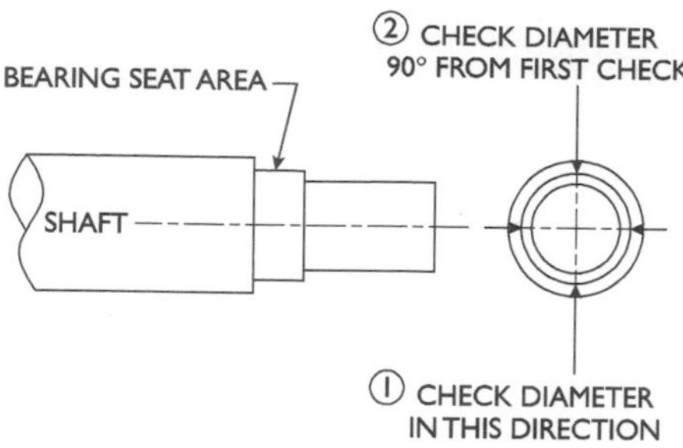

Figure 10-32 Check the shaft for proper dimensions.

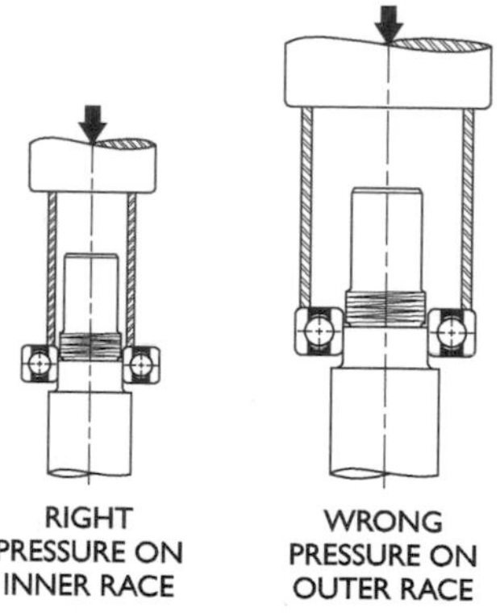

Figure 10-33 Correct and incorrect methods of applying force.

Thermal mounting uses a technique of heating the bearing (causing expansion), cooling the shaft (causing it to shrink), or a combination of the two to mount the bearing onto the shaft. In most applications, the inner ring should be uniformly heated to a temperature not to exceed 250°F. *Never* use a direct flame or hot plate. Heating a bearing to a temperature in excess of 250°F for extended periods of time will anneal the bearing metal and reduce the hardness—a good way to shorten the bearing life drastically.

A bearing can be heated using a cone heater (Figure 10-34), an electric oven, an induction heater, a hot oil bath, or even an electric light bulb centered in the bearing inner race. Using a cone heater has the advantage that the heater is quick, portable, and easy to operate. It is not uncommon to see the cone heater as the tool of choice in a maintenance shop.

Some cone heaters have a magnetic thermocouple that attaches to the bearing and senses the actual temperature—either alarming when a temperature of 250°F is reached or controlling to keep the temperature from exceeding this limit. A less expensive cone heater makes use of a temperature crayon (temp-stick) that is applied to the inner race. When the crayon just starts to melt and flow, the bearing is at the correct temperature.

Figure 10-34 Heating the inner race using a cone heater. *(Courtesy Maintenance Troubleshooting)*

When the bearing reaches the proper temperature, it should fit easily on the shaft. The example that follows shows the expected clearance by heating a 25-mm bore bearing to install it onto a shaft.

Example A 50-mm bore bearing must be assembled onto a shaft using heating. The maximum allowable temperature for heating the bearing is 250°F. The measurements of the bearing show the bore to be 1.9680 in., and the shaft is checked with a micrometer and has a diameter of 1.9690 in. in the area of the bearing seat. Obviously, the assembly will give an interference fit. The temperature in the maintenance shop is 70°F.

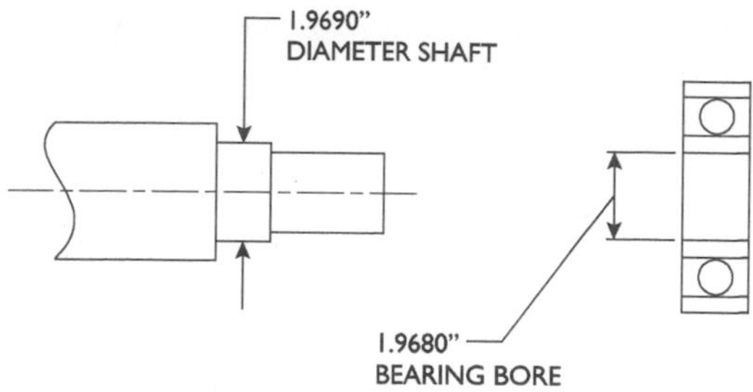

Figure 10-35 Thermal fitting.

1. The thermal linear coefficient of expansion of steel = .0000063 in/in/°F.
2. The formula for expansion of the bore of a bearing due to heating is

   Inner Ring Bore Expansion = (Diameter of bore) × (Change in temperature) × (Coefficient of expansion of steel)

   Inner Ring Bore Expansion = (1.9680 in.) × (250°F − 70°F) × (.000063 in/in/°F)

   Inner Ring Bore Expansion = (1.9680 in.) × (180°F) × (.0000063 in/in/°F) = 0.0022 in.

   Heated Inner Ring Bore Diameter = Inner Ring Bore Diameter Cold + Inner Ring Bore Expansion

   Heated Inner Ring Bore Diameter = (1.9680 in.) + (0.0022 in.) = 1.9702 in.

   Clearance = (Heated Inner Ring Bore Diameter) − (Shaft Diameter)

   Clearance = (1.9702) − (1.9690) = 0.0012 in.
3. The inner ring will slide over the shaft with a 0.0012-in. clearance (an easy job).
4. After cooling, the inner race will have an interference fit to the shaft.

Installation of the Split Pillow Block Spherical Roller Bearing

Quite a few pieces of machinery use spherical-roller bearings with split pillow block housing. Usually one bearing is fixed (held), and the other is allowed to float (free) to give axial movement of the shaft under temperature and load. The fixed bearing should be mounted first to allow for the proper amount of shaft extension. The floating bearing is then positioned in relation to the fixed bearing. The steps to install these types of bearings are as follows:

1. Slide the inner (inboard) seal ring onto the shaft. It should slide freely into position (Figure 10-36).

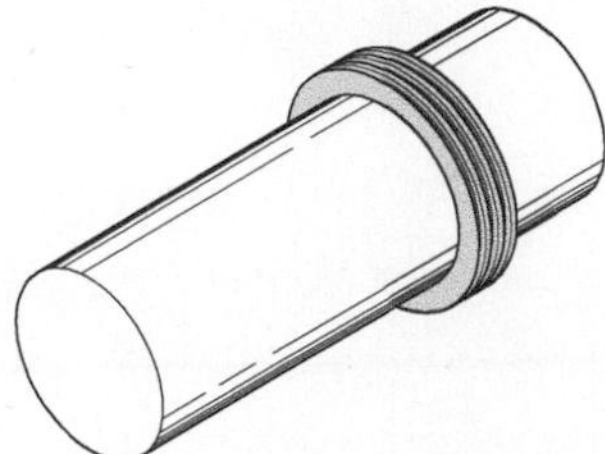

Figure 10-36 Inner seal.

2. Position the adapter sleeve onto the shaft, threads facing outboard, to the approximate location on the shaft to provide the proper center-to-center bearing distance (Figure 10-37).

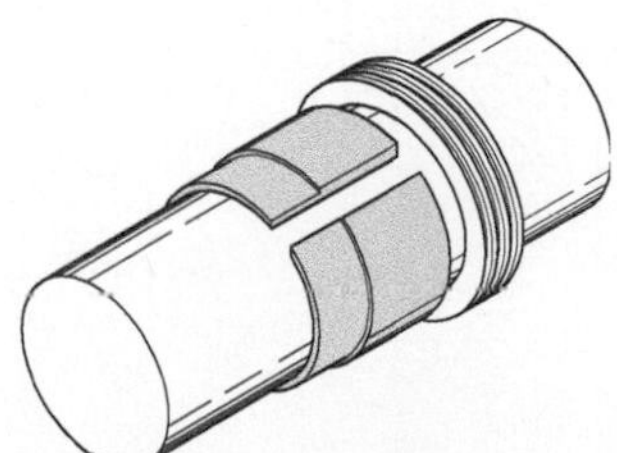

Figure 10-37 Adapter sleeve.

3. Measure the unmounted internal clearance in the bearing by inserting progressively larger feeler gauge blades the full length of the roller between the most vertical unloaded roller and the outer ring sphere. Do not roll the feeler through the clearance; slide it through. Record the measurement of the largest feeler that will slide through. This is the unmounted internal clearance (Figure 10-38).
4. Mount bearing on adapter sleeve, starting with the large bore of the inner ring to match the tape of the adapter. With the bearing hand tight on the adapter, locate the bearing to the

proper position on the shaft. *Do not apply lockwasher at this time because the drive-up procedure may damage the locknut* (Figure 10-39).

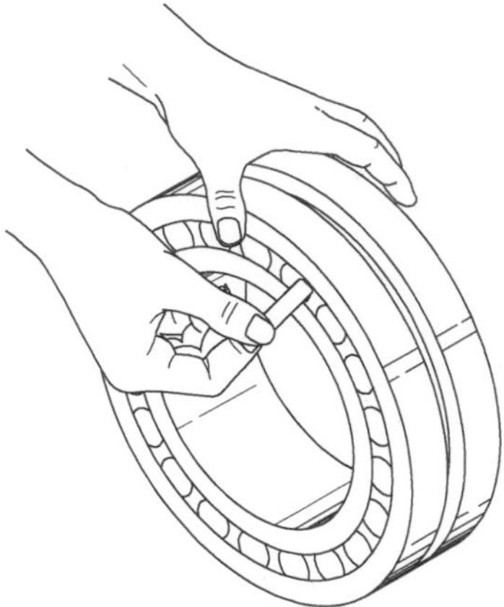

Figure 10-38 Unmounted clearance.

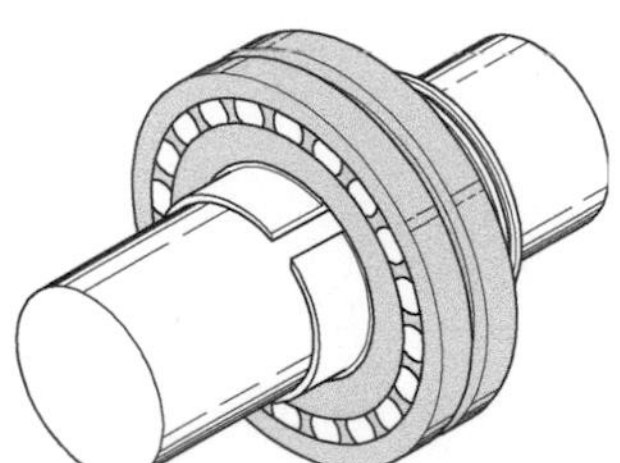

Figure 10-39 Bearing.

5. Apply the locknut with the chamfered face toward the bearing. Use a light coating of oil on the face of the locknut where it contacts the inner-ring face of the bearing. (Use a spanner wrench and a hammer to tighten the locknut.) Large bearings will require a heavy-duty spanner wrench and sledgehammer to obtain the required reduction in internal clearance. Do not attempt to tighten the locknut with a hammer and drift pin. The locknut will be damaged, and chips can enter the bearing. Tighten the locknut until the internal clearance is less than the figure measured in step 3 by the amount shown in Table 10-2.

Table 10-2 Internal Reduction Clearance for Spherical-Roller Bearings

Shaft Diameters (in.)		
Over	***Including***	***Clearance Reduction***
1¼	2¼	.001
2¼	3½	.0015
3½	4¼	.002
4¼	5	.0025
5	6½	.003
6½	7¼	.0035
7¼	8¼	.0045
8¼	9¼	.0045
9¼	11	.005
11	12½	.006
12½	14	.007

Measure the internal clearance with the feeler gauge between the most vertical unloaded roller and the outer ring (Figure 10-40).

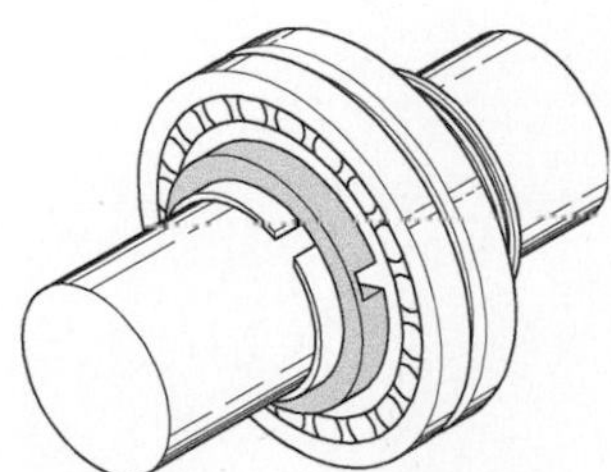

Figure 10-40 Locknut.

For example, to determine the amount to reduce the internal fit for a 3⁷⁄₁₆-in.-diameter shaft mounting:

a. Unmounted internal clearance was measured as 0.004 in.

b. Reduction in internal clearance from Table 10-1 equals .0015 in.

c. Final mounted internal clearance: (.004 in.) − (.0015 in.) = .0025 in.

6. Remove locknut and mount lockwasher on adapter sleeve with the inner prong of the lockwasher toward the face of the bearing and located in the slot of the adapter sleeve. Replace and retighten the locknut with the spanner wrench. Hit the spanner wrench one time with the hammer to seat the locknut firmly. Check to make certain that clearance has not changed (Figure 10-41).

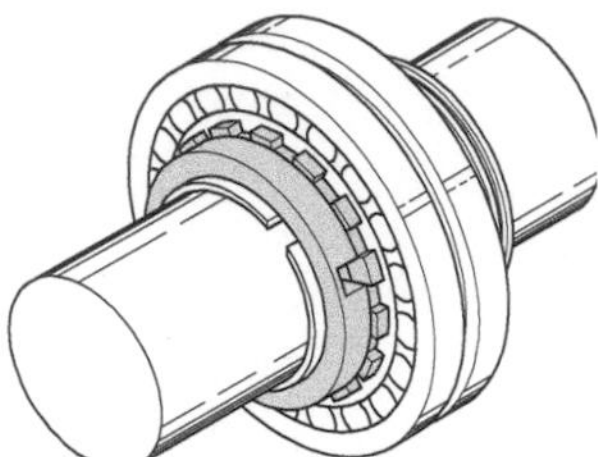

Figure 10-41 Locknut and lockwasher.

7. Slide the outer seal onto the shaft. Locate both the inner and outer seals to match the labyrinths in the housing (Figure 10-42).

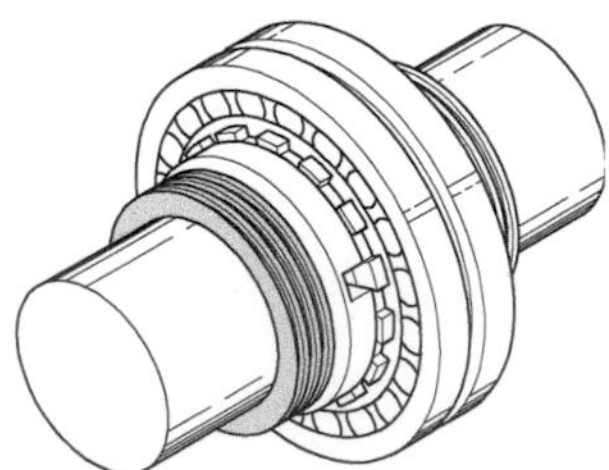

Figure 10-42 Outer seal.

8. Remove any paint and burrs from the mating surface at the split, and thoroughly clean the pillow block housing. The vertical hole at the bottom of each enclosure groove must be free of foreign matter. Place the shaft with the bearings into the lower half of the housing while carefully guiding the seal rings on the shaft into the seal grooves. Bolt the fixed bearing into place (Figure 10-43).

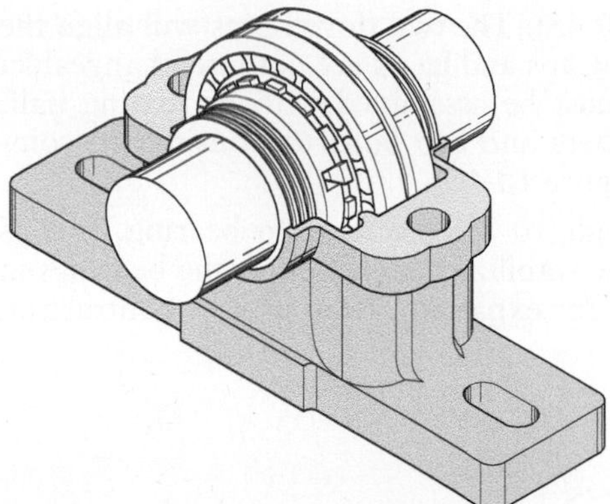

Figure 10-43 Lower half of housing.

9. Move the shaft axially so that the stabilizing ring can be inserted between the fixed bearing outer race and the housing shoulder on the locknut side of the bearing. Center all other bearings on the shaft in their housing seats. If a bearing has to be repositioned after being tightened, it has to be loosened on the adapter sleeve before being moved. Steps 4 through 6 must be repeated. There must be only one fixed bearing per shaft. Other bearings must be free to permit shaft expansion (Figure 10-44).

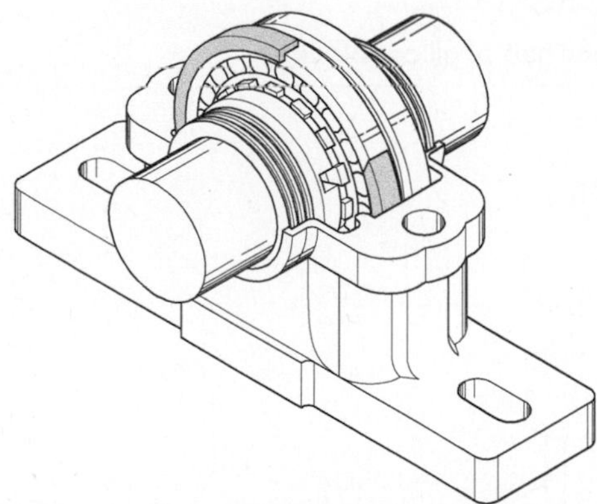

Figure 10-44 Stabilizing ring (held or fixed bearing).

10. The bearing seat in the upper half of the pillow block housing (cap) should be deburred, thoroughly cleaned, and placed over

the bearing (Figure 10-45). The two dowel pins will align the upper half of the cap. Caps and bases are not interchangeable. Each cap and base must be assembled with its mating half. Then, apply lockwashers and cap bolts to housings to complete the assembly (Figure 10-46).

11. Follow steps 1 through 10 for the floating bearing. In this case, do not install the stabilizer rings. Center the bearings in the housings to allow for expansion from heat or contraction from cold.

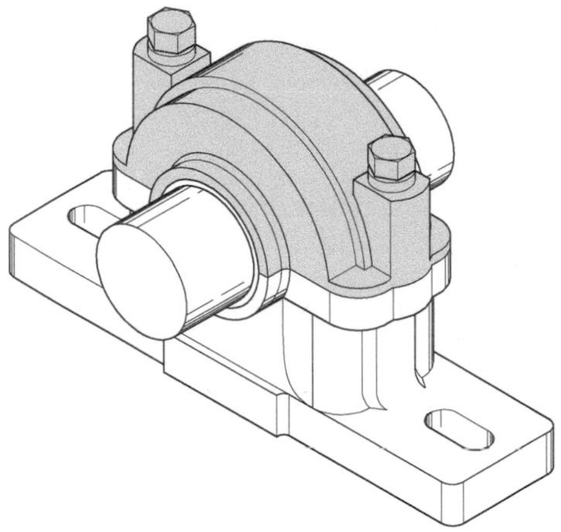

Figure 10-45 Mounting upper half of pillow block.

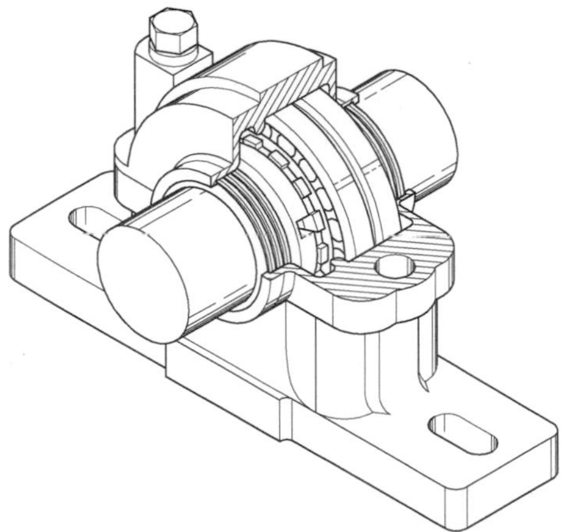

Figure 10-46 Cutaway illustration.

Installation of the One-Piece Pillow BlockRoller Bearing

Roller bearings are secured to the shaft either with setscrews (or screws) in the inner ring or by an eccentric collar.

Setscrew Installation

Secure these bearings to the shaft simply by tightening the setscrews in the inner-ring extension. In some cases, the shaft can be filed slightly or shallow drilled to provide more secure locking of the setscrews (Figure 10-47).

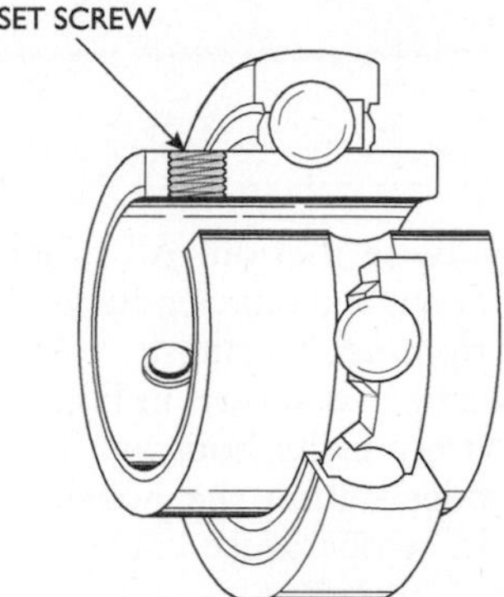

Figure 10-47 Setscrew installation.

Eccentric Locking Collar

The eccentric locking collar has a recess that is eccentric in relation to the bore. On the inner ring, there is a corresponding external eccentric section (Figure 10-48).

Place the eccentric locking collar in position on the inner ring, and turn it sharply in the same direction that the shaft will rotate in until it locks. Tighten the setscrew to keep the collar from loosening.

If the direction of rotation is not known, then lock the bearings at either end of the shaft in opposite directions.

Fixed and Floating Bearings Temperature variations will expand and contract the components of any machine. Because of this, it is essential that such parts be permitted to expand and contract without restriction. For this reason, only one bearing on any one shaft should be fixed axially in the bearing housing. This is called a *fixed*,

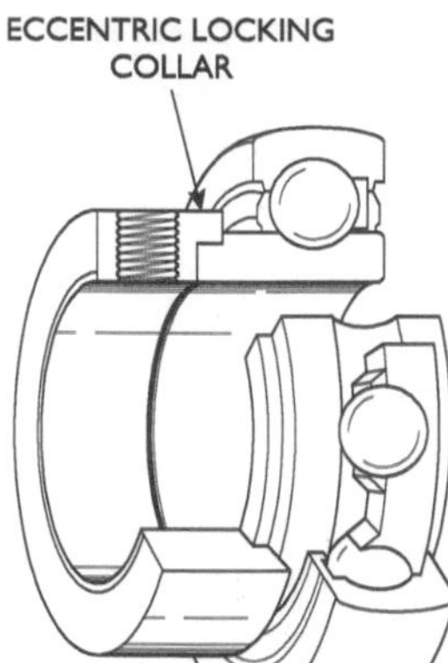

Figure 10-48 Eccentric locking collar.

or *held*, bearing and prevents axial or endwise motion. All other bearings on that same shaft should have adequate axial clearance in the housing and are referred to as *free*, or *floating*, bearings.

A typical *fixed* and *floating* bearing assembly is shown in Figure 10-49. The fixed bearing is clamped securely in its housing. The floating bearing has clearance on both sides within the housing. Thus, movement is allowed as changes in temperature cause the shaft to increase or decrease in length.

Generally it is preferred to hold the bearing at the drive end of a rotating assembly. However, consideration must sometimes be given to bearing loading; the bearing carrying the smaller radial load is fixed. This is usually in cases where the fixed bearing is subjected to a thrust load of some magnitude. The distribution of the load in this manner tends to equalize it between the two units.

Another very important consideration in the determination of which bearings should be fixed is operating clearance. When the clearances of components rotating with the shaft must be held to close tolerances, the closest bearing to the critical clearance point should be fixed.

The herringbone gear reducer is a special case using fixed and floating bearings. A common practice is to fix the large gear position with a fixed bearing on one end of the gear shaft. In this assembly, the V shape of the gear teeth will locate the mating gear and shaft assembly in proper axial position. The mating gear and shaft assembly is unrestrained, or full floating.

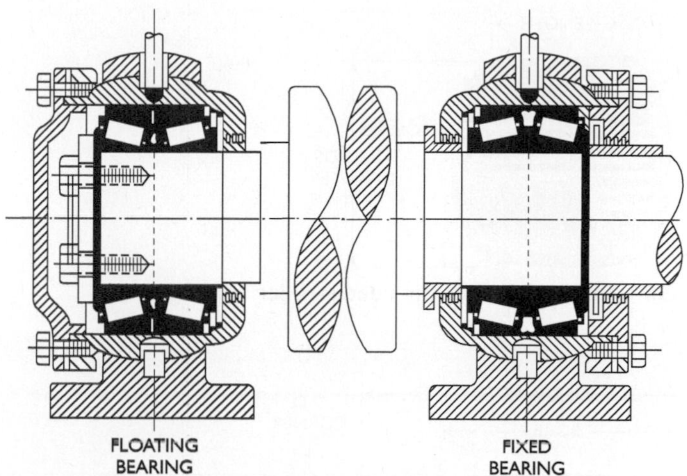

Figure 10-49 Fixed and floating bearings.

Bearing Handling

The two most important rules in handling antifriction bearings are as follows:

1. *Keep bearing and parts clean.* The principal reason for antifriction bearing failure is the entrance of dirt or grit.
2. *Apply force to the tight ring only.* Transfer of force from one ring through the rolling elements to the other ring can cause indentation of the races, which will cause early bearing failure.

The following practices and procedures for mounting and dismounting antifriction bearings are recommended by the Anti-Friction Bearing Manufacturers Association:

- Clean shafts and bearing housing thoroughly.
- Clean dirt out of keyways, splines, and grooves.
- Remove burrs and slivers.
- Clean and oil bearing seats.
- Press bearing on *straight* and *square.*
- Press only on the ring that takes the tight fit.
- Press bearings until they are seated against the shaft or housing shoulder.

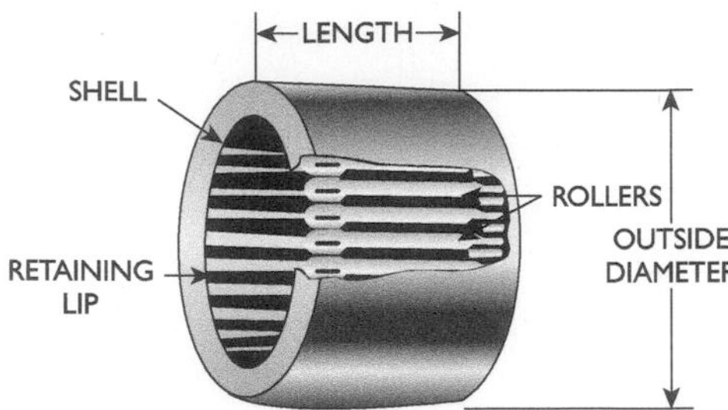

Figure 10-50 Various parts of the needle-roller bearing.

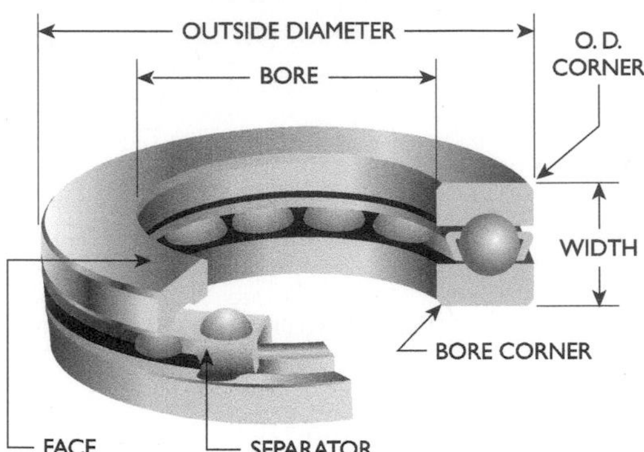

Figure 10-51 Various parts of the ball-thrust bearing.

The preferred mounting procedure is to press the bearing on in an arbor press, as this is the best way to control alignment, support, and force. Care must be exercised so that the supporting blocks are under the tight ring to ensure that the force is not transferred through the rolling elements. Figure 10-52 illustrates a shaft being pressed into a properly supported bearing in an arbor press.

If the distance from the bearing seat to the end of the shaft is fairly short, hold the shaft in a vise and press the bearing onto the shaft with a clean tube. This method of mounting may also be accomplished by tapping with a hammer evenly around the tube as illustrated in Figure 10-53. Tap lightly at first to make sure the bearing goes on square and does not scrape or burr the seat. Be sure that the bearing is tapped to a firm seat against the shaft shoulder.

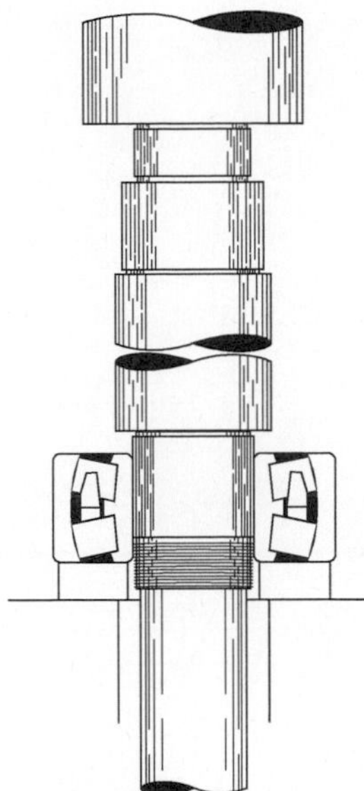

Figure 10-52 Pressing a bearing assembly on a shaft.

Do not use a hammer directly on a bearing under circumstances other than an emergency. Bearings are precision assemblies of very hard, highly finished parts that can be easily ruined by such an impact. See Figure 10-54.

Installation or mounting of bearings can be made easy in some circumstances by the use of heat. Generally, the use of heat is limited to a single fit, i.e., mounting a bearing on a shaft outside the housing. The bearing must then be allowed to cool before installation into the housing is attempted.

The use of heat is a rather simple operation consisting of heating the bearing to a temperature of 200 to 250°. This expands the inner ring sufficiently so that it should slip over the shaft to the bearing seat. Do not overheat the bearing or its hardness will be affected. If expanding the ring is not enough to get it on, the shaft may be contracted by cooling it in dry ice.

Care must be exercised in the heating of the bearing to ensure that it is done uniformly. A torch should never be used, as the risk of spot overheating is great. An easy and safe method is to place the bearing in a thermostatically controlled oven. This not only gives accurate heat control but also allows the bearing to be kept clean.

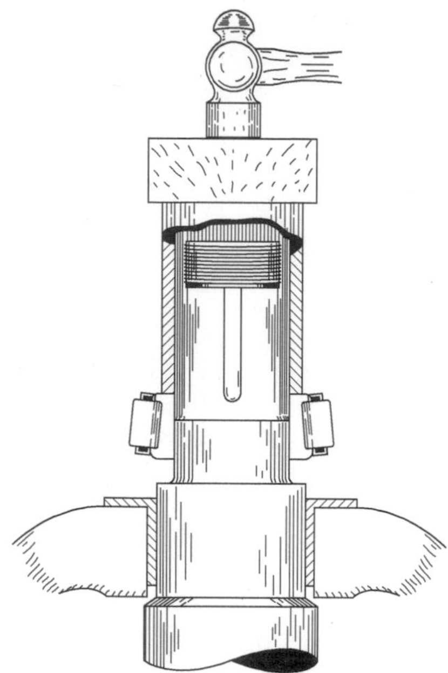

Figure 10-53 One proper way of installing a bearing on a short shaft.

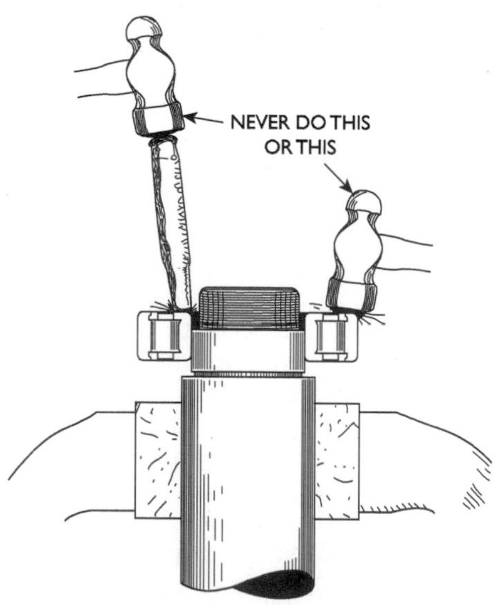

Figure 10-54 Avoid this type of installation.

The use of steam should be avoided because this introduces moisture into the bearing directly onto the metal surfaces while the slushing oil is washed away by the steam. Oil baths with controlled temperature are also an excellent way of heating; however, these are seldom available in small repair shop operations. When no facilities are available for heating, the bearing may be wrapped carefully and placed on hot steam lines or apparatus. Another simple and easy way to heat small bearings is to carefully place them on a lighted incandescent electric lamp. Covering the bearing will help to retain the heat, and turning it occasionally will ensure even heating. If the bearing is heated until it is just slightly too hot to touch with the fingers, its temperature will be in the area of 180 to 200°.

The practices and precautions previously discussed when mounting bearings must also be followed to avoid damage when dismounting bearings. Figure 10-55A shows the correct method of removing a bearing in an arbor press. The press ram is pushing down against the shaft on which the bearing inner ring is tightly fitted. Note that the stripping is being done by the support blocks under the rings.

In Figure 10-55B, an incorrect method is illustrated; here, the support blocks are spaced too far out. If done in this manner, pressure will be against the outer ring as force is applied by the ram. This will in turn place a heavy stress on the balls or rollers, which can damage them and cause indentation of both races.

Another important consideration in bearing removal is to prevent the shaft from falling when released from the bearing. Some means must be provided to keep the shaft from falling to the floor and being damaged.

In bearing removal, as in mounting, it is not always possible or practical to use the arbor press. This operation may also be performed quite safely with improvised methods if the right tools and proper care are used. Direct hammer blows, as shown in Figure 10-56, must not be employed if the bearing is to be safely removed without damage.

Shown in Figure 10-57 are several devices and precautions that can assist in removal of bearings without damage. However, when using brass or other soft metal as a driver, care must be taken that the ends are properly dressed. Soft metal will mushroom and chip from repeated blows. This can result in a safety hazard from flying chips and possible bearing damage if chips enter the bearing and are not noticed.

In some cases it may be extremely difficult to remove a tight bearing that is tightly held against a shoulder. This may require that the bearing be removed in pieces with a torch. To do this, first cut the outer ring, which will allow it and the rolling elements and

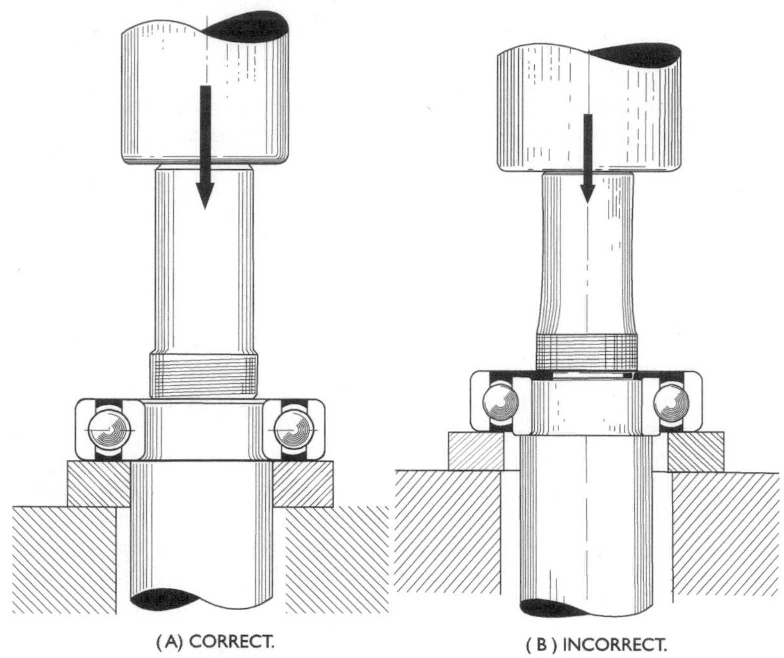

Figure 10-55 Correct and incorrect way of removing a bearing from a shaft.

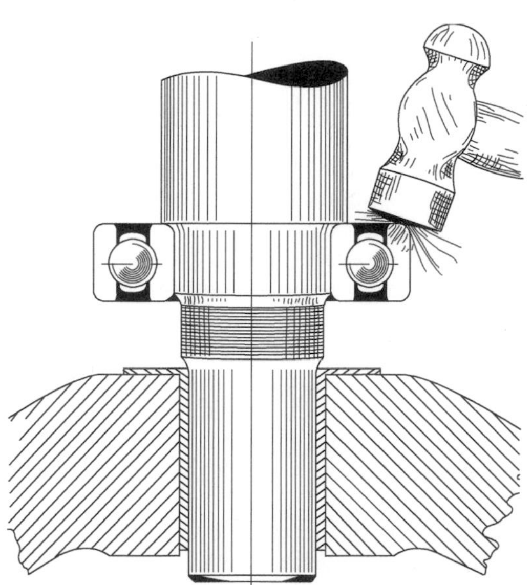

Figure 10-56 This method of bearing removal must never be used.

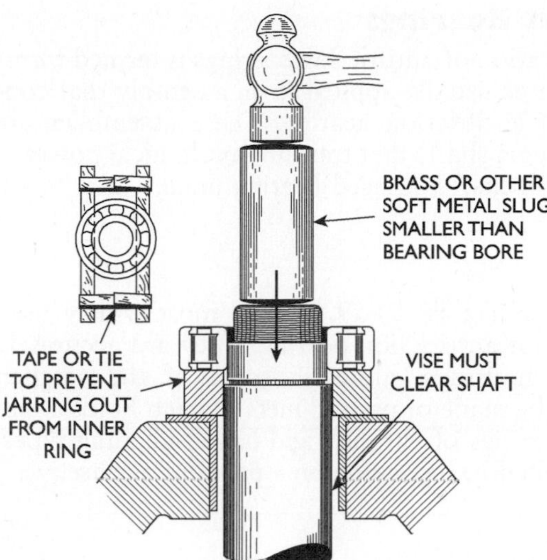

Figure 10-57 Proper method used to remove bearings from shaft.

separator to be removed. To remove the inner ring, carefully cut a notch in it with a torch as shown in Figure 10-58. Do not allow the flame to cut through and damage the shaft surface. Pinching the ring in a vise and striking it sharply with a hammer will cause it to fracture at the weakened point.

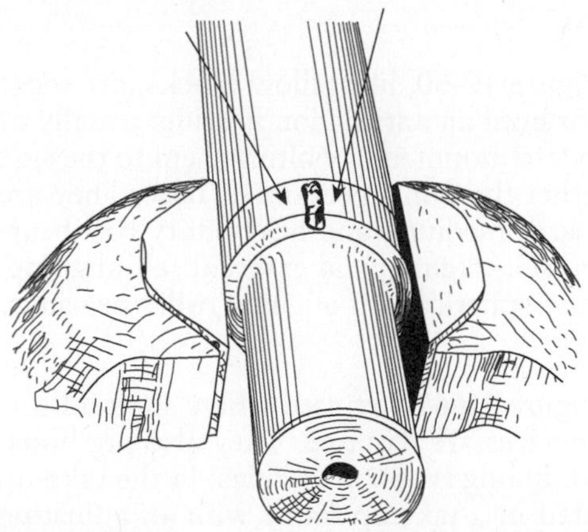

Figure 10-58 One method of removing the bearing inner ring from a shaft.

Transmission Unit Bearings

Another broad classification of antifriction bearings is termed *transmission units.* The term is usually applied to an assembly that contains and supports an antifriction bearing. These assemblies are commonly used to support shafts that transmit mechanical power—for example, "pillow blocks," "flanged bearing units," and "take-up units."

Pillow Blocks

Pillow blocks, shown in Figure 10-59, are the most widely used type of transmission unit antifriction bearing. They are accurately constructed cast-iron housings, although some of the smaller, lightweight types may be made of pressed metal, which holds a ball or roller bearing or bearings of various rigid or self-aligning types. The pillow block is bolted into position on a pedestal or base.

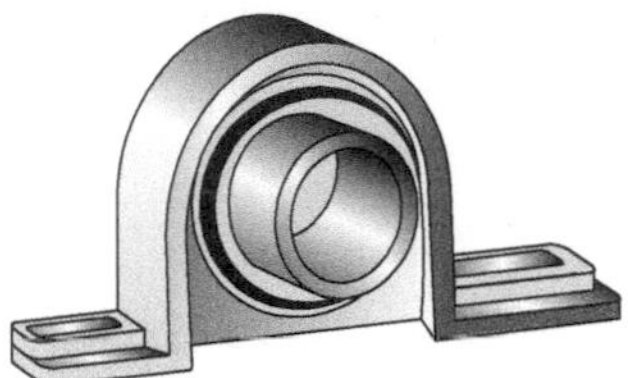

Figure 10-59 A typical pillow block.

Flanged Bearing Units

Flanged units, shown in Figure 10-60, like pillow blocks, are separate housings or covers that hold an antifriction bearing, usually of the self-aligning type. They are mounted by bolting them to the side of a piece of machinery rather than on a pedestal or base. They are available with light, standard, medium, and heavy-duty ball bearings. The types are internal self-aligning and external self-aligning. Also, there are types that use spherical self-aligning roller bearings.

Take-Up Units

Take-up units, shown in Figure 10-61, are used where shaft adjustment and belt tightening devices are required. They also are housings or covers to hold self-aligning types of bearings. In the take-up unit, the housing is mounted in a take-up frame with an adjusting screw, which permits moving the bearing housing to obtain movement as required.

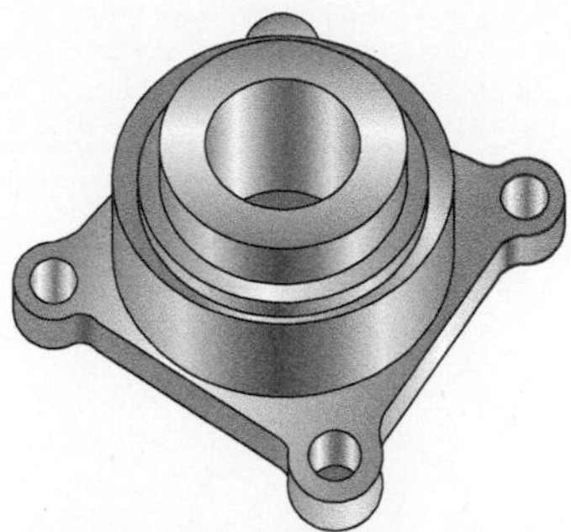

Figure 10-60 A flanged bearing unit.

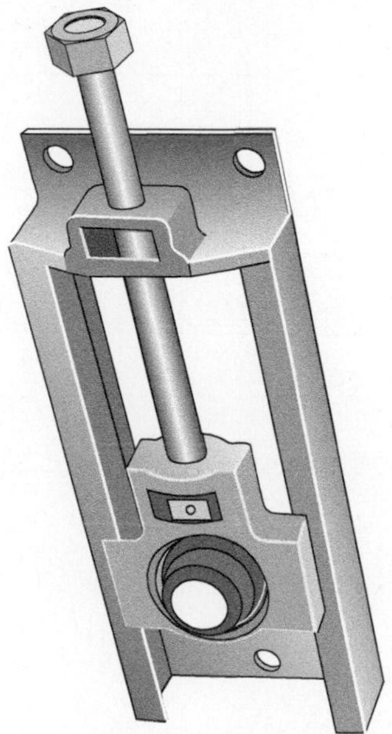

Figure 10-61 A take-up bearing assembly.

Chapter 11

Principles of Mechanical Power Transmission

Practically all equipment that moves in today's industry does so by means of power-transmission equipment. The installation, adjustment, and maintenance of this vital equipment is the millwright's or mechanic's job. To perform this job well, he or she must have an understanding of the basic principles and the concept of operation, as well as a practical knowledge of work procedures. Although there are many systems and an infinite number of design variations, the terms and concepts that apply to all are relatively simple.

The term *power-transmission equipment* applies to specific parts of a machine assembly which functions as a mechanical transmission of power. By far the most common form is the transmission of power from shaft to shaft. This can be broken into two types of rotary-motion power transmission: shaft to shaft axially and shaft to adjacent shaft. To transmit rotary motion axially, some form of coupling or clutch is used. These will be discussed later. There are three major systems in use for transmission of rotary motion between adjacent shafts: belts, chains, and gears.

Certain words, terms, and concepts apply equally to these three systems. An understanding of them is a prerequisite to an understanding of the systems.

Pitch

The word "pitch" is commonly used in connection with many kinds of machinery and types of mechanical operations and calculations. Its definition, as applied to mechanical power transmission, is simple yet very important: *the distance from a point to a corresponding point.* In Figure 11-1 are several examples of this measurement of distance from point to corresponding point.

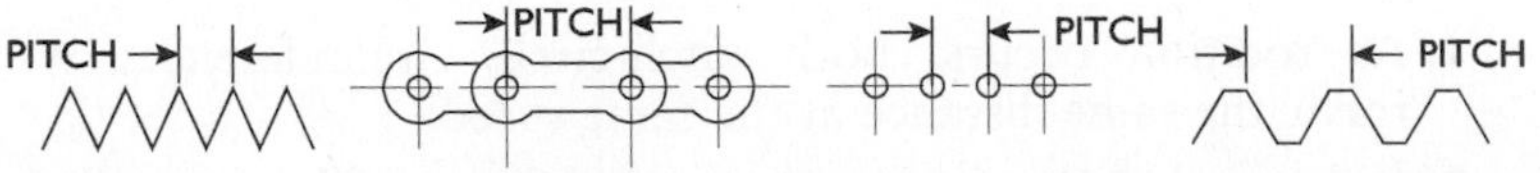

Figure 11-1 The measurement of distance from point to corresponding point.

Pitch Diameter

The diameter of a circle is the dimension across its center; hence, "pitch diameter" is the dimension across the center of a *pitch circle*. Although pitch circles are usually imaginary, pitch diameters are specific dimensions.

Pitch Circle

Although the "pitch circle" is not visible, its dimension can be stated specifically as the pitch diameter of a gear, sheave, sprocket, etc. These dimensions are a necessary part of all rotary power-transmission calculations. These calculations are based on the concept of discs or cylinders in contact, as illustrated in Figure 11-2.

The rotation of one disc causes the disc with which it is in contact to rotate. This concept assumes that no slippage occurs between the surfaces of the discs. The surfaces then travel equal distances at equal surface speeds.

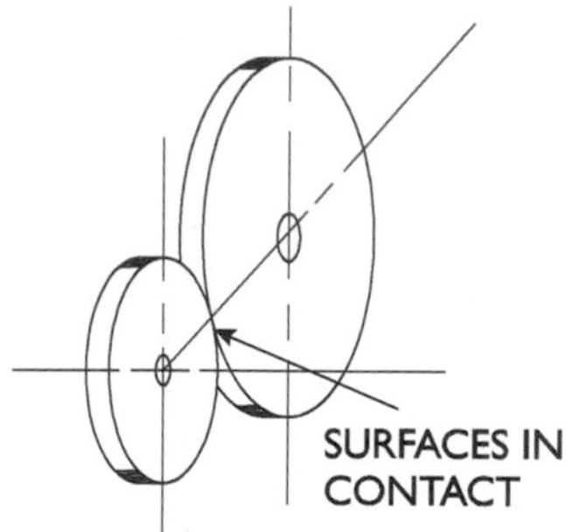

Figure 11-2 Showing the surface contact of a disc or cylinder.

Figure 11-3 is a graphic illustration of the concept. The discs are the pitch circles, and their specific sizes are their pitch diameters. This concept may also be applied to belt and chain drives, although their discs are actually separated. This is true because the belt or chain is in effect an extension of the pitch-circle surface. The belt or chain, as well as the surfaces of the discs, travel equal distances at equal speeds. Based on this concept, the following statements can be made:

1. As rotation occurs, both pitch-circle circumferences will travel the same distance at the same speed.
2. If the pitch diameters are unequal, the circumferences will be unequal, and although surface speeds are the same, the rotational or shaft speeds will be unequal.

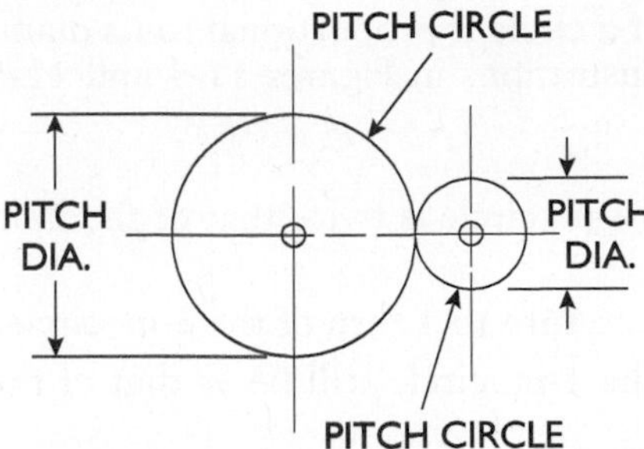

Figure 11-3 A graphic representation of surface contact.

The relationship of pitch diameters to shaft speeds may be visualized by the illustration in Figure 11-4, as it applies to gears, and in Figure 11-5, as it applies to belts.

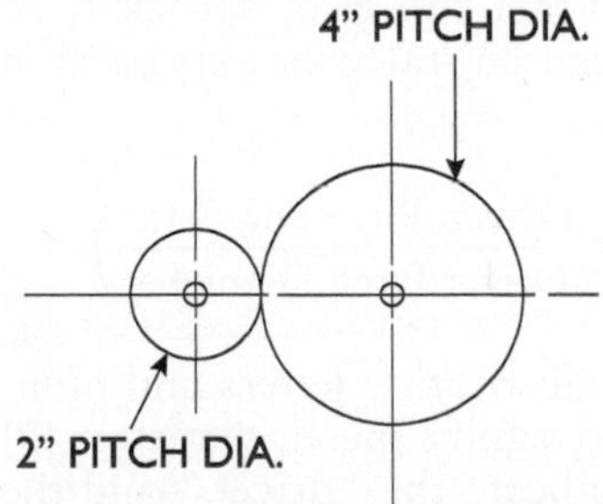

Figure 11-4 One revolution of a 2-in. circle will result in a ½ revolution of a 4-in. circle.

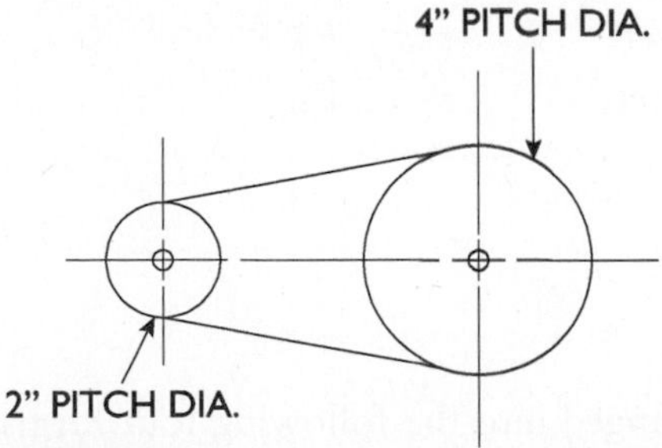

Figure 11-5 One revolution of a 4-in. circle will result in two revolutions of a 2-in. circle.

Because the circumference of a circle is proportional to its diameter, the relationships of the illustrations in Figures 11-4 and 11-5 are as follows:

1. The circumference of the 4-in. circle is twice that of the 2-in. circle.
2. The 4-in. circle will make ½ turn to 1 turn of the 2-in. circle.
3. The rotational speed of the 4-in. circle will be ½ that of the 2-in. circle.

Speed and pitch-diameter calculations for gears, belts, and chains are based on the relationship of pitch diameters to rotational speeds. This relationship may be stated as follows:

Shaft speeds are inversely proportional to pitch diameters.

Using the term "driver" to indicate the driving gear, sprocket, or sheave and the term "driven" to indicate the gear, sprocket, or sheave that is being driven, the ratio can be stated or expressed in the form of an equation, as follows:

$$\left(\frac{\text{Driver Rotational Speed}}{\text{Driven Rotational Speed}} = \frac{\text{Driven Pitch Diameter}}{\text{Driver Pitch Diameter}}\right)$$

The equation can be simplified by substituting letters and numbers for the words. Use the letter (S) to signify speed, the letter (P) to indicate pitch, the number (1) to indicate the "driver," and the number (2) to indicate the "driven."

Driver rotational speed	=	$S1$
Driven rotational speed	=	$S2$
Driver pitch diameter	=	$P1$
Driven pitch diameter	=	$P2$

The equation then becomes

$$\frac{S1}{S2} = \frac{P2}{P1}$$

The equation may be rearranged into the following four forms, one for each of the four values.

$$\text{Driver rotational speed} = S1 = \frac{P2 \times S2}{P1}$$

$$\text{Driven rotational speed} = S2 = \frac{P1 \times S1}{P2}$$

$$\text{Driver pitch diameter} = P1 = \frac{S2 \times P2}{S1}$$

$$\text{Driven pitch diameter} = P2 = \frac{S1 \times P1}{S2}$$

These equations may be used to find unknown values by simple substitution of known values into the appropriate equation. The following is an example of the use of each of these equations.

Example (1)

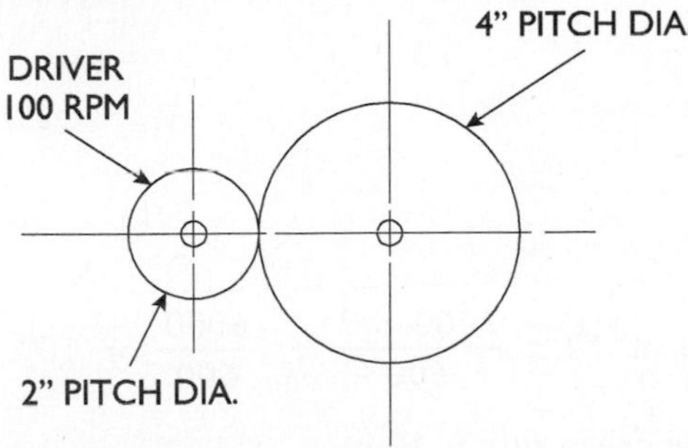

The pitch diameter of the driver unit that is turning at 100 rpm is 2 in. What will be the speed of the driven unit if its pitch diameter is 4 in.?

Known values:

$S1 = 100$
$P1 = 2$
$P2 = 4$

$$\text{Unknown } S2 = \frac{P1 \times S1}{P2} \text{ or } S2 = \frac{2 \times 100}{4} \text{ or } \frac{200}{4} \text{ or } 50$$

Speed of driven unit is 50 rpm.

Example (2)

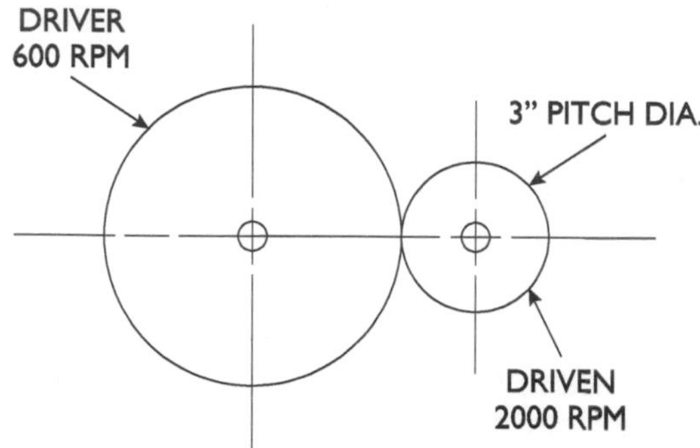

The driver unit is turning at a speed of 600 rpm. The driven unit is turning at 2000 rpm and its pitch diameter is 3 in. What is the pitch diameter of the driver unit?

Known values:

$$S1 = 600$$
$$S2 = 2000$$
$$P2 = 3$$

$$\text{Unknown } P1 = \frac{S2 \times P2}{S1} \text{ or } P1 = \frac{2000 \times 3}{600} \text{ or } \frac{6000}{600} \text{ or } 10$$

The pitch diameter of the driver unit is 10 in.

Example (3)

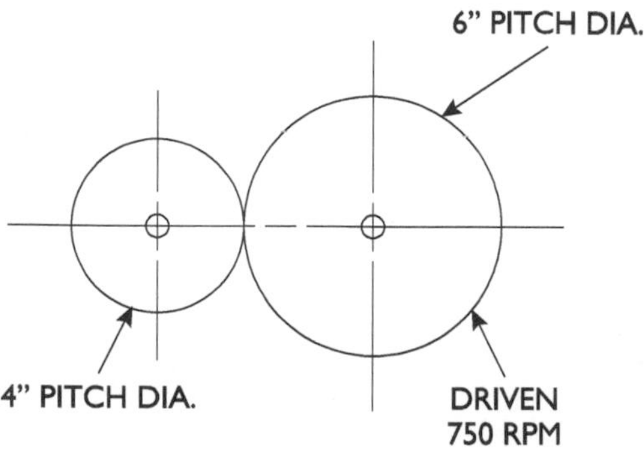

The pitch diameter of the driver unit is 4 in. The pitch diameter of the driven unit is 6 in., and it is turning at a speed of 750 rpm. What is the speed of the driver unit?

Known values:

$$P1 = 4$$
$$P2 = 6$$
$$S2 = 750$$

$$\text{Unknown } S1 = \frac{P2 \times S2}{P1} \text{ or } S1 = \frac{6 \times 750}{4} \text{ or } \frac{4500}{4} \text{ or } 1125 \text{ rpm}$$

The speed of the driver unit is 1125 rpm.

Example (4)

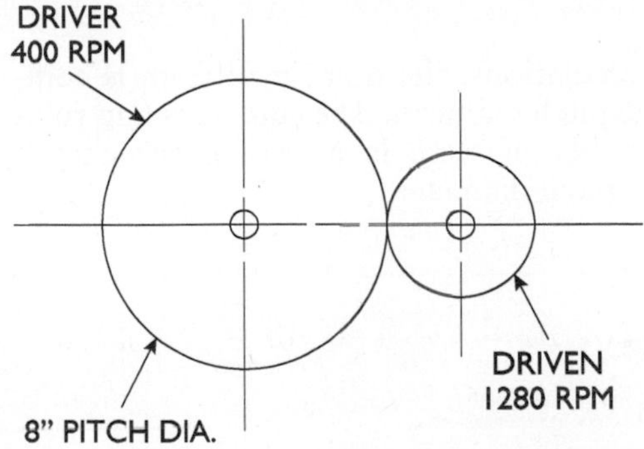

The pitch diameter of the driver unit, which is turning at 400 rpm, is 8 in. The speed of the driven unit is 1280 rpm. What will be the pitch diameter of the driven unit?

Known values:

$$P1 = 8$$
$$S1 = 400$$
$$S2 = 1280$$

$$\text{Unknown } P2 = \frac{S1 \times P1}{S2} \text{ or } P2 = \frac{400 \times 8}{1280} \text{ or } \frac{3200}{1280} \text{ or } 2.5$$

The pitch diameter of the driven unit is 2.5 in.

Another convenient way to find unknown shaft speeds and pitch diameters is to use written rules. These rules are based on the relationship previously stated—that shaft speeds are inversely proportional to pitch diameters. For some, it may be easier to follow these rules than to substitute values in the preceding equations.

> To find *driving shaft speed*—Multiply *driven pitch diameter* by *speed of driven shaft* and divide by *driving pitch diameter.*
>
> To find *driven shaft speed*—Multiply *driving pitch diameter* by *speed of driving shaft* and divide by *driven pitch diameter.*
>
> To find *driving pitch diameter*—Multiply *driven pitch diameter* by *speed of driven shaft* and divide by *speed of driving shaft.*
>
> To find *driven pitch diameter*—Multiply *driving pitch diameter* by *speed of driving shaft* and divide by *speed of driven shaft.*

In gear and sprocket calculations, the number of teeth is commonly used rather than the pitch diameter. The equations and rules hold true because the number of teeth in a gear or sprocket is directly proportional to its pitch diameter.

Chapter 12

Shafting

Machine shafting accomplishes three purposes:

- Provides an axis of rotation
- Transmits power
- Helps position or mount gears, pulleys, bearings, and other working elements

Shafting is rated on its ability to accommodate bending and torsional loads. Shaft loads result from external forces caused by misalignment, belt drives, chain drives, flywheels, brakes, or other accessories. Depending on customer preference, inch and metric shaft extensions are available from most manufacturers.

Shafting Specifications

Although machines can make use of the many different materials used for shafting, the most common shaft is made of steel. In the chemical or petrochemical industries, shafts for pumps or other machinery (where the wetted surfaces are in contact with the rotating element), stainless steel and other exotic steel alloys for shafting materials might be appropriate.

Purchasing fan shafts, pump shafts, and other replacement parts directly from the manufacturer guarantees that the shaft is exactly what is required for the application. The cost of replacement shafting from the original equipment manufacturer (OEM) is expensive. Where the shaft material is known, it is a common practice in many maintenance shops to purchase one shaft from the OEM, accurately measure it, and then make a detailed mechanical drawing. This drawing can be sent to a local machine shop to allow spare part shafts to be made, often at a greatly reduced cost.

In particular, when a machine has been in service for many years beyond its manufacturer's warranty period, there is little or no reason why replacement shafts need to be purchased from the OEM.

General Shaft Dimensions

Table 12-1 shows the typical dimensions for machine shafting from ½ in. through 8 in. in diameter.

Table 12-1 General Shaft Dimensions

Diameter	*Tolerance Range*	
Inches	*Maximum, in.*	*Minimum, in.*
½	.500	.498
9/16	.562	.560
5/8	.625	.623
11/16	.687	.685
¾	.750	.748
13/16	.812	.810
7/8	.875	.873
15/16	.937	.935
1	1.000	.998
1 1/16	1.062	1.059
1 1/8	1.125	1.122
1 3/16	1.187	1.184
1¼	1.250	1.247
1 5/16	1.312	1.309
1 3/8	1.375	1.372
1 7/16	1.437	1.434
1½	1.500	1.497
1 9/16	1.562	1.559
1 5/8	1.625	1.622
1 11/16	1.687	1.684
1¾	1.750	1.747
1 13/16	1.812	1.809
1 7/8	1.875	1.872
1 15/16	1.937	1.934
2	2.000	1.997
2 1/16	2.062	2.058
2 1/8	2.125	2.121
2 3/16	2.187	2.183
2¼	2.250	2.246
2 5/16	2.312	2.308
2 3/8	2.375	2.371
2 7/16	2.437	2.433
2½	2.500	2.496
2 5/8	2.675	2.671
2¾	2.750	2.746

2⅞	2.875	2.871
3	3.000	2.996
3⅛	3.125	3.121
3¼	3.250	3.246
3⅜	3.375	3.371
3½	3.500	3.496
3⅝	3.675	3.671
3¾	3.750	3.746
3⅞	3.875	3.871
4	4.000	3.996
4¼	4.250	4.245
4¾	4.750	4.745
5	5.000	4.995
5¼	5.250	5.245
5½	5.500	5.495
5¾	5.750	5.745
6	6.000	5.995
6¼	6.250	6.244
6½	6.500	6.494
6¾	6.750	6.744
7	7.000	6.994
7¼	7.250	7.244
7½	7.500	7.494
7¾	7.750	7.744
8	8.000	7.994

Emergency Shaft Repair

In an emergency, you might look to used parts to repair a piece of production equipment. If a bearing, seal, or key unexpectedly fails, there may not be adequate time to locate a spare shaft or even to machine one. With today's cost-efficiency concerns, a maintenance department cannot purchase and store a shaft for each piece of equipment in its area of responsibility.

If the used shaft is worn in the area of the bearing seat, where the bearing inside diameter would spin on the seat if the old shaft were used, some emergency techniques may save the day. The shaft can be chucked up in a lathe, and the area of the bearing seat can be knurled. Knurling throws up metal and expands the diameter of the shaft to make up for the undersized area. The knurl is then turned to obtain the correct diameter required for press fit (or shrink fit) of the bearing. Keep in mind that bearing manufacturers do not recommend this procedure at all. For the most part, the knurl will not support the inner ring of a bearing properly for a long time, so the manufacturer is correct. Knurling might be good enough, however, to allow the machine to operate for a few weeks until a spare shaft is obtained (Figure 12-1).

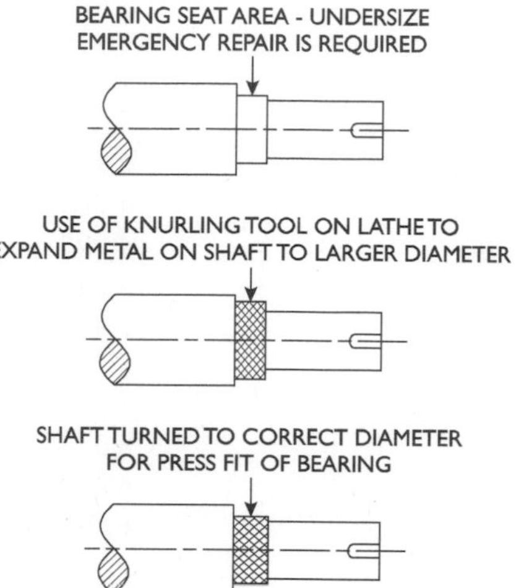

Figure 12-1 Although not recommended, knurling can buy some time.

Another emergency shaft repair method is spray metallizing. This is a perfectly acceptable repair on the sealing area and other non-load-carrying portions of a shaft, but it is not recommended for use in the area of the bearing journal. Again, the bearing manufacturer advises against this technique as a permanent fix, but spraying can keep a machine in operation while you locate a new spare or build a proper replacement shaft at the machine shop.

It is important to restate that both of these procedures are *not* recommended by machinery or bearing manufacturers as permanent solutions, but plenty of mechanics have saved the day using these bootstrap methods and following up by replacing the temporary fix with a proper repair as soon as possible.

Permanent Shaft Repair

Obviously, the most permanent shaft repair would be to obtain a replacement part from the OEM or to machine a duplicate at a local shop. Often a shaft can be salvaged with a perfectly good quality repair that "makes the shaft as good as new."

Submerged Arc Welding

Shaft repair at the site of a bearing journal can be done with *submerged arc welding*—the highest-quality repair method that eliminates the possible flaking and peeling problems of other, lesser-quality repair techniques.

This special process involves welding new material on worn shaft areas under tightly controlled conditions. As the shaft slowly turns in a lathe, heat generated during welding is evenly distributed over the shaft surface. This heat distribution provides the necessary "stress relief" to prevent warping or distortion of the shaft itself.

Once the appropriate amount of buildup is achieved, the shaft is remachined to specifications, and the repaired shaft is considered as good as new.

Metal Spraying

Although not considered a suitable repair for the bearing journal of a shaft, metal spraying is a quick and cost-effective way to reclaim shaft surfaces in the area for seal journals or mechanical seals.

Metal spraying can also be used to improve the wear resistance of a shaft. Wear resistance coating can protect against fretting and abusive wear, and it is impact abrasion-, erosion-, and cavitation-resistant (Figure 12-2). Other sprayed coatings can improve the corrosion resistance of a shaft. These coatings protect less noble materials—predominantly low-alloy steels—from chemical attack. This type of coating also protects against oxidation, sulfidation, and galvanic corrosion. The chemical and petrochemical processing industries often use this type of thermal spray coating.

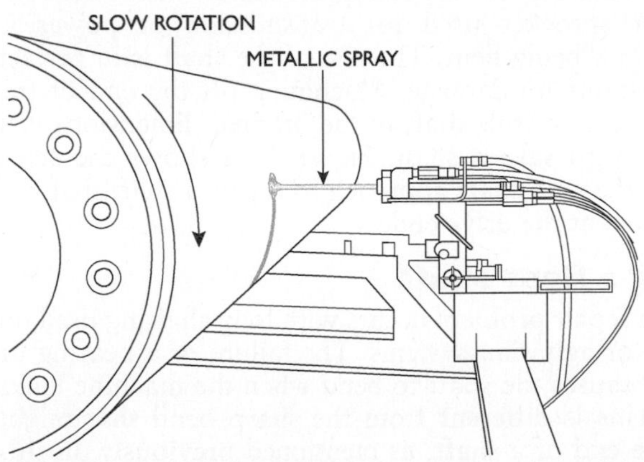

Figure 12-2 Metal spraying to build up a worn area in the shaft.

TIG Welding

Another permanent shaft repair technique for the seal and key slot areas is tungsten inert gas (TIG) welding. TIG welding, although slower than MIG or stick welding, offers better control of the heat-affected zone normally responsible for shaft breakage next to a weld.

Hard Chrome Plating

Hard chrome plating is used for hard surfacing specific areas of a shaft to produce a wear-resistant surface. *Hard* chrome is a misnomer. The electroplating industry refers to any deposition of chrome with a thickness of 0.005 in. or more as hard chrome; deposits of less than that are called "decorative" plating. The area where wear resistance is required is turned down, chrome plated to an oversized OD, and then ground down to produce the correct dimension and a smooth, almost mirror-like surface for wear resistance. Shaft areas that are prone to packing or seal rubs are good candidates for hard chroming.

Use of a Shaft Sleeve

A shrink-fit metal sleeve is an acceptable shaft repair for a bearing seat (Figure 12-3). Here, the shaft is turned down in the area where the damage occurred. A suitable sleeve is machined where the ID of the sleeve is bored to be shrunk-fit to the shaft and the OD of the sleeve is made to be larger than the original shaft dimension. The sleeve is heated and shrunk onto the shaft. When the sleeve has cooled, the shaft is again chucked up in the lathe, and the OD of the sleeve is turned down to the proper final dimension.

Use of a Stub Shaft

In many machine failures, the very end of the shaft—where the coupling, sheave, or sprocket used for transmission of power is located—will end up being bent. The rest of the shaft is in perfect condition and sustains no damage. Machining off the end of the shaft and constructing a stub shaft in the original dimensions is a perfectly good way to save a shaft. Figure 12-4 shows the steps involved in using the stub shaft approach to repair a shaft that has been bent or broken at the drive end.

Straightening a Bent Shaft

Another common repair problem occurs with long shafting used on deep well pumps or agitation systems. The failure of a bearing or other component causes the shaft to bend when the machine locks up under load. This is different from the sharp bend that might occur at the drive end of a shaft, as mentioned previously. In this

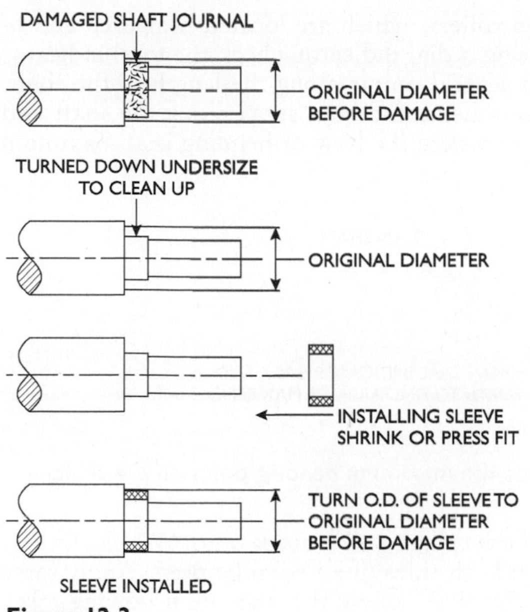

Figure 12-3

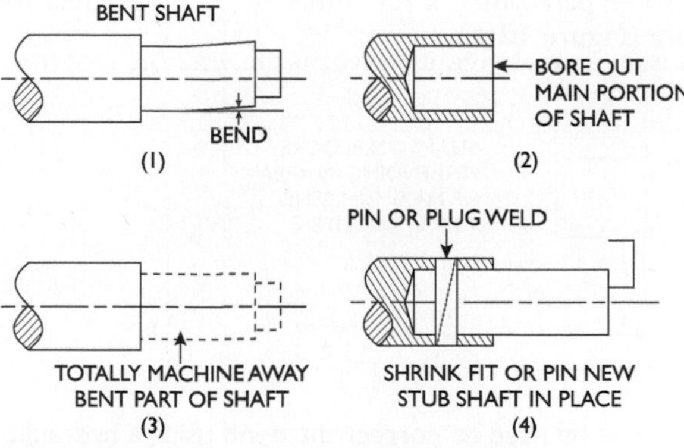

Figure 12-4 Using a stub shaft for repair.

case, the shaft may be 4 to 8 ft or more in length and has been bowed. Straightening a shaft is accomplished using a hydraulic press and a dial indicator. The following procedure gives the steps required to straighten and reuse a bowed shaft:

1. Spin the shaft on rollers, which are located near each end of the shaft and, using a dial indicator, check the total indicator reading (TIR) at several points along the length of the shaft. Use a marker to indicate the high spots along the shaft and determine the area where the bow or bending is at maximum (Figure 12-5).

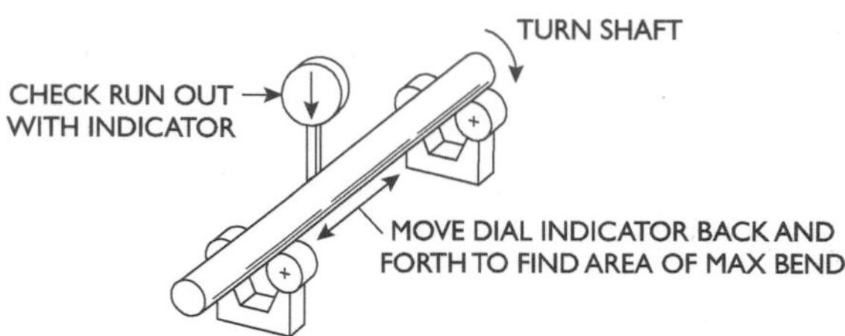

Figure 12-5 Determine the maximum bending point on the shaft.

2. Lift the shaft off the rollers and move it onto "V" blocks that are positioned on both sides of a hydraulic press. Apply force at the point on the shaft where the maximum bowing takes place, and use the press to straighten the shaft. Care must be taken not to "overbend" the shaft. Bending and checking might need to be performed a few times to develop a feel for the procedure (Figure 12-6).

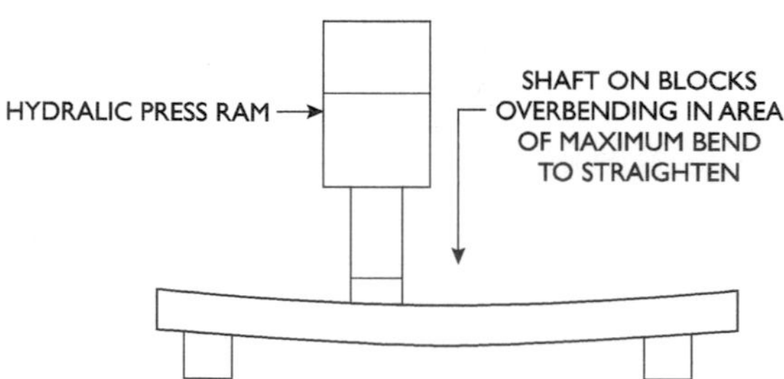

Figure 12-6 Care must be used to correct the bend using a hydraulic press.

3. Remove the shaft, and recheck it with a dial indicator on the rollers. A rule of thumb is that the maximum allowable TIR is about 0.0005 in./ft of shaft length but not more than 0.001 in. within and 1 ft of length. For instance, a 6-ft shaft should not have more than $6 \times 0.0005 = 0.003$ in. of runout.

4. If the TIR is still too large, additional marking must be done on the shaft, depicting the results of the new dial indicator readings, and the shaft must be transferred to the press for additional straightening.

Shaft straightening is a straightforward process. Careful work and measurement, combined with slow and steady control of the pressure to the ram on the hydraulic shop press, will result in a shaft that is as good as new.

Chapter 13

Flat Belts

Although there are few new installations of flat-belt drives in industry, there are still many installations in operation. Flat belting can be made from leather, rubber, or canvas. Leather belts are by far the most commonly used. Rubber belting is usually used where there is exposure to weather conditions or moisture because rubber belts do not stretch under these conditions. Canvas or similar fabrics, usually impregnated with rubber, are used when the materials (such as liquids) in contact with the belt would have an adverse effect on leather or rubber.

Leather belting is specified by thickness and width. The two general thickness classifications are single and double. These are further divided into light, medium, and heavy. The thickness specifications for first-quality leather belting are as follows:

Medium Single—5/32 to 3/16 in.
Heavy Single—3/16 to 7/32 in.
Light Double—15/64 to 17/64 in.
Medium Double—9/32 to 5/16 in.
Heavy Double—21/64 to 23/64 in.

In the installation of leather belts, precautions should be taken to put the flesh side on the outside and the grain, or smooth, side toward the pulley. It has been found by experience that when the belt is installed in this manner it will wear much longer and deliver more power than if put on in the reverse way.

Belt Speed

The maximum speed of a belt should not exceed a mile per minute. If the speed is raised above this limit, insufficient contact friction between belt and pulley will cause the effect of centrifugal force, which will tend to *throw* the belt. The most economical speed has been found to be from 3000 to 5000 ft per minute, though practical limitations such as speed of shafting and pulley diameters often limit belt speeds to much lower values.

If for the sake of simplicity such factors as slip and creep are omitted, the belt speed will obviously be the same as that of the pulley surface on which it travels. The circumferential speed (V) of the pulleys shown in Figure 13-1 may be written as

$$V = \pi \times D \times N \tag{1}$$

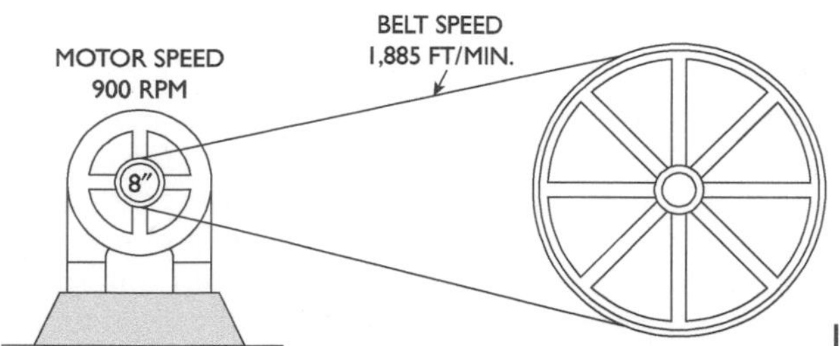

Figure 13-1 Relation between motor speed, pulley size, and belt speed.

where

V = speed in ft per min.
D = pulley diameter in ft
N = revolutions per min. (rpm) of pulley shaft
π = 3.1416

It will now be an easy matter to determine the speed of any belt, provided that the rpm and the diameter of the pulley are known. The former may be determined by a reliable speed indicator, and the latter by direct measurement.

Example
What is the belt speed when the full-load speed of a motor is 900 rpm and the diameter of the pulley is 8 in.?

Solution
With reference to Equation (1),

$$\text{The belt speed } V = \frac{\pi \times 8 \times 900}{12} = 1885 \text{ ft. per min.}$$

Example
Compute the belt speed of a motor having a full-load speed of 860 rpm when the diameter of the pulley is 10 in.

Solution
Inserting numerical values in the previously discussed equation, then

$$V = \frac{\pi \times 860 \times 10}{12} = 2255 \text{ ft. per min. approx.}$$

Pulley Diameters

There is no need to absorb a large quantity of formulas in order to determine pulley diameters and shaft speeds. The only way to master anything completely is to understand how it works.

With reference to Figure 13-2, it is obvious that if minor factors such as slip and creep are omitted, the speed of the belt connecting the two pulleys is the same throughout the loop. Any point on the driven pulley surface will travel at exactly the same speed as a corresponding point on the driver pulley surface regardless of the difference in their respective diameters.

As the circumferential speeds (V) of the pulleys are equal, it may be written

$$\pi \times D \times N = V = \pi \times d \times n, \text{ or} \qquad (2)$$

$$D \times N = d \times n \qquad (3)$$

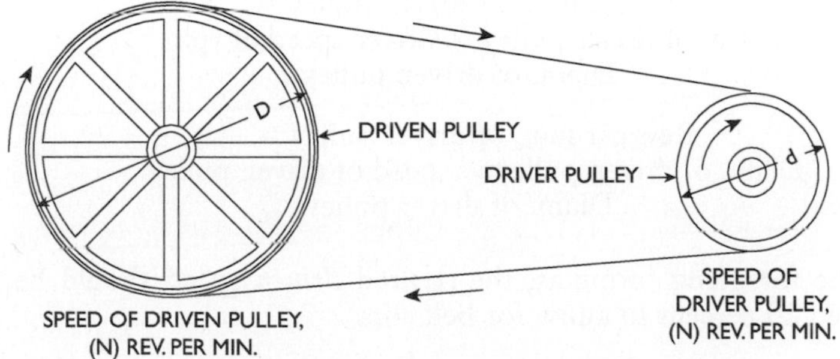

Figure 13-2 Relation between shaft speed and pulley diameter.

where

D = diameter of *driven* pulley
d = diameter of *driver* pulley
N = revolutions per min. (rpm) of *driven* pulley
n = revolutions per min. (rpm) of *driver* pulley

From equation (3), the following formulas may be written:

$$D = \frac{d \times n}{N} \tag{4}$$

$$d = \frac{D \times N}{n} \tag{5}$$

$$N = \frac{d \times n}{D} \tag{6}$$

$$n = \frac{D \times N}{d} \tag{7}$$

If the formulas are written down without abbreviations,

Diam. of driven pulley =
Diam. of driver pulley × driver speed in rpm
÷ Rev. per min. (rpm) of driven pulley

Diam. of driver pulley =
Diam. of driven pulley × rpm of driven pulley
÷ Rev. per min. (rpm) of driver pulley

Rev. per min. of driven pulley =
Diam. of driver pulley × driver speed of rpm
÷ Diam. of driven pulley

Rev. per min. of driver pulley =
Diam. of driven pulley × speed of driven pulley
÷ Diam. of driver pulley

In the foregoing formulas, the desired *driven* speed should be increased 20 percent to allow for belt slip.

To Find Diameter of Driven Pulley

Example
What size pulley is required on a 200-rpm line shaft if it is driven by a 1200-rpm motor that has a pulley diameter of 4 in.?

Solution
Substituting values in equation (4),

$$\text{Diam. of driven pulley } D = \frac{d \times n}{N} = \frac{44 \times 1200}{200} = 24 \text{ inches}$$

To Find Diameter of Driver Pulley

Example

What size pulley must be used on an 860-rpm motor employed to drive a fan at 215 rpm when the fan pulley diameter is 24 in.?

Solution

Substituting values in equation (5),

$$\text{Diam. of driver pulley, } d = \frac{D \times N}{n} = \frac{24 \times 215}{860} = 6 \text{ inches}$$

To Find Speed of Driven Pulley

Example

At what speed will a generator run if it has an 8-in. pulley and is driven by an engine having a 48-in. pulley and running at 180 rpm?

Solution

Substituting values in equation (7),

$$\text{Rev. per min. of driver pulley, } n = \frac{D \times N}{d} = \frac{60 \times 170}{12} = 850 \text{ rpm}$$

Creep in Belts

It is a well-known fact that slip causes a decrease in driven pulley speed below that which would be expected from the speed of the driver and the ratio of diameters of the two pulleys. The result of *creep* similarly affects the speed of the driven pulley, and *slip* and *creep* jointly cause power loss.

The effect of slip can be greatly reduced by adhering to conservative pulley ratios and correct alignment, but creep, being a physical characteristic of the belt itself, is inherent in all power-transmitting operations where belting is employed as the transmitting medium.

In order to obtain a clear concept regarding this phenomenon, the diagrams shown in Figure 13-3 will be of assistance.

Figure 13-3B shows a belt drive in which the driving and driven pulleys are of equal diameters. In order to facilitate the understanding of creep, the belt thickness has been greatly enlarged. When the belt is first applied to the pulleys, it must be given a certain initial tension, which is the same in both strands. It is this tension that causes friction between the belt and pulley surfaces and in this manner permits power to be transmitted from one pulley to the other.

When this transmission takes place, the driving pulley rotates in the direction of the arrow by action of its prime mover, which causes a pull on the lower strand of the belt. Due to this increase in tension in the lower strands, there will be a decrease in tension in the upper strand. This causes the former to stretch and become thinner and the latter to contract and grow thicker, as indicated.

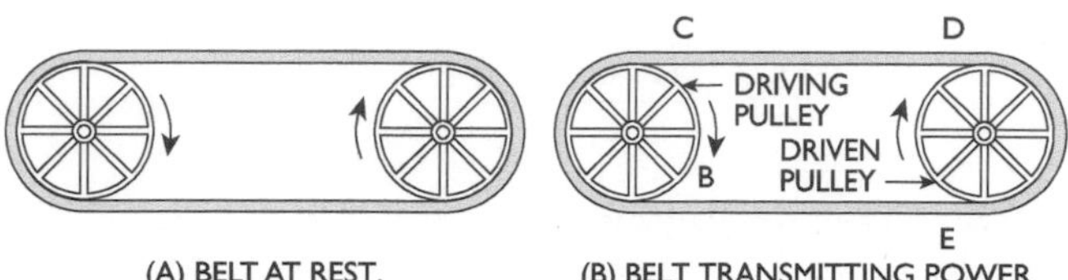

Figure 13-3 Belt thickness and condition of belt when at rest and when transmitting power.

As a result, a long section of belt (B at the driving pulley) is delivered to C. This is due to the elasticity of the belting material. In a manner similar to that already described, a short section of belt (D) is delivered to the pulley as it comes off the long section (E). The belt on the slack side, in other words, is being delivered to the pulley at a slower rate than it comes off the tight side. The belt is therefore creeping ahead on the pulley, and its belt surface runs more slowly than the belt itself.

Because the creep, as already pointed out, is caused by the elasticity of the belt, it increases with the load and is usually somewhat less than one percent at the load the belt is designed to carry under normal conditions. However, if the load on the belt is constantly increased above normal, a point will be reached where the friction between pulley and belt is not sufficient to drive the load, and the belt will slip on either or both pulleys, finally causing the belt to be thrown off.

Belt Slippage

The effect of *creepage* and *slip* is to reduce the speed of the driven pulley and, consequently, the power transmitted. For commercial applications, it will be close enough to multiply the calculated speed of the driven shaft by 98 percent, assuming 2 percent slip. The diameter of the driven pulley should be slightly smaller than that calculated on the basis of no slip, in order to obtain a certain speed at the driven shaft. In other words, the speed of the driven shaft is reduced by slip unless compensated for by reducing the driven-pulley diameter a proportionate amount.

How to Figure Belt Slip

Example
A synchronous motor having a full load speed of 900 rpm and a pulley of 8 in. diameter is required to drive a machine at a speed of 300 rpm. What should be the diameter of this pulley if the belt slip is estimated to be 2 percent?

Solution
If D denotes the diameter of driven pulley, d the diameter of motor pulley, n the motor speed in rpm, and N the speed of driven pulley, then

$$D = \frac{d \times n}{N} = \frac{8 \times 900}{300} = 24 \text{ in.}$$

Finally, taking the belt slip into account, it is evident that in order to compensate for this the pulley should be slightly smaller (2 percent), which is a reduction of 24 × 0.98 or 23.52 in.

A small amount of belt slip is not harmful, but excessive slip is. Severe slippage burns the belt quickly, destroying its usefulness, while at best the belt surface is polished so that the grip on the pulley is materially reduced. Belt slip can be detected by noting the condition of the pulley surface. When the belt is slipping, the pulley will have a very shiny appearance, as contrasted with the smooth but rather dull appearance it should have. Excessive slip is caused by poorly designed drives in which the driving pulley is too small or the load too great, by running the belt too loose, or by not giving proper attention to the care of the belt.

Balancing Pulleys

It is of the utmost importance, especially in high-speed drives, that the pulleys be correctly balanced. This is necessary because, if the centrifugal forces generated by the pulleys' rotation are greater at one side than at the other, it will cause the pulley shaft to vibrate whenever the amount of unbalanced centrifugal force becomes sufficient to bend the shaft or deflect the frames holding the shaft. There are two methods of testing the balance of a pulley:

1. The standing balance (static).
2. The running balance (dynamic).

A *standing balance* does not balance a pulley but merely corrects the want of balance to a limited degree. A *running balance* correctly balances a pulley when running up to a speed at which the balance was made but does not balance for greater speed.

A common method of balancing is to set the pulley in slow rotation several times in succession, and if the same part of the pulley circumference comes to rest in each case at the bottom, then this particular part of the pulley is the heaviest and its weight should be reduced, or weight should be added to the diametrically opposed side.

Table 13-1 will facilitate the calculation of belt speed in feet per minute when the corresponding motor speed and pulley diameter are known.

Example

What is the corresponding speed of a motor having a pulley of 11 inches in diameter and assuming a full-load speed of 1750 rpm?

Solution

On the 1750-rpm line in Table 13-1, read the figure corresponding to a pulley diameter of 11 in., which is 5040. In other words, when an 11-in. diameter pulley rotates at 1750 revolutions per minute, a belt on this pulley travels at a speed of 5040 feet per minute. In a similar manner, belt speeds of motors whose rpm and pulley diameters are given in the table may be read directly without calculation.

Table 13-1 is based on the formula

$$\text{Belt speed in feet per minute} = \frac{3.1416 \times \text{pulley diameter in inches} \times \text{rpm}}{12}$$

Belt Length

The length of belts is found by various methods, such as

1. Direct measurement.
2. Calculation.
3. Scale method.

Table 13-1 Belt Speeds (in feet per minute)

Full-Load Speed of Motor (rpm)	*Pulley Diameters in Inches*										
	3½	4	4½	5	5½	6	7	8	9	10	11
1750	1605	1835	2060	2290	2520	2750	3205	3670	4125	4580	5040
1450	1330	1520	1710	1900	2090	2275	2660	3040	3420	3800	4170
1150	1055	1205	1355	1505	1655	1810	2110	2410	2710	3010	3315
860	790	900	1015	1130	1240	1350	1575	1800	2030	2255	2480
690		725	815	905	995	1085	1265	1445	1625	1810	1990
575			680	755	830	905	1055	1205	1355	1505	1655

Full-Load Speed of Motor (rpm)	*Pulley Diameters in Inches*										
	12	13	14	15	16	17	18	19	20	21	22
1450	4560	4930									
1150	3615	3915	4220	4520	4820	5120					
860	2705	2930	3160	3380	3610	3835	4060	4280	4510	4740	4960
690	2170	2350	2530	2710	2890	3075	3250	3435	3615	3800	3980
575	1805	1955	2105	2260	2410	2560	2710	2860	3010	3160	3310

Full-Load Speed of Motor (rpm)	*Pulley Diameters in Inches*										
	23	24	25	26	27	28	29	30	31	32	33
690	4160	4340	4520	4700	4880	5060					
575	3460	3610	3760	3915	4060	4210	4360	4515	4665	4815	4965

(Table gives belt speed in ft per min. corresponding to various motor speeds and pulley diameters.)

Direct Measurement

This is the simplest and undoubtedly the safest method. A steel tape carefully adjusted over the pulley crowns will give the length of belting required. To this should be added a reasonable amount of material for jointing as required in each individual case.

Calculation

The calculation method is useful in locations where a direct measurement cannot readily be obtained. In many instances, it may be necessary to obtain the length of belting from *blueprints* where only shaft center distances and pulley diameters are given. This is most often the case, especially on new constructions, where the entire plant machinery is laid out on drawings according to a carefully engineered plan.

In belting up such a plant, it is imperative that the millwright has the knowledge necessary to intelligently cut and assemble belts without undue waste of material.

Generally, belts may be divided into two groups:

- Belts driving their connecting pulleys in the *same* direction are *open belts.*
- Belts driving their connecting pulleys in the *opposite* direction are *crossed belts.*

Length of Crossed Belts

With reference to Figure 13-4, the exact length (L) of a crossed belt may be written

$$L = (\pi + \frac{\pi}{90} \times \phi)\,(R + r) \; + (2 \times C\,cos\phi) \qquad (8)$$

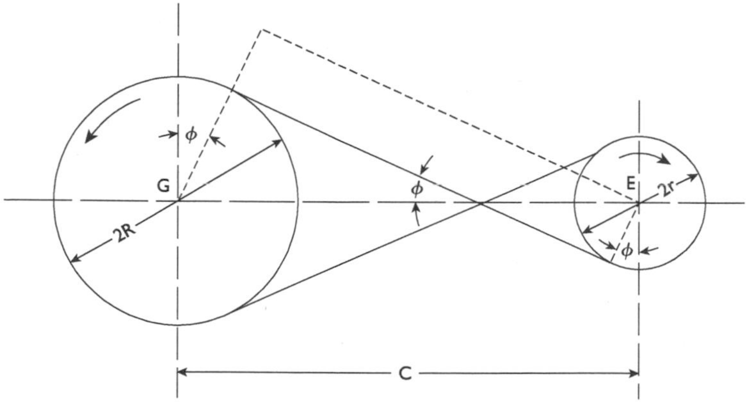

Figure 13-4 Calculating the length of a crossed belt.

where

R = radius of the larger pulley
r = radius of the smaller pulley
C = distance between shaft centers
ϕ = angle whose sine is $(R + r)/C$

Although this formula is found in most handbooks in one form or another, it should perhaps not be used unless the reader clearly understands its derivation. The length of the belt with reference to the illustration is simply found by adding up the various elements involved. Thus, starting with the larger of the two pulleys, the length of the pulley arc indicated by angle ϕ is

$$\frac{2\pi R}{360} \times \phi$$

but inasmuch as there are two such elements, their combined lengths are

$$\frac{4\pi R}{360} \times \phi$$

In a similar manner, the arcs on the smaller pulley are obtained as

$$\frac{4\pi r}{360} \times \phi$$

Finally, to the four arcs should be added twice the distance EF and half the circumference of each pulley. The length (L) of the belt is hence

$$L = \frac{\pi R\phi}{90} + \frac{\pi r\phi}{90} + \pi R = \pi r + (2C\ cos\phi)$$

which after rearrangement of terms becomes

$$L = (\pi + \frac{\pi\phi}{90})\ (R + r) + (2C\ cos\phi)$$

as previously written.

If the radii of the pulleys are substituted for their diameters, respectively, the formula becomes

$$L = \left(\pi + \frac{\pi\phi}{90}\right)\left(\frac{D + d}{2}\right) + (2C\ cos\phi) \qquad (9)$$

A somewhat simpler formula giving the approximate length (L) of a crossed belt is

$$L = \frac{\pi}{2}(D + d) + \left(2\sqrt{C^2 + \left(\frac{D + d}{2}\right)^2}\right). \quad (10)$$

Length of Open Belts

Contrary to the case of a crossed belt, the exact length of an open belt is somewhat difficult to obtain. An approximate formula in which the errors do not exceed practical limitations is derived from Figure 13-5. If the length between the pulley centers (C) is given, and the pulley diameters are respectively (D) and (d), then the length of the belt is

$$L = \frac{\pi D}{2} + \frac{\pi d}{2} + \left(2\sqrt{C^2 + \left(\frac{D - d}{2}\right)^2}\right).$$

or

$$L = \frac{\pi}{2}(D + d) + \left(2\sqrt{C^2 + \left(\frac{D - d}{2}\right)^2}\right). \quad (11)$$

Example

The shaft-center distance between two pulleys in an *open belt* drive is 12 ft. If the pulley diameters are 4 ft and 1 ft, respectively, what is the length of belting required?

Solution

Substituting values in equation (11),

$$L = \frac{\pi}{2}(4 + 1) + 2\sqrt{12^2 + 1.5^2}$$

or

$$L = \frac{5\pi}{2} + \left(2\sqrt{146.25}\right) = 7.854 + (2 \times 12.1) = 32.054 \text{ ft.}$$

Example

If in a *cross-belt* transmission drive the pulley-shaft distance is 10 ft and the pulley diameters are 54 in. and 9 in., what length of belt should be used?

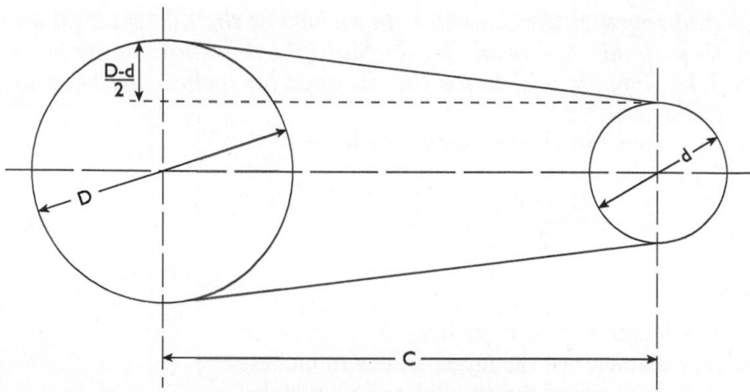

Figure 13-5 Calculating the length of an open belt.

Solution

Inserting values in equation (9) and remembering that the angle ϕ in Figure 13-1 is that angle whose sine is $(R + r)/C$, then sine $\phi = 31.5/120 = 0.2625$; from a table of natural sines and cosines, the corresponding angle ϕ is 15.2°. Substituting numerical values,

$$L = \left[3.14 + (0.035 \times 15.2)\right]\left[\frac{4.50 + 0.75}{2}\right] + \left[2 \times 10 \times 0.9650\right]$$

from which

$$L = 28.93 \text{ ft.}$$

Scale Method

This method consists of laying out the pulleys set at their proper distance apart, either full size or in a suitable scale, measuring the length of the side of the belt, and then adding to this one-half the circumference of each pulley, assuming the belt to envelop one-half the circumference of each pulley.

Length of Belts for Equal Size Pulleys

When the pulleys are of the same diameter, or nearly so, the length of an *open belt* may be obtained as follows:

Add together the diameter (in inches) of the two pulleys, and then divide the result by 2. Multiply the quotient by π, or 3.14. Finally, add twice the distance (in inches) between the pulley centers.

If this is written algebraically, we have

$$L = \frac{(D + d)\pi}{2} + 2C \tag{12}$$

where

L = length of belt in inches
D = diameter of the larger pulley in inches
d = diameter of the smaller pulley in inches
C = distance between pulley shaft centers in inches

Example

How many feet of belting are necessary to connect two pulleys with diameters of 28 and 21 in., respectively, when their shaft center distance is 18 ft 6 in.?

Solution

By employing the previously enumerated rule or formula,

$$L = \frac{(28 + 21)}{2} 3.14 + (2 \times 12 \times 18.6) = 520.93 \text{ inches}$$

If this is divided by 12 to obtain feet, the result is 43.41 ft, or 43 ft 5 in. This is the net length of the belt without the lap, for which allowance should be made.

Belt Strength

With reference to Figure 13-6, the power transmitted from one pulley to the other is directly proportional to the difference in tension $(T_1 - T_2)$ and the speed of the belt. If $(T_1 - T_2)$ is measured in pounds and the belt speed (V) in ft per minute, the horsepower (hp) transmitted may be written

$$hp = \frac{(T_1 - T_2)V}{33{,}000} \tag{13}$$

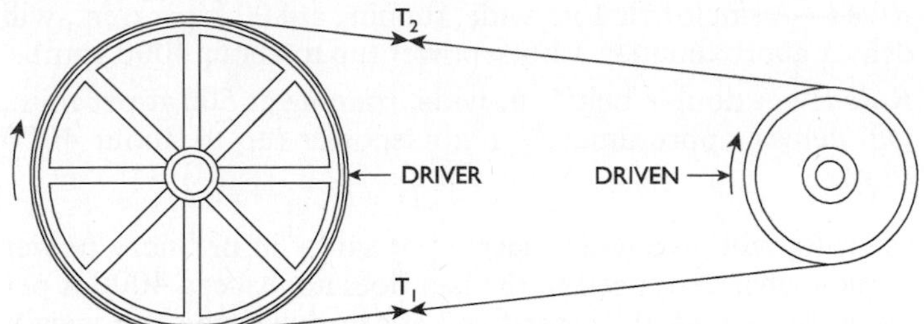

Figure 13-6 Belt tension in typical belt drive.

but since the speed $V = \pi \times D \times N$, it follows that

$$hp = \frac{(T_1 - T_2)\pi \times D \times N}{33{,}000} \tag{14}$$

where

D = pulley diameter in feet
N = revolutions per min. of pulley shaft
$\pi = 3.1416$

Example

The full-load speed of a motor that has a 9-in.-diameter pulley is 850 rpm. If the high- and low-tension pulls are 150 lb and 30 lb, what is the horsepower transmitted to the belt?

Solution

Substituting numerical values into the formula, the horsepower is

$$hp = \frac{(150 - 30)\pi \times 850}{33{,}000} = 7.28 \text{ hp}$$

The customary method of expressing the amount of slack side tension (T_2) necessary for a successful drive is in the form of a ratio commonly called the *tension ratio*. This ratio is written T_1/T_2 and usually varies over rather large limits, depending on pulley diameter ratios, center distances, pulley gripping quality, type of belting employed, etc.

Determination of Belt Sizes

For a given condition, the minimum belt width is determined by the horsepower to be transmitted and the speed of the belt. The millwright's rules are as follows:

Rule I—A single belt 1 in. wide, running at 800 ft per min., will deliver approximately 1 horsepower (up to about 4000 rpm).

Rule II—A double belt 1 in. wide, running at 500 ft per min., will deliver approximately 1 horsepower (up to about 4000 rpm).

These rules will give wide margins of safety in ordinary power transmission when the speed of the belt does not exceed 4000 ft per min. Above this speed, the centrifugal effects due to the belt weight begin to be rather appreciable and should therefore be included.

Pulley Diameters versus Belt Thickness

The smallest diameters of pulleys over which belts of various thicknesses will work satisfactorily vary with the care with which the belt is manufactured, as well as the grade of material involved.

The thickness of various types of belting in relation to minimum pulley diameters is given in Table 13-2.

Table 13-2 Various Belting Thicknesses

Type of Belt	*Thickness (in.)*	*Min. Dia. of Pulleys (in.)*
Light single	1/8–5/32	3
Medium single	5/32–3/16	4
Heavy single	3/16–7/32	5
Light double	1/4–9/32	6
Medium double	5/16	10
Heavy double	3/8	16
Medium three ply	1/2	30
Heavy three ply	9/16	36

Belt Joints

There are several methods by means of which belting is made endless, such as

1. Cementing.
2. Lacing, which may be subdivided into various classes such as *lap*, *butt*, and *apron* joints.
3. Patented hook and plates.

However, before proceeding with the various methods of joint making, it may be worth repeating that the safest method to determine the length of a belt is by direct measurement over the pulleys with a steel tape. If the old belt is present, measure over it and deduct six times the belt's thickness from the tape measurement.

Cement Joint

After the required length has been measured off, a sufficient amount of length should be added to prepare for the necessary laps. These will vary with the width of the belt to be installed. For single belts, the practice is to use a 4-in. lap for belts up to 3 in. in width and a 6-in. lap for belts exceeding this width. It should be remembered that extra-heavy belts or extra-light belts may call for longer or shorter laps. See Figure 13-7.

The laps in either single or double belting should be tapered carefully with a special plane or drawknife on a level surface to ensure a good and even performance. Oil and grease should be cleaned from leather before application of cement. After the two edges have been thoroughly cleaned, the cement is spread over both pieces and they are put together. A suitable pressure is applied until the cement has set. The lap should be allowed to dry undisturbed for ten to fifteen minutes and should not be put in use for one to two hours after sticking.

Cement Joints on Double Belts

When making joints in double belts, the same general procedure is followed as that already described for single belts. With reference to Figure 13-8, the belts and lap are made up of one male and one female and prepared in such a manner as to eliminate any possibility of uneven thickness throughout. See Figure 13-9 for direction of overlap and belt travel.

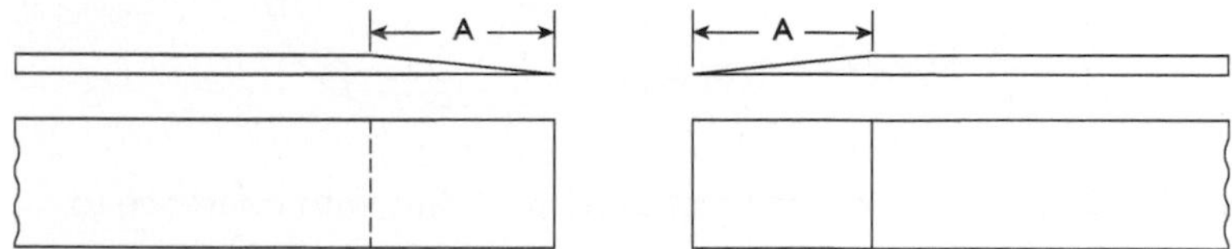

Figure 13-7 Method of splicing a single belt.

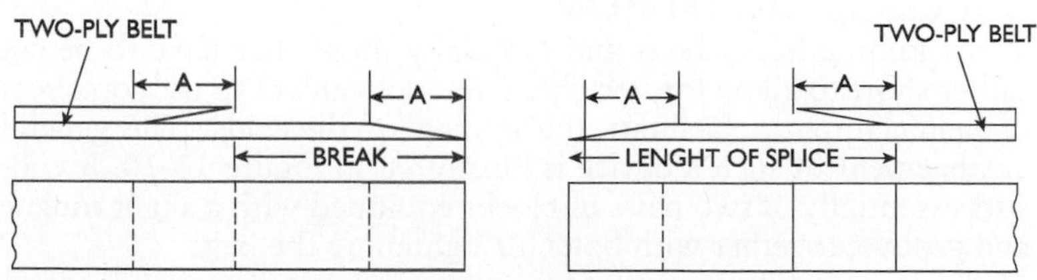

Figure 13-8 Method of splicing a double belt.

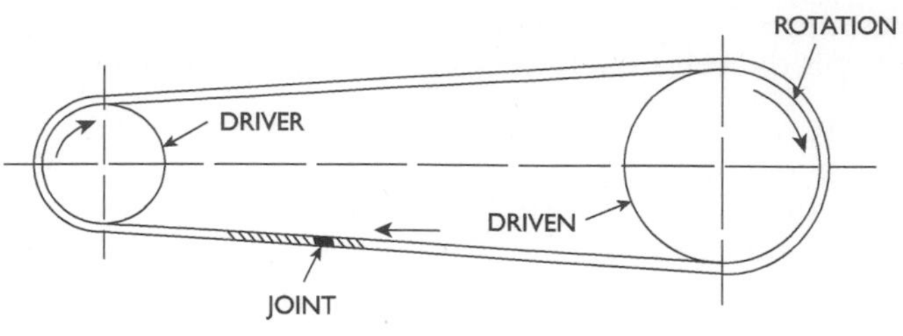

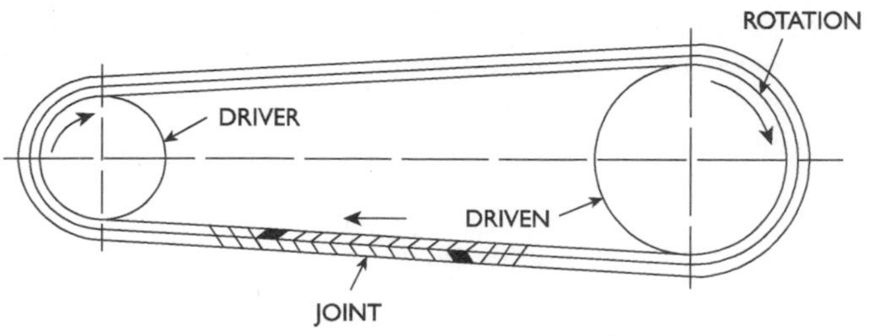

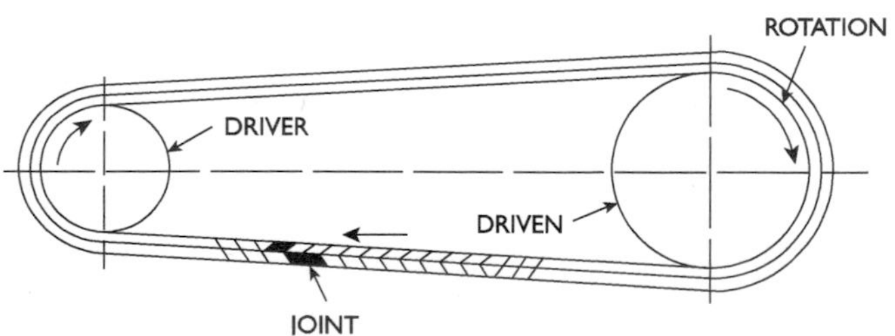

Figure 13-9 Proper direction of overlap in joint with reference to direction of belt travel.

Belt Clamps and Their Use

When joining heavy belts and especially those that have to be cut rather short to allow for stretching, it is convenient to use some sort of belt clamps to facilitate the joining of the ends. The general arrangement of such a device is illustrated in Figure 13-10. It consists essentially of two pairs of blocks equipped with a slight tongue and groove, together with bolts for tightening the belt.

After insertion of the belt in such a manner as to allow for joining clearance, the clamps are securely fastened and pulled together by means of the two main rods (4 to 10 ft long) equipped with right- and left-hand threads with large nuts on one end and vise handles on the other. When fastening the clamps on the belt, care should be taken that the clamps are at right angles to the belt and are correctly centered on it. The main rods should be taken up both evenly and simultaneously so that the ends to be joined may fit together correctly. A press block of suitable size consisting of two blocks with flat surfaces and take-up bolts will facilitate the application of pressure on the cemented joint.

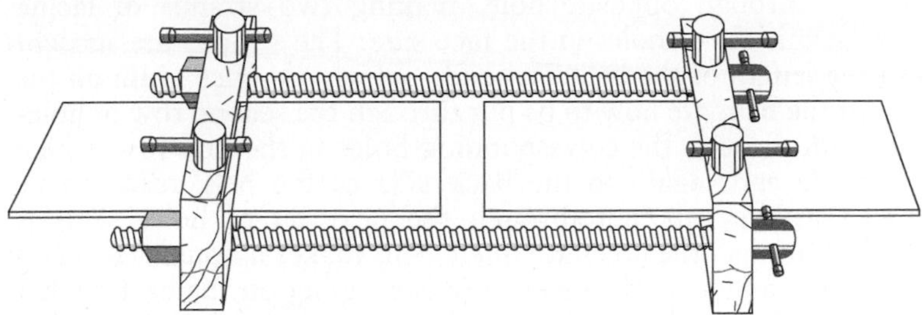

Figure 13-10 Arrangement of belt clamps used to hold belt securely in position when making up joints.

Making a Laced Joint

The best lacing material consists of *rawhide* or *Indian tanned lace.* Laced joints are used very frequently when leather belting is being employed. Their advantages over other types are the following:

1. They are made more rapidly than other types.
2. If correctly made, they are very strong.
3. The absence of a metallic surface against the pulley increases the efficiency of the belt.
4. They prevent injury to the hands of the operator when shifting is necessary.

Although there are no specific rules for making a laced joint for various sizes, a good method of lacing a leather belt is as follows. The first row of holes is punched one inch from the end of the belt, and the second row two inches from the end. The number of holes across the belt should equal the number of inches in the width, with the holes on the side 3⁄8 in. from the edge. The lacing should always be made with the flesh side out and the grain side in or toward the pulley.

Commencing the lacing from the side of the belt, the ends being passed from the inside, each end is then passed through the hole in the opposite end of the belt, making two thicknesses of lacing on the inner side. See Figure 13-11. Each lacing end is then passed through the hole in the back row and again through the same hole as before in the first row, thus filling the four holes at the side and leaving the lacing straight with the lengths of the belt.

With both ends of the lacing now on the inside of the belt, they must be put through the second hole in the first row on the opposite end, thus causing the lacing to cross on the back or inside of the belt. The ends are now put through these same holes again, opposite end through opposite hole, making two strands of lacing between this set of holes in the face side. The strands are straight with the length of the belt. The ends, which are once again on the back of the belt, are now to be put through the second row of holes and again through the corresponding holes in the first row, which brings the ends again to the back side of the belt, ready to be crossed as they are put through the next set of holes. This is repeated until all the holes are filled. This makes a double lacing on the outside and smooth single strands crossing on the back with a straight pull on each strand on the face as it passes through each of the four holes in line. See Figure 13-12.

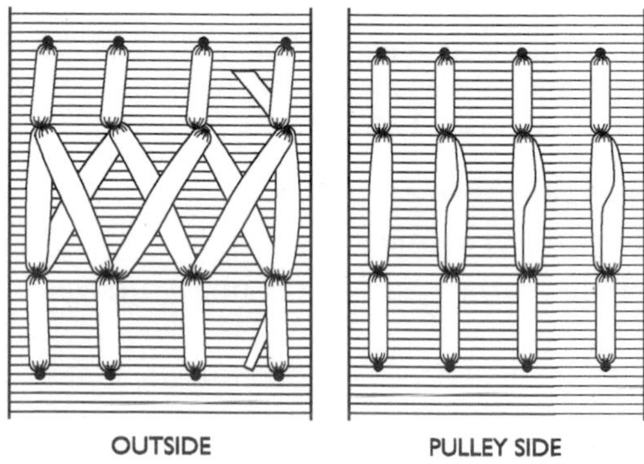

Figure 13-11 A good method of lacing a 4-inch belt.

As the lacing is drawn through each of the four holes, the ends will be found securely fastened and will bear a strand or two being cut. It is necessary to use pliers to draw the laces through the holes.

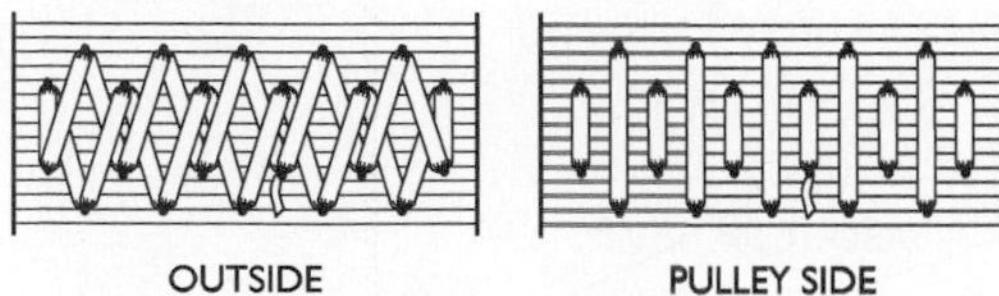

Figure 13-12 A simple and reliable method of lacing a belt joint for use with Indian rubber belting.

Installing Belts on Pulleys

Belts should under no circumstances be cut and spliced too short, thereby requiring undue strain to force them on the pulleys. Because it is difficult to determine just how much a belt will stretch over a period of time, the belt should be assembled so that only a moderate amount of strain will be required to install it on the pulleys.

If the belt is spliced too short, it will almost invariably become dented at the point where it is unduly forced on, which is due to the unusual force on one edge, the result of which will be an unsatisfactory and weak belt. This will also put an excessive strain on shafts and bearings. In all cases where conditions are such that belts cannot readily be spliced off the pulleys, an arrangement of belt clamps such as that shown in Figure 13-10 will have to be employed.

Drive Arrangements

Horizontal drives are always the best. The proper procedure here is to arrange the belt as shown in Figure 13-13C, that is, with the *lower side* of the belt *driving*; the sag of the upper side tends to increase the *arc* of contact on the pulley. This sag should be about 1½ in. for every 10 ft of center distance between the shafts. If too loose, the belt will have an unsteady flapping motion that will injure both belt and machinery. Again, if too tight, the bearing will be put under a severe strain, and the belt will be quickly destroyed.

Vertical drives arranged as shown in Figure 13-13B should be avoided whenever possible. This is particularly true where the lower pulley is the smaller of the two. The reason why vertical drives prove less satisfactory than horizontal drives is that in the vertical drive the effective tension and arc contact are substantially reduced, and as a result, the slip will increase in extreme cases to the point where the normal load cannot be carried.

In group drives where several belts transmit power from a line shaft, it is advantageous, wherever possible, to locate the drive shaft so that the bearing pressure can be equalized and reduced by alternating the direction of drive first on one side and then on the other.

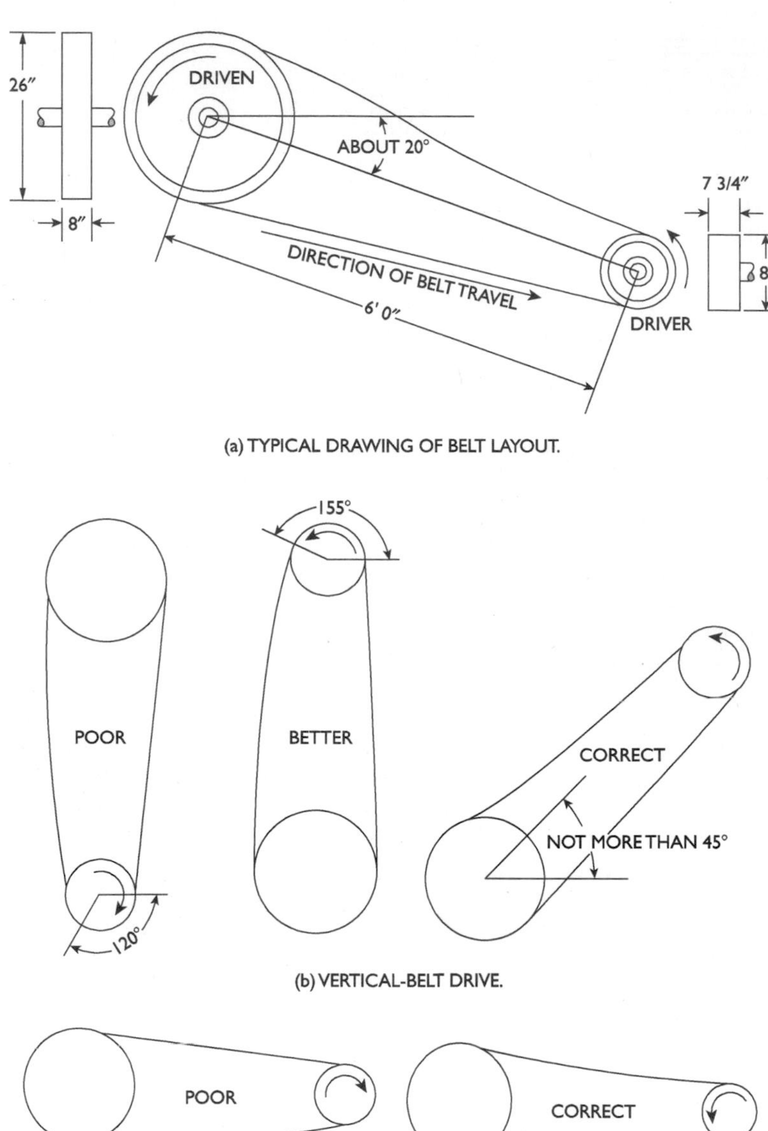

(a) TYPICAL DRAWING OF BELT LAYOUT.

(b) VERTICAL-BELT DRIVE.

(c) HORIZONTAL-BELT DRIVE.

Figure 13-13 Typical belt layout and various drive arrangements.

Belt Maintenance

It is of utmost importance in order to secure the best possible economy that a proper maintenance schedule be observed. This consists of numerous details, the most important of which are

1. Keeping belts tight.
2. Taking up slack of belts.
3. Running in new belts.
4. Cleaning dirty belts.
5. Dressing belts.
6. Maintaining belt guards.

In locations where idlers or tension-base drives are employed, the tightness of belts is automatically adjusted to a desired value. Because of the absence of the necessary space requirements, however, this method is not always feasible, and the surplus length will therefore have to be removed at certain intervals of time. Some belts stretch more than others under similar conditions and loads. The temperature and humidity are two factors that greatly influence the tension of belts. Locations of high humidity and moisture are known to cause a belt to expand more quickly.

Putting New Belts in Service

Before a new belt is put in service, it is a good practice to run the belt under light load to make it flexible. This holds true for all kinds of machinery where power is transmitted from one part to another. Because a belt is usually a component part of one machine, it follows that this rule also applies when the belt is utilized as a means of transferring power between two separate machines.

Cleaning

In locations where there is a considerable amount of dust of one kind or another, and especially where there are particles of oil present, the belt has a tendency to lose its smoothness and become stiff and glazy, causing excessive slip and unsatisfactory operation. This accumulation of dirt should be scraped off with a suitable tool (square edge or a block of wood), taking care that the scraping operation is performed in the direction of the point of the lap to protect against damage of the leather.

Belt Dressing and Application

There are many belt dressings on the market, some of a very high quality, although others may actually cause damage to the belt instead of resurrecting it. The surest method of obtaining a desirable belt dressing is to purchase it from a reputable manufacturer. The object of the belt dressing is to maintain the surface in a pliable condition to ensure a gripping effect on the pulley crown.

A good belt dressing consists of tallow and cod oil melted together to form a paste that is worked into the belt by means of a suitable brush or, if the belt is running, as shown in Figure 13-14. When applying the dressing, care should be taken not to apply an excessive amount because this makes the belt soggy, thus decreasing its lateral stiffness.

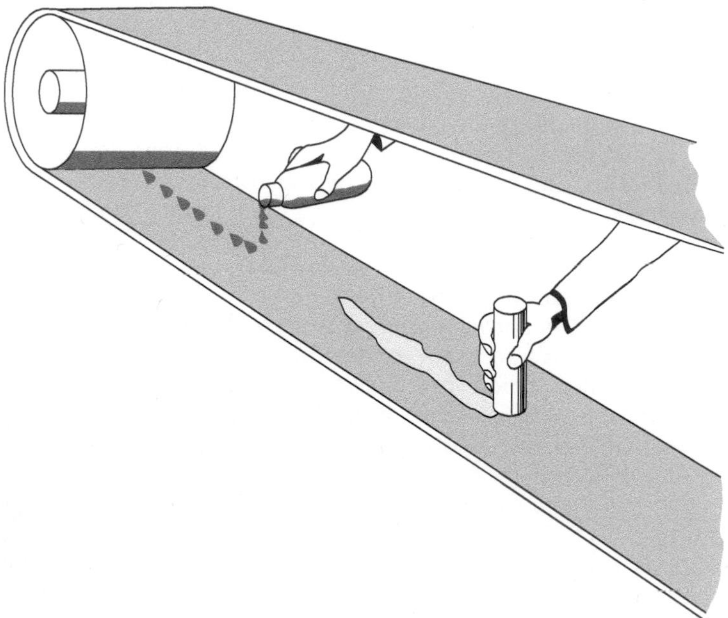

Figure 13-14 Method of applying dressing to a running belt.

Belt Notes

After the elastic limit of a leather belt is reached, it may be stretched to the breaking point without making it any tighter.

The friction increases regularly with the pressure; the more elastic the surface is, the greater the friction will be.

To obtain the greatest amount of power from belts, the pulleys should be covered with leather; this will allow the belts to run very slack and give more wear.

The leather in a belt should be pliable; of fine, close fiber; solid in its appearance; and of smooth, polished surface.

Belts derive their power to transmit motion from the friction between the surface of the belt and the pulley and from nothing else.

The thickness, as well as the width of belts, must be considered; consequently, a double belt must be used where it is necessary to transmit a greater power than possible with a single belt.

To increase the driving power of a belt, the pulleys may be enlarged in circumference, thus increasing the speed of the belt. This is usually an advantage, provided that the speed is not increased above the safe limit.

Pulleys and Their Arrangement

Pulleys are named from their construction or for the uses to which they are put, as in the following examples:

1. *Compound pulley*, by which the power to raise heavy weights is gained at the expense of speed.
2. *Cone pulley*, which is usually stepped to different diameters to facilitate the change of speed imparted to a machine.
3. *Dead pulley*, i.e., a pulley running idle on a shaft. It is placed beside a fast or keyed pulley to receive the belt when it is thrown off the latter to stop the machine or shaft. This pulley is alternately called *loose pulley.*
4. *Two-speed pulley.* A combination of two loose pulleys and toothed gearing with one fast-driven pulley, whereby two different speeds of rotation may be obtained with pulleys of the same diameter by shifting the belt from the fast pulley to one of the loose pulleys. This pulley is sometimes called a *double-speed pulley.*
5. *Sliding pulley.* A driving pulley to which a friction clutch is attached and placed so as to slide backward and forward on a shaft. It is used to couple and disengage machinery.

Materials Used

Pulleys are usually manufactured from cast iron, fabricated steel, paper, fiber, or various kinds of wood. They are made solid or split, and in either case the hub may be split so as to facilitate the assembly of the pulley with the shaft.

Comparison of Various Types of Pulleys

The chief objection to wood or pulp pulleys, as compared with cast iron, is that all wood pulleys are necessarily made of built-up segments glued together. These segments, due to temperature and moisture changes, may come apart where the service is unusually severe. However, in ordinary use, this objection need not be considered.

The advantage of the wooden pulley is that it has a more advantageous coefficient of friction, especially when new. Sometimes wood and iron pulley surfaces are covered with leather, fastened by various methods. The best fastening method is to connect a smooth leather covering on the pulley, hair side outward. The leather should be uniform in thickness, of the width of the pulley surface, and of equal length to the circumference plus the amount of lap.

Shape and Width of Pulley Surfaces

Ordinarily, the width of the pulley face should slightly exceed that of the belt it is to carry. It has been found good practice for the pulley face to be approximately *one inch wider* than the belt, for belts less than 12 in. wide. For belts from 12 to 24 in. in width, the pulley should be *two inches wider*, and for belts exceeding 24 in. in width, the pulley should be *three inches wider* than the belt.

Flat Pulley Surfaces

All driving pulleys that require belt shifting should have a flat surface; all other pulleys should have a crowned surface.

Crowned Pulley Surface

The main reason for the use of crowned pulleys is that belts *adjust* or *center themselves* better when the pulley is given a slight crown. The crown of the belt is effective only when the slip of the belt is at a minimum because a slipping belt will run off a crowned pulley more easily than a flat-faced pulley.

Leather belts require higher crowns than rubber or cotton belts. The form of the crown used may be either a convex or a very slight V form. See Figure 13-15. A good figure for the amount of crowning is that the height of the pulley crown should be 1/96 of the width, or 1/8 in. per foot of pulley width.

Idle Belts

It is good practice when belts are out of service for any length of time not to let them rest on the shafting but to throw them off the pulleys. This is best accomplished by a suspension hook, as shown in Figure 13-16.

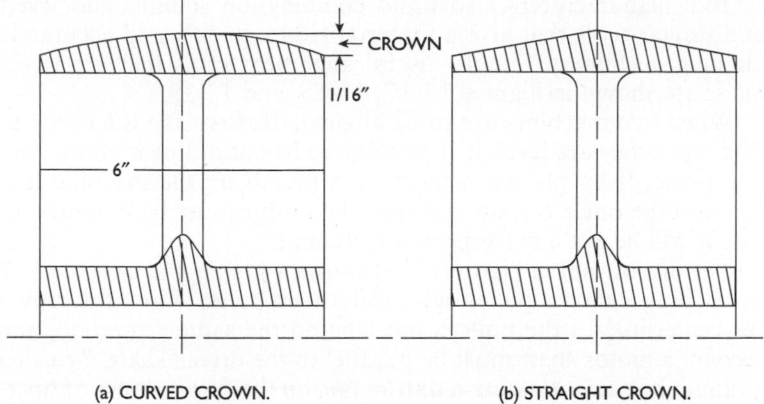

Figure 13-15 Typical cross section of curved and straight-sided crowning of pulleys.

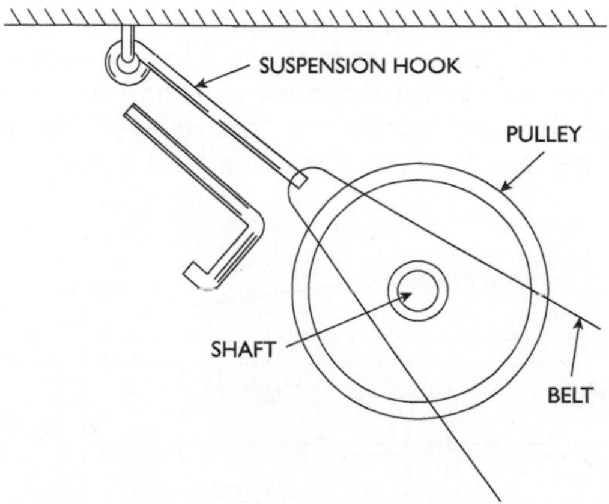

Figure 13-16 Belt suspension hook to remove belts from shaft and pulley when not in service.

Alignment of Pulleys and Shafting

After the motor has been properly located and the right kind of mounting has been provided, the next step is to align the machine with its drive. The tools usually employed for alignment of machinery or line shafting are the square, plumb bob, and level.

Tool manufacturers also build combination squares and levels and similar tools that give a material aid in quickly and accurately aligning machinery. Simple, useful alignment tools that are easily made are shown in Figures 13-17, 13-18, and 13-19.

When two machines are to be aligned, the first step is a check to find out if they are level. It is possible to be out of line in more than one plane. If by placing a level on a plumb on the machine it is found to be out a certain amount, the motor must be mounted so that it will be out a corresponding amount.

The illustration in Figure 13-18 shows a simple and easy method of aligning a motor pulley with the driven pulley. First, the crown (or centerline) of the pulleys must be on the same centerline, and second, a motor shaft must be parallel to the driven shaft. By using a plumb bob and drawing a datum line on the floor, a base of operation is established. Next, drop a plumb line from the center of the driven pulley to the floor. With a square, draw a line perpendicular to the datum line. Drop a plumb line from the center of the motor pulley, and move the motor up or back until the plumb bob rests on the centerline of the driven pulley. From the pulley centerline, perpendiculars may be drawn through the centers of the holes in the motor feet. A level should be used to verify that the line shafting is level. If it is not level, then the motor feet must be shimmed up so that the motor shaft and the line shaft will be "out of level" by the same amount. Chain drive may be aligned in a similar manner.

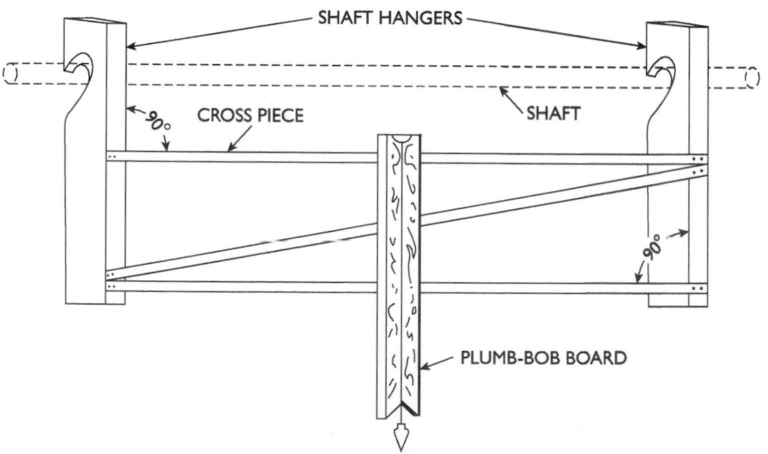

Figure 13-17 Method of shaft leveling or alignment.

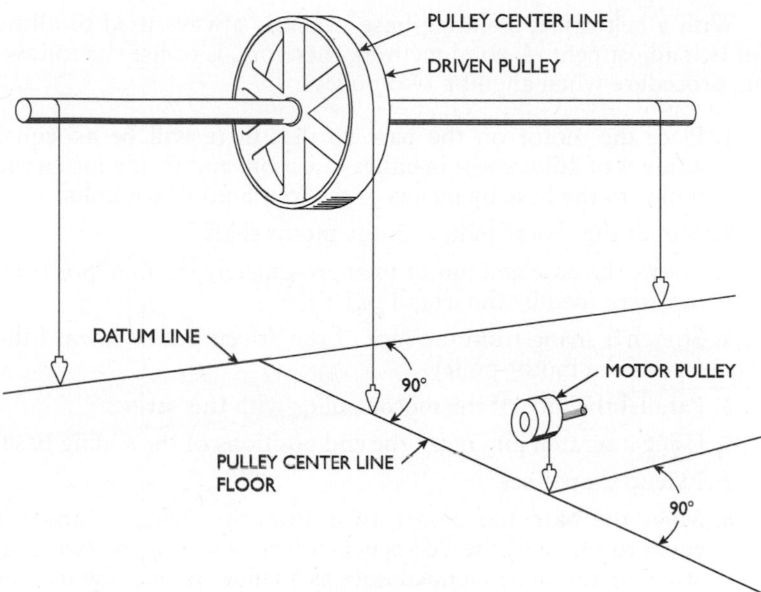

Figure 13-18 Method used to align two pulleys.

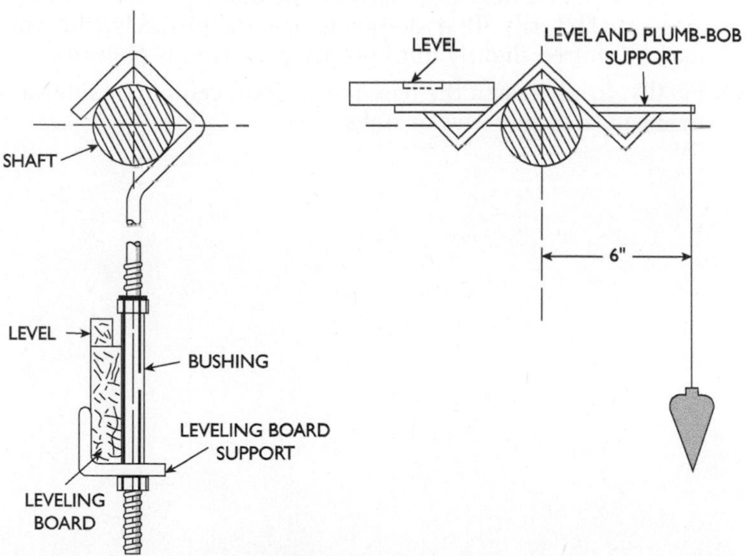

Figure 13-19 One method to align shaft and locate exact point on the floor in relation to shaft center.

With a belt drive, a sliding base is nearly always used to allow for belt adjustment. A good method, therefore, is to use the following procedure when aligning two pulleys:

1. Place the motor on the base so that there will be an equal amount of adjustment in either direction, and firmly fasten the motor to the base by means of the four hold-down bolts.
2. Mount the motor pulley on the motor shaft.
3. Locate the base and motor in approximately the final position, as determined by the length of belt.
4. Stretch a string from the face of the driven pulley toward the face of the motor pulley.
5. Parallel the face of the motor pulley with this string.
6. Using a scratch pin, mark the end positions of the sliding base.
7. Extend these lines.
8. Move the base and motor away from the string an amount equal to one-half the difference in face width of the two pulleys. Use the two extended lines as a guide to keep the base in its proper position.
9. The belt should now be placed on the pulleys to see if it operates satisfactorily. If it does not operate properly, the base may be shifted slightly until proper operation is obtained.
10. Finally, firmly fasten the base to the floor, ceiling, or sidewalls by means of lag screws or bolts.

Chapter 14

V-Belt Drives

Belt drives provide a quiet, compact, and resilient form of power transmission. They are employed widely throughout industry in many forms and styles. The so-called rubber belt, manufactured with a combination of fabric, cord, and/or metal reinforcement vulcanized together with natural or synthetic rubber compounds, is today's basic form of power transmission belt. The most widely used style of rubber belt is the V-belt. Its tapered cross-sectional shape causes it to wedge firmly into the groove of the sheave under load. Its driving action takes place through frictional contact between the sides of the belt and the sheave groove surfaces. Normally the V-belt does not contact the bottom of the sheave groove.

Because V-belts are so vital to industrial operations, and because installation and maintenance are the duty of the millwright or mechanic, he or she must be thoroughly familiar with them. This includes a working knowledge of the types of belts as well as how to install, adjust, and maintain them. There are three general classifications of V-belts:

1. Fractional horsepower.
2. Standard multiple.
3. Wedge.

Fractional-Horsepower Belts

Fractional-horsepower belts are principally used as single belts on fractional-horsepower drives. These belts are designed for intermittent and relatively light use rather than continuous operation. They are intended for service with small-diameter sheaves on drives and machinery, generally within the capacity of a single belt.

Standard-Multiple Belts

Standard-multiple belts are designed for the continuous service usually encountered in industrial applications. As the name indicates, more than one belt provides the required power-transmission capacity. The standard belt is intended for the majority of industrial drives that have normal loads, speeds, center distances, sheave diameters, and operating conditions. A premium-quality belt is manufactured for severe conditions. This is recommended when the drive is subjected to repeated shock loads, pulsation or vibration, high speeds, extremes of temperature and humidity, and substandard sheave diameters.

Wedge Belts

The wedge belt is an improved-design V-belt that makes possible a reduction in size, weight, and cost of V-belt drives. Utilizing improved materials, these multiple belts have a smaller cross section per horsepower and use smaller-diameter sheaves at shorter center distances than is possible with standard multiple belts. Because of the premium-quality, heavy-duty, balanced construction, only three cross sections of belt are used to cover the duty range of the five sizes of standard multiple belts.

V-Belt Lengths and Matching

Satisfactory operation of multiple-belt drives requires that each belt carry its share of the load. To accomplish this, all belts in a drive must be essentially the same length. Use of belts of unequal length results in the shorter-length belts carrying the load and the longer belts getting a free ride. However, because of the variation in stretch, shrink, etc., of the cords and fabric in rubber belts, it is not economically practical to manufacture them to exact length. Most manufacturers of V-belts therefore follow a practice of code marking to indicate exact length.

Each belt is measured under specific tension and marked with a code number to indicate its variation from nominal length. The number 50 is commonly used as the code number to indicate a belt within tolerance of its nominal length. For each 1⁄10 in. over nominal length, the number 50 is increased by 1. For each 1⁄10 in. under nominal length, 1 is subtracted from the number 50. The code number is separated from the regular belt number that specifies the belt cross-section size and its nominal length, as shown in Figure 14-1. For example, a 60-in. B-section belt manufactured 3⁄10 of an inch long will be code-marked 53; a 60-in. B-section belt manufactured 3⁄10 of an inch short will be code-marked 47. Although both of these belts are B60's, they could not be used satisfactorily in a set because of the difference in their actual length.

It is possible for the length of a belt to change slightly during storage. Under satisfactory storage conditions, however, such changes should not exceed measuring tolerances. Therefore, properly stored belts may be combined by matching code numbers. Ideally, sets should be made up of belts with the same code numbers, although the resiliency of the belts allows some length variation. Table 14-1 lists the maximum recommended variations for standard-multiple belts when matching.

Figure 14-1 Typical code marking.

In operation where only a small number of spare belts are carried in stock, matching may not be possible. In this case, it is advisable that belts be purchased in matched sets, as shown in Figure 14-2, to ensure proper load sharing when installed. A properly matched set of belts on an operating drive will all be fairly even on the tight side of the drive—that is, the side on which the belts are approaching the motor sheave. Sagging belts on the tight side are an indication of variations in belt length. If the belts are all new, it is a sure sign of mismatched belts.

Figure 14-2 Matched set of belts.
(Courtesy Emerson Power Transmission/Browning)

Table 14-1 Recommended Variations for Standard-Multiple Belts

Matching Number Range	*A*	*B*	*C*	*D*	*E*
2	26–180	35–180	51–180		
3		195–315	195–255	120–255	144–240
4			270–360	270–360	270–360
6			390–420	390–660	390–660

Note

In recent years, certain top-shelf manufacturers have improved their process of making V-belts. Tighter tolerances on the rubber, fiber, carbon, and other raw ingredients, plus much closer process control in making and curing the belts, have led to a whole new class of "matchless" belts. This quality of belt does not require length matching because each belt is exactly the same as any other for a given size. Obviously, purchasing belts from different manufacturers for use on the same machine is not a good idea because the standards of each manufacturer for "matchless" belts will vary.

V-Belt Drive Alignment

The life of a V-belt is, of course, dependent on the quality of materials and manufacture. However, high-quality belts cannot give good service life unless they are correctly installed and properly maintained. Frequently, unsatisfactory service can be traced to improper installation and maintenance practices. One of the most important installation factors influencing operating life is belt alignment. In fact, excessive misalignment is probably the most frequent cause of shortened belt life.

Although V-belts, because of their inherent flexibility, can accommodate themselves to a degree of misalignment not tolerated by other types of power transmission, they still must be held within reasonable limits. Maximum life will, of course, be attained only with true alignment. As misalignment increases, belt life is proportionally reduced. If the misalignment reaches an amount greater than ¹⁄₁₆ in. for each 12 in. of center distance, very rapid wear will result. The safest way is not to bother measuring the belts. If sheaves are out of line, correct them before trouble develops.

Misalignment of belt drives results from shafts being out of angular or parallel alignment or from the sheave grooves being out of axial alignment. These three types of misalignment are illustrated in Figure 14-3.

Because the shafts of most V-belt drives are in a horizontal plane, angular shaft alignment is easily obtained by leveling the shafts. In cases where shafts are not horizontal, a careful check must be made to ensure that the inclination angle of both shafts is the same.

Checking and adjusting for parallel-shaft alignment and axial-groove alignment of most drives may be done simultaneously. The best accuracy can be attained with a straightedge, as shown in Figure 14-4. The second choice is a taut line. In either case, the shaft and sheave must first be checked for runout. Any shaft runout or sheave wobble will cause proportionate inaccuracies in alignment.

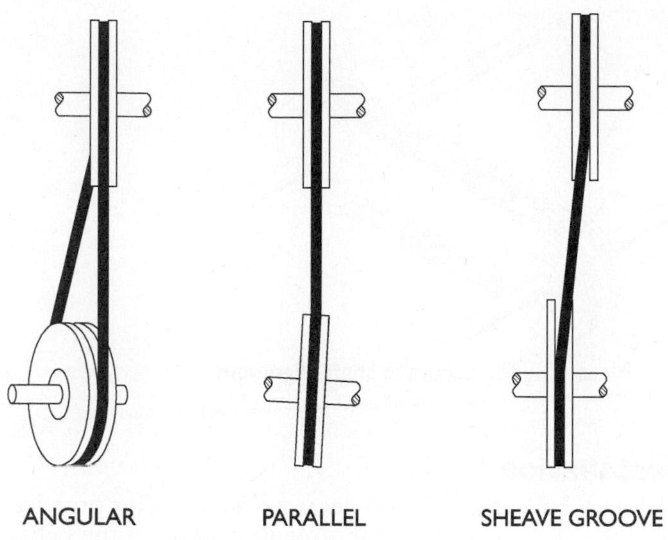

Figure 14-3 V-belt misalignment.

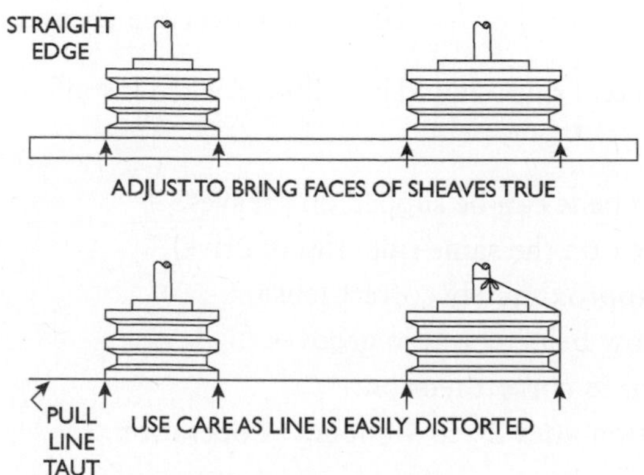

Figure 14-4 Checking shaft and sheave alignment.

In cases where very accurate parallel-shaft alignment is required, as with heavily loaded positive-drive belts, shaft parallelism should be checked by measurement, as shown in Figure 14-5.

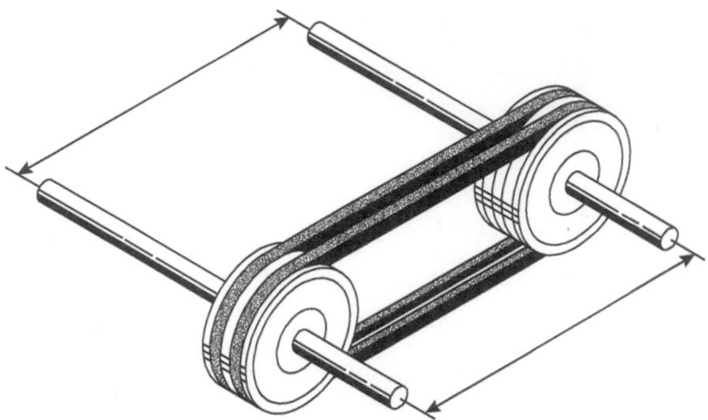

Figure 14-5 Measuring for accurate shaft alignment.

V-Belt Installation

When installing V-belts, they should never be *run on*. This practice places excessive stress on cords in the strength section of the belts, usually straining or breaking some of them. A belt damaged in this manner will flop under load and turn over in the sheave groove. Proper method of installation is to loosen the motor-mount bolts, slide the motor forward, and slip the belts loosely over the sheave, as shown in Figure 14-6.

The following six general rules should be followed when installing V-belts:

1. Reduce centers so belts can be slipped on sheaves.
2. Have all belts slack on the same side (top of drive).
3. Tighten belts to approximately correct tension.
4. Start unit and allow belts to seat in grooves.
5. Stop and retighten to correct tension.
6. Recheck belt tension after 24 to 48 hours of operation.

Figure 14-6 The incorrect and correct methods of installing V-belts.

Checks for Runout

All installed sheaves should be checked for eccentricity (outside diameter runout) or wobble (face runout), as shown in Figure 14-7.

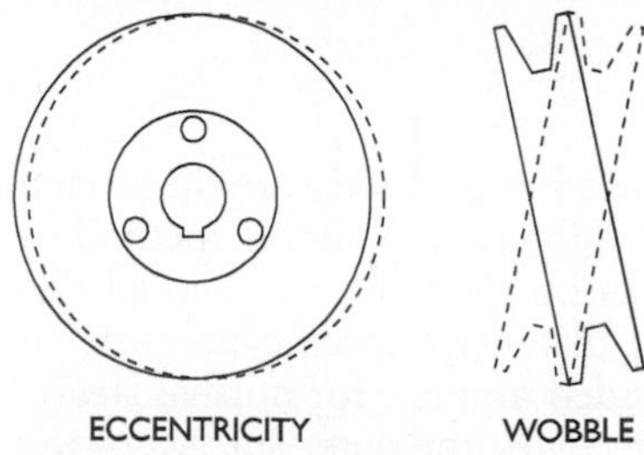

Figure 14-7 Check for runout.

Improperly bored hubs, bent shafts, or improperly installed bushings are the common causes for these conditions. The following standards should be used when checking for wobble or eccentricity using a dial indicator:

1. Radial runout (eccentricity) limits for sheaves:
   Up through 10-in. diameter—0.010 in. maximum.
   For each additional inch of diameter, add 0.005 in.
2. Axial runout (wobble) limits for sheaves:
   Up through 5-in. diameter—0.005 in. maximum.
   For each additional inch of diameter, add 0.001 in.

Checking Belt Tension

Belt tension is a vital factor in obtaining maximum operating efficiency and belt life. V-belt drives must be under proper tension during operation to produce the wedging action in the sheave grooves that give them their pulling power. If belt tension is too low, there will be slippage and rapid wear of both belts and sheave grooves. Slippage is usually apparent by the squealing or howling noise that accompanies it. Correction of slippage should not be attempted by application of belt dressings. These usually contain chemicals that tend to soften V-belts. Although immediate improvement or slippage reduction may be accomplished, eventually deterioration of the rubber compounds will shorten belt life.

The tensioning of fractional horsepower and standard-multiple belts may be accomplished satisfactorily by tightening until the proper "feel" is attained. The proper "feel" might be described as a lively, springy action of the belt. A way to check this is to strike the belts with the hand. They will feel alive and springy if properly tightened. If there is insufficient tension, the belt will feel loose or dead when struck. Too much tension will cause the belts to feel taut, as there will be no give to them.

Percent Elongation Method

A more precise way of checking belt tension is to measure the stretch or elongation of the belts as they are tightened. A simple method of stretch measurement is to place marks on the belts a measured distance apart and tighten to increase the distance a specific amount. An increase of 2 percent is the recommended amount for normal steady drives. For example, before mounting, mark the outer surface of the untensioned belt with two gauge lines, as shown in Figure 14-8. These lines should be as far apart as possible to obtain maximum accuracy. After the belt is mounted, tighten it until the distance between marks has increased by the required percentage. As indicated in Figure 14-9, the lines that were 100 in. apart on the untensioned belt should become 102 in. apart by tightening and stretching the belt.

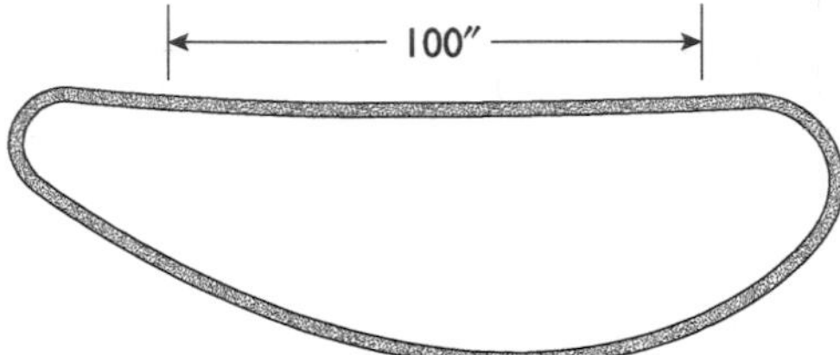

Figure 14-8 Measuring a specified distance on untensioned belts.

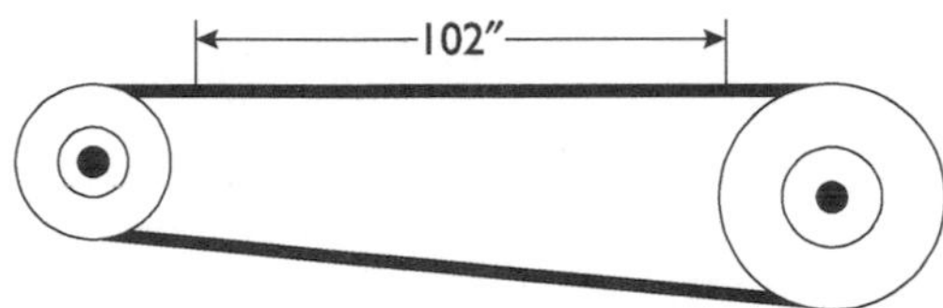

Figure 14-9 Measuring the same distance after belt has been tightened. The measured increase should be 2 percent of the original measurement.

The recommended 2 percent stretch is adequate for normal steady drives. However, additional tension is required for more severe service conditions. The following list of service conditions may be used as a guide to the amount of additional stretch required:

Normal steady drives	2 percent stretch
High starting loads or sudden peak loads	Increase of ½ percent
Extreme temperature or humidity variations	Increase of ½ percent
Extreme dust conditions	Increase of ½ percent
Extreme oily or greasy conditions	Increase of ½ percent
Vertical shafts	Increase of ½ percent
Belt surface speed over 6000 rpm	Increase of ½ percent
Belt surface speed over 8000 rpm	Increase of 1 percent
Total stretch not to exceed 5 percent	

Force-Deflection Method

To assist the mechanic in correctly tensioning V-belts, the use of a belt-tension measuring tool is recommended. Through the use of such a tool, accurate setting can be made and measured.

Belt tension is a vital factor in operating efficiency and service life. Too low a tension results in slippage and rapid wear of both belts and sheave grooves. Too high a tension stresses the belts excessively and unnecessarily increases bearing loads. The following instructions (provided by Maintenance Troubleshooting) show the force-deflection method of tensioning and give an example of proper use of a tension tester for correctly setting belt tension.

Here is an example showing the nine steps for setting the tension of a V-belt using the force-deflection method.

1. Measure the belt span (tangent to tangent) as shown in Figure 14-10.

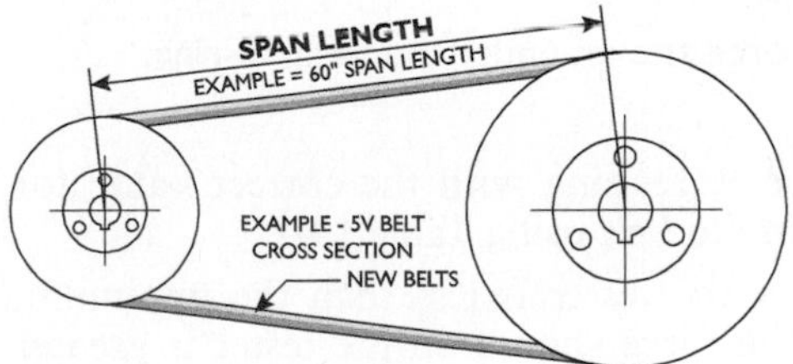

Figure 14-10 Checking span length.

2. Set the large O-ring on the span scale as shown in Figure 14-11.

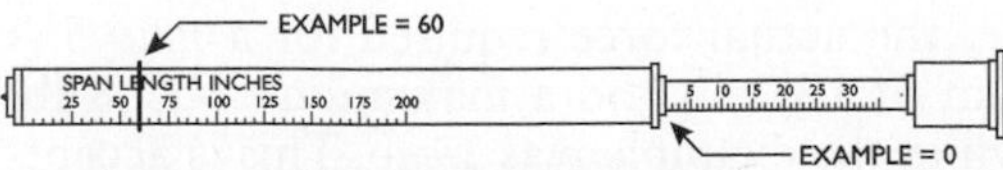

Figure 14-11 Tension tester showing settings of large and small O-rings.

3. Set the small O-ring at zero on the force scale as shown in Figure 14-11.
4. Place the metal end of the tester on one belt at the center of the span length as shown in Figure 14-12.

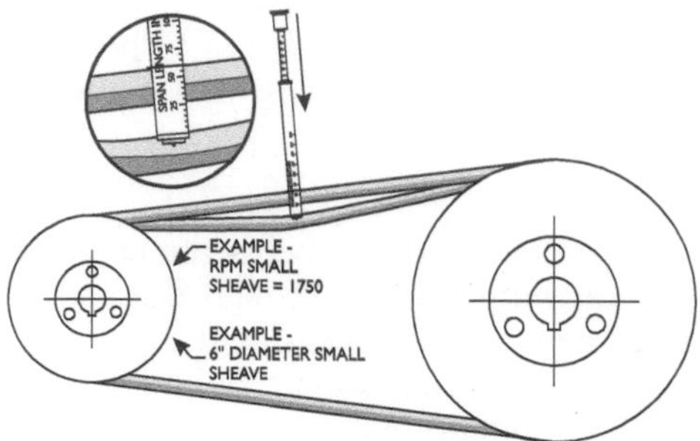

Figure 14-12 Checking tension using the force-deflection method.

5. Apply force to the plunger until the bottom of the large O-ring is even with the top of any adjacent belt as shown in Figure 14-12.
6. Read the force scale under the small O-ring to determine the force used to give the deflection as shown in Figure 14-13.

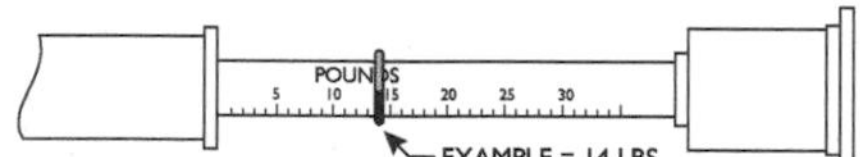

Figure 14-13 Reading the force shown under the small O-ring.

7. Compare the force scale reading with the correct value for either a new belt or an old belt using Table 14-2.
8. If the force shown on the tester is less than the minimum, tighten the belts. If the force shown on the tester is greater than the maximum, loosen the belts.
9. If the chart calls for more tension than one tension tester can deliver, use two or more testers side by side and add the forces together.

Note: In this example, the actual force required for a *new* 5V-belt would be a minimum of 13.2 lb and a maximum of 19.8 lb. The actual reading shown in the example was 14 lb. This is acceptable because it is within the range as shown.

Table 14-2 Force-Deflection Table

Belt X-Section	Smaller Sheave Range: Diameter, in.	Smaller Sheave Range: rpm	Maximum/Minimum Recommended Deflection Force: New Belt Maximum	New Belt Minimum	Old Belt Maximum	Old Belt Minimum
A	3 to 3⅝	1000–2500	8.3	5.5	5.5	3.7
		2501–4000	6.3	4.2	4.2	2.8
	3⅞ to 4⅞	1000–2500	10.2	6.8	6.8	4.5
		2501–4000	8.6	5.7	5.7	3.8
	5 to 7	1000–2500	12.0	8.0	8.0	5.4
		2501–4000	10.5	7.0	7.0	4.7
B	3⅜ to 4¼	850–2500	10.8	7.2	7.2	4.9
		2501–4000	9.3	6.2	6.2	4.2
	4⅜ to 5⅝	850–2500	11.9	7.9	7.9	5.3
		2501–4000	10.0	6.7	6.7	4.5
	5⅞ to 8⅝	850–2500	14.1	9.4	9.4	6.3
		2501–4000	13.4	8.9	8.9	6.0
C	7 to 9	500–1740	25.5	17.0	17.0	11.5
		1741–3000	20.7	13.8	13.8	9.4
	9½ to 16	500–1740	31.5	21.0	21.0	14.1
		1741–3000	27.8	18.5	18.5	12.5
D	12 to 16	200–850	55.5	37.0	37.0	24.9
		851–1500	47.0	31.3	31.3	21.2
	18 to 20	200–850	67.8	45.2	45.2	30.4
		851–1500	57.0	38.0	38.0	25.6
E	21⅝ to 24	100–450	71.0	47.0	47.0	31.3
		451–900	48.0	32.0	32.0	21.3
3V	2¼ to 2⅜	1000–2500	7.4	4.9	4.9	3.3
		2501–4000	6.5	4.3	4.3	2.9
	2⅝ to 3⅝	1000–2500	7.7	5.1	5.1	3.6
		2501–4000	6.6	4.4	4.4	3.0
	4⅛ to 6⅞	1000–2500	11.0	7.3	7.3	4.9
		2501–4000	9.9	6.6	6.6	4.4
5V	4⅜ to 6⅝	500–1740	22.8	15.2	15.2	10.2
		1741–3000	19.8	13.2	13.2	8.8
	7⅛ to 10⅞	500–1740	28.4	18.9	18.9	12.7
		1741–3000	25.0	16.7	16.7	11.2
	11⅞ to 16	500–1740	35.1	23.4	23.4	15.5
		1741–3000	32.7	21.8	21.8	14.6
8V	12½ to 17	200–850	74.0	49.3	49.3	33.0
		851–1500	59.8	39.9	39.9	26.8
	18 to 22⅜	200–850	88.8	59.2	59.2	39.6
		851–1500	79.0	52.7	52.7	35.3

V-Belt Selection

Millwrights and mechanics are seldom required to make initial belt selections; this is usually done by the person who designs the drive. They are, however, frequently required to choose replacement belts. Because the top widths of some of the various sizes in the three V-belt types are essentially the same dimension, errors in choice might be made. In Figure 14-14, the dimension of the various sizes and types of V-belts are charted. Corresponding fractional-horsepower and standard-multiple sizes are also listed.

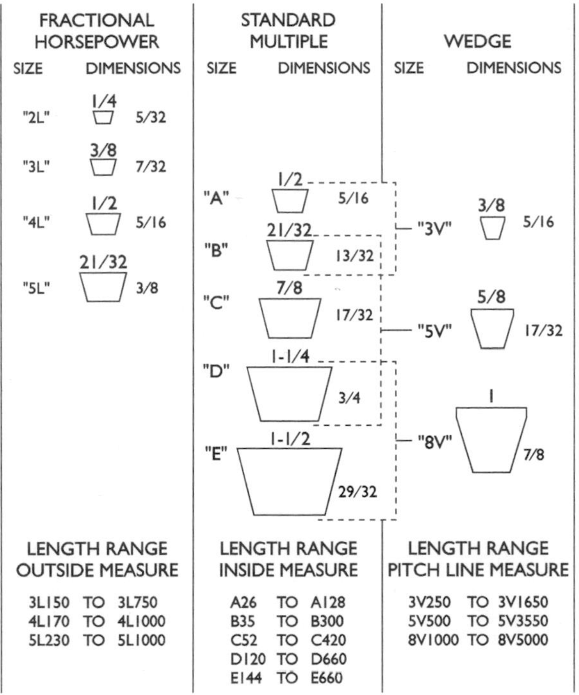

Figure 14-14 Size, dimension, and comparison of V-belts.

The three wedge belt sizes are related to the five standard-multiple sizes by dashed line brackets. The fractional-horsepower belt size designations 2L, 3L, 4L, and 5L are now used by most manufacturers that formerly used the old size designations 1000–2000–3000 or the 0–1–2 size numbers. The length range of standard belts in each size and type is also noted. Length measurements are made on a different surface or line for each type of belt. The fractional-horsepower belt length is measured on the outside surface of the belt. The standard-multiple type is measured on the inside surface, and the wedge type is measured along the pitch line. Actual dimensions of V-belts of various manufacturers may vary somewhat from nominal dimensions. Because of this, belts of different manufacturers should not be mixed on the same drive.

Having to determine the belt length that is required is another situation a mechanic may frequently be faced with. For most drives, it is not necessary to be exact when calculating V-belt lengths. This is because of the adjustment built into the drive and the fact that belt selection is limited to the standard lengths available. As standard lengths vary in steps of several inches, an approximate length calculation is usually adequate. Therefore, the following method of calculating belt length can be used for most V-belt drives:

1. Add the two sheave pitch diameters and multiply by 1.5.
2. To this figure, add twice the distance between centers.
3. Select the nearest longer standard belt length.

In Figure 14-15, an example of belt-length calculation is illustrated. This method of calculation is more exact when 1.57, or $\frac{1}{2}\pi$, is used as a multiplier rather than 1.5.

In cases where an accurate length must be determined, the following formula is recommended. (This might be when centers are fixed or when extreme pitch-diameter differences and short center distances are involved.)

$$\text{Belt Pitch Length} = 2C + 1.57\,(D + d) + \frac{(D - d)}{4C}$$

where

C = center distance

D = large sheave pitch diameter

d = small sheave pitch diameter

An example of the use of this formula is shown in Figure 14-16.

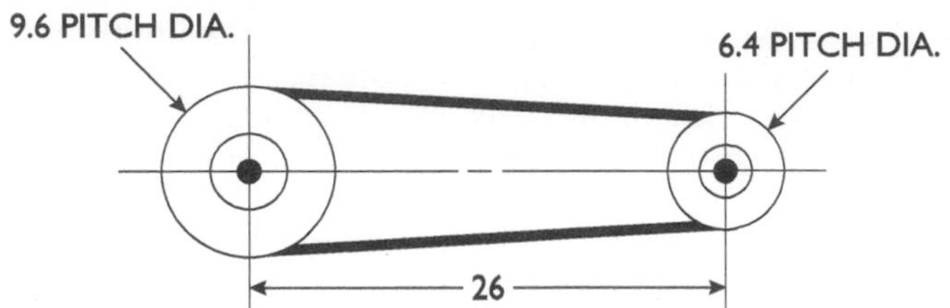

Figure 14-15 Calculating belt size.

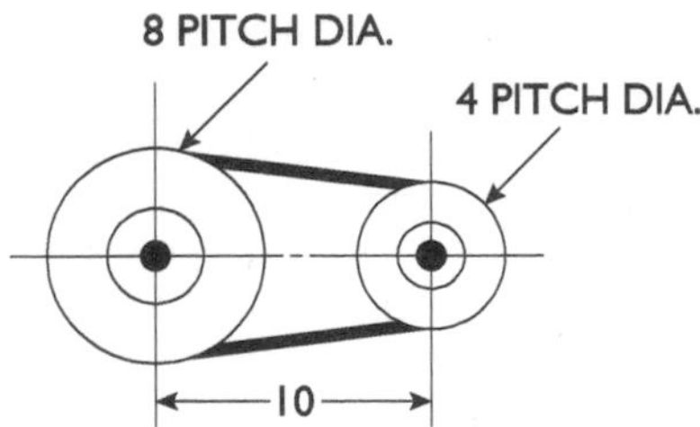

Figure 14-16 A more accurate system of determining belt length.

Standard Belt Lengths

Fractional-horsepower standard belt lengths vary by one-inch increments between the minimum and maximum length for each size. In addition to the belts of even-inch lengths, some manufacturers' stock lists include belts made to fractional-inch lengths.

The belt numbers indicate the nominal *outside length* of the belt; the last digit of the number indicates *tenths of an inch*.

The actual pitch-line length of standard-multiple belts may be from one to several inches greater than the nominal length indicated by the belt number. This is because belt numbers indicate the length of the belt along its inside surface. Belt-length calculations are in terms of a belt-length measurement on the pitch line. Table 14-3 gives actual pitch-line lengths and should be used when choosing a belt. The table lists pitch-line lengths for only the most commonly used belt sizes. However, some longer and/or shorter lengths than those listed are available from some manufacturers.

Wedge-belt numbers indicate the effective pitch-line length of the belt, so the nominal belt number can be used when choosing belts. Table 14-4 lists the wedge-belt lengths that are commonly available. Again, some longer and/or shorter lengths than those listed are available from some manufacturers.

Table 14-3 Standard V-Belt Lengths, Standard-Multiple Type—Standard Pitch Lengths

Belt Number Indicating Nom. Length	A	B	C	D	E
26	27.3				
31	32.3				
33	34.3				
35	36.3	36.8			
38	39.3	39.8			
42	43.3	43.8			
46	47.3	47.8			
48	49.3	49.8			
51	52.3	52.8	53.9		
53	54.3	54.8			
55	56.3	56.8			
60	61.3	61.8	62.9		
62	63.3	63.8			
64	65.3	65.8			
66	67.3	67.8			
68	69.3	69.8	70.9		
71	72.3	72.8			
75	76.3	76.8	77.9		
78	79.3	79.8			
80	81.3				
81		82.8	83.9		
83		84.8			
85	86.3	86.8	87.9		
90	91.3	91.8	92.9		
96	97.3		98.9		
97		98.8			
105	106.3	106.8	107.9		
112	113.3	113.8	114.9		
120	121.3	121.8	122.9	123.3	
128	129.3	129.8	130.9	131.3	
136		137.8	138.9		
144		145.8	146.9	147.3	
158		159.8	160.9	161.3	
162			164.9	165.3	

(Continued)

Belt Number Indicating Nom. Length	A	B	C	D	E
173		174.8	175.9	176.3	
180		181.8	182.9	183.3	184.5
195		196.8	197.9	198.3	199.5
210		211.8	212.0	213.3	214.5
240		240.3	240.9	240.8	241.0
270		270.3	270.9	270.8	271.0
300		300.3	300.9	300.8	301.0
330			330.9	330.8	331.0
360			360.9	360.8	361.0
390			390.9	390.8	391.0
420			420.9	420.8	421.0
480				480.8	481.0
540				540.8	541.0
600				600.8	601.0
660				660.8	661.0

Table 14-4 Standard V-Belt Lengths, Wedge Type

Effective Pitch Length	*Belt No. 3V*	*Belt No. 5V*	*Belt No. 8V*
25	3V250		
26½	3V265		
28	3V280		
30	3V300		
30½	3V305		
33½	3V335		
37½	3V375		
40	3V400		
42½	3V425		
45	3V450		
47½	3V475		
50	3V500	5V500	
53	3V530	5V530	
56	3V560	5V560	
60	3V600	5V600	

Effective Pitch Length	**Belt No. 3V**	**Belt No. 5V**	**Belt No. 8V**
63	3V630	5V630	
67	3V670	5V670	
71	3V710	5V710	
75	3V750	5V750	
80	3V800	5V800	
85	3V850	5V850	
90	3V900	5V900	
95	3V950	5V950	
100	3V1000	5V1000	8V1000
106	3V1060	5V1060	8V1060
112	3V1120	5V1120	8V1120
118	3V1180	5V1180	8V1180
125	3V1250	5V1250	8V1250
132	3V1320	5V1320	8V1320
140	3V1400	5V1400	8V1400
150		5V1500	8V1500
160		5V1600	8V1600
170		5V1700	8V1700
180		5V1800	8V1800
190		5V1900	8V1900
200		5V2000	8V2000
212		5V2120	8V2120
224		5V2240	8V2240
236		5V2360	8V2360
250		5V2500	8V2500
265		5V2650	8V2650
280		5V2800	8V2800
300		5V3000	8V3000
315		5V3150	8V3150
335		5V3350	8V3350
355		5V3550	8V3350
375			8V3750
400			8V4000
425			8V4250
450			8V4500

Maintenance of V-Belts

Satisfactory service from V-belts requires that the installation be performed properly and, equally important, that the drives be properly maintained. Proper maintenance involves not just the drive components but the surrounding conditions as well. Most V-belt operating problems can be avoided or alleviated when the basic cause is determined.

Heat is the most destructive enemy of rubber. Most belts will operate efficiently in ambient temperatures as high as 140°. Specially constructed belts are designed for service up to 180°. Above these temperatures, the life expectancy of the belt decreases at a rapid rate. Although V-belts will operate satisfactorily at temperatures as high as 300°, their life expectancy will only be a fraction of normal.

Excessive heat causes cracking or checking on the bottom surface of the belt. This characteristic cracked appearance is the first sign of eventual failure. When this type of damage is detected, replacement should be made at the first opportunity.

Belt drives, to avoid overheating, must be adequately ventilated. The use of solid guards should be avoided because they do not allow sufficient air circulation. Guards should be made of perforated or expanded metal as shown in Figure 14-17. This style of guard can perform the important guarding function with minimum interference with heat dissipation.

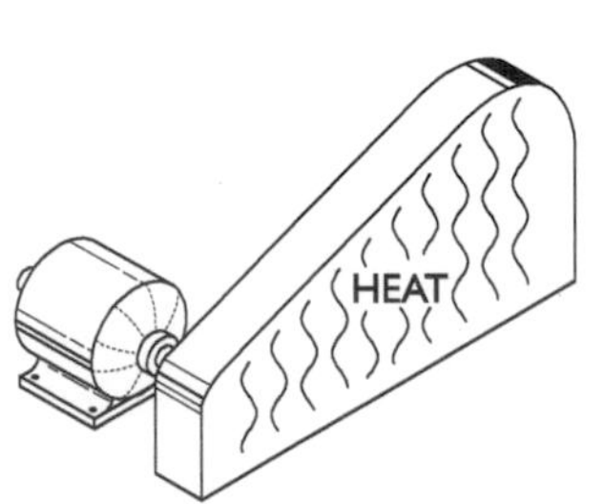

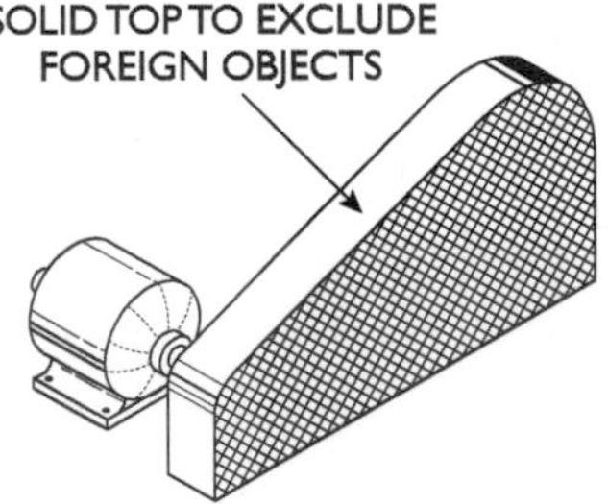

Figure 14-17 Proper ventilation for belt operation.

Oil and grease cause the rubber in regular V-belts to soften, swell, and deteriorate very rapidly. Because of this, belts must be protected against these dangers. On those applications where contact with oil or grease cannot be avoided, special belts that are resistant to oils should be used. Some mechanics feel that a drop or two of oil or belt dressing on belts will give them greater grip and make them last longer. This is not true. Oil and belt dressing should not be used because foreign substances of any kind weaken belt grip. Because their grip depends on friction, belts operate best when they, and the sheave grooves, are smooth and dry.

All belts and sheaves will wear to some degree with use. As wear occurs, the belts will ride lower in the grooves. When inspecting sheaves for wear, a general rule is that a new belt should not seat more than 1⁄16 in. below the top of the sheave groove, as shown in Figure 14-18.

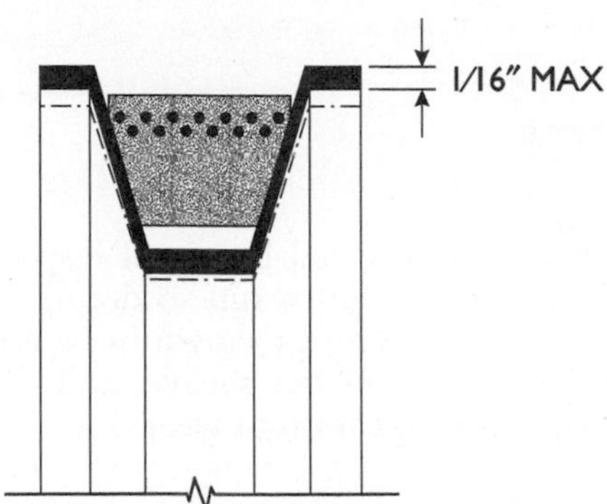

Figure 14-18 Illustrating sheave wear.

As wear occurs at the contact surface on the sides of the grooves, a dished condition develops. When this happens, the belt's wedging action is reduced, gripping power is lost, slippage occurs, and wear of the belts and the sheave grooves is accelerated. Before wear on the sidewalls of the sheave grooves reaches 1⁄32 in., as shown in Figure 14-19, the sheave should be repaired or replaced. If continued in operation, the bottom shoulders on the sidewalls will quickly chew off the bottom corners of the new belts.

In a well-engineered and well-maintained belt drive, sheave-groove sidewall wear is extremely slight. However, wear can be quite rapid if abrasive dusts are present, particularly when belt slippage occurs. Sheave wobble (which results in belt whip) and misalignment produce rapid wear of both sheave grooves and belts. If moisture or chemicals are present, corrosion of the highly polished sidewalls can also contribute to wear. Because corrosion is polished away by the action of the belts, it may be unnoticed. Excessive

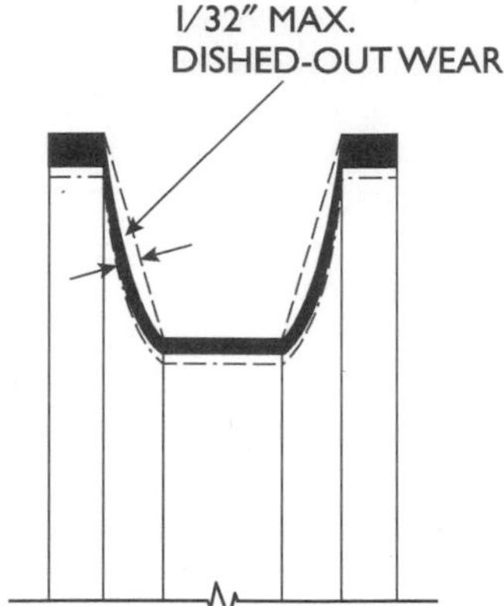

Figure 14-19 Side wear on a sheave.

sheave-groove wear may sometimes be detected by visual inspection. Extreme groove wear will allow the belt to run so deeply in the groove that it contacts the sheave bottom, as shown in Figure 14-20. If this occurs, it is usually on the smaller sheave, and the resultant shiny groove bottom is a sure indication of wear.

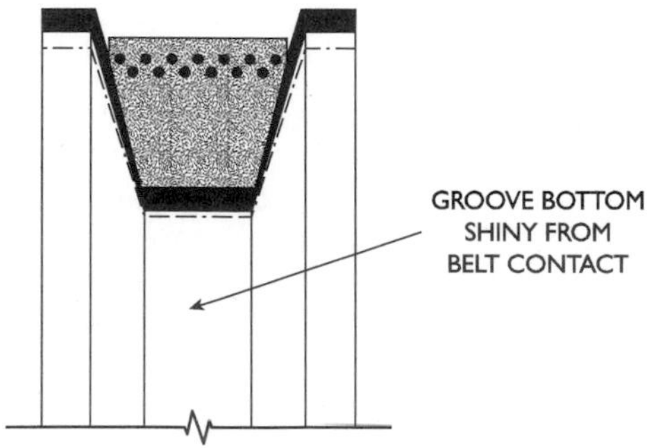

Figure 14-20 Groove bottom shiny because of belt wear.

Unequal groove wear has the same effect on a drive as unmatched belts. A new belt in a worn groove will be loose and will not pull its share of the load. When changing belts, an accurate sidewall-wear check should be made. This is done with a sheave-groove gauge or template as shown in Figure 14-21. Some belt manufacturers provide inexpensive cardboard templates for this purpose.

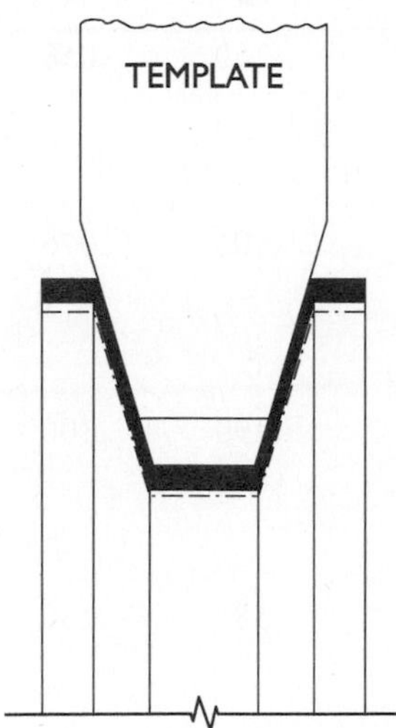

Figure 14-21 Checking sheave walls for wear.

When checking sheave grooves, care must be taken that the correct gauge or template is used. It must be correct not only with respect to the type and size of belt but also for the pitch diameter of the sheave being checked.

As the belt bends around the sheave, the inside, or compression, section is compressed and a slight bulging occurs. This causes a small change in the side angle of the belt. The smaller the diameter of the sheave, the greater the bulge and angle change. Sheave-groove side angles are designed to conform to this change. The small-diameter sheave will have a smaller angle on the groove side-walls than the larger sheaves.

It is very important that the correct dimensions and angles be maintained when worn sheave grooves are repaired. Tables 14-5 and 14-6 give sheave-groove dimensions and angles for fractional-horsepower, multiple, and wedge belts.

Mechanical interference can also cause serious damage to belts. This condition is easily detected because it is accompanied by a ticking noise (if the contact is slight) or a heavy slapping or rubbing

Table 14-5 Sheave-Groove Dimensions—Fractional Horsepower

Size	Outside Diameter	Groove Angle	W	D	X
2L	Under 1.5	32°	.240	.250	.050
	1.5 to 1.9	34°	.240		
	2.0 to 2.5	36°	.246		
	Over 2.5	38°	.250		
3L	Under 2.2	32°	.360	.406	.075
	2.2 to 3.2	34°	.364		
	3.2 to 4.2	36°	.368		
	Over 4.2	38°	.372		
4L	Under 2.6	30°	.485	.490	.100
	2.6 to 3.2	32°	.490		
	3.2 to 5.6	34°	.494		
	Over 5.6	38°	.504		
5L	Under 3.9	30°	.624	.580	.150
	3.9 to 4.9	32°	.630		
	4.9 to 7.3	34°	.637		
	Over 7.3	38°	.650		

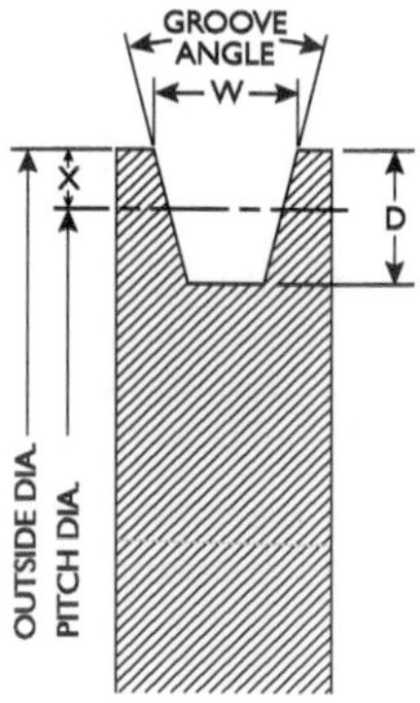

noise (where contact is heavier). In situations where rubbing is severe, heat is generated and actual burning may result. Other mechanical deficiencies to check for are wobbling sheaves (with resultant belt whip), chips out of cast iron sheaves, or dents and bends in pressed metal sheaves.

A troubleshooting checklist (at the end of this chapter) will give causes and corrections of basic troubles. By using this list, trouble causes can be recognized and the appropriate corrective action determined. In those situations where belt failure occurs without opportunity for prior troubleshooting, the failure cause should be determined before belts are replaced. The first step is inspection of the belts. The failure cause can usually be spotted by the characteristic effect it has on the belts. The troubleshooting chart will help in determining the cause of failure and the probable reason. Inspection of the drive and a check of operating conditions will indicate what, if any, action is necessary.

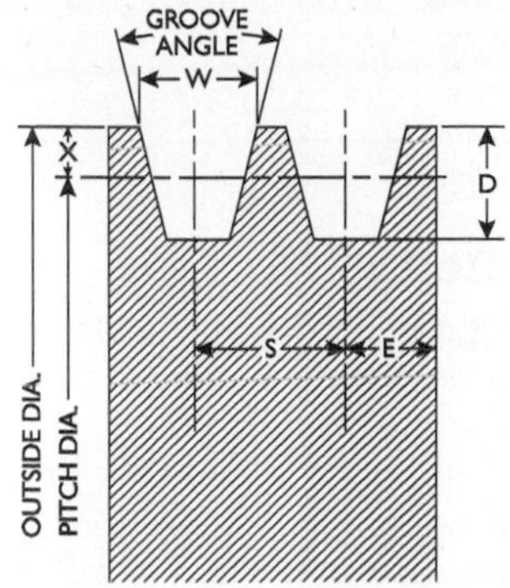

Table 14-6A Sheave-Groove Dimensions—Standard Multiple

Size	*Outside Diameter*	*Groove Angle*	*W*	*D*	*X*	*S*	*E*
A	2.6 to 5.4	34°	.494	.490	.125	5/8	3/8
	Over 5.0	38°	.504	.490	.125	5/8	3/8
B	4.6 to 7.0	34°	.637	.580	.175	3/4	1/2
	Over 7.0	38°	.650	.580	.175	3/4	1/2
C	7.0 to 7.99	34°	.879	.780	.200	1	11/16
	8.0 to 12.0	36°	.887				
	Over 12.0	38°	.895				
D	12.0 to 12.99	34°	1.259	1.050	.300	1 1/16	7/8
	13.0 to 17.0	36°	1.271				
	Over 17	38°	1.283				
E	18.0 to 24.0	36°	1.527	1.300	.400	1 3/4	1 1/8
	Over 24.0	38°	1.542	1.300	.400	1 3/4	1 1/8

Table 14-6B Sheave-Groove Dimensions—Wedge

Size	*Outside Diameter*	*Groove Angle*	*W*	*D*	*X*	*S*	*E*
3V	Under 3.5	36°	.350	.350	.025	13⁄32	11⁄32
	3.5 to 6.0	38°					
	6.0 to 12.0	40°					
	Over 12.0	42°					
5V	Under 10.0	38°	.600	.600	.050	11⁄16	½
	10.0 to 16.0	40°					
	Over 16.0	42°					
8V	Under 16.0	38°	1.000	1.000	.100	1⅛	¾
	16.0 to 22.4	40°					
	Over 22.4	42°					

Micro-V and Common-Backed Belts

In the past few years, the micro-V-belt has gained popularity. Often used as a drive on an automobile engine to power the alternator, air conditioner, and other accessories from the crankshaft pulley, the micro-V-belt appears as a tiny row of "V" shapes molded to a flat common backing.

Wedge-type and conventional V-belts are also made in a banded configuration (Figure 14-22). These belts provide more uniform power transmission on multiple belt drives. Two or more belts are connected at the top by a layer of rubber and fabric. The connecting layer helps to stabilize the belt.

Figure 14-22 Banded belts.

Troubleshooting V-Belt Problems

A properly designed and maintained V-belt drive should provide a belt life of three to five years and sheave service of 10 to 20 years, depending on the application. Service life that falls far short of these applications warrants investigation. Table 14-7 will help prevent a recurrence of many V-belt problems.

Table 14-7 Troubleshooting V-Belt Problems

Symptom	*Cause*	*Correction*
Broken belt	Shock loads or heavy starting load	Check tension. It may be necessary to redesign the drive for increased belt capacity.
Excessive heat	Belt slip	Check tension.
	Misalignmen	Check alignment of sheaves and shafts.
	Objects falling on belt or getting wedged into sheave groove	Protect the drive with a fine-wire mesh guard.
	Broken tension cords	Probably caused by prying belt onto sheave. Follow correct procedure for belt installation.
Belt squeal	Not enough tension	Increase tension. Check with tension tester.
	Insufficient arc of contact (belt does not have enough wraparound sheave to transmit the load)	Investigate the possibility of redesigning the drive for a decreased sheave ratio.
Premature sheave wear	Dirty environment	Use a closed-at-the-top belt guard.
Irregular belt vibration	Insufficient tension	Tension using tension tester.
	Unbalanced shaft or rotors	Use vibration analysis equipment to test for unbalance. Send out for balancing or perform field balance in place.
	Sheaves not balanced	Purchase dynamically balanced sheaves.
	Framework or shafting too light	Investigate possibility of replacing with heavier components or beefing up existing ones.
Belt turnover	Broken tension cords	Probably because of poor installation where prying was done. Follow correct installation procedures.
	Overloaded drive	Consider redesign.
	Misalignment of sheave or shaft	Perform corrective alignment with straightedge or string.
	Worn sheave grooves	Replace sheaves.

Positive-Drive Belts

Positive-drive belts, also called timing belts and gear belts, combine the flexibility of belt drives with the advantages of chain and gear drives. Power is transmitted by positive engagement of belt teeth with pulley grooves, as in chain drives, rather than by friction as in belt drives. This positive engagement of belt teeth with pulley grooves eliminates slippage and speed variations. There is no metal-to-metal contact and no lubrication required. See Figure 14-23.

Figure 14-23 A positive-drive belt, sometimes called a gear belt. *(Courtesy Emerson Power Transmission/Browning)*

The positive-drive belt is constructed with gear-like teeth that engage mating grooves in the pulley. Unlike most other belts, they do not derive their tensile strength from their thickness. Instead, these belts are built thin, their transmission capacity resulting from the steel-cable tension members and tough molded teeth shown in Figure 14-24.

Figure 14-24 Exploded view of gear belt. *(Courtesy Emerson Power Transmission/Browning)*

High-strength synthetic materials are used in positive-belt construction. Figure 14-25 shows the four basic components: neoprene backing, steel tension members, neoprene teeth, and nylon facing. The nylon covering on the teeth is highly wear-resistant and, after a short "run-in" period, becomes highly polished with resultant low friction. Belts are constructed so that tooth strength exceeds the tensile strength of the belt when 6 or more teeth are meshed with a mating pulley.

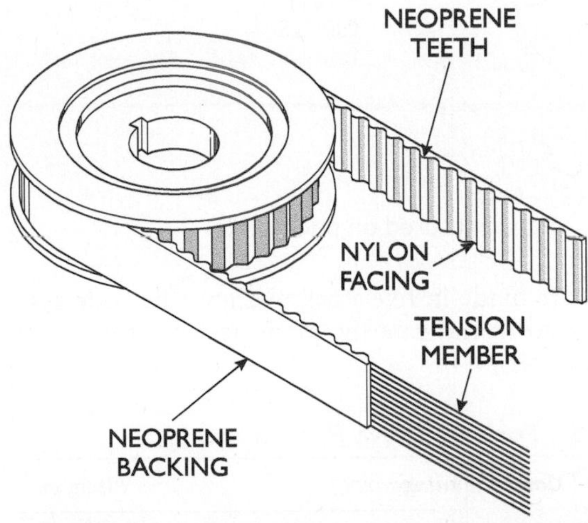

Figure 14-25 The four basic components used in the construction of the gear belt.

With positive-drive belts, as with gear and chain drives, pitch is a fundamental consideration. In this case, *circular pitch* is the distance between tooth centers (measured on the pitch line of the belt) or the distance between groove centers (measured on the pitch circle of the pulley) as indicated in Figure 14-26.

The *pitch line* of a positive-drive belt is located within the cable tension members. The *pitch circle* of a positive-drive pulley coincides with the *pitch line* of the belt mating with it. The pulley pitch diameter is *always greater* than its face diameter. All positive-drive belts must be run with pulleys of the same pitch. A belt of one pitch *cannot* be used successfully with pulleys of a different pitch.

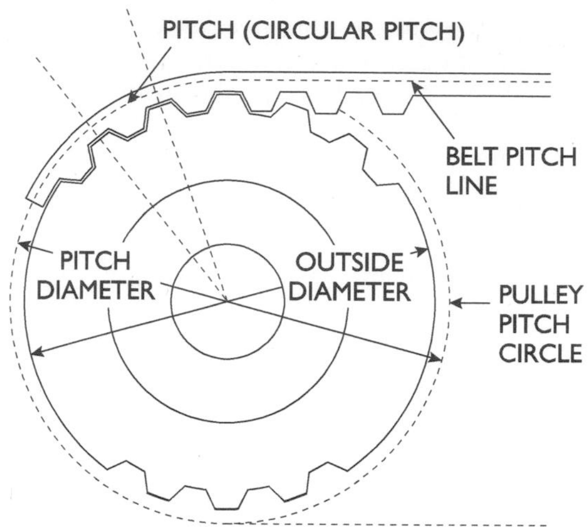

Figure 14-26 Circular pitch, measured on pitch circle.

Positive-drive belts are made in five stock pitches. The code system in Table 14-8 is used to indicate the pitch of a positive-drive system.

Table 14-8 Positive Drive Pitch Spacing

Code	*Code Meaning*	*Pitch, in.*
XL	Extra Light	1/5
L	Light	3/8
H	Heavy	1/2
XH	Extra Heavy	7/8
XXH	Double Extra Heavy	1 1/5

The standard positive-drive belt numbering system is made up of three parts: (1) the pitch length of the belt, which is the actual pitch length multiplied by 10; (2) the code for the pitch of the drive; and (3) the belt's width multiplied by 100.

For example, 390 L 100 (for the belt shown in Figure 14-2) is made up of the following parts:

390—pitch length (39.0 in. multiplied by 10)

L—pitch of drive (light, 3/8-in. pitch)

100—belt width (1 in. multiplied by 100)

Table 14-9 shows the stock sizes of positive-drive belts carried by most manufacturers. Much wider and longer drives can be furnished on special order.

Table 14-9 Positive Drive Lengths

Code	*Pitch, in.*	*Stock Widths, in.*	*Length Range, in.*
XL	1/5	1/4–1/16–3/8	6–26
L	1/8	1/2–3/4–1	12–60
H	1/2	3/4–1–1 1/2–2–3	24–170
XH	7/8	2–3–4	50–175
XXH	1 1/4	2–3–4–5	70–180

Because of a slight side thrust of positive-drive belts in motion, at least one pulley in a drive must be flanged. When the center distance between shafts is eight or more times the diameter of the small pulley, or when the drive is operating on vertical shafts, both pulleys should be flanged.

High-Torque Drive Belts

A new device used in the positive-drive arena is the high-torque drive (HTD) or curvilinear drive. Similar to a timing belt, an HTD drive uses a circular tooth design to allow teeth to enter and leave the mating sprocket grooves in a smooth rolling manner, functioning in much the same manner as the teeth in a gear. The circular tooth design improves stress distribution and allows higher overall loading (Figure 14-27). These belts are available in 8-, 14-, and 20-mm widths.

Belt Installation

Satisfactory operation of positive-drive belts requires that shafts and pulleys be carefully aligned. This is especially important on heavily loaded drives because the belt has a tendency to move to one side of the pulley. Shaft parallelism should be carefully checked with a feeler bar or a millwright's measuring stick. Pulley alignment may be checked with a straightedge, with care being taken to avoid any part of the pulley flanges that may be bent or dented. On long center drives, if the belt tends to crowd to one side, it may be necessary to offset the driven pulley to compensate. Shafts should not be moved out of parallel to cause the belt to climb away from the side it crowds. This can result in sudden reversal of the crowding tendency and cause serious damage to the drive.

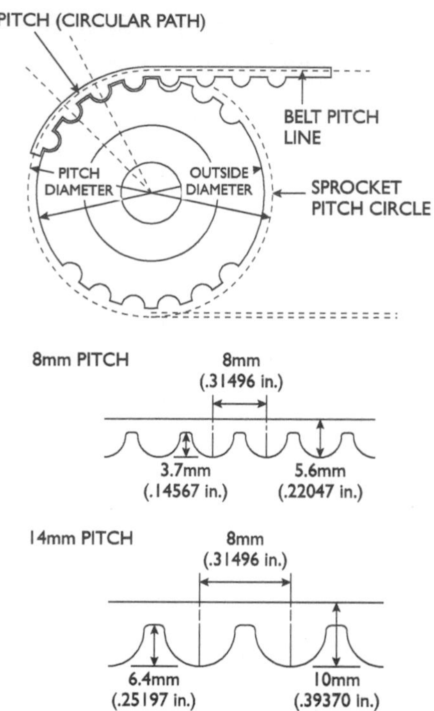

Figure 14-27 Typical HTD drive.

When installing positive-drive belts, the center distance should be reduced and the belt placed loosely on the pulleys. Belts should not be forced in any way over the pulley or flange, as damage to the belt will result. The belt should be tightened to a snug fit. Because the positive-drive belt does not rely on friction, there is no need for high initial stress. However, if torque is unusually high, a loose belt may "jump grooves." In such cases, the tension should be increased gradually until satisfactory operation is attained. Care must be exercised so that shaft parallelism is not disturbed while doing this. On heavily loaded drives with wide belts, it may be necessary to use a tension-measuring tool to accurately tighten the belt. Belt manufacturers should be consulted for their recommendations of equipment and procedures to follow for special situations of this type.

Chapter 15

Chain Drives

Chain drives are used to transmit mechanical power in practically all industrial plants. They consist of a driving sprocket, one or more driven sprockets, and an endless chain whose links mesh with the sprocket teeth. Chain drives do not slip or creep, and so they maintain a positive speed ratio between the driving and driven sprockets. Their principal advantages are their simplicity, economy, efficiency, and adaptability. Probably the most widely used style of chain for power transmission is the *roller chain*. It is available in single- and multiple-strand construction and in a wide range of sizes. The standard single-strand roller chain is illustrated in Figure 15-1 and the multiple-strand construction in Figure 15-2.

Figure 15-1 A single-strand roller chain.
(Courtesy Rexnord Industries, Inc.)

Figure 15-2 A multiple-strand roller chain.
(Courtesy Rexnord Industries, Inc.)

Roller Chain

A roller chain is composed of an alternating series of *roller links*, as shown in Figure 15-3, and *pin links*, as shown in Figure 15-4. Each link is an assembly of precision parts. A roller link consists of two rollers and two bushings. The bushings are press-fitted into the side members, which are called *link plates*. A pin link consists of two link plates into which are press-fitted two pins.

Figure 15-3 Roller-link section.
(Courtesy Rexnord Industries, Inc.)

Figure 15-4 Pin-link section.
(Courtesy Rexnord Industries, Inc.)

When the parts are assembled, the pins move freely inside the bushings that are surrounded by the freely turning rollers. Figure 15-5 shows the parts in various stages of assembly and the relation of one part to another.

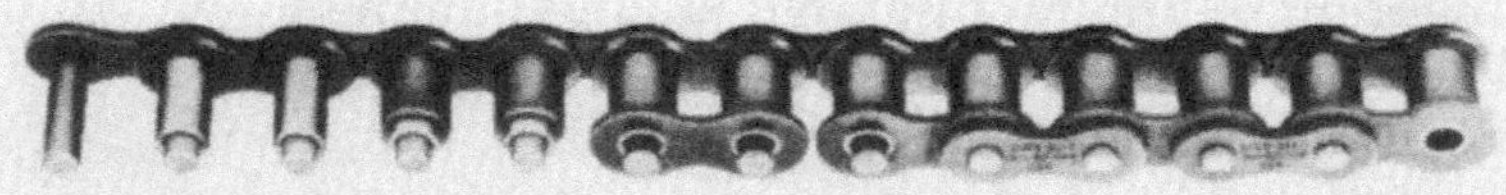

Figure 15-5 Roller links and pin links connected.
(Courtesy Rexnord Industries, Inc.)

Standard roller chain is manufactured to the specifications for "transmission roller chains and sprocket teeth" (ANSI Standard B29.1). Because of this standardization, chains and sprockets of different manufacturers are interchangeable. Also, identification is standardized and replacements may easily be selected from any manufacturer's stock. Although there will be differences in different manufacturers' products, these differences will not affect interchangeability.

Roller-Chain Dimensions

The three principal dimensions used to identify roller chain are *pitch*, *chain width*, and *roller diameter.* Pitch is the distance from a point on one link to a corresponding point on an adjacent link (usually from center to center). Chain width, also called nominal width, is the minimum distance between link plates of a roller link. This dimension determines the width of the sprocket teeth with which the roller links must mesh. The roller diameter is the outside diameter of the roller and is approximately ⅝ of the pitch. Sprocket-tooth form is determined by this dimension. The dimensions of standard roller-chain parts are pitch proportioned, that is, based on a ratio to the pitch dimension. The approximate standard proportions for roller-chain dimensions are as follows:

1. Roller diameter is approximately ⅝ of the pitch.
2. Chain width is approximately ⅝ of the pitch.
3. Pin diameter is approximately 5⁄16 of the pitch.
4. Thickness of link plates is approximately ⅛ of the pitch.

Standard-series chains range in pitch from ¼ to 3 in. Included in the standard are heavy-series chains from ¾-in. pitch and up. They differ from the standard series in the thickness of link plates. Their value is in their capability to carry higher loads at lower speeds. The letter *H* following the number of a chain denotes a chain of the heavy series.

Standard Roller-Chain Numbers

The standard roller-chain numbering system provides a complete identification of a chain by number. The right-hand digit in the chain number is 0 for chain of the usual proportions, 1 for lightweight chain, and 5 for a rollerless bushing chain. The number to the left of the right-hand digit denotes the number of ⅛ in. in the pitch. The letter *H* following the chain number denotes the heavy series. The hyphenated 2 suffixed to the chain number denotes a double-strand chain, 3 a triple-strand chain, etc.

For example, the number 60 indicates a chain with a 6⁄8-in., or ¾-in., pitch of basic proportions; number 41 indicates a narrow, lightweight ½-in.-pitch chain; number 25 indicates a ¼-in.-pitch rollerless chain; and number 120 indicates a chain of basic proportions with a pitch of 12⁄8 in. or 1½ in. In multiple-strand chains, 50-2 designates two strands of 50 chain, 50-3 represents a triple strand, etc. General chain dimensions for standard chain from ¼-in. pitch to 3-in. pitch are tabulated in Table 15-1.

A length of roller chain before it is made endless will normally be made up of an even number of pitches. At either end will be an unconnected roller link with an open bushing. A special type of pin link called a "connecting link" is used to connect the two chain ends. The partially assembled connecting link consists of two pins press-fitted and riveted in one link plate. The pin holes in the free link plate are sized for either a slip fit or a light press fit on the exposed pins, and the plate is secured in place either by a cotter pin, as shown in Figure 15-6, or by a spring clip, as shown in Figure 15-7.

Figure 15-6 Cotter pins secure the plate in place.

Figure 15-7 A spring clip secures the plate in place.

If an odd number of pitches are required, an "offset link," as shown in Figure 15-8, may be substituted for the roller link at one end of the open chain and connected to that of the other end to make the chain endless. By the use of a removable pin, the offset link can also serve as a connecting link.

Figure 15-8 An offset link used to connect chain ends on odd-number pitches.

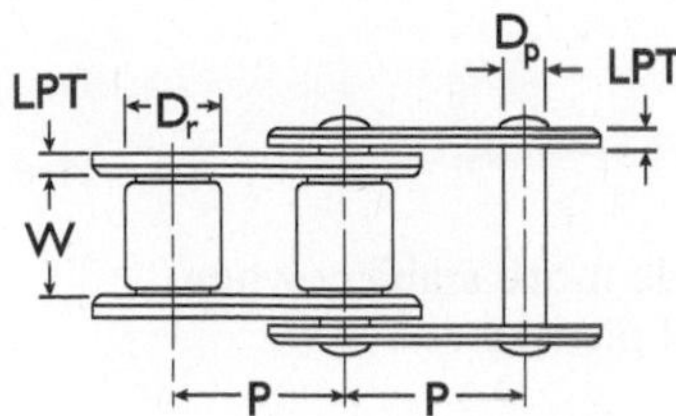

Table 15-1 General Chain Dimensions (in inches)

ANSI Standard Chain No. Std.	ANSI Standard Chain No. Heavy	Pitch (P)	Max. Roller Diameter (D_r)	Width (W)	Pin Diameter (D_p)	Link Plate Thickness (LPT) Std.	Link Plate Thickness (LPT) Heavy	Meas. Load, lbs
25[a]	H	1/4	0.130[a]	1/8	0.0905	0.030		18
35[a]	H	3/8	0.200[a]	3/16	0.141	0.050		18
41[b]	H	1/2	0.306	1/4	0.141	0.050		18
40	H	1/2	5/16	5/16	0.156	0.060		31
50	H	5/8	0.400	3/8	0.200	0.080		49
60	H	3/4	15/32	1/2	0.234	0.094	.125	70
80	80H	1	5/8	5/8	0.312	0.125	.156	125
100	100H	1 1/4	3/4	3/4	0.375	0.156	.187	195
120	120H	1 1/2	7/8	1	0.437	0.187	.219	281
140	140H	1 3/4	1	1	0.500	0.219	.250	383
160	160H	2	1 1/8	1 1/4	0.562	0.250	.281	500
180	180H	2 1/4	1 13/32	1 13/32	0.687	0.281	.312	633
200	200H	2 1/2	1 9/16	1 1/2	0.781	0.312	.375	781
240	240H	3	1 7/8	1 7/8	0.937	0.375	.500	1125

[a] *Without rollers.*

[b] *Lightweight chain.*

Another method of providing an odd number of pitches is by use of the "offset section" shown in Figure 15-9. Its use provides greater stability than the single offset link. Because the pin is riveted in place, a standard connecting link must be used with the offset section.

Figure 15-9 An offset section to provide more stability when connecting chain ends on odd number of pitches.

Roller-Chain Sprockets

Roller-chain sprockets may be described as toothed wheels whose teeth are shaped to mesh with roller chain. So-called cut-tooth sprockets, which have precisely machined teeth, are most suitable for use with roller chain. The dimensions, tolerances, and tooth form for standard precision cut-tooth sprockets are specified in "transmission roller chains and sprocket teeth" (ANSI Standard B29.1). The standard includes four types: Type A is a plain sprocket without hub, type B has a hub on only one side, type C has a hub on both sides, and type D has a detachable hub. The four types are illustrated in Figure 15-10.

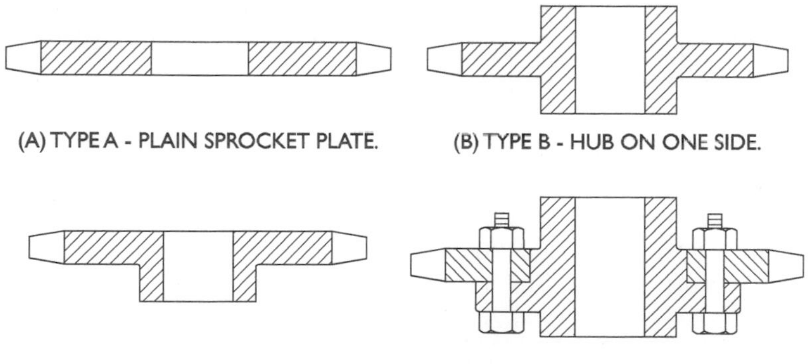

Figure 15-10 The four standard sprocket hubs.

The shear-pin type of sprocket is a special design that is widely used to prevent damage caused by overloads. It consists of a hub, keyed to the shaft, and a sprocket that is free to rotate either on the shaft or on a hub when the driving pin shears. The two types are shown in Figure 15-11.

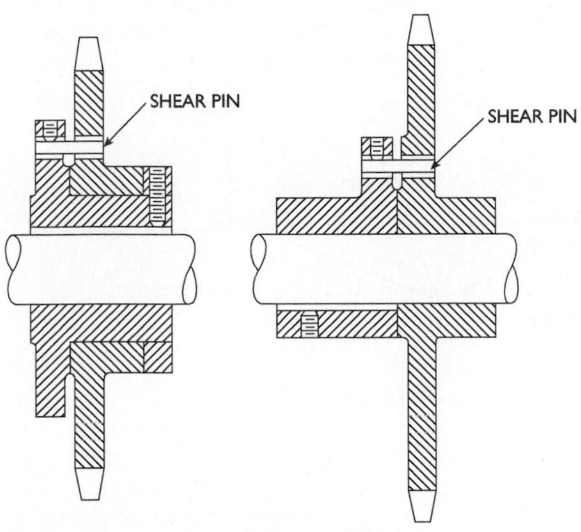

Figure 15-11 Shear-pin sprocket.

The number of teeth in sprockets, rather than their pitch diameters, is used for speed calculations. Although this is similar to gear-drive speed calculations, and many of the same terms are used, their meanings may not be identical. For example, in Figure 15-12, note that the pitch is measured on a straight line between the centers of adjacent teeth; thus, the chain-pitch lines form a series of chords of the pitch circle. The pitch diameter therefore is a function of the chain pitch and the number of teeth in the sprocket; its value may be determined by use of the following equation:

$$PD = \frac{P}{\sin\left[\frac{180°}{N}\right]}$$

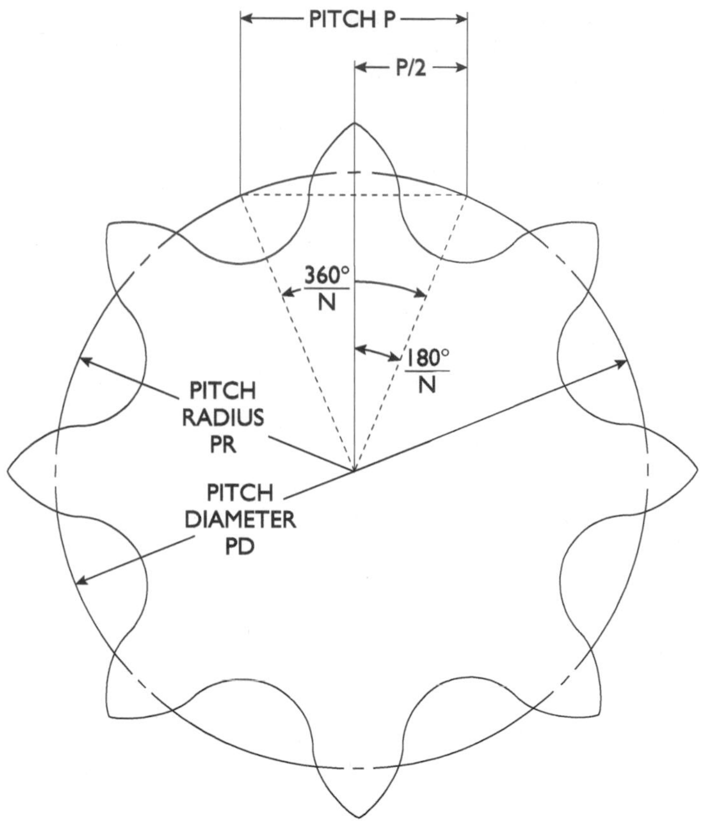

Figure 15-12 Measuring the pitch of sprocket teeth.

Other terms and dimensions are shown in Figure 15-13. The caliper diameter, which applies to sprockets only, is an important measurement used to check the precision to which the bottom diameter is machined. For sprockets with an even number of teeth, it is obviously equal to the bottom diameter. For sprockets with an odd number of teeth, it is the distance from the bottom of one tooth gap to that of the nearest opposite tooth gap.

Roller-Chain Installation

Correct installation of a roller-chain drive requires that the shafts and the sprockets be accurately aligned. Shafts must be set level, or if inclined from a level position, both shafts must be at exactly the

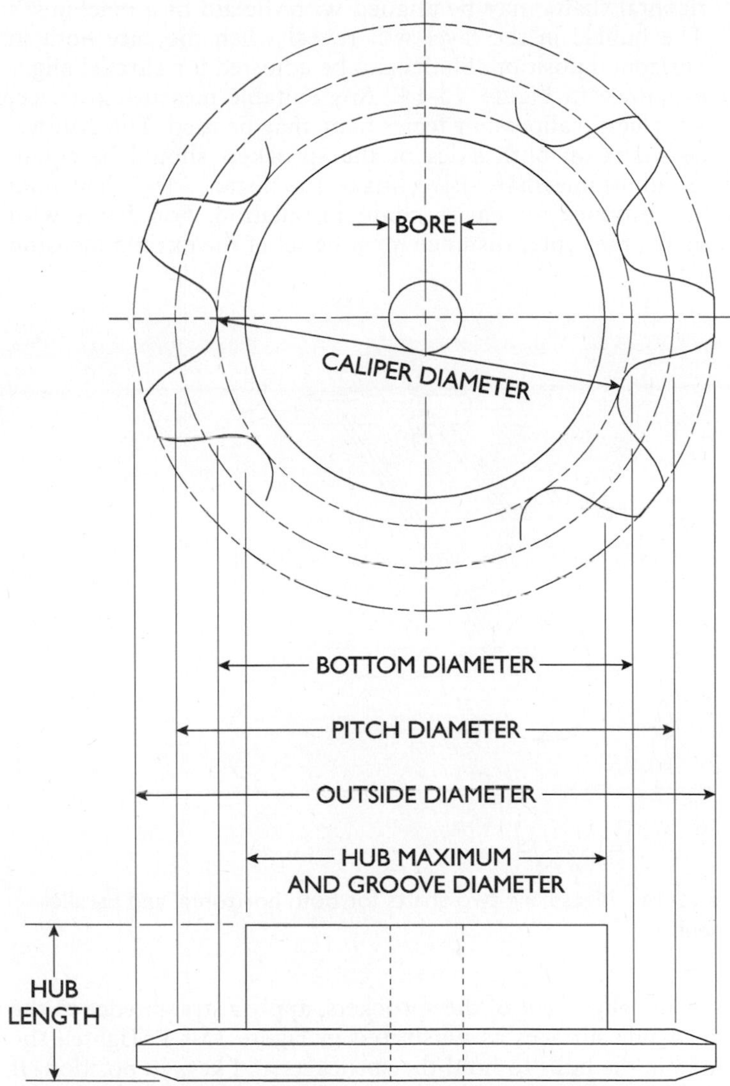

Figure 15-13 Other measurements made on a sprocket.

same angle. They must also be positioned parallel within very close limits. The sprockets must be in true axial alignment for correct sprocket-tooth and chain engagement.

Horizontal shafts may be aligned with the aid of a machinist's level. The bubble in the level will reveal when they are both in exact horizontal position. Shafts can be adjusted for parallel alignment as shown in Figure 15-14. Any suitable measuring device, such as verniers, calipers, or feeler bars, may be used. The distance between shafts on both sides of the sprockets should be equal. With an adjustable shaft drive, make the distance less than final operating distance for easier chain installation. For drives with fixed shafts, the center distance must be set at the exact dimension specified.

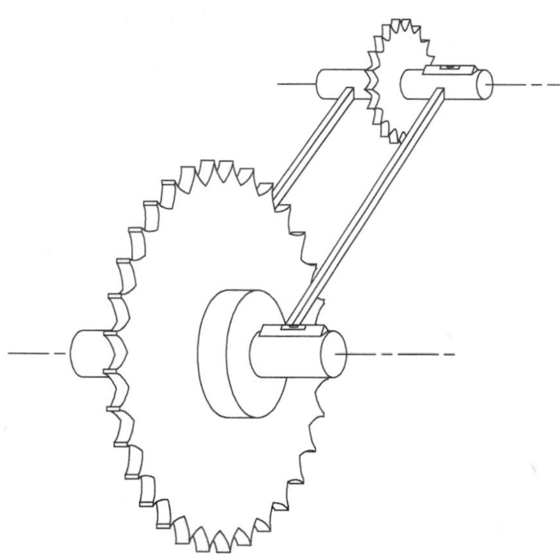

Figure 15-14 Measuring two shafts for both horizontal and parallel alignment.

To ensure alignment of the sprockets, apply a straightedge to the machined side surfaces as illustrated in Figure 15-15. Tighten the setscrews in the hubs to hold the sprockets and keys in position. If one of the sprockets is subject to end float, locate the sprocket so that it will be aligned when the shaft is in its normal running position. If the center distance is too great for the available straightedge, a taut piano wire may be used.

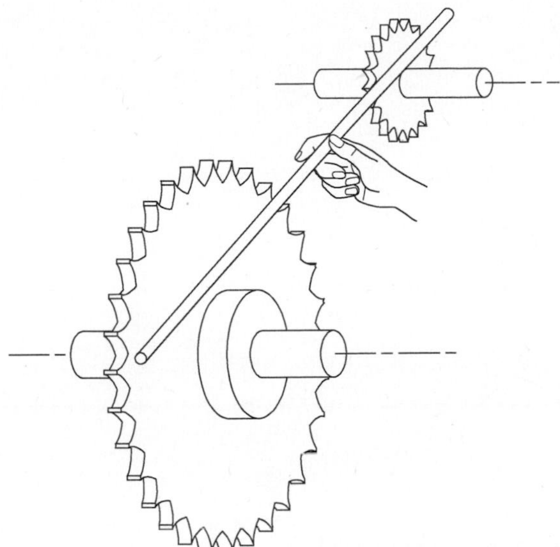

Figure 15-15 Measuring sprocket for alignment.

Before installing the chain, recheck all the preceding adjustments and correct any that may have been disturbed. If a used chain is being reinstalled, inspect it to be sure it is free from dirt or grit. If it is dirty, thoroughly clean it in a solvent. When dry, relubricate it, making sure the oil reaches the pin and bushing surfaces. To install the chain, fit it on both sprockets, bringing the free chain ends together on one sprocket. Insert the pins of the connecting link in the two end links of the chain as shown in Figure 15-16, and then install the free plate of the connecting link. Fasten the plate with the cotters or spring clip, depending on the type used. After the fastener is in position, tap back the ends of the connecting link pins until the outside of the plate rests snugly against the fastener. This will prevent the connecting link from squeezing the sprocket teeth, which might interfere with free flexing of the joint and proper lubrication.

On drives with an adjustable shaft, the shaft must be adjusted to provide proper chain tension. Horizontal and inclined drives should have an initial sag equal to 2 percent of the shaft centers.

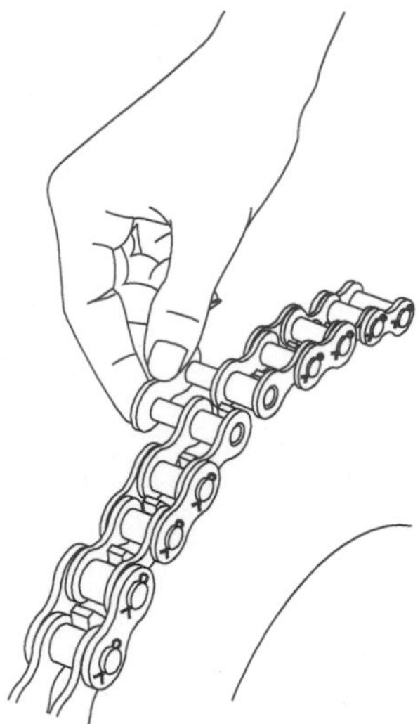

Figure 15-16 Installing the connecting link.

Measurements are made as shown in Figure 15-17. Table 15-2 shows the measurements for various center distances to obtain approximately 2 percent sag.

To determine the amount of sag, pull the bottom side of the chain taut so that all of the "excess" chain will be in the top span. Pull the top tension-free part of the chain down at its center and measure the sag as illustrated in Figure 15-17; then, adjust the centers until the proper amount is obtained. Make sure the shafts are rigidly supported and securely anchored to prevent deflection or movement that would destroy alignment.

Adequate lubrication is one of the most important elements in chain operation. With few exceptions, it is the prime influence on chain life, assuming the chain has been properly selected and installed. The primary purpose of chain lubrication is to reduce wear between moving parts. In addition, the lubricant retards corrosion and rust, reduces sprocket impact, and dissipates heat.

There are a variety of ways to lubricate chain, ranging from manual and gravity drop to semiautomatic and automatic lubrication.

Table 15-2 Sag for Various Center Distances

Shaft Centers, in.	*Sag, in.*	*Shaft Centers, in.*	*Sag, in.*
18	3/8	60	1 1/4
24	1/2	70	1 1/2
30	5/8	80	1 5/8
36	3/4	90	1 7/8
42	7/8	100	2
48	1	125	2 1/2
54	1 1/8		

Whatever means are employed, they should be directed toward accomplishment of the following:

1. Lubrication should be regular and planned.
2. The lubricant should be directed into the chain joints.
3. Chains should be kept clean to allow flow of lubricant in and out of joint clearances.
4. Protect chains from abrasive and corrosive contaminants.
5. The higher the speed is, the more critical lubrication becomes.

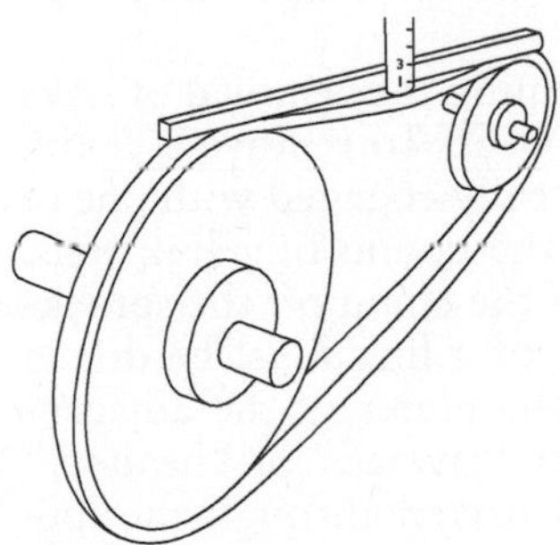

Figure 15-17 Measuring for proper tension.

Roller-Chain Maintenance

A roller-chain drive, like any other mechanical operating equipment, requires maintenance attention to obtain satisfactory performance and long service life. Periodic inspection to discover any incipient faults before major problems develop pays dividends in extended chain life and freedom from shutdowns. Lubrication and cleanliness have already been discussed and are important items to check during inspection. Another important inspection procedure is to check for evidence of wear.

If wear is apparent on the inner surface of the roller-link sidebars, the drive is probably misaligned. Companion wear will probably also be apparent on the side of the sprocket teeth. Alignment of shaft and sprocket should be checked in any case, even if wear is not evident. It may reveal misalignment before wear on chain and sprockets becomes apparent. Excessive tooth wear will change the tooth shape, resulting in a "hooked" appearance. The life of the sprockets in such cases may be extended by reversing them on the shaft to present the less worn tooth surfaces to the chain.

When performing maintenance work on chain drives, certain procedures can cause rapid failure rather than improve operation. An example of this is the installation of a new link in an old chain that has been appreciably elongated by wear; the pitch of the new link will be shorter than that of the other links, and the resulting shock, each time the link engages the sprocket, will destroy the chain. The installation of a new chain on badly worn sprockets will do more damage to the chain in a few hours of operation than would years of normal use.

Maintenance repairs to chain drives require the removal of links to shorten a chain elongated by normal wear. To remove a chain, turn the drive until the connecting link is fully engaged with one of the sprockets to relieve the tension on the connecting link pins. Then, remove the connecting link and lift the chain off the sprockets. To cut the roller chain, the two pins of a link must be driven out of its link plate. At the same time, the plates of the adjacent links must not be distorted. If the pins are "riveted" or "headed," they should be ground off before being driven through the link plates.

The chain detacher, shown in Figure 15-18, is a convenient bench tool for uncoupling precision roller chain. It is designed to prevent damage to the chain during disassembly. It consists of two pieces, a fork and an anvil block. The fork is slotted to fit the chain pitch and roller diameter with tines that slip through the roller link sidebars. The block is of proper size to hold the chain and fork, allowing the pins to be driven squarely out of the sidebar. A detacher is required for each pitch of chain.

Figure 15-18 A chain detacher, used to separate chain. *(Courtesy Rexnord Industries, Inc.)*

There is, of course, a limit to the wear and elongation beyond which the chain must be replaced. Use of an excessively worn chain will result in a tendency for the chain to jump the teeth. In most cases, if the chain is worn to such a degree, quite probably the sprockets are also badly worn and must be replaced to avoid damage to the new chains.

Roller-Chain Replacement

During operation, chain pins and bushings slide against each other as the chain engages, wraps, and disengages from its sprockets. Even when the parts are well lubricated, some metal-to-metal contact does occur, and these parts eventually will wear. This progressive joint wear elongates chain pitch, causing the chain to lengthen and ride higher on the sprocket teeth. The number of teeth in the large sprocket determines the amount of joint wear that can be tolerated before the chain jumps or rides over the ends of the sprocket teeth. When this critical degree of elongation is reached, the chain must be replaced.

Chain manufacturers have established tables of maximum elongation to aid in the determination of when wear has reached a critical point and replacement should be made. By placing a certain number of pitches under tension, elongation can be measured. When elongation equals or exceeds the limits recommended in Table 15-3, the chain should be replaced.

Table 15-3 Maximum Chain Elongation Limits

Chain Number			*Measuring Length*		*Length of Chain When Replacement Is Required*[a]							
					Number of Teeth in Largest Sprocket (T)							
Link-Belt PTC	*ANSI*	*Pitch (in.)*	*Number of Pitches*	*Nominal Length (in.)*	*Up to 67*	*68–73*	*74–81*	*82–90*	*91–103*	*104–118*	*119–140*	*141–173*
RC 35	35	3/8	32	12	12 3/8	12 11/32	12 5/16	12 9/32	12 1/4	12 7/32	12 3/16	12 5/32
RC 40	40	1/2	24	12	12 3/8	12 11/32	12 5/16	12 9/32	12 1/4	12 7/32	12 3/16	12 5/32
RC 50	50	5/8	20	12 1/2	12 7/8	12 19/32	12 13/16	12 25/32	12 3/4	12 23/32	12 11/16	12 21/32
RC 60	60	3/4	16	12	12 3/8	12 11/32	12 5/16	12 9/32	12 1/4	12 7/32	12 3/16	12 5/32
RC 80	80	1	24	24	24 3/4	24 11/16	24 5/8	24 9/16	24 1/2	24 7/16	24 3/8	24 5/16
RC 100	100	1 1/4	20	25	25 3/4	25 11/16	25 5/8	25 9/16	25 1/2	25 7/16	25 3/8	25 5/16
RC 120	120	1 1/2	16	24	24 3/4	24 11/16	24 5/8	24 9/16	24 1/2	24 7/16	24 3/8	24 5/16
RC 140	140	1 3/4	14	24 1/2	25 1/4	25 3/16	25 1/8	25 1/16	25	24 15/16	24 7/8	24 13/16
RC 160	160	2	12	24	24 3/4	24 11/16	24 5/8	24 9/16	24 1/2	24 7/16	24 3/8	24 5/16
RC 180	180	2 1/4	11	24 3/4	25 1/2	25 7/16	25 3/8	25 5/16	25 1/4	25 3/16	25 1/8	25 1/16
RC 200	200	2 1/2	10	25	25 3/4	25 11/16	25 5/8	25 9/16	25 1/2	25 7/16	25 3/8	25 5/16
RC 240	240	3	8	24	24 3/4	24 11/16	24 5/8	24 9/16	24 1/2	24 7/16	24 3/8	24 5/16

[a] *Valid for drives with adjustable centers or drives employing adjustable idler sprockets.*

The recommended measuring procedure is to remove the chain and suspend it vertically. If this is not possible, it may be laid on a smooth horizontal surface and a load applied to remove all slack. In cases when the chain must be measured while on sprockets, remove slack on a span of chain and apply sufficient tension to keep the chain taut. See Figure 15-19.

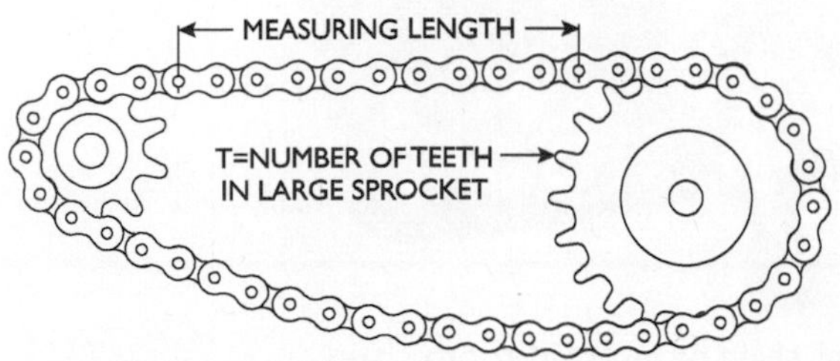

Figure 15-19 Measuring chain to check for possible wear.

Silent-Chain Drives

Silent chain, also called "inverted-tooth" chain, is constructed of leaf links with inverted teeth designed to engage cut-tooth wheels in a manner similar to the way a rack engages a gear. Silent chain combines the flexibility and quietness of a belt and the positive action and durability of gears with the convenience and efficiency of a chain. The links are alternately assembled with either pins or a combination of joint components.

Silent chain is manufactured to specification for "inverted-tooth (silent) chains and sprocket teeth" (ANSI Standard B29.2). This standard is intended primarily to provide for interchangeability between chains and sprockets of different manufacturers. It does not provide for a standardization of joint components and link plate contours, which differ in each manufacturer's design. However, all manufacturers' links are contoured to engage the standard sprocket tooth so that joint centers lie on the pitch diameter of the sprocket. Figure 15-20 shows the style of joint design used in Link-Belt silent chain, using pins and bushings.

Silent chain is manufactured in a wide range of pitches and widths in various styles. Chain under ¾-in. pitch has outside guide links, shown in Figure 15-21, which engage the sides of the sprocket. (Some manufacturers use the term "wheel" rather than sprocket.) Outside-guide-link chain is also termed "side-flange" silent chain.

Figure 15-20 Silent chain or inverted-tooth chain.
(Courtesy Rexnord Industries, Inc.)

Figure 15-21 The side-flange silent chain.
(Courtesy Rexnord Industries, Inc.)

The most widely used style of silent chain is the middle-guide design with one or more rows of guide links that fit guide grooves in wheels. The single-guide or center-guide style of silent chain is illustrated in Figure 15-22.

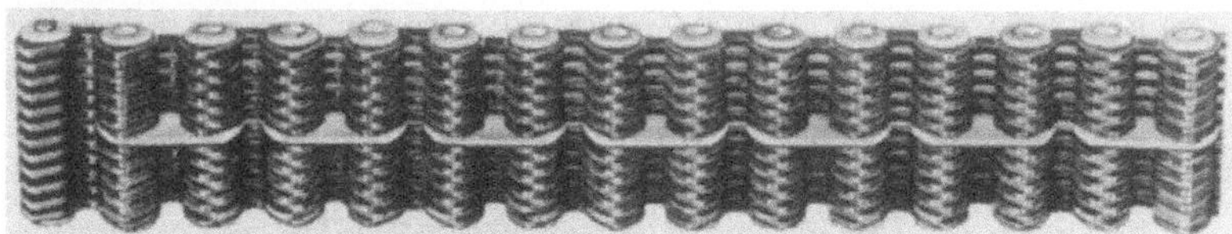

Figure 15-22 The silent chain with a middle guard.
(Courtesy Rexnord Industries, Inc.)

Standard Silent-Chain Numbers

In the United States, standard silent chains are designated by a combined letter-and-number symbol as follows:

1. A two-letter symbol: SC
2. One or two digits indicating the pitch in eighths of an inch
3. Two or three digits indicating the chain width in quarter inches

For example, SC302 designates a silent chain of ⅜-in. pitch and ½-in. width. SC1012 designates a silent chain of 1¼-in. pitch and 3-in. width. Link plates of silent chain manufactured to the ANSI Standard B29.2 are stamped with a symbol indicating the pitch. For example, SC6, or simply 6, indicates a chain with a ¾-in. pitch. Link plates of silent chains not conforming to the ANSI Standard B29.2 are generally stamped with a number of the manufacturer's choice. Standard silent-chain pitches are listed in Table 15-4.

Table 15-4 Silent-Chain Identification Stampings

Chain Number (width in ¼ in.)	*Chain Pitch*	*Stamp*
SC3	⅜	SC3 or 3
SC4	½	SC4 or 4
SC5	⅝	SC5 or 5
SC6	¾	SC6 or 6
SC8	1	SC8 or 8
SC10	1¼	SC10 or 10
SC12	1½	SC12 or 12
SC16	2	SC16 or 16

Offset couplers may be used where a chain length of an odd number of pitches is required or when an adjustment in chain length is desirable and other means of adjustment are not available. Offset couplers are made in two styles, the three-pitch shown in Figure 15-23 and the five-pitch shown in Figure 15-24.

Silent-Chain Sprockets

Silent-chain sprockets are called "wheels" by some manufacturers and may be described as toothed wheels whose teeth are shaped to mesh with the leaf links of silent chain. All teeth are precision-machined to conform to established standards. The dimension, tooth form, and tolerances are specified to "inverted-tooth (silent) chains and sprocket teeth" (ANSI Standard B29.2).

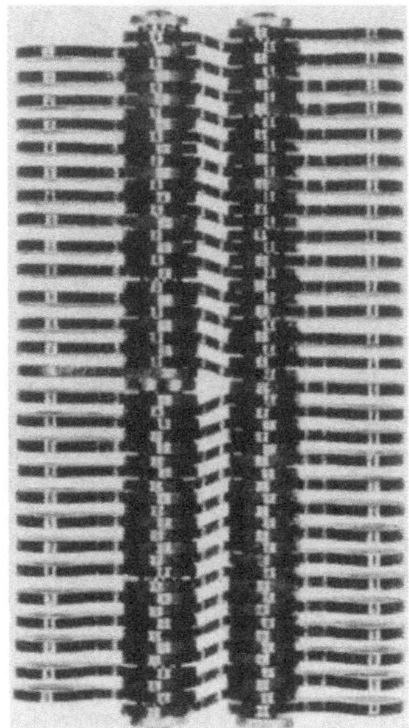

Figure 15-23 Three-pitch silent-chain offset coupler. *(Courtesy Rexnord Industries, Inc.)*

Silent-chain sprockets, or wheels, are manufactured in three basic types identified by the hub arrangement. These are classified as types A, B, and C, similar to the roller-chain sprockets. The type A wheels do not have any hub. Type B wheels have a hub on only one side. Pinions are usually furnished as type B and are machined from steel bar stock. Type C wheels have hubs on both sides. The hub projections are usually equidistant from the centerline of the wheel. Large-diameter wheels are usually furnished as type C. Split- or clamp-hub wheels are used to facilitate installation or replacement. Cross sections of the various types are shown in Figure 15-25.

Silent-Chain Installation

Accurate alignment of silent-chain drives is an important requirement to maximize chain and sprocket wheel life. The same basic alignment steps covered in detail for roller chain must also be followed for silent chain.

Figure 15-24 Five-pitch silent-chain offset coupler. *(Courtesy Rexnord Industries, Inc.)*

To obtain shaft alignment, mount the sprocket wheels on their respective shafts. As illustrated in Figure 15-26, align the shafts horizontally with a machinist's level and adjust the shafts for parallel alignment with a vernier, caliper, or feeler bar. The distance between shafts on both sides of the wheel must be equal. When shafts have been accurately aligned, the motor, bearings, etc., should be bolted securely in place so that alignment will be maintained during operation.

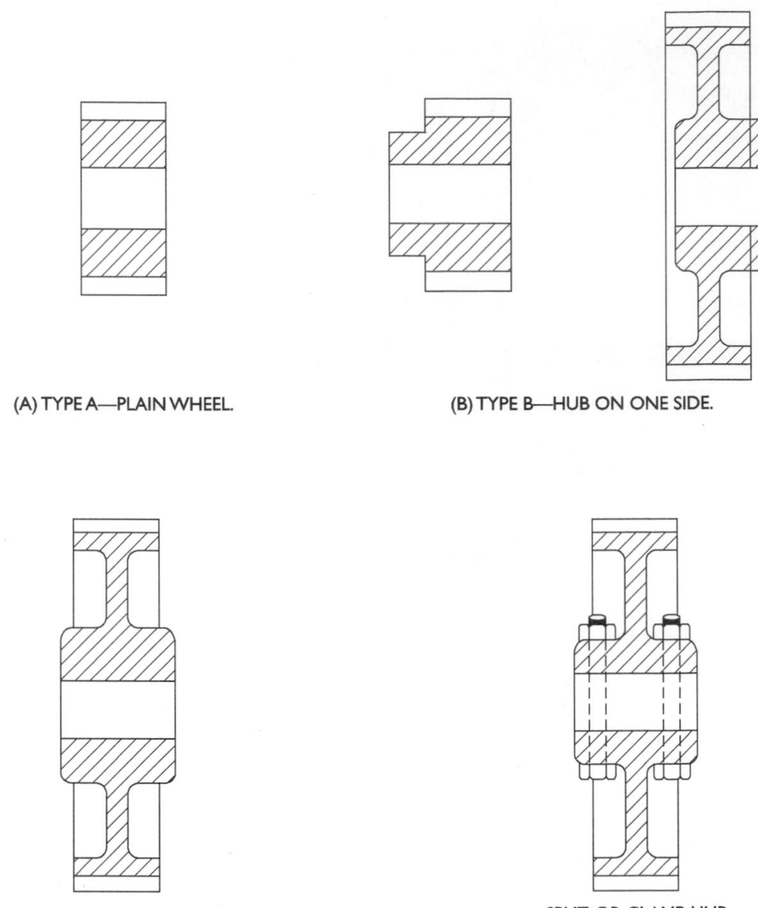

Figure 15-25 Silent-chain sprocket hubs.
(Courtesy Rexnord Industries, Inc.)

Wheels must be in axial alignment for correct chain and sprocket tooth engagement. Apply a straightedge or taut wire to the wheel surfaces as shown in Figure 15-27. When a shaft is subjected to end float, the wheel should be aligned for normal running position. Tighten setscrews in wheel hubs to guard against lateral wheel movement and to hold the key in position.

Figure 15-26 Aligning shafts both horizontally and vertically. *(Courtesy Rexnord Industries, Inc.)*

Figure 15-27 Aligning sprockets. *(Courtesy Rexnord Industries, Inc.)*

Before installing the chain, inspect it to be sure it is free from dirt and grit. Clean and relubricate if necessary. Fit the chain around both wheels, bringing the free ends together on one wheel as shown in Figure 15-28. Since joint components vary with each manufacturer, the method of assembly will vary. In general, the components are assembled and the pin is peened over the washer to rivet it in place.

Figure 15-28 Installing the connecting link.
(Courtesy Rexnord Industries, Inc.)

When it is not convenient to couple the chain on a wheel, a coupling tool as shown in Figure 15-29 may be necessary. This device provides a convenient means for drawing the ends of the chain together, which otherwise would be very difficult with a heavy chain.

The adjustment of drive centers for proper tension must be done while the drive is stationary. Normally, horizontal and inclined drives should be installed with an initial sag equal to approximately 2 percent of wheel centers. Vertical center drives and those subject to shock loading, reversal of rotation, or dynamic braking should be operated with both spans of chain almost taut.

To determine the amount of sag in a horizontal drive, pull one side of the chain taut, allowing all the excess chain to accumulate in the opposite span. As illustrated in Figure 15-30, place a straightedge over the slack span and, pulling the chain down at the center, measure for proper sag as listed in Table 15-5.

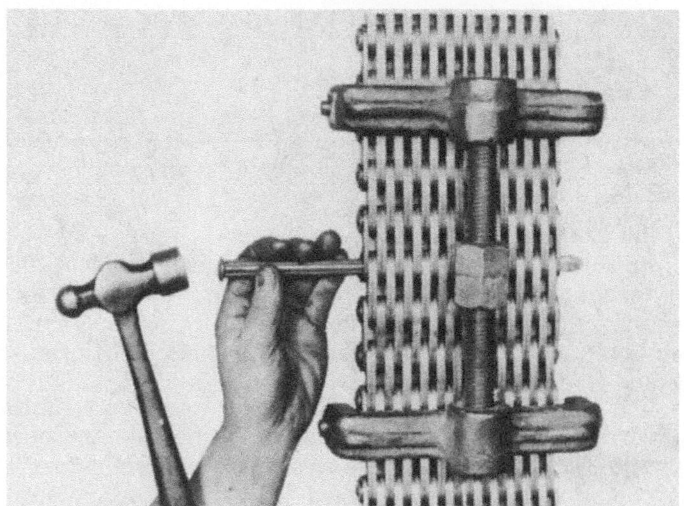

Figure 15-29 Showing the use of a coupling tool. *(Courtesy Rexnord Industries, Inc.)*

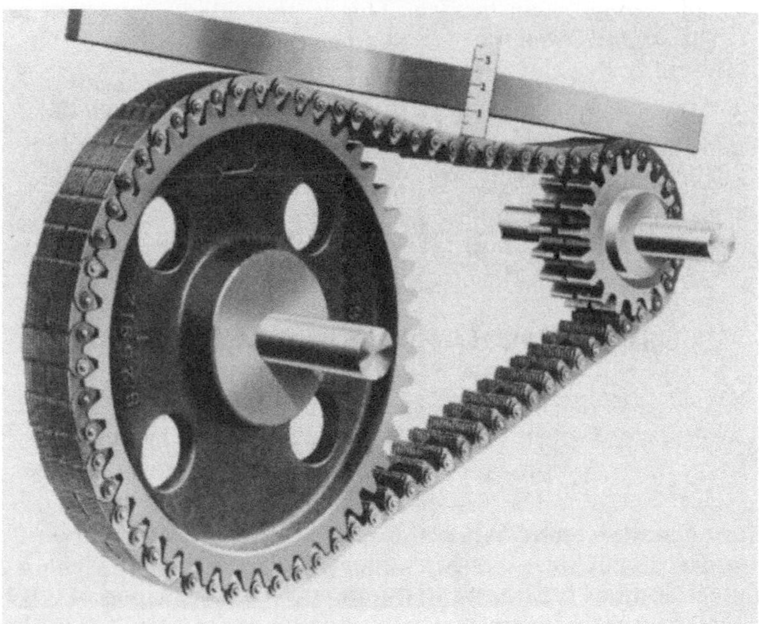

Figure 15-30 Checking for proper chain adjustment. *(Courtesy Rexnord Industries, Inc.)*

Table 15-5 Sag for Various Silent Chains

Shaft Center (in.)	***20***	***30***	***40***	***50***	***60***	***70***	***80***	***90***	***100***	***125***	***150***
Sag (in.)	½	⅝	⅞	1	1¼	1½	1⅝	1⅞	2	2½	3

Silent-Chain Maintenance

The establishment of periodic inspection is a basic requirement for satisfactory silent-chain operation. The first inspection should be after 100 hours of operation to find and correct any improper conditions before serious problems can develop. The following inspection procedure is recommended:

1. Check wheel alignment. If wear is apparent on only one side of the guide links or guide grooves, there is misalignment. This condition must be corrected immediately to prevent undue wear of chain and wheel teeth. Check and tighten setscrew.
2. Adjust wheel centers to take up accumulated slack and provide proper chain tension. This is particularly important at the initial 100-hour check, as some slack will be apparent as a result of the seating of the chain joints.
3. After the first 500 hours of operation, the lubricant in the drive casing should be drained and refilled with fresh oil. Oil changes and inspections should be made at approximately 2500 hours of operation thereafter. If at any time the lubricant is found to be contaminated, the casing should be flushed and the chain carefully cleaned. The chain should be immersed in oil and thoroughly dried before being reinstalled.

Wear of the chain and the wheel teeth should be checked at each inspection. Reduced tooth sections are signs of excessive wear. When chain and teeth are found in this condition, they should be replaced. It is never advisable to operate new chain on badly worn wheels, but it is sometimes possible to reverse the wheels so that relatively unworn tooth surfaces contact the chain.

Silent-Chain Replacement

Progressive joint wear elongates chain pitch, causing the chain to lengthen and ride higher on the sprocket teeth. The number of teeth in the large sprocket determines the amount of joint wear that can be tolerated before the chain jumps or rides over the ends of the sprocket teeth. When this critical degree of elongation is reached, the chain must be replaced. The "maximum chain elongation limits" (Table 15-3 and Figure 15-19) for roller chain may also be used for silent chain. Select the appropriate pitch in column three of the table. The length of chain when replacement is required can be found in the appropriate sprocket-tooth-number column.

Troubleshooting Chain Drive Problems

Table 15-6 will help you troubleshoot common chain drive problems.

Table 15-6 Chain Troubleshooting Guide

Problem	*Cause*	*Solution*
Excessive noise	Sprocket misalignment	Realign.
	Loose bearings	Tighten setscrews.
	Chain or sprocket wear	Replace components.
	Chain pitch size too large	Recalculate drive selection.
	Too much or too little slack	Adjust take-up.
Vibrating chain	High load fluctuation	Make use of fluid coupling or torque converter.
	Resonance	Change speed of machine.
Chain climbs sprockets	Heavy overload	Reduce load or install stronger chain.
	Excessive slack	Adjust take-up.
	Worn chain	Replace chain.
Excessive wear of link plate and one side of sprocket teeth	Misalignment	Realign shafts and sprockets.
Broken pins, roller, or bushings	Speed too high for sprocket size and pitch	Use shorter-pitch chain or install larger-diameter sprockets.
	Shock loading	Reduce source of shock load or use stronger chain.
	Inadequate lubrication	Perform proper PM lubrication at correct intervals.

(Continued)

Problem	*Cause*	*Solution*
	Sprocket or chain corrosion	Install stainless or other corrosion-resistant chain and sprockets.
	Material buildup in sprocket teeth	Install material relief sprockets that allow material to spill out.
Chain clings to sprocket	Center distance too long or high load fluctuation	Adjust center distance or install idler take-up.
	Excessive chain wear	Replace chain.
	Excessive chain slack	Adjust center distance or install idler take-up.
Breakage of link plates	Vibration	Install vibration dampener.
	Shock load	Install a shock absorber.
Stiff chain	Inadequate lubrication	Perform proper PM lubrication at correct intervals.
	Misalignment	Realign sprockets and shafts.
	Corrosion	Replace with noncorrosive components.

Chapter 16

Gears

A gear is a form of disc, or wheel, that has teeth around its periphery for the purpose of providing a positive drive by meshing the teeth with similar teeth on another gear or rack.

Spur Gears

The *spur gear* might be called the basic gear, as all other types have been developed from it. Its teeth are straight and parallel to the center bore line, as shown in Figure 16-1. The spur gear may run together with other spur gears or parallel shafts, with internal gears on parallel shafts, and with a *rack*. A rack such as the one illustrated in Figure 16-2 is in effect a straight-line gear. The smaller of a pair of gears (Figure 16-3) is often called a *pinion*.

The involute profile or form is the one most commonly used for gear teeth. It is a curve that is traced by a point on the end of a taut line unwinding from a circle. The larger the circle, the straighter the curvature; for a rack, which is essentially a section of an infinitely large gear, the form is straight or flat. The generation of an involute curve is illustrated in Figure 16-4.

Figure 16-1 One example of a spur gear.
(Courtesy Emerson Power Transmission/Browning)

Figure 16-2 A rack, which is a straight-line gear.
(Courtesy Emerson Power Transmission/Browning)

The involute system of spur gearing is based on a rack having straight, or flat, sides. All gears made to run correctly with this rack will run with each other. The sides of each tooth incline toward the center top at an angle called the *pressure angle*, shown in Figure 16-5.

The 14.5° pressure angle was standard for many years. In recent years, however, the use of the 20° pressure angle has been growing, and today 14.5° gearing is generally limited to replacement work. The principal reason for this is that a 20° pressure angle results in a gear tooth with greater strength and wear resistance and permits the use of pinions with a few less teeth. The effect of the pressure angle on the tooth of a rack is shown in Figure 16-6.

Figure 16-3 Typical spur gears.
(Courtesy Emerson Power Transmission/Browning)

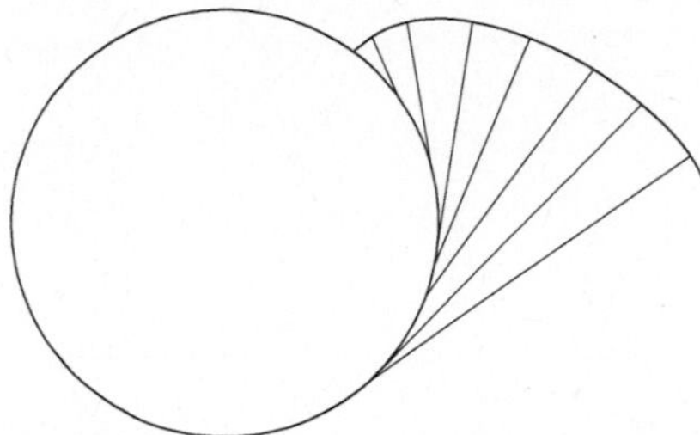

Figure 16-4 Illustrating the involute curve.

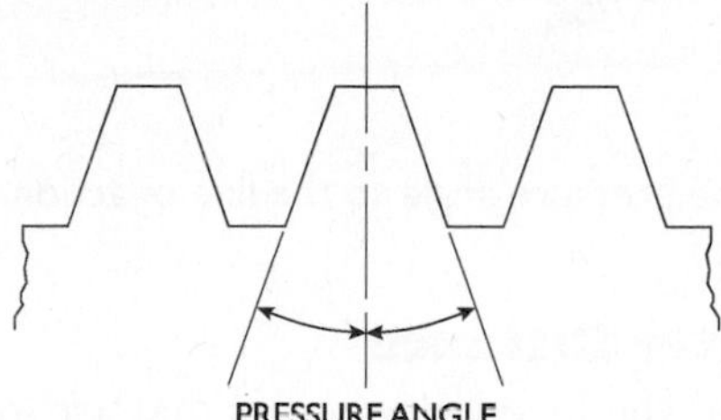

Figure 16-5 Illustrating pressure angle.

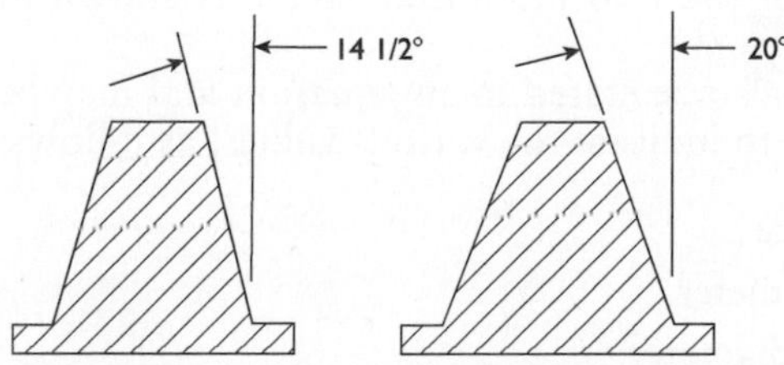

Figure 16-6 Illustrating different pressure angles on gear teeth.

It is extremely important that the pressure angle be known when gears are mated, as all gears that run together must have the same pressure angle. The pressure angle of a gear is the angle between the line of action and the line tangent to the pitch circles of mating gears. Figure 16-7 illustrates the relationship of the pressure angle to the line of action and the line tangent to the pitch circles.

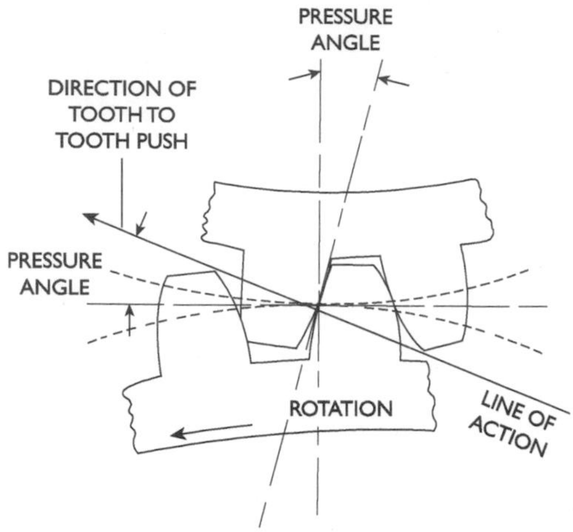

Figure 16-7 The relationship of the pressure angle to the line of action.

Pitch Diameter and Center Distance

Pitch circles have been defined as the imaginary circles that are in contact when two standard gears are in correct mesh. The diameters of these circles are the pitch diameters of the gears. The center distance of the two correctly meshed gears, therefore, is equal to one half of the sum of the two pitch diameters, as shown in Figure 16-8.

This relationship may also be stated in an equation and may be simplified by using letters to indicate the various values, as follows:

C = Center distance
D_1 = First pitch diameter
D_2 = Second pitch diameter

$$C = \frac{D1 + D2}{2} \qquad D1 = 2C - D2 \qquad D2 = 2C - D1$$

Example

Illustration in Figure 16-9.

$$C = \frac{D1 + D2}{2} = \frac{4 + 8.5}{2} = \frac{12.5}{2} = 6.25$$

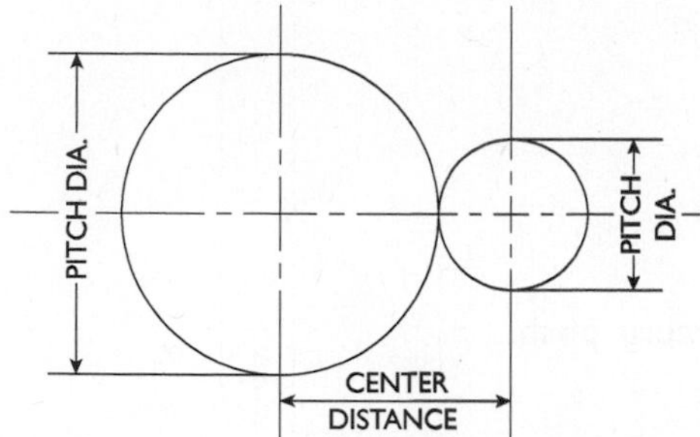

Figure 16-8 Pitch diameter and center distance.

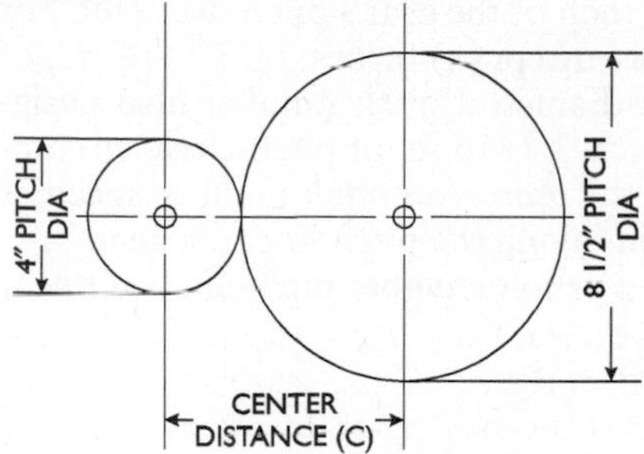

$$(C) = \frac{\text{PITCH DIA (D1)} + \text{PITCH DIA (D2)}}{2}$$

$$C = \frac{4 + 8\ 1/2}{2} = \frac{12\ 1/2}{2} = 6\ 1/4''$$

Figure 16-9 The center distance can be found if the pitch diameters are known.

Circular Pitch

The size and proportion of gear teeth are designated by a specific type of pitch. In gearing terms, there are two specific types of pitch: *circular pitch* and *diametral pitch*. Circular pitch is simply the distance from a point on one tooth to a corresponding point on the next tooth, measured along the pitch line or circle, as illustrated in Figure 16-10. Large-diameter gears are frequently made to circular-pitch dimensions.

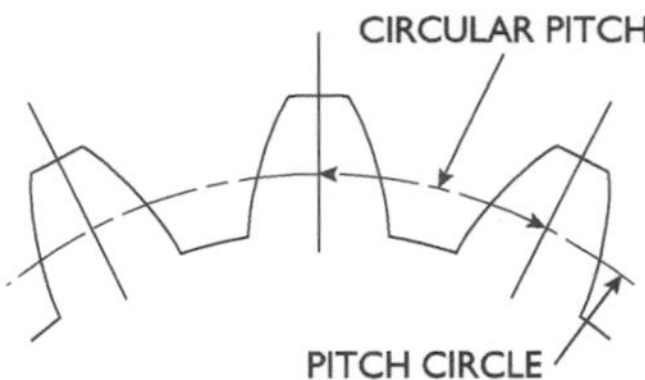

Figure 16-10 Illustrating circular pitch.

Diametral Pitch

The diametral pitch system is the most widely used, as practically all common-size gears are made to diametral pitch dimensions. It designates the size and proportions of gear teeth by specifying the number of teeth in the gear for each inch of the gear's pitch diameter. For each inch of pitch diameter, there are pi (π) inches, or 3.1416 in., of pitch-circle circumference. The diametral pitch number also designates the number of teeth for each 3.1416 in. of pitch-circle circumference. Stated in another way, the *diametral pitch* number specifies the number of teeth in 3.1416 in. along the pitch line of a gear.

For simplicity of illustration, a whole-number pitch-diameter gear (4 in.) is shown in Figure 16-11.

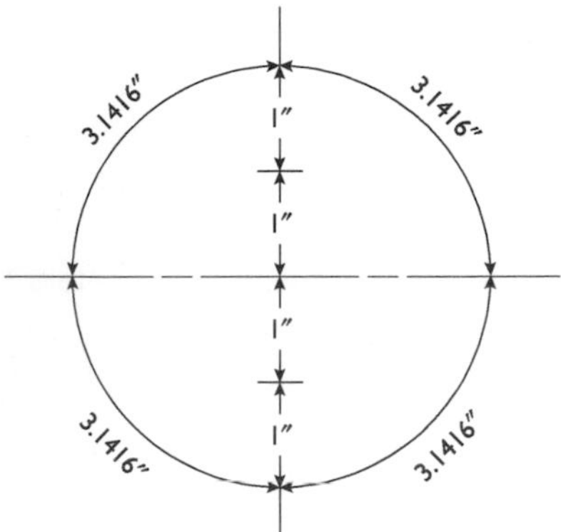

Figure 16-11 The pitch diameter and the diametral pitch.

Figure 16-11 illustrates that the *diametral pitch* number specifying the number of teeth per inch of pitch diameter must also specify the number of teeth per 3.1416 in. of pitch-line distance. This may be more easily visualized and specifically dimensioned when applied to the rack in Figure 16-12.

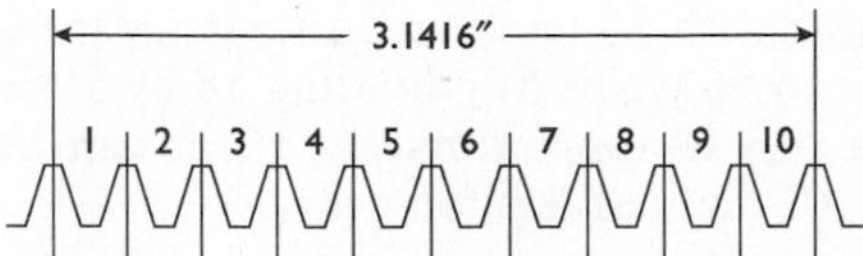

Figure 16-12 The number of teeth in 3.1416 inches.

Because the pitch line of a rack is a straight line, a measurement can be easily made along it. In Figure 16-12, it is clearly shown that there are 10 teeth in 3.1416 in.; therefore, the rack illustrated is a 10-diametral-pitch rack.

A similar measurement is illustrated in Figure 16-13, along the pitch line of a gear. As the diametral pitch is the number of teeth in 3.1416 in. of pitch line, the gear in this illustration is a 10-diametral-pitch gear.

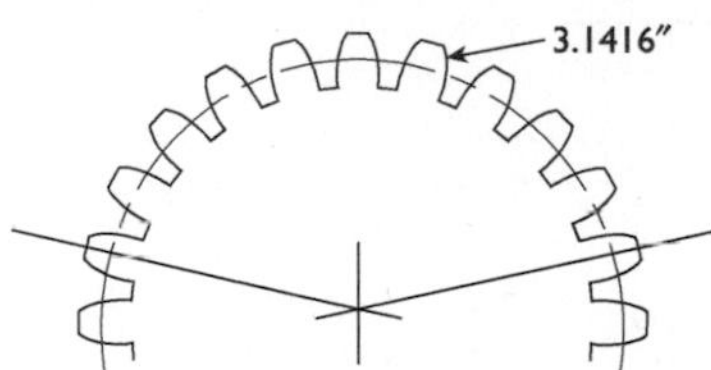

Figure 16-13 The number of teeth in 3.1416 in. on the pitch circle.

Diametral Pitch Approximation

In many cases, particularly on machine repair work, it may be desirable for the mechanic to determine the diametral pitch of a gear. This can be done very easily without the use of precision measuring tools, templates, or gauges. Measurements need not be exact because diametral pitch numbers are usually whole numbers. Therefore, if an approximate calculation results in a value close to a whole number, that whole number is the diametral pitch number of the gear.

The following three methods may be used to determine the approximate diametral pitch of a gear. A common steel rule, preferably flexible, is adequate to make the required measurements.

Method 1

Count the number of teeth in the gear, add 2 to this number, and divide by the outside diameter of the gear. Scale measurement of the gear to the closest fractional size is adequate accuracy.

Figure 16-14 illustrates a gear with 56 teeth and an outside measurement of 5 13⁄16 in. Adding 2 to 56 yields 58; dividing 58 by 5 13⁄16 gives an answer of 9 31⁄32. As this is approximately 10, it can be safely stated that the gear is a 10-diametral-pitch gear.

Method 2

Count the number of teeth in the gear and divide this number by the measured pitch diameter. The pitch diameter of the gear is measured from the root or bottom of a tooth space to the top of a tooth on the opposite side of the gear.

Figure 16-15 illustrates a gear with 56 teeth. The pitch diameter measured from the bottom of the tooth space to the top of the opposite tooth is 5 5⁄8 in. Dividing 56 by 5 5⁄8 gives an answer of 9 15⁄16 in. or approximately 10. This method also indicates that the gear is a 10-diametral-pitch gear.

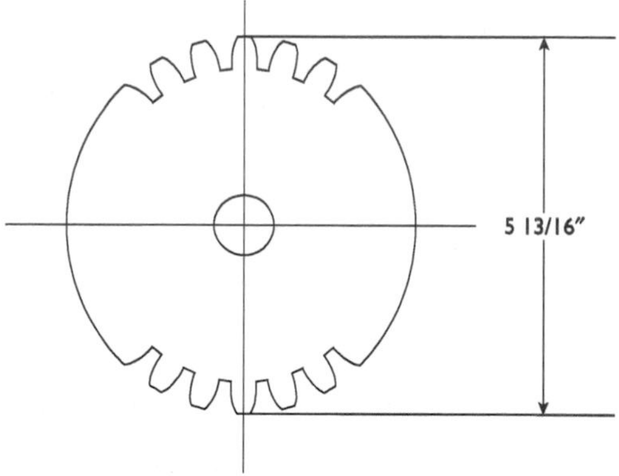

Figure 16-14 Using method 1 to approximate the diametral pitch. In this method, the outside diameter of the gear is measured.

Method 3

The third method was described previously when the term *diametral pitch* was discussed. Using a flexible scale, measure approximately 3 1⁄8 in. along the pitch line of the gear. To do this, bend the scale to match the curvature of the gear, being careful to position the scale about midway between the base and top of the teeth. This is the location of the imaginary pitch line. If the gear can be rotated, it is helpful to draw a pencil line on the gear to indicate the pitch line. The line can then be used to bend and position the flexible scale. Count the number of teeth in 3 1⁄8 in. on the scale to determine the diametral pitch of the gear.

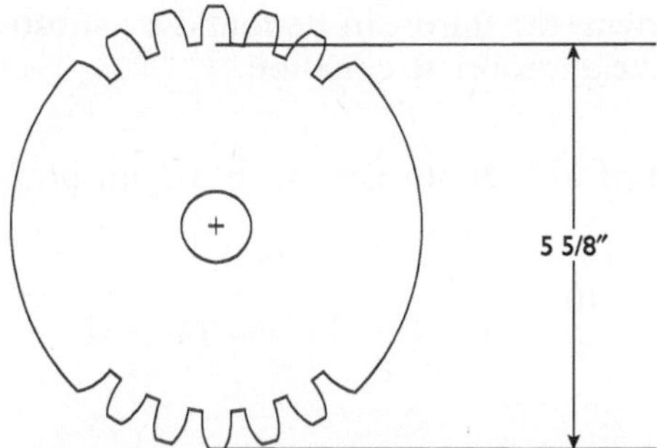

Figure 16-15 Using method 2 to approximate the diametral pitch. This method uses the pitch diameter of the gear.

Figure 16-16 illustrates the section of a gear with an approximate length of $3\frac{1}{8}$ in. along the pitch line of the gear. Counting the number of teeth in this distance indicates that it is a 10-diametral-pitch gear.

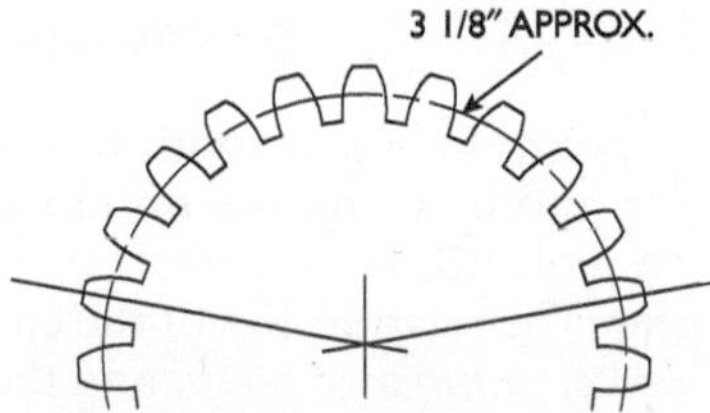

Figure 16-16 Using method 3 to determine the number of teeth in approximately $3\frac{1}{8}$ in. on pitch circle.

Pitch Calculations

Diametral pitch, usually a whole number, denotes the ratio of the number of teeth to a gear's pitch diameter. Stated another way, it specifies the number of teeth in a gear for each inch of pitch diameter. The relationship of pitch diameter, diametral pitch, and number of teeth can be stated mathematically as follows:

$$P = \frac{N}{D} \qquad D = \frac{N}{P} \qquad N = D \times P$$

where

D = pitch diameter
P = diametral pitch
N = number of teeth

If any two values are known, the third can be found by substituting the known values in the appropriate equation.

Example 1
What is the diametral pitch of a 40-tooth gear with a 5-in. pitch diameter?

$$P = \frac{N}{D} \quad \text{or} \quad P = \frac{40}{5} \quad \text{or} \quad P = 8 \text{ diametral pitch}$$

Example 2
What is the pitch diameter of a 12-diametral-pitch gear with 36 teeth?

$$D = \frac{N}{P} \quad \text{or} \quad D = \frac{36}{12} \quad \text{or} \quad D = 3'' \text{ pitch diameter}$$

Example 3
How many teeth are there in a 16-diametral-pitch gear with a pitch diameter of 3¾ in.?

$$N = D \times P \quad \text{or} \quad N = 3\tfrac{3}{4} \times 16 \quad \text{or} \quad N = 60 \text{ teeth}$$

Circular pitch is the distance from a point on a gear tooth to the corresponding point on the next gear tooth as measured along the pitch line. Its value is equal to the circumference of the pitch circle divided by the number of teeth in the gear. The relationship of the circular pitch to the *pitch-circle circumference*, *number of teeth*, and the *pitch diameter* may also be stated mathematically as follows:

$$\text{Circumference of pitch circle} = \pi D$$

$$p = \frac{\pi D}{N} \qquad D = \frac{pN}{\pi} \qquad N = \frac{\pi D}{p}$$

where

D = pitch diameter
N = number of teeth
p = circular pitch
π = pi, or 3.1416

If any two values are known, the third can be found by substituting the known values in the appropriate equation.

Example 4
What is the circular pitch of a gear with 48 teeth and a pitch diameter of 6 in.?

$$p = \frac{\pi D}{N} \quad \text{or} \quad \frac{3.1416 \times 6}{48} \quad \text{or} \quad \frac{3.1416}{8} \quad \text{or} \quad p = .3927 \text{ inches}$$

Example 5
What is the pitch diameter of a .500 in. circular-pitch gear with 128 teeth?

$$D = \frac{pN}{\pi} \quad \text{or} \quad \frac{.5 \times 128}{3.1416} \qquad D = 20.371 \text{ inches}$$

The list that follows includes just a few names of the various parts given to gears. These parts are shown in Figures 16-17 and 16-18.

Addendum. The distance the tooth projects above, or outside, the pitch line or circle.

Dedendum. The depth of a tooth space below, or inside, the pitch line or circle.

Clearance. The amount by which the dedendum of a gear tooth exceeds the addendum of a matching gear tooth.

Whole Depth. The total height of a tooth or the total depth of a tooth space.

Working Depth. The depth of tooth engagement of two matching gears. It is the sum of their addendums.

Tooth Thickness. The distance along the pitch line or circle from one side of a gear tooth to the other.

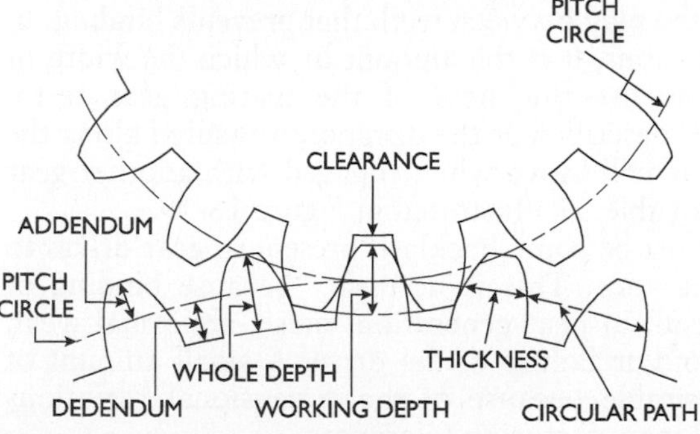

Figure 16-17 Names of gear parts.

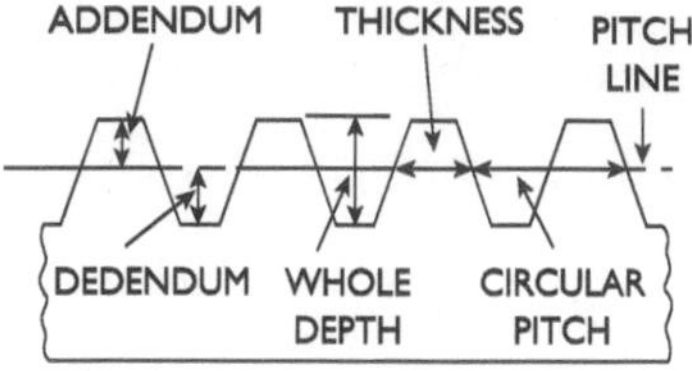

Figure 16-18 Names of rack parts.

Tooth Proportions

The *full-depth involute system* is the gear system in most common use. The formulas (with symbols) shown below are used for calculating tooth proportions of full-depth involute gears. Diametral pitch is given the symbol *P* as before.

Addendum $$a = \frac{1}{P}$$

Whole Depth $$Wd = \frac{2.2}{P} + .002 \quad \text{(20}p\text{ or smaller)}$$

Dedendum $$Wd = \frac{2.157}{P} \quad \text{(Larger than 20}p\text{)}$$

Whole Depth $$b = Wd - a$$

Clearance $$c = b - a$$

Tooth Thickness $$t = \frac{1.5708}{P}$$

Backlash

Backlash in gears is the play between teeth that prevents binding. In terms of tooth dimensions, it is the amount by which the width of tooth spaces exceeds the thickness of the mating gear teeth. Backlash may also be described as the distance, measured along the pitch line, that a gear will move when engaged with another gear that is fixed or unmovable, as illustrated in Figure 16-19.

Normally there must be some backlash present in gear drives to provide running clearance. This is necessary because binding of mating gears can result in heat generation, noise, abnormal wear, possible overload, and/or failure of the drive. A small amount of backlash is also desirable because of the dimensional variations involved in practical manufacturing tolerances.

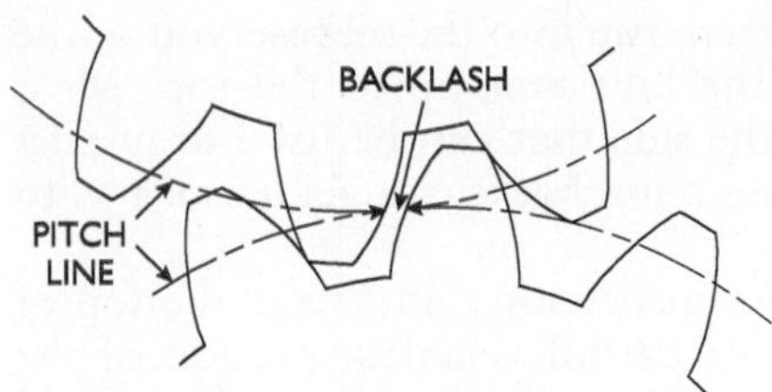

Figure 16-19 Illustrating backlash.

Backlash is built into standard gears during manufacture by cutting the gear teeth thinner than normal by an amount equal to one-half the required figure. When two gears made in this manner are run together, at standard center distance, their allowances combine, provided the full amount of backlash is required.

On nonreversing drives or drives with continuous load in one direction, the increase in backlash that results from tooth wear does not adversely affect operation. However, on reversing drives and drives where timing is critical, excessive backlash usually cannot be tolerated.

Table 16-1 lists the suggested backlash for a pair of gears operating at the standard center distance.

Table 16-1 Backlash Suggestions

Pitch	*Backlash*
3 P	.013
4 P	.010
5 P	.008
6 P	.007
7 P	.006
8–9 P	.005
10–13 P	.004
14–32 P	.003
33–64 P	.0025

Gearboxes

In most industrial locations, gears are supplied—rather than exposed—within a gearbox. Boxes can be purchased that are single-, double-, or triple-reduction, depending on the number of gears and shafts. In most cases, the gearbox is filled to a certain level with a high-quality gear oil, and the teeth run into the oil reservoir at the bottom of the box and carry the lubricant up to the top. Most boxes include a sight glass on the side that can be used to inspect the level of oil, and in some cases a mechanic can get an idea as to whether the oil looks dirty.

Most large gearboxes have an inspection panel near the top of the box that can be removed for partial visual inspection of the gears. Most boxes are built so that the top half can be removed and a complete inspection can be done, including backlash checking.

Other Gear Types

Many styles and designs of gears have been developed from the spur gear. Although they are all commonly used in industry, many are complex in design and manufacture. Only a general description and explanation of principles will be given, as the field of specialized gearing is beyond the scope of this book. Commonly used styles will be discussed sufficiently to provide the millwright or mechanic with the basic information necessary to perform installation and maintenance work.

Bevel and Miter Gears

Two major differences between bevel gears and spur gears are their shape and the relation of the shafts on which they are mounted. The shape of a spur gear is essentially a cylinder, whereas the shape of a

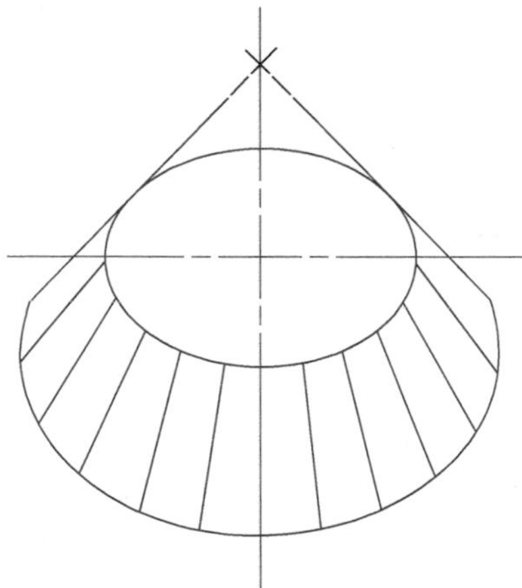

Figure 16-20 The basic shape of bevel gears.

bevel gear is a cone. Spur gears are used to transmit motion between parallel shafts, whereas bevel gears transmit motion between angular or intersecting shafts. The diagram in Figure 16-20 illustrates the bevel gear's basic cone shape. Figure 16-21 shows a typical pair of bevel gears.

Figure 16-21 A typical set of bevel gears.
(Courtesy Emerson Power Transmission/Browning)

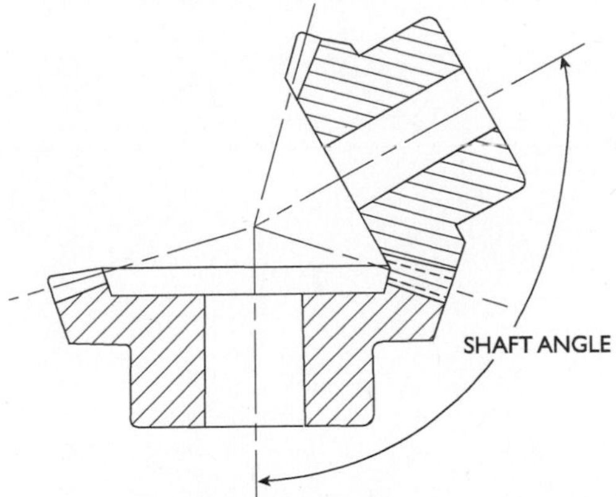

Figure 16-22 The shaft angle, which can be at any degree.

Special bevel gears can be manufactured to operate at any desired shaft angle, as shown in Figure 16-22. Miter gears are bevel gears with the same number of teeth in both gears operating on shafts at right angles or at 90°, as shown in Figure 16-23.

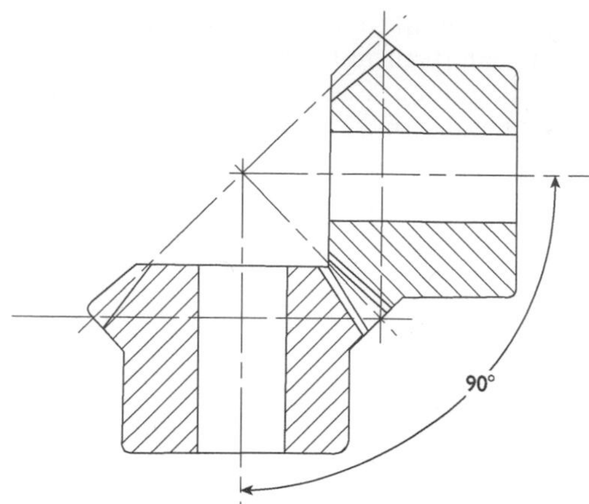

Figure 16-23 Miter gears, which are shown at 90°.

A typical pair of straight miter gears is shown in Figure 16-24. Another style of miter gear having spiral rather than straight teeth is shown in Figure 16-25. The spiral-tooth style will be discussed later.

Figure 16-24 A typical set of miter gears.
(Courtesy Emerson Power Transmission/Browning)

The tooth size of bevel gears is established by the diametral pitch number, as with spur gears. Because the tooth size varies along its length, it must be measured at a given point. This point is the outside part of the gear where the tooth is the largest. Because each gear in a set of bevel gears must have the same angles and tooth lengths, as well as the same diametral pitch, they are manufactured and distributed only in mating pairs. Bevel gears, like spur gears, are manufactured in both the 14.5° and 20° pressure-angle designs.

Figure 16-25 Miter gears with spiral teeth. *(Courtesy Emerson Power Transmission/Browning)*

Helical Gears

Helical gears are designed for parallel-shaft operation like the pair in Figure 16-26. They are similar to spur gears except that the teeth are cut at an angle to the centerline. The principal advantage of this design is the quiet, smooth action that results from the sliding contact of the meshing teeth. A disadvantage, however, is the higher friction and wear that accompanies this sliding action. The angle at which the gear teeth are cut is called the *helix angle* and is illustrated in Figure 16-27.

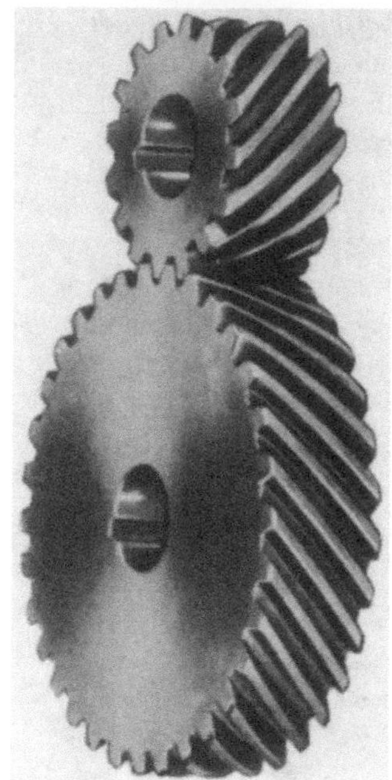

Figure 16-26 A typical set of helical gears. *(Courtesy Emerson Power Transmission/Browning)*

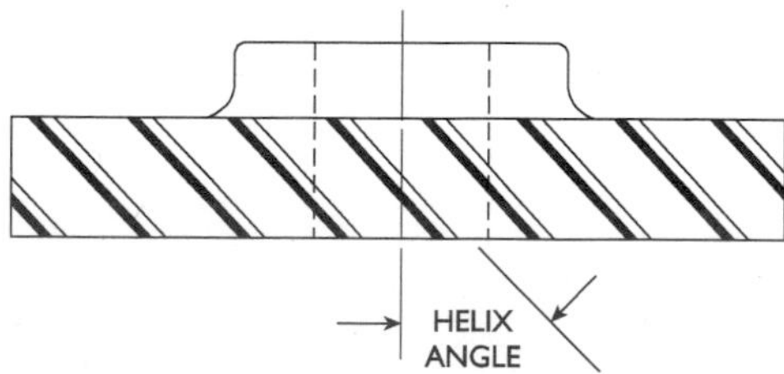

Figure 16-27 Illustrating the angle at which the teeth are cut.

It is very important to note that the helix angle may be on either side of the gear's centerline. In other words, if compared to the helix angle of a thread, it may be either a "right-hand" or a "left-hand" helix. The hand of the helix is the same regardless of how it is viewed. Figure 16-28 illustrates a helical gear as viewed from opposite sides; the hand of the tooth's helix angle cannot be changed by altering the position of the gear. A pair of helical gears, as illustrated in Figure 16-26, must have the same pitch and helix angle but must be of opposite hands (one right hand and one left hand).

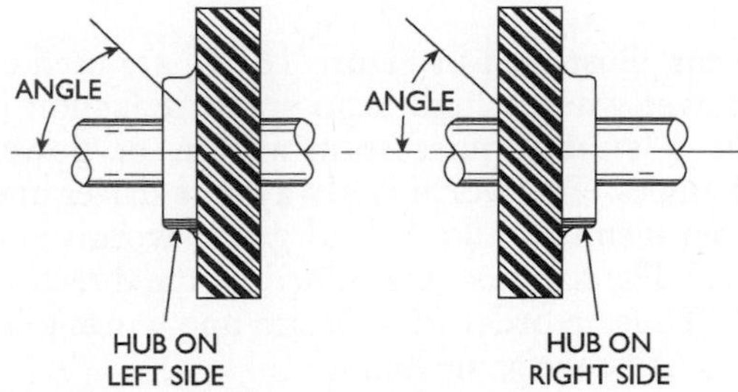

Figure 16-28 The helix angle of the teeth is the same no matter from which side the gear is viewed.

Helical gears may also be used to connect nonparallel shafts. When used for this purpose, they are often called "spiral" gears or crossed-axis helical gears. This style of helical gearing is shown in Figure 16-29.

Figure 16-29 A typical set of spiral gears. *(Courtesy Emerson Power Transmission/Browning)*

Worm Gears

The worm and worm gear, illustrated in Figure 16-30, are used to transmit motion and power when a high-ratio speed reduction is required. They provide a steady, quiet transmission of power between shafts at right angles. The worm is always the driver and the worm gear the driven member. Like helical gears, worms and worm gears have "hand." The hand is determined by the direction of the angle of the teeth. Thus, in order for a worm and worm gear to mesh correctly, they must be the same hand.

Figure 16-30 A typical set of worm gears.
(Courtesy Emerson Power Transmission/Browning)

The most commonly used worms have one, two, three, or four separate threads and are called single, double, triple, and quadruple thread worms. The number of threads in a worm is determined by counting the number of starts or entrances at the end of the worm. The thread of the worm is an important feature in worm design, as it is a major factor in worm ratios. The ratio of a mating worm and worm gear is found by dividing the number of teeth in the worm gear by the number of threads in the worm.

Herringbone Gears

To overcome the disadvantage of the high end thrust present in helical gears, the herringbone gear, illustrated in Figure 16-31, was developed. It consists simply of two sets of gear teeth, one right hand and one left hand, on the same gear. The gear teeth of both hands cause the thrust of one set to cancel out the thrust of the other. Thus, the advantage of helical gears is obtained, and quiet, smooth operation at higher speeds is possible. Obviously, they can only be used for transmitting power between parallel shafts.

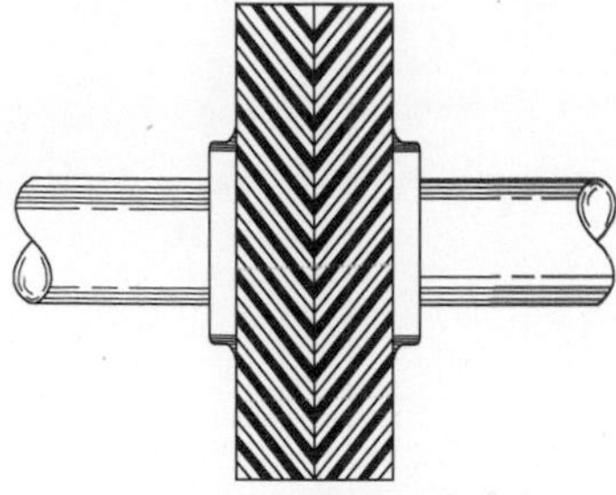

Figure 16-31 Herringbone gear.

Troubleshooting Gear Problems

Gear problems such as excessive wear are easy to spot on exposed gearing, but they are not so obvious when the gears operate with a closed gearbox. Of course, the box inspection plate may be removed or the box may be disassembled for inspection, but these procedures take time and may result in excessive machinery downtime for the production department.

Spectrographic oil analysis is an excellent tool for checking potential internal gear or bearing problems on a gearbox. A small amount of oil is properly sampled and sent to a vendor laboratory for analysis. Results may show high ferrous content (indicating gear wear), high silicon levels (indicating dirty oil), or high levels of chrome or nickel (indicating potential bearing wear). Pulling an oil sample every six months and checking for nonproblems until a problem occurs that is worthy of shutting down the box and opening it for a full investigation can save time in the long run.

Chapter 17

Couplings

Power-transmission couplings are the usual means of connecting coaxial shafts so that one can drive the other—for example, connecting an electric motor to a pump shaft or to the input shaft of a gear reducer or connecting two pieces of shafting together to obtain a long length, as with line shafting. Power-transmission couplings for such shaft connections are manufactured in a great variety of types, styles, and sizes. They may, however, be divided into two general groups or classifications: "rigid" couplings, which are also called "solid" couplings, and a group called "flexible" couplings.

Rigid Couplings

A rigid (or solid) coupling connects shaft ends together rigidly, making the shafts so connected into a single, continuous unit. Rigid couplings provide a fixed union that is equivalent to a shaft extension. They should be used only when *true* alignment and a solid, or rigid, coupling is required, as with line shafting, or where provision must be made to allow parting of a rigid shaft. A feature of rigid couplings is that they are self-supporting and automatically align the shafts to which they are attached when the coupling halves on the shaft ends are connected. Two basic rules should be followed to obtain satisfactory service from rigid couplings:

- A force fit must be used in assembly of the coupling halves to the shaft end.
- After assembly, a runout check of all surfaces of the coupling must be made, and any surface found to be running out must be machined true.

Checking surfaces is especially necessary if the coupling halves are assembled by driving rather than pressing. Rigid couplings should *not* be used to connect the shafts of independent machine units that must be aligned at assembly. The sketch in Figure 17-1A illustrates a typical rigid coupling, and the self-aligning surfaces can be seen in the cross section in Figure 17-1B.

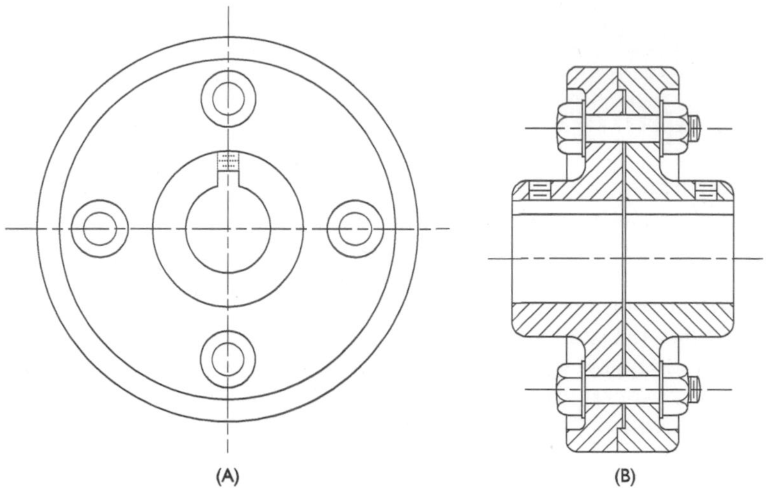

Figure 17-1 Typical rigid couplings.

Flexible Couplings

The transmission of mechanical power often requires the connection of two independently supported coaxial shafts so that one can drive the other. Prime movers, such as internal combustion engines and electric motors, connected to reducers, pumps, variable speed drives, etc., are typical examples. For these applications, a flexible coupling is used because perfect alignment of independently supported coaxial shaft ends is practically impossible. In addition to the impracticality of perfect alignment, there is always wear and damage to the connected components and their shaft bearings, as well as the possibility of movement from temperature changes and external forces.

In addition to allowing coaxial shafts to operate satisfactorily with slight misalignment, flexible couplings allow some axial movement, or end float, and, depending on the type, may allow torsional movement as well. Another benefit that may result when flexible couplings have nonmetallic connecting elements is electrical insulation of connected shafts. In summary, four conditions exist when coaxial shaft ends are connected. Flexible couplings are designed to compensate for some or all of these conditions:

- Angular misalignment
- Parallel misalignment
- Axial movement (end float)
- Torsional movement

Figure 17-2 shows the four types of flexibility that may be provided by flexible couplings.

Flexible couplings are intended to compensate for only the slight, unavoidable misalignment inherent in the design of machine components and the practices followed when aligning coaxial shafts of connected units. If the application is one where misalignment must exist, universal joints or some style of flexible shafting may be necessary. Flexible couplings should not be used to compensate for deliberate misalignment of connected components. Couplings designed for offset operation are discussed later in the chapter.

Almost all flexible couplings are made up of three basic parts: two hubs that attach to the shaft ends to be connected and a flexing member or element that transmits power from one hub to the other. There are a great variety of flexible coupling designs, all of which have characteristics and features to meet specific needs.

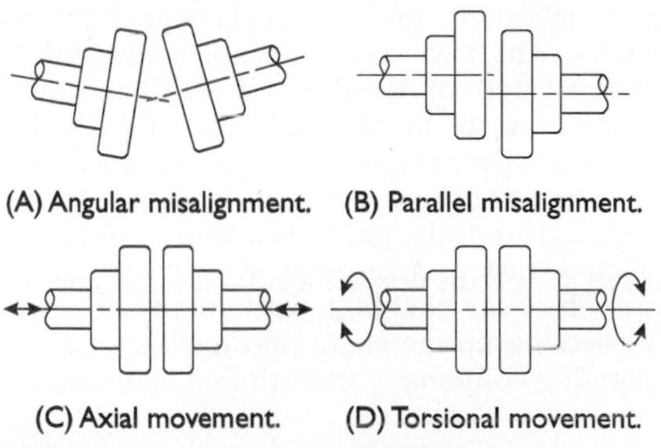

(A) Angular misalignment. (B) Parallel misalignment.

(C) Axial movement. (D) Torsional movement.

Figure 17-2 Four types of flexibility.

Chain and Gear Couplings

Angular and parallel misalignments are allowed by couplings in this group, as well as end float and limited torsional flexibility. Lubrication is usually necessary. Examples of chain-style couplings are the *roller chain* shown in Figure 17-3 and the *silent chain* shown in Figure 17-4. Shown in Figure 17-5 is a gear type manufactured by the Falk Corp. The hub teeth are 20° full depth and barrel-shaped, with three curved surfaces—top, flank, and root—in combination with straight-sided sleeve teeth to provide load distribution and eliminate tooth top loading and some axial movement.

Figure 17-3 Roller-chain couplings.
(Courtesy Emerson Power Transmission Manufacturing, L.P. and affiliates)

A chain coupling is essentially two sprockets with hubs for attachment to the shafts. The two sprockets are connected by assembling around them a length of double chain that has a number of links that correspond to the number of sprocket teeth. The chain may be roller chain or silent chain. Some manufacturers use standard roller chain, whereas others use chain that has special features to suit the manufacturer's design. Lubrication is necessary and is retained in a close-fitting cover that wraps around the sprocket and chain section of the coupling. This cover also serves as a safety device by covering up the chain and teeth so that the outer surface of the rotating coupling is smooth and cannot catch or snag.

Gear couplings are made up of meshing internal and external gears or splines. Single-engagement types use one of each. Double-engagement types, which are the ones most widely used, use two of each. Flexibility results from both the fit of the teeth and from a special shaping of the teeth that permits them to pivot. In general, these couplings permit more angular deflection, parallel misalignment, and end float than chain-type couplings do.

With few exceptions, gear couplings must be lubricated; they are designed with a complete housing, seals, and provisions for relubrication. The exceptions have one or more gears made of nonmetallic material. The materials used include plastic, molded fiber, and special rubber compounds. Couplings using nonmetallic gears offer additional torsional resilience compared with those using metallic gears, in addition to providing electrical insulation.

Figure 17-4 Silent-chain couplings.
(Courtesy Emerson Power Transmission Manufacturing, L.P. and affiliates)

Figure 17-5 Gear-type coupling.
(Courtesy Falk Corp. Steelflex® Grid, Torus® Elastomer, and Lifelign® Gear are registered trademarks of Falk Corp.)

Jaw and Slider Couplings

Couplings in this category may be described as having *tightly* fitted hard parts that are constructed to allow sliding action. In one style, two flanged hubs are connected by a square, nonmetallic slider member. The C-shaped flange jaws engage opposite sides of the slider, permitting relative sliding motion of the two flanges at right angles to each other. In another style, called a jaw coupling, both hubs have protrusions ("jaws"), and a removable leather or fiber member cushions the driving and driven surfaces. In still another style, the hubs have openings, and the slider, also called a spider, is disc-shaped with lugs to engage the openings in the hubs. Couplings in this group compensate for angular and parallel misalignment and generally have zero torsional flexibility.

In general, slider couplings have two hubs, which attach to the shaft ends and are constructed with surfaces designed to receive a sliding member. This sliding element, which provides the coupling's flexibility, is referred to as the "slider." The slider is driven by jaws, keys, or openings, and it in turn drives the other hub by the same method. Usually, slider couplings have to be lubricated. The slider may be made of self-lubricating material or oil-impregnated sintered metal to accomplish this. Some slider couplings incorporate lubricant reservoirs. Replaceable bearing strips are also used on some sliders.

Flexible or Resilient Element Couplings

This style of coupling is probably made by more manufacturers and in more design variations than the other two styles combined. The flexible elements used in these couplings may be metal, rubber, leather, or plastic. The most widely used are metal and rubber. Typical couplings using metallic flexible elements, as well as some of the many designs using nonmetallic flexible elements, are shown in Figures 17-6 through 17-9. Couplings in this category will allow angular and parallel misalignment and end float. The torsional flexibility covers a wide range. Some have virtually none, whereas some that use rubber flexible elements have more torsional flexibility than any other coupling. Most couplings in this group do not need lubrication. Exceptions are those using flexible metallic elements. An extremely wide range of speed and horsepower applications is covered by these couplings.

Coupling Selection

Although coupling selection is usually the responsibility of the engineer rather than the mechanic, it is a good idea for the mechanic to

Figure 17-6 Sure-Flex rubber gear.
(Courtesy T. B. Wood's Incorporated)

have an understanding of the factors involved. One important benefit of having such an understanding is that it aids in the recognition of possible selection errors when coupling failure occurs.

Figure 17-7 Steelflex metal grid.
(Courtesy Falk Corp. Steelflex® Grid, Torus® Elastomer, and Lifelign® Gear are registered trademarks of Falk Corp.)

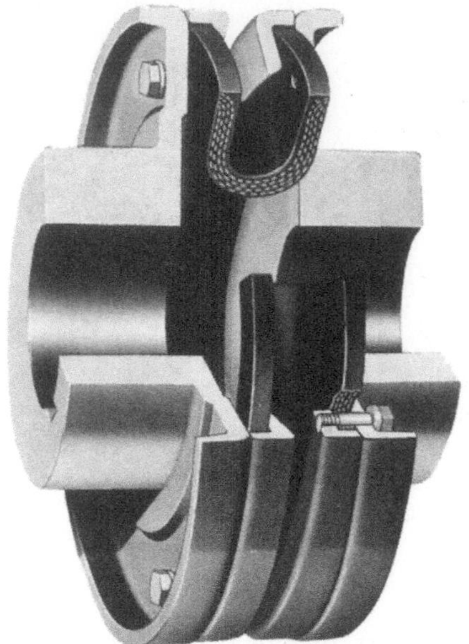

Figure 17-8 Torus rubber element.
(Courtesy Falk Corp. Steelflex® Grid, Torus® Elastomer, and Lifelign® Gear are registered trademarks of Falk Corp.)

Figure 17-9 Moreflex rubber bushing.
(Courtesy Emerson Power Transmission Manufacturing, L.P. and affiliates)

Flexible coupling selection is primarily a matter of matching job requirements with coupling specifications. Manufacturers supply detailed product information and data to enable this to be accomplished with a high assurance of satisfactory operation. For example, if the application is one that combines high speed and horsepower ratings with appreciable end float, the gear coupling has these characteristics. On the other hand, if a high degree of torsional flexibility is needed, this is found only in couplings with resilient flexible elements.

Most manufacturers' selection recommendations place heavy emphasis on the type or nature of the load. For example, the load on a coupling driving a fan or a centrifugal pump would be of a different nature than if it were driving a machine tool, a punch press, or a rock crusher. Horsepower or torque rating of the coupling to be selected must be modified to allow for service conditions. To do this, the actual horsepower of the job is multiplied by a "service factor" to find the rating of the coupling to use. There are wide variations in service factors for different flexible couplings, so it is important that only factors for the make and type of coupling selected be used.

There are many applications where the job requirements might be met by almost any available coupling. In these cases, other factors that affect installation and maintenance may have a strong influence on selection. Some additional factors to consider are the following:

1. Method and ease of installation and removal
2. Lubrication consideration (feasibility, etc.)
3. Dimension and weight limits
4. Cost
5. Vibration, quietness, etc.

Coupling Alignment

The designation "coupling alignment" reflects the accepted use of the term to describe the operation of bringing coaxial shaft ends into alignment. The most common situation is one where a coupling is used to connect and transmit mechanical power from shaft end to shaft end. Although the term implies that the prime function in performing this operation is to align the surfaces of the coupling, the principal objective is actually to bring the centerlines of the coaxial shafts into alignment. Although this may seem a fine distinction, it is important that the difference be understood. This point plays a large part in one's understanding of the procedures that are followed in correctly aligning a coupling.

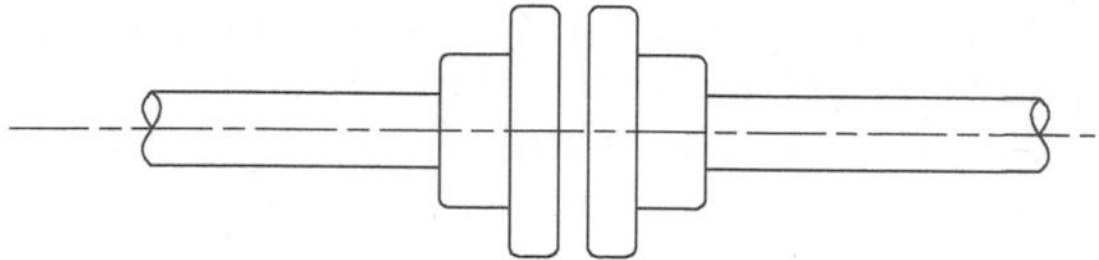

Figure 17-10 Alignment of centerlines.

To illustrate this important point, Figure 17-10 shows two coupling halves upon whose surfaces the alignment operations are performed. When these coupling surfaces are aligned, the centerlines of the shafts are also aligned, with this very important proviso: The surfaces of the coupling halves must run *true* with the centerlines of the shafts. If the surfaces do not run true with the shaft centerlines, *alignment* of the untrue surfaces will result in *misalignment* of the shaft centerlines. True running surfaces, therefore, are a basic requirement if the alignment procedure that follows is to be successful.

As the purpose is to accomplish alignment, it may also be described as an operation to correct or eliminate misalignment. We must then define and locate the misalignment if we are to eliminate it. In respect to the shafts to be coupled, they may be misaligned in two ways. They may be at an angle rather than in a straight line (angular misalignment), or they may be offset (parallel misalignment), as shown in Figure 17-11.

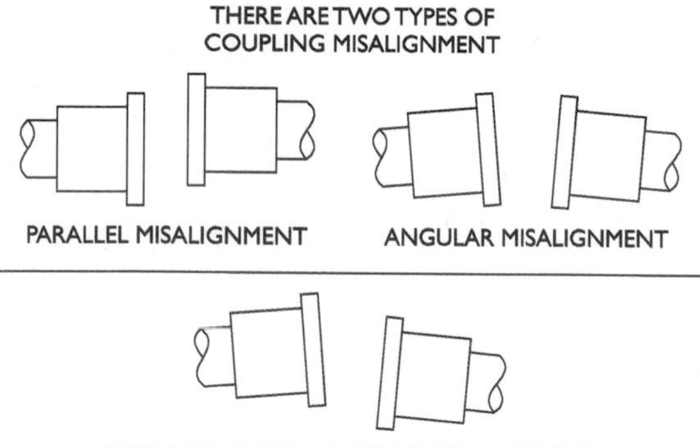

Figure 17-11 Angular and parallel misalignment.

The misalignment of the shaft's centerlines may be located in any plane within the full circle of 360°. Angular misalignment may be a tilt up, down, or to the sides of one shaft with respect to the other. Parallel misalignment may be one shaft high, low, or to the sides of the other. A practical method of correcting misalignment is to align the centerlines in two planes at right angles. The practice commonly followed, because it is convenient, is to check and adjust in the vertical and horizontal planes. Misalignment in these two planes is illustrated in Figure 17-12.

Aligning the centerlines of coaxial shafts in two planes, the vertical and the horizontal, with respect to both angularity and parallelism, requires *four* separate operations—an angular and parallel alignment in the vertical plane and the same alignments in the horizontal plane. Too frequently, these operations are performed in random order, and adjustments are made by trial and error. This results in a time-consuming series of operations; often, some of the adjustments disturb settings made during prior adjustments.

An organized procedure can eliminate the need to repeat operations. When a definite order is established, one operation is completed before another is started. In many cases, when the correct order is followed, it is possible to satisfactorily align a coupling by going *once* through the four operations. When extreme accuracy is required, the operations may be repeated but should always be performed in the correct order.

It was previously stated that alignment operations are performed on the coupling surfaces because they are convenient to use. An extremely important requirement when this is done is that these surfaces, and the shaft, run true. If there is any runout of the shaft or the coupling surfaces, a proportionate error in alignment will be assembled into the coupling connection. Although measurements may indicate true alignment, the alignment will be out of true if the surfaces from which measurements are made are out of true. Therefore, prior to any alignment measurements and adjustments, the condition of the shaft and coupling surfaces should be checked and runout eliminated if present. If runout correction cannot be made, the alignment of the coupling must be done with an indicator, and both shafts must be rotated while making measurements.

Four basic steps are required to correct misalignment:

1. Correct for angular misalignment in the side-view plane (Figure 17-12).
2. Correct for parallel misalignment in the side-view plane (Figure 17-13).

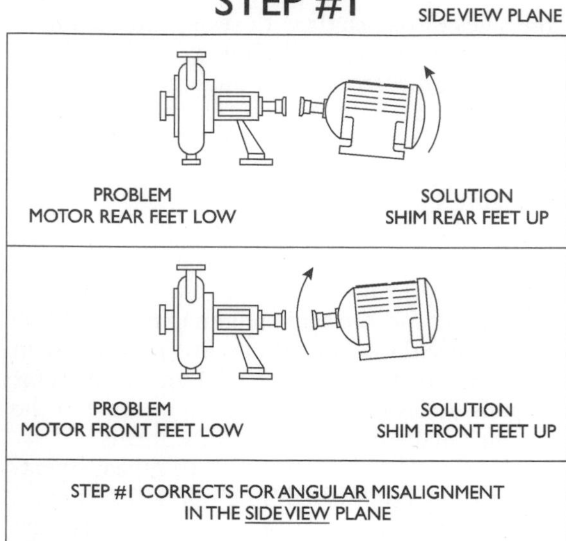

Figure 17-12

3. Correct for angular misalignment in the top-view plane (Figure 17-14).

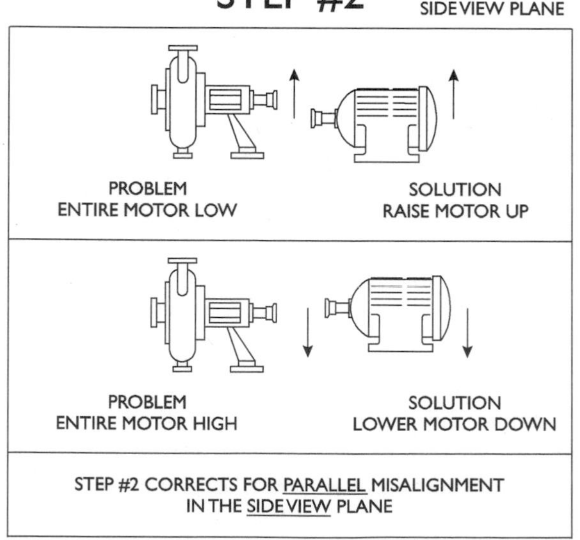

Figure 17-13

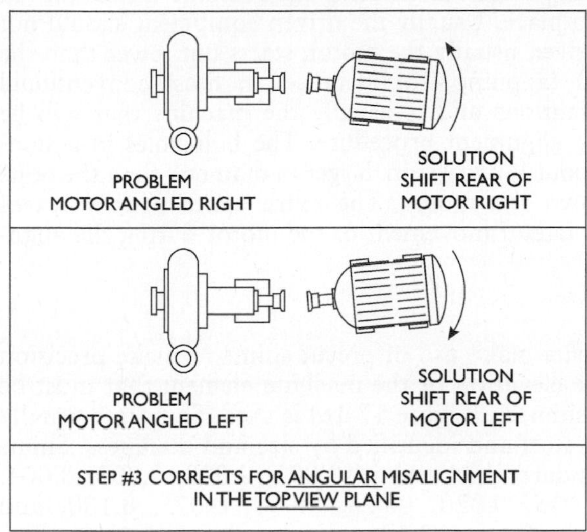

Figure 17-14

4. Correct for parallel misalignment in the top-view plane (Figure 17-15).

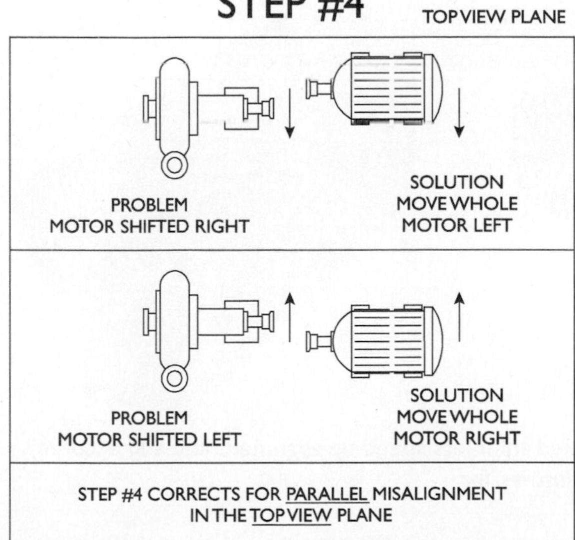

Figure 17-15

When performing a shaft alignment, one machine will be moved and shimmed into place. Usually the driven equipment should not be moved. The driver, usually the motor, starts out lower than the driven equipment (a pump, for instance) in most conventional machinery combinations and is usually the machine that will be moved during the alignment procedure. The bolt holes in a standard motor are about 1⁄16 to 3⁄16 in. larger in diameter than the bolts used to hold down the motor. The extra space usually leaves enough room for lateral movement of the motor during the alignment procedure.

Shimming

Accurate alignments make use of precut shims to make precision movements in the elevation of the machine element that must be moved. Often, a shim kit (Figure 17-16) is used. Shims are usually made of stainless steel and identified by size and thickness. Shims come with a standard thickness of 0.001, 0.002, 0.003, 0.004, 0.005, 0.010, 0.015, 0.020, 0.025, 0.050, 0.075, 0.100, and 0.120, as shown in Figure 17-17, and are selected from a chart (Table 17-1).

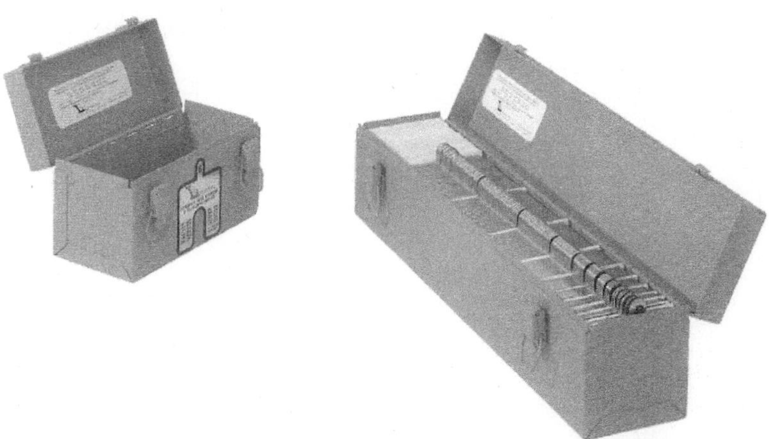

Figure 17-16 Boxed shim kits speed up alignment and add accuracy. *Courtesy Lawton Industries, Inc.*

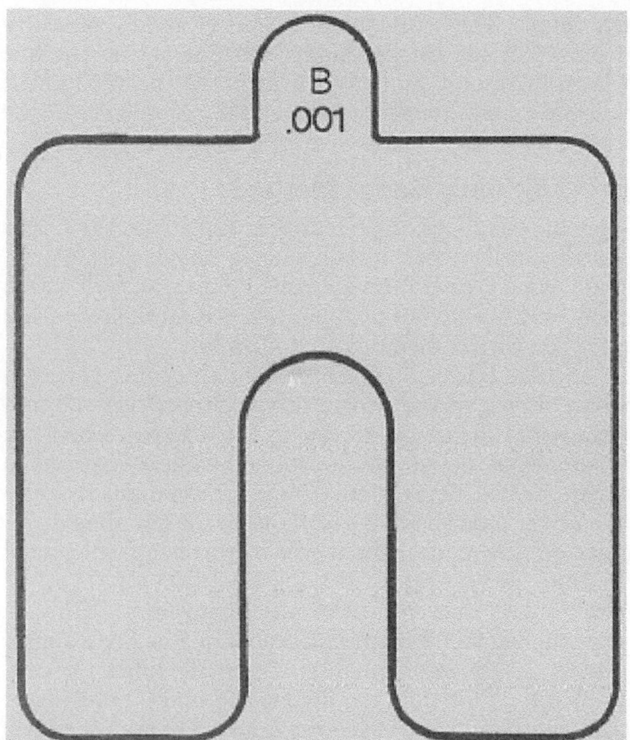

Figure 17-17 Typical shim footprint.
Courtesy Lawton Industries, Inc.

Table 17-1 Sizing Chart for Shim Selection

Motor Horsepower	*.25–15*	*10–60*	*50–200*	*150–1000*	*Over 1000*	*Over 1000*
Overall shim dimensions	2 × 2	3 × 3	4 × 4	6 × 5	7 × 7	8 × 8
Shim slot opening	5/8	13/16	1 1/4	1 5/8	1 3/4	2 1/4
Correct shim size	A	B	C	D	G	H

It is important to use shims that are corrosion resistant and die-cut to make sure that the shim is uniform. There must be a sufficient land area on each shim to prevent shim shifting or crushing under heavy machinery loads. The better shims are cut with "double-entry dies." This means that the thicker shims are flat at the cut line because each die shears only halfway through the metal. Because double-entry dies are used, the shim is not "fat" or thicker at the edges of the metal, which can be the case with single-entry dies.

Straightedge—Thickness Gauge Method

Practically all flexible couplings on drives operating at average speeds will perform satisfactorily when misaligned as much as 0.005 in. Some will tolerate much greater misalignment. Alignment well within 0.005 in. is easily and quickly attainable using the straightedge and thickness gauge when correct methods are followed.

For vertical angular (face) alignment, use a thickness gauge to measure the width of the gap at the top and bottom between the coupling faces as shown in Figure 17-18. Using the difference between these two measurements, determine the shim thickness required to correct alignment. (It will be as many times greater than the misalignment as the driver base length is greater than the coupling diameter.) Shim under the low end of the driver to tilt it into alignment with the driven unit.

For example, assume that measured misalignment is 0.160 in. minus 0.152 in., equaling 0.008-in. misalignment in 5 in. See example illustrated in Figure 17-19. Base length is about 2½ times the coupling diameter; therefore, the shim required is 2½ times 0.008 in., or 0.020 in. A 0.020-in. shim placed under the low end of the driver will tilt it into approximate angular alignment with the driven unit. (Slight error is well within tolerance of flexible coupling alignment.)

For vertical parallel (height) alignment, use a straightedge and thickness gauge to measure the height difference between driver and driven unit on the OD surface of the coupling. Place shims at all driver-support points equal in thickness to the measured height difference. This is illustrated in Figure 17-19.

For horizontal angular and parallel (face and OD) alignment, use a straightedge as shown in Figure 17-20; check alignment of OD's at sides of coupling. Using a thickness gauge as in Figure 17-20, check the gap between coupling faces at the side of the coupling. Adjust driver as necessary to align the OD's at the side and set the gap between faces equal at the sides. Do not disturb shims.

The standard "thickness gauge," also called a "feeler gauge," is a compact assembly of high-quality heat-treated steel leaves of various thickness, as shown in Figure 17-21. The leaves usually vary in thickness by .001 in., and the exact thickness of each leaf is marked on its surface.

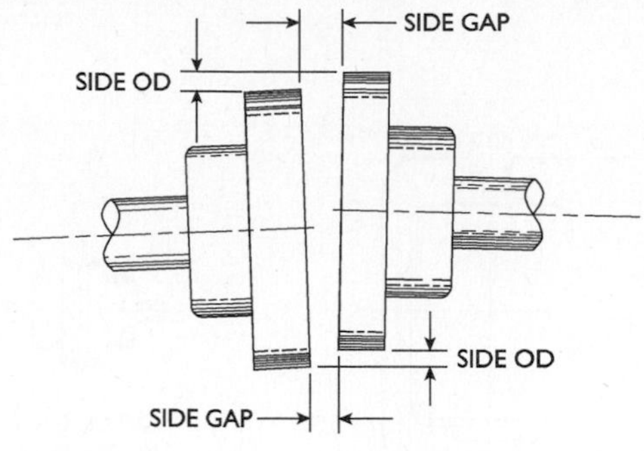

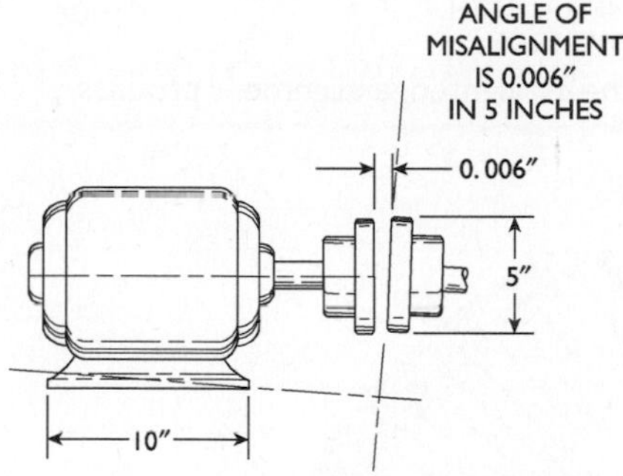

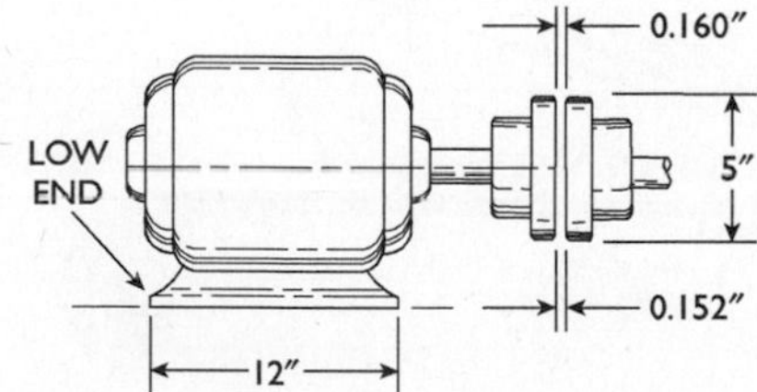

Figure 17-18 Step 1 in the straightedge alignment process.

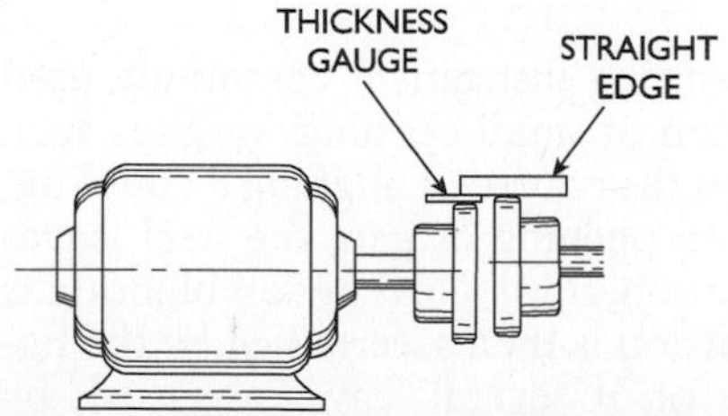

Figure 17-19 Step 2 in the straightedge alignment process.

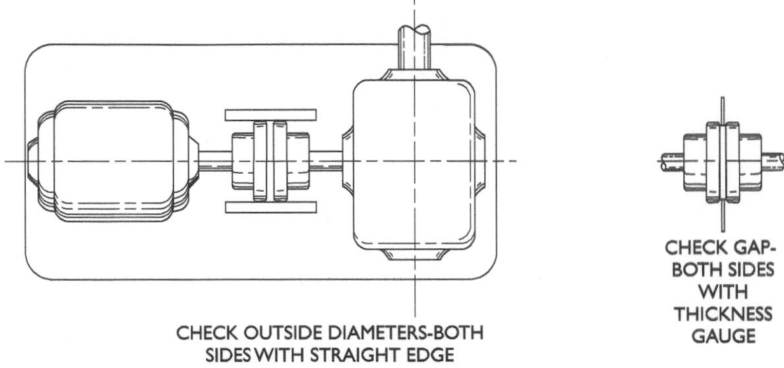

Figure 17-20 Steps 3 and 4 in the straightedge alignment process.

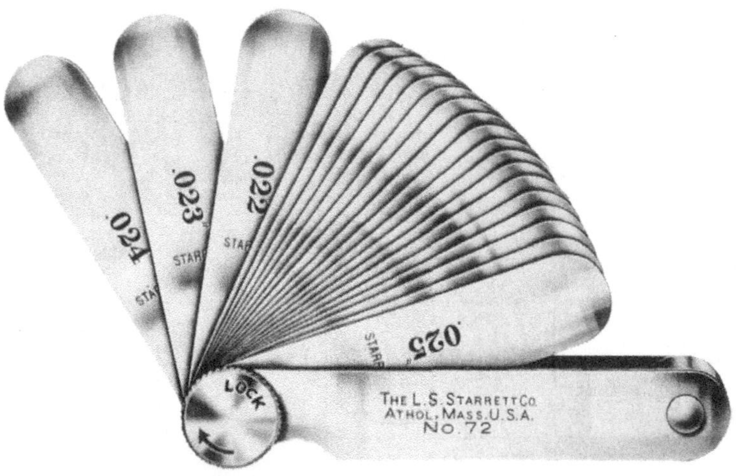

Figure 17-21 Thickness gauge.
Courtesy The L. S. Starrett Co.

A thickness gauge is the measuring instrument commonly used to determine the precise dimension of small openings or gaps such as those that must be measured in the course of aligning a coupling. To determine the dimension of an opening or gap, the steel leaves are inserted, singly or in combination, until a leaf or combination is found that fits snugly. The dimension is then ascertained by the figure marked on the leaf surface or, if several leaves are used, by totaling the surface figures.

Another precision measuring tool, not as widely known or used but ideally suited for coupling alignment, is shown in Figure 17-22. This is the "taper gauge," sometimes called a "gap gauge" by mechanics. Its principal advantage for coupling alignment is that it gives a direct reading and does not require trial-and-error "feeling" to determine a measurement. The tool end is inserted into an opening, or gap, and the opening size is read on the graduated face. Two measurement systems, inch and metric, are shown in the illustration.

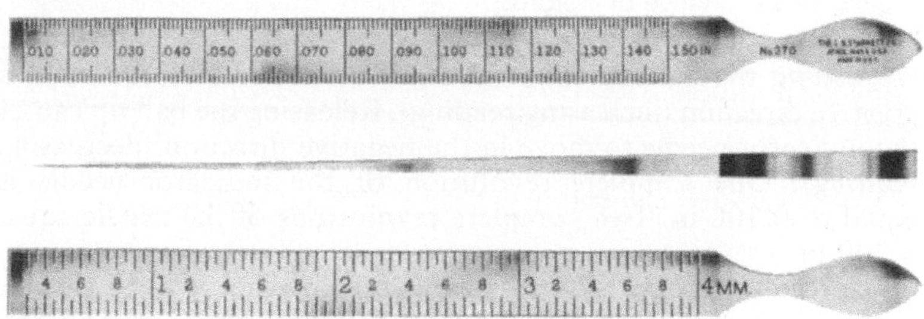

Figure 17-22 Taper gauge.
Courtesy The L. S. Starrett Co.

Coupling Alignment Using Dial Indicator

A dial indicator is a precision tool that can be used in the alignment of equipment (Figure 17-23).

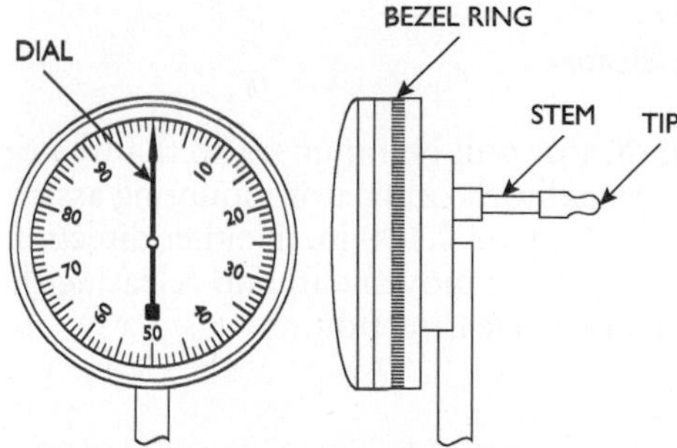

Figure 17-23 The dial indicator.

A common indicator used for shaft alignment has a range of zero to 0.200 in. with 0.001-in. graduations (Figure 17-24).

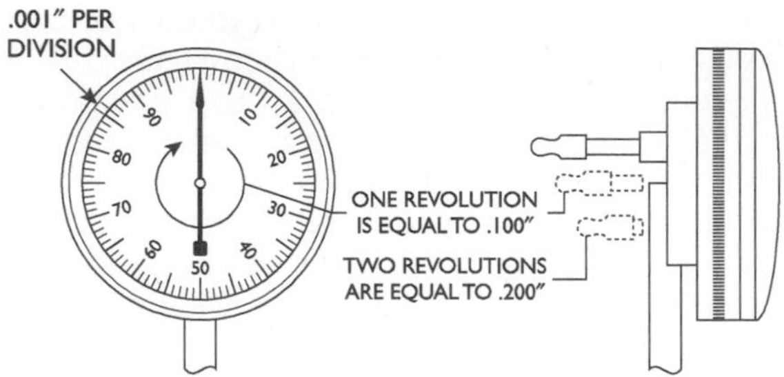

Figure 17-24 Range of indicator.

Pushing the ball tip causes the indicator needle to move in a positive direction (increasing reading). Releasing the ball tip causes the indicator needle to move in the negative direction (decreasing reading). One complete revolution of the indicator needle is equal to 0.100 in. Two complete revolutions of the needle equal 0.200 in.

To zero the indicator, adjust the bezel ring on the outside edge. You can zero the indicator by rotating the bezel ring (Figure17-25).

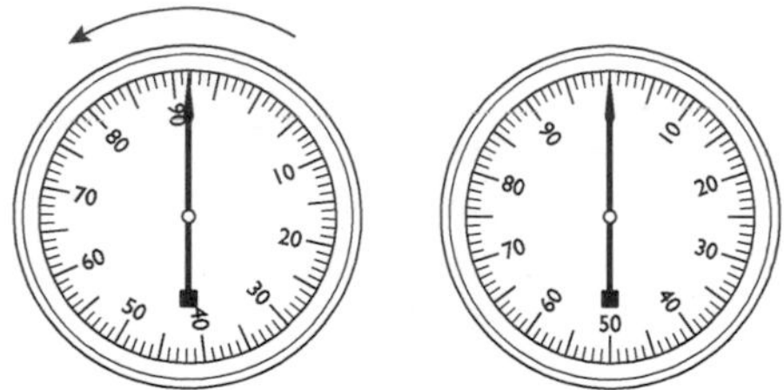

Figure 17-25 Zeroing the indicator.

During normal use, the indicator will be mounted so that the tip can travel in either direction. Usually, the indicator mounting assembly is tightened so that the tip can travel 0.100 in. in either direction. Pushing the tip in will cause a positive movement, and releasing the tip will move the needle in the negative direction.

Conventions

Alignment readings are usually represented on a circle. The position of the indicator is shown in the center of the circle. This is usually represented as M-P, indicating that the readings are being taken from the motor to the pump, or P-M, indicating that the readings are being taken from the pump to the motor. Rim readings are recorded on the outside of the circle, and face readings are recorded on the inside of the circle (Figure 17-26).

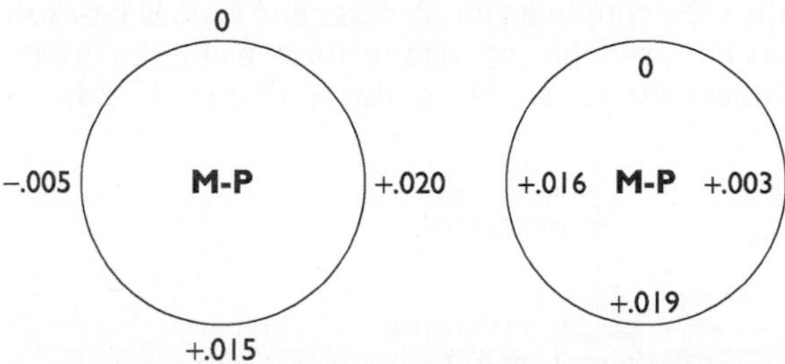

Figure 17-26 Recording rim and face readings.

Using the degrees of a circle as a reference, the data positions on the circle are often referred to as the 0°, 90°, 180°, and 270° readings. Another convention in use refers to the hour positions on a clock face. The corresponding data positions are then referred to as the 12 o'clock, 3 o'clock, 6 o'clock, and 9 o'clock readings, as shown in Figure 17-27.

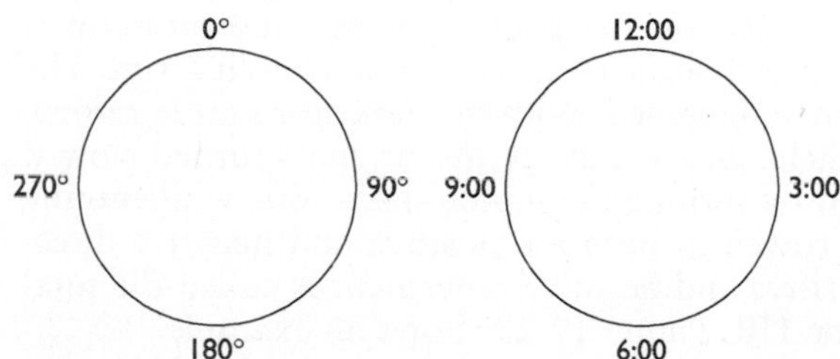

Figure 17-27 Recording readings using a clock-face convention.

Pump and Motor Example

Working through an example of a pump and a motor coupled with a flexible coupling will provide an understanding of proper alignment. To perform an alignment, follow eight important steps.

1. Lock out and tag the machinery for safety.

Make sure the power to the equipment is shut off and locked out with a padlock. Press the start button on the machinery before beginning work to ensure that the power is turned off.

2. Check and eliminate runout due to a bad coupling or a bent shaft. To check for coupling runout, a dial indicator can be mounted on an immovable surface, such as the machinery base. Next, depress the tip of the indicator against the coupling rim. A magnetic base is best for this purpose. If this is not available, an arrangement using the driver and the driven machinery shafts may be employed (Figure 17-28).

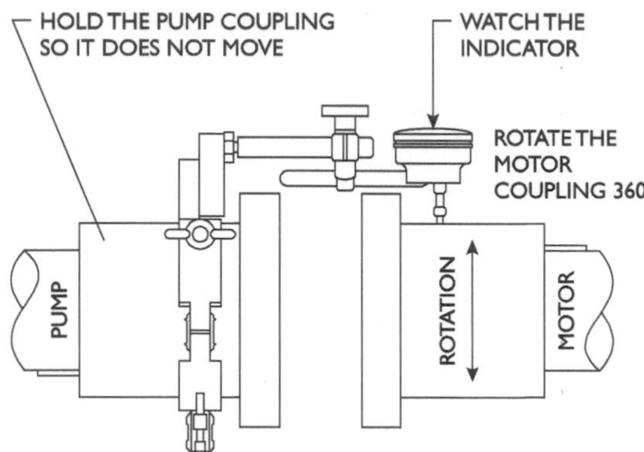

Figure 17-28 Checking runout.

The clamp is mounted on the pump, and the indicator tip is depressed on a machined surface on the motor coupling rim. The whole assembly is then tightened. With the indicator needle zeroed, the pump shaft is held steady and the motor shaft turned slowly. Runout is measured by turning the motor shaft 360° while noting the total indicator travel, in both the positive and negative directions. The total positive and negative movement is called the total indicator reading, or TIR. Figure 17-29 shows an example.

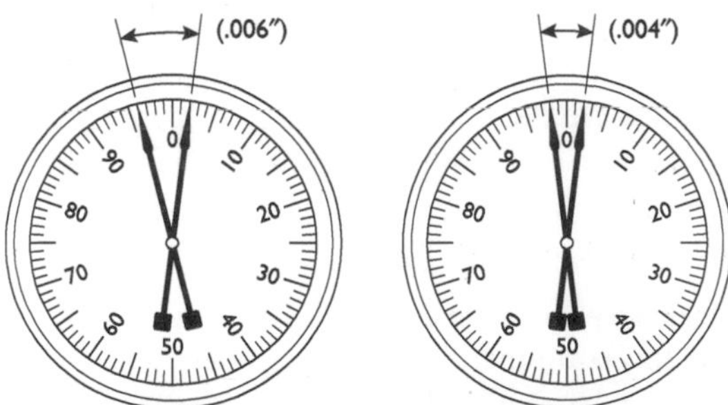

Figure 17-29 Example of runout.

A problem exists if the TIR is greater than 0.005 in. The cause of the problem is either a bent shaft or a bad coupling. To isolate the problem, the coupling half should be removed and the indicator tip depressed on the motor shaft (Figure 17-30). The motor shaft is turned 360° while noting the indicator needle travel in both the positive and negative directions. If the TIR is greater than 0.005 in., then the shaft is bent and must be repaired or replaced prior to attempting an alignment procedure. If the TIR is less than or equal to 0.005 in., the coupling half may be bored off center and should be replaced.

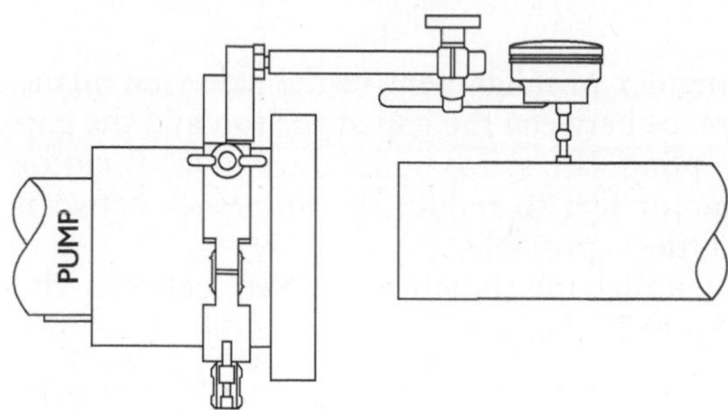

Figure 17-30 Checking runout on the motor shaft.

Next, check the pump shaft by mounting the clamp on the motor shaft and repeating the steps that were described to check the motor shaft and motor coupling. The same standards used to judge the motor shaft apply to replacing or repairing the pump shaft or pump coupling.

3. Find and correct the soft foot.
A machine with four legs tends to distribute its weight on three feet more than on the fourth. This phenomenon is called a soft foot. One foot may actually lift off the base. This condition must be corrected before two shafts can be aligned.

To find and correct the soft foot, clean the area under all the motor feet and the base plate mating surfaces. Starting with a 0.002-in. feeler gauge, attempt to slide the gauge under each foot of the motor (Figure 17-31). If the gauge slides freely under any foot, a soft foot exists. Increase the size of the feeler gauge, and determine the shim size required to correct the soft foot.

4. Perform a rough alignment.
Rough alignment helps eliminate errors and speeds up the use of the dial indicator. A very close rough alignment will result in relatively small corrections measured with the dial indicators.

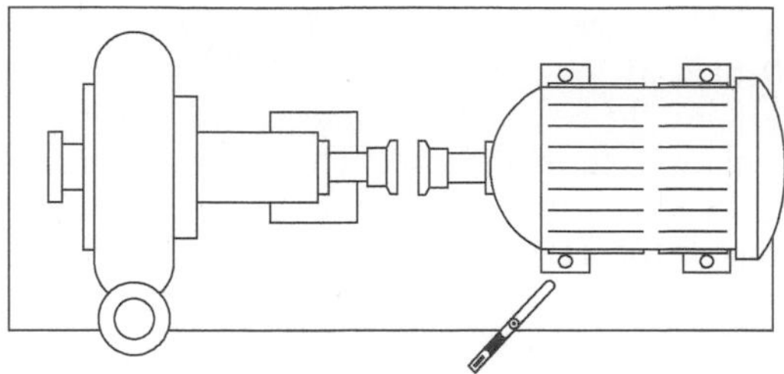

Figure 17-31 Feeler gauge check for soft foot.

First, measure the angular misalignment in the side-view plane by measuring the difference between the gap at the top and the gap at the bottom of the coupling. Use shims under the two front motor feet or the two back motor feet to reduce the difference between both measurements as little as possible.

Next, measure the parallel misalignment (offset) between the two coupling halves (Figure 17-32).

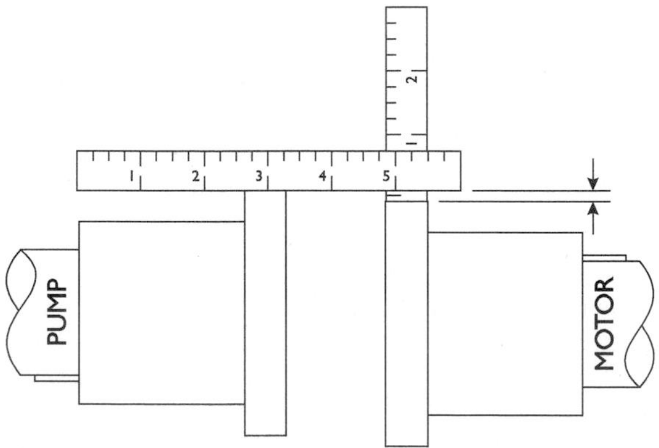

Figure 17-32 Checking for parallel misalignment.

Insert shims under all four motor feet to reduce this offset.

Repeat the procedure in the top-view plane. In this case, use a soft-blow mallet to reduce the difference in the angular and parallel (offset) measurements (Figure 17-33).

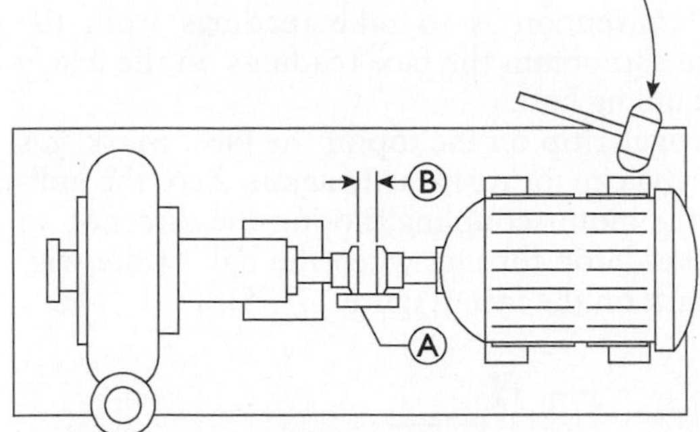

Figure 17-33 Moving using soft blow technique.

5. Correct angular misalignment in the side-view plane.

To correct angular misalignment in the side-view plane, determine the difference remaining in the gap between the top and bottom faces of the two coupling halves. The gap between the two faces is either wider at the top or wider at the bottom. Mount the dial indicator to measure face readings on the coupling halves. This is often easier to accomplish with the indicator mounted on a "wiggler" (Figure 17-34).

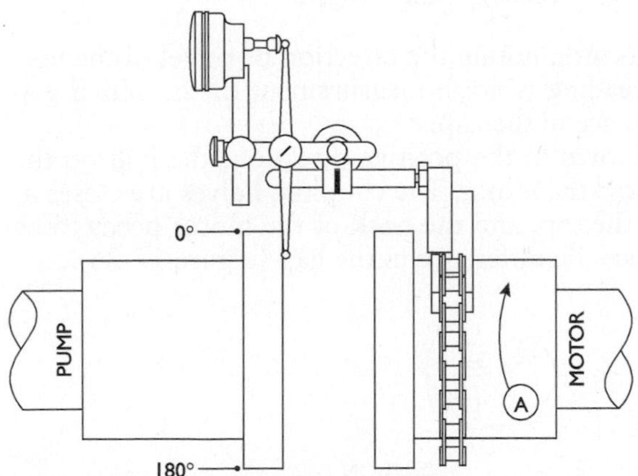

Figure 17-34 Use of the wiggler.

The most common convention is to take readings from the motor to the pump. You can obtain the face readings on the inside or outside face of the coupling half.

Set the indicator or wiggler tip on the top of the face. Mark this point, called 0°, so that you can locate it easily again. Zero the indicator, and slowly turn the motor coupling, noting the direction in which the indicator moves. Stop turning when the ball of the wiggler reaches the 180° mark on the face (Figure 17-35).

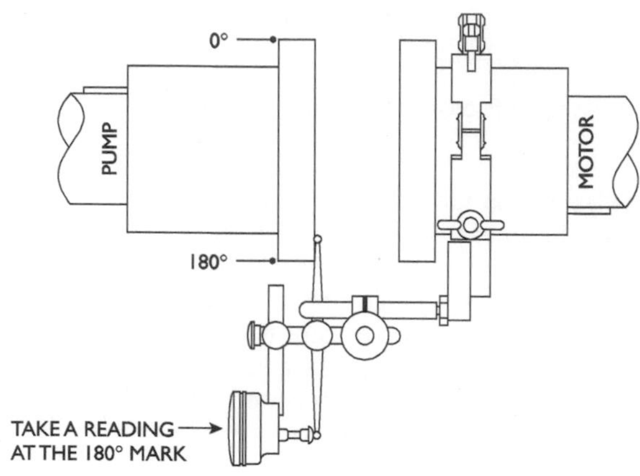

Figure 17-35 Example of reading using a wiggler.

Read the dial indicator, noting the direction of travel of the needle. The indicator reading is not a measurement of the actual gap but rather the difference in the gap.

If the indicator moves in the positive direction, the ball on the wiggler moved *toward* the motor. The coupling halves are closer at the bottom than at the top, and the back of the motor needs to be shimmed up to reduce the difference in the gap (Figure 17-36).

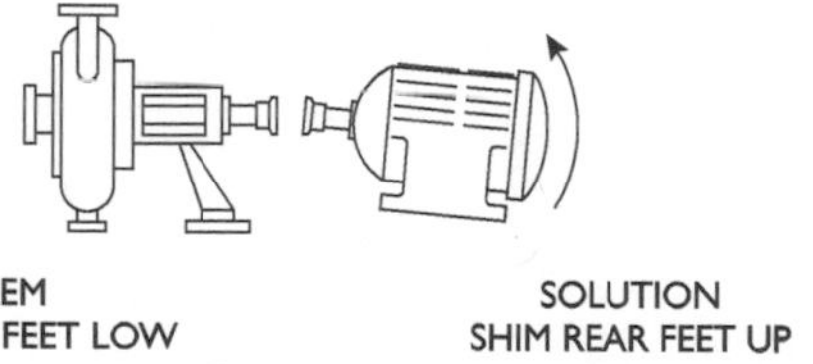

Figure 17-36 Proper shimming—back of motor.

If the indicator needle moves in a negative direction, the ball on the wiggler moved *away* from the motor. The coupling halves are further away at the bottom than at the top, and the front of the motor needs to be shimmed up to reduce the difference in the gap (Figure 17-37).

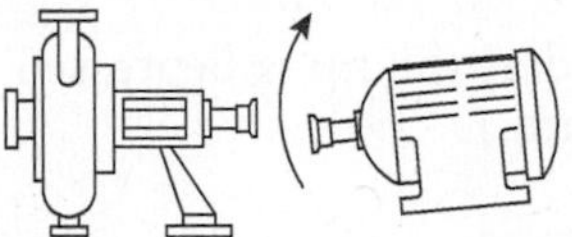

PROBLEM
MOTOR FRONT FEET LOW

SOLUTION
SHIM FRONT FEET UP

Figure 17-37 Proper shimming—front of motor.

The shims required to correct the angular misalignment can be calculated from two dimensions: the *diameter of the coupling* (or, more accurately, the diameter of the circle drawn by the indicator tip) and the *distance from the front bolt hole to the back bolt hole of the motor feet.* Figure 17-38 shows the use of the proper formula and where the data must be collected.

$$\text{Shim Thickness} = \text{Indicator Reading} \times \frac{\text{Motor Bolt Distance}}{\text{Coupling Diameter}}$$

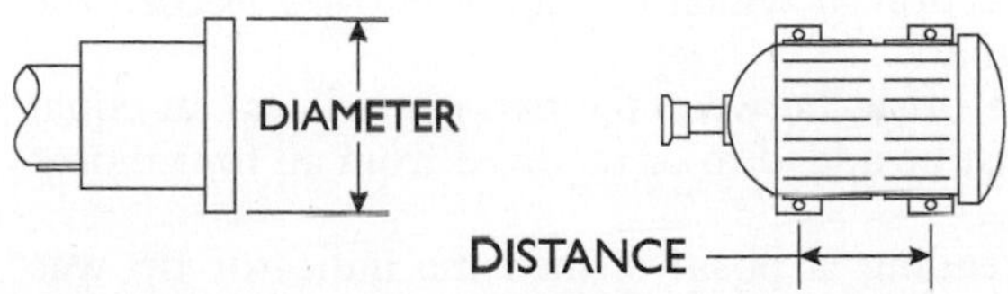

Figure 17-38 Formula for shim calculation.

For example, if the indicator reading is 0.016 in. in the positive direction, the coupling diameter is 4 in., and the motor bolt distance is 12 in., then

$$\text{Shim Thickness} = .016'' \times \frac{12''}{4''} = .048$$

The back end of the motor must be raised 0.048 in. to correct for the 0.16 in. difference between the two coupling halves.

Make the necessary correction, and recheck the readings. Angular misalignment in the side plane is corrected when the reading of the dial indicator is as close to 0.000 in. as possible.

6. Correct parallel misalignment (offset) in the side-view plane. When angular misalignment has been corrected, the two shaft centerlines are parallel but are not necessarily the same in the side-view plane. Make an accurate measurement to determine the offset between the two centerlines.

Mount the dial indictor so that the tip is depressed against the rim of the pump coupling (Figure 17-39).

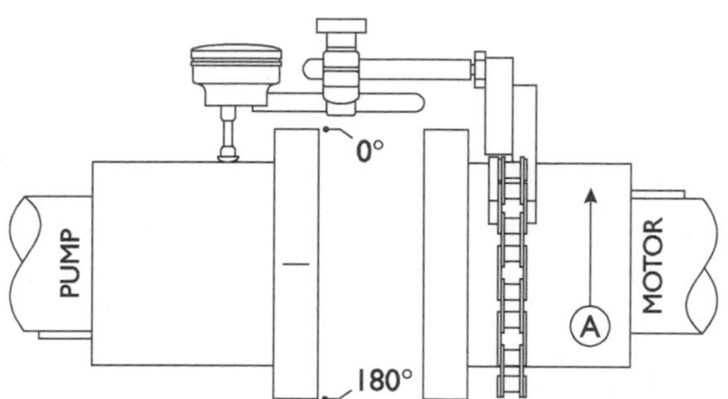

Figure 17-39 Using the dial indicator in the side-view plane.

Zero the needle, and slowly rotate the motor shaft from 0 to 180°, noting the direction in which the needle moves. Record the reading at the 180° mark.

To correct for the offset between the two centerlines, an equal number of shims must be added to or removed from all four motor feet.

If the indicator reading is positive, then the indicator tip was pushed in, so shims must be removed to lower the motor (Figure 17-40).

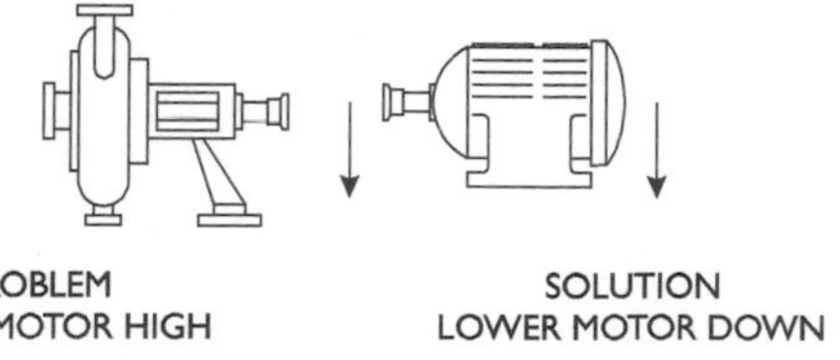

Figure 17-40 Removing shims.

If the indicator reading is negative, the indicator tip was let out, so shims must be added to raise the motor (Figure 17-41).

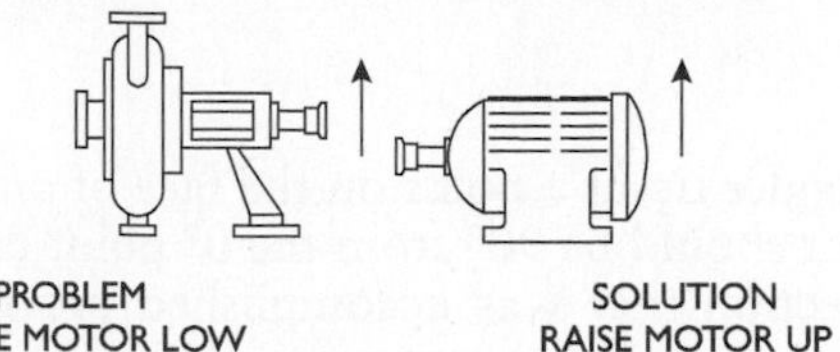

Figure 17-41 Adding shims.

Calculate the shims that must be added or removed from under the motor using the following formula:

$$\text{Shim Thickness} = \tfrac{1}{2} \times \text{Dial Indicator Reading}$$

As an example, assume the indicator had a reading of (negative) 0.064 in. at 180°:

$$\text{Shim Thickness} = \tfrac{1}{2} \times 0.064 \text{ in.} = 0.032 \text{ in.}$$

So, 0.032 in. of shims must be added under each motor foot to bring the motor up and correct for the offset between the two shafts.

After making the required correction, check the readings again. The idea is to make the indicator reading as close to 0.000 in. as possible.

7. Correct angular misalignment in the top-view plane.
After angular and parallel corrections are made in the side-view plane, corrections in the top-view plane should be made. The motor needs to be moved from side to side.

To correct for angular misalignment in the top-view plane, determine the difference remaining in the *gap between either side of the coupling halves*. Mount the dial indicator to measure face readings on the coupling halves using the wiggler (Figure 17-42).

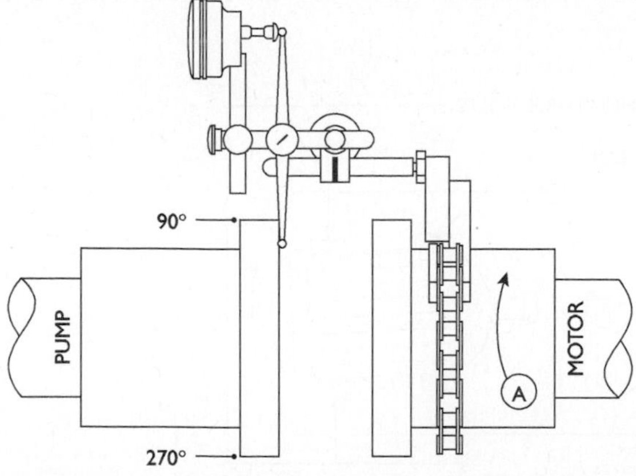

Figure 17-42 Using an indicator to make a face reading.

Set the indicator or wiggler tip at a point on the face of one side of the coupling. This point should be 90° from the 0° point used in the side-view plane correction that was accomplished previously. Zero the indicator, and slowly turn the motor coupling, noting the direction of movement of the indicator. Stop turning when the wiggler ball reaches 270° on the face (Figure 17-43).

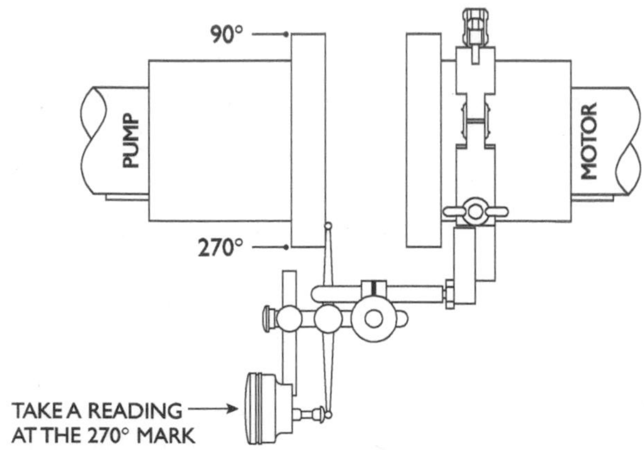

Figure 17-43 Wiggler check at the 280° position.

If the indicator needle moved in the positive direction, the ball on the wiggler has moved toward the motor. The coupling halves are closer at 270° than at 90°, and the back of the motor needs to be moved over to the right to reduce the difference in the gap (Figure 17-44).

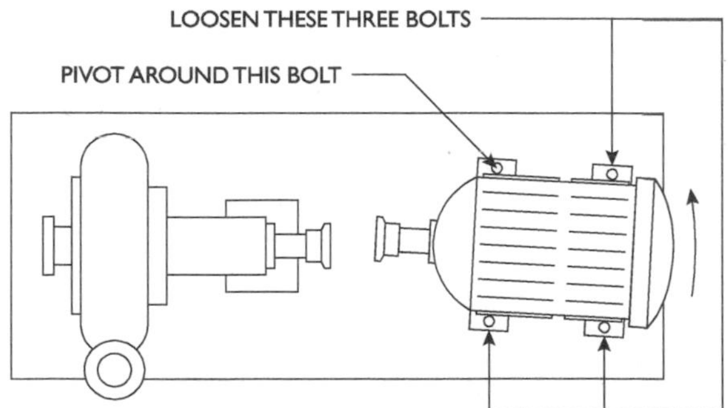

Figure 17-44 Move back of motor right.

If the indicator needle moved in a negative direction, the ball on the wiggler has moved away from the motor. The coupling halves are further apart at 270° than at 90°, and the back of the motor needs to be moved over to the left to reduce the difference in the gap (Figure 17-45).

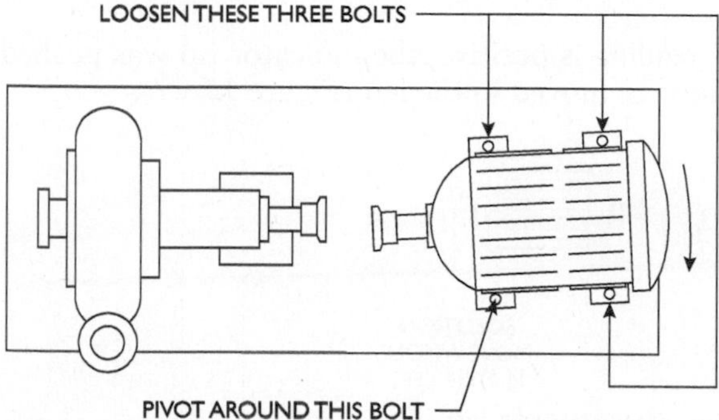

Figure 17-45 Move front of motor left.

Use a soft-blow mallet to move the back of the motor. Make the necessary corrections, and recheck the readings. Angular misalignment in the top-view plane is corrected when the dial indicator reading is as close to 0.000 in. as possible.

8. Correct parallel misalignment in the top-view plane.
When angular misalignment has been corrected in the top-view plane, the two shaft centerlines are parallel but not necessarily the same. Make accurate measurements to determine parallel misalignment (offset) between the two centerlines.

Mount the dial indicator so that the tip is depressed against the rim of the pump coupling (Figure 17-46).

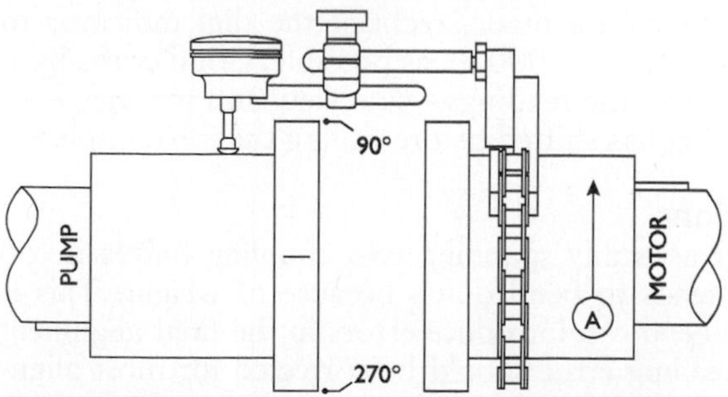

Figure 17-46 Depress tip and zero the needle.

Zero the needle, and slowly rotate the motor shaft from 90° to 270°, noting the direction in which the needle moves. Record the reading at 270°.

To correct for the parallel misalignment (offset) between the two centerlines, the whole motor must be moved to one side or the other. Take care to move the motor in a straight line perpendicular to its axis of rotation.

If the indicator reading is positive, the indicator tip was pushed in, so the motor must be moved to the left (Figure 17-47).

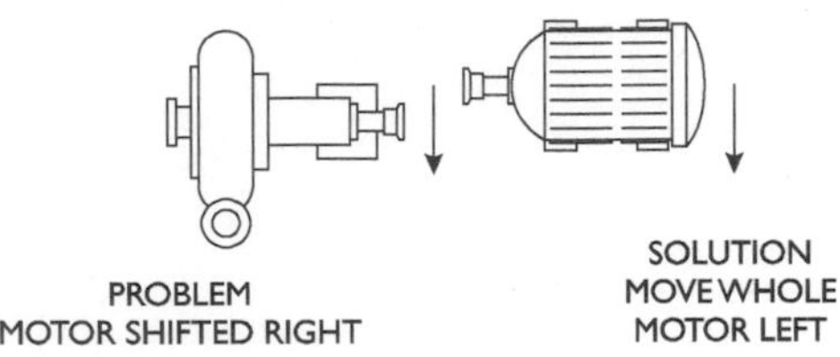

Figure 17-47 Move whole motor left.

If the indicator reading is negative, the indicator tip was let out, so the motor must be moved to the right (Figure 17-48).

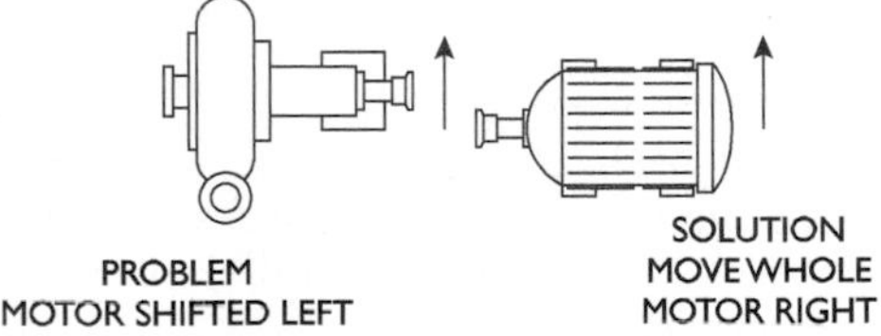

Figure 17-48 Move whole motor right.

Use a soft-blow mallet to move the motor over carefully. When the necessary corrections are made, recheck the dial indicator to produce a reading as close to 0.000 in. as possible. Good craftsmanship dictates checking all the readings—side view and top view—to make sure that nothing has shifted before calling the job complete.

Bar Sag Corrections

When an alignment assembly spanning two coupling halves is too long, the assembly tends to bend down because of weight. This is called bar sag. This bend can introduce errors in the final alignment readings. The rim reading error should be corrected for most alignment assembly spans longer than 3 in. Face sag usually can be ignored.

Figure 17-49 shows the effect of bar sag on two perfectly aligned coupling halves.

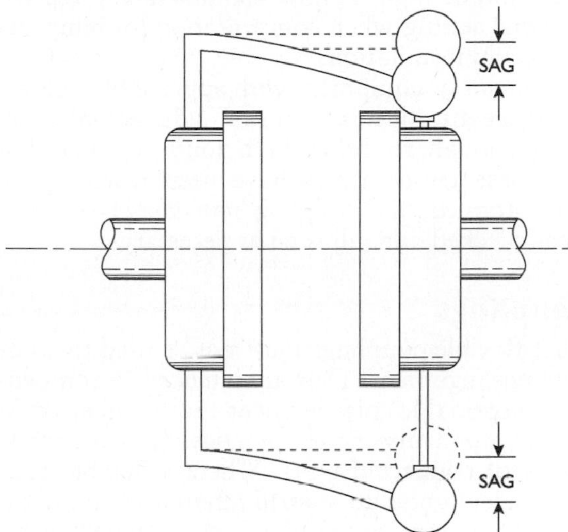

Figure 17-49 Note the effect of bar sag.

The indicator assembly sags when on the top. With the indicator zeroed at the 0° point, the assembly is turned to the 180° point. The indicator assembly now sags in the other direction. The total indicator reading will be the sum of the two sags.

Bar sag is easy to determine. After taking the parallel readings in the side-view plane, remove the assembly and remount it on a shaft or piece of pipe. With the indicator zeroed, turn the rig upside down and read the sag indicated. As an example, if the indicator read -0.008 in., the sag value is 0.008 in. To correct the readings for bar sag, simply add the sag value to the 180° reading, or

Corrected 180° Rim Reading − 180° Rim Reading (TIR) + Sag Value

Temperature Change Compensation

To compensate for temperature differences between installation conditions and operating conditions, it may be necessary to set one unit high or low when aligning. For example, centrifugal pumps handling cold water that are directly connected to electric motors require a low motor setting to compensate for expansion of the motor housing as its temperature rises. If the same units were handling liquids hotter than the motor operating temperature, it might be necessary to set the motor high. Follow manufacturers' recommendations for the initial setting when compensation for temperature change is made at cold installation.

Make the final alignment of equipment with appreciable operating temperature difference after it has been run under actual operating conditions long enough to bring both units to operating temperatures. When these temperatures have been reached, the equipment should be stopped, the coupling immediately disconnected, and alignment checked and adjusted as necessary.

Offset Shaft Couplings

It has been stated that flexible couplings must not be used to compensate for deliberate misalignment. They are intended to compensate for only slight, unavoidable misalignment that is inherent in the design of machine components and the practices followed when aligning coaxial shafts of connected units. Whereas flexible couplings are designed to accommodate a small amount of misalignment, offset shaft couplings are designed to accommodate large amounts.

If a coupling is subjected to shaft misalignment, then undesirable side loads may be introduced by the coupling. These side loads may be the result of frictional loads or loads caused by flexing or squeezing material. Such side loads exert excessive forces that overheat bearings and shorten the life of machine components, as well as increase vibration and noise, which can also be detrimental. The offset coupling is designed to avoid these side loads and transmit power through relatively large misaligned shafts.

Offset shaft couplings were developed to provide for wide ranges of misalignment in all four categories of basic shaft movements: axial, parallel, angular, or torsional. One of the great advantages of these couplings is that they allow the drive components to be installed in fixed position, thus simplifying design. Heavy motors, reducers, etc., can be installed in fixed position, making possible movement of driven components only. This can be an enormous advantage for feed drives, embossing rolls, straighteners, and all manner of operations where offset shaft capability is an advantage.

Illustrated in Figure 17-50 are several styles of offset couplings designed to provide large parallel shaft offset at high accuracy with no side loads. These couplings provide features that are advantageous for applications in industries such as papermaking and converting, woodworking, textile operations, machine tools, printing, steelmaking, and many others. They are available in a wide range of sizes and capabilities. Shaft displacement for the smaller sizes can be in fractions of an inch and up to 24 inches in the very large special sizes. As shown in the illustration, the coupling's capabilities are made possible by a design that incorporates two pairs of parallel links installed 90° out of phase with each other. This allows for transmission of torque and constant angular velocity through relatively large misaligned parallel shafts.

Another offset coupling design that incorporates an angular misalignment capability, as well as that of parallel misalignment, is illustrated in Figure 17-51. This is a modification of the parallel offset coupling shown in Figure 17-50. It is designed to accommodate all four basic shaft movements, thus allowing adjustment to any possible misaligned shaft position. Because of its large misalignment capabilities, this style of coupling can replace spindle drives and save important floor space for steel mills, mining applications, and similar installations. The capability for angular misalignment is accomplished by installing spherical-roller bearings in the parallel links rather than the needle- or straight-roller bearings utilized in the coupling designed for offset misalignment.

Figure 17-50 Offset couplings.
Courtesy Schmidt Couplings Inc.

Figure 17-51 Modified offset coupling.

Chapter 18

Gaskets, Packings, and Seals

A *gasket* is a material that is used to seal two faces of a machine. Gaskets can be made of soft materials, such as asbestos or elastomers, or they can be made of harder materials, such as metal ring gaskets made of iron, steel, and other materials. Combinations of materials—such as spiral-wound metal/asbestos-filled gaskets—are also common.

Gaskets are designed for compressibility and sealability. Compressibility is a measure of the gasket's ability to deflect and conform to the faces being sealed. Gasket compressibility compensates for surface irregularities, such as minor nicks, nonparallelism, corrosion, and variations in groove depth. Sealability is the measure of fluid leakage through and across both faces of a gasket. Most of the leakage for a properly installed face-to-face connection will occur through the gasket.

Types of Gaskets

Gaskets are classified usually either by the material from which they are made and the type of construction or design or in some cases—both.

Flat Gasket

Flat gaskets (Figure 18-1) are gaskets cut from flat stock of gasket material. The material can be an elastomer, either natural (rubber) or synthetic (hypalon, viton, SBR). Fiber material is often formed with a binder into flat-stock gasket material. Flat stock comes in standard thickness, ranging from 1/64 to 1/4 in. Flat-stock gaskets are often cut into full-face gaskets, in which the gasket design incorporates the appropriate bolt-hold pattern and the flange bolts are used to center the gasket in place. The full-face design is most often used with the flat-face flange design. Ring-face gaskets are cut so that the outer diameter of the gasket rests inside the bolt pattern, and the gasket is centered by resting it on the flange bolts.

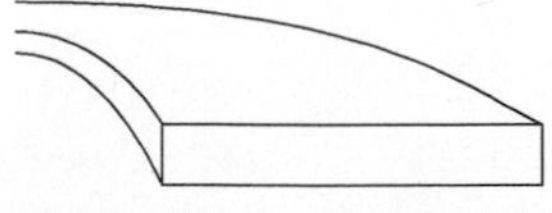

Figure 18-1 Flat gasket.

Envelope Gasket

Envelope gaskets (Figure 18-2) consist of an elastomeric material protected by another material as a jacket. TFE is a common envelope material because of its resistance to many chemicals. Envelope gaskets are usually ring-face sized so that the gasket is centered by resting on the flange bolts. Different metals can be used as the envelope material.

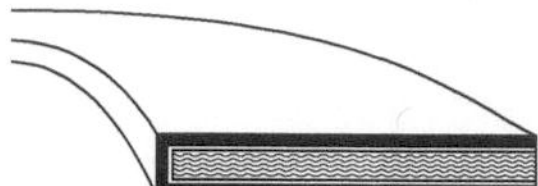

Figure 18-2 Envelope gasket.

Spiral-Wound Metal-Filled Gaskets

Spiral-wound metal-filled gaskets (Figure 18-3) are the most common and popular. They consist of a thin metal spiral separated by different filler materials, depending on service conditions. They are available in full-face design, but they are most commonly employed as ring-face design. The spiral-wound section of the gasket exists only where the flange faces meet. The remainder of the gasket consists of a backing ring. Special spiral-wound gaskets with no backing rings are used in some tongue-and-groove and male-and-female joint designs.

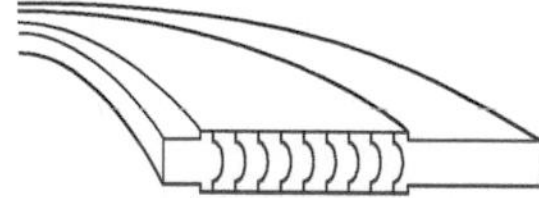

Figure 18-3 Spiral-wound metal-filled gasket.

Grooved Metal and Solid Flat Metal Gaskets

Grooved metal and *solid flat metal gaskets* (Figure 18-4) are most often employed in tongue-and-groove and male-and-female joint design.

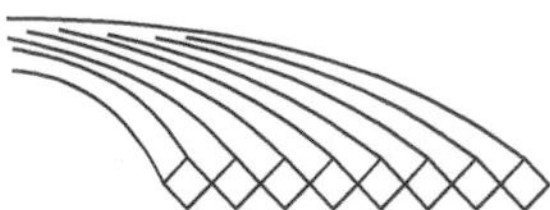

Figure 18-4 Grooved metal and solid flat metal gaskets.

Metal Ring Joint Gaskets

Metal ring joint gaskets (Figure 18-5) are either oval or octagonal in cross section. They are used exclusively in ring joint flange configurations.

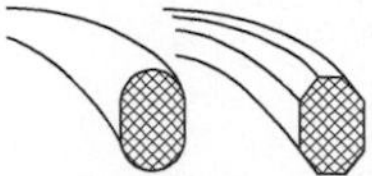

Figure 18-5 Metal ring gasket.

Choosing a Gasket

When choosing a gasket, remember that it must perform under the system temperature and pressure conditions. Gasket manufacturers commonly list the maximum temperature and maximum pressure ratings of their gasket materials. They also give the maximum pressure times temperature ($P \times T$) rating. Table 18-1 gives P and T data for several common gaskets.

Table 18-1 Gasket Materials' Physical Properties

Material	*Temperature Max. (°F)*	*Pressure Max (psi)*	*P × T, Max.*
Natural rubber	200	100	15,000
SBR	200	100	15,000
Neoprene	250	150	20,000
Nitrile	250	150	20,000
EPDM	300	150	20,000
Asbestos/rubber binder	900	3000	350,000
Asbestos/SBR binder	750	1800	350,000
Asbestos/neoprene binder	750	1500	350,000
Asbestos/nitrile binder	750	1500	350,000

A *seal* may be described as a device for controlling the movement of fluids across a joint or opening in a vessel or assembly. *Packings* are a form of seal commonly associated with a condition where relative motion occurs—for example, the sealing of moving parts, either rotating or reciprocating, at the point of entry to a pressure chamber. Various types of packing material are also used to seal stationary parts, such as cylinder heads and flanges, and internally between sections.

Stuffing Boxes

The oldest and still one of the most widely used seals is the mechanical arrangement called a *stuffing box*. It is used to control leakage at the point where a rod or shaft enters an enclosed space that is at a pressure above or below that of the surrounding area. See Figure 18-6.

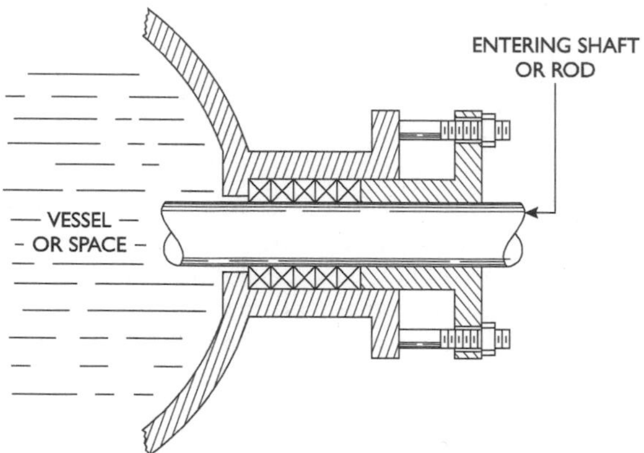

Figure 18-6 A typical stuffing box.

A plain stuffing box is composed of three parts: (1) the packing chamber, or box; (2) the packing rings; and (3) the gland follower, or stuffing gland (Figure 18-7).

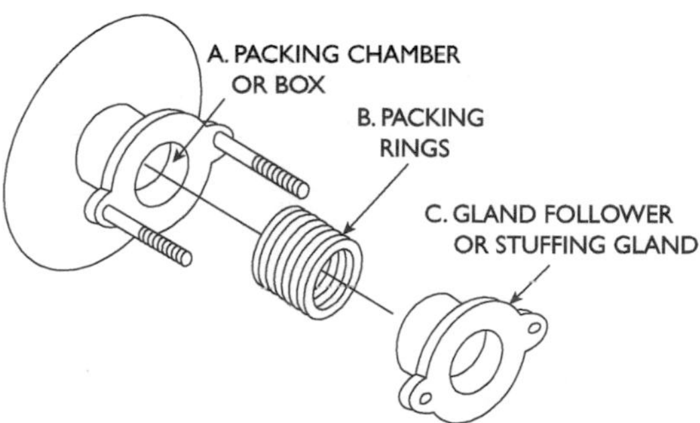

Figure 18-7 The three basic components of the stuffing box.

Sealing is accomplished by squeezing the packing between the throat of the box and the gland. The packing is subjected to compressive forces that cause it to flow outward to seal against the bore of the box and inward to seal against the surface of the moving shaft or rod. See Figure 18-8.

Leakage is controlled by the intimate contact between the packing and the surface of the rod shaft or sleeve and between the outside surface of the packing and the inside bore surface of the stuffing box. Leakage through the packing material is prevented by the lubricant contained in the packing. As the packing is compressed, it must have the ability to deform in order to seal. It must also have a certain ruggedness of construction so that it may be readily cut into rings and assembled into the stuffing box without serious breakage or deformation.

Packing used in stuffing boxes is referred to as soft or compression packing. It requires frequent adjustment to compensate for the wear and loss of volume that occurs continuously while it is subjected to operating conditions. A fundamental rule for satisfactory operation of a stuffing box is that *there must be controlled leakage.* This is necessary because, in operation, a stuffing box is a form of braking mechanism that generates heat. The frictional heat is held to a minimum by the use of smooth, polished shaft surfaces and a continuous supply of lubricant from the packing to the shaft-packing interface. The function of stuffing-box leakage is to assist in lubrication and to carry off the generated heat. Maintaining packing pressures at the lowest possible level helps to keep heat generation to a minimum.

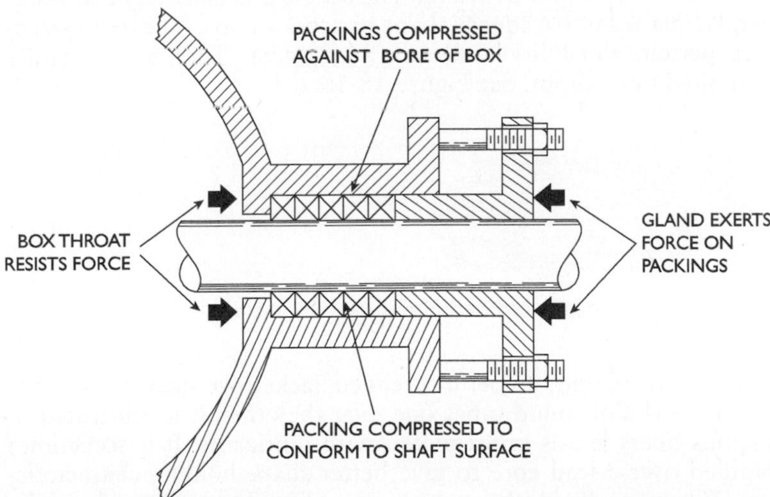

Figure 18-8 The principal operating features of the stuffing box.

Packing Materials and Construction

Compression packing is manufactured from various forms of fibers, such as vegetable, animal, mineral, or synthetic, combined with binders and lubricants. The four principal types of construction are twisted, square-braided, interlocked-braided, and braid-over-braid.

The twisted type, which is the simplest and a widely used type of packing, is usually made from twisted strands of asbestos or cotton lubricated with mineral oil and graphite. Twisted packing is not as strong as the braided type, but it is quite easy to install. One size of packing may be adapted to many different stuffing-box sizes. A twisted packing of a large diameter may be adapted to small stuffing boxes by removing a few strands. See Figure 18-9.

Figure 18-9 Twisted fibers used in packing construction.

Square-braided packing is made of asbestos, cotton, plastic, or leather and sometimes includes metal wires such as lead or copper. It is usually grease- or oil-impregnated. It is made by gathering a number of yarns into a strand. The strands, usually eight in number, are plaited into a square shape or cross section. The loose structure permits flexibility and easy adjustment. They are generally graphited throughout. See Figure 18-10.

Figure 18-10 Square-braided fibers used in packing.

Braid-over-braid, sometimes called jacket-over-jacket, is made up of a series of round tubes one over the other. It is fabricated of various fibers and is impregnated with lubricants. It is sometimes braided over a lead core to give better shape-holding characteristics. When graphited, the graphite is ordinarily applied only to the outside. See Figure 18-11.

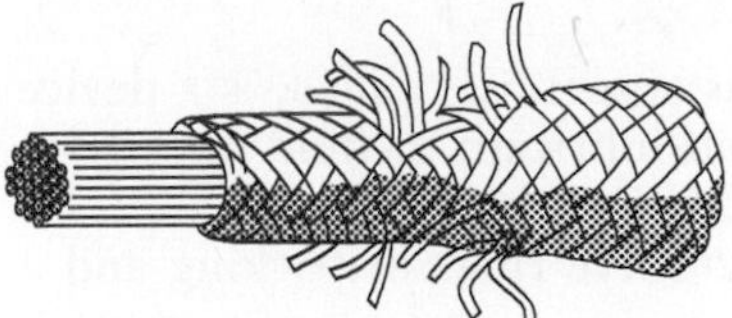

Figure 18-11 Braid-over-braid packing construction.

Interlocked-braided packing has some of the characteristics of both braided and braid-over-braid. All yarns in this type of construction are interlocked to form a solid square braid that is extremely resistant to unraveling. It is most frequently supplied with oil lubrication and is available in asbestos, cotton, plastic, and other fibers. See Figure 18-12.

All compression packing in the various types discussed above is impregnated with a variety of lubricants. Approximately 30 percent of the total volume of compression packing is lubricant.

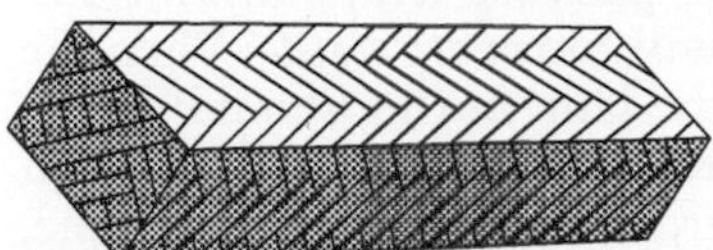

Figure 18-12 Interlocking fiber packing construction.

Stuffing Box Arrangement

The function of the multiple rings of packing in a stuffing box is to break down the pressure of the fluid being sealed so that it approaches zero gauge pressure, or atmosphere, at the follower end of the box. Theoretically, a single ring of packing, properly installed, will seal. In practice, the bottom ring in a properly installed set of packing does a major share of the sealing operation. Because the bottom ring is the farthest from the follower, it has the least pressure exerted upon it. Therefore, to perform its important function of a major pressure reduction, it is extremely important that it be properly installed. Correct installation of compression packing will be discussed later.

Experience indicates that a practical stuffing-box packing arrangement is five rings of packing and a lantern ring. Usually three rings of packing are placed in the bottom of the box, followed by a lantern ring and the two remaining packing rings.

Lantern Rings

The lantern ring, or seal cage as it is sometimes called, is a device that allows introduction of additional lubricant or fluid directly to the interface of packing and shaft. Illustrated in Figure 18-13 is a newly packed stuffing box containing five rings of packing and a lantern ring.

A common practice with a pump that has a suction pressure below atmospheric pressure is to connect the pump's discharge to the lantern ring. Fluid introduced through the lantern ring acts as both a seal to prevent air from being drawn into the pump and as a lubricant and cooling medium for the packing. Lantern rings are also commonly used in pumps handling slurries, particularly if the slurries contain abrasive materials. In this case, clear liquid from an external source, at a higher pressure than the slurry, is introduced into the stuffing box through the lantern ring.

Stuffing boxes incorporating lantern rings require special attention when packing. The lantern ring must be positioned between the packing rings so that its front edge is in line with the inlet port, as illustrated in Figure 18-13. As the packing wears and the follower is tightened, the lantern ring will move forward. When the packing has been fully compressed, the lantern ring should still be in line with the inlet port, as illustrated in Figure 18-14. Although lantern rings may, on occasion, be troublesome to the mechanic, they should not be discarded because they are an important part of the stuffing-box assembly.

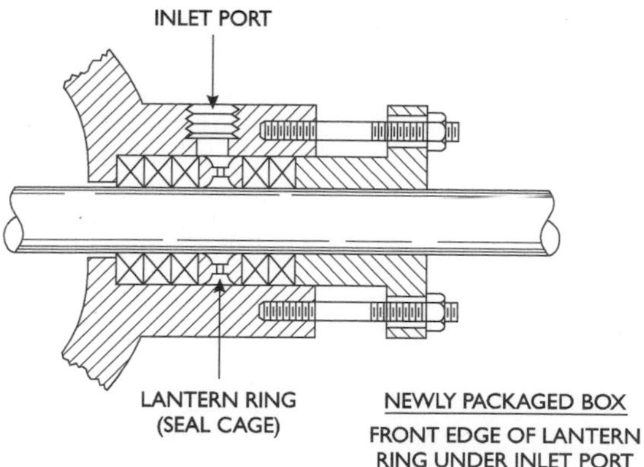

Figure 18-13 Position of the lantern ring when installing new packing.

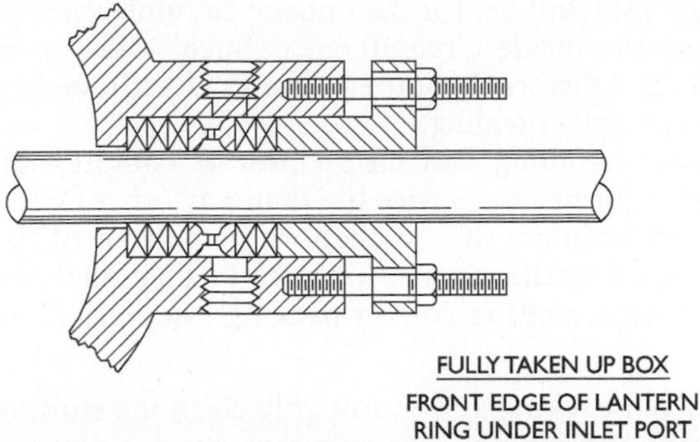

Figure 18-14 Position of the lantern ring when packing is compressed.

Packing Installation

Although the packing of a stuffing box appears to be a relatively simple operation, it is often done improperly. It is generally a hot, dirty, uncomfortable job that is completed in the shortest possible time with the least possible effort. Short packing life and damage to shaft surfaces can usually be traced to improper practices rather than deficiencies in materials and equipment. For example, a common improper practice is to lay out packing material on a flat surface and cut it to measured length with square ends.

When lengths of packing with square ends are formed into rings, they have a wedge-shaped void at the ends, as illustrated in Figure 18-15. The thicker the packing in relation to the shaft size, the

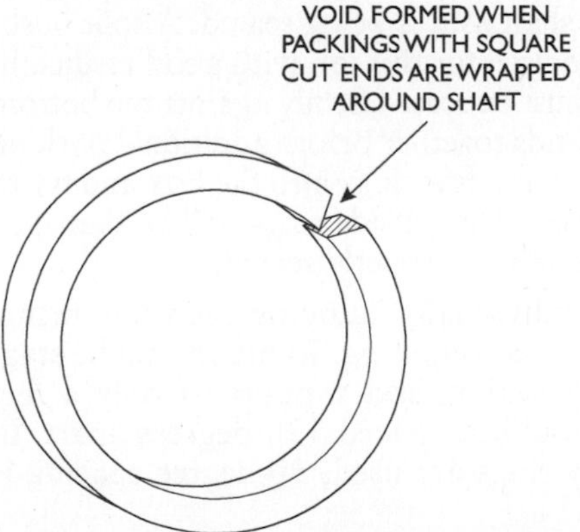

Figure 18-15 Cutting ends at angle will eliminate the pie-shaped void shown in the illustration.

more pronounced the void will be, for the outside circumference of a ring is greater than the inside circumference. Such voids cause unequal compression and distortion of the packing. Overtightening is then required to accomplish sealing.

Actually, the proper handling and installation of compression packing has a greater influence on service life than any other factor. Packing manufacturers estimate that as high as 75 percent of the packing life is determined by the manner in which it is handled and installed. The basic critical steps in correct packing installation are as follows:

1. Remove all the old packing and thoroughly clean the stuffing box. Check ports and piping to be sure they are not plugged and are free from obstruction. Remove all accumulations inside the box and on the surface of the rod or shaft.
2. Cut packing rings on the shaft or on a mandrel of the same diameter as illustrated in Figure 18-16. This is very important, as each ring must fully fill the packing space with no gap at the ends.
3. Keep the job simple and cut rings with plain butt joints. There is no advantage to a diagonal or skive cut.
4. Form the first packing gently around the shaft and enter the ends first into the box. The installation of this first packing ring is probably the most critical step in packing installation. If this operation is not properly performed, a satisfactory seal cannot be attained. The first ring should be gently pushed forward using tamping tools with flexible shafts, which will aid in keeping the packing square with the shaft as it is being seated. A split bushing of proper size may be used for this job with good results. In any event, the first ring must be seated firmly against the bottom of the box with the butt ends together before additional packing rings are installed. Never put a few rings into the box and try to seat them with the follower. The outside rings will be damaged and the bottom rings will not be properly seated.
5. Insert additional rings individually, tamping each one firmly into position against the preceding ring. Joints should be staggered to provide proper sealing and support. If only a few rings are used, joints should be spaced 120 degrees apart. In installation where many rings are used, 90-degree spacing is satisfactory.
6. Position the lantern ring correctly as previously discussed and illustrated in Figure 18-13.

7. Insert individually the remaining rings required to fully pack the stuffing box.
8. Install the gland follower. In a properly designed stuffing box, the gland follower, when the box is newly packed, will enter only a small amount. The gland portion should extend out of the box an amount equal to ⅓ of the packing depth. When the follower is fully tightened, this will allow compressing the packing to ⅔ of its original volume.
9. Tighten the follower snugly while rotating the shaft by hand. When done in this manner, it is immediately apparent if the follower becomes jammed as a result of cocking or if the packing is overtightened. Slack off and leave finger tight.
10. Open valves to allow fluid to enter equipment. Start equipment; fluid should leak from the stuffing box. If leakage is excessive, take up slightly on gland follower. Do not eliminate leakage entirely; slight leakage is required for satisfactory service. During the first few hours of run-in operation, the equipment should be checked periodically, as additional adjustment may be required.

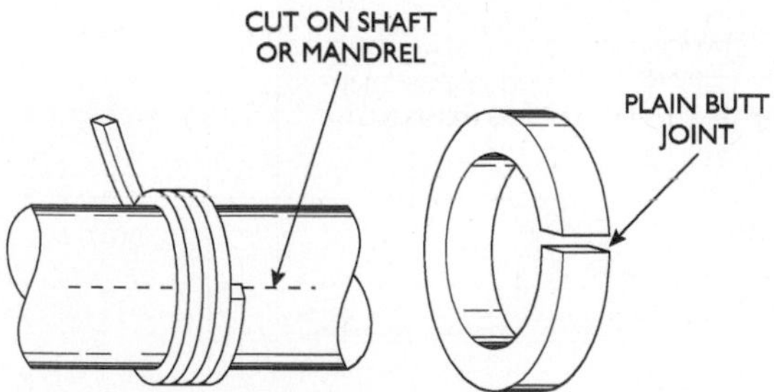

Figure 18-16 One example of cutting packing material.

Mechanical Seals

The sealing of rotating shafts at the point of entrance to a pressure chamber has for years been accomplished with a stuffing-box seal. A serious deficiency of the stuffing-box type of seal is its leakage requirement for satisfactory operation. Although still widely used because of its simplicity and ability to operate under extremely adverse conditions, the mechanical seal has replaced it in many applications.

The mechanical seal is an end-face seal designed to provide rotary seal faces that can operate with practically no visible leakage. The principle of the mechanical seal is to use two replaceable antifriction mating rings—one rotating, the other stationary—to provide sealing surfaces at the point of relative movement. These mating rings are statically sealed, the rotating ring to the shaft and the stationary ring to the housing.

Because the parts are stationary in respect to the surfaces in contact, sealing at this junction point is accomplished easily through the use of gaskets, O-rings, V-rings, cups, etc.

The mechanical seal assembly therefore incorporates three sealing points; two are static, having no relative movement, and the third is the rotary seal at the faces of the mating rings. The mechanical seal also incorporates some form of self-contained force to hold the mating rings in contact.

This force is usually provided by a spring-loading apparatus such as a single coil spring, multiple springs, or in some cases wave springs, which are thin spring washers into which waves have been formed. The illustration in Figure 18-17 is a cross-section diagram of the principle, showing the various parts and their relative positions.

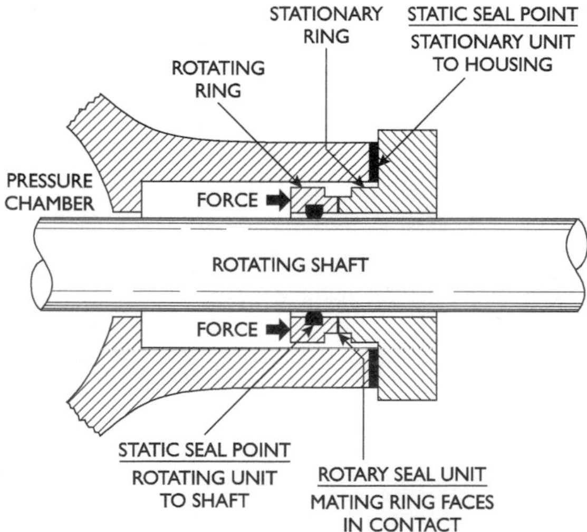

Figure 18-17 Operating principle of the inside mechanical seal.

Construction

A mechanical seal functions in a stuffing box as a spring-loaded check valve. There are two basic subassemblies in all seal designs:

1. A rotating assembly, which is attached to the shaft or the shaft sleeve
2. A stationary assembly, which is fastened to the stuffing box

The rotating assembly of the seal illustrated in Figure 18-18 comprises a collar, springs, compression ring, seal ring, and packing. The collar is attached by means of setscrews to the shaft and revolves with it. The collar drives the spring-loaded compression ring, which compresses the packing against the shaft and in turn drives the seal ring, which rotates against a face on the stationary unit.

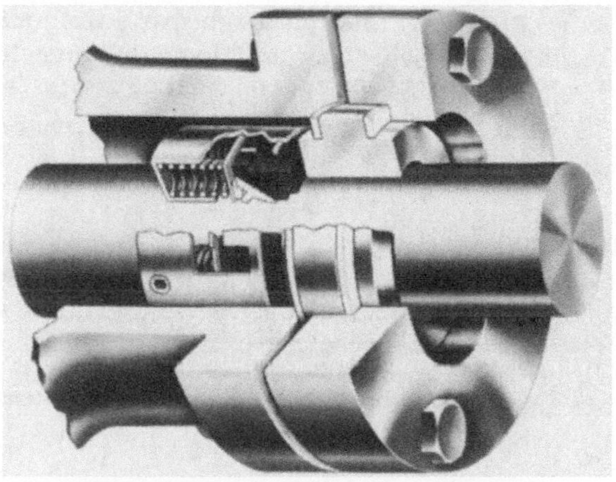

Figure 18-18 Cutaway view of a spring-loaded mechanical seal. *(© John Crane)*

The stationary unit consists of a gland insert and gland. The gland insert is held tightly in the gland, which is firmly held in place by bolting to the stuffing box. The face of the gland insert provides a running surface for the face of the rotating assembly. The surfaces of both faces are lapped to an extremely fine surface finish and flatness.

These two mated faces provide the running seal of the stuffing box. Leakage along the shaft and under the seal-rotating element is prevented by the compressed shaft packing. A simple gasket seals the gland; thus, with the exception of a very thin film that works its way between the seal faces, a liquid can be completely sealed within the stuffing box.

Basic Design

Although there are many design variations and numerous adaptations, there are only three basic designs of mechanical seals. These are the inside mechanical seal, the outside mechanical seal, and the double mechanical seal.

Inside Seal

The principle of the inside mechanical seal is shown in Figure 18-17. The rotating unit of an inside seal is inside the stuffing box. Because the fluid pressure inside the box acts on the parts and adds to the force holding the faces together, the total force on the faces of an inside seal will increase as the pressure of the fluid increases. If the force built into the seal plus the hydraulic force is high enough to squeeze out the lubricating film between the mating faces, the seal will fail. Although the mechanical seal is commonly considered to be a positive seal with no leakage, this is not true. Its successful operation requires that a lubricating film be present between the mating faces. For such a film to be present, there must be a very slight leakage of the fluid across the faces. Although this leakage may be so slight that it is hardly visible, the seal will fail if it does not occur. This is why a mechanical seal must never be run dry. The stuffing box must be completely filled and the seal submerged in fluid before the equipment is started and always while operating.

Outside Seal

This style of seal is also named for the location of the rotating unit, in this case outside the stuffing box, as shown in Figure 18-19. Because all the rotating parts are removed from the liquid being handled, it is superior for applications where corrosive or abrasive materials are present. Because the hydraulic pressure of the fluid is imposed on the sealing faces and tends to open them by overcoming the self-contained seal force, it is limited to moderate pressures.

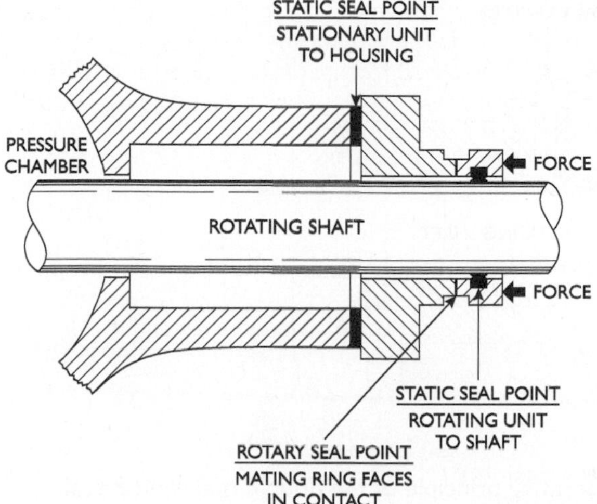

Figure 18-19 Operating principle of the outside mechanical seal.

Double Seal

A double seal is basically an arrangement of two single seals placed back to back inside a stuffing box. The double seal provides a high degree of safety when handling hazardous liquids. This is accomplished by circulating a nonhazardous lubricating liquid inside the box at higher pressure than the material being sealed. Any leakage therefore will be the nonhazardous lubricating liquid inward rather than the hazardous material in the pressure chamber outward. Illustrated in Figure 18-20 is the basic principle of the double mechanical seal.

Lubrication of Seal Faces

The major advantage of the mechanical seal is its low leakage rate. This is so low that there is virtually no visible leakage. To operate satisfactorily in this manner requires that the film of lubricant between the seal faces be extremely thin and uninterrupted. This is accomplished by machining and finishing these faces to very high tolerances in respect to flatness and surface finish. Maintaining this high quality of precision and surface finish requires that seal parts be carefully handled and protected. Mating faces should never be placed in contact without lubrication. To operate satisfactorily, this film must be maintained at all times. If there is insufficient leakage to provide this film, it will quickly overheat and fail. Liquid must *always* be present during the operation because running dry for a matter of seconds can destroy the seal faces.

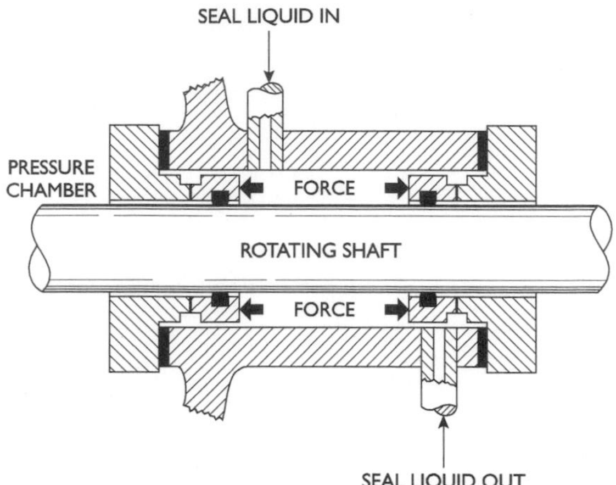

Figure 18-20 Operating principle of the double mechanical seal.

Installation

No one method or procedure for installation of mechanical seals can be outlined because of the variety of styles and designs. In cases where the seal is automatically assembled into correct position by the shape and dimensions of the parts, installation is relatively simple and straightforward. In many cases, however, the location of the seal parts must be determined by the mechanic at installation. In such cases, the location of these parts is critical because their location determines the amount of force that will be applied to the seal faces. This force is a major factor in seal performance because excessive face pressure results in early seal failure. Parts must therefore be located to apply sufficient force to hold the mating rings in contact without exerting excessive face pressure.

The procedure for location and installation of outside seals is usually relatively obvious and easily accomplished. This is a desirable feature of the outside seal with its exposed parts, allowing critical positioning to be done with the equipment in the assembled state.

The inside mechanical seal is more difficult to install because some parts must be located and attached while the equipment is disassembled. The location of these parts must be such that the proper force will be applied to the seal faces when the assembly is complete.

Seal designs and styles vary with manufacturers; however, the same basic principles apply to all when locating and installing inside mechanical seals. The following general procedure is applicable to most styles in common use.

Step 1

Determine the *compressed length* of the seal component incorporating the force mechanism. This is its overall length when it is in operating position (springs properly compressed).

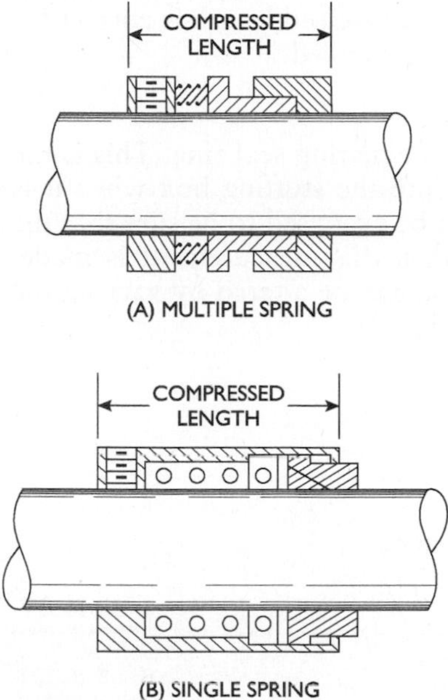

Figure 18-21 Typical inside-seal designs.

Figure 18-21 shows two widely used seal designs, one incorporating multiple springs and the other a single helical spring. In either case, the spring or springs must be compressed the amount recommended by the seal manufacturer before the measurement to determine compressed length is made. This is vitally important because the force exerted on the seal faces is controlled by the amount the springs are compressed. Insufficient compression could result in too low a force on the seal faces, which could lead to leakage if this force were low enough to be overcome by the system pressure force. The other extreme, overcompression of the springs, would involve so high a force on the faces that a lubricating film could not be maintained and failure of the seal faces would result.

Manufacturers' practices vary in the method of determining correct spring compression. In some cases, the springs should be compressed to obtain a specific gap or space between sections of the seal assembly. In other cases, it is recommended that the spring or springs be compressed until alignment of lines or marks is accomplished. In any case, consult the manufacturer's instructions and be sure the method used to determine compressed length is correct for the make and model of the seal being installed.

Step 2

Determine the *insert projection* of the mating seal ring. This is the distance the seal face will project into the stuffing box when it is assembled into position. Care must be exercised to be sure that the static seal gasket is in position when this measurement is made. Obviously, the amount of projection can be altered by varying the thickness of the gasket. See Figure 18-22.

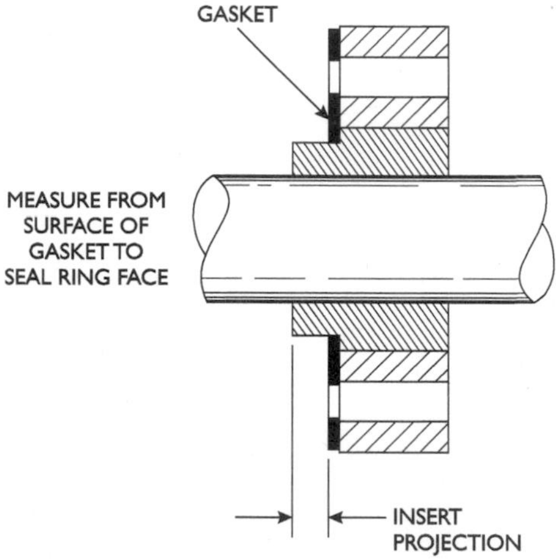

Figure 18-22 Illustrating the insert projection of a seal.

Step 3

Determine the *location dimension.* This is done by simply adding the compressed-length dimension (found in step 1) to the insert-projection dimension found in step 2. Figure 18-26 shows the rotary unit assembled into position in a stuffing box with springs compressed. This illustration shows that the location dimension is the distance from the face of the box to the back face of the rotary unit, which is the compressed length plus the insert projection.

Step 4

Witness-mark the shaft in line with the face of the stuffing box. A good practice is to blue the shaft surface in the area where the mark is to be made. A flat piece of hardened steel, such as a tool bit ground on only one side to a sharp edge, makes an excellent marking tool. The marker should be held flat against the face of the box and the shaft rotated in contact with it. This will provide a sharp, clear witness-mark line that is exactly in line with the face of the box. Figure 18-23 illustrates the relation of the witness mark to the box face and the use of the marking tool.

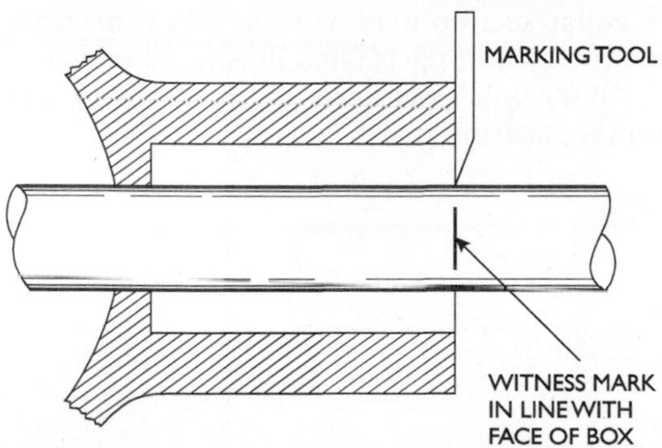

Figure 18-23 Locating the witness mark.

At this point, the equipment must be disassembled in such a manner to expose the area of the shaft where the rotary unit of the seal is to be installed. The amount and method of disassembly will vary with the design of the equipment. In some cases, it may be necessary to completely remove the shaft. In other cases, as with back-pullout design pumps, the stuffing-box section of the equipment is removed, leaving the shaft exposed.

Step 5

With the shaft either removed or exposed, blue the area where the back face of the rotary unit will be installed. From the witness mark placed on the shaft in step 4, measure the location dimension distance and place a second mark on the shaft. This is called the *location mark*, as it marks the point where the back face of the rotary unit is to be located. Figure 18-24 shows the relationship of the location mark to the witness mark.

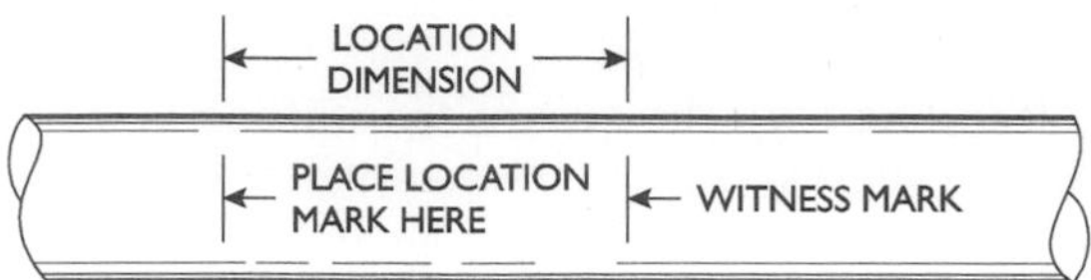

Figure 18-24 Placing the location mark on the shaft.

Step 6

Assemble the rotary unit on the shaft with its back face on the location mark. Fasten the unit securely to the shaft in this position. Some seal designs allow separation of the rotary unit components. In such cases, the back collar section may be installed at this time and the remainder of the rotary unit parts later. Illustrated in Figure 18-25 is a single-spring rotary unit assembled on the shaft with the rotary unit back face on the location mark.

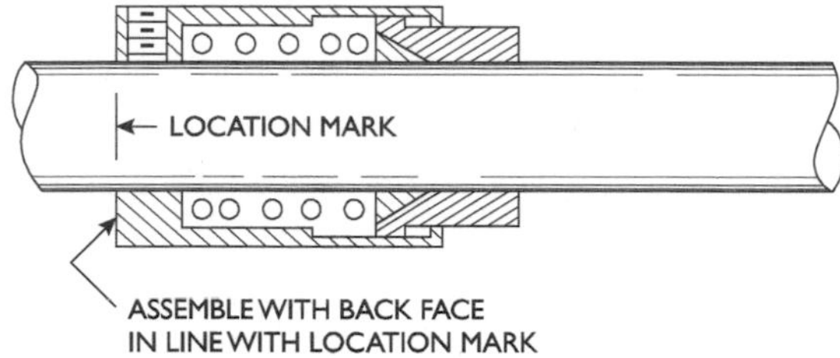

Figure 18-25 Locating a single-spring rotary member on the location mark.

Step 7

Reinstall the shaft into the equipment and complete the seal assembly. *Be sure the seal faces are lubricated.* The sequence of parts assembly depends on the type and design of the seal and the equipment. Figure 18-26 illustrates a completely installed inside seal of the multiple-spring design.

The final assembly operation will be the tightening of the gland-follower bolts or nuts. When this is done, the lubricated seal faces should be brought into contact very carefully. When the faces initially contact, there should be a space between the face of the box and the follow-gland gasket. The amount of space should be the same as the amount the springs were compressed in step 1 when the compressed length of the rotary unit was being determined. This should be very carefully observed because it is a positive final check on correct location of an inside mechanical seal.

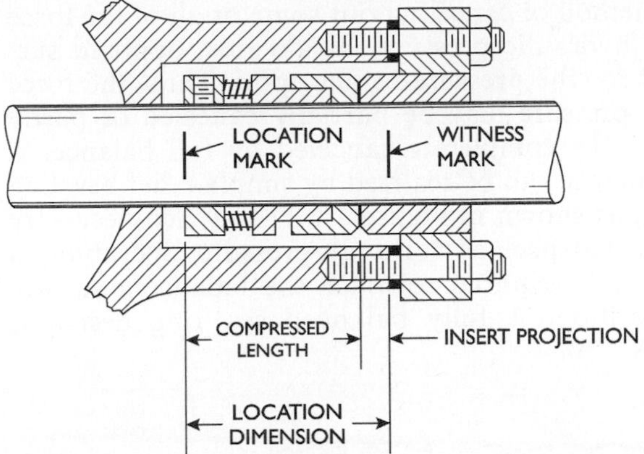

Figure 18-26 A multiple-spring inside seal completely installed.

Installation Precautions

It is very important to make some checks as you install a seal. The following list will help to keep mistakes from happening:

- Check the shaft with an indicator for runout and end play—maximum allowable is .005 in.
- All parts must be clean and free of burrs and sharp edges.
- All parts must fit properly without binding.
- Inspect seal faces carefully. There should be no nicks or scratches.
- Never allow faces to make dry contact. Lubricate them with a good grade of oil or with the liquid to be sealed.
- Protect all static seals, such as O-rings, V-rings, V-cups, or wedges, from damage on sharp edges during assembly.
- Before operating, *be sure* proper valves are open and seal is submerged in liquid. If necessary, vent box to expel air and allow liquid to surround the seal rings.

Balancing

Pressure inside the stuffing box tends to separate the faces of outside seals and to increase the force exerted on the faces of inside seals. As pressures are increased, the friction and resultant wear are increased on the faces of inside seals. Hence, it is advisable to reduce the load on the faces whenever possible. This is accomplished by balancing the forces that act upon the seal.

Balancing is a method of canceling out some or all of the force due to the system hydraulic pressure. To do this, opposed surfaces are provided for the pressure to act upon. Thus, the force due to the system pressure may be partially canceled or partly balanced, or it may be completely canceled by full balance. A small amount of balance can be realized by simple relief beveling of the mating ring, as shown in Figure 18-27A. When necessary to provide greater balance, as for high-pressure operation, a stepped-shaft design is required to place the balancing surface area in proper position. A fully balanced seal-ring design is shown in Figure 18-27B.

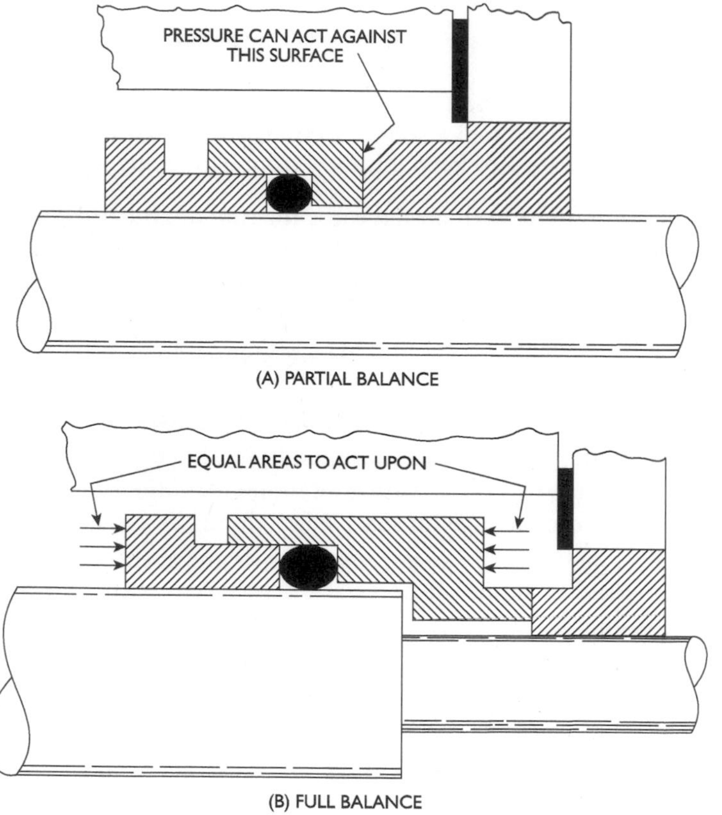

Figure 18-27 Balancing areas in high-pressure sealing.

O-Rings

The O-ring, shown in Figure 18-28, is a squeeze-type packing made from synthetic rubber. It is manufactured in several shapes, the most common being the circular cross section from which it derives its name. It is a very simple device and probably was not developed earlier because suitable materials and precision molding were not available. The principle of operation of the O-ring can be described as controlled deformation. A slight deformation of the cross section, called a mechanical squeeze, illustrated in Figure 18-29, deforms the ring and places the material in compression. The deformation squeeze flattens the ring into intimate contact with the confining surfaces, and the internal force squeezed into the material maintains this intimate contact.

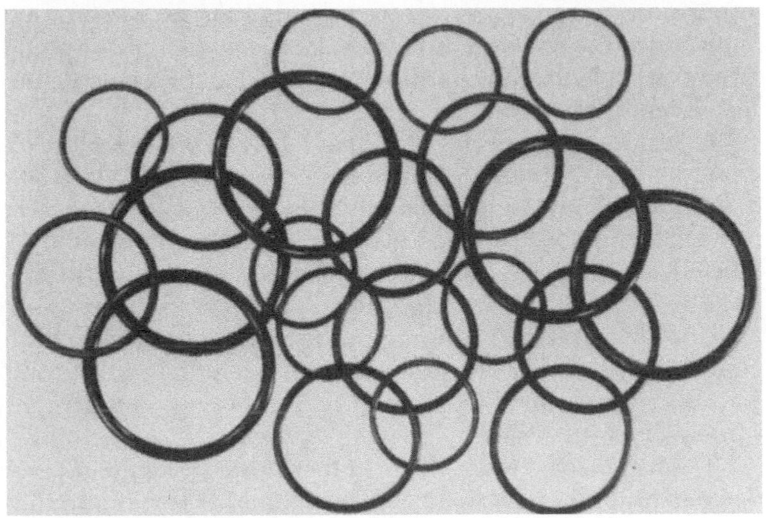

Figure 18-28 Various sizes of O-rings.

Additional deformation results from the pressure the confined fluid exerts on the surface of the material. This in turn increases the contact area and the contact pressure, as shown in Figure 18-30.

The principal applications for O-rings are for static sealing and installations involving reciprocating motion. In reciprocating applications, the sealing action is similar to that of a piston ring. In some cases, they may be used for oscillating and rotary motion if speeds and pressure are low.

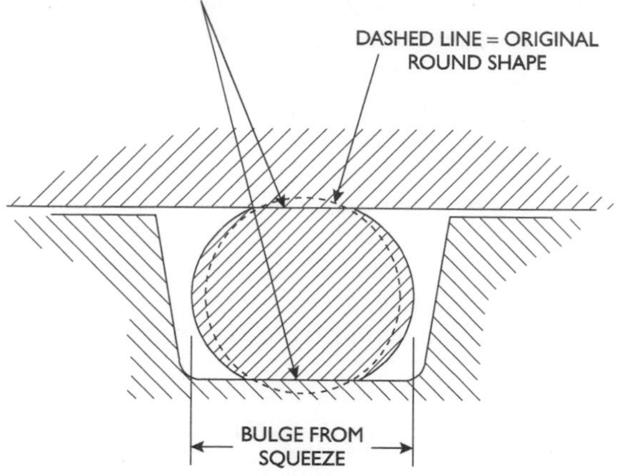

Figure 18-29 The mechanical squeeze, which forms the sealing effect of the O-ring.

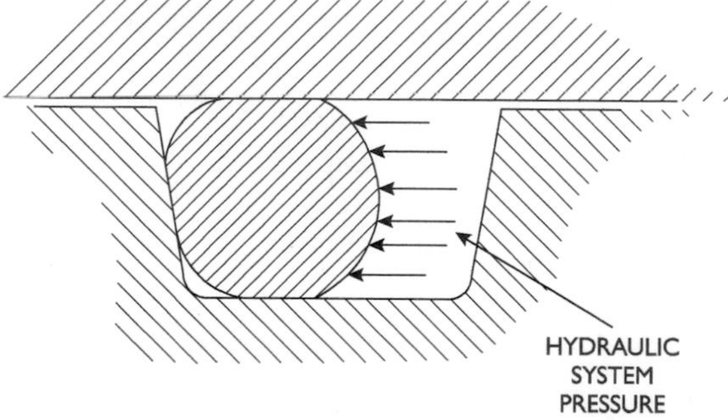

Figure 18-30 Pressure from the confined fluid develops a seal.

The squeeze-type O-rings are generally fitted to a rectangular groove in the hydraulic or pneumatic mechanism. Although other groove shapes are sometimes used, the rectangular groove has proved the simplest from the manufacturing standpoint and is as efficient as, if not superior to, other shapes. It is usually about one and one-half to twice as wide as the diameter of the O-ring cross section. This allows the O-ring to slide and roll to the side of the groove away from the pressure. This rolling and sliding distributes the wear, thus extending the life of the O-ring, and assists in lubrication of its surface.

The initial mechanical squeeze of the O-ring at assembly should be equal to approximately 10 percent of its cross-sectional dimension. The general-purpose industrial O-ring is made to dimensions that, in effect, build the initial mechanical squeeze into the product. This is done by manufacturing the O-ring to a cross-sectional dimension 10 percent greater than its nominal size. Table 18-2 lists the nominal and actual dimensions of the standard O-ring in common use. Note that the actual dimensions are the nominal plus 10 percent.

Table 18-2 Standard O-Ring Dimensions

Nominal Cross-Section Diameter	*Actual Cross-Section Diameter*
1⁄32	.040
3⁄64	.050
1⁄16	.070
3⁄32	.103
1⁄8	.139
3⁄16	.210
1⁄4	.275

Because the cross-sectional dimension of the O-ring is 10 percent oversize, the inside and outside diameter dimensions must be changed proportionally. Thus, the actual dimension of an O-ring to fit a groove with an inside diameter of 2 in. and an outside diameter of 2⅜ in., as shown in Figure 18-31, would be .210 × 1.977 × 2.398.

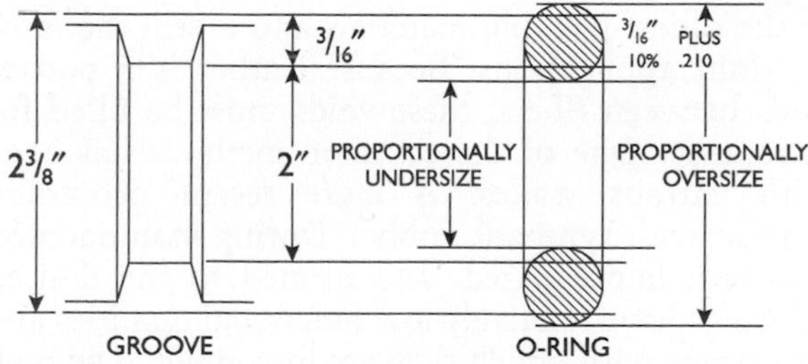

Figure 18-31 Actual O-ring dimensions.

Many O-ring standards have been established by various agencies, societies, industrial groups, etc. The dimensions used in all of them are comparatively the same; however, the numbering systems are not. In wide use by many manufacturers is what is called the *uniform dash* number system. In this system, the numbers from -001 to -475 are used to identify the specific dimensions of O-rings. Using a uniform dash number table, one can select an O-ring of correct size from the nominal dimensions. Simply select the dash number that corresponds to the nominal dimensions of the installation. The dimensions of an O-ring so selected will be proper to ensure the required 10 percent mechanical squeeze at installation. Included in the appendix is a table of uniform dash numbers with the corresponding nominal ring dimensions.

Formed and Molded Packings

The principle of operation of formed and molded packings is quite different from compression packings in that no compressing force is required to operate them. The pressure of the fluid being sealed provides the force that seats the packings against the mating surface. They are therefore often classed as *automatic* or *hydraulic packings*. Packings formed or molded in the shape of a cup, flange, U-shape, or V-shape are classed as lip-type packings. These styles are usually produced with the lips slightly flared to provide automatic preload at installation. The fluid being sealed then acts against the lips, exerting the force that presses them against the mating surface. Lip-type packings are used almost exclusively for sealing during reciprocating motion. They must be installed in a manner that will allow the lips freedom to respond to the fluid forces. They may be grouped into classifications based on the material from which they are manufactured.

Leather Packings

Leather is one of the oldest packing materials and is still the most satisfactory for certain applications. Because leather is a porous material with voids between fibers, these voids must be filled for satisfactory performance. One of the common methods has been impregnation with various waxes. A more recent procedure involves impregnation with synthetic rubber. During manufacture, the leather is softened, impregnated, and formed to the desired shape. Leather packings have relatively low functional qualities and tend to burnish a surface with which they are in contact. The high tensile strength of the fibers makes leather packings resistant to extrusion; therefore, they do not require as close clearances as other types. They are also resistant to mechanical damage and can tolerate moderately rough surfaces and, to some degree, gritty materials. Because of this ruggedness, they are recommended for use in what might be called rough and difficult applications. They also perform well when subjected to high pressures and in situations where clearances cannot be held to minimum values. Because leather absorbs fluids, it tends to be self-lubricating.

Fabricated Packing

Fabricated packing is made by molding woven duck, asbestos cloth, and more recently synthetic fibers with rubber or synthetic rubber. The fabric used may be compared to the fibers in leather, as it acts as a reinforcement and provides additional strength to the packing. This reinforcement enables this type of packing to contain relatively high pressures and to be resistant to extrusion. Fabricated packings have a wider temperature range than leather and can be made of materials resistant to acids and alkalies. Although metal surface finish is not too critical, the material will abrade or rub away in contact with a rough surface. They are used chiefly on heavy industrial equipment with common metal finishes where clearances are held to minimum values. Fabricated packings tend to wipe drier than leather packings, although the fabric will absorb some fluid and has a slight lubricating action.

Homogenous Packings

Homogenous packings are compounded from various types of synthetic rubber. They are made in many degrees of hardness depending on the application and pressure. They do not contain any fabric or other reinforcement materials. Like the fabricated packings, they will operate satisfactorily over a wide temperature range and are resistant to acids and alkalies. They lack resistance to extrusion, as they lack the strength of reinforced packings. Clearance must therefore be held to a minimum, particularly at higher pressures. When clearances cannot be held to close limits, and when pressures are above 1500 psi, antiextrusion rings should be used. Antiextrusion rings, which might be called back-up rings, are flat rings, usually made of leather, that are placed under the packing. The back-up ring bridges the clearance, preventing the homogenous packing from extruding into the clearance opening. Metal surface finish for successful operation of homogenous packings is critical. A fine finish is essential because the material will quickly abrade and wear away if mated with common metal-finish surfaces. Homogenous packings will not pick up or absorb fluid, have no self-lubricating qualities, and usually wipe contact surfaces quite dry.

Plastic Packings

Many styles of packings are molded from various kinds of plastics for special cases. One of the most widely used is packing made of fluorocarbon resins. Although plastic-lip packings are inert to most chemicals and solvents, they have little elasticity or flexibility. Because of this, these packings are difficult to install and are susceptible to leakage in applications where conditions vary and some resiliency is required. Although fluorocarbons have a slippery feel and resist adhesion to most metals, their friction under pressure is high compared to packings made of other materials. Because of their lack of resilience, plastic packings are used primarily in chemical applications where a material inert to chemicals and solvents is required.

Packing Installation and Operation

The principle of operation is the same for all lip-type packings regardless of the type of material. They must be installed in a manner that will allow them to expand and contract freely. They should not be placed under high mechanical pressure because this transforms them to compression packings. Overtightening a lip-type packing improperly preloads it. Although slight preloading is needed for a tight fit and sealing at low pressure, it should occur automatically as a function of the size and shape of the packing.

A packing will take the shape of the back support surface after a few applications of pressure. This will occur whether or not the packing has a radius-shape shoulder. The greatest wear occurs at the shoulder of a packing; therefore, proper support at this point is essential. Excessive clearance will result in extrusion of the packing into the clearance gap. The greater the clearance or pressure, the more readily the packing will extrude into it. Lip-type packings also require inside support. This is usually in the form of filler materials such as flax and rubber or a metallic filler ring. Inside support rings should have clearance, and corners should be rounded to avoid cutting of the packings. Because most packing materials swell to some extent, depending on the material and the fluid, the clearance is necessary to allow the packing freedom of movement. A packing that is too tight at installation will lead to binding and premature failure. Fillers such as flax and rubber, however, may be packed tightly inside of the packing, as they are porous and pliable. These fillers lend support to the packing and prevent its collapse. They do not interfere with the free movement of the fluid to all inside surfaces of the packing.

Cup Packing

Cup packings are one of the most widely used styles of packing. They are simple to install and highly satisfactory for plunger-end

Figure 18-32 Molded rubber cup packings.

applications. They are considered to be unbalanced packing because they have only one lip; see Figure 18-32. U- and V-shaped packings, shown in Figures 18-33 and 18-34, are considered balanced because they have double lips, which result in the pressure and friction being equal on both sidewalls.

Figure 18-33 Molded rubber U-shaped packings.

Figure 18-34 Molded rubber V-shaped packings.

Cup packings inside follower plates, shown in Figure 18-35, should not be overtightened because the bottom of the packing may be crushed and cut through. The inside follower plate need only be carefully snugged tight to prevent initial leaking of the fluid. The pressure will then force the packing against the back support surface, preventing any leakage from occurring past the center hole and by the bottom of the cup.

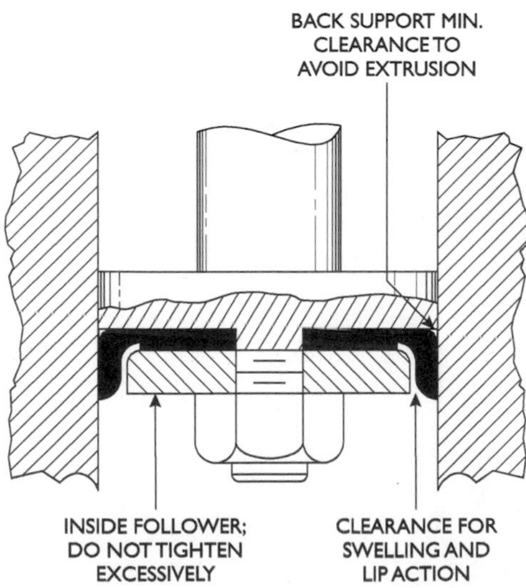

Figure 18-35 Inside-packed single-acting cup packing.

A cup pack expands from the force acting on the inside and is remolded as the heel expands to the diameter of the cylinder wall. When this occurs, the packing loses depth or becomes thinner on the bottom or both. The heel or shoulder is the point of greatest wear and is usually the failure point. In most cases, the lip will show very little wear when the packing fails at the heel. For satisfactory operation of cup packings on plunger applications, diametral clearance between the back supporting plate and the cylinder wall must be held to a minimum. Table 18-3 shows the clearance allowances that may be used as a guide.

Table 18-3 Cylinder Clearance Allowances

Cylinder Diameter	*Up to 500 psi*	*Over 500 psi*
Under 8 in.	.006 in.	.004 in.
Over 8 in.	.010 in.	.006 in.

Packing manufacturers should be consulted for specific clearance allowances. Clearance must also be provided between the inside of the follower plate and the packing lips. This clearance allows free movement of the lips and space for swelling of the packing material. The tips of the lips must also be protected from bumping. This protection is frequently provided by stops or by an inside follower plate thicker than the height of the lips.

U-Packings

U-packings are balanced, as they seal on both the inside- and outside-diameter surfaces. Properly installed and supported, they are automatic in action. Proper support includes provision to keep the lips from collapsing as well as adequate back support. To support the lips, the recess of the U is filled with a supporting material called a filler. Such materials as flax, rubber, hemp, or fiber are used for this purpose. Metal rings are also used for lip support and are called pedestal rings. Clearance between the side of a pedestal ring and the inside walls of the packing is necessary to accommodate swelling of the packing and avoid binding. The nose of a pedestal ring should be made with a slight radius to prevent cutting the packing. The ring must also be high enough to give clearance between the bevel lips of the packing and the base of the ring to prevent the lips from striking the bottom, as shown in Figure 18-36. Bottoming of the lips will cause them to turn inward and will destroy their ability to seal. A pedestal ring should clamp the packing firmly at the base, leaving the sidewalls and lips free to respond to the actuating fluid forces. It must be drilled crosswise to allow equalization of actuating fluid pressure on all inside surfaces of the packing.

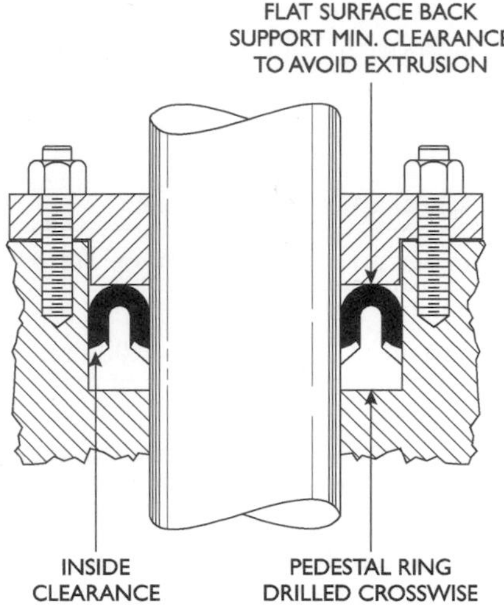

Figure 18-36 Outside-packed U-shaped packing assembly.

The function of all fillers, whether metal or soft material, is to support the sidewalls and prevent their collapse. If fillers are made of hard, nonporous materials, such as fiber, they must be provided with pressure equalization openings similar to the holes in metal pedestal rings. This is not necessary for porous materials because they allow passage of the fluid and equalization of pressures and accommodate any swelling that might occur. When rubber fillers are used, they should be in strip form equal in thickness to the width of the packing recess. Strips should be cut slightly longer than required, with a butt joint, and forced into the recess to prevent them from falling out.

Figure 18-36 shows an outside-packed U-packing assembly used for a gland-sealing application. This design is commonly used in place of a compression-packing stuffing box. A good inside-packed design for plunger applications is shown in Figure 18-37. A serious drawback to this design is that complete dismantling is required to repack.

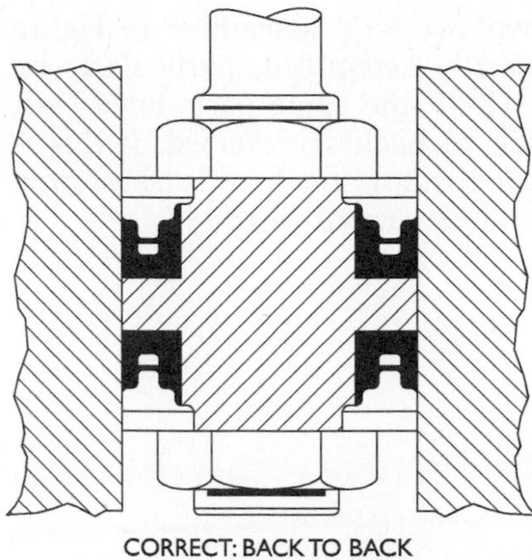

Figure 18-37 Inside-packed double-acting U-shaped packing.

A more practical design from the standpoint of packing replacement is shown in Figure 18-38. This is a compromise design that allows repacking from one side. It is a compromise because there is some trapping of pressure between packings. This disadvantage is more than offset by the convenience of repacking without having to dismantle and remove the plunger.

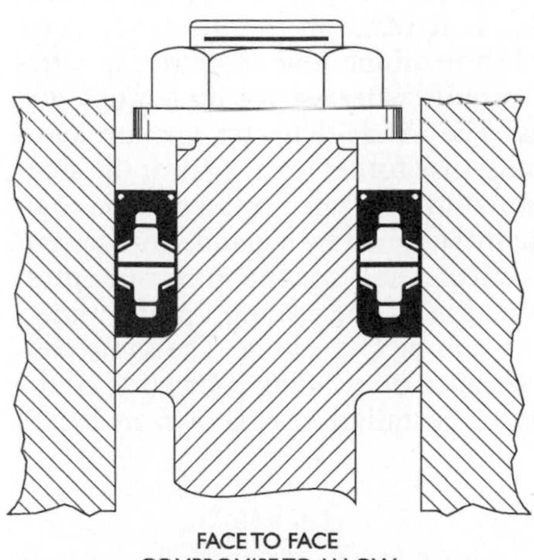

Figure 18-38 Another type of inside-packed double-acting U-shaped packing.

The compromise design, shown correctly assembled in Figure 18-38, may very easily be incorrectly assembled, particularly by inexperienced people. In Figure 18-39 the same parts have been assembled with the packings back to back or reversed. If this is done, the packings do not have support and the load is transferred through them, causing crushing and very rapid wear.

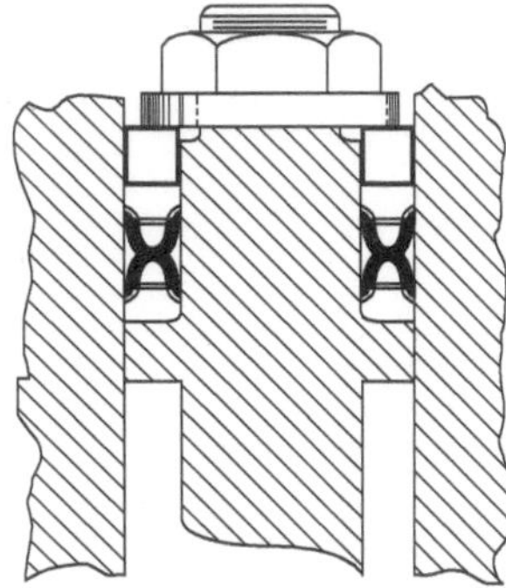

Figure 18-39 Incorrect packing of Figure 18-38.

Flange Packing

The flange packing is probably the least used of the lip-style packings. It is an unbalanced style, as it seals on the inside diameter only, and is generally restricted to retaining low pressures. It is frequently used for outside applications where space limitations will not allow other packing types. The inside-diameter lips are automatically actuated by the pressure that forces them against the shaft or rod. The base or flange section must be sealed by an outside compressing force. This force is usually supplied by a gland arrangement and must be strong enough to effect a seal for the maximum pressure that will be applied. The action of the sealing lip of a flange packing is the same as that of a cup or U-packing. The same considerations must be given to clearance or swelling protection of the lip edge. A typical flange-packing installation is shown in Figure 18-40.

V-Shaped Packing

The V-packing is one of the most widely used of all lip-type packings. V-packing installations can be used for low or high pressure and rotating or reciprocating motion; they may be installed on pistons (inside-packed) or installed in glands (outside-packed). It is the one

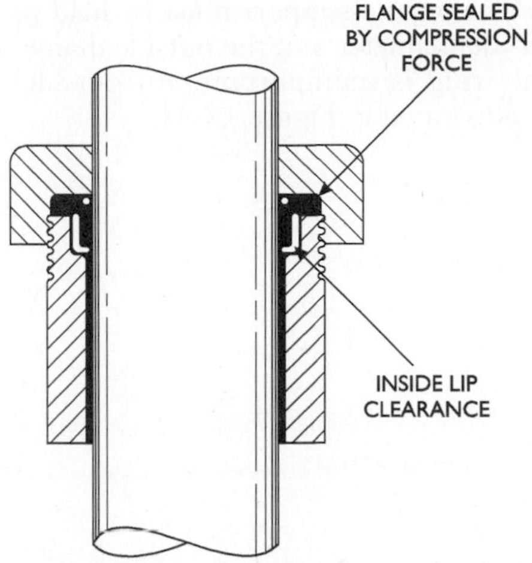

Figure 18-40 Outside-packed flange packing.

type of installation that can be used for practically all designs. Onc of the principal advantages of V-packings is that a small cross section can be used, thus reducing the stress on the gland ring. V-packing installations, therefore, need not be as bulky as that required for other types. It has been found that V-packing installations will far outlast and outperform any other type, especially at high pressures. Manufacturers have determined that a 45° angle (90° included angle) is the most satisfactory for V-packings and have standardized on this angle. V-packings operate as automatic packings but offer the distinct advantage of permitting taking up on the gland ring when excessive wear develops, thereby opening the V and effecting a seal.

All lip-type packings discussed previously have been unit seals. The V-packing is a multiple packing and is installed in stacks or nested sets. It is customary to use a minimum of three rings to a maximum of six rings per set. They are stacked with a support ring or adapter at the top and bottom. These conform to the V shape on one side and are flat on the other. This provides square flat ends for the V-packing stack. Male and female support rings can be made from metal, leather, or homogenous synthetic rubber. When made from metal, they are referred to as support rings. When made from materials other than metal, they are referred to as adapters. Although the metal support rings can be used over and over again with each new installation of packing, the female ring will wear in time and should be checked regularly for excessive clearance so as to prevent packing extrusion. The female support must be held to close tolerances on both the inside diameter and the outside diameter. The clearance on the male ring is unimportant. An outside-packed V-packing assembly is illustrated in Figure 18-41.

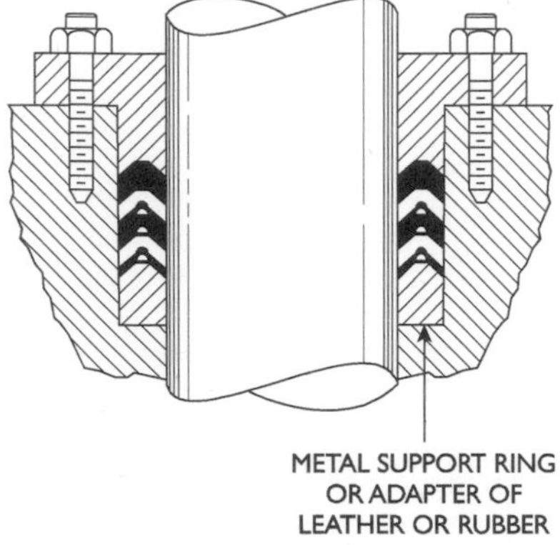

Figure 18-41 Outside-packed V-shaped packing.

Packing Installation

Lip-type packings may be installed for one-directional sealing in glands (outside-packed) or on pistons (inside-packed). For one-directional sealing, they should be installed with the inside of the packing exposed to the actuating fluid. Proper installation of packings in this style of application is obvious if the principle of the lip-type automatic packing is understood. Double-acting applications require greater care and attention to operating principles if satisfactory service life is to be obtained. When mounted with the inside of the packings facing together, they are referred to as face-to-face mounted. When mounted with the bottoms toward one another and the inside facing away, they are referred to as back-to-back mounted. The ideal double-acting packing arrangement is back-to-back with solid shoulders for back support, as illustrated in Figure 18-37. With this arrangement, each packing is fully supported, and no trapping of pressure between packings can occur. Because such an arrangement requires that packings be installed from both sides of the plunger head or end, it frequently is not practical. In such cases, face-to-face mountings are used. A typical assembly to accommodate packing from one end is illustrated in Figure 18-38. Although this allows repacking from one end, it also introduces an undesirable condition. When packings are mounted face-to-face, the actuating fluid must pass the first packing and open the second packing as it is facing toward the fluid. When the second packing is expanded by the fluid, pressure tends to back up into the first packing, expanding it and locking pressure between the two packings. From a practical standpoint, it is frequently necessary to pack in this manner. However, service life from this type of installation cannot compare with properly mounted back-to-back installation.

Improper Installation

The importance of proper backup support has been discussed previously. The position of the packing in relation to the flow and buildup of pressure in the actuating fluid is also vitally important. Reversing a packing, whether done inadvertently or through misunderstanding, can shorten the life of a packing or, in some cases, cause malfunction of the equipment. The most frequent mistake is to reverse the assembly, that is, to install the packing face-to-face in an assembly designed for back-to-back operation, or vice versa. Figures 18-39 and 18-42 illustrate two examples of incorrect assembly. In Figure 18-42 the packing has been installed face-to-face. This will cause trapping of pressure between the packings with excessive friction and wear.

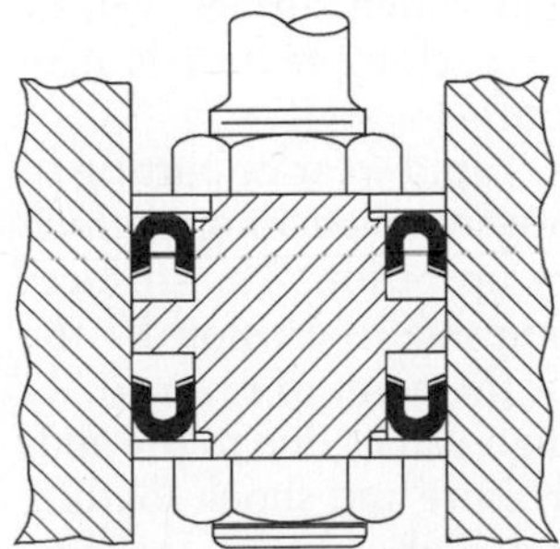

Figure 18-42 Incorrect back-to-back packing.

If assembled in this manner, the equipment may appear to operate satisfactorily, but the packing life will be greatly reduced. In some cases, as illustrated in the face-to-face assembly in Figure 18-38, this is done deliberately to make possible packing from one side. However, this is not necessary in the Figure 18-42 illustration, and face-to-face assembly is incorrect. Figure 18-39 illustrates what is probably the most serious packing installation error. If packings are installed in the manner illustrated, the load is transferred from one packing through the other. This results in crushing of the compressed packing. Under these conditions, the packings are soon crushed out of shape and the assembly becomes free to move with each reversal of pressure. Rolling and jamming of the packing develops, and complete failure soon occurs.

Packing Procedure

Before any work is undertaken, the equipment must be properly locked out. This includes the disconnecting or closing of pipelines to the equipment. In some cases, it may be necessary to block equipment to prevent its moving when pressure is interrupted. Consideration should also be given to the possibility of pressure buildup when valving off. Starting or movement of the equipment might cause the buildup of high and possibly hazardous pressure under these circumstances.

The first step in replacing packings in equipment is the removal of the old packing. Frequently, the condition of the old packing is an indication of the cause of failure or other malfunction. Excessive wear may be an indication of misalignment. Extruded packings indicate excessive clearance. Extrusion may also occur if there is an out-of-round condition, probably due to wear or cramping of the components. Packings that are jammed or rolled over may indicate improper assembly. Such things as extremely rough surface, grit or abrasive material in the fluid, or improper packing materials may also be determined from packing inspection.

Worn, grooved, or cut surfaces will cause rapid wear or cutting of packings. This is especially true with homogenous packings, which require fine surface finishes for satisfactory operation. The surfaces that are most subject to cutting or grooving due to erosion by the fluid are located at the ends of the stroke or the point of reversal. At this point, the pressure may build to very high values due to the sudden reversal of direction. This excessive pressure and shock loading tends to cause failure of the packing and high-velocity leakage. Surfaces may also become badly damaged if foreign material gets into the system.

The importance of clean surfaces and clean fluids cannot be overemphasized. Although leather packings can tolerate some foreign material, most packings made of synthetic materials are damaged very quickly if the system is not clean. The system, including feed and discharge lines, should be thoroughly cleaned if there is any indication of foreign matter. Fluid should be drained and the system refilled if gritty substances are present. Screens should be checked and cleaned before refilling the system.

Formed and molded packings must be of the proper size for satisfactory operation. Because they are not adjustable and are automatic in action, the fit determines success or failure of the packing. Each installation should be checked to determine proper dimensions. Leather packing, being quite sensitive to climate conditions, may change dimension if stored in a warm, dry location. This can be determined by measurement, and some correction is possible if the packings are soaked before installation. Fabricated and homogenous packings do not present this problem because they are not affected by air moisture.

The proper packing for the equipment should be obtained. Consideration must be given to the material of which the packings are constructed as well as their shape or style. Where fillers or adapters are required, these should be obtained and installed in some cases. For example, flax or rubber fillers in U-packings should be cut to size and installed in the packing. Where hard fillers are used, provisions to allow equalization of pressures must be made. Where metal rings, pedestal rings, etc., are used, their condition should be checked, and they should be checked for dimension with respect to the packing.

Before installing any packings, the assembly should be studied so that the function is completely understood. The packings must be installed so that the fluid will properly act on the lips to produce automatic sealing. A practice is sometimes followed of replacing packings in the same manner as they are removed. This can lead to continuing an assembly error if someone has previously repacked it improperly. If the principle of operation of lip packings is understood, a brief check will soon determine the proper order of assembly.

Packings should be installed in the order that has been determined by a study of the equipment. The lips of the packing should be in contact with the mating surface with some preload. There should be inside clearances to allow freedom of the packing lips to respond to the fluid pressures. The tips of packing should be protected from bumping by stops or other mechanical features. Packings should rest on proper backup support surfaces.

Lip-type packings are automatic in operation and should not require any compression. In some cases, they should be tightened to ensure an initial seal. In no case, however, should they be tightened excessively, as this can crush and possibly damage the packing.

Radial Lip Seals

The *radial lip seal*, frequently called an *oil seal*, is a device that applies a sealing pressure to a mating cylindrical surface to retain fluids and in some cases exclude foreign matter. Although the rotary shaft application of the radial seal is most common, it is also used when shaft motion is oscillating or reciprocating. A typical radial lip seal is shown in Figure 18-43.

Figure 18-43 Typical radial lip seal.
(Courtesy of Klozure® Oil Seals and Bearing Isolators, a division of Garlock Sealing Technologies)

The radial lip seal is manufactured in a wide variety of types and sizes to meet the needs of particular applications. Three broad categories of these seals are (1) *cased seals*, where the sealing element is retained in a manufactured metal case (Figure 18-44); (2) *bonded seals*, where the sealing element is bonded to a washer or a formed metal case (Figure 18-45); and (3) *dual-element seals*, where the sealing element is double (Figure 18-46). The double element may be used for applications where liquids are present on both sides of the seal or where the service is unusually severe and a double element is necessary to hold leakage to a minimum.

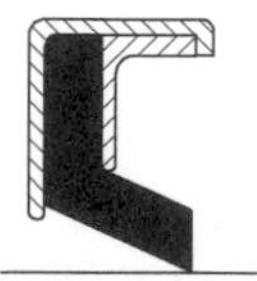
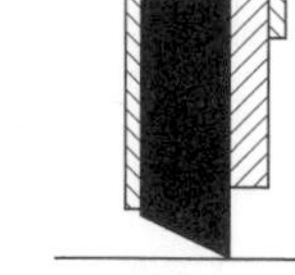
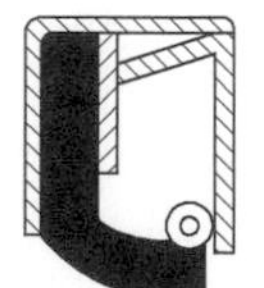

Figure 18-44 Cased seals.

Figure 18-45 Bonded seals.

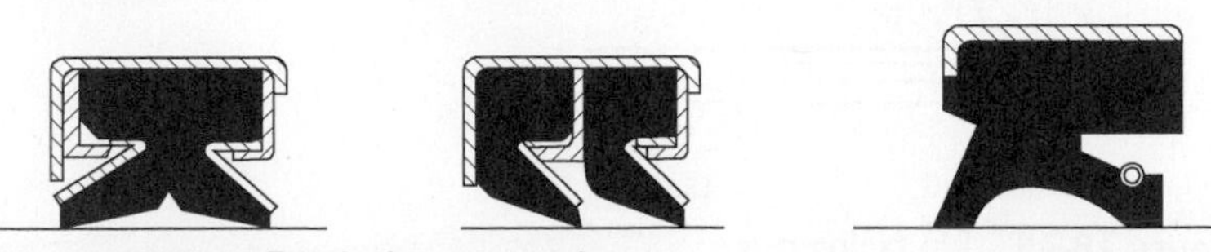

Figure 18-46 Dual-element seals.

Radial lip seals have two specific functions: retention of fluids and exclusion of foreign matter. The single-lip seal, although usually used for liquid retention, can also be used for exclusion (i.e., to exclude destructive contaminants). When used for retention, it is installed with the lip facing inside, as shown in Figure 18-47. When used for exclusion, it is installed with the lip facing away from the housing, as shown in Figure 18-48. One lip of the dual-element seal shown in Figure 18-49 retains lubricant, whereas the other excludes contaminants.

In some seal designs, various types of springs—coiled, garter-like, or finger—are used to squeeze the sealing element more intimately against the mating surface. In operation, the sealing pressure is a result of an interference fit between the flexible sealing element and the mating component. This pressure is augmented when springs are used.

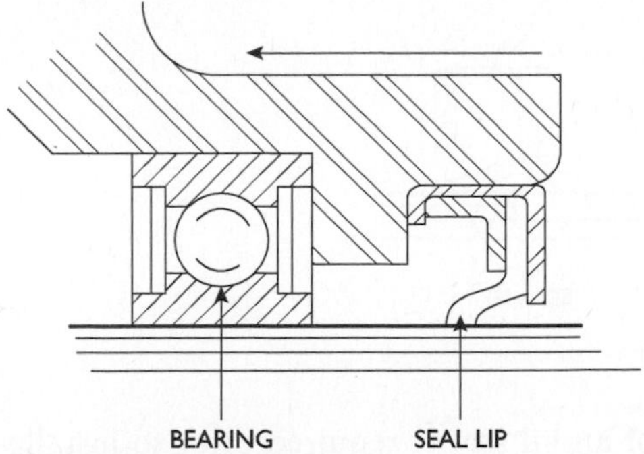

Figure 18-47 Lip facing in.

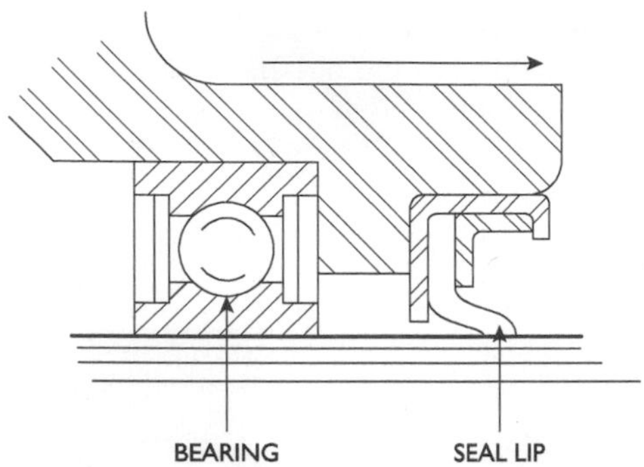

Figure 18-48 Lip facing out.

Like any precision-made machine part, radial lip seals should be handled with reasonable care when being installed in equipment. Although these seals are compactly built and under ordinary circumstances are not fragile, rough handling may mar or damage the flexible lip of the sealing member, resulting in an imperfect seal when placed in operation.

Before proceeding with the installation of an oil seal, the following dimension checks should be made:

1. Is the seal of correct inside size to squeeze-fit the shaft?
2. Is the seal's outside diameter correct for the bore in which it is to be installed so that a leakless press fit will be obtained?

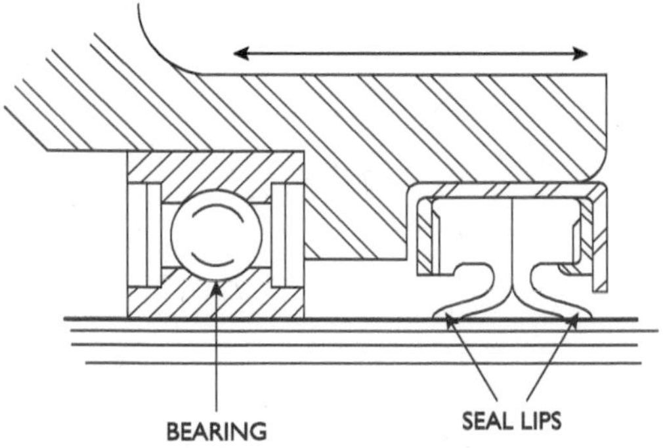

Figure 18-49 Dual-lip seal.

No special treatment of an oil seal is required prior to installation; however, a film or coating of suitable lubricant makes it easier to apply the seal over the shaft.

The shaft surface over which the seal will slide must be smooth and free of burrs, nicks, or scratches, which may damage the sealing lip. Ideally, the end of the shaft should be chamfered or rounded to prevent lip damage and to ease installation of the seal. In those cases where the shaft shoulder is not chamfered or rounded, a suitable mounting tool or device should be used when applying the seal. This is probably the area where the mechanic encounters the most difficulty when installing radial lip seals. Suitable tools or thimbles are usually provided to the assemblyman when factory installations are made. This is seldom the case for field installations. To aid in safe assembly when a thimble is not available, a simple but effective practice is to cover the edge or shoulder with a protective covering. This may be done by using thin sheet material such as brass shim stock, aluminum, or stiff plastic sheeting; well-lubricated kraft paper can even be used. The protective sheet should be wrapped around the shaft, over the end that otherwise might damage the sealing lip. Care must also be exercised to cover any keyways or holes the seal might pass over. The seal is assembled over the sheet material, protecting the sealing lip from damage on the shoulder edge or keyway as it passes over them while being slipped into position.

Because seals are an interference fit in the housing bore, they must be either pressed or driven into position. Ideally, they should be press-assembled, using a suitable installation tool, thimble, or sleeve that has a flat face to contact the back of the seal. In field assembly, these ideal conditions seldom exist, and the mechanic must use alternative methods. A tube or short length of pipe, slightly smaller in outside diameter than the housing bore, may be placed against the seal and tapped with a mallet. If such an article is not available, a block of hard wood may be placed squarely on the back of the seal surface and tapped lightly with a mallet as it is moved around the circumference of the seal. For large-diameter seals, two blocks should be used simultaneously at opposite points on the circumference of the seal. *Never* strike the seal directly with a hammer or apply force to the inner area of the seal.

Chapter 19

Lubrication and Oil Analysis

Most lubricants used in industry are mineral-based and obtained from petroleum by refining processes and further purification and blending.

Functions of a Lubricant

Lubricants have three major functions: limit friction, minimize wear, and dissipate heat.

Limit Friction

Friction is defined as the resistance to motion of contacting surfaces. Even smooth metal surfaces have microscopic rough spots called *asperities*. Friction is increased by the presence of asperities on surfaces. Attempts to overcome the force of friction will increase the localized heat generated between contacting surfaces. This heat can actually create temperatures high enough to weld two surfaces together.

Lubrication prevents peaks of asperities from touching each other through what is called film strength. Molecules of lubricants are naturally bonded together, often in chains. Any attempt to break the chain creates an opposite tension that prevents separation.

Minimize Wear

Wear is the removal of material from one or more moving surfaces in contact with each other. The material removed becomes the source of additional friction and increased wear on the surfaces involved.

A quality lubricant will fill the valleys of the asperities and provide an additional film over the peaks of the asperities. The asperities of the two surfaces are prevented from contacting each other, and wear will be minimized.

Dissipate Heat

Even well-lubricated parts will heat up as a result of friction and external heat. One advantage of liquid lubricants is their ability to absorb and dissipate point sources of heat.

Types of Industrial Lubricants

Lubricants can be divided into two types: *solid* and *liquid.*

Solid Lubricants

Solid lubricants are materials such as graphite, molybdenum disulfide, and PTFE (polytetrafluoroethylene); they are used in smaller equipment or on surfaces where just a minor amount of movement is expected. Lead, babbitt, silver, gold, and some metallic oxides are solid lubricants that can provide for more movement or pressure between surfaces. Some machine designers use ceramics or intermetallic alloys to coat the surfaces of moving parts; they can also be considered lubricants.

Liquid Lubricants

Liquid lubricants, in industry, fall into two categories: greases and oils. We will look at both of these types of lubricants in the following sections.

Measuring the Properties of Greases and Oils

Criteria for measuring the properties of grease are these:

Hardness. Greases range from hard to soft. Table 19-1 shows this range. The NLGI is the National Lubricating Grease Institute.

Table 19-1 Hardness of Grease

NGLI Number	*Consistency*	*ASTM Worked Penetration at 25°C (77°F) 10^{-1}mm*
000	Very fluid	445–475
00	Fluid	400–430
0	Semi-fluid	355–385
1	Very soft	310–340
2	Soft	265–295
3	Semi-firm	220–250
4	Firm	175–205
5	Very firm	130–160
6	Hard	85–115

Greases with an NLGI of #000 are like a liquid, whereas #6 greases are almost solid. The most frequently used greases are #0, #1, and #2.

Dropping point. This is the temperature at which the grease will change from semisolid to liquid—basically the melting point.

Water resistance. This determines whether a grease will dissolve in water. This is a very important quality if there is a chance that water will come in contact with the lubricant.

Stability. This property determines the ability of a grease to retain its characteristics with time.

Criteria for *measuring the properties of oil* are these:

Viscosity. This is the most important characteristic; it refers to the thickness of the fluid and can also be described as the resistance to flow. Viscosity is affected by temperature and decreases as temperature increases. There are many ways of measuring viscosity, but they are all based on the time taken for a fixed volume of oil to pass through a standard orifice under laboratory conditions.

There are three commonly used terms for viscosity: Saybolt Universal Seconds (SUS), centipoise (cP), and centistokes (cSt).

Saybolt Universal Seconds is an indication of the time it takes 60 millimeters of fluids to flow through a calibrated Saybolt Universal tube (also called a viscosimeter). This is an old method used to describe viscosity, but a lot of American companies still use it.

A *centipoise* is 0.01 poise. A poise is an absolute viscosity unit in the metric system. A *centistoke* is 0.01 stoke. A one-stoke fluid has an absolute viscosity of 1 poise and a density of 1 gram per cubic centimeter. In other words, centistokes differ from centipoise by a density factor.

Some industrial organizations classify oils by the type and viscosity rating:

- International Standards Organization—Viscosity Grade (ISO VG)
- Society of Automotive Engineers (SAE)—Viscosity Number
- SAE—Gear Viscosity Number
- American Gear Manufacturers Association (AGMA)—Lubricant Number

Viscosity Index. This is the rate of change of viscosity with temperature. A high viscosity index shows that the oil will remain the same over a wide range of temperatures, whereas a low index indicates that the oil will thin out rapidly with an increase in temperature.

Flash Point. This is the temperature at which the vapor of a lubricant will ignite.

Fire Point. This is the temperature (higher than the flash point) that is required to form enough vapor from the lubricant to cause it to burn steadily.

Pour Point. This is the low temperature at which the lubricant becomes so thick it will not flow.

Oxidation Resistance. If oil is exposed to the atmosphere, especially at a higher temperature, oxygen is absorbed into the oil. A chemical change takes place in the oil that drastically reduces its lubricating properties.

Emulsification. This is the measure of the tendency for oil and water to mix together.

Synthetic oils are being used more frequently today. Current uses include situations of high pressure or vacuum, very high or low temperature, nuclear radiation, and chemical contamination. They have a better film strength than mineral oils and are much better at inhibiting oxidation.

Regreasing Interval for Antifriction Bearings

Figure 19-1 shows lubrication frequency for rolling element bearings.

Example

A 320 series bearing is turning 1800 rpm at 90°C. This bearing is a radial ball bearing with a bore of 100 mm.

1. Find 1800 rpm at the bottom of the chart.
2. Move up the chart to the curve for 100 mm.
3. Move to the left from this point, and read the hours for radial ball bearing = 4000 hours.
4. Cut this time in half for the higher temperature.
5. Convert the hours to months or weeks of operation: 2000 hours = 12 weeks or 3 months.

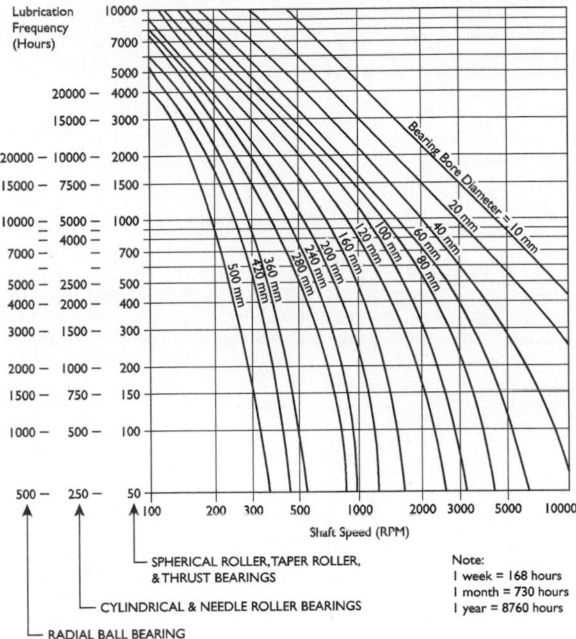

Figure 19-1

Oil Viscosity Classification Chart

Figure 19-2 compares oils rated by different standards. Place a straightedge horizontally across the chart to compare one type of reading with another.

For instance, SAE gear oil of viscosity #90 is the same viscosity as SAE 50 weight used in a car. A middle-of-the-night maintenance tip is that ISO, AGMA, and SAE gear oils are, for the most part, interchangeable. An SAE #90 gear oil can be substituted for an AGMA #5 or #6.

Chain Drive Lubrication

Most chain drive applications are not lubricated beyond the manufacturer's initial lubrication. This is unfortunate because chain life can be increased by three times the unlubricated life. Table 19-2 suggests viscosity to lubricate chain drives.

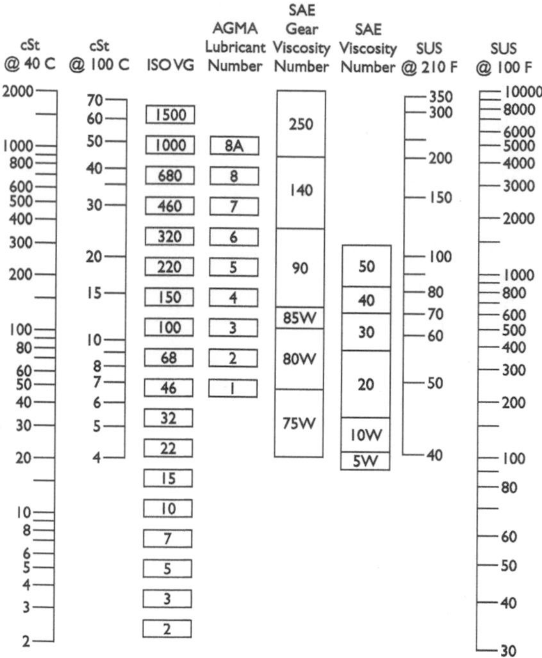

Figure 19-2 Viscosity of oils using different standards.

Table 19-2 Chain Drive Lubricant

Ambient Temperature	*Viscosity at 40°C*
-5–5°C	30–80 cSt
5–40°C	80–120 cSt
40–50°C	120–180 cSt
50–60°C	180–300 cSt

The most common oil used is SAE 30 because it meets the temperature requirements most often encountered in industry.

Regreasing Schedule for AC Electric Motors

Although it is a good idea to check with the manufacturer for regreasing intervals, Table 19-3 can be used for ball-bearing motors if no better information exists.

Table 19-3 AC Motor Regreasing Schedule (Ball Bearing Only)

Type of Service	*Typical Examples*	*½ – 7 ½ Horsepower*	*10–40 Horsepower*	*50–200 Horsepower*
Easy	Motor operating infrequently—1 hour per day	10 years	7 years	5 years
Standard	Machine tools, fans, pumps, textile machinery	7 years	5 years	3 years
Severe	Motors for continuous operation in key or critical locations	4 years	2 years	1 year
Very severe	Dirty and vibrating applications where end of shaft is hot from high ambient temperature	9 months	4 months	4 months

Regreasing intervals for motors are usually longer than most people assume they will be. Overgreasing can present a problem, resulting in grease entering the motor. It is particularly important to follow the manufacturer's recommendation for motors.

Oil Analysis

The most commonly used technique to monitor oil and use the information to indicate both the serviceability of the oil and the internal wear of components is spectrographic oil analysis. Often, this service is provided free by the oil vendor to the plant or purchased separately at a cost of about $10.00 to $15.00 for each analysis. Large hydraulic oil reservoirs, gearbox sumps, and internal combustion equipment are prime candidates for lubrication testing.

Taking a sample is not hard but requires some thinking. The correct sample desired is a *representative* sample of the oil in the reservoir. The best time to take a sample is with the machinery running, if possible. The oil fill port is opened, and a clean tube connected to a small portable hand-powered vacuum pump is inserted into the oil fill port. The tube is lowered into the oil reservoir, and a sample is "pulled" using the vacuum pump. The sample bottle is capped, identified, labeled, and sent out for testing. The tube and pump are cleaned with acetone or other fast-drying solvent and made ready for the next sample to be taken. In many cases, however, taking a sample while the equipment is running is not so easy. Inserting the sample tube into the gearbox and catching the tube in the gear mesh is a possibility. In these cases, the equipment needs to be shut down and the sample taken as soon as possible to ensure that the oil is still mixed well and any particles suspended in the oil are captured in the sample.

Imagine that the oil being sampled is from an automobile engine or diesel engine that might be attached to the emergency generator at a plant location. The results back from the analytical lab show the oil is serviceable but loaded with chrome. The report shows what would be expected as a normal chrome level and indicates the chrome is much higher that the norm. The piston rings in an internal combustion engine are often chrome plated. In this case, ring wear might be suspected from this simple test. A compression test on the cylinders could be run to confirm this analysis of the condition. Obviously, it takes a great amount of time to run a compression test on all cylinders as opposed to a few minutes to get an oil sample and send it to the lab. Oil analysis makes more sense as a screening type of tool. If a problem is found, then the additional check of equipment components makes good economic sense.

Other good candidates for oil analysis are large gearboxes and large hydraulic reservoirs. Spending a few dollars on an oil analysis versus draining all the oil and removing inspection plates or covers to physically inspect the critical components is money well spent.

The report back from the laboratory now indicates the normal amounts of metal in the oil but also shows the amount of dirt and/or a change in viscosity. Again, if the oil is found to be in good condition and the metal levels are normal, there is no reason to change the oil. Spending $10.00 to eliminate an unnecessary $700 oil change is a wise expenditure of funds. Manufacturers' recommendations for oil changes are based on an average usage of their equipment. If the machinery used is lightly loaded and used infrequently, oil analysis shows that the oil is serviceable and should not be changed. If the machinery is located in a dirty environment and is heavily loaded, the oil analysis shows that the optimum change-out is more frequent than the manufacturer advised. In either case, the use of oil analysis to determine the real conditions gives the right answers.

Chapter 20

Vibration Measurement

Vibration measurement is a simple but effective method of looking at the condition or "health" of rotating machinery. Periodically checking machine conditions by trending is easy to do with vibration meters. Vibration meters help mechanics spot deteriorating machine conditions before they become critical (Figure 20-1). By identifying and quantifying a vibration problem, you can take corrective action before the problem becomes significant. Trending with meters allows mechanics or maintenance personnel to plan repairs during normal work hours instead of scheduling expensive overtime or shutting down production.

Almost all mechanical rotating equipment vibrates when it is running. Excessively high vibration is a symptom of a problem, such as unbalance, looseness, misalignment, worn shafting, and other faults, just as a high temperature for a human being might indicate infection, organ failure, or inflammation in the human body.

Simple Vibration Meter

A simple vibration meter usually measures vibration velocity in units of inches per second (often shown as in./sec or ips) and can measure the bearing condition to determine flaws inside an antifriction bearing. Figure 20-2 shows a simple vibration meter with a magnetic pickup used to take a reading on an electric motor. Most simple vibration meters are battery powered and may use an analog or a digital scale. Many experienced maintenance persons prefer the analog scale using a needle because the swing or variation of the needle can be used to help determine the cause of the vibration.

Types of Readings

The two most useful readings are velocity and bearing G's. The destructive forces generated in rotating machinery are proportional to vibration velocity. Checking the vibration velocity at key points on a machine is a very good indicator of machinery severity. The bearing G is a measurement of the pulses caused by impacts between bearing parts that have defects and flaws. In most cases, the vibration instrument is capable of obtaining both velocity and bearing-G readings by flipping a selector switch, thereby changing the circuitry in the instrument to obtain the appropriate reading.

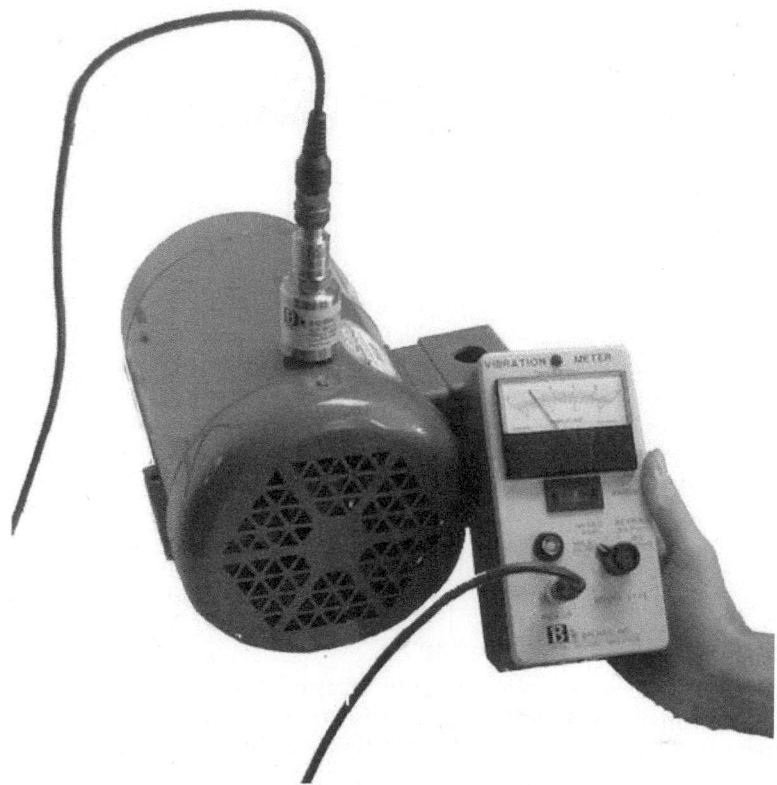

Figure 20-1 Vibration meter and components.
(Courtesy Balmac, Inc.)

Figure 20-3 shows the front panel and face of a typical vibration instrument. Notice the selector switch, which allows readings of velocity and bearing G's. Other vibration quantities, such as displacement and acceleration, can also be obtained but are less useful for simple machinery "health" determination.

Reading the Instrument

Most simple vibration meters have an amplitude meter with two scales on the meter face. The top scale is marked from 0 to 10 and the bottom scale from 0 to 3, as shown in Figure 20-4. When obtaining a reading using the instrument, the scale switch should be set to allow the needle to read in the upper two-thirds of the scale

Figure 20-2 Simple vibration meter in use.
(Courtesy Balmac, Inc.)

without pegging the meter (going off the scale). Because vibration can be any value on a particular machine, it is all right to allow the meter to peg until the correct scale setting can be obtained.

Setting the range of the instrument determines where to place the decimal point when taking a reading. If the range switch is on any of the ranges that use a "1" (0.1, 1, 10, or 100), then the top portion of the amplitude meter scale should be read. If the range switch is on any of the ranges that use a "3" (.03, 0.3, 3, or 30), then the bottom portion of the amplitude meter scale should be read.

The examples shown in Figure 20-5 give visual explanations of various measurements and their correct readings.

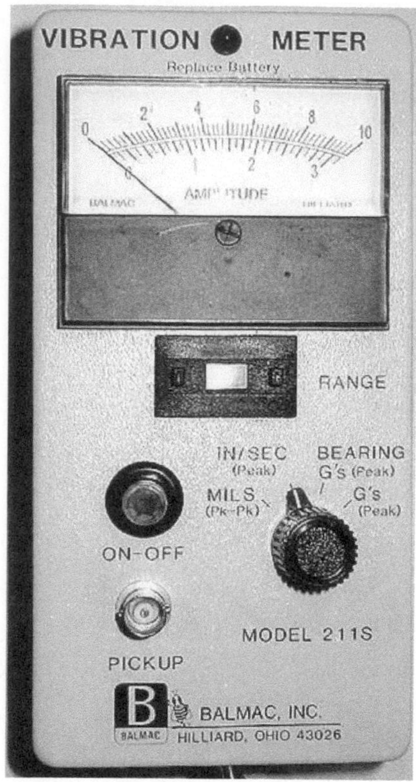

Figure 20-3 Front panel of vibration instrument. *(Courtesy Balmac, Inc.)*

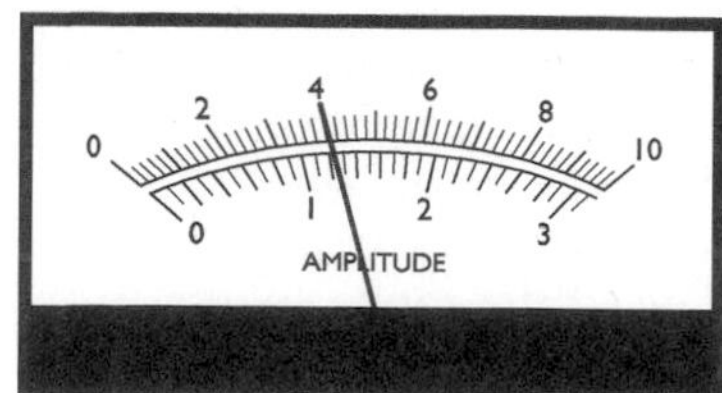

Figure 20-4 Meter face and scales.

Obtaining the Readings

Vibration forces can be measured in horizontal, vertical, and axial directions, as shown in Figure 20-6. The vibration pickup may be hand-held using the extension probe or affixed using the

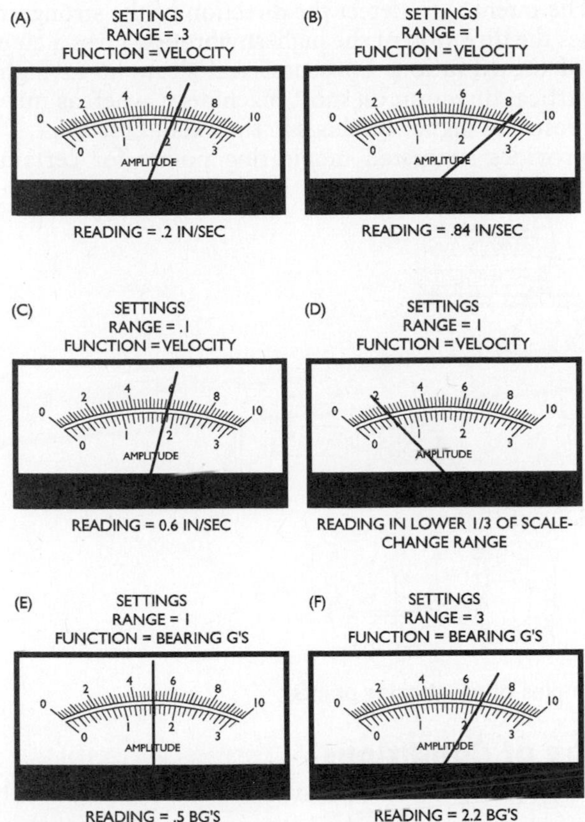

Figure 20-5 See if you can read the meter correctly.

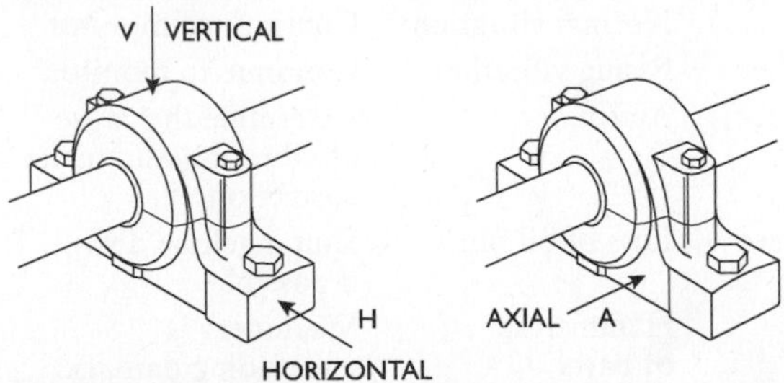

Figure 20-6 Measurements can be taken in the horizontal, vertical, and axial directions.

magnetic base. The intent is to detect the direction of the strongest signals. Sometimes the direction of the highest vibration gives a clue as to the cause of the vibration. Looseness will show up as high readings in the vertical direction on most machinery, whereas misalignment might result in higher readings in the axial directions.

Figure 20-7 provides suggested monitoring points for certain generic classes of machinery.

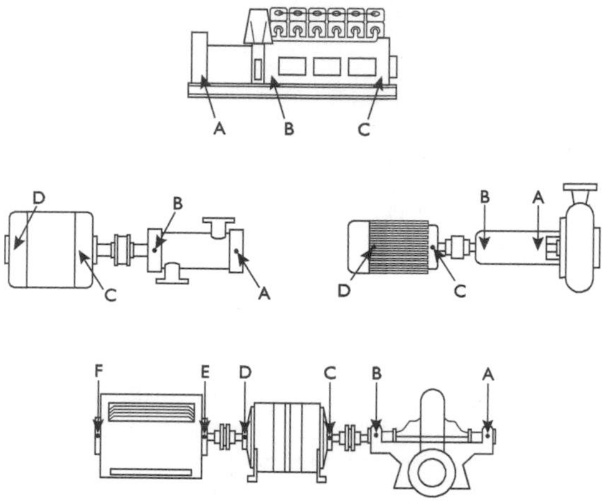

Figure 20-7 Examples of monitoring points.

Determination of Conditions

Use Table 20-1 to judge the severity of vibration levels for general machinery.

Table 20-1 Machinery Severity Table

Vibration Reading	*Severity Level*	*Action Needed*
.15 in./sec or less	Normal vibration	Continue to monitor.
.15 in./sec to .29 in./sec	Rising vibration	Continue to monitor.
.30 in./sec to .59 in./sec	Alarm	Determine the cause of vibration and make correction.
.60 in./sec to .89 in./sec	Loss of oil film	Shut machine down if possible.
.90 in./sec or more	Hammering of parts	Machinery is undergoing damage. Continued operation will produce early failure.

The bearing-G readings in Table 20-2 can be used to judge the condition of antifriction bearings for general machinery.

Table 20-2 Bearing Condition Table

Bearing Reading	*Bearing Condition*
.00 to .49 bearing G's	No cause for alarm. Bearing is good.
.50 to .99 bearing G's	Spalling apparent in bearing. Minor damage.
1.0 to 1.49 bearing G's	Minor to major damage to rolling elements or raceways.
Above 1.50 bearing G's	Seriously consider bearing replacement before bearing failure occurs.

Predicting Failure

There is no absolute level of vibration that indicates failure for a machine. It is possible to measure vibration levels above .90 in./sec on a machine for weeks and still not have a catastrophic failure. When the machine is disassembled, however, most parts will have to be replaced, and machine work will probably be required to repair worn journals or housing fits. In short, the machine will be taken out of service while it is still running, but the components will be in need of major replacement or repair. Finding the source of the vibration and correcting it when it just starts to be considered excessive will reduce the level of vibration and save a piece of equipment from costly repair.

Vibration Analysis

Determining the cause of vibration with a simple hand-held meter is not always easy. More complex instruments, called vibration analyzers, are often brought to the machine site to assist in determining the specific cause. Figure 20-8 shows a typical vibration analyzer.

Vibration analysis makes use of a tunable filter to allow the maintenance technician to look at the amount of vibration at specific frequencies. Much like a radio can be tuned to frequencies (stations) where news, music, police calls, and fire calls might be found, a vibration analyzer allows a look at a machine to determine if unbalance, looseness, misalignment, bad gears, bad bearings, or other machinery faults might exist. The filter is tuned to frequencies expressed in cpm (cycles per minute), and the association between cpm and rpm (revolutions per minute) gives the mechanic insight into the nature of the problem.

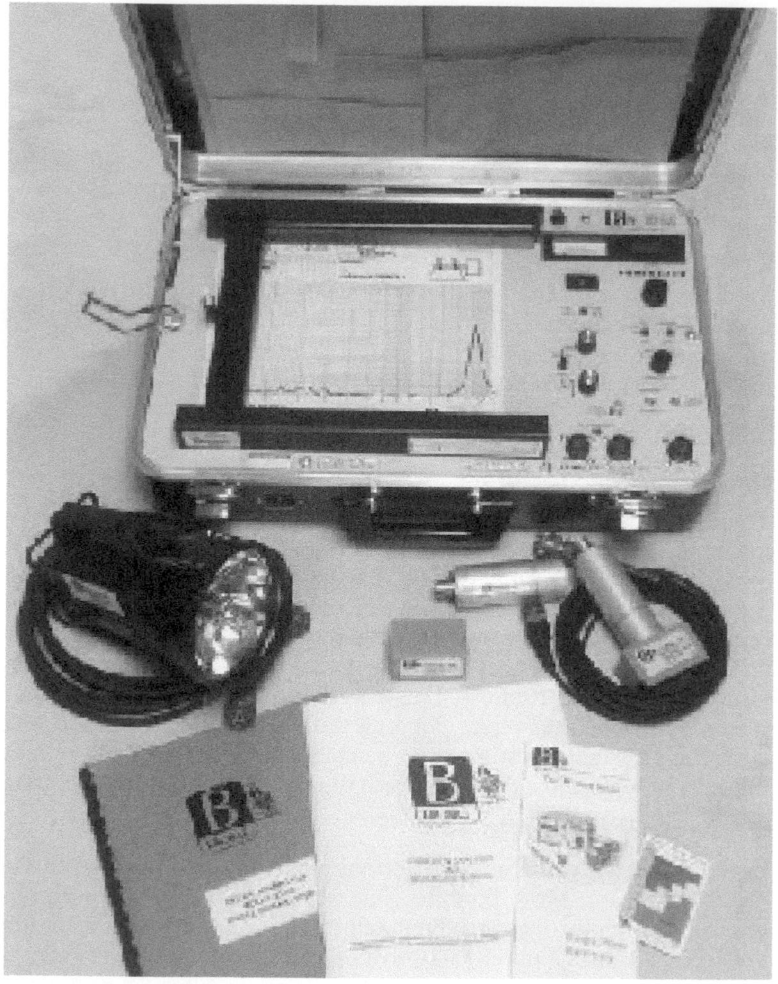

Figure 20-8 Vibration analyzer.

Imagine that a fan is driven by a motor using a V-belt and sheaves. The motor is operating at a speed of 1750 rpm, the belt at 675 rpm, and the fan at 1200 rpm. If a frequency scan of a problem vibration showed that most of the vibration was occurring at 1200 cpm, then the fan would be the offending component. If the problem vibration occurred at 675 cpm or 1750 cpm, then the offending component would be the belt or the motor, respectively.

Detailed vibration analysis is beyond the scope of this book and usually requires advanced training by the instrument manufacturer. Often an outside service is called on to determine and help correct the problem after the local mechanic has found that a problem exists using the simple hand-held vibration meter.

Chapter 21

Preventive and Predictive Maintenance

Preventive maintenance (PM) is a vital and necessary step in keeping equipment in good operating condition. Preventive maintenance is time based. Certain tasks (such as lubrication, inspection, or adjustment) are made at a required time or frequency on a piece of machinery. Predictive maintenance (PDM) makes use of physical measurements on a machine (such as vibration levels, bearing condition, noise emission readings, oil sample results, or temperature) and compares these readings to a norm and also trends the readings to determine impending failure in a machine or components.

The installation of both PM or PDM programs follows some logical steps:

1. Get a process drawing, building equipment location drawing, or other "map" of the plant or process that lists all the equipment and shows the location.
2. Make a list of equipment that will require PM or PDM. It is beneficial to group the equipment by name, number, location, and type. If equipment has never been numbered, it might be a smart idea to "invent" an intelligent numbering system that helps in sorting or locating the equipment.

   Examples of intelligent numbering include:

   PU-2004. A PUmp located on the second deck and the fourth one in the process train.

   100-EX-405. A heat EXchanger in building 100 on the fourth floor and the fifth item in the process train.

   100-AC-405. An AC motor used to drive the gear pump on the heat exchanger shown in the previous entry.

3. Perform a field visit to each piece of equipment. Make sure it exists and that it corresponds to the information previously obtained. Take down all nameplate information to determine the vendor, model number, serial number, and other identification characteristics. A digital camera can speed up this process.

4. Locate all the maintenance manuals for the equipment. Be thorough—check the engineering offices, foreman's desk, and mechanics' benches. If there are no manuals available, use the nameplate information to find the local sales office or manufacturer's location to obtain a contact person or phone number. The Internet is also a good source of manufacturer information. Contact the manufacturer, supply the nameplate information, and request a maintenance manual for each piece of equipment. In some cases, there may be a nominal charge. Make sure to request manuals with "cut sheet" or "exploded diagrams" to make spare-part identification easier when working on the equipment.
5. Create a set of equipment files. Use the number for the equipment, and set up an empty manila folder for each piece of equipment. Use this file to store nameplate information, vendor contact information, maintenance manuals, or other reference materials for each particular piece of equipment.
6. Use a copy machine to copy the PM page for the particular equipment from the equipment manual. Look over the manufacturer's recommended PM activities. Mentally group the activities into lubrication tasks, visual observation tasks, shutdown inspection tasks, teardown inspection tasks, and any other grouping that puts similar activities together.
7. Set up a meeting with key personnel who are responsible for the operation of the equipment—both maintenance and operations. Review the manufacturers' information with others and see if the recommended PM makes sense. Some manufacturers will use an overkill approach and suggest much more PM than is necessary. In addition, pay attention to the suggested frequencies of PM that are mentioned. Read closely. A manufacturer may suggest that the equipment be serviced every month, but such a recommendation is based on continuous usage. If the equipment is run for only one week in the month, then the PM frequency needs to be adjusted. In addition, consider which PDM tools might be brought into play to inspect the equipment. Should oil analysis be run on a large gearbox? Is vibration measurement the right method for checking the fan bearings?
8. Begin setting up the PM and PDM activities. Pay attention to common types of equipment (all the gearboxes, all the motors) and common activities (greasing, oil changes, chain inspection).

It may be more expeditious to assign PM and PDM by task—for example, one person follows a route through the facility and performs all the lubrication or all the vibration measurement.

It may be more expeditious to assign PM and PDM by equipment type—for example, a work order for PM calls for PM inspection of all gearboxes during an outage, or all fans are checked for vibration just before a shutdown for cleaning.

It may be more expeditious to assign PM or PDM by area—for example, all the PM or PDM for a certain floor or process area is done by one person.

Usually a combination of these methods will produce a workable PM/PDM system.

9. Assign frequencies to the PM/PDM activities (daily, weekly, hours of operation, etc.).
10. Set up the PM/PDM routes. Schedule things to be accomplished so that people use their time properly. Make sure personnel do not have to retrace their steps or back up to accomplish the PM/PDM tasks.
11. Begin the PM/PDM system. In some cases, it is easier to take a small section of the plant or building as a model to see how well the route is laid out. If the model meets with success, then expand it. If the model has problems, correct them before going further.
12. Incorporate the entire PM/PDM system into the daily planning and scheduling of personnel.
13. Develop a feedback loop to adjust tasks, frequency, and duration of activities. This feedback loop should allow tailoring of the PM/PDM system to adjust for changes in production, changes in machinery loading, or changes from aging of equipment.
14. Continue to audit the PM/PDM program every three months. Look for problems. Look for solutions. Effect changes.
15. Modify the program to change tasks, change frequencies, add equipment, remove equipment, change routes, and change personnel based on results of the PM/PDM audit.

PM Visual Inspection Checklist

Visual inspection of machinery or machine components should be quantified. Using a checklist with specific activities and some method of measurement is preferred.

Example
Gearbox inspection (partial list shown):

1. Make sure the gearbox is shut down and locked out.
2. Clean off the area around the inspection cover.
3. Remove the inspection cover and, using a flashlight, check for gear wear.

    [ ] Gears show no wear.
    [ ] Gears show some signs of pitting.
    [ ] Gears show mushrooming and have sharp edges when touched.
    [ ] Gears have cracked teeth.
    [ ] Gears have missing teeth.

4. Inspect each oil seal on the gearbox.

    [ ] Input shaft seal on drive side leaking
        [ ] small [ ] medium [ ] extreme
    [ ] Input seal on drive side not leaking
    [ ] Input shaft on opposite drive side leaking
        [ ] small [ ] medium [ ] extreme
    [ ] Input shaft on opposite drive side not leaking
    [ ] Output shaft on drive side leaking
        [ ] small [ ] medium [ ] extreme
    [ ] Output shaft on drive side not leaking
    [ ] Output shaft on opposite drive side leaking
        [ ] small [ ] medium [ ] extreme
    [ ] Output shaft on opposite drive side not leaking

Quantifying things is important. A well-designed checklist gives the mechanic an exact procedure and a place to record observations easily without a lot of writing, which may be difficult to do in the field.

PM Instrument or Gauge Inspection

Some basic instruments or checking gauges make PM inspections more meaningful. A strobe light can be used to perform a slow-motion study of a shaft or coupling while it is rotating. Cracks, loose parts, and movement are readily observable using this method (Figure 21-1).

Figure 21-1 Using a strobe light to find defective parts.

A mechanical stethoscope is inexpensive and a great asset to a PM mechanic. If the PM checklist calls for listening for noises, it makes good sense to give the mechanic a tool to pinpoint the source.

Sheave inspection on the V-belt drive can be done with a sheave inspection tool as shown in Figure 21-2. Belt tensioning can be checked using a tension tester.

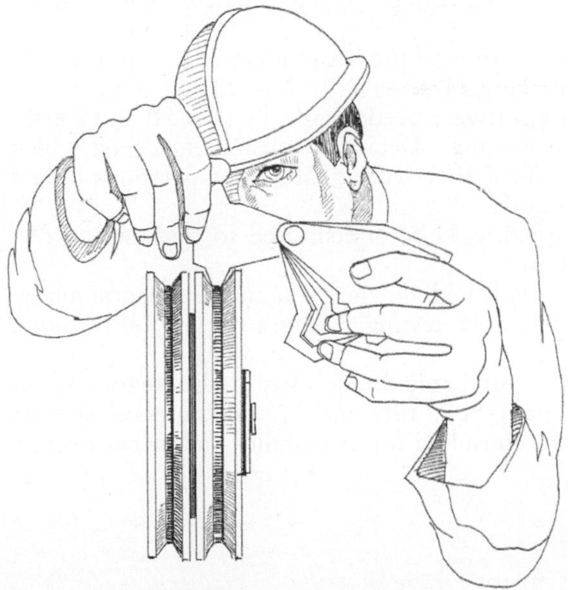

Figure 21-2 Sheave inspection using a sheave gauge.
(Courtesy Maintenance Troubleshooting)

A pocket scale or tape measure allows the check of chain elongation to determine wear as shown in Figure 21-3.

Figure 21-3 Checking for chain elongation.

Any device that allows a mechanic to measure for wear or obtain information without shutting the equipment down can greatly enhance a PM system.

Predictive Maintenance

Predictive maintenance (PDM) compares the trend of measured physical parameters against known engineering limits for the purpose of detecting, analyzing, and correcting problems before failure occurs.

A classical implementation of predictive maintenance follows the flowchart shown in Figure 21-4.

The hardware and software needed must be selected and a critical equipment list generated. Decisions must be made on which predictive test is the most effective for catching the various failure modes that might occur.

The PDM cycle (Figure 21-5) is dedicated to find and correct problems before failure.

Typical predictive tools include vibration and lubrication analysis, electrical time/resistance testing (Megger testing), ultrasound, infrared scanning, etc.

Table 21-1 helps the mechanic decide what predictive tool works the best for a particular type of machine. Manufacturers of specific equipment need to be consulted for availability and instruction on the proper tooling.

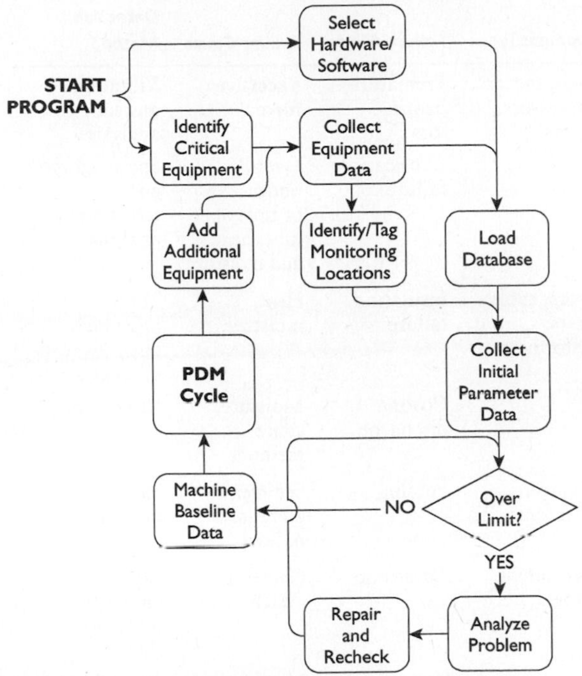

Figure 21-4 PDM implementation.

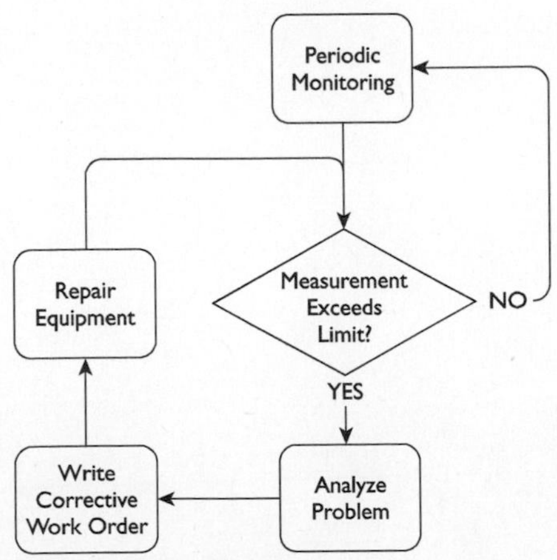

Figure 21-5 PDM cycle.

Table 21-1 Choices for Predictive Maintenance Tools

Equipment Category	*Equipment Types*	*Failure Mode*	*Failure Cause*	*Detection Method*
Rotating machinery	Pumps, motors, compressors, blowers	Premature bearing loss	Excessive force	Vibration and lube analysis
		Lubrication failure	Over-, under-, or improper lube; heat and moisture	Spectrographic and ferrographic analysis
Electrical equipment	Motors, cable, starters, transformers	Insulation failure	Heat, moisture	Time/ resistance tests, I/R scans, oil analysis
		Corona discharge	Moisture, splice methods	Ultrasound
Heat transfer equipment	Exchangers, condensers	Fouling	Sediment/ material buildup	Heat transfer calculations
Containment and transfer equipment	Tanks, piping, reactors	Corrosion	Chemical attack	Corrosion meters, thickness checks
		Stress cracks	Metal fatigue	Acoustic emission

Chapter 22

Electricity

In many large industrial plants, the millwright and/or mechanical repairman is not required to perform work on the electrical system or equipment. In smaller plants, where repairmen are responsible for all types of maintenance coverage, they may perform some work that requires a general understanding of electricity. Such an understanding is essential even for minor electrical repairs or simple parts replacement if it is part of the mechanic's duties. All mechanics, whether they do the actual electrical work or not, should have such an understanding if they must do troubleshooting. This is very important because a check of electrical systems and controls is a major consideration in any troubleshooting operation. Therefore, the objective in this chapter is to present the practical basic electrical information useful to a mechanic. Detailed technical explanations and involved theory will be avoided.

Electrical Safety

Because of the ever-present dangers of electrical energy, a basic requirement when working with electricity is that there is no guesswork or risk taking. Activities in this area should be restricted to those things with which you have experience or about which you have specific knowledge or understanding.

First and especially important is a full appreciation by the mechanic of the hazard involved in working with electricity. This is an area where a little knowledge can be a dangerous thing if used improperly. The potential danger, ever present with electrical energy, cannot be overemphasized. A basic requirement for electrical work safety is that there be no guesswork or chance taking. If one does not know what conditions are present or the possible consequence of the action being considered, no action should be taken. Competent assistance should be obtained. On the other hand, we are constantly using electricity and electrical equipment safely all the time both in the home and at work. Therefore, if one restricts his activities in the area of electrical work to those things he understands and/or has specific knowledge of or experience with, these things can be done safely.

Electrical Terms

An understanding of the words and terms that make up the language of electricity is a first requirement. Although these definitions may not be technically correct in all cases, they are generally accurate and descriptive.

Electromotive Force

Electromotive force is the force that causes electricity to flow when there is a difference of potential between two points. It is abbreviated emf, and the unit of measurement is the volt.

Current Flow

The movement of electrons is commonly referred to as the flow of electricity, or *current flow*. It is measured in amperes.

Voltage

Voltage is the value of the electromotive force in an electrical system. It may be compared to pressure in a hydraulic system.

Amperage

The quantity and rate of flow in an electrical system is referred to as the *amperage*. It may be compared to the volume of movement in a hydraulic system.

Resistance

Resistance refers to the ability of materials to impede the movement of electrons or the flow of electricity. It is commonly stated in terms of ohms.

Ohm's Law

This is the universally used electrical law stating the relationship between current, voltage, and resistance. This is done mathematically by the formula shown here. Current is stated in *amperes* and abbreviated *I*, resistance is stated in *ohms* and abbreviated *R*, and voltage is stated in *volts* and abbreviated *E*.

$$\text{Current} = \frac{\text{voltage}}{\text{resistance}} \quad \text{or} \quad I = \frac{E}{R}$$

The arrangement of values gives two other forms of the same equation:

$$R = \frac{E}{I} \quad \text{and} \quad E = I \times R$$

Example

An ammeter placed in a 110-V circuit indicates a current flow of 5 amps. What is the resistance of the circuit?

$$R = \frac{E}{I} \quad \text{or} \quad R = \frac{110}{5} \quad \text{or} \quad R = 22 \text{ ohms}$$

Direct Current

Current flow in one direction is referred to as *direct current.* This is commonly associated with continuous direct current, which is nonpulsating, as from a storage battery.

Alternating Current

Alternating current refers to current flow that is continuously reversing or alternating in direction, resulting in a regularly pulsating flow.

Cycle

A *cycle* is the interval or period during which alternating current starts at zero and increases to maximum force in a positive direction, reverses to a maximum force in the negative direction, and returns to zero value. One cycle of such a current flow is illustrated in Figure 22-1.

Frequency

Frequency refers to the number of complete cycles per second of the alternating current flow. The most commonly used alternating-current frequency is 60 cycles per second. This is the number of complete cycles per second; thus, the pulsation rate is twice this, or 120 pulses per second. Frequency is now specified as so many hertz. The term *hertz* is defined as cycles per second and is abbreviated Hz.

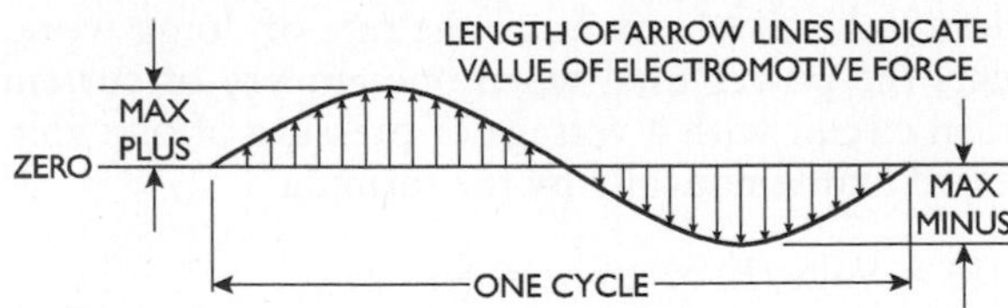

Figure 22-1 Illustrating one cycle.

Phase

The word *phase* applies to the number of current surges that flow simultaneously in an electrical circuit. The illustration of alternating current in Figure 22-1 is a graphic description of a single-phase alternating current, the single line indicating a current flow that is continuously increasing or decreasing in value.

Three-phase current has three separate surges of current flowing together. In any given instant, however, their values differ, as the peaks and valleys of the pulsations are spaced equally apart. The three-phase current wave, shown in Figure 22-2, illustrates that the peaks and valleys of each phase are uniformly spaced. If each cycle were considered to be a full 360° rotation, each of the phases would be spaced 120° apart.

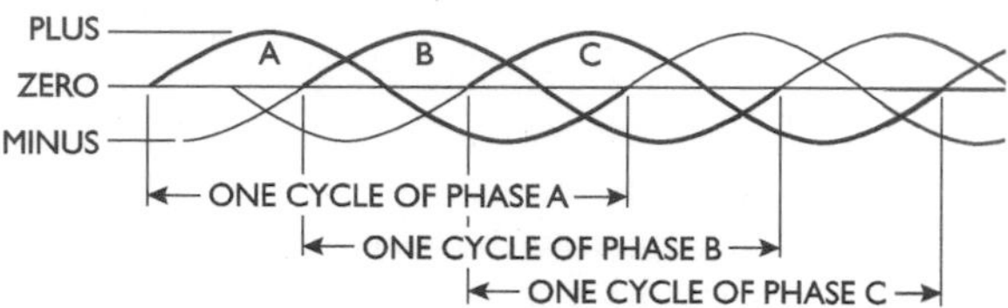

Figure 22-2 Three-phase sine wave.

The waveforms in Figure 22-2 are lettered A, B, and C to represent the alternating current flows for each phase during a complete cycle. They graphically illustrate the pulsation from plus to minus values and the relationship between the three phases in which any one current pulse is always one-third of a cycle out of step with another.

Single-phase current is used for lighting and small motors requiring small amounts of power and comparatively light loads. The single-phase motor is used in the home, on the farm, and in industry whenever possible because it is less expensive than the three-phase motor. Three-phase motors, by contrast, are more expensive and require more expensive three-phase service and auxiliary equipment.

Watt

The *watt* is the electrical unit of power, or the rate of doing work. One watt represents the power used when one ampere of current flows in an electrical circuit with a voltage or pressure of one volt. This can be expressed mathematically by the formula

$$\text{Watts } (P) = \text{Volts } (E) \times \text{Amperes } (I)$$
$$P = E \times I$$

This formula for electrical power indicates the rate at any given instant at which work is being done by the current moving through the circuit. The name on a piece of electrical equipment usually indicates the wattage rating as well as the voltage and current. The power limits are given as maximum safe voltages and currents to which the device may be subjected. In those cases where a device is not limited to a specific operating voltage, its wattage indicates its maximum operating limit.

Electrical apparatus such as soldering irons, motors, heaters, and lights are rated in watts. This rating indicates the rate at which electrical energy is converted into heat, light, or motion by the device. If the normal rating of a piece of electrical equipment is exceeded, it will overheat and probably be damaged. Although the immediate damage can be serious, the potential for fire resulting from the overheating may be of even greater concern. Overload devices, which will be discussed later, are designed to protect against these hazards.

Watt-Hour

The *watt-hour* expresses watts in time measurement of hours. For example, if a 100-W lamp is in operation for a two-hour period, it will consume 200 watt-hours of electrical energy.

Kilowatt-Hour

One kilowatt is equal to 1000 W. One kilowatt-hour is the energy expended at the rate of one kilowatt over a period of one hour.

Overload

An overload is the electrical load or demand for electrical energy that results in a flow of current above the maximum rating of the circuit or any of its components. The electrical device the mechanic may most frequently be concerned with is that which provides protection against electrical overloads. This may be a fuse, a circuit breaker, or any other overload protective device. Such devices may be located near the equipment they protect, or they may be incorporated in the wiring system. Protective devices are installed to guard against overloading of the electrical equipment and the wiring system.

Although overloads may occur for many reasons, a common situation is one involving overloading of motors. This may result from mechanical problems, such as bearing failure or parts breakage, or from operating practices, such as jamming or overfeeding. It might be that the work being attempted is beyond the capacity of the equipment. In any case, a high flow of electrical current results, causing the fuse element to overheat and melt or the circuit breaker to open. Overloads might also be produced in the electrical system itself. These might be caused by such things as high or low voltage, wires improperly connected, or short circuits resulting from insulation damage or other conditions that allow wires to make direct contact.

Fuse

A fuse is an overload device consisting of a wire or strip of fusible metal usually contained in a plug or cartridge. If the electrical current flowing through the system exceeds the rating of the fuse, the wire or metal element will overheat and melt, which opens the circuit. The function of the fuse is to melt when overloaded, thus interrupting the flow of electrical current before damage can occur to other parts of the system.

The two general forms of fuse are the screw-plug and the cartridge. Screw-plug fuses include the Edison-base style, which has a threaded base and is the size of a standard incandescent lamp, and the newer type-S style, which has a base of smaller thread and is designed for use in circuits of 125 V or less.

More widely used in industrial plants are cartridge fuses, which are made in two voltage classifications. One class is designed for circuits with a 250-V maximum and the other for circuits with a 600-V maximum. Standard fuses are made in the following ampere ratings, each rating having a different cartridge size: 15, 20, 25, 30, 35, 40, 45, 50, 60, 70, 80, 90, 100, 110, 125, 150, 175, 200, 225, 250, 300, 350, 400, 450, 500, 600, 700, 800, 1000, 1200, 1600, 2000, 2500, 3000, 4000, 5000, and 6000. The 30- and 60-amp cartridges have ferrules at each end, which are the contact surface and mate in the clips of the fuse holder. Larger-size cartridge fuses have protruding blades at each end called knife-blade contacts, which are held firmly by the clips in this style of fuse holder. Some cartridge-type fuses are renewable, with ends that can be unscrewed to allow replacement of a removable fuse link.

Circuit Breaker

Another common form of overload protection is the circuit breaker, which opens, or disconnects, automatically to protect branch circuits and equipment from harmful overloads. Its function is the same as that of a fuse—opening when there is a short circuit or overload on the line. In addition, it may function as a switch in that it can be opened and closed by hand. When a circuit breaker has opened automatically because of an overload, it may be reclosed by hand without replacing any part, as is required with a fuse. However, the circuit breaker will not stay closed until the overload or other trouble has been removed from the circuit. There are many kinds and types of circuit breakers, some of them quite complicated. The most widely used types may be classified as air breakers and oil breakers.

One of the principal applications for air circuit breakers is lighting branch circuits where the 15-amp size is in common use. Larger breakers are rated to match the current ratings of various conductors, for example, 20, 25, 35, and 50 amps. This classification of breaker is not limited to such applications, however, as they are available in ratings from a few amps to thousands of amps. Breakers are manufactured as complete, self-contained units in metal cabinets that provide a neat appearance and safe operation. The oil-filled circuit breaker is generally found where large amounts of power are to be controlled, as in substations and industrial applications.

A circuit breaker looks like a toggle switch and has a handle to turn power on and off (Figure 22-3). Inside of a circuit breaker is a mechanism that trips the breaker if an electrical overload exists. When the breaker trips, the load is disconnected. In most cases, resetting the breaker consists of forcing the handle beyond the "off" position and then returning it to the "on" position. All circuit breakers have a time delay built into them. The breaker will carry 80 percent of its rated load indefinitely. The breaker can carry a small overload for a short period of time before tripping, but it trips quickly if a large overload is present. A circuit breaker installed in a motor circuit has enough time delay that the motor can start without tripping the breaker.

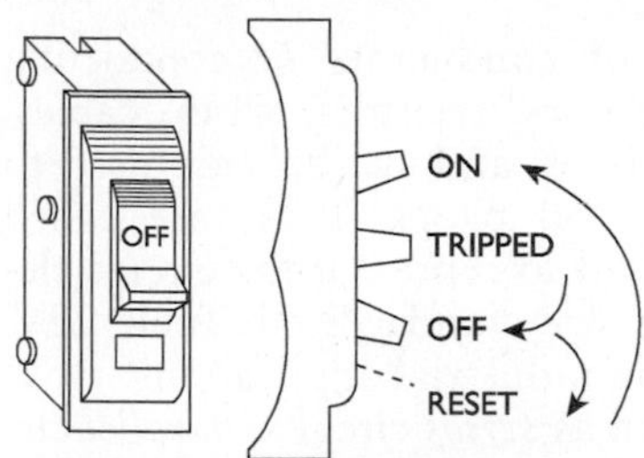

Figure 22-3 Circuit breaker.

Power Formula

This formula indicates the rate at any given instant at which work is being done by current moving through a circuit. Voltage and current are abbreviated *E* and *I* as in Ohm's law, and watts are abbreviated *W*.

$$\text{Watts} = \text{voltage} \times \text{amperes} \quad \text{or} \quad W = E \times I$$

The two other forms of the formula, obtained by rearrangement of the values, are these:

$$E = \frac{W}{I} \quad \text{or} \quad I = \frac{W}{E}$$

Example

Using the same values as those used in the preceding example, 5 amps flowing in a 110-V circuit, how much power is consumed?

$$W = E \times I \quad \text{or} \quad W = 110 \times 5 \quad \text{or} \quad W = 550 \text{ watts}$$

Example

A 110-V appliance is rated at 2000 W. Can this appliance be plugged into a circuit fused at 15 amps?

$$I = \frac{W}{E} \quad \text{or} \quad I = \frac{2000}{110} \quad \text{or} \quad I = 18.18 \text{ amperes}$$

Obviously, the fuse would blow if this appliance were plugged into the circuit.

Circuit Basics

An electrical circuit is composed of conductors or conducting devices such as lamps, switches, motors, resistors, wires, cables, batteries, or other voltage sources. Lines and symbols are used to represent the elements of a circuit on paper. These are called *schematic diagrams*. The symbols used to represent the circuit elements, including the voltage source, are standardized. Table 22-1 shows the symbols commonly used in industrial applications.

Electrical circuits may be classified as *series* circuits, *parallel* circuits, or a combination of series and parallel circuits. A series circuit is one where all parts of the circuit are electrically connected end-to-end. The current passes through each element as it flows from one terminal of the power source to the other. The same amount of current flows in each part of the circuit. Figure 22-4 shows an example of a series circuit.

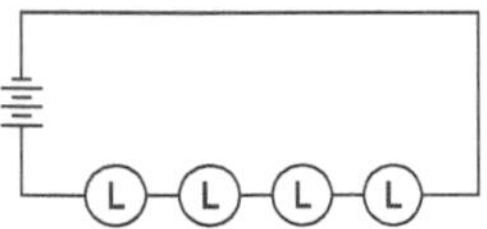

Figure 22-4 Series circuit.

In a parallel circuit, each element is connected in such a way that it has direct flow to both terminals of the power source. The voltage across any element in a parallel circuit is equal to the voltage of the

Table 22-1 Electrical Symbols

Symbol	*Meaning*	*Symbol*	*Meaning*
	Crossing of conductors not connected		Knife switch
	Crossing of conductors connected		Double-throw switch
	Joining of conductors not crossing		Cable termination
	Grounding connection		Resistor
	Plug connection		Reactor or coil
	Contact normally open		Transformer
	Contact normally closed		Battery
	Fuse		Ammeter
	Air circuit breaker		Voltmeter
	Oil circuit breaker		

source, or power supply. Figure 22-5 shows an example of a parallel circuit.

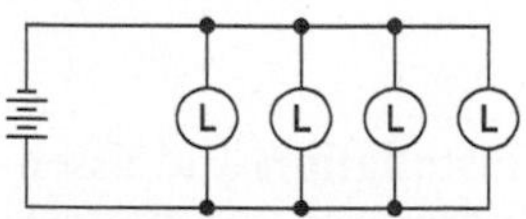

Figure 22-5 Parallel circuit.

The relationship between values in series and parallel circuits using Ohm's law and the power formula is illustrated in Figure 22-6 and compared in the following examples:

Current flow through series circuit $I = \frac{E}{R} = \frac{120}{400} = .3$ amp

Current flow through parallel circuit $I = \frac{E}{R} = \frac{120}{25} = 4.8$ amps

Voltage across one lamp $E = IR = .3 \times 100 = 30$ volts

Current flow through one lamp $I = \frac{E}{R} = \frac{120}{100} = 1.2$ amps

Current flow through one lamp $I = \frac{E}{R} = \frac{30}{100} = .3$ amp

Voltage across one lamp $E = IR = 1.2 \times 100 = 120$ volts

Power used by one lamp $W = EI = 30 \times .3 = 9$ watts

Power used by one lamp $W = EI = 120 \times 1.2 = 144$ watts

Power used by circuit $W = EI = 120 \times .3 = 36$ watts

Power used by circuit $W = EI = 120 \times 4.8 = 576$ watts

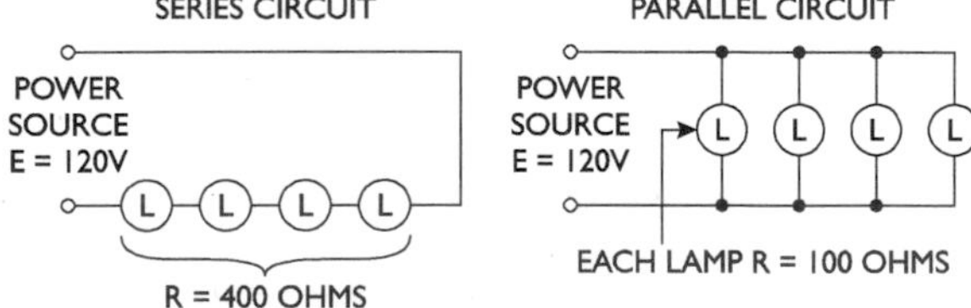

Figure 22-6 Parallel and series circuits.

Electrical Wiring

The term *electrical wiring* is applied to the installation and assembly of electrical conductors. The size of the wire used for electrical conductors is specified by gauge number according to the American Wire Gauge (AWG) system. The usual manner of designation is by the abbreviation AWG. Table 22-2 lists the AWG numbers and the corresponding specifications using the *mil* unit to designate a 0.001-in. measurement.

Table 22-2 AWG Table

Size of Wire, AWG	*Diameter of Wire, mils*	*Cross Section, Circular, mils*	*Resistance, & OHgr/1000 ft at 68°F (20°C)*	*Weight, lf/pounds per 1000 ft*
0000	460	212,000	0.0500	641
000	410	168,000	0.062	508
00	365	133,000	0.078	403
0	325	106,000	0.098	319
1	289	83,700	0.124	253
2	258	66,400	0.156	201
3	229	52,600	0.197	159
4	204	41,700	0.248	126
5	182	33,100	0.313	100
6	162	26,300	0.395	79.5
7	144	20,800	0.498	63.0
8	128	16,500	0.628	50.0
9	144	13,100	0.792	39.6
10	102	10,400	0.998	31.4
11	91	8,230	1.26	24.9
12	81	6,530	1.59	19.8
13	72	5,180	2.00	15.7
14	64	4,110	2.53	12.4
15	57	3,260	3.18	9.86
16	51	2,580	4.02	7.82
17	45	2,050	5.06	6.20
18	40	1,620	6.39	4.92
19	36	1,290	8.05	3.90
20	32	1,020	10.15	3.09
21	28.5	810	12.80	2.45
22	25.3	642	16.14	1.94
23	22.6	509	20.36	1.54
24	20.1	404	25.67	1.22
25	17.9	320	32.37	0.970
26	15.9	254	40.81	0.769
27	14.2	202	51.47	0.610
28	12.6	160	64.90	0.484
29	11.3	127	81.83	0.384
30	10.0	101	103.2	0.304
31	8.9	79.7	130.1	0.241
32	8.0	63.2	164.1	0.191

Table 22-3 Current Capacities

Wire Size	In Conduit or Cable		In Free Air		Weatherproof Wire
	*Type RHW**	*Type TW, R**	*Type RHW**	*Type TW, R**	
14	15	15	20	20	30
12	20	20	25	25	40
10	30	30	40	40	55
8	45	40	65	55	70
6	65	55	95	80	100
4	85	70	125	105	130
3	100	80	145	120	150
2	115	95	170	140	175
1	130	110	195	165	205
0	150	125	230	195	235
00	175	145	265	225	275
000	200	165	310	260	320

**Types RHW, TW, or R are identified by markings on outer cover.*

Table 22-4 Adequate Wire Sizes

Load in Building, A	*Distance, in ft, from Pole to Building*	*Recommended* Size of Feeder Wire for Job*
Up to 25 A,	Up to 50	No. 10
120 V	50–80	No. 8
	80–125	No. 6
	Up to 80	No. 10
20–30 A,	80–125	No. 8
240 V	125–200	No. 6
	200–350	No. 4
	Up to 80	No. 8
30–50 A,	80–125	No. 6
240 V	125–200	No. 4
	200–300	No. 2
	300–400	No. 1

**These sizes are recommended to reduce "voltage drop" to a minimum.*

Table 22-5 Circuit Wires for Single-Phase Motors

Horse-power of Motor	Volts	Appro-ximate Starting Current, A	Appro-ximate Full Load Current, A	Length of Run, in ft, from Main Switch to Motor								
				Feet	25	50	75	160	150	200	300	400
¼	120	20	5	Wire Size	14	14	14	12	10	10	8	6
⅓	120	20	5.5	Wire Size	14	14	14	12	10	8	6	6
½	120	22	7	Wire Size	14	14	12	12	10	8	6	6
¾	120	28	9.5	Wire Size	14	12	12	10	8	6	4	4
¼	240	10	2.5	Wire Size	14	14	14	14	14	14	12	12
⅓	240	10	3	Wire Size	14	14	14	14	14	14	12	10
½	240	11	3.5	Wire Size	14	14	14	14	14	12	12	10
¾	240	14	4.7	Wire Size	14	14	14	14	14	12	10	10
1	240	16	5.5	Wire Size	14	14	14	14	14	12	10	10
1½	240	22	7.6	Wire Size	14	14	14	14	12	10	8	8
2	240	30	10	Wire Size	14	14	14	12	10	10	8	6
3	240	42	14	Wire Size	14	12	12	12	10	8	6	6
5	240	69	23	Wire Size	10	10	10	8	8	6	4	4
7½	240	100	34	Wire Size	8	8	8	8	6	4	2	2
10	240	130	43	Wire Size	6	6	6	6	4	4	2	1

Table 22-6 Extension Cord Table

	Type	*Wire Size*	*Use*
Ordinary Lamp Cord	POSJ, SPT	No. 16 or 18	In residences for lamps or small appliances.
Heavy duty—with thicker covering	S, SJ, or SJT	No. 10, 12, 14, or 16	In shops, and outdoors for larger motors, lawn mowers, outdoor lighting, etc.

Table 22-7 Cord-Carrying Capacity

Wire Size	*Type*	*Normal Load*	*Capacity Load*
No. 18	S, SJ, SJT or POSJ	5.0 A (600 W)	7 A (840 W)
No. 16	S, SJ, SJT or POSJ	8.3 A (1000 W)	10 A (1200 W)
No. 14	S	12.5 A (1500 W)	15 A (1800 W)
No. 12	S	16.6 A (1900 W)	20 A (2400 W)

Table 22-8 Length of Cord Set

Light Load (to 7 A)	*Medium Load (7–10 A)*	*Heavy Load (10–15 A)*
To 25 ft—Use No. 18	To 25 ft—Use No. 16	To 25 ft—Use No. 14
To 50 ft—Use No. 16	To 50 ft—Use No. 14	To 50 ft—Use No. 12
To 100 ft—Use No. 14	To 100 ft—Use No. 12	To 100 ft—Use No. 10

Note: As a safety precaution, be sure to use only cords that are listed by Underwriters' Laboratories. Look for the Underwriters' seal when you make your purchase.

The *circular mil* unit used in the table is a measurement of cross-sectional area based on a circle one mil in diameter.

Switching

Switches are the most widely used of all electrical wiring devices. They are connected in series with the devices they control, and they allow current to flow when closed and interrupt current flow when open. One of the most common of switch applications is the control of one or more lamps from a single location. The schematic diagram for such a circuit is illustrated in Figure 22-7, and a sketch of actual wiring connections is shown in Figure 22-8.

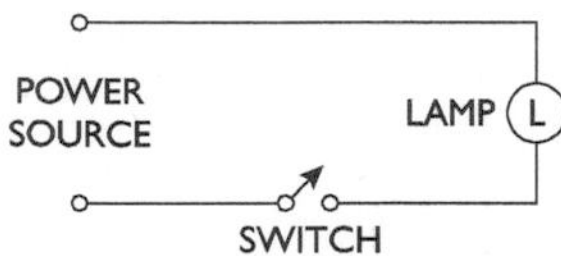

Figure 22-7 Schematic diagram—lighting circuit.

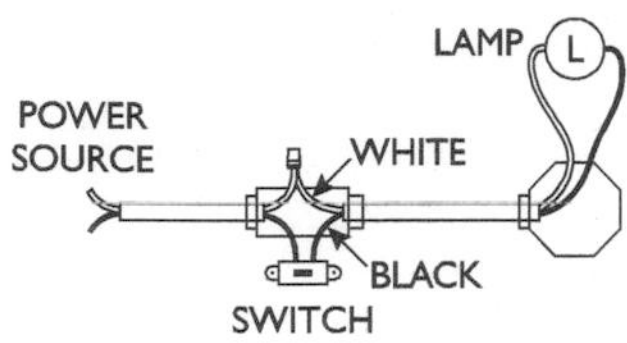

Figure 22-8 Actual wiring—lighting circuit.

Controlling lamps from two points requires a switch called a three-way switch. It has three terminals, one of which is arranged so that current is carried through it to either of the other two. Its function is to connect one wire to either of two other wires. Figure 22-9 shows a lamp circuit controlled from two points using three-way switches. The actual connection boxes and the wires in the cable between the boxes for the three-way-switch circuit are shown in Figure 22-10.

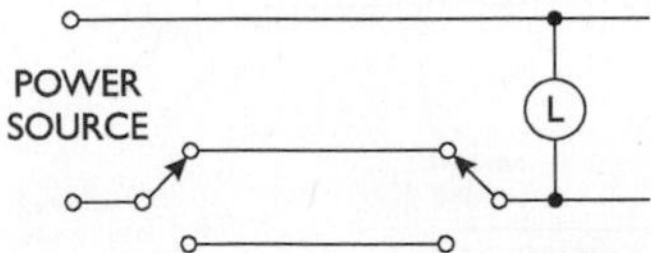

Figure 22-9 Schematic—three-way switching.

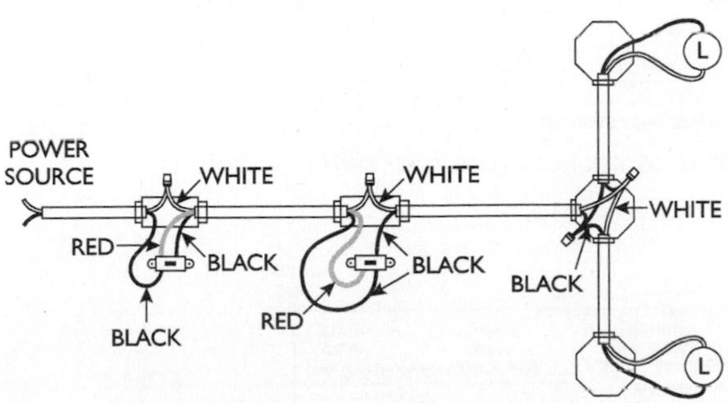

Figure 22-10 Actual—three-way switching.

In factory or residential circuits, it is often important to control lighting (or other devices) from more than one switching location. Use of two-way, three-way, and four-way switches can be a bit confusing. Although some simple schematics were shown previously in this section, the following diagrams are useful to compare the circuits and the connections. Figure 22-11, Figure 22-12, and Figure 22-13 depict schematic diagrams for the use of these switching devices and show the differences in the connections required to effect the various types of control.

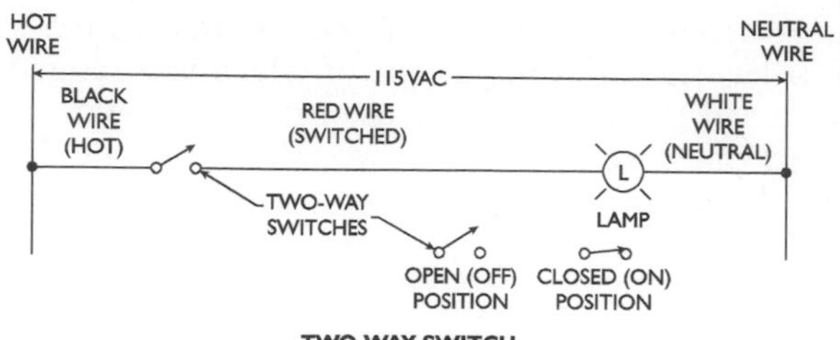

Figure 22-11 Light is controlled from one switch location.

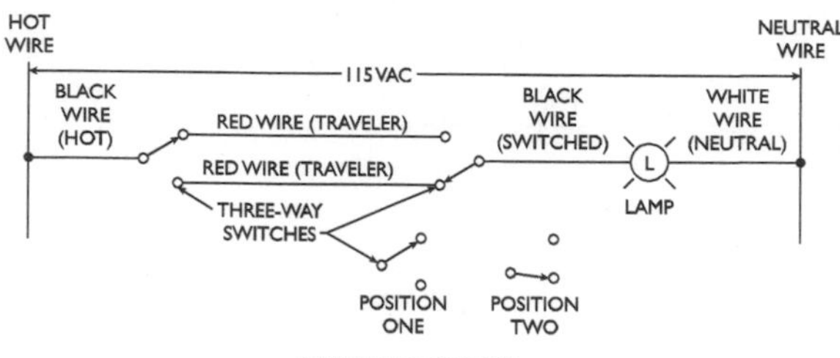

Figure 22-12 Light is controlled from two switch locations.

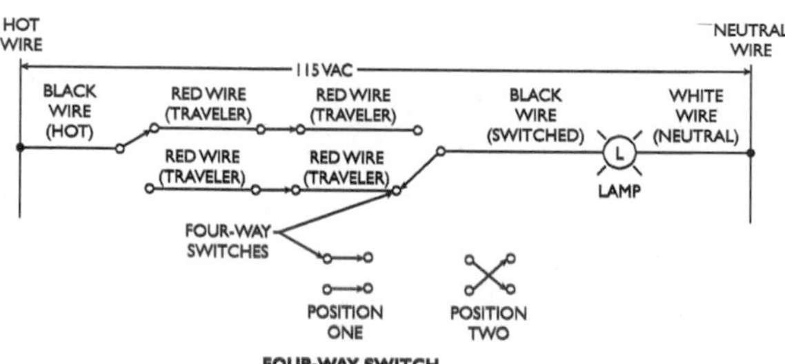

Figure 22-13 Light is controlled from three or more switch locations.

Chapter 23

AC Motors

Magnetism makes motors work. All motors include a series of electromagnets that are energized to produce a force that allows the motor to run. Electromagnets are simple devices. They consist of a coil of wire wound around an iron core (Figure 23-1). When a current passes through the wire, a magnetic field is produced. The north and south poles are at opposite ends of the iron core, and the poles can be reversed by changing the direction of current in the coil.

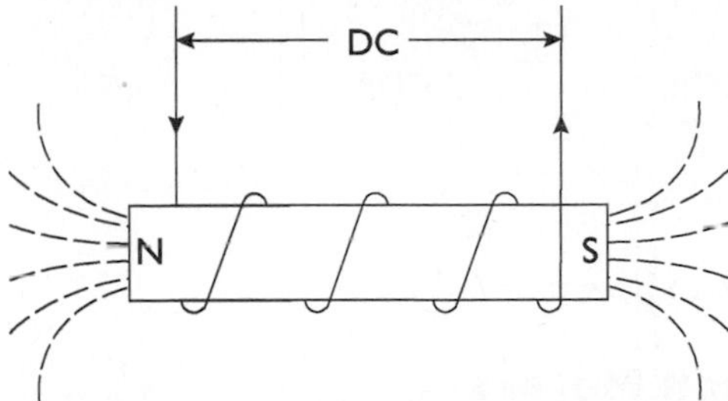

Figure 23-1 Coil and iron core.

Although there are many types of electric motors, the basic principle of all motors can be shown using a permanent magnet and two electromagnets. Figure 23-2A shows a current passed in coil A and coil B in a direction to cause north and south poles to occur next to the permanent magnet. In this simple example, the permanent magnet represents the moving portion of the motor, spinning around an axis. The laws of physics show that unlike poles attract and like poles repel. The permanent magnet starts to spin because the poles at each end are repelling. As the permanent magnet gets halfway around, the force of attraction between unlike poles keeps the permanent magnet rotating. In Figure 23-2B the rotating magnet turns all the way around until the unlike poles line up (attraction), and the rotor would be expected to stop. Suddenly, the current is reversed, and the poles of the electromagnets also change and reverse. Now there is a condition where the poles are the same and repulsion takes place. The rotor continues to spin, as shown in Figure 23-2C. In short, if the current in the coils is reversed every time the permanent magnet turns halfway around, the magnet (rotor) will continue to operate. This simple example shows the principle by which all motors work. Motors may be more complex, but the idea is the same.

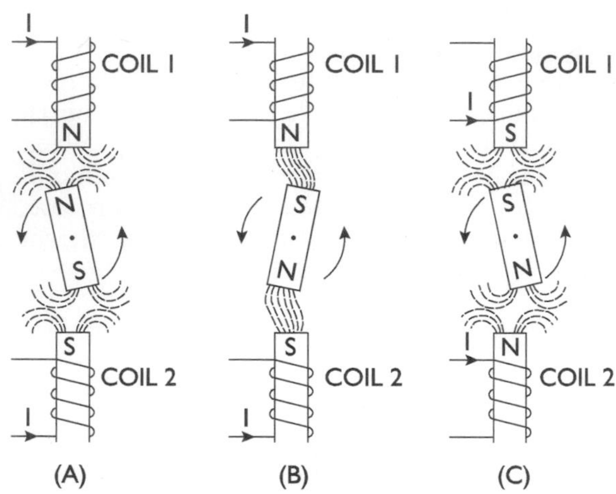

Figure 23-2 Simple example of how motors work.

Single-Phase Induction Motors

Alternating current alters its direction (or reverses) many times in each second. The amplitude of current increases to a maximum in one direction, drops off to zero, and then increases to a maximum in the opposite direction, as shown in Figure 23-3. This complete process is called a *cycle*. The *frequency* of a current is the number of times that this process (cycle) occurs in one second and is normally expressed in cycles per second, or hertz (Hz). Most ac power in the United States operates at a frequency of 60 cycles per second, or 60 Hz.

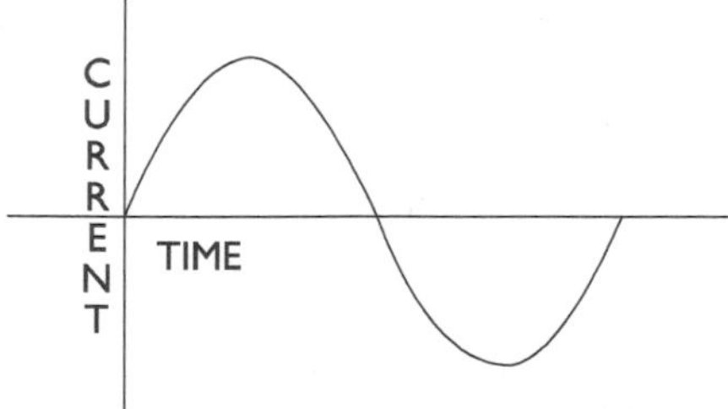

Figure 23-3 Single-phase alternating current.

Single-phase current does not set up a rotating magnetic field in the stator. Single-phase current flow causes the polarity of the stator field to reverse and the poles to alternate back and forth as the direction of the current changes. This alternation sets up a pulsating magnetic field in the stator, which can maintain rotation of the squirrel-cage rotor but cannot start it rotating. Providing a rotating magnetic field in the stator during the starting period starts single-phase inductor motors. Two widely used methods of doing this are by phase splitting and by capacity.

Split-Phase Motor

Split-phase motors range in size from ⅛ to ¾ horsepower, are relatively low in cost, and are widely used on appliances and other light applications. The split-phase motor uses a squirrel-cage rotor with two separate sets of windings in the stator. One set of windings is for running, the other set for starting only. The starting-coil windings are displaced 45° from the running-coil windings, as shown in Figure 23-4.

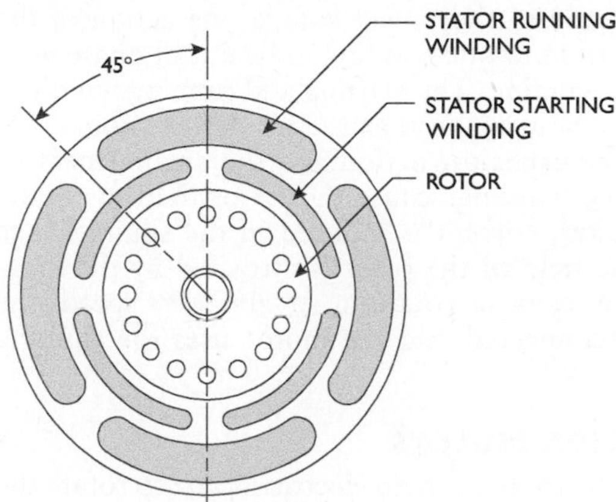

Figure 23-4 Split-phase motor.

The starting winding has finer wire, offering more resistance than the heavy wire on the running windings. The greater resistance in the starting windings delays the current enough to let it peak at a different time from the current in the running windings. The alternating current flowing through coils of differing resistance sets up rotation of the stator magnetic field. Similar to the three-phase motor operation, the rotating magnetic field of the stator induces current in the rotor. This induced current sets up a magnetic field in the rotor, and it begins to rotate. When the rotor reaches about ¾ speed, the magnetic field of the starting windings is no longer needed, and the power source is automatically disconnected. The pulsating field of the stator, with its reversal of polarity, provides the necessary force of rotation for the rotor, and the motor maintains a constant running speed.

Capacitor-Start Motors

The *capacitor-start* motor ranges from ⅛ to about 7½ horsepower and is used to run small equipment such as compressors that require high starting power. The capacitor-start motor also uses a squirrel-cage rotor and extra starting windings in the stator. As with the split-phase motor, once the motor is running, the starting winding is not needed. To start the motor, a capacitor is put in the circuit between the power source and the starting winding. An electric charge that is out of phase with the power source is continually built up and discharged from the capacitor. As the running windings are connected directly to the power source, the action of the capacitor causes the current in one winding to be out of phase with the current in the other winding. The starting and running windings are placed in the stator housing so that their poles are about 45° apart (Figure 23-4). The capacitor action and the arrangement of the starting and running windings cause the stator field to rotate. As the stator field rotates, current is induced in the squirrel-cage rotor, and the magnetic field of the rotor is attracted by the magnetic field of the stator, causing rotation. At about ¾ speed, the starting winding is disconnected and the motor uses the running winding to operate.

Polyphase Induction Motors

The effects of magnetism are used in an electric motor to rotate the shaft and convert electrical energy to mechanical power. Most widely used for industrial application is the three-phase induction-type electric motor. Some of its advantages are fairly constant speed under load, simple construction, a wide range of horsepower, and an ability to start under load. The two basic parts of an induction motor are the *stator* and the *rotor* (Figure 23-5). The stator is a set of electromagnets contained in a frame and arranged to form a cylinder. The rotor is an assembly of conductor bars in the form of a cage (called a *squirrel cage*) embedded in an iron core and free to rotate inside the stator.

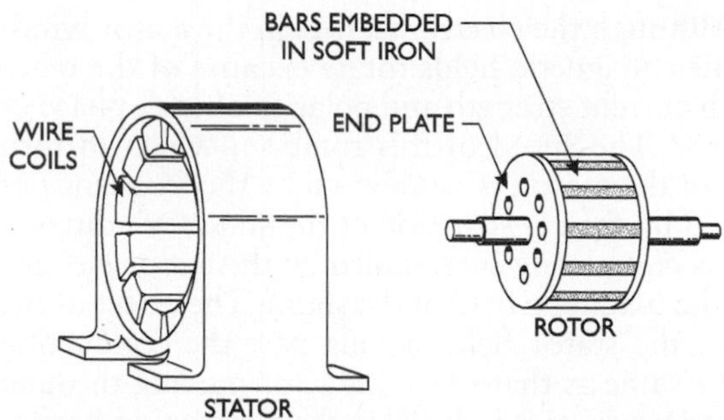

Figure 23-5 Induction motor—stator and rotor.

The ends of the cage of conductor bars in the rotor are connected to plates at each end of the soft iron core. This core, which is usually constructed of many thin iron plates called laminations, acts to concentrate the strength of the magnetic field of the rotor. The end plates act as conductors and connect the bars to form the cage within the iron core. Power to energize the electromagnets in the stator is supplied from an outside source.

When three-phase electric current flows to the stator magnets arranged in a cylinder, magnetic fields are set up that are increasing, decreasing, and reversing as the current alternates. This alternation of current causes the polarity of the magnetic fields in the stator to move or rotate (Figure 23-6).

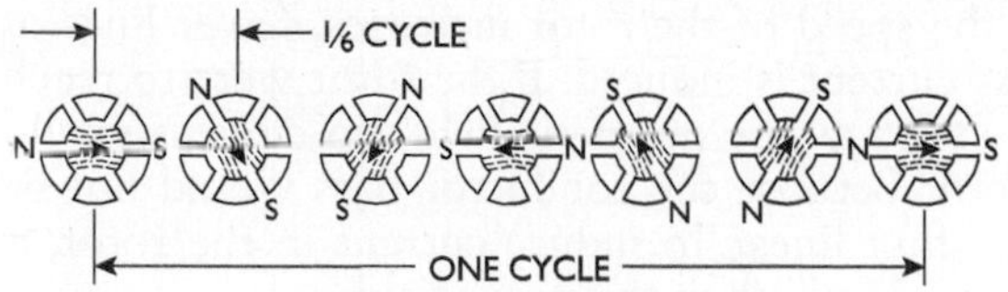

The polarity of the magnetic fields in the stator as these fields rotate.

Figure 23-6 Rotating magnetic fields.

This rotation of the magnetic field does not occur when single-phase alternating current is supplied to the stator. A pulsation of the magnetic field is all that occurs because there is only a single flow of power. With three-phase current, there are three separate surges of power, the maximum intensity of each (either positive or negative) being separated by one-sixth of a cycle. This results in the two magnetic poles shifting steadily or, in effect, rotating always in the same direction. Although the electromagnets in the stator windings are stationary, their magnetic fields rotate because of the regular alternation of both current strength and polarity of each phase of the three-phase current. The speed of this rotation, which in turn determines the speed of the motor, is established by the frequency of the electric current and by the construction of the stator windings.

Because the rotor is completely surrounded by the stator electromagnets, it is within the magnetic field of the stator. The effect of the magnetic flux lines of the stator field moving past the conducting bars of the rotor is the same as that of a conductor moving through a magnetic field. Electric current is induced in the conducting bars of the rotor, which in turn sets up magnetic fields in the rotor. A north pole in the stator field sets up a south pole in the rotor, and a stator south pole sets up a rotor north pole. Because unlike poles attract, the moving north pole of the stator attracts the south pole of the rotor, and vice versa. In effect, the rotor becomes one large bar magnet whose poles are attracted by the rotating poles of the stator, and so it rotates to follow the rotating stator field.

At starting, when the rotor is stationary and an outside power source is connected to the stator windings, the stator's magnetic field immediately starts to rotate at a constant speed. The difference in speed between the stator rotating field and the stationary rotor is at a maximum. The number of lines of force being cut as the stator field crosses the rotor bars is at a maximum, and the current induced in the rotor is also at a maximum. Therefore, during the starting period, an induction motor develops a strong rotational force or torque. As the speed of the rotor increases, fewer lines of force are cut and less current is induced. If the rotor were to reach the same rotational speed as the magnetic field, no current would be induced in the rotor because the conductor bars would not be cutting any magnetic flux lines. To induce current in the rotor, it must always turn more slowly than the stator field.

This required difference in speed between the stator field and the rotor is called *slip*. Without this slip, a squirrel-cage induction motor cannot run. As loads increase on an induction motor, slip increases, which induces higher rotor current and creates a stronger magnetic field in the rotor. The increase in the strength of the magnetic field increases horsepower output to handle the increased load. Although under different loads there are different amounts of slip, the overall speed of squirrel-cage induction motors is fairly constant. The direction of rotation of the rotor in the squirrel-cage induction motor is the same as that of the stator's rotating magnetic field. The direction of a three-phase motor can be reversed by interchanging the connections of any two supply leads. This interchange will reverse the sequence of phases in the stator and the direction of the rotor rotation. The interchanging of leads may be made at any point before the power reaches the motor, and the effect will be the same. For this reason, it is important that care be taken not to interchange leads when disconnecting and reconnecting supply voltage to a three-phase motor. If this is not practical, check the motor direction carefully after reconnection.

Motor Control

Some form of control equipment is required to operate all electric motors. In some cases, this need only be a device to connect the motor to the power source and to disconnect it as needed. It may also be a more complicated control to stop, start, reverse, or vary the rotational speed. Other controls may be necessary to start and stop automatically at time intervals or in synchronization with related equipment. Two general terms are used to describe motor controls—*starter* and *controller.* The term "starter" is commonly used when referring to simple start-and-stop devices, and "controller" is used when referring to more complex equipment that may perform several functions.

For fractional-horsepower, low-voltage motors, a simple on-off contact switch may adequately serve as a starter. In this case, the line fuses provide the only overload protection. The electromagnetic full-voltage, or across-the-line, starter is widely used in industry for stop-start control of larger ac motors. Generally, this type of starter employs a solenoid to close the contacts electromagnetically when the control start button is pushed. A simplified diagram of a simple full-voltage starter is shown in Figure 23-7.

When the control start button is pushed, current flows to the solenoid, which closes the line switch and connects the power source to the motor. In this simplified diagram, the power to energize the solenoid comes from the same power source as the motor. For safety, the control circuit for motors operating at high voltage is supplied with a lower-voltage current. The control switches touched directly by the operator do not carry the high voltage supplied to the motor.

A starter for control of a three-phase motor is, in effect, three simultaneously acting switches, as all three power lines are controlled. In addition, a fourth switch to supply current to a holding coil is part of the starter mechanism. The three main sets of contacts control the current flow to the motor. The fourth switch keeps current flowing to the holding coil after the start button is released. When the stop button is pushed, the current flow to the holding coil

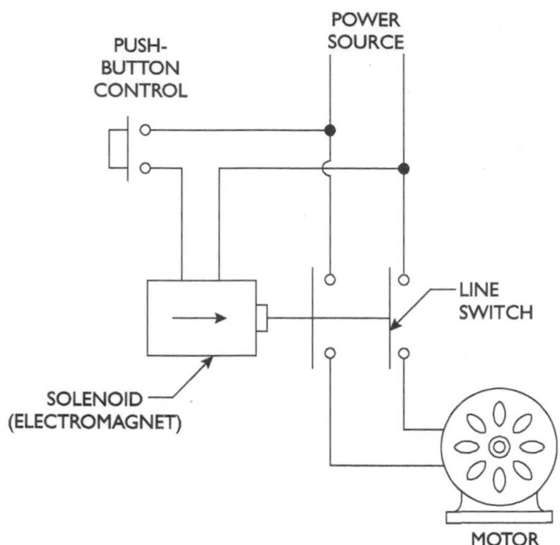

Figure 23-7 Full-voltage starter circuit.

is interrupted, the electromagnet loses its magnetism, and the starter opens. Figure 23-8 shows a three-phase starter and push-button stations with the single-phase-control circuit power supplied from the main power line.

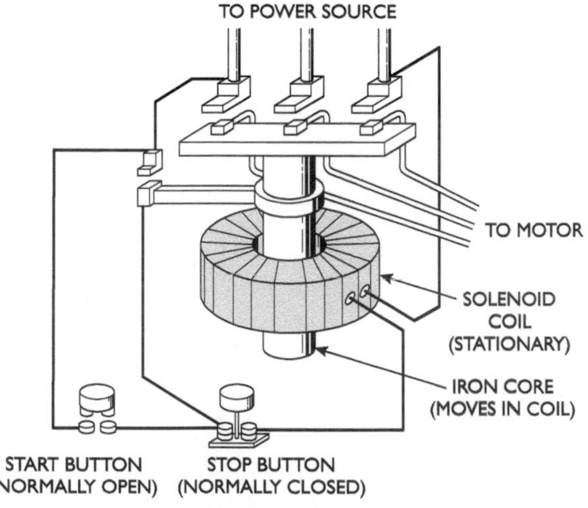

Figure 23-8 Three-phase starter—single-phase control.

The solenoid used to close the contacts on a starter electromagnetically is made up of a stationary coil, called a holding coil, and a movable soft-iron core. Because of the weight of the core and the parts attached to it, the core drops down out of the center of the coil. Many solenoids employ a spring to supply the force to move the core out of the center of the coil. When current is supplied to the holding coil, its core is magnetically pulled into the center of the coil, thus closing a switch. When current is cut off, gravity or force supplied by the spring causes the core to move out of the center of the coil and open the switch.

Overload devices to protect the motor from damage are also placed in the control circuit. These are usually actuated by the heat that results from excessive current flow. The heat causes the overload contacts to open, interrupting the current in the control circuit in the same manner as when the stop button is pushed. Because the protective device can be made ineffective if the start button is held in contact, it should be pressed only momentarily. The current to the holding coil will then flow through the magnetically operated contacts, and the circuit can be interrupted by either the stop button or the overload device.

Troubleshooting Polyphase AC Motors

Troubleshooting motors can be done in a systematic way using the following chart. First find the appropriate symptom and then isolate the most probable cause and solution.

Table 23-1 Polyphase Motor Troubleshooting Chart

Symptom	*Possible Cause*	*Solution*
Motor will not start	Overload tripped	Wait for overload to cool. Reset and try again.
	Power not connected	Connect power to control and control to motor. Check clip contacts.
	Faulty fuses	Test fuses.
	Low voltage	Check motor nameplate values against power supply. Also make sure motor wire size is adequate.
	Wrong control connections	Check connections against wiring diagram.
	Loose terminals	Tighten connections.
	Driven machine locked	Disconnect motor from load. If motor starts properly, check driven machine for blockage or damage.

(Continued)

Symptom	*Possible Cause*	*Solution*
	Open circuit in stator or rotor winding	Check for open circuits.
	Short circuit in stator winding	Check for shorted coil.
	Winding grounded	Test for grounded winding.
	Bearings stiff	Free bearings or replace.
	Grease too stiff	Use different lubricant.
	Faulty control	Check control wiring.
	Overload	Reduce load on motor.
Motor noisy	Motor running single phase	Stop motor, try to start again. If it is single phasing, it will not start.
	Electrical load unbalance	Check current balance with ammeter.
	Shaft bumping (sleeve bearing motor)	Check alignment and condition of belt drive. Check end play and axial centering of motor.
	Vibration	Driven machine may be unbalanced. Remove motor from load. If motor is still noisy, rebalance motor rotor.
	Air gap not uniform or rotor rubbing on stator	Center the rotor, and if necessary replace bearings.
	Noisy ball bearings	Check lubrication. Replace bearings if necessary.
	Loose parts or loose rotor on shaft	Tighten all holding bolts.
	Object caught between fan and end shields	Disassemble motor and clean.
	Motor loose on foundation	Tighten hold-down bolts. Motor might have to be realigned.
	Coupling loose	Tighten coupling bolts. Realign if necessary.
Motor temperature too high	Overload	Measure motor loading with ammeter. Reduce load.
	Electrical load unbalance	Check for voltage unbalance or single phasing. Check for an opening in one of the lines or circuits.

(Continued)

Symptom	*Possible Cause*	*Solution*
	Restricted ventilation	Clean air passages and windings.
	Incorrect voltage and frequency	Check motor nameplate values with power supply. Also check voltage at motor terminals under load.
	Motor stalled by driven machine or tight bearings	Remove power from motor. Check machine for cause of stalling.
	Stator windings with loose connections	Test windings for short circuit or ground.
	Motor used for rapid reversing service	Replace with motor designed for this service.
	Belt too tight	Remove tension and excessive pressure on bearings.
Bearings hot	End bells loose or not replaced properly	Make sure end bells fit squarely and are properly tightened.
	Excessive belt tension or excessive gear slide thrust	Reduce belt tension or gear pressure, and realign shafting.
	Bent shaft	Straighten shaft or replace.
Sleeve bearings hot	Insufficient oil	Add oil. If oil supply is very low, drain, flush, and refill.
	Foreign material in oil or poor grade of oil	Drain oil, flush, and relubricate with proper lubricant.
	Oil rings rotating slowly or not rotating at all	Oil too heavy; drain and replace.
	Motor tilted too far	Level motor or reduce tilt and realign, if necessary.
	Oil rings bent or otherwise damaged in reassembling	Replace oil rings.
	Oil ring out of slot	Adjust or replace retaining clip.
	Motor tilted, causing end thrust	Level motor, reduce thrust, or use motor designed for thrust.
	Defective bearings or rough shaft	Replace bearings. Resurface shaft.
Ball bearing hot	Too much grease	Remove relief plug and let motor run. If excess grease does not come out, flush and relubricate.

(Continued)

Symptom	Possible Cause	Solution
	Wrong grade of grease	Add proper grease.
	Insufficient grease	Remove relief plug and regrease bearing.
	Foreign material in grease	Flush bearings, relubricate; make sure the grease supply is clean. Keep can covered when not in use.
	Bearings misaligned	Align motor and check bearing housing assembly. See that bearing races are exactly 90° with the shaft.
	Bearings damaged	Replace bearings.

Table supplied by Maintenance Troubleshooting—Newark, Delaware, USA.

Motor Nameplate Information

The organization that sets the standards for electric motor manufacture is the National Electrical Manufacturers Association (NEMA). This group was formed in 1926 and has specifications covering nomenclature, composition, construction, dimensions, tolerances, safety, operating characteristics, performance, quality, rating, testing, and service for polyphase induction motors.

The nameplate on a motor contains a wealth of information. Some specific items on the nameplate are extremely beneficial to understand.

Voltage

Each motor is designed for optimum performance at a specific line voltage. If a motor is operated at its rated voltage, it will provide the most efficient service and longest life.

Line voltage does vary, and a motor must be able to cope with some voltage variation. A standard polyphase induction motor will tolerate voltage variations of plus or minus 10 percent.

If a nameplate call for 460 V, then the motor will give satisfactory but not necessarily ideal performance when running at a voltage between 416 V (minus 10 percent) and 506 V (plus 10 percent).

This explains why "brown outs" or undervoltage conditions are such a concern to industry. If the power company drops voltage by 15 or 20 percent, the motor will continue to run, but the life of the insulation gets very short. A few months later, your plant may experience quite a few inexplicable motor failures. In many large facilities, there is a daily check of the incoming voltage at each motor control center to check for brown-out conditions. In many cases, the power company is under no obligation to inform the customer of brown outs. Sometimes it is wiser to shut down large motors if brown-out conditions continue rather than subject the motor insulation to the higher heat loading imposed by reduced voltage.

Full-Load Current (FLA or FLC)

As the load on a motor increases, the line current increases. The wiring, starter, circuit breaker, and thermal overloads are all sized based on the full-load current shown on the nameplate. Some quick rules of thumb that can be useful are the following:

- At 460 V, a three-phase motor draws 1.25 amps per horsepower.
- At 230 V, a three-phase motor draws 2.5 amps per horsepower.
- At 230 V, a single-phase motor draws 7 amps per horsepower.
- At 115 V, a single-phase motor draws 14 amps per horsepower.

These values are only "rules of thumb" and should never be used to size switchgear, protective devices, or wire for a motor. The actual nameplate information should always be used.

Insulation Class

There are four classes of insulation that are used in a motor. Each has an absolute temperature that cannot be exceeded without loss of insulation life. Table 23-2 shows the values.

Table 23-2 Insulation Class and Temperature

Insulation Class	*Temperature Limit*
Class A	105°C
Class B	130°C
Class C	155°C
Class D	180°C

Class A and B insulation are hydroscopic, i.e., they absorb water. A motor with this type of insulation will rapidly degrade in outside service. In addition, when the motor is shut down, condensation can occur on the windings. Because both class A and B insulations absorb moisture, the life will be shortened if condensation occurs.

Class F and H insulation are nonhydroscopic, i.e., they do not absorb moisture. If a motor is to be used in an outside location or in an area of high ambient temperature, then class F or H should be used.

A "rule of thumb" for obtaining the actual motor winding temperature is as follows:

1. Using a thermometer, infrared thermometer gun, or other suitable means, check the temperature of the motor under load on the outside of the motor at the junction where the end bell is fitted to the housing (Figure 23-9).

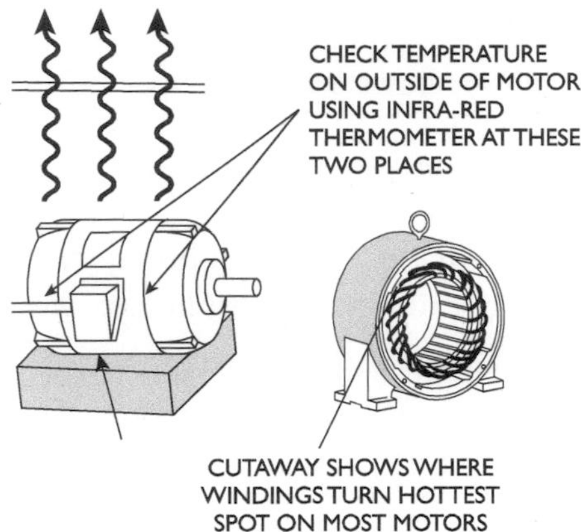

Figure 23-9 Correct place for taking motor temperature on housing.

2. Add 10° to this number, and this will be the approximate temperature of the windings. For example,

   Using a thermometer or infrared gun, a motor was measured at 125°C under load. Adding 10°C gives 135°C for the winding temperature. The nameplate shows insulation class "B"; a check of the chart (Table 23-2) shows that "B" insulation is rated for a maximum of 130°C. The motor is operating above the limit, and the motor life will be diminished if this situation is not corrected.

Chapter 24

Fans and Blowers

Any type of device that moves air or a gas is called a *fan* or *blower.* The shop name for a centrifugal fan is a "squirrel cage" fan. A fan is usually considered for low-pressure applications below 1 psi, whereas blowers are in use for pressures up to 10 psi.

Types of Fans

Fans are divided into two major classes: *centrifugal* and *axial.* Airflow through a centrifugal fan is circular. Air enters along the axis of rotation of a centrifugal fan and discharges in a radial direction, as shown in Figure 24-1. Airflow in an axial fan is straight. A centrifugal fan builds pressure by moving the air in a circular motion, whereas the axial fan adds energy to the air by pushing it through the fan.

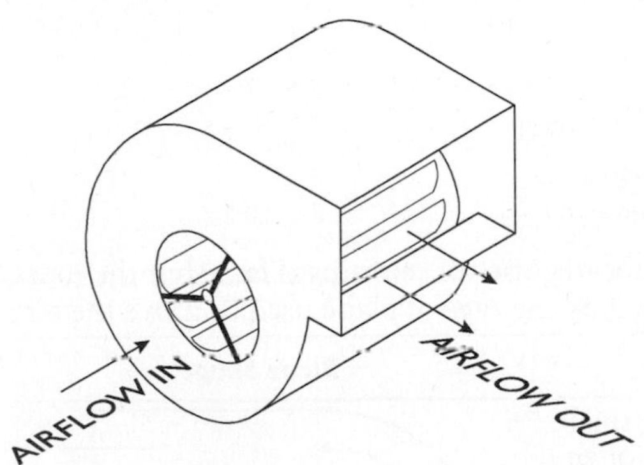

Figure 24-1 Air enters along the axis and discharges radially.

Centrifugal Fans

The *fan housing* or *scroll* is the metal casing that encloses the fan. The *cut-off* is a piece of metal that keeps air that is discharged from the wheel from reentering the scroll and reducing fan efficiency. The *fan wheel* includes the blades, backpiece, and any framework that holds these components in position (Figure 24-2).

Blade design varies and is dependent on the design purpose of the fan. The *inlet* or *vortex* is usually a conical component, which allows efficient draw of the intake air into the rotating wheel.

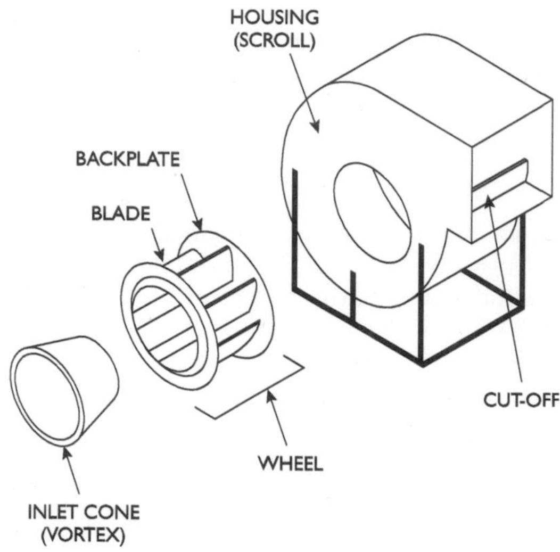

Figure 24-2 Fan components.

There are numerous subcategories of centrifugal fans, but the most common ones are identified by the type of blade used to move the air:

Type of Fan	***Fan Use***	***Blade Shape***
Forward-curved	Used for small systems, often in residences.	FORWARD-CURVED-BLADE FAN

Backward-curved	Blades slant away from the direction of rotation—curved blades.	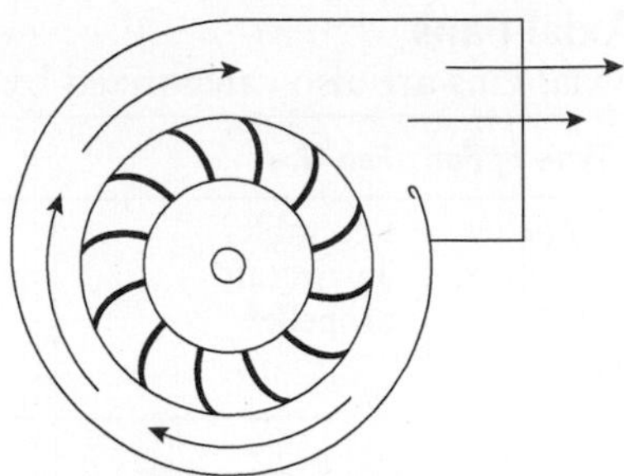 BACKWARD-CURVED BLADES
Backward-inclined	Blades slant away from the direction of rotation—straight blades.	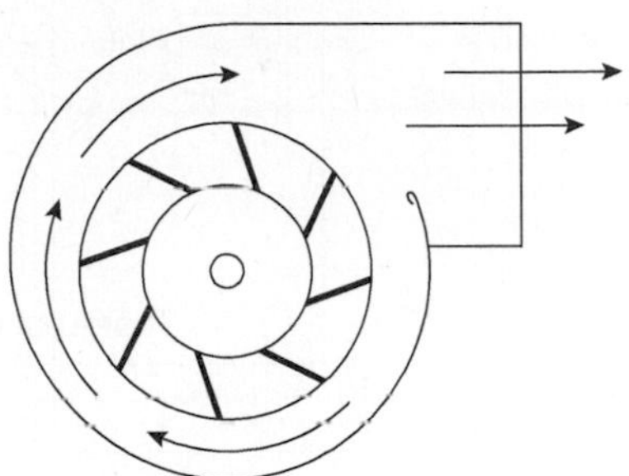 BACKWARD-INCLINED BLADES
Radial blade	Used for material handling.	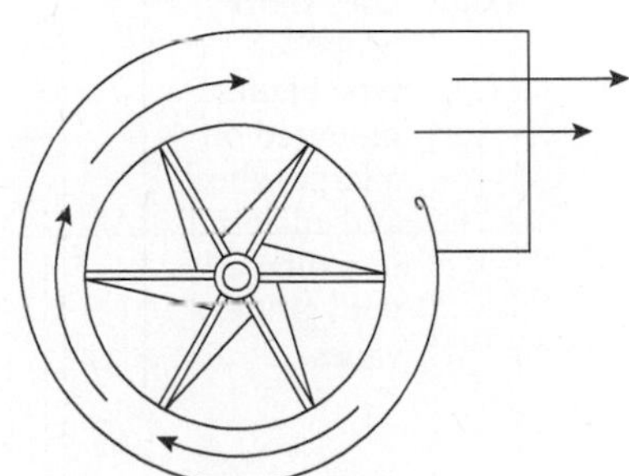 RADIAL-BLADE FAN
Airfoil	Cutaway of blade looks like airplane wing—most efficient fan.	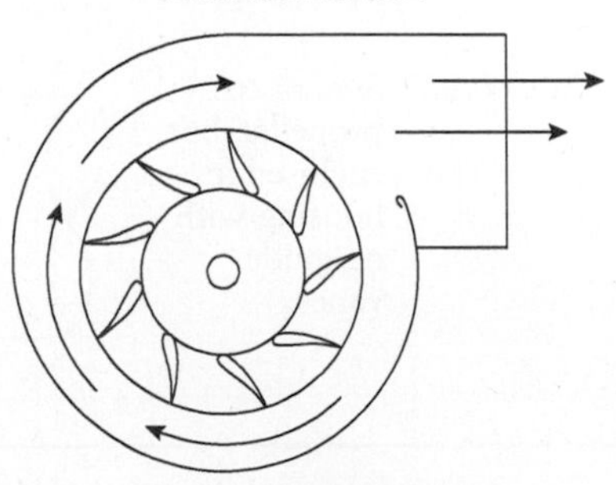 AIRFOIL-BLADE FAN

Axial Fans

Axial fans are also categorized by blade style:

Type of Fan	*Fan Use*	*Blade Shape*
Propeller	Resembles an aircraft propeller.	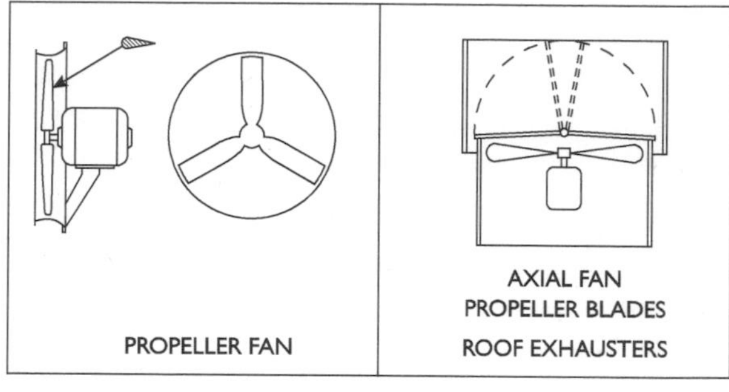
Vane-axial	Uses short propeller-type blades mounted on a large wheel and enclosed in a tube with guide vanes.	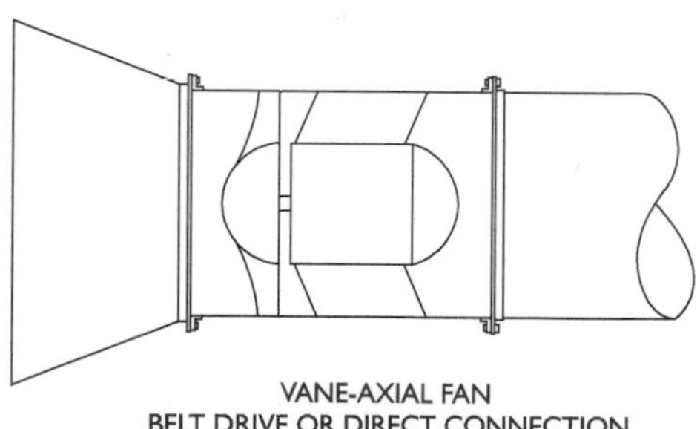
Tube-axial	Similar to propeller, but enclosed in housing with no guide vanes.	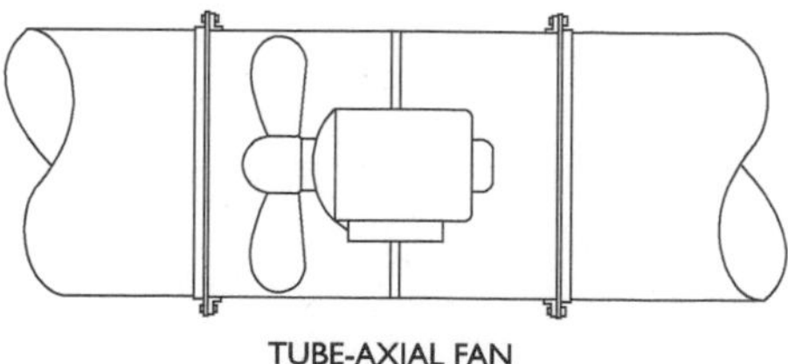

Inspection of Fans

A checklist for fan maintenance and inspection would include the following items:

V-Belt Drive

- Inspect for sheave wear. Look at sidewall of sheave for grooving and curving.
- Inspect for belt wear. Look for fraying, cracking, or slapping belts.
- Check for V-belt alignment and tension. Use a straightedge or string to check sheave alignment.

Fan Wheel

- Look for corrosion, rust buildup, and corrosion cracking on blades.
- Check for loss of balancing weights or clips.
- Look for dirt buildup on blades. Use a wire brush or water blast to clean.
- Check for missing bolts or setscrews that attach wheel to shaft.
- Look for hairline cracks at welds.

Lubrication

- Clean fan bearing grease fittings, and lubricate properly.
- Clean motor bearing grease fittings, and lubricate properly.
- Lubricate damper assembly.

Noises or Knocks

- Listen for unusual noises, rattles, rubbing, grinding, or knocking.
- Feel and listen for air escaping—whistling—indicating leakage.
- Listen for fluctuating air noise. Check for possible loose damper.

Vibration

- Test for excessive vibration using vibration meter.
- Look at any baseplate spring isolators and see if any are broken or completely collapsed.
- Check for broken welds or cracks in expansion joints in ductwork.

Fan Laws

It is very helpful to understand the relationship between speed and other fan characteristics. Equations showing these relationships are referred to as the *fan laws*. These equations are developed from the science of fluid mechanics. The fan laws can be expressed as shown here:

$$\frac{CFM_2}{CFM_1} = \frac{RPM_2}{RPM_1}$$

$$\frac{TP_2}{TP_1} = \left(\frac{RPM_2}{RPM_1}\right)^2$$

$$\frac{SP_2}{SP_1} = \left(\frac{RPM_2}{RPM_1}\right)^2$$

$$\frac{VP_2}{VP_1} = \left(\frac{RPM_2}{RPM_1}\right)$$

$$\frac{BHP_2}{BHP_1} = \left(\frac{RPM_2}{RPM_1}\right)^3$$

where

CFM = flow rate in cubic feet per minute
SP = specific pressure in pounds per square inch (psi)
VP = velocity pressure in pounds per square inch (psi)
TP = total pressure in pounds per square inch (psi)
BHP = brake horsepower—that is, the horsepower requirement for the fan

Taking a closer look at these equations, you also see the following:

- CFM varies directly with the fan speed. If the fan speeds up by 10 percent, then the CFM will increase by 10 percent, assuming that the speeding-up operation is within the allowable rpm limits for the particular fan.
- TP (total pressure) and SP (static pressure) vary as the square of the fan speed. For example, if the fan speed is doubled (2x), then the total pressure and static pressure are increased fourfold ($2^2 = 4$).
- BHP (horsepower of the motor) varies as the cube of the fan speed. For example, if the fan speed were to be doubled, then the horsepower required would be increased eightfold ($2^3 = 8$).

Another useful equation that gives rapid estimates for the amount of horsepower required for a fan is as follows:

$$\text{BHP} = \frac{\text{CFM} \times 7\text{P}}{6356 \times \text{ME}_{\text{OF FAN}}}$$

ME is the *mechanical efficiency* of the fan. It can be obtained from the manufacturer. Most centrifugal fans show an ME between 0.50 and 0.65.

Troubleshooting Fan Problems

The following troubleshooting guide may be helpful in solving problems with fans that have an appearance as shown in Figure 24-3.

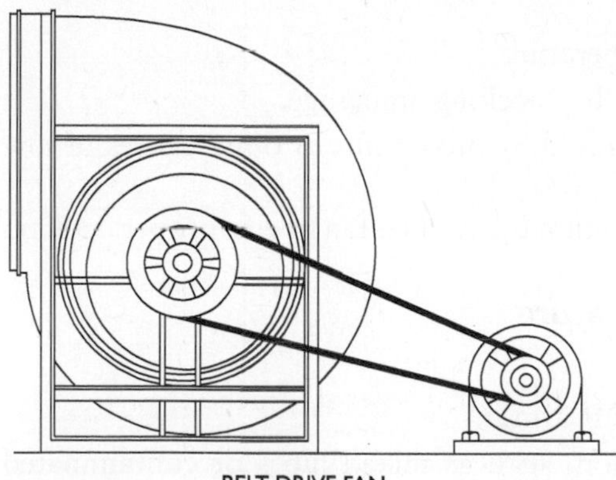

Figure 24-3 Typical fan in service.

Vibration

Check for the following:

1. Loose bolts in bearings and pedestals or improper mounting.
2. Poor alignment of bearings or shaft couplings.
3. Excessively worn or defective bearings.
4. Unbalanced wheel due to dirt buildup or abrasive wear to blades.
5. Cracked welds.
6. Improper wheel clearance between wheel and inlet vortex.
7. Loose or slapping V-belts.

8. Reversed wheel rotation.
9. Loose setscrews on wheel to shaft hub.
10. Bent shaft due to high-temperature shutdown without proper cooling.
11. Beat frequency with other fans on common base.
12. Motor or fan is causing structural base to resonate. If so, base must be stiffened or motor/fan speed must be changed.

Hot Motor

Check for the following:

1. Improper ventilation to motor or blocking of the cooling air.
2. High ambient temperature.
3. High current draw by checking amperage.
4. Power problems caused by brownouts or other causes of low voltage.
5. Motor is wrong rotation for cooling fan to give proper cooling.

High Bearing Temperature

Check for the following:

1. Overlubrication of bearings.
2. Improper lubrication, such as mixed lubes or contaminated lubrication.
3. V-belts too tight.
4. Defective or misaligned bearings after recent overhaul.
5. Lack of lubrication to bearings.
6. Heat flinger missing on fan shaft.
7. Floating bearing endplay is restricted.

Excessive Starting Time

Check for the following:

1. Improper sizing of motor to fan.
2. Failure to close inlet dampers during startup.
3. Low voltage at motor terminals.
4. Improper selected time-delay starting circuit.

Air or Duct Noise

Check for the following:

1. Duct thinner than housing.
2. Flattened, cracked, or compressed expansion joints.
3. Poor duct design.
4. Rusted or cracked ducting.
5. Foreign material in fan housing.

Dynamic Balancing of a Fan

If a fan vibrates excessively, the bearing caps should be checked with a vibration meter. See Chapter 20 for instruction on how to use the meter and take accurate readings.

If the fan exhibits high vibration levels, there is a good chance that it has become unbalanced. First, clean the fan; solvent wash or water blast is preferred. Then, check the vibration again. If the fan is still shaking, it will most likely need to be balanced. Although other types of equipment may have problems of misalignment, looseness, rubbing, or the like, high vibration on fans is most commonly due to unbalance.

The following procedure outlines a three-point method for balancing a fan using a simple vibration meter. This procedure assumes that the source of the major vibration in the fan is due to unbalance alone and not other faults such as looseness, misalignment, bad bearings, and so on.

The tools required are a simple vibration instrument, a compass, a protractor, and a ruler.

Basically, the procedure involves obtaining an original reading of vibration on the bearing cap of the fan rotor and recording it. Then, a trial weight is added to the fan wheel or blade in three different positions, called "A," "B," and "C." These positions are approximately 120° apart. The amount of vibration is measured and recorded as the weight is placed at each position. After the data are obtained, the amount of weight needed to correct the unbalance is calculated and the correct position on the rotor is determined. Adding the right weight at the right spot will balance the rotor.

1. With the fan operating at normal speed, measure and record the original amount of vibration as "O." Let's assume an offending vibration of 0.6 in./sec.

2. Draw a circle with a radius equal to "O," as shown in Figure 24-4. In the example, ¼ in. = 0.1 in./sec. The radius would be set at 6⁄4 in., or 1¼ in., to equal an amount of 0.60 in./sec.

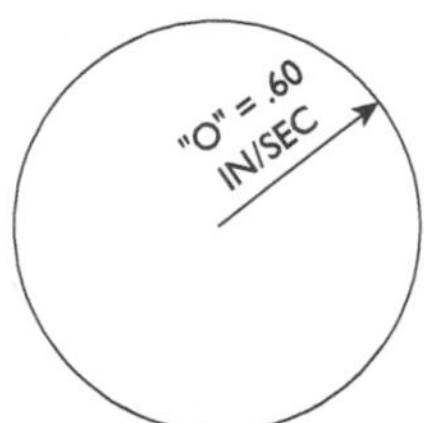

Figure 24-4 Construct a circle with a radius scaled to equal the original ("O") vibration reading.

3. Stop the fan. Mark the fan wheel at three points, "A," "B," and "C," approximately 120° apart. These three points on the fan wheel need not be exactly 120° apart; however, the precise angles of separation, whatever they may be, must be known. In this example, the weights will not be equally spaced to illustrate the way this may be done.

   As shown in Figure 24-5, point "A" is the starting point and is considered 0°. Mark the respective positions of points "A," "B," and "C" on the original circle. The correct angles can be laid out using a protractor or other angle-measuring device.

4. Select a suitable trial weight and attach it to position "A" on the fan. In many cases, a good guess can be made by looking at other weights that are already on the fan and picking a trial weight similar in size.

   In the example, a trial weight of 10 grams was chosen.

5. Start the fan and bring it to operating speed. Measure and record the new vibration at the same place on the bearing cap as "O + T_a." In the example, "O + T_a" = 0.40 in./sec.

6. Using point "A" on the original circle as the center point, draw a circle with a radius equal to "O + T_a."

   In the example, this circle will have a radius of "O + T_a" = 0.40 in./sec, or 4⁄4 in. or 1 in., as shown in Figure 24-6.

7. Stop the fan and move the trial weight to position "B."

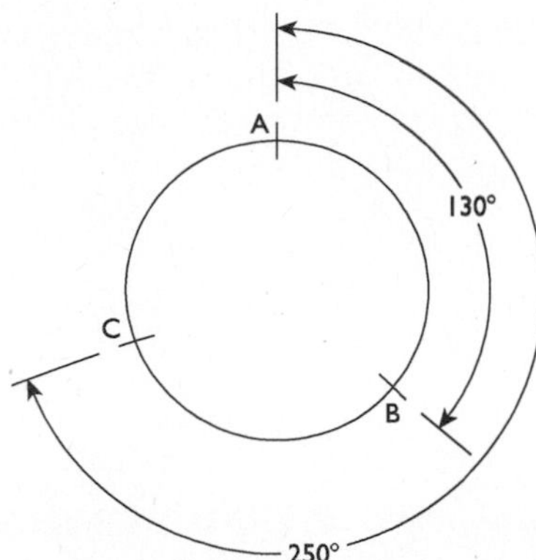

Figure 24-5 Marking the respective positions for adding a test weight.

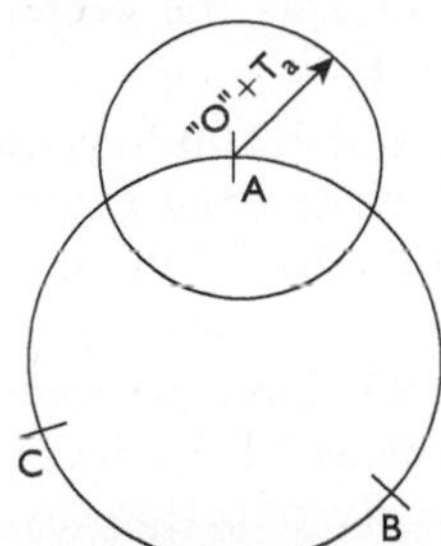

Figure 24-6 Drawing the "O + T_a" circle.

8. Start thc fan and bring it to normal operating speed. Use the vibration meter to record the new vibration amount as "O + T_b."

In the example, "O + T_b" = 0.80 in./sec.

9. Using point "B" on the original circle ("O") as the center point, draw a circle with radius equal to "O + T_b."

In the example, this circle will have a radius of "O + T_b" = 0.80 in./sec, or 8/4 in. or 2 in., as shown in Figure 24-7.

10. Stop the fan and move the trial weight to the third position, position "C."

11. Start the fan and bring it up to operating speed. Measure the new vibration amount and record it as "O + T_c."

In the example, "O + T_c" = 1.1 in./sec.

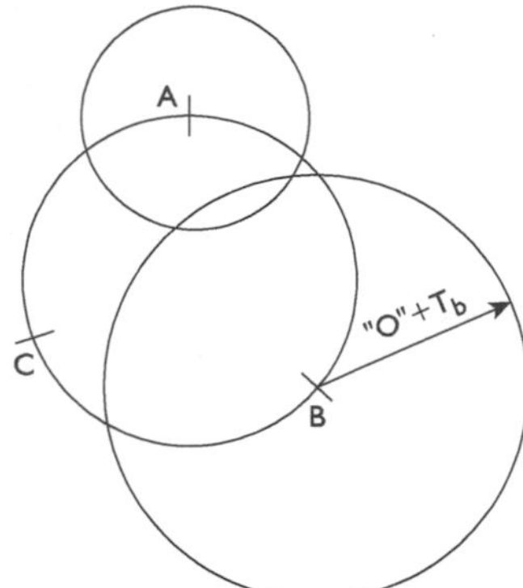

Figure 24-7 Drawing the "O + T_b" circle.

12. Using point "C" on the original circle ("O") as the center point, draw a circle of radius equal to "O + T_c."

In the example, this circle will have a radius of 1.1 in./sec, or 11⁄4 in. or 2¾ in., as shown in Figure 24-8. Note from Figure 24-8 that the three circles drawn from points "A," "B," and "C" intersect at a common point "D."

13. Draw a straight line from the center of the "O" circle to point "D," as shown in Figure 24-9. Label this line as "T."

14. Measure the length of line "T" using the same scale that was chosen to draw the circles. Figure 24-9 is used to do this.

In this case, ¼ in. = 0.10 in./sec.

Line "T" measures 1⁵⁄16 in., or 21⁄16 in., divided by (¼ in. per 0.10 in./sec) = 0.525 in./sec.

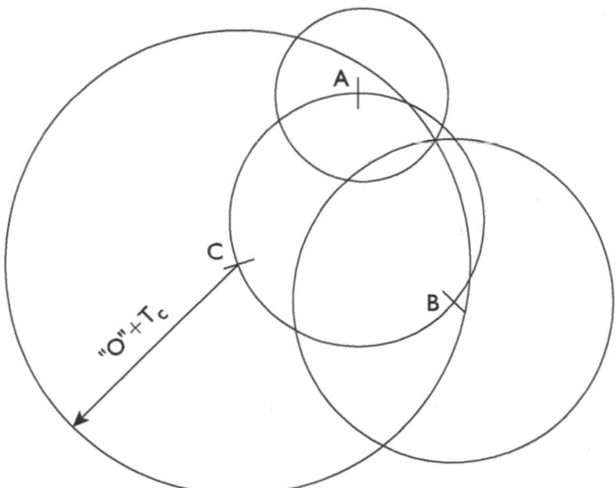

Figure 24-8 Drawing the "O + T_c" circle.

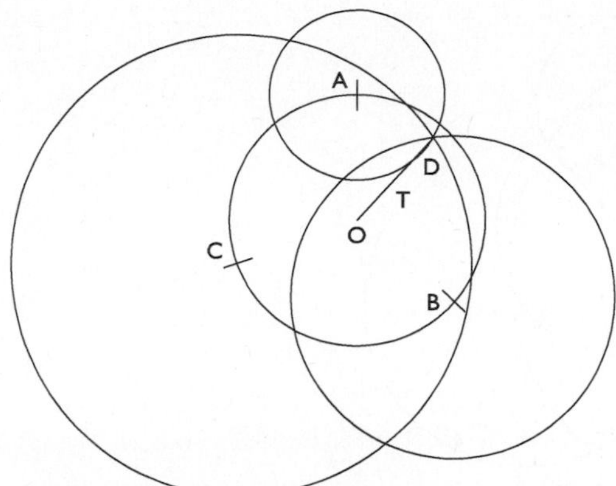

Figure 24-9 Constructing "T."

15. Calculate the amount of correct weight to balance the rotor from the formula

CW = TW (O/T), where
CW = Correct Weight
TW = Trial Weight
O = "O"riginal Unbalanced Reading
T = Measured length "T" converted to vibration units.

For the example, the solution is found as follows:

CW = TW (O/T)
CW = 10 grams × (0.60 in./sec / 0.525 in./sec)
CW = 11.4 grams

16. Using the protractor, measure the angle between line "T" and line "OA," as shown in Figure 24-10. This measured angle is the angular location of the correct weight, located relative to point "A" on the rotor.

In the example, the angle equals 41°.

17. Stop the fan and remove the original trial weight from point "C."

18. Attach the correct weight calculated in step 15 to the fan wheel at the angular position determined in step 16.

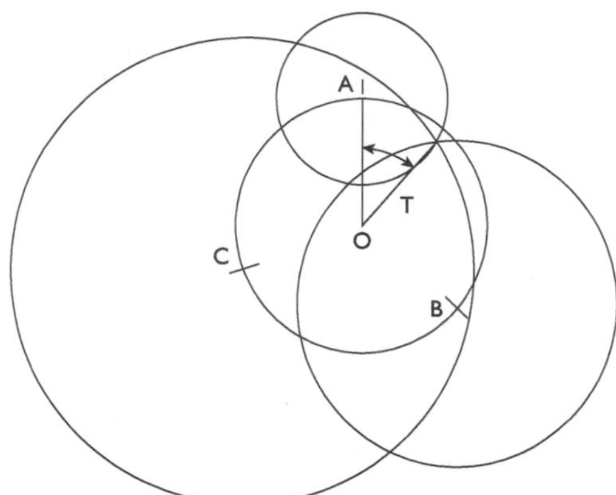

Figure 24-10 Measuring the angular location for the correct weight.

In the example, the calculated correct weight of 11.4 grams is added to the fan at an angular position 41° clockwise from position "A" on the fan wheel, as shown in Figure 24-11.

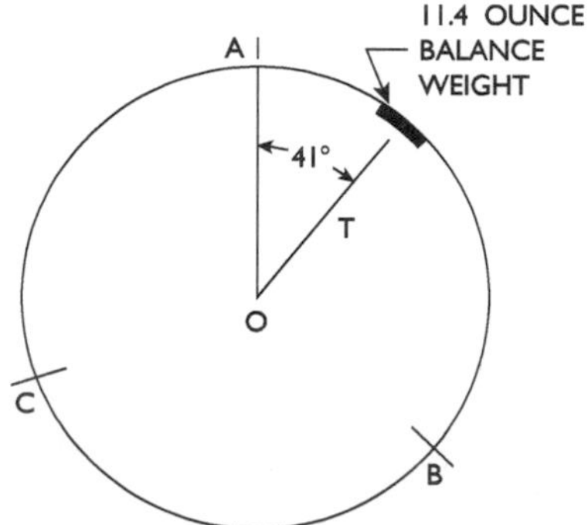

Figure 24-11 Adding the correct weight to the proper location.

With the balance-correction weight calculated and located on the fan in accordance with the above instructions, the fan should now be balanced.

Chapter 25

Pumps

Pumps may be defined as mechanical devices or machines that lift or convey liquids against the force of gravity or remove exhaust gases from closed vessels. A pump for liquid may be used to lift the liquid from a source below the pump to a higher elevation, to move a liquid from one location to another, to move liquid from a low-pressure area to a higher-pressure area, or to increase the rate of liquid flow. The machine used to withdraw air or other gases from a closed vessel is also classified as a pump. However, the machine used to compress air is known as a compressor, and the machine used to convey air is known as a blower. These classifications for machines that handle air and other gases frequently overlap.

Pumps are broadly classified with respect to their construction or the service for which they are designed. The three groups into which most pumps in common use fall are centrifugal, reciprocating, and rotary.

Centrifugal Pumps

The centrifugal pump takes its name from the type of force (centrifugal force) that this design of pump utilizes. Centrifugal force acts on a body moving in a circular path, tending to force it farther from the center of the circle. Inside the body of a centrifugal pump, the liquid is forced to revolve and so generates the force that enables this style of pump to develop pressure and move liquids. Two essential parts of the centrifugal pump are shown in Figure 25-1: the blades, or vanes, called the *impeller*; and the surrounding case, or housing, called the *volute*.

The rotation of the vanes of the impeller moves the liquid in a circular path, generating centrifugal force, which forces the liquid to the outer tips of the impeller vanes. Liquid that has moved to the outer rim of the impeller is forced against the inside of the volute and follows it to the discharge nozzle. The volute surface is a spiral or a scroll-like shape that constantly recedes from the center of the impeller, thus forming a progressively expanding passageway for the liquid forced from the rim of the impeller. Some of the centrifugal pump terms in common use are illustrated in Figure 25-2.

Liquid enters the centrifugal pump casing through the *suction* inlet in the center of the casing, passing into the impeller at its center,

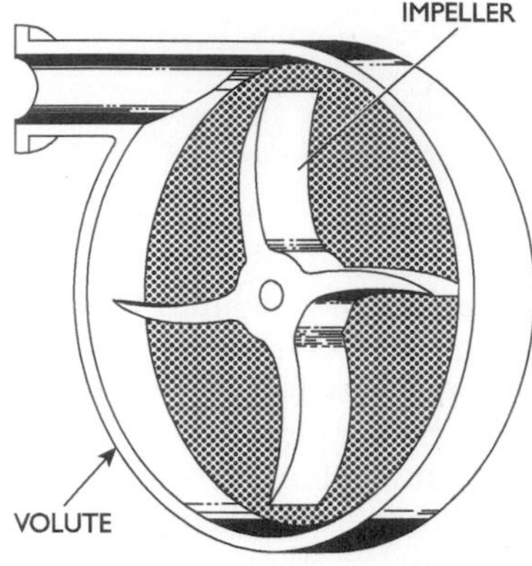

Figure 25-1 The main parts of a centrifugal pump.

or *eye*, and is moved by the impeller vanes as previously explained. As the liquid is forced to the outer rim of the impeller and out the *discharge* nozzle, a lowering of the pressure or partial vacuum is created at the suction inlet of the pump. The supply liquid flows into the suction inlet because of the relatively lower pressure, and operation is continuous. A centrifugal pump, therefore, must have the impeller vanes in the correct relation to the volute shape, and the rotation of the impeller must be in the direction of the volute's expanding passageway. Centrifugal pumps are *not* reversible.

The impeller of a centrifugal pump is attached to a shaft, which in turn is rotated by an outside power source or driver. At the point where the shaft passes through the casing, a packing or seal is used to prevent leakage of the liquid. This portion of the pump casing is usually called the packing or stuffing box. Installation and maintenance of stuffing-box packings and mechanical seals are covered in Chapter 18.

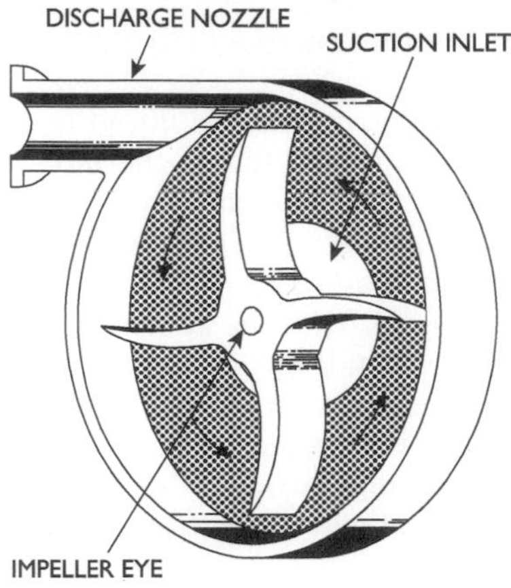

Figure 25-2 Common terms used in centrifugal pumps.

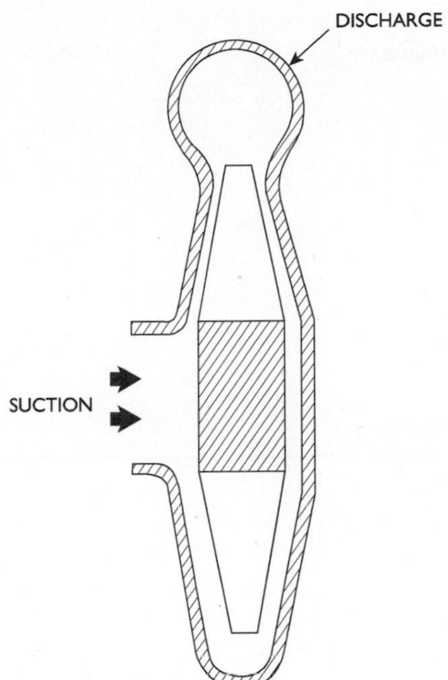

Figure 25-3 A single-admission pump.

Centrifugal pumps are made in a great variety of types, styles, and sizes to suit application requirements and operating conditions. A basic design difference between pumps is the number of stages. Pumps with one impeller are classified as single-stage pumps, and those with two or more as multistage pumps. A multistage pump may be compared to two or more single-stage pumps joined in series. For example, a two-stage pump may be compared to two single-stage pumps joined together, with the discharge from the first pump connected to the suction of the second pump. Although each stage is a separate pump, all are in the same casing with impellers mounted on a common shaft. The liquid enters the first stage, where pressure is increased to its capacity, and is then directed into the second stage. This process continues throughout the entire succession of stages, with the pressure increasing in each stage.

The manner of liquid entrance into the casing of either single-stage or multistage pumps may be from one side or from both sides of the impeller. Pumps with single entrance are classified as *single-admission*, or *end-suction*, pumps. A diagram of this design is shown in Figure 25-3. The double-entrance design is diagrammed in Figure 25-4. This style of pump is classified as *double-admission*, or *double-suction*.

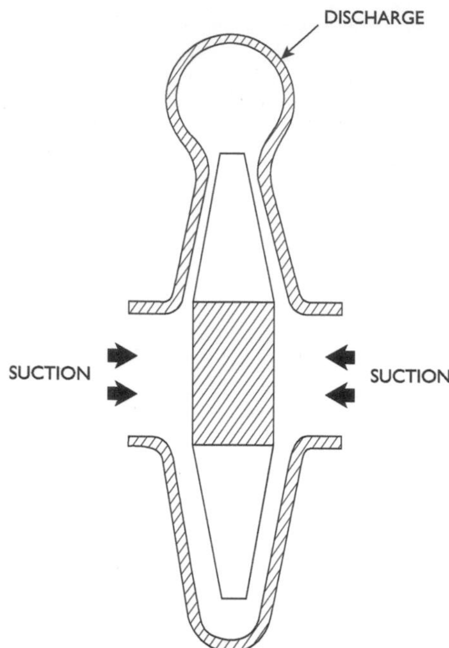

Figure 25-4 A double-admission pump.

To meet the wide range of service conditions under which centrifugal pumps must operate, manufacturers design and build a great variety of impeller types and styles. These are all derived from three basic impeller types, broadly classified as *open*, *semi-open*, and *enclosed*. The simplest impeller, both in construction and operation, is the open design shown in Figure 25-5A. It is most suitable for handling liquids containing a limited amount of solids and grit. The semi-open design, illustrated in Figure 25-5B, is better suited to handle liquids containing sediment and other foreign matter in suspension. The enclosed design, illustrated in Figure 25-5C, is usually used where high-efficiency pump operation is a prime consideration.

Centrifugal pump casings and impellers may be equipped with *wear rings*. These rings are located at the point where the liquid from the high-pressure discharge area leaks back into the suction area of the pump. The wear ring in the casing is stationary, and the impeller wear ring rotates. Figure 25-6 shows the location of wear rings on an enclosed impeller pump with end suction, and Figure 25-7 shows their location on an enclosed impeller pump with double suction.

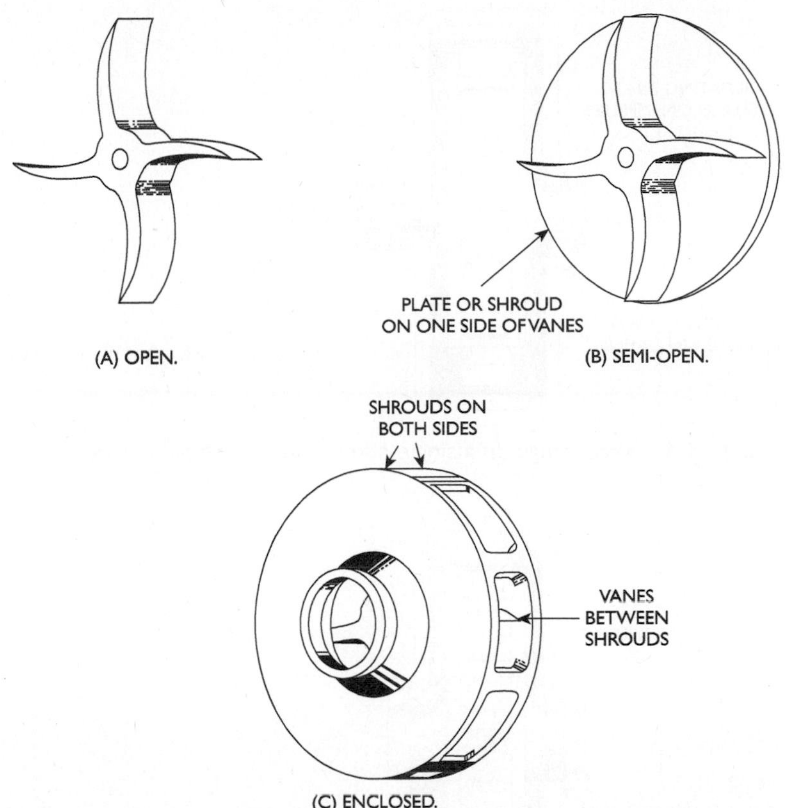

Figure 25-5 Three basic designs of centrifugal pump impellers.

The leakage of liquid from the high-pressure area to the low causes wear, increasing the clearance between impeller and casing and reducing the efficiency of the pump. By the use of wear rings, an initial close fit is possible at this point to hold leakage to a minimum. When wear reaches the point where excessive leakage occurs, the wear rings may be replaced much more quickly and economically than a pump casing or impeller. Because there is relative motion between the surfaces of the stationary and rotating wear rings, there must be clearance between them. Some leakage, therefore, must occur during operation. This leakage is necessary to act as a lubricant and coolant, to keep the surfaces separated, and to prevent rubbing and seizing of the surfaces. For this reason, a centrifugal pump should never be started unless it is filled with liquid.

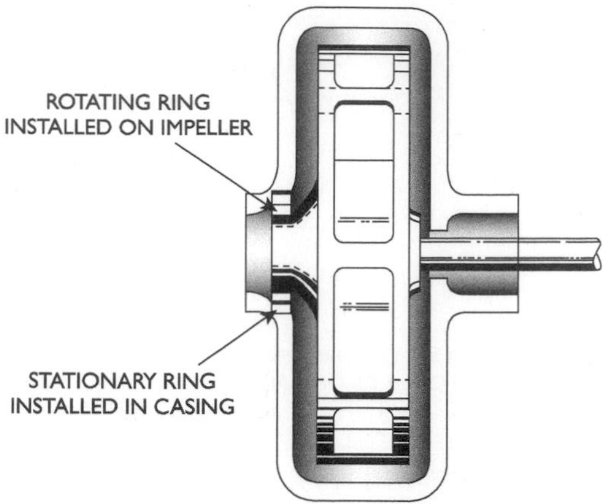

Figure 25-6 Wear rings on a single-admission centrifugal pump.

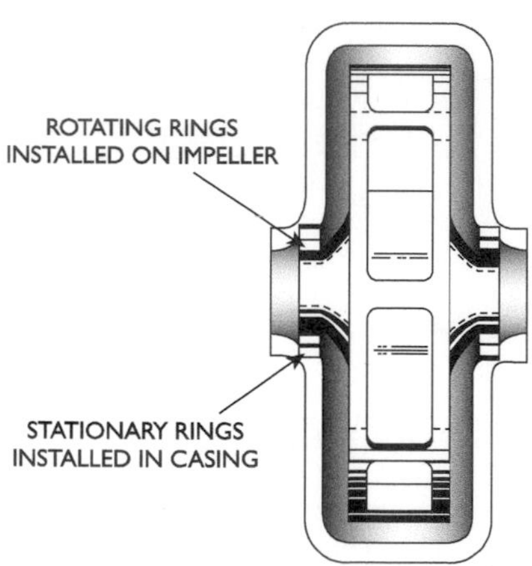

Figure 25-7 Wear rings on a double-admission centrifugal pump.

Centrifugal Pump Features

The centrifugal pump is simple in design and construction and relatively low in cost. It is most efficient for relatively large capacities and low pressures. Liquids containing foreign matter can be handled by centrifugal pumps without clogging. The flow of liquid can be stopped without building excessive pressures in the casing, as the impeller is free to rotate in the liquid. However, even though there is no excessive buildup of pressure, there is generation of heat if the impeller is allowed to rotate with the discharge closed off. Centrifugal pumps deliver a smooth, nonpulsating flow, have no valves or reciprocating parts, and are suitable for high-speed operation by direct connection to motors, steam turbines, or engines.

Troubleshooting Centrifugal Pumps

Table 25-1 will assist in troubleshooting centrifugal pump problems.

Table 25-1 Centrifugal Pump Troubleshooting Chart

Problem or Symptom	*Cause*	*Correction*
No liquid being pumped	Suction line plugged or clogged	Clear obstructions.
	Pump not primed	Reprime pump.
	Impeller clogged	Try back-flushing to clear impeller.
	Incorrect rotation	Reverse motor leads to achieve proper rotation.
Pump not producing rated flow or pressure	Air leak at stuffing box	Tighten packing or replace mechanical seal.
	Air leak at gasket	Replace with new gasket and proper gasket sealant if required.
	Impeller partly clogged	Back flush.
	Low suction head	Open suction valve all the way and check suction piping.
Hot bearings	Improper lubrication	Check for proper grease or oil and relubricate.
	Improper alignment	Realign pump and motor.
Pump vibrates	Poor alignment	Align motor and pump shafts.
	Worn bearings	Replace bearings.
	Pump is cavitating	Recalculate pump or check system attributes.
		Replace impeller.
	Unbalanced impeller	If due to foreign material, back flush; if impeller is worn, replace.

(Continued)

Problem or Symptom	Cause	Correction
	Broken parts on impeller or shaft	Replace impeller.
Pump begins to pump and then stops	Air pocket in suction line	Change piping to eliminate air pocketing.
	Air leak in suction line	Plug the leak.
	Pump not primed properly	Reprime pump.
Motor amperage draw is excessive	Rotating parts are binding	Check rebuilding procedures and correct interfering parts.
	Pump is pumping too much liquid	Machine impeller to smaller size.
	Stuffing box packing is too tight	Readjust.
Leakage at stuffing box	Worn mechanical seal parts	Replace seal.
	Shaft sleeve (if used) is scored or cut	Replace sleeve with new, or remachine and replace.
	Packing improperly adjusted	Tighten or replace packing material.

Reciprocating Pumps

Reciprocating pumps take their name from the back-and-forth motion of the pumping element. Whereas the pumping element (impeller) of a centrifugal pump moves in a continuous circular path, the pumping element in a reciprocating pump alternately moves forward and backward. The principle of operation also differs entirely from that of the centrifugal pump. In the centrifugal pump, liquid is moved by the centrifugal force developed in the liquid. The reciprocating pump moves liquid by displacing the liquid with a solid. This principle of operation is called *positive displacement* and is illustrated in Figure 25-8.

Figure 25-8A shows a container of liquid, filled level with the overflow. Under the overflow pipe is a smaller container that is empty. In Figure 25-8B, a solid has been placed in the larger container, causing the liquid to overflow into the smaller container. The volume of liquid displaced into the smaller container is equal to the volume of that portion of the solid that is submerged in the liquid. As the solid is further submerged in the liquid, it will continue to displace, and flow will

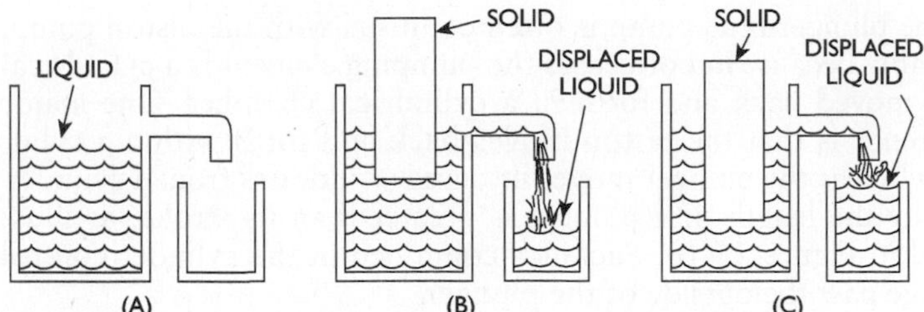

Figure 25-8 The basic operation of a positive-displacement reciprocating pump.

continue until the condition in Figure 25-8C is reached, where the solid is against the bottom of the container. This displacing action is what occurs in a positive-displacement pump when the pumping element enters the liquid-filled pump chamber.

Three basic types of positive-displacement reciprocating pumps are in common use. Each is named for the style of its pumping element. They are classified as *piston*, *plunger*, and *diaphragm*. A piston-type pump is illustrated in Figure 25-9.

The pumping element, or piston, is a relatively short cylindrical part that is moved back and forth in the pump chamber, or cylinder. Generally the distance that the piston travels back and forth, called the *stroke*, is greater than the piston length. Leakage past the outside of the piston is usually controlled by packing or piston rings.

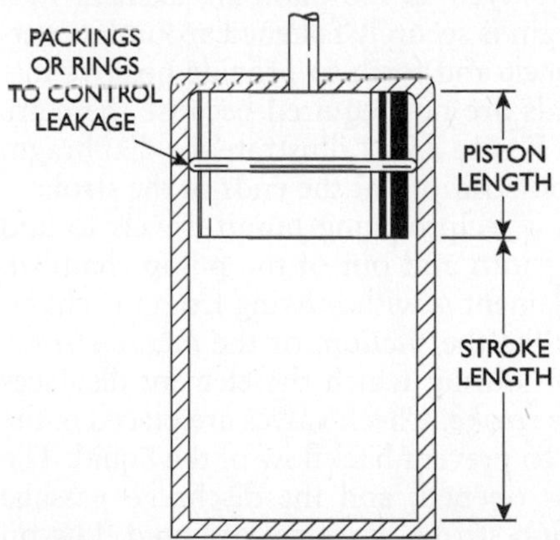

Figure 25-9 The operating principle of the piston-type reciprocating pump.

The plunger-type pump is often confused with the piston pump, probably because in both cases the pumping element is a cylindrical solid moved back and forth in a cylindrical chamber. One major difference is that the piston moves back and forth within a cylinder, whereas the plunger moves into and withdraws from a cylinder. Usually the length of a plunger is greater than its stroke, as illustrated in Figure 25-10. Packings contained in the cylinder control leakage past the outside of the plunger.

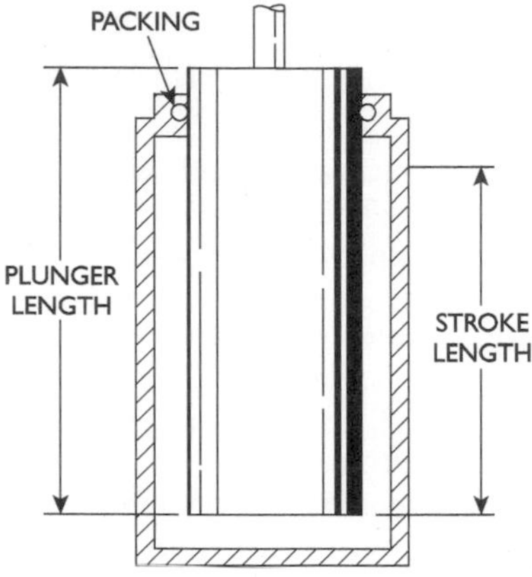

Figure 25-10 The basic parts of the plunger-type reciprocating pump.

A flexible diaphragm is employed as the pumping element in a diaphragm pump. The diaphragm is securely fastened around its outside, and its center is moved back and forth to provide positive displacing action. Packings or seals are not required because there are no sliding fits. The diagrams in Figure 25-11 illustrate the diaphragm pumping element in the extreme positions at the ends of the stroke.

As the pumping element in a reciprocating pump travels to and fro, liquid is alternately moved into and out of the pump chamber. The period during which the element is withdrawing from the chamber and liquid is entering is called the *suction*, or the *intake stroke*. Travel in the opposite direction during which the element displaces the liquid is called the *discharge stroke*. Check valves are placed in the suction and discharge passages to prevent backflow of the liquid. The valve in the suction passage is opened, and the discharge passage valve is closed during the suction stroke. Reversal of liquid flow on the discharge stroke causes the suction valve to close and the discharge valve to open. The diagram in Figure 25-12A illustrates the position of the valves during the suction stroke of a piston pump. The suction valve is open to admit liquid, and the discharge valve is closed to prevent backflow from the discharge passage. In Figure 25-12B the diagram shows the piston on the discharge stroke. The suction valve is closed to prevent backflow into the suction passage, and liquid is moving out through the open discharge valve.

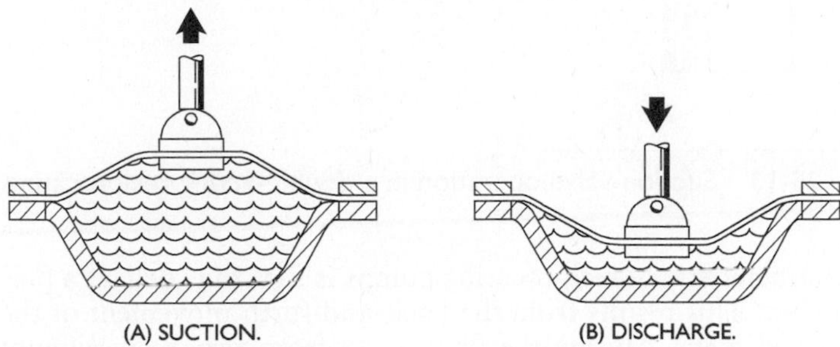

Figure 25-11 The basic parts of the diaphragm-type reciprocating pump.

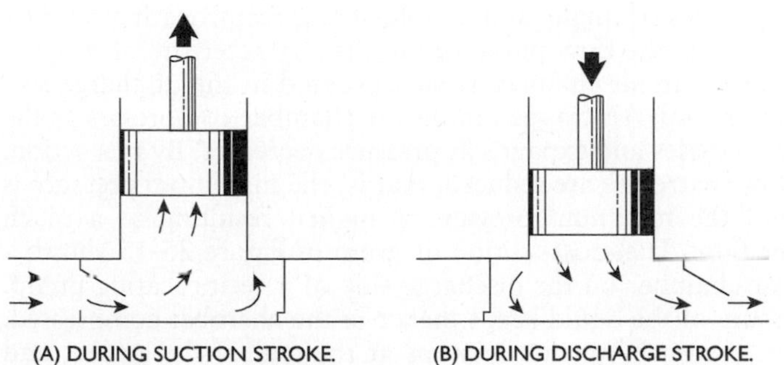

Figure 25-12 Suction-valve operation.

Pumps constructed as shown in Figure 25-12 are called *single-acting* pumps because liquid flows from the pump during only one direction of movement of the pumping element. Pumps constructed as shown in Figure 25-13 are called *double-acting* pumps because liquid flows during movement of the pumping element in both directions. The double-acting pump is more efficient and discharges a more continuous flow. Valve action is the same as in the single-acting pump; however, valves must be provided at both ends of the cylinder of a double-acting pump.

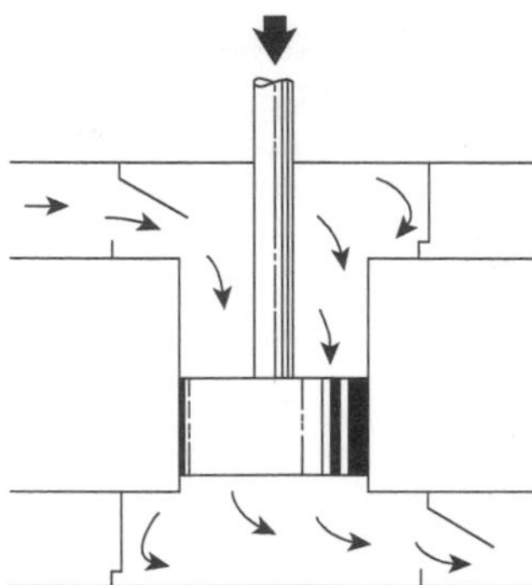

Figure 25-13 Suction-valve operation in a double-acting reciprocating pump.

A characteristic of reciprocating pumps is that they deliver a pulsating flow. This results from the back-and-forth movement of the pumping element, causing the flow to go from zero to maximum and then back to zero. Single-acting pumps discharge during one direction only, whereas double-acting pumps discharge during movement in both directions. The graphs in Figure 25-14 show the pressure profiles of single- and double-acting reciprocating pumps.

To overcome the flow pulsation that is characteristic of reciprocating pumps, an air chamber is incorporated in the discharge section of the pump. Air trapped in the air chamber compresses as the pressure increases and expands as pressure decreases. By this action, the pressure extremes are reduced; that is, the maximum pressure is lower and the minimum pressure is higher, resulting in a much smoother flow. The cross-section diagram in Figure 25-15 shows a typical air chamber on the discharge side of a reciprocating pump. The pressure of the liquid keeps the air in the chamber compressed. When the piston momentarily stops at the end of the stroke, and during the return stroke when liquid is not being pumped, the air expands and tends to maintain the discharge pressure.

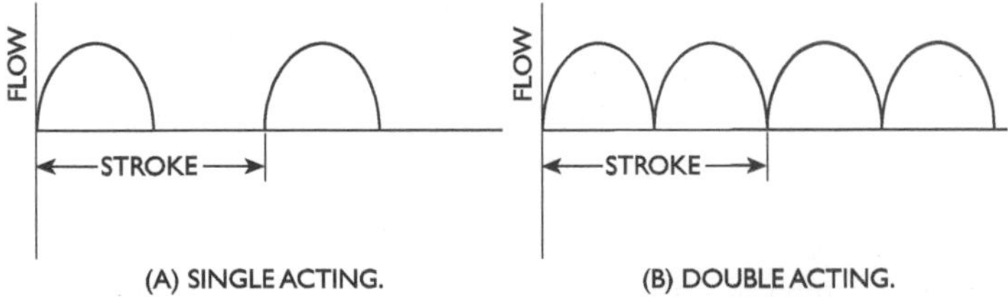

Figure 25-14 Pressure flow from reciprocating pumps.

The pressure profile of a reciprocating pump equipped with an air chamber graphically illustrates the smoothing out of the flow. In Figure 25-16 the normal graph of a double-acting reciprocating pump is shown, with the profile of the air-chamber-equipped pump represented by the dashed line.

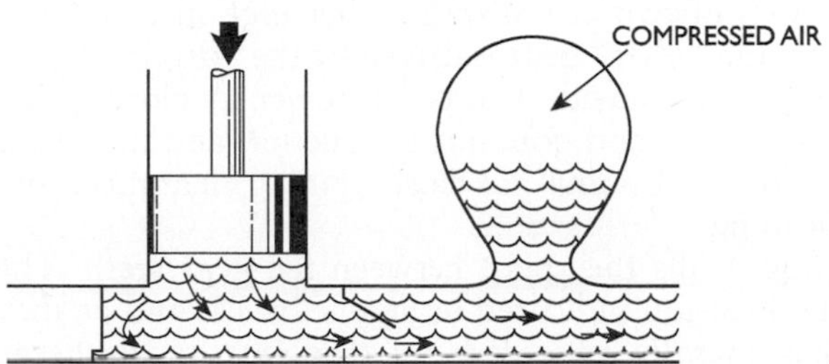

Figure 25-15 A typical air chamber connected to a reciprocating pump.

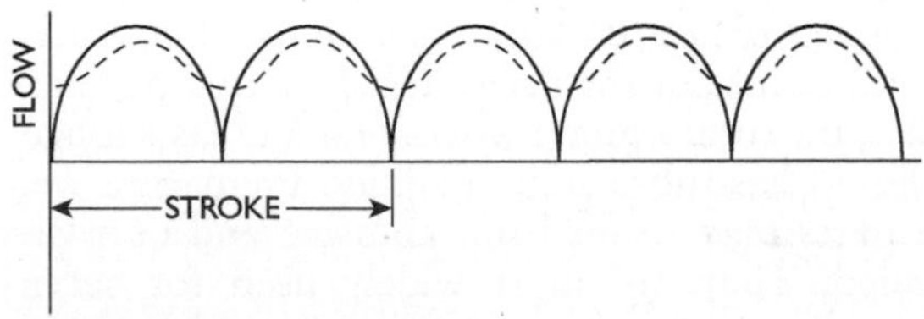

Figure 25-16 The pressure flow from a reciprocating pump connected to an air chamber.

A problem associated with an air chamber is the absorption of the air by the liquid. The result of this absorption is that the air chamber gradually fills up with liquid and is rendered ineffective in cushioning shock and maintaining uniform pressure. For effective operation, therefore, air-chamber-equipped pumps must also have provisions for makeup air. A variety of devices are used for this purpose, from simple air check valves similar to those used in pneumatic tire valve stems to more complex devices employing floats and ball valves. The function of any of these devices is to admit small amounts of air during the suction stroke, which will automatically accumulate at the highest point, or within the air chamber.

Rotary Pumps

Rotary pumps, like reciprocating pumps, are positive displacement in operation. However, their design is much simpler because no check valves are required, and their flow is continuous. Different designs make use of such elements as vanes, gears, lobes, and cams to move the material being pumped. A common style, illustrated in Figure 25-17, consists of two gears, spur or herringbone, in mesh. In its simplest form, the "idler" gear is driven by the "driving" gear, which is rotated from an outside source of power. A close-fitting casing surrounds the gears and contains the suction and discharge connections. This design of pump has close running clearances and generally is self-priming.

In operation, liquid fills the space between the gear teeth. The gears rotate in the direction indicated in Figure 25-18, and as they mesh, the liquid is literally squeezed out through the discharge opening. As the teeth move into the suction area, they separate, creating a partial vacuum, and the liquid flows into the spaces between the teeth from the suction. This operation produces a very continuous flow with almost no pulsation.

As in the case of the reciprocating pump, the capacity delivery is constant regardless of the pressure, within the limits of operating clearances and power. Because of the necessary close clearance and metal-to-metal contact, the rotary pump works best and lasts longest when the pumping liquid has lubricating qualities. Premature wear of the closely fitted parts may result from abrasive and corrosive material being handled. They are most widely used for service requiring relatively small capacities, medium heads, and high suction lifts.

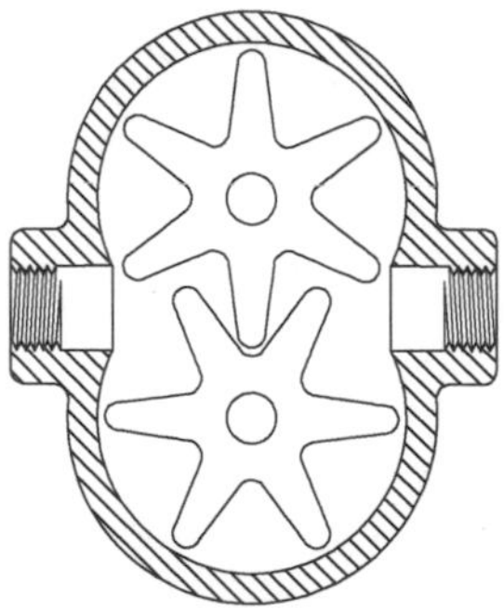

Figure 25-17 The basic parts of the rotary pump.

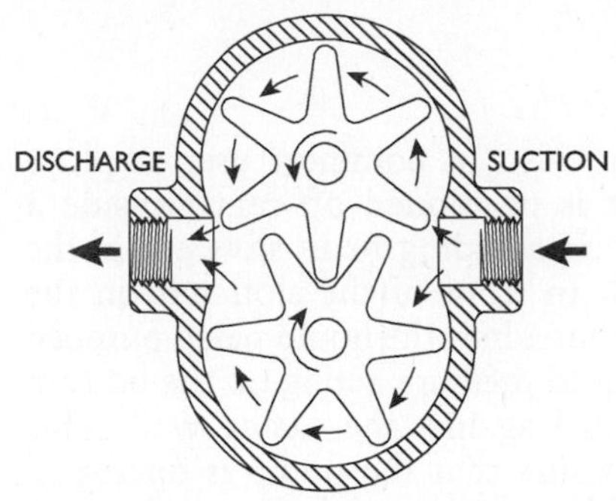

Figure 25-18 The operating principle of the rotary pump.

Another style of rotary pump in wide use is the internal gear type shown in Figure 25-19. The internal gear is the rotor of the pump and drives the smaller idler gear. The idler is offset inside the rotor to mesh the teeth of the two gears. A stationary crescent-shaped section projects from the surface of the pump cover to fill the space between the outside surface of the idler gear and the inside of the rotor. With the gear teeth being out of mesh and completely separated at this point, this filler piece is necessary to seal the discharge from the suction.

In operation, the material entering through the suction of the pumps fills the openings of both the rotor and the idler gears. It stays in these spaces between the teeth of the rotor and idler until the teeth mesh and the liquid is forced from the spaces and out the discharge of the pump. The *crescent seal*, as the projecting filler piece is called, seals the suction area from the discharge area and prevents leakage between the two.

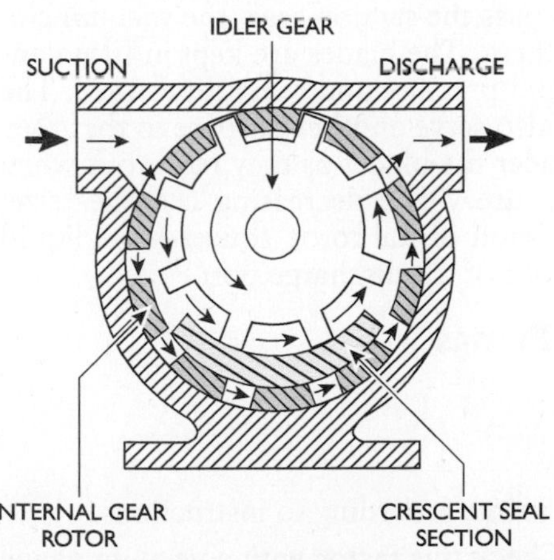

Figure 25-19 The internal-gear rotary pump.

A third style of rotary pump, also in common use, employs blade-type vanes in a rotor that is positioned off-center inside a cylindrical chamber. The blades are a sliding fit in slots cut in the rotor, as shown in Figure 25-20. In front of the slots and in the direction of rotation are grooves that admit the liquid being pumped by the blades. The force of the liquid pressure acting on the bottom of the blades moves them outward against the casing wall. This force varies directly with the pressure that the pump is operating against. Centrifugal force caused by rotation of the rotor also acts to hold the blades in contact with the casing wall. The centrifugal force is an important factor, as it maintains the blades in contact while they are traveling through the low-pressure area, thus preventing leakage between the suction and discharge areas.

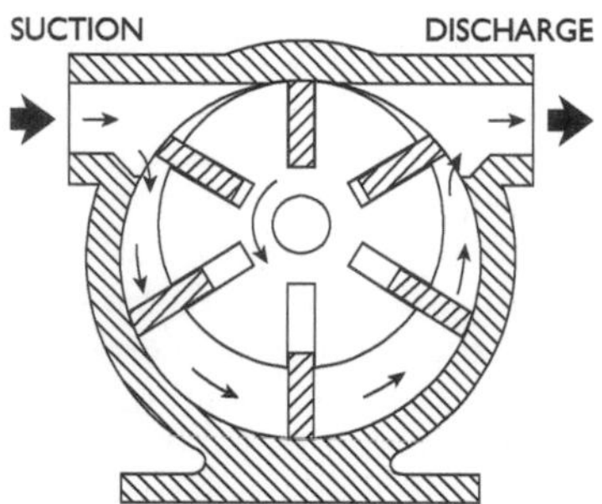

Figure 25-20 Blade-type vanes in the rotor of a rotary pump.

In operation, as the vanes pass the suction area, the vacuum created fills the space between them. The blades are kept in firm contact by a combination of liquid pressure and centrifugal force. The grooves ahead of the blades also serve another purpose in that they break the vacuum formed under the blades as they move out when they go through the suction area. The decreasing chamber size, because of the off-center position of the rotor, squeezes the liquid trapped ahead of the blades out of the discharge port.

Troubleshooting Rotary Pumps

No Liquid Delivered

1. Stop pump immediately.
2. If pump is not primed, prime according to instructions.
3. Lift may be too high. Check this factor with a vacuum gauge on the inlet. If the lift is too high, lower the position of the

pump and increase the size of the inlet pipe; check the inlet pipe for air leaks.

4. Check for incorrect direction of rotation.

Insufficient Liquid Delivered

1. Check for air leak in the inlet line or through stuffing box. Oil and tighten the stuffing-box gland. Paint the inlet pipe joints with shellac or use RTV rubber to seal.
2. Speed is too slow. Check the rpm with manual tach or strobe light. The driver may be overloaded, or the cause may be low voltage or low steam pressure.
3. Lift may be too high. Check with vacuum gauge. Small fractions in some liquids vaporize easily and occupy a portion of the pump displacement.
4. There is too much lift for hot liquids.
5. Pump may be worn.
6. Foot valve may not be deep enough (not required on many pumps).
7. Foot valve may be either too small or obstructed.
8. Piping is improperly installed, permitting air or gas to pocket inside the pump.
9. There are mechanical defects, such as defective packing or damaged pump.

Pump Delivers for Short Time and Quits

1. There is a leak in the inlet.
2. The end of the inlet valve is not deep enough.
3. There is air or gas in the inlet.
4. Supply is exhausted.
5. Vaporization of the liquid in the inlet line has occurred. Check with vacuum gauge to be sure the pressure in the pump is greater than the vapor pressure of the liquid.
6. There are air or gas pockets in the inlet line.
7. Pump is cut by the presence of sand or other abrasives in the liquid.

Rapid Wear

1. Grit or dirt is in the liquid that is being pumped. Install a fine-mesh strainer or filter on the inlet line.

2. Pipe strain on the pump casing causes working parts to bind. The pipe connections can be released and the alignment checked to determine whether this factor is a cause of rapid wear.
3. Pump is operating against excessive pressure.
4. Corrosion roughens surfaces.
5. Pump runs dry or with insufficient liquid.

Pump Requires Too Much Power

1. Speed is too fast.
2. Liquid is either heavier or more viscous than water.
3. Mechanical defects occur, such as bent shaft, binding of the rotating element, stuffing-box packing too tight, misalignment of pump and driver, misalignment caused by improper or sprung connections to piping.

Noisy Operation

1. Supply is insufficient. Correct by lowering pump and increasing size of inlet pipe.
2. Air leaks in inlet pipe cause a cracking noise in pump.
3. There is an air or gas pocket in the inlet.
4. Pump is out of alignment, causing metallic contact between rotor and casing.
5. Pump is operating against excessive pressure.
6. Coupling is out of balance.

Pump Capacity, Pressure, and Head

The amount of material, usually liquid, that a pump moves in a given length of time determines its capacity. The usual units of measure used in industrial applications are the gallon and the minute. The standard pump rating in general use is *gallons per minute*, abbreviated gpm. Although this states the volume delivered by the pump, another very important consideration is the pressure at which the liquid is moved. Two systems of pressure specification are used: *pounds per square inch* and *head.*

A force acting on a unit area is called pressure. In pump operation, the units commonly used are the square inch of area and the pound of force. Pump pressures, therefore, are usually stated in pounds per square inch, abbreviated psi. Pressure is developed within the fluid by the weight of the fluid above it. A tank of water

23 ft high, as shown in Figure 25-21, has pressure within the tank, due to the weight of the water, of 0 psi at the top of the tank and approximately 10 psi at the bottom. This pressure may be calculated by multiplying the height of the column of water by its weight and dividing the square-inch area upon which the weight acts.

One cubic foot of water weighs 62.4 lb; therefore, the column of water in Figure 25-21 would weigh 62.4 times 23, or 1435.2 lb. The pressure at the bottom of the tank, therefore, is 1435.2 lb per square foot. To reduce this to the terms commonly used, i.e., pounds per square inch (psi), requires division of the total weight force by the number of square inches upon which it is acting—in this case, division by 144, the number of square inches in a square foot. This calculates to 9.666, or ⅓ lb under 10 psi.

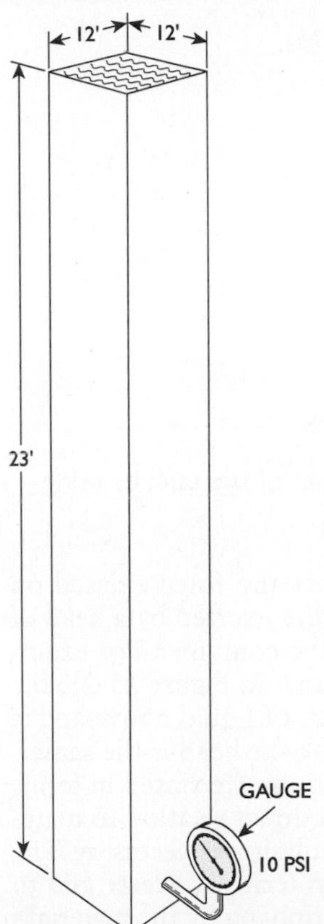

Figure 25-21 Calculating the pressure of water in a tank in psi.

A similar column of water may be used to explain the meaning of *head*, which is simply the height of a liquid. The head of the water in the column in Figure 25-22 is 10 ft. To calculate the pressure at the bottom of the column, multiply the unit water weight (62.4) by the height of the column (10), and divide by 144 square inches of supporting surface. The pressure at the base of the water column, therefore, is 4.333 psi, or it may be stated that a 10-ft head of water exerts a force of 4.333 psi. The unit most commonly used is the pressure exerted by a 1-ft head of water, which is of course ¹⁄₁₀ that of the column in Figure 25-22, or .433 psi. The pressure exerted by any column of water may be determined by multiplying its head, or height, by 0.433. For example, a 15-ft head of water exerts 6.49 psi (15 × 0.433).

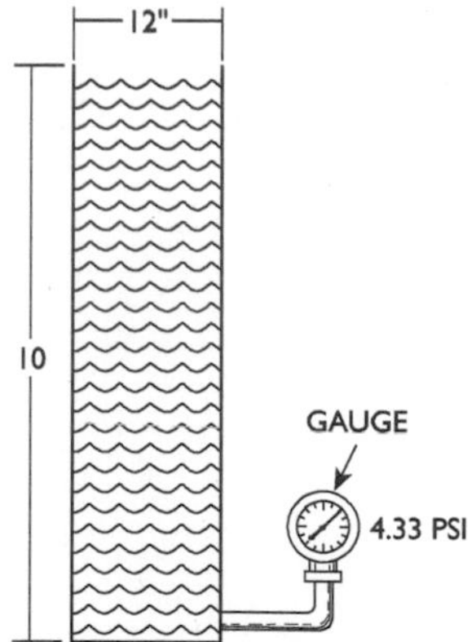

Figure 25-22 Calculating the psi at the bottom of the tank by using the head, or height, of the liquid.

In each case, the stated pressure describes the force exerted on one square inch of area; therefore, the pressure exerted by a head of liquid does not depend on the diameter of the container. For example, at any point on the bottom of the container in Figure 25-23, the pressure depends on only the head, or height, of liquid above and is the same at the bottom of both tanks because the head is the same.

Because pump specifications and calculations are stated in terms of head pressures, and head pressures are in direct relation to material weight, a standard system of weight calculation is necessary. The system used is to calculate head pressures in terms of water and to adjust this figure to suit the material by multiplying by the material's specific gravity. The specific gravity is a number indicating how many times a certain volume of a material is heavier or lighter than an equal volume of water. For example, the two tanks in Figure 25-24 are filled to equal height, a head of 15 ft. The 15-ft head pressure at the bottom of tank A, which is filled with water, is 6.49 psi (15 × 0.433). Tank B is filled with a material weighing only one-half that of water, its specific gravity being 0.5. To determine the head pressure at the bottom of tank B, the same calculation is made, and the result is multiplied by the material's specific gravity (0.5). The pressure at the bottom of tank B is 3.24 psi (15 × 0.433 × 0.5).

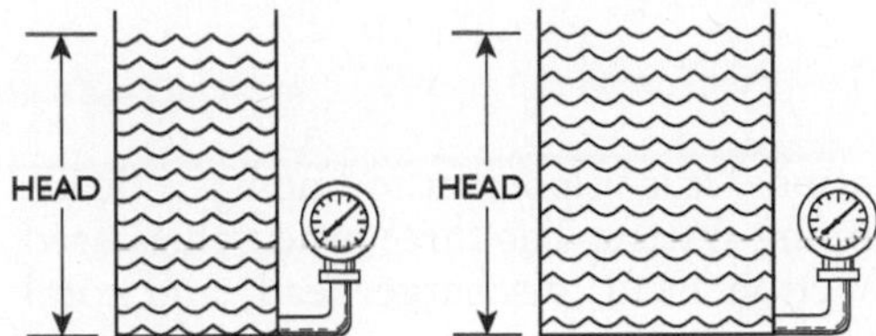

Figure 25-23 The pressure at the bottom depends on the head, or height, of the liquid.

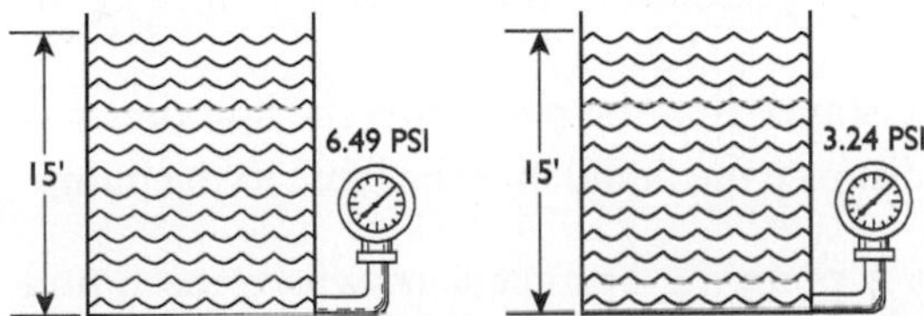

Figure 25-24 The pressure will also depend on the specific gravity of the liquid.

The calculation of pressure resulting from the head of a liquid may be stated in a formula as follows:

$$\text{Pressure} = \text{head} \times 0.433 \times \text{spec. gravity}$$

The formula may be rearranged to determine head when the pressure is known:

$$\text{Head} = \frac{\text{pressure}}{0.433 \times \text{specific gravity}}$$

For example, given an 18-ft column of material with a specific gravity of 0.78, what would be the pressure? Using the first formula,

$$\text{Pressure} = 18 \times 0.433 \times 0.78$$

$$\text{Pressure} = 6 \text{ psi}$$

Using the same figures, what would be the head of a column of material with a specific gravity of 0.78 if the pressure at the base were 6 psi?

Using the second formula,

$$\text{Head} = \frac{6}{0.433 \times 0.78}$$

$$\text{Head} = 18 \text{ feet}$$

In making pump calculations, it is important to know the heads at various levels in the pumping system. The three head values used in such calculations are suction head, discharge head, and total head. A fourth value that is also a vital factor in pump calculations is suction lift.

Suction head is the pressure at the suction, changed to head.

Discharge head is the pressure at the discharge, changed to head.

Total head is the distance that the pump lifts the liquid.

Suction lift is the distance the liquid must be lifted to the pump.

Figure 25-25 shows a pumping arrangement where the liquid supply to the pump is above the suction, and the vessel into which the pump discharges is at a higher elevation than the supply. In this case, the total head is less than the discharge head.

The liquid rises to the supply level in the Figure 25-25 system without being pumped. It is lifted from this level to the discharge tank by the pump. Therefore, in this case, the difference in level between the supply and discharge is the total head. The total head in such a system may be calculated by subtracting the suction head from the discharge head.

When the pump location is above the supply level, as shown in Figure 25-26, the liquid must be lifted to the pump. The total head in this case is the suction lift plus the discharge head.

In each case, to convert head in feet to pressure in pounds, we have calculated the weight of the liquid in the column above the level at which the pressure is being exerted. If the liquid is contained in a system that is under pressure, the liquid pressure will be increased, and

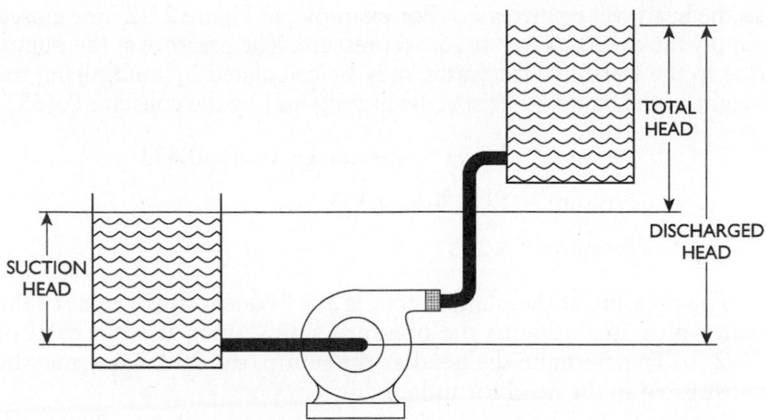

Figure 25-25 Calculating the total head in a pumping system where the pump is in line with the suction head.

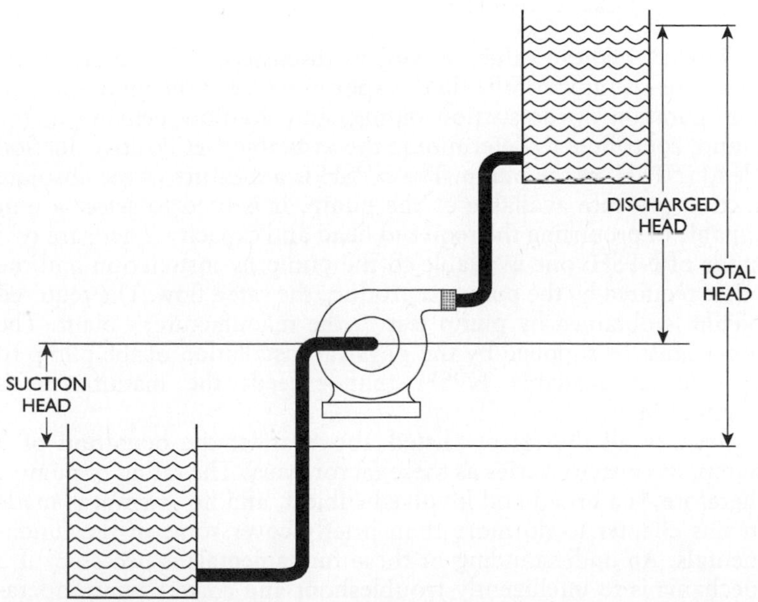

Figure 25-26 Calculating the total head when the pump is located above the suction lift.

so the head will be increased. For example, in Figure 25-27 the closed supply tank holds water at 15 psi pressure. The pressure at the pump, due to the weight of the water, may be calculated by multiplying the height of the liquid by its specific gravity and by the constant 0.433.

$$\text{Pressure} = \text{height} \times \text{specific gravity} \times 0.433$$

$$\text{Pressure} = 12 \times 1 \times 0.433$$

$$\text{Pressure} = 5.2 \text{ lb}$$

The pressure at the pump, then, is 5.2 lb due to the weight of the water plus 15 lb due to the pressure acting upon it, or a total of 20.2 lb. To determine the head at the pump, the 20.2 value may be substituted in the head formula.

$$\text{Head} = \frac{\text{pressure}}{\text{specific gravity} \times 0.433}$$

$$\text{Head} = \frac{20.2}{1 \times 0.433}$$

$$\text{Head} = 46.6 \text{ feet}$$

The several head values previously discussed, and such other factors as temperature of the fluid, vapor pressure of the fluid, size and configuration of the suction piping, and the flow demand of the pump, contribute to determining the available Net Positive Suction Head (NPSH) at the pump. The NPSH is a measure of the absolute suction pressure available at the pump. It is used to select a unit capable of producing the required head and capacity. There are two kinds of NPSH: one available to the pump by installation and the other required by the pump to produce the rated flow. The required NPSH is obtained by pump test in the manufacturer's plant. The other must be supplied by the physical installation of the pump to provide an available NPSH that exceeds the manufacturer's required figure.

Because all the factors listed above affect the operation of a pump, its capacity varies as these factors vary. The rating of pumps, therefore, is a broad and involved subject, and no attempt is made in this chapter to do more than briefly cover some of the fundamentals. An understanding of these fundamentals is necessary if a mechanic is to intelligently troubleshoot and correct pump operation problems. Because questions often arise regarding NPSH, this area requires further discussion.

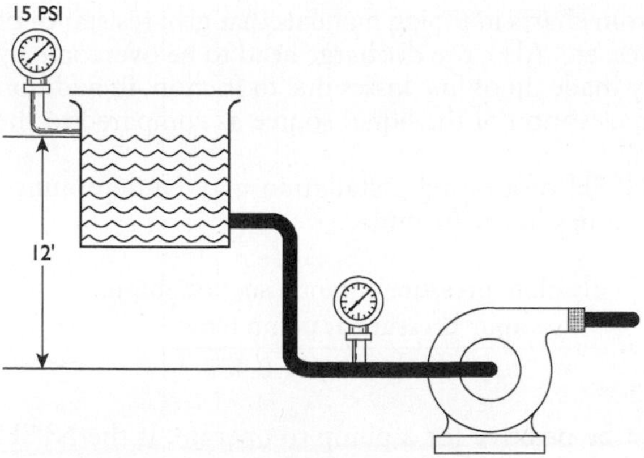

Figure 25-27 Calculating pump pressure.

As previously stated, available NPSH is the absolute pressure available at the suction nozzle of the pump. Thus, it is the amount of pressure on the surface of the liquid in the suction tank plus the liquid head and minus the pressure lost through friction in moving the liquid to the suction of the pump. To better understand some of these factors, let's discuss the meaning of some of the terms, starting with "absolute pressure."

The pressure of air, gases, or fluids is generally measured either in absolute pressure or in gauge pressure. When measured in absolute pressure, the pressure of the atmosphere is included; the gauge pressure is the pressure above that of the atmosphere. The pressure of air at sea level is 14.7 lbs per square inch. The absolute pressure, therefore, is gauge pressure plus 14.7. If a gauge reads 15, the gauge pressure is 15 psig (pounds per square inch gauge) and the absolute pressure is 29.7 psia (pounds per square inch absolute).

The vapor pressure of a liquid is the pressure that the liquid's vapor exerts when confined or under pressure. A liquid vaporizes rapidly, or boils, when pressure is reduced below a minimum level; therefore, if the suction pressure is low enough, liquid entering a pump will vaporize or boil. If the pump were to fill up with vapor rather than liquid, it would be vapor-locked and could not operate. To keep the liquid from vaporizing, the absolute pressure at the pump suction must be higher than the vapor pressure of the liquid at the pumping temperature.

The pressure lost through friction in moving a liquid to the suction of a pump is determined by the size of pipe, its length, the number and type of fittings, and the volume of liquid to be pumped. These values may be obtained from charts in piping manuals that give resistance of pipes, fittings, valves, etc. Also, the discharge head to be overcome by the pump is usually made up of line losses due to friction, in addition to the difference in elevation of the liquid source as compared to the discharge elevation.

The available NPSH of a pump installation may be determined by substitution of values in the formula.

$$\text{NPSH} = \frac{\text{absolute pressure at pump suction minus vapor pressure at pump temp.}}{\text{specific gravity} \times 0.433}$$

The NPSH must be positive for a pump to operate. If the NPSH were negative, the liquid would vaporize in the pump suction.

Manufacturers provide performance curves for their pumps that show the relationship of capacity to NPSH, efficiency, total head, and horsepower. The values shown in these graphs are obtained by pump tests conducted by the manufacturer. Typical curves of this type are shown in Figure 25-28. The graph is arranged so that capacity is read at the bottom and efficiency, horsepower, and total head are read at the left.

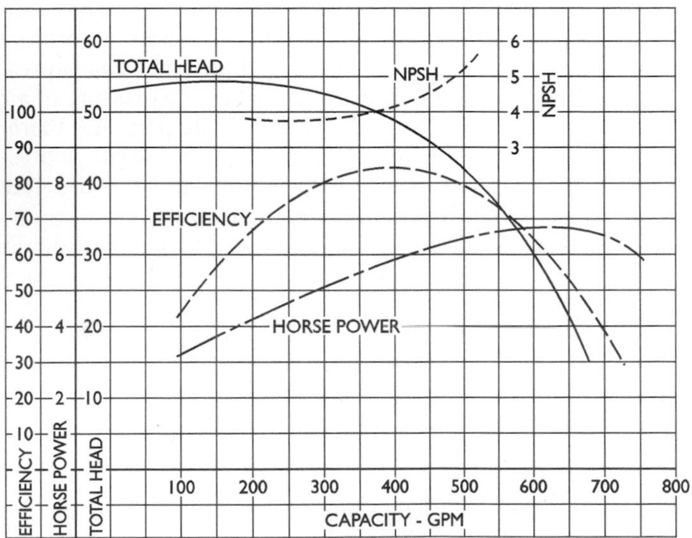

Figure 25-28 Capacity versus efficiency, horsepower, and total head.

The suction lift has already been defined as the distance the liquid must be lifted to the pump. Also, there is pressure lost in moving a liquid to the suction of a pump. The total suction lift, therefore, is made up of two factors:

1. The vertical distance the pump is above the liquid (known as static lift)
2. The frictional resistance in the suction line

Pumps do not actually suck up liquid. The liquid is forced into the pump's suction by atmospheric pressure when a partial vacuum is built up in the pump. This vacuum is created by the piston or plunger in the reciprocating pump, by the cam or gear action in rotary pumps, and in centrifugal pumps by the liquid forced out of the pump discharge. The total suction lift of a pump, therefore, is limited to the distance the atmospheric pressure can lift the liquid minus the frictional resistance in the line. The distance that atmospheric pressure can lift a liquid depends on the weight of the liquid. Because 1 ft of water exerts 0.433 lb per square inch of pressure, exerting 1 lb per square inch of pressure requires a height, or head, of 2.31 ft of water.

Given that 1 lb of water pressure is equal to 2.31 ft of head in a water column, and atmospheric pressure is 14.7 lb per square inch at sea level, the theoretical distance the atmospheric pressure can lift a column of water is 2.31 × 14.7, or 33.9 ft. However, various losses reduce this theoretical lift: losses through pistons, valves, rotary-pump cams, bypassing, stuffing-box leakage, etc. Losses in positive-acting pumps (reciprocating and rotary) are less than in centrifugal pumps. The maximum practical suction lift is generally recognized as 22 ft on positive-acting pumps and 15 ft on centrifugal pumps.

These are maximum total suction lifts, which include both factors: vertical distance the pump is above the liquid and frictional resistance in the suction line. For example, on a reciprocating pump where the maximum lift is 22 ft, if it is found from a table of friction losses that there is a 5-ft frictional loss in the suction line, it means that the vertical distance of the pump above the liquid cannot exceed 22 ft minus 5 ft, or 17 ft.

Pump Operating Problems

Probably more cases of pump- and motor-bearing failures and excessive coupling wear and failure are caused by misalignment of the pump and driver during installation than by any other single cause. The importance of motor and driver alignment at the time of installation cannot be overemphasized. Alignment procedures are covered in Chapter 17. A common misconception about factory-assembled pumps is in connection with the rigidity of the bedplates or baseplates supplied. Commercially speaking, there is no such thing as a perfectly rigid bedplate. Because of the weight of the pump and the driver and the distribution of the weight on the bedplate, all bedplates will distort in handling and shipping. Therefore, no matter how carefully the pump and driver have been aligned at the factory, they must be realigned at the time of installation. If manufacturers were to supply rigid bedplates that would not distort, they might cost more than the pump because they would have to be so thick and heavily ribbed that the weight would be excessive.

Chapter 26

Air Compressors

Compressed air is *air forced into a smaller space than it originally occupied*, thus increasing its pressure. The power available from compressed air is used in many applications as a substitute for steam or other force, as in operating rock drills, shop tools, and engines.

A compressor is *a machine (driven by any prime mover) that compresses air into a receiver to be used at a greater or lesser distance.* The system is not subject to loss by condensation in the pipes, as is the case when carrying steam in pipes for long distances. Air stored under pressure in a reservoir can be used expansively in an ordinary steam engine, returning an amount of work equivalent to that which was required to compress it, minus the friction.

The Compression of Air

When the space occupied by a given volume of air is changed, both its pressure and temperature are changed in accordance with the following laws:

> Boyle's law: *At constant temperature, the absolute pressure of a gas is inversely proportional to its volume.*
>
> Charles' law: *At constant pressure, the volume of gas is proportional to its absolute temperature.*

In the ordinary process of air compression, therefore, two elements contribute toward the production of a higher pressure:

1. The reduction of volume.
2. The increasing temperature due to the increasing pressure corresponding to the reduced volume. See Figure 26-1.

The application of the two laws is illustrated in Figure 26-2, which shows a cylinder fitted with an airtight piston. If the cylinder is filled with air at atmospheric pressure (14.7 lbs per sq. in. absolute), represented by volume A, and the piston is moved to reduce the volume, say to ⅓ A, as represented by B, then according to Boyle's law the pressure will be tripled: $14.7 \times 3 = 44.1$ lbs absolute, or $44.1 - 14.7 = 29.4$ gauge pressure. *In reality, however, a pressure gauge on the cylinder would at this time show a higher pressure than 29.4 gauge pressure because of the increase in temperature produced in compressing the air.*

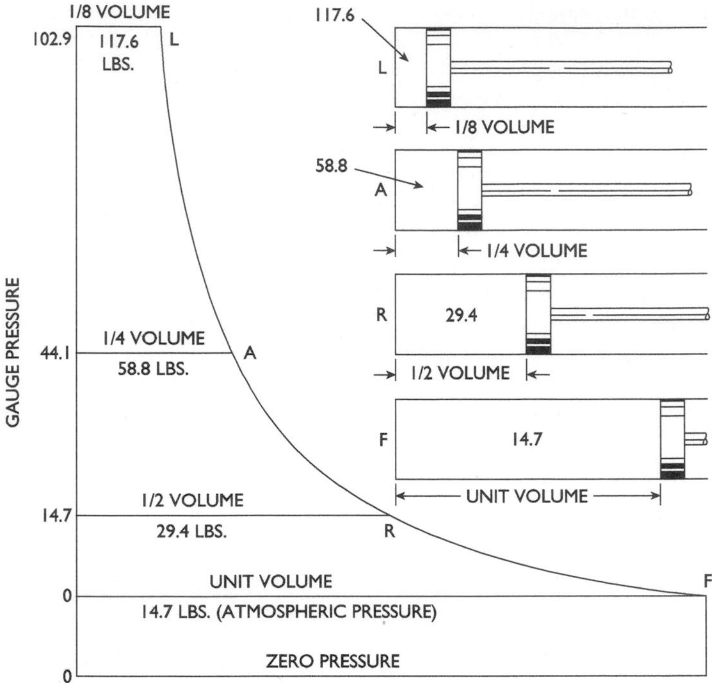

Figure 26-1 Diagram showing volume versus pressure.

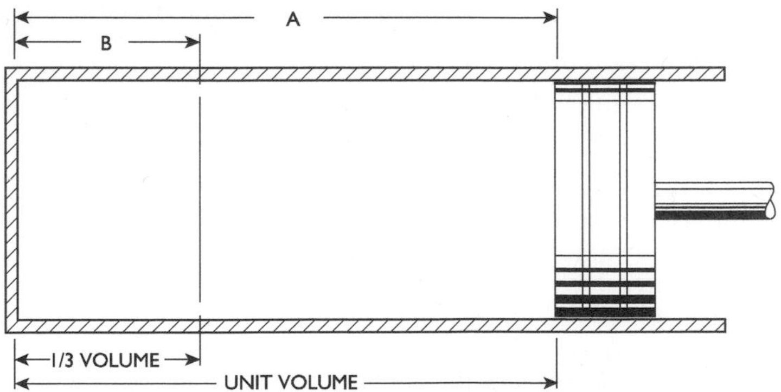

Figure 26-2 Elementary air compressor illustrating the phenomena of compression stated in Boyle's and Charles' laws.

Now, in the actual work of compressing air, it should be carefully noted that *the extra work that must be expended to overcome the excess pressure due to rise of temperature is lost because the compressed air cools after it leaves the cylinder, and the pressure drops to what it would have been if compressed at constant temperature.*

Accordingly, in the construction of air compressors, where working efficiency is considered, some means of cooling the cylinder is provided, such as projecting fins or jackets for the circulation of cooling water.

Heat of Air Compression

This subject has probably received more consideration in air-compressor design than any other. The principal losses in the earlier compressors were traceable to this source. Figure 26-3 shows why the heat of compression results in a loss.

It should be noted that the heat of compression, as already explained, represents work done upon the air for which there is usually no equivalent obtained, given that the heat is all lost by radiation before the air is used. The selection of an air cylinder lubricant is, of course, governed to a considerable extent by knowledge of the cylinder temperature it must withstand.

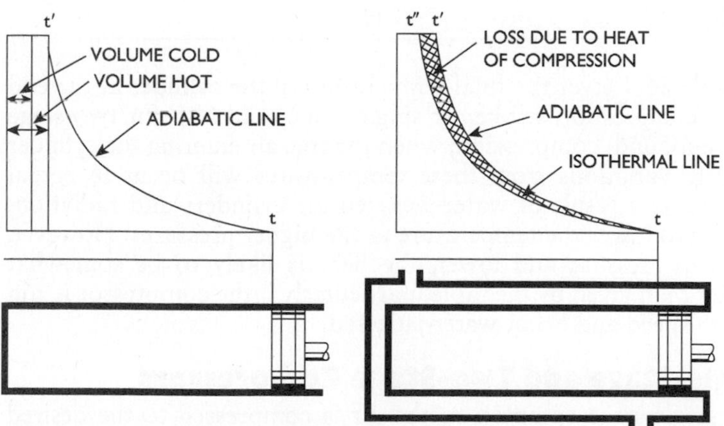

Figure 26-3 Loss due to heat of compression.

When the air pressures are known, the corresponding temperatures are ascertained fairly accurately, as shown in Table 26-1.

Table 26-1 Cylinder Temperatures at End of Compression

Air Pressure (lb gauge)	*Final Temperature, Single Stage (°F)*	*Final Temperature, Two Stage (°F)*
10	145	
20	207	
30	255	
40	302	
50	339	188
60	375	203
70	405	214
80	432	224
90	459	234
100	485	243
110	507	250
120	529	257
130	550	265
140	570	272
150	589	279
200	672	309
250	749	331

Table 26-1 gives the final temperature in the cylinder at the end of the compression stroke, or single stage, and also for two-stage (or compound) compression, when the free air entering the cylinder is 60°F. Variations from these temperatures will occur in actual practice as a result of water-jacketed air cylinders and radiation, tending to lower the temperature at the higher pressures. However, at 50 lb pressure and lower, the heat is likely to be somewhat greater than given by the table, particularly if the compressor is run at high speed and is not water-jacketed.

Single-Stage and Two-Stage Compressors

In a single-stage compressor, the air is compressed to the desired pressure in one operation. The air is taken into the air cylinder at zero gauge pressure (atmospheric pressure, or 14.7 lb per sq. in.) and compressed with one stroke of the piston to the desired pressure. It is then discharged directly into the air receiver.

In a two-stage compressor, the desired pressure is reached in two operations, and two separate cylinders are required. The air is taken into the low-pressure (large) cylinder and compressed to an intermediate pressure. Then, it is passed through an intercooler to the high-pressure (small) cylinder, in which the compression to the desired pressure is completed. The principal advantage of compound compression over simple compression is the reduction of the loss due to the heat of compression. This is due to the fact that more time is taken to compress a certain volume of air and that this air, while being compressed, is brought into contact with a larger percentage of jacketed surfaces.

A typical air-cooled single-stage compressor unit is shown in Figure 26-4. This is a self-contained unit consisting of compressor, motor, and regulation equipment all mounted on a receiver tank. This style of unit is for very light duty, usually in the ½ to 1 horsepower range, with a capacity of 1½ to 2½ cubic feet of air per minute (cfm) at a maximum of 150 pounds gauge pressure (psig).

Figure 26-4 An air-cooled single-stage air compressor.
Courtesy Ingersoll-Rand Co.

The compressor unit shown in Figure 26-5 is similar in appearance but is a two-stage design. This unit is very widely used in industrial and commercial applications. It is manufactured in a wide range of sizes, up to about 25 horsepower, with capacities of 75 to 80 cfm at a maximum of 250 psig.

Figure 26-5 A typical two-stage air-cooled air compressor.
Courtesy Ingersoll-Rand Co.

Other important advantages of the two-stage compressor may be enumerated as follows:

1. Cooler intake air
2. Better lubrication
3. Reduction of clearance losses
4. Lower maximum strains and nearer uniform resistance

If the temperature of the air leaving the intake cylinder is low, the cooling influence of the jacket is better, the cylinder walls are cooler between strokes, and the air enters the cylinder cooler than in a single-stage compressor. The lubricant for cylinders and valves is not subject to the pernicious influence of high temperatures, and the clearance losses, or losses due to dead spaces, are less in a compound compressor than in a simple compressor.

Clearance loss in an air compressor is principally a loss in capacity and therefore affects only the intake cylinder. It increases with the terminal pressure, but because the terminal pressure of the intake or low-pressure cylinder of a compound compressor is much less than the terminal pressure of a simple compressor, the volumetric efficiency of the compound compressor is greater than that of the simple compressor. The life of a compound compressor is longer than that of a simple compressor for like duty, due to better distribution of pressures.

More heat is generated in compressing to a high pressure than to a low pressure. Up to a pressure of about 60 lb per sq. in., it is not practical to remove this heat, and single-stage compressors should be used. Above 60 lb, a two-stage compressor will not only deliver more air than a single-stage compressor of equal size but also consume less power.

Intercoolers

By definition, an intercooler is *a species of surface condenser placed between the two stages of a compound air compressor* so that the heat of compression liberated in the first cylinder may be removed from the air as it passes to the second, or high-pressure, compression cylinder. The cooling surface usually consists of small brass or copper tubes through which water circulates.

Aftercoolers

Moisture in compressed air or gas is costly and annoying. Carried into the lines in the form of vapor, it condenses when cooled and has many harmful effects. In compressed air, it washes away lubricant from the tools and machines through which it passes. It freezes in valves, ports, and other openings because of the sudden expansion of the air. It hastens corrosion of all metal that it reaches and hastens the decay of rubber hose.

In the case of compressed gas for distribution, it is one of the leading causes of line and meter troubles. Removal of moisture before the air or gas is introduced into the lines is the best method of protection. This can be done effectively by an aftercooler, which cools the air (or gas) to a point where most of the moisture and oil condense and can be removed. See Figure 26-6.

This is accomplished by bringing the air (or gas) into contact with pipes through which cooling water is constantly circulated. This not only eliminates the difficulties caused by moisture at points where the air is used but also ensures more effective distribution. The air leaves the aftercooler at a uniform and relatively low temperature, thus obviating the alternate expansion and contraction of lines previously referred to.

With efficient aftercoolers, sufficient moisture is removed to satisfy most applications of air, and no further care need be exercised. When the air is used for such purposes as paint spraying, enameling, and food preparation, further drying of the air can be effected by passing it through special separators immediately before it is used. This removes the moisture that may condense as a result of further cooling of the air in pipelines after it leaves the aftercooler.

Figure 26-6 A typical horizontal aftercooler.
Courtesy Ingersoll-Rand Co.

Piston Displacement

The displacement of a compressor is *the volume displaced or swept through by the piston during the compression stroke.* It is not a measure of the amount of air that the compressor will actually deliver.

Actual Air Delivered

The amount of air actually delivered by the compressor is *always less than the piston displacement* and is the *amount of air available for useful work.* It is expressed in cu. ft per minute of free air.

Volumetric Efficiency

By definition, volumetric efficiency is the *ratio of the actual air delivered to the piston displacement.* For instance, if a compressor has a displacement of 20 cu. ft per minute and the actual air delivered is 16 cu. ft per minute, its volumetric efficiency is

$$\frac{16}{20} = 80\%$$

Control Methods

Because the power-driven compressor is almost always a constant-speed machine, various methods of regulation are employed. Constant speed means constant piston displacement. The problem of delivering a variable volume of air with constant piston displacement becomes one of making a portion of that displacement noneffective in the compression and delivery of air.

The following methods should be noted:

First method—This is really one of unloading rather than of regulating. *A pressure-controlled mechanism is arranged so that, when pressure exceeds normal, a communication is opened between the two sides of the compressor piston.* Usually this is accomplished by opening and holding open one or several of the discharge valves at both ends of the cylinder; the air is then swept back and forth from one side of the piston to the other through the open valves and the air discharge passage. When normal pressure is restored, the valves are automatically closed, and compression and delivery are resumed. Obviously this is practically a total unloading of the machine for a longer or shorter period—a sudden release from load and a sudden resumption of load. Moreover, the air swept back and forth by the piston in its travels is air under full pressure; when the discharge valves suddenly close, the piston at once encounters a full cylinder of air at maximum pressure. These facts limit regulators of this class to machines of comparatively small capacity.

Second method—By means of a pressure-operated device, the partial or total closing of the compressor intake under reduced load is accomplished. To avoid the dangers and sudden action upon such an operation, these devices are provided with some damping mechanism so that they are compelled to operate slowly, making the release or resumption of the load gradual.

Third method—This is very similar to the first, *except that here the inlet valves, instead of the discharge valves, are held open when the machine is unloaded.* The piston simply draws in and forces out air at atmospheric pressure. It is open to the same criticism (though in somewhat less degree) as the first method, namely, undue shock and strain on release and resumption of load.

Fourth method—A pressure-controlled valve is used on the compressor discharge of a single-stage machine, combining the functions of a check valve to limit the escape of air from the receiver or air line. Excessive pressure blows the discharge to atmosphere instead of into the line. This arrangement is also used on two-stage machines by placing it on the low-pressure discharge to the intercooler. When the governor valve is opened by excess pressure, the low-pressure cylinder discharges to atmosphere, and the high-pressure cylinder acts simply as a low-pressure cylinder with intake at atmospheric pressure.

This device is more of a relief valve than an unloader, for the piston must continue to compress to a pressure that will open the discharge valves; this volume of compressed air is wasted.

Fifth method—This provides auxiliary clearance spaces, or pockets, at each end of the cylinder, which are successively "cut in" as load diminishes. The excess air is simply compressed into these clearance spaces and expanded on the back stroke. The capacity of the cylinder is reduced without any appreciable waste of power. The energy used in compressing the clearance air is given back by its expansion.

Air-Compressor Types

Air compressors can be classified into either of two general categories: *positive-displacement* or *dynamic units.* The positive-displacement type includes reciprocating and rotary compressors. In the dynamic type, centrifugal and axial-flow designs are used.

Cylinders in reciprocating machines may be either single or double acting. The single-acting arrangement is one in which the compression of the air takes place in only one end of the cylinder as the piston moves back and forth. In this design, the inlet and discharge ports are located at the compression end of the cylinder. Such a compressor may be a single- or a two-stage unit using single-acting cylinders in either or both stages of compression.

Compressors defined as units having double-acting compression cylinders compress air at both ends of the cylinder as the piston moves back and forth. The cutaway view in Figure 26-7 shows the internal arrangement in a typical double-acting horizontal compressor. This design requires that both inlet and discharge ports be located at each end of the cylinder. The double-acting compression cylinder is used on single-stage and multistage compressors. In machines with more than one cylinder, the cylinders can be arranged in a vertical/horizontal combination, "V" or "L" fashion, radial, and similar patterns in tandem and parallel.

The three principal styles of rotary positive-displacement compressors are the sliding-vane, impeller, and liquid-piston. The rotor in the sliding-vane compressor is positioned off-center to the cylinder bore as shown in Figure 26-8. In operation, the rotor blades slide in and out in the rotor slots in accordance with the varying clearance in the cylinder. Air is trapped ahead of the vanes as they move through the intake area. Because of the off-center position of the rotor with respect to the cylinder, the size of the chambers rapidly decreases as they move toward the discharge port. During this movement from intake to discharge, there is a smooth, continuous compression of the confined air.

Figure 26-7 Cutaway view of a horizontal double-acting compressor. *Courtesy Ingersoll-Rand Co.*

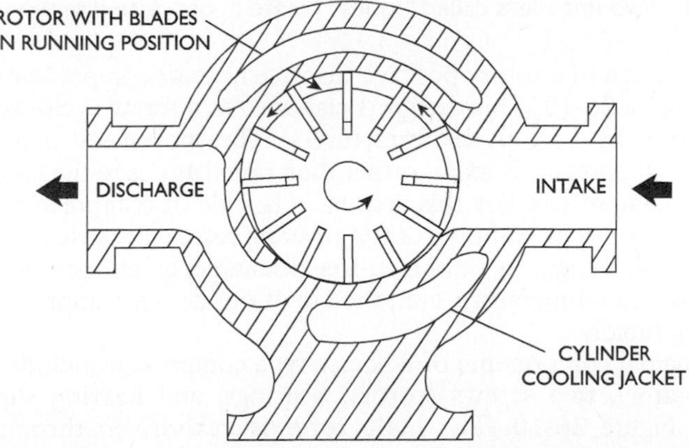

Figure 26-8 The off-center rotor in a sliding-vane compressor.

In the impeller design of a rotary positive-displacement air compressor, shown in Figure 26-9, two impellers called *lobes* are rotated in opposite directions within a housing or cylinder. The lobes are so designed that they mesh as they rotate. Connecting or timing gears connect the two shafts, maintaining proper mating relationship. A minimum of operating clearance prevents leakage of the confined air during compression. Such units can handle large volumes of air per minute at a relatively low pressure. They are widely used for dry conveying systems of all types, where large-volume, steady, low-pressure airflow is required. This style of compressor is commonly designated as a blower when used for conveyor applications.

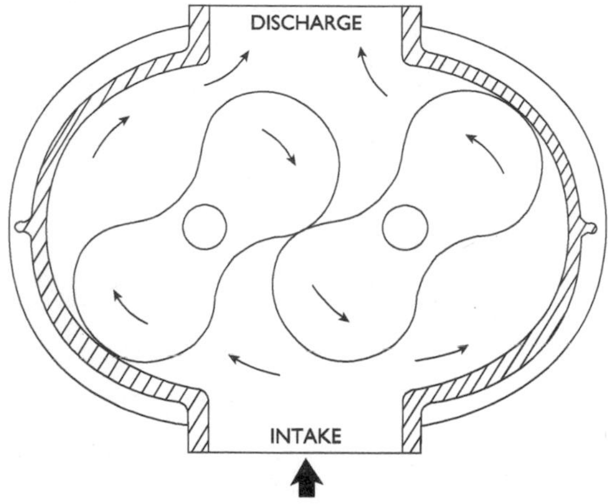

Figure 26-9 Two impellers called "lobes" rotate in opposite directions.

Another design of a rotary positive-displacement air compressor is shown in Figure 26-10. This design is classified as a screw cycloidal, or twist type, by various manufacturers. The movement of air through this compressor is axial, rather than radial as it is in the lobe-type rotary compressor. For this reason, this style of compressor is also classified as an axial-flow rotary air compressor. Because of its relatively simple design, its pulsation-free discharge, quiet operation, low discharge air temperature, etc., the use of this design compressor is increasing rapidly.

The principal components of a screw-type compressor include a housing (stator), two screws (rotor), bearings, and bearing supports. See Figure 26-10. The male rotor is gear-driven through speed-up gears by the motor. The male rotor has fewer lobes than

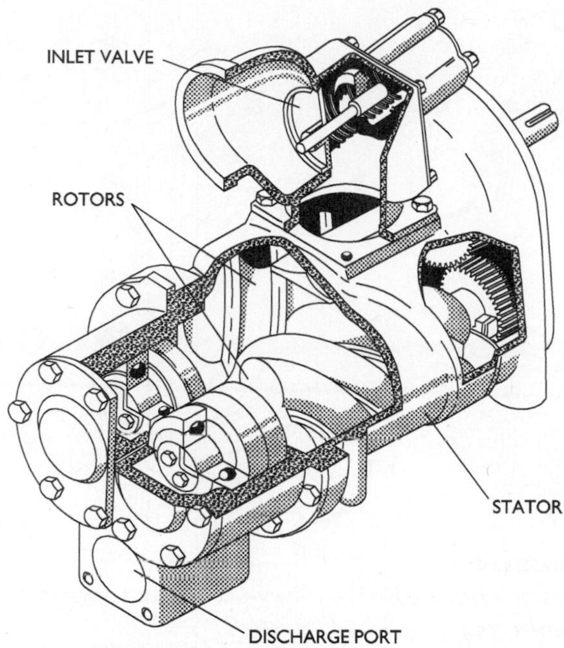

Figure 26-10 A cycloidal or twist-type rotary-positive air compressor. *(Courtesy Gardner Denver Inc.)*

the female, the relationship generally being four lobes on the male rotor and six on the female.

The principle of operation of the compressor is illustrated in Figure 26-11. It will be noted that the initial volume (displacement) of air is obtained by the trapping of air in pockets between the threads or lobes of the two rotors. As the rotors turn, the lobes mesh, reducing the volume in the pockets and thus compressing the trapped air. Compressed air leaves the compressor unit through a discharge port, which is located at the proper volume ratio to give the desired discharge pressure. An oil pump, built into the compressor, ensures lubrication of bearings, gears, and rotors. This oil also serves as a cooling agent to maintain low discharge air temperatures and helps to seal running clearances. As oil passes through the compressor, it mixes with the air being compressed and is discharged with the compressed air into the oil sump. Nearly all of the oil in the air drops out in the sump because of impingement and velocity change. The air then passes through a separator where most of the remaining oil is removed.

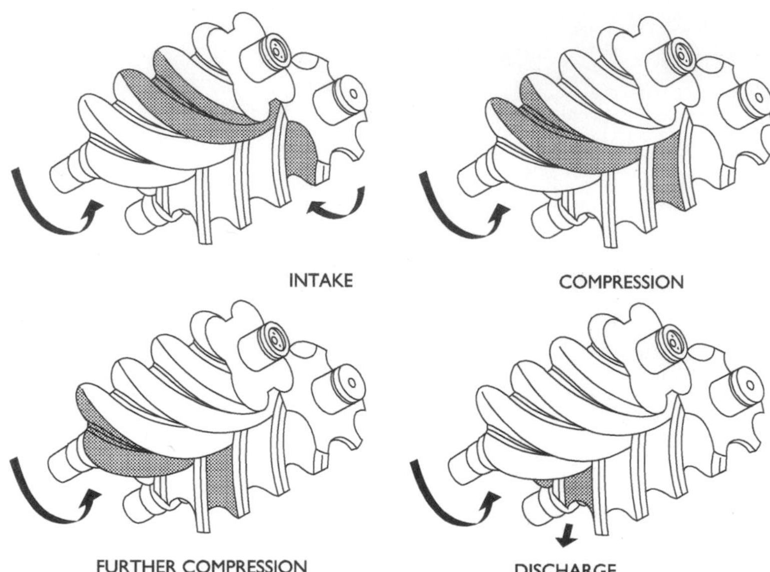

Figure 26-11 Rotors or lobes used in the rotary-positive air compressor. *(Courtesy Gardner Denver Inc.)*

The liquid-piston design compressor uses water to apply a compressing force. Figure 26-12 is a cutaway diagram that illustrates the principle of operation of this type of compressor. The actual disassembled parts are shown in Figure 26-13.

The rotor revolves freely without contact in an elliptical casing containing a liquid, usually water. The rotor is a circular casting consisting of a series of blades, which project from a cylindrical hub to form pockets or chambers. Ports are arranged at the bottom of each chamber. A cone-shaped casting consisting of two inlet and two outlet ports fits without contact into the rotor hub. Starting at the discharge side, the chambers are full of water. The water, turning with the rotor and constrained to follow the casing by centrifugal force, alternately recedes from, and is forced back into, the rotor twice in a revolution. As the water recedes from the rotor, it draws air from the inlet into the cone, through the cone inlet port, and into the rotor by means of the port in the bottom of the rotor chambers. When the water is forced back into the rotor by the converging casing, the air is discharged through the ports at the bottom of the rotor chambers. The air travels through the cone outlet ports and out the discharge.

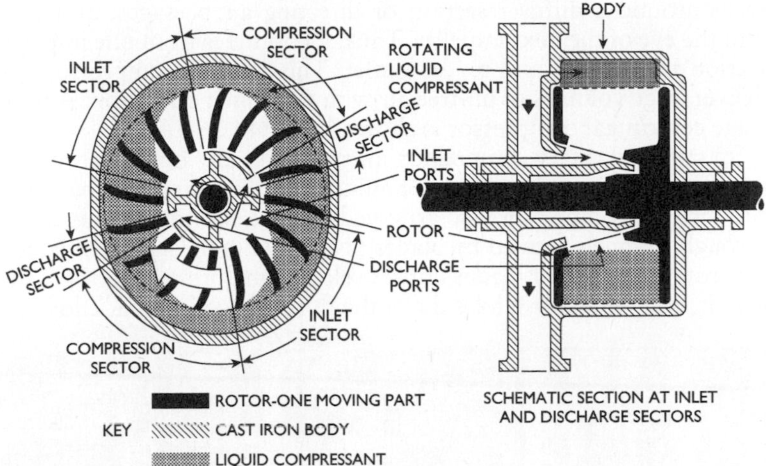

Figure 26-12 Internal view of the liquid-piston air compressor. *(Courtesy nash–elmo industries)*

Figure 26-13 The disassembled parts of the liquid-piston air compressor. *(Courtesy nash–elmo industries)*

A small amount of seal water must constantly be supplied. Most of the water stays in the compressor. Excess water is carried over with the air and is usually run to waste.

Dynamic-type air-compressor units depend principally on centrifugal force and high-speed movement of air to effect compression. In the centrifugal compressor, air moves at high velocity because of the rotational speed of the impeller. Air moves radially outward, turns around a diffuser section or directing air passages, and goes into the eye of the next impeller. Thus, it continues from one impeller section to the next and to the outlet. This kind of compressor can deliver large volumes of oil-free air with very quiet operation. A four-stage centrifugal compressor is shown in Figure 26-14.

In the axial rotary compressor, air travels axially from inlet to outlet. It is a high-speed compressor suitable for constant-volume applications. The compressor is so arranged that the air into the unit passes through channels formed by blades positioned in the casing or stator. The rotating vanes or blades impart velocity and pressure to the flowing air, and the stator blades direct the flow and control backflow.

Figure 26-14 A four-stage centrifugal air compressor.
Courtesy Ingersoll-Rand Co.

Horizontal Compressors

The horizontal-type reciprocating air compressor, so called because of the horizontal arrangement of the cylinder, has been widely used in industry for many years. The unit shown in Figure 26-15 is a single-stage, double-acting type that is belt-driven, usually at constant speed, by an electric motor. Air enters the compression chambers through the inlet port at the bottom of the cylinder and is discharged through the outlet directly above on top of the cylinder.

Figure 26-15 A typical horizontal reciprocating air compressor. *Courtesy Ingersoll-Rand Co.*

Vertical Compressors

Very similar to the horizontal compressor, the vertical style is designed primarily for use where space is at a premium. The unit shown in Figure 26-16 is also single-stage and double-acting; however, the

Figure 26-16 A typical vertical reciprocating air compressor. *Courtesy Ingersoll-Rand Co.*

complete unit is supported by the frame in a vertical position. Inlet and outlet ports are located identically in respect to the cylinder, requiring horizontal pipe connections on either side of the cylinder.

To generate air power for an industrial plant, more is needed than just an air compressor. A complete system includes an inlet filter and silencer, an aftercooler, a moisture separator, condensate trap, air receiver, motor, motor starter, and control. In addition, it may be necessary to have an air dryer and an automatic control system. To efficiently supply this equipment, engineered package units of matched components are manufactured. Figures 26-17 and 26-18 are examples of horizontal and vertical single-stage complete packaged units. The complete unit is factory assembled and tested, requiring only pipe and electrical connections.

Figure 26-17 A complete horizontal compressor for industrial use. *Courtesy Ingersoll-Rand Co.*

L-Frame Compressors

The L-frame cylinder arrangement is used for heavy-duty, two-stage, double-acting types of air compressors. Figure 26-19 shows a package unit incorporating a flange-mounted electric motor connected directly to the compressor shaft. The internal design and construction is illustrated in the cutaway view in Figure 26-20. An intercooler is built into the unit, and air flows from the first stage discharge, through the intercooler to the second stage inlet, through passages within the frame and cylinders.

The trend in industry is to the packaged air plant, which provides a single, pretested operating unit requiring only service connections.

These units are engineered and built to provide the maximum efficiency and dependability to suit the service requirements. They are manufactured in both positive-displacement and dynamic styles and in a great variety of designs within these two broad classifications. Examples of the range of complete packaged plants have already been mentioned. Figure 26-15 shows the horizontal positive-displacement single-stage reciprocating air-compressor package incorporating compressor, motor, starter, inlet filter, aftercooler, moisture separator, condensate trap, air receiver, wiring, safety devices, and controls. Figure 26-21 shows a "radial"-style packaged unit complete with controls and intercooler.

Figure 26-18 A complete vertical air compressor used in industrial plants.
Courtesy Ingersoll-Rand Co.

A rotary-type packaged unit is shown in Figure 26-22. This is a stationary rotary compressor unit with two meshing helical rotors to deliver air without pulsations—a self-contained unit incorporating all the components to make up a complete packaged air plant.

Figure 26-23A shows the two meshing helical rotors that interact to compress the air smoothly without pulsation. The male rotor, driven by a speed-increasing gear, imparts its energy to the entrapped air; the female rotor acts as a pressure container. The illustration in Figure 26-23B shows an end view with the male and female rotors in meshed position.

Figure 26-19 An L-frame compressor with a directly coupled motor. *Courtesy Ingersoll-Rand Co.*

Another rotary-style packaged unit is shown in Figure 26-24; this is a two-stage rotary compressor of the sliding-vane type directly coupled to an electric motor. Included are all components generally accepted as essential in a well-engineered air power plant.

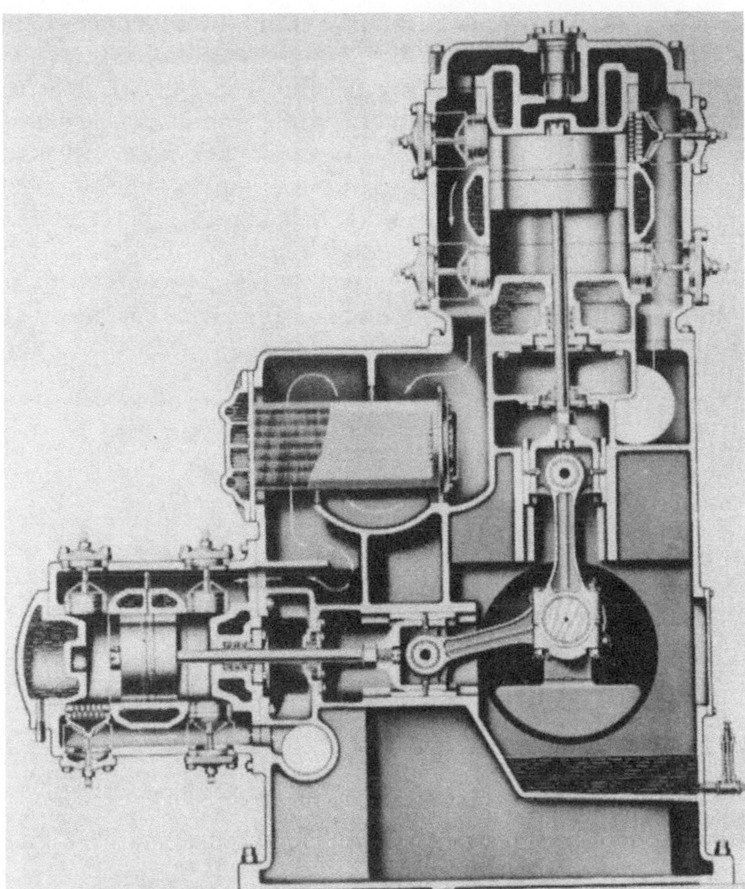

Figure 26-20 Cutaway view of an L-frame compressor.
Courtesy Ingersoll-Rand Co.

Centrifugal compressor packaged units are also furnished, as shown in Figure 26-14. The unit illustrated employs four stages of compression and incorporates intercoolers between stages and a built-in aftercooler. Centrifugal compressor units are outstanding for quiet, smooth operation and clean, oil-free air.

Reciprocating Compressor Construction

All reciprocating air compressors—horizontal, vertical, or other configuration in either single- or multiple-stage design—are constructed from certain basic sections that are relatively standard in general design. The major sections are the frame, distance piece with crosshead guide, and cylinder. Incorporated into these sections are the many components of varying design that provide the features of the specific compressor type.

Figure 26-21 A radial packaged unit ready for connection. *(Courtesy Gardner Denver Inc.)*

The compressor frame is the base or bed upon or around which the machine is built. It is usually a one-piece casting, well ribbed to ensure accurate and permanent alignment of running parts. The main bearings for the crankshaft are in the frame and support the heavy forces developed as the crank moves the rod and piston back and forth. The reciprocating on-off type of loading tends to cause vibration. The frame must be heavy enough to absorb a major portion of this vibration.

The distance piece is the intermediate section between the frame and the cylinder. The crosshead guide is usually contained in the distance piece, although in some styles it is an integral part of the frame casting. Supports for rod wipers are also part of the distance piece.

Figure 26-22 A typical rotary packaged unit completely assembled and shipped ready for connections.
Courtesy Ingersoll-Rand Co.

The cylinder section, either finned for air cooling or jacketed for water cooling, is mounted on the distance piece. The piston travels back and forth in the bore of the cylinder. The inboard head, through which the rod enters the cylinder, contains the stuffing box for rod sealing. The valves are also incorporated in the cylinder casting, as are the inlet and discharge pipe connections.

Figure 26-23 The meshing helical rotors.
Courtesy Ingersoll-Rand Co.

Figure 26-24 Two-stage rotary compressor packaged unit.
Courtesy Ingersoll-Rand Co.

Crankshaft

The rotary motion provided by the driving power unit is converted into reciprocating motion by the crankshaft. Crankshafts are generally forged of steel in one solid piece. The stroke of the piston in the compressor cylinder is determined by the offset distance of the crank on the shaft. The piston stroke is twice the crank arm, or crank offset. The most widely used style of crankshaft is the center-throw design with the crank section in the center, which provides bearing support on both sides of the crank. Crankshafts of this design are made for single-throw and multiple-throw applications to operate single- and multiple-cylinder compressors. A typical single-throw crankshaft with counterweights is shown in Figure 26-25. The function of the counterweights is to act against the centrifugal forces tending to distort the crankshaft and to balance the compression forces exerted on the piston.

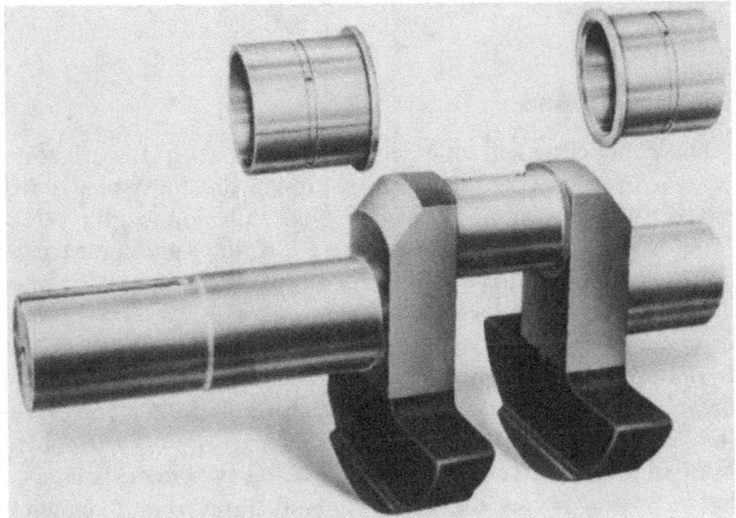

Figure 26-25 Typical crankshaft and crankshaft bearings.
Courtesy Ingersoll-Rand Co.

Connecting Rod

The connecting rod connects the crankshaft to the crosshead, transforming the rotating reciprocation at the crank to a straight-line reciprocation at the crosshead. Generally, connecting rods are steel forgings of I or similar section for rigidity and low weight. Some are designed to provide a means of adjustment to compensate for wear. A common design uses liners or shims on the crank pin and a wedge block with adjusting screw on the crosshead end. By tightening the screw, the wedge is drawn in against a mating surface to reduce the bore size and reduce the running clearance. The shims between the cap and the rod on the crank end may be changed for adjustment purposes. Floating-type bearing liners, as shown in Figure 26-26, are used by Ingersoll-Rand in the connecting rods of many of their reciprocating compressors. No adjustment is necessary, as the bearings are full-floating and pressure-lubricated, free to rotate both in the bearing journal and within the bearing housing. Rubbing is reduced, and bearing wear is distributed around the entire bearing surface, both inside and out.

Crosshead

The crosshead is a casting or forging that acts as a sliding joint between the connecting rod and the piston rod. In small air compressors, no

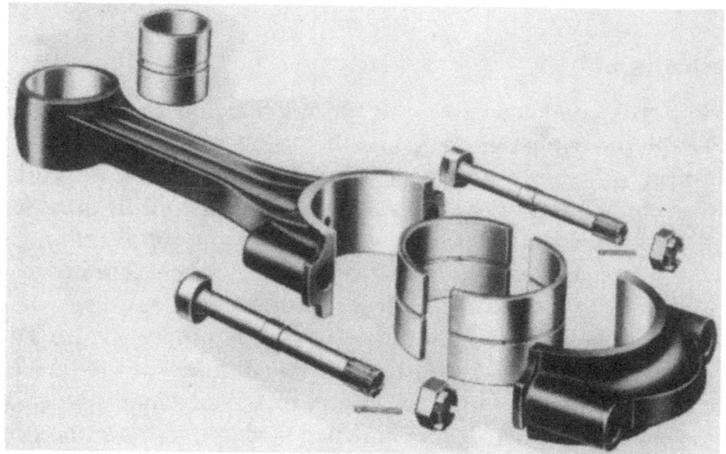

Figure 26-26 Typical connecting rod and bearing assembly.
Courtesy Ingersoll-Rand Co.

crosshead is necessary because the connecting rod is a direct connection between crankshaft and piston. In most large compressors, this is not possible. A shaft or rod is rigidly connected to the piston and extended out to meet the connecting rod at the crosshead. The wearing surfaces of the crosshead are usually cylindrical in shape so that they may be finished more easily. These finished surfaces fit against the crosshead guide surfaces, which are either in the compressor frame or the distance piece. The crosshead guides are also cylindrical in shape to mate the crosshead and also for easy machining. The guides are made slightly shorter than the piston stroke so that they will wear evenly and shoulders will be avoided as wear progresses.

Although provisions are made in many compressors to take up crosshead wear by adjusting wedge-shaped shoes, the trend is to replaceable wearing surfaces. Figure 26-27 shows a crosshead with special alloy wearing surfaces and floating bushings.

Pistons

The compression of air in an air compressor is accomplished as the air trapped between the end of the cylinder and the end of the piston is reduced in volume by the force of the advancing piston. The piston material and construction must be capable of withstanding this continual application of force. In the low-pressure stage of two-stage compressors, aluminum pistons are often used because of

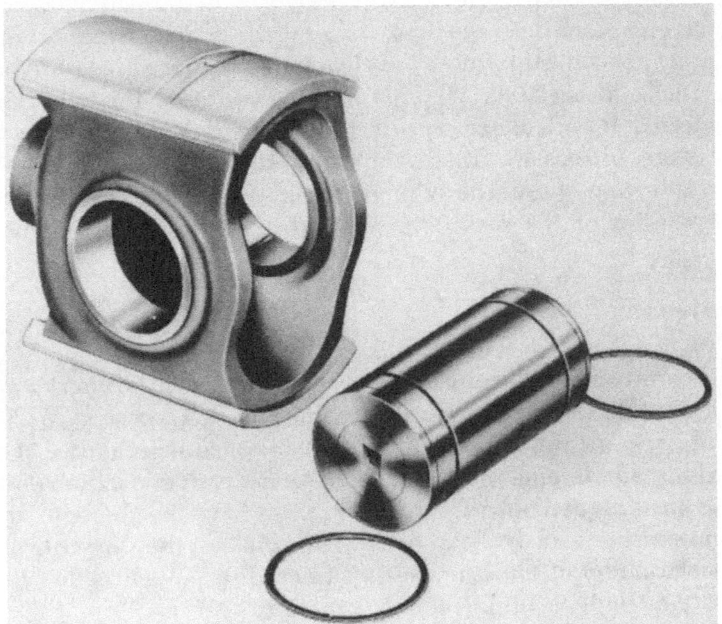

Figure 26-27 Typical crosshead assembly.
Courtesy Ingersoll-Rand Co.

their low weight. A stronger cast-iron or alloy-steel piston is used for high-pressure stages when greater strength is required. Pistons are manufactured in a great variety of designs, depending on the compressor type and the service application.

Generally in the smaller compressors with a connecting rod link directly from the crankshaft to the piston, a long-skirted piston called a trunk piston is used. This piston is similar to that used in internal combustion engines; also, the applied forces are similar. The principal reason for using this style is that the extra length provided by the skirt helps to evenly distribute the side thrust of the piston in the cylinder, due to the rotational throw effect from the crankshaft. In crosshead compressors, the pistons are made as short as possible, consistent with adequate strength and support surface. In some lightweight high-performance compressors, the pistons are extremely light to reduce the inertia loads to a minimum. Any weight reduction that can help to lower inertia loads means less unbalanced force and smoother operation.

Pistons must slide freely back and forth in the cylinder; therefore, clearance between the cylinder wall and the piston surface is required. Although the amount of clearance in terms of measurement is small—about 0.001in. per inch of bore diameter—this gap is large in terms of air leakage past the piston. To control or seal the air leakage, pistons are fitted with piston rings. Pistons are grooved for insertion of the sealing rings and may have two to eight rings, depending on the service for which the compressor is designed.

Valves

Although each part of any machine is important to the operation of the complete unit, there is usually some part that is critical. With air compressors, this critical part is the valves. Although some of the lightest and smallest of the machine's parts, valves perform one of the most rugged operations. Valves must have the ability to function satisfactorily for long periods of time, so they are critical to the economy of the compressor's operation as well as the continuity of both air supply and air pressure.

Air-compressor valves are one-way, or check-type, valves. They are placed in the cylinder to permit air to flow in one direction, either into or out of the cylinder. Old-style valve designs were of the positive-acting, mechanically operated type, using valve lifting mechanisms, timing mechanisms, etc. Today, all air compressors utilize some type of automatic valve, operated by the action of the air it controls. They are opened by the difference in pressure across the valve; no positive mechanical device is involved. Closing of the valves, however, is usually assisted by some form of spring action.

Valve action must be very rapid because valves open and close once for each revolution of the crankshaft. The valves in a compressor turning at 600 rpm, therefore, have only 1⁄10 of a second to open, let the air pass through, and close. To do this, valves must be free-acting without slamming or flutter. They must be tight under severe pressure and temperature conditions, and they must be durable under many kinds of abuse. Valves in a continuously operating compressor at 600 rpm must open and close over 25,000,000 times in a month.

Four basic valve designs are in general use in various types of air compressors. These may be classified as finger, leaf, channel, and annular ring. Within these classifications, there are a variety of detail differences in design to suit operating speeds, capacities, etc. For example, annular-ring valve assemblies may have one or a number of separate concentric rings or valves. Examples of these valve designs are shown in Figures 26-28 through 26-32. A finger-valve assembly used on small compressors in the ½- to 5-hp range is shown in Figure

26-28. These are quick-acting valves of heat-treated stainless steel. The concentric-ring valve design shown in Figure 26-29 is used on larger compressors. This design features large opening area for longer life and is also made of stainless steel.

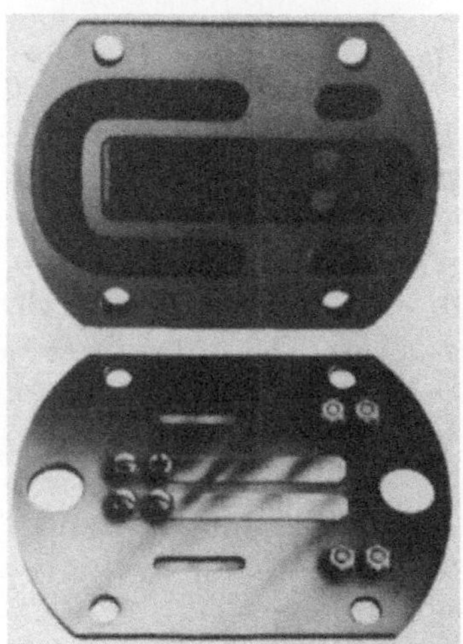

Figure 26-28 Finger-valve assembly used on small compressors.
Courtesy Ingersoll-Rand Co.

Figure 26-29 The concentric-ring valve used on medium-size compressors.
Courtesy Ingersoll-Rand Co.

The channel valve is actually a combination of several small valves. Each channel valve with its bowed leaf spring operates individually, closing over a corresponding slot-shaped port. With this valve, a cushioning is effected when a small volume of air is trapped between the channel and its spring, causing the channel to float silently to a stop. This positive cushioning action, prior to the critical point of valve impact with the stop plate, reduces valve breakage. Figure 26-30 shows a channel valve assembly with the stop plate tilted up to show the channels, springs, guides, etc., assembled on the valve seat.

Figure 26-30 The channel valve assembly.
Courtesy Ingersoll-Rand Co.

The exploded view in Figure 26-31 shows the various parts of an intake and a discharge channel valve in their relative positions. Each channel with its bowed leaf spring operates individually, closing over a corresponding port. The seal plate is made of stainless steel and fits on top of the valve seat. The slots in the seat plate are made slightly larger than those in the valve seat. The seat plate is so designed that it can be turned over for continued use, thus doubling its life, or it can be replaced by a new one. On top of the seat plate are the channels and the valve guides. The guides are used to retain the channels in their proper position above the seat plate. The valve springs fit within the channels and the stop plate on top of the assembly to retain the valve springs with the correct applied pressure. Flat-head machine screws are used to hold the assembly together securely.

The cutaway views in Figure 26-32 show a single-channel valve with the valve guide in position at one end and a segment of the

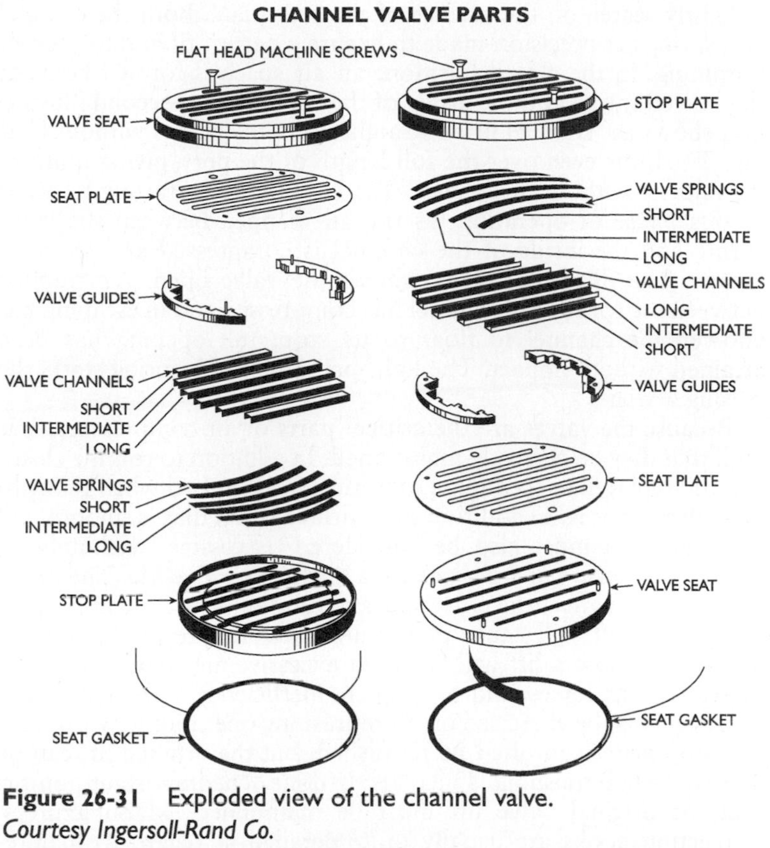

CHANNEL VALVE PARTS

FLAT HEAD MACHINE SCREWS
VALVE SEAT
STOP PLATE
SEAT PLATE
VALVE SPRINGS
SHORT
INTERMEDIATE
LONG
VALVE GUIDES
VALVE CHANNELS
LONG
INTERMEDIATE
SHORT
VALVE CHANNELS
SHORT
INTERMEDIATE
LONG
VALVE GUIDES
VALVE SPRINGS
SHORT
INTERMEDIATE
LONG
SEAT PLATE
STOP PLATE
VALVE SEAT
SEAT GASKET
SEAT GASKET

Figure 26-31 Exploded view of the channel valve.
Courtesy Ingersoll-Rand Co.

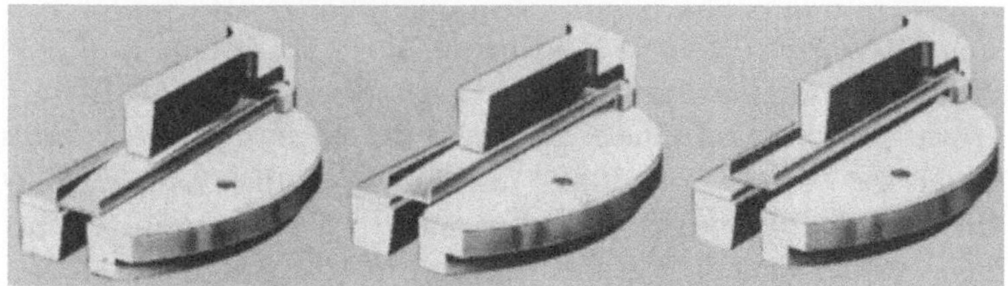

Figure 26-32 A cutaway view of a single-channel valve assembly.
Courtesy Ingersoll-Rand Co.

stop plate in position. The three illustrations show various stages of valve operation. In the first view, the valve is closed and the channel is tightly seated on the surface of the seat plate. Both the channel and spring are precision-made to ensure a perfect fit and to prevent slamming. In the closed position, an air space is formed between the bowed spring and the inside of the channel. The second illustration shows the channel lifted straight up in the guides without flexing. The lift is even over the full length of the port, giving uniform air velocity without turbulence. The cushioning is starting to occur at this phase of operation, as the air trapped between the valve spring and the inside of the channel is compressed and starts to escape. The third illustration shows the valve open. Air trapped between the spring and channel has compressed and in escaping has allowed the channel to float to its stop; full opening has been attained without impact. The light pressure of the spring starts the closing action.

Because the valves are the critical parts of air compressors, it is vital that they be correctly maintained. In addition to routine cleaning and replacement, extreme care must be exercised when refinishing valve surfaces. In addition to a surface finish that is smooth and flat, valve lift must also be considered. Excessive machining or grinding of valve parts may cause a change in valve life. The lift of the valves can affect the efficiency and durability of the compressor. Too high a lift will cause premature valve failure due to impact fatigue. Too low a lift will result in excessive air velocity through the valve, high losses, and consequent inefficiency. Care and understanding must be exercised on all refinishing operations. Worn seats and stop plates can often be refinished, but the original lift can be altered if the refinishing is incorrectly done. The important point is that the original valve lift must be maintained. Manufacturers' instruction books are usually quite detailed in regard to maintenance and refinishing of air compressor valves. Their instructions should be carefully followed.

Nonlubricated Compressors

The nonlubricated air compressor was developed for applications where the air must be completely free of oil, mist, and vapors. In such cases, oil lubrication cylinder and rod packings must be eliminated. When this is done, there are three areas where high temperature and metal-to-metal sliding contact can cause serious trouble. These are the valves, the pistons and piston rings, and the piston-rod packing. Metal-to-metal sliding contact results in rapid wear without oil lubrication. To reduce this wear, compressors may be fitted with carbon or Teflon piston rings. Carbon rings depend to some degree on the condensation of moisture in the air for part of their lubrication. Self-lubricating materials may also be used in the valve assemblies and the rod packing to prevent any metal-to-metal rubbing contact at these points.

Highly polished piston rods and cylinder walls are essential for nonlube service. Where clean air is required, the design of the compressor may be altered. The distance piece and piston rod may be lengthened so that the cylinder packing and the crankcase oil seal do not touch the same part of the piston rod. Thus, no part of the rod contacted by the oil will enter the cylinder stuffing box containing self-lubricating packing material. Another major alteration used on large cylinder compressors involves supporting the piston free of the cylinder wall by means of tail-rod construction. The tail rod is supported at the outer end of the cylinder by a second crosshead support. Special precautions are necessary to prevent the oil lubrication to the second crosshead from getting on the rod and working into the cylinder. Two serious disadvantages of this construction are that a second stuffing box is required and that there is a reduction in capacity because of the cross-sectional area of the rod, which is usually of heavier and stiffer construction to limit deflection.

Air Compressor Troubleshooting

The ability to accurately troubleshoot mechanical equipment is often the difference between an outstanding mechanic and a mediocre one. This is particularly true in air compressor maintenance because minor troubles, if recognized and corrected, can avert major problems and breakdowns. There are various symptoms for which corrective action may be taken. Because in most cases the remedy is apparent once the cause of the trouble has been determined, the following troubleshooting guide is set up in terms of problems and causes.

Delivery Below Rated Capacity

- Excessive system leakage
- Restriction of intake
- Intake filters clogged
- Broken or worn valves
- Valves not seated in cylinder
- Leaking gaskets
- Unloader defective or improperly set
- Worn or broken piston rings

- Cylinder or piston worn or scored
- Leaking rod packing or safety valve
- Slipping belts
- Speed too low
- Excessive system demand

Below-Normal Discharge Pressure

- Excessive system leakage
- Restriction of intake
- Intake filter clogged
- Broken or worn valves
- Valves not seated in cylinder
- Leaking gaskets
- Unloader defective or improperly set
- Worn or broken piston rings
- Cylinder or piston worn or scored
- Leaking rod packing or safety valve
- Slipping belts
- Speed too low
- Excessive system demand

Noisy or Knocking Compressor

- Valves worn or broken
- Piston rings worn, stuck, or broken
- Cylinder or piston worn
- Pulley or flywheel loose
- Anchor bolts loose or foundation uneven
- Piston-to-head clearance too small
- Excessive crankshaft endplay
- Piston or piston nut loose
- Bearings loose
- Liquid in cylinder
- Inadequate lubrication

Extra-Long Operating Cycle

- Excessive system demands
- Intake filter clogged
- Valve worn or broken
- Valves not seated in cylinder
- Leaking gaskets
- Defective unloader or control
- Piston rings worn, stuck, or broken
- Excessive system leakage

Motor Overheats

- Motor too small, wrong connections, low voltage
- Excessive number of starts, belts too tight, speed too high
- High ambient temperature, poor ventilation
- Inadequate lubrication or incorrect lubricant
- Unloader setting incorrect, pressure setting above rating
- Compressor mechanical trouble, valves, piston, packings, gaskets

Compressor Overheats

- Pressure setting above rating
- Restricted intake or clogged filter
- Broken or worn internal parts, piston, rings, cylinder, valves
- Insufficient water or clogged jackets
- Inlet water temperature too high
- Speed too high, belts misaligned
- Bearings need adjustment or renewal
- Inadequate or incorrect lubricant
- Poor ventilation
- Wrong rotation

Excessive Valve Wear and Breakage

- Insufficient or incorrect lubrication
- Liquid carryover

- Dirt or foreign material entering cylinder
- Incorrect assembly or installation
- Springs broken
- New parts combined with worn parts

Excessive Compressor Vibration

- Speed too high
- Belt misalignment
- Pulley or flywheel loose
- Foundation bolts loose, base inadequately supported
- Improperly piped or piping unsupported
- Pressure setting above rating

Excessive Wear: Piston and Cylinder

- Piston rings worn, stuck, or broken
- Cylinder or piston worn or scored
- Inadequate or incorrect lubricant
- Dirt or foreign material entering cylinder
- Defective air filter

Excessive Wear: Rod and Packings

- Inadequate or incorrect lubricant
- Defective air filter
- Dirt or foreign material reaching wearing surfaces
- Packing rings worn, stuck, or broken
- Rod worn, pitted, or scored

Excessive Oil Pumping

- Intake filter clogged or intake restricted
- Piston rings worn, scored, or broken
- Cylinder or piston worn or scored
- Too much or wrong type of lubricant
- Oil pressure too high
- Piston ring gaps not staggered or ring drain holes plugged

Low Oil Pressure

- Defective gauge
- Insufficient or wrong type of lubricant
- Defective oil-relief valve, pipe leaks
- Pump suction air leaks
- Worn or defective oil pump
- Excessive clearances

Low Intercooler Pressure

- Excessive system demand or system leakage
- Restricted intake or clogged intake filter
- Valve worn or broken in low-pressure cylinder
- Piston rings worn, scored, or broken in low-pressure cylinder
- Cylinder or piston worn or scored in low-pressure cylinder
- Unloader defective or incorrectly set

High Discharge Temperature

- Discharge pressure setting above rating
- Intake filter clogged or intake restricted
- Valves worn, broken, or not seated in cylinder
- Unloader or control defective
- Piston rings worn, stuck, or broken
- Cylinder or piston worn or scored
- Excessive speed
- Too much or incorrect lubricant
- Poor ventilation, airflow blocked
- Insufficient cooling water, jacket clogged
- Water temperature too high, cooling lines restricted

High Intercooler Pressure

- Discharge pressure set above rating
- Valves worn, broken, or not seated in high-pressure cylinder
- Cylinder or piston worn or scored in high-pressure cylinder
- Unloader or controls defective or improperly set

- Intercooler passage clogged
- Water insufficient, temperature too high, or passages clogged

High Outlet Water Temperature

- Discharge pressure set above rating
- Dirty cylinder head or intercooler
- Water temperature too high, water jackets or intercooler dirty
- Air discharge temperature too high, intercooler pressure too high
- Speed too high

Chapter 27

Pipe Fittings

The work of installing and maintaining pipe systems is called pipe fitting. Those who work as pipe fitters, steam fitters, and plumbers specialize in specific types of pipe-fitting work. Those in other occupations, such as millwrights and mechanics, also perform pipe fitting as part of their duties. As with all mechanical operations, an understanding of basic principles and procedures is essential to satisfactory performance of pipe-fitting work.

Pipe is manufactured from metals such as cast iron, wrought iron, steel, brass, copper, and lead. Various kinds of plastic pipe are also in wide use, as well as special materials such as glass and cement. Steel, wrought iron, brass, and stainless steel are probably the most commonly used materials and are available commercially in what are known as standard pipe sizes. Standard pipe is made in different wall thicknesses and to standardized dimensions. The actual outside diameter in these various wall thicknesses does not change; the increased thickness of the wall decreases the internal diameter. Except for the large sizes, the size of pipe is specified in terms of the nominal inside diameter. The actual inside diameter of standard pipe is usually greater than the nominal, especially in the smaller sizes. For example, the inside diameter of ⅜-in. pipe may vary from 0.423 in. for a heavy-wall, extra-strong pipe to as large as 0.545 inch for a light-wall stainless-steel pipe.

Pipe has, for many years, been specified as standard, extra strong, and double extra strong, and while this terminology is not used in the ANSI B36.10 standard, it is still in common use throughout industry. In the same nominal size, the standard, extra strong, and double extra strong all have the same outside diameter. The added wall thickness of the extra- and double-extra-strong (XX) pipe is on the inside. Thus, the outside diameter for 1-inch pipe in all three weights is 1.315 in. The inside diameter of the standard 1-in. pipe is 1.049 in., of the 1-in. extra strong is 0.957 in., and of the 1-in. double extra strong is 0.599 in.

The ANSI B36-10 standard gives a means of specifying wall thicknesses by a series of "schedule numbers," which indicate the approximate strength of the pipe. The lower numbers have the least strength, and the strength increases as the number increases. The schedule numbers indicate approximate values for the expression

$$1000 \times \frac{P}{S}$$

where P is the pressure and S is the allowable stress. This permits an approximation of the wall thickness if the service pressure and the value of the allowable stress for the material and service conditions are known. Although this is of primary concern to the designer, the mechanic should also have an understanding of schedule numbers.

Although the ANSI B36.10 standard is for steel and wrought-iron pipe, brass, copper, stainless steel, and aluminum pipe are made to the same nominal diameters. Table 27-1 gives pipe dimensions for those schedule numbers that are in common use, listing the nominal size and actual outside and inside diameters.

Pipe Threads

The method used to connect pipe depends on the materials, service, and work practices. Steel, wrought iron, and brass pipe, in the small and medium sizes, are normally threaded and made up with screwed connections. Therefore, the bulk of the pipe-fitting work performed by millwrights and mechanics is joining with screwed connections. The American Standard pipe thread (formerly known as the National Standard and the American Briggs Standard) is the standard thread for pipe connections. The form of the American Standard pipe thread is similar to the United Screw Thread form in that it has an angle of 60° and a slight flat at crest and root. The ANSI B2.1 standard for the American Standard pipe threads (except Dryseal) provides two types of pipe thread: tapered and straight. The type normally used for pipe connections employs a taper internal thread and taper external thread. The threads are cut on a taper of $\frac{1}{16}$ in. per inch, measured on the diameter, as illustrated in Figure 27-1.

Table 27-1 Commercial Pipe Dimensions—Inside Diameter

Nominal Size	Outside Diameter	Stainless Steel Only Schedule 5	Stainless Steel Only Schedule 10	Standard Schedule 40	Extra Strong Schedule 80	Schedule 160	Double Extra Strong
1/8	.405		.347	.269	.215		
1/4	.540		.410	.364	.302		
3/8	.675		.545	.493	.423		
1/2	.840		.674	.622	.546	.466	.252
3/4	1.050	.920	.884	.824	.742	.614	.434
1	1.315	1.185	1.097	1.049	.957	.815	.599
1 1/4	1.660	1.530	1.442	1.380	1.278	1.160	.896
1 1/2	1.900	1.770	1.682	1.610	1.500	1.338	1.100
2	2.375	2.245	2.157	2.067	1.939	1.689	1.503
2 1/2	2.875	2.709	2.635	2.469	2.323	1.885	1.771
3	3.500	3.334	3.260	3.068	2.900	2.625	2.300
3 1/2	4.000	3.834	3.760	3.548	3.364		
4	4.500	4.334	4.260	4.026	3.826	3.438	3.152
5	5.562	5.345	5.295	5.047	4.813	4.313	4.063
6	6.625	6.407	6.357	6.065	5.761	6.187	4.897
8	8.625	8.407	8.329	7.981	7.625	6.813	6.875

Thread Designations

In addition to the widely used taper pipe thread, there are several other types of standard pipe threads for specific applications. All American Standard pipe threads are designated by specifying, in sequence, the nominal size, number of threads per inch, and the symbols for thread series.

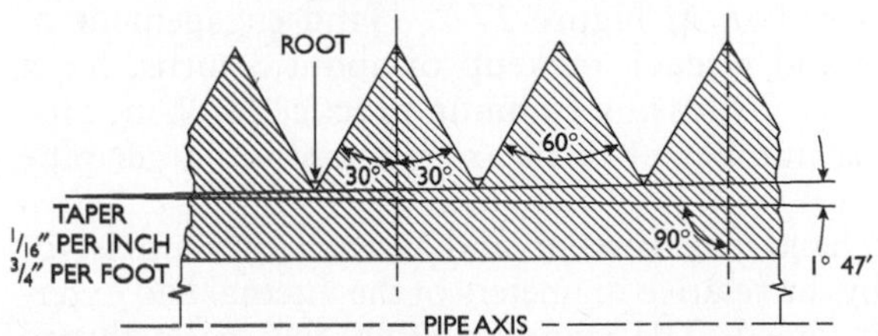

Figure 27-1 Standard taper pipe thread form.

Nominal Size	No. of Threads	Symbols
3/8	18	N.P.T

Each of the letters in the symbols has the following significance:

N—American (National) Standard
P—Pipe
T—Taper
C—Coupling
S—Straight
L—Locknut
R—Railing Fitting
M—Mechanical

The following are examples of thread designation:

3/8—18—NPT Am. Std. Taper Pipe Thread
3/8—18—NPSC Am. Std. Straight Coupling Pipe Thread
1—11½—NPTR Am. Std. Taper Railing Pipe Thread
½—14—NPSM Am. Std. Straight Mechanical Pipe Thread
1—11½—NPSL Am. Std. Straight Locknut Pipe Thread

Taper-Thread Connections

The joining or "makeup" of a taper-threaded pipe connection is performed in two distinct operations termed "hand engagement" and "wrench makeup." Hand engagement is the amount the connection can be turned freely by hand. Wrench makeup is the additional turning required to intimately join the tapered internal and external threads as shown in Figure 27-2. Hand engagement of about 3 or 4 turns and wrench makeup of about 3 turns, for a total of approximately 7 turns, is common practice for 1-in. pipe and under. A greater number of turns are required on larger pipe sizes.

The amount of hand engagement and wrench makeup is, of course, controlled by the relative diameters of the internal and external members being joined. The standard establishes basic thread dimensions and values for hand engagement and wrench makeup. These are listed in Table 27-2, which gives the nearest fractional dimensions rather than the decimals given in the standard.

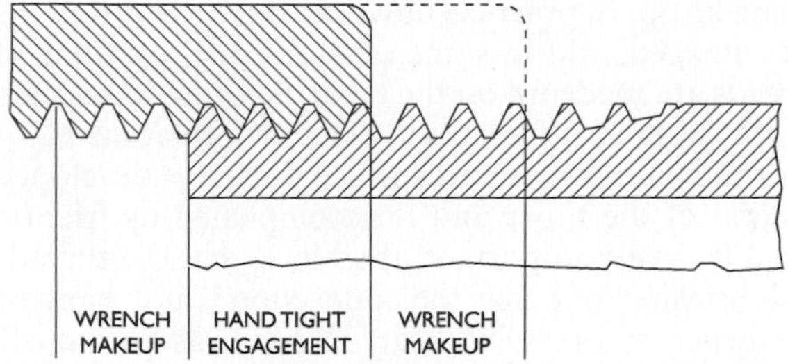

Figure 27-2 Hand engagement and wrench makeup.

Because these are basic dimensions and the allowable variation in the commercial product is one turn, large or small, from the basic dimension, there can be considerable variation in makeup of pipe and fittings threaded to commercial tolerances. For example, mating of the extremes could result in a connection as much as two turns short or, in the other direction, as much as two turns long. Although the probability of this occurring is remote, it points up the fact that makeup of threaded pipe connections requires knowledge and care on the part of the mechanic to obtain accurate results.

Table 27-2 Taper Pipe Thread Makeup

Nominal Pipe Size	Threads per Inch	Hand Tight Dimension	Hand Tight Threads	Wrench Makeup Dimension	Wrench Makeup Threads	Total Makeup Dimension	Total Makeup Threads
1/8	27	3/16	4 1/2	3/32	2 1/2	9/32	7
1/4	18	7/32	4	3/16	3	13/32	7
3/8	18	1/4	4 1/2	3/16	3	7/16	7 1/2
1/2	14	5/16	4 1/2	7/32	3	17/32	7 1/2
3/4	14	5/16	4 1/2	7/32	3	17/32	7 1/2
1	11 1/2	3/8	4 1/2	1/4	3 1/4	5/8	7 3/4
1 1/4	11 1/2	13/32	4 3/4	9/32	3 1/4	11/16	8
1 1/2	11 1/2	13/32	4 3/4	9/32	3 1/4	11/16	8
2	11 1/2	7/16	5	1/4	3	11/16	8
2 1/2	8	11/16	5 1/2	3/8	3	1 1/16	8 1/2
3	8	3/4	6	3/8	3	1 1/8	9

The makeup of taper-threaded pipe connection also requires care and proper procedure to obtain a pressure-tight assembly. As both the internal and external threads have the same form, theoretically all surfaces should make up. In practice, however, there are variations in form as well as damaged, and in some cases missing, portions of threads. As the threads are made up on the taper, deformation occurs and matching on the surfaces takes place. This deformation is, of course, the result of the high forces and surface pressures developed by the wedging action of the taper and is accompanied by friction and heat generation. It is very important, therefore, that the threads be clean and well lubricated and that the connection is not screwed up fast enough to generate excessive heat. Friction, as previously stated, produces heat, which in turn causes expansion. As the cross-sectional area of pipe and fitting are not equal, expansion and contraction on cooling are not equal. Therefore, a connection screwed up fast enough to generate heat may be tight when initially made up but may leak under pressure when cool.

Many substances are available for use as pipe-joint lubricants, and these are generally referred to as "dope." The prime purpose of a pipe dope is to provide lubrication and allow thread surfaces to deform and mate without galling and seizing. In practice, however, the dope does help plug openings resulting from improper threads and in some cases acts as a cement to prevent leakage of solvent and other hard-to-seal liquids. The important points, therefore, in makeup of tight pipe connections are good-quality threads, clean threads, proper dope for the application, and slow final makeup to avoid heat generation.

Pipe Measurement

Pipe systems are assemblies of pipes, fittings, and accessories, all joined by pipe connections. They are represented by two general systems of pipe drawings. These may be described as scale layout drawings and diagrammatic pipe drawings. Scale layout is principally used for large piping or critical work such as assemblies cut and threaded at one location for installation at some removed location. On small-scale drawings, sketches, etc., the diagrammatic system is used. In this system, a line is used to represent pipe, and fittings are shown by symbol. A typical example of scale layout is shown in Figure 27-3, and diagrammatic drawing is shown in Figure 27-4.

In either case, the dimensions on pipe drawings are principally location dimensions, all of which are made to centerlines. Actual pipe lengths are not specified; therefore, the mechanic must determine the length of pipe required to give the desired center-to-center dimension. When calculating pipe lengths for threaded connections, allowances must be made for the length of the fitting and the distance the threaded pipe is screwed into the fitting. The method of doing this is to subtract an amount called the "takeout" from the

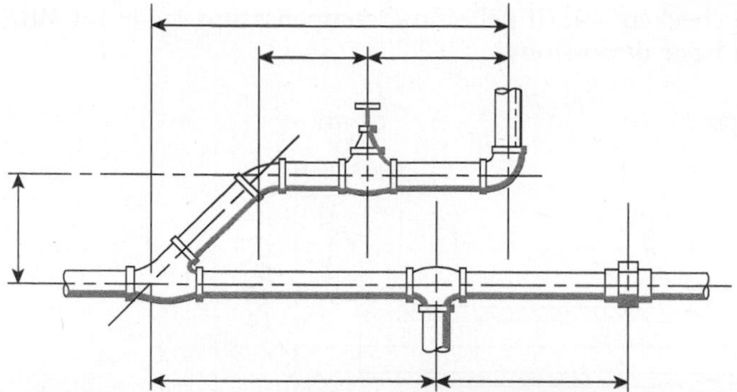

Figure 27-3 A scale layout of pipe and fittings.

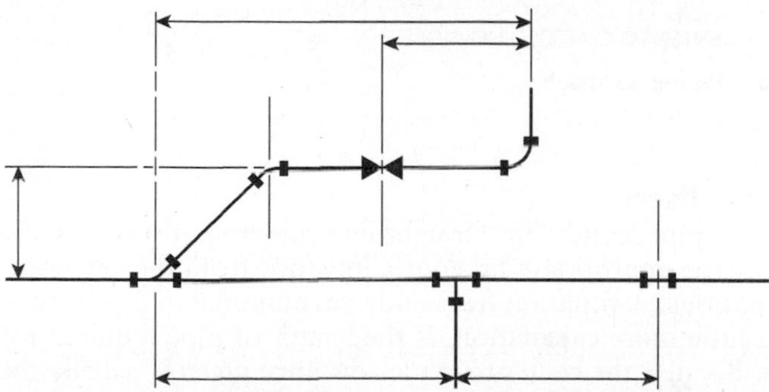

Figure 27-4 A diagrammatic drawing representing pipe and fittings.

center-to-center dimension. The relationship of the takeout to other pipe connection distances, termed *makeup*, *center-to-center*, and *end-to-end*, is illustrated in Figure 27-5.

To determine the actual pipe length (end-to-end), the takeout is subtracted from the center-to-center dimension. Many mechanics memorize takeout allowances for the various sizes and types of fittings they frequently use. Another common practice is to measure each fitting and figure its takeout allowance by subtracting the makeup for the pipe size from the center-to-face dimension of the fitting.

Standard takeout dimensions, as shown in Table 27-3, are calculated from basic thread and fitting dimensions and may be used for this purpose. However, because the commercial product may vary as much as one turn, large or small, these tables must be used with judgment. Where accurate pipe connections are required, materials should be checked and, if necessary, compensation made for variance from basic dimensions.

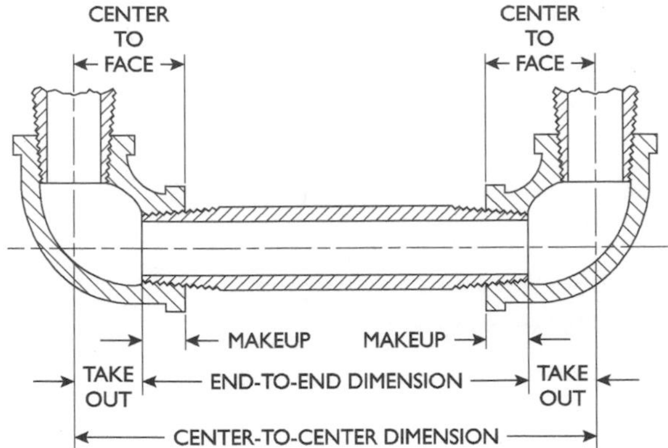

Figure 27-5 Piping terms.

45-Degree Offsets

Calculation of pipe lengths for straight-line runs requires simply the deduction of the appropriate takeout allowance from the center-to-center dimensions. A situation frequently encountered in pipe fitting, requiring a little more calculation, is the length of pipe required for 45° offsets. Because the center-to-center distance given is usually the "run" or the "offset" dimension (see Figure 27-6), mechanics must calculate the "travel" dimension before they can deduct the appropriate takeout allowance.

The offset and the run dimensions for a 45-degree offset are equal, as shown in Figure 27-6. The travel dimension, therefore, is the diagonal of a square or the distance across corners of a square, which is determined by multiplying the distance across flats by 1.414. The problem may also be considered a simple right-angle triangle, with the run and offset being the sides and the travel the hypotenuse. In this case, the hypotenuse is found by multiplying the side by the cosecant of the angle. As the cosecant of 45° is 1.414, the travel is found by multiplying the offset or the run by 1.414. This may be stated in a simple equation as follows:

$$\text{Travel} = \text{Offset} \times 1.414$$
$$\text{Travel} = \text{Run} \times 1.414$$

There may also be occasions where the travel is known and the dimension of the offset and run may be wanted. Again, this may be done by substitution of values into a simple right-angle equation. In this case, the length of the side may be found by multiplying the hypotenuse by the cosine of 45°, which is the travel multiplied by 0.707. This may also be stated in a simple equation as follows:

$$\text{Offset} = \text{Travel} \times 0.707$$
$$\text{Run} = \text{Travel} \times 0.707$$

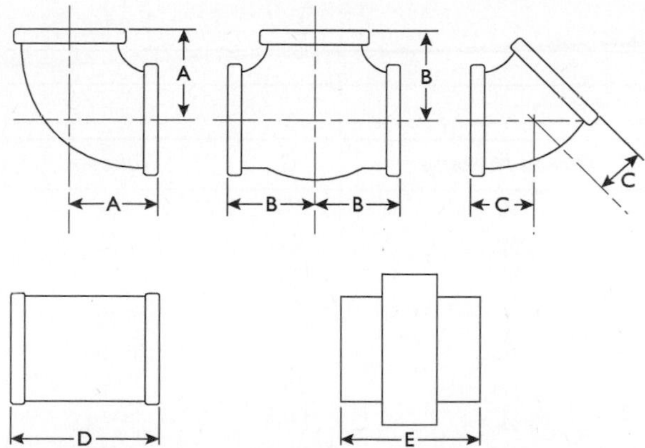

Table 27-3 Takeout Dimensions

90° Elbow

Pipe Size	*Thread Makeup*	*A*	*Takeout*
1/8	1/4	11/16	7/16
1/4	3/8	13/16	7/16
3/8	3/8	15/16	9/16
1/2	1/2	1 1/8	5/8
3/4	9/16	1 5/16	3/4
1	9/16	1 1/2	7/8
1 1/4	5/8	1 3/4	1 1/8
1 1/2	5/8	1 15/16	1 1/4
2	11/16	2 1/4	1 5/8

Tee			
Pipe Size	***Thread Makeup***	***B***	***Takeout***
1/8	1/4	11/16	7/16
1/4	3/8	13/16	7/16
3/8	3/8	15/16	9/16
1/2	1/2	1 1/8	5/8
3/4	9/16	1 5/16	3/4
1	9/16	1 1/2	7/8
1 1/4	5/8	1 3/4	1 1/8
1 1/2	5/8	1 15/16	1 1/4
2	11/16	2 1/4	1 5/8

45° Elbow			
Pipe Size	***Thread Makeup***	***C***	***Takeout***
1/8	1/4	9/16	1/4
1/4	3/8	3/4	3/8
3/8	3/8	13/16	7/16
1/2	1/2	7/8	3/8
3/4	9/16	1	7/16
1	9/16	1 1/8	9/16
1 1/4	5/8	1 5/16	11/16
1 1/2	5/8	1 7/16	3/4
2	11/16	1 11/16	1

Coupling			
Pipe Size	***Thread Makeup***	***D***	***Takeout***
1/8	1/4	1	1/4
1/4	3/8	1 1/8	3/8
3/8	3/8	1 1/4	3/8
1/2	1/2	1 3/8	3/8
3/4	9/16	1 1/2	3/8
1	9/16	1 3/4	1/2
1 1/4	5/8	2	3/4
1 1/2	5/8	2 1/8	7/8
2	11/16	2 1/2	1 1/4

Union			
Pipe Size	***Thread Makeup***	***E***	***Takeout***
1⁄8	1⁄4	1½	¾
¼	3⁄8	1 5⁄8	7⁄8
3⁄8	3⁄8	1¾	1
½	½	1 7⁄8	1
¾	9⁄16	2 1⁄8	1 1⁄16
1	9⁄16	2 3⁄8	1¼
1¼	5⁄8	2 5⁄8	1 3⁄8
1½	5⁄8	3	1½
2	11⁄16	3¼	1¾

Dimensions of fittings may vary with manufacturers, as well as with material. The dimensions given in the tables have been rounded off to the closest 1⁄16 in.

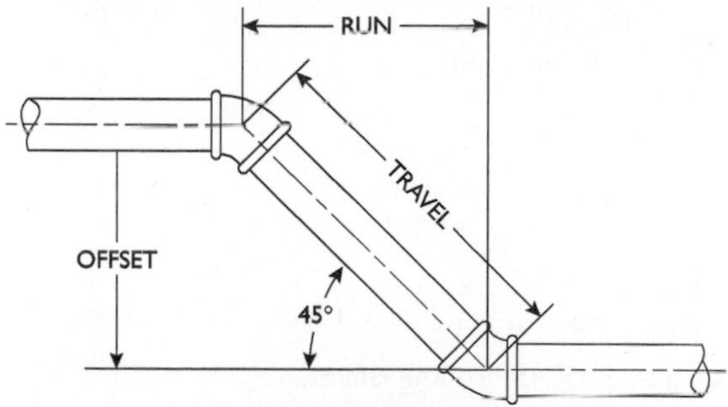

Figure 27-6 Figuring 45° offsets.

Example: What is the travel for a 16-in. offset? Using the formula,

Travel = Offset × 1.414
Travel = 16 × 1.414
Travel = 22.624
Travel = 22 5⁄8 inches

Example: What is the run for a 26-in. travel? Using the formula,

Run = Travel × 0.707
Run = 26 × 0.707
Run = 18.382
Run = 18 3⁄8 inches.

In each case, the calculated distances are center-to-center distances, and the appropriate takeouts must be deducted to find the end-to-end dimension of the pipe.

45-Degree Rolling Offset

The 45° offset is often used to offset a pipeline in a plane other than the horizontal or vertical. This is done by rotating the offset out of the horizontal or vertical plane. It is known as a *rolling offset* and can best be visualized by the three-dimensional pictorial sketch in Figure 27-7.

The run and offset distances are equal because of the 45° angle; however, two other dimensions have been added: the *roll* and the *height.* These are the centerline offset distances in the horizontal and vertical planes.

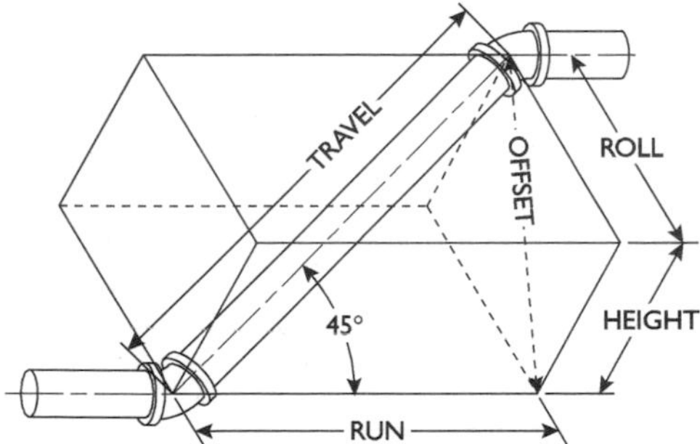

Figure 27-7 Pictorial sketch of 45° rolling offset.

Figure 27-7 shows that two right-angle triangles must now be considered, the original one with the offset and run as the sides and the travel as the hypotenuse and a new one with the roll and height as the sides and the offset as the hypotenuse.

The equations for the simple 45° offset previously discussed can still be used for calculating rolling offset distances; however, additional equations are needed because of the two added dimensions, roll and height.

These additional equations are as follows:

$$\text{Offset} = \sqrt{\text{Roll}^2 + \text{Height}^2}$$

$$\text{Run} = \sqrt{\text{Roll}^2 + \text{Height}^2}$$

$$\text{Roll} = \sqrt{\text{Offset}^2 - \text{Height}^2} \quad \text{or} \quad \sqrt{\text{Run}^2 - \text{Height}^2}$$

$$\text{Height} = \sqrt{\text{Offset}^2 - \text{Roll}^2} \quad \text{or} \quad \sqrt{\text{Run}^2 - \text{Roll}^2}$$

Depending on what the known values are, it may sometimes be necessary to solve two equations to find the distance wanted.

Example: What is the travel for a 6-in. roll with 7-in. height? Using the formula,

$$\text{Offset} = \sqrt{\text{Roll}^2 + \text{Height}^2}$$
$$\text{Offset} = \sqrt{(6 \times 6) + (7 \times 7)} \quad \text{or} \quad \sqrt{36 + 49} \quad \text{or} \quad \sqrt{85}$$
$$\text{Offset} = \sqrt{85} \quad \text{or} \quad 9.22 \quad \text{or} \quad 9\tfrac{7}{32} \text{ inches.}$$

Then,

$$\text{Travel} = \text{Offset} \times 1.1414$$
$$\text{Travel} = 9\tfrac{7}{32} \times 1.1414 \quad \text{or} \quad 13.035 \quad \text{or} \quad 13\tfrac{1}{32} \text{ inches.}$$

Example: What is the roll for a 10-in. offset with 6-in. height? Using the formula,

$$\text{Roll} = \sqrt{\text{Offset}^2 - \text{Height}^2}$$
$$\text{Roll} = \sqrt{11 \times 11 - (8 \times 8)} \quad \text{or} \quad \sqrt{121 - 64} \quad \text{or} \quad \sqrt{57}$$
$$\text{Roll} = \sqrt{57} \quad \text{or} \quad 7.55 \quad \text{or} \quad 7\tfrac{9}{16} \text{ inches.}$$

Example: What is the height for a 16-in. offset with 12in. roll? Using the formula,

$$\text{Height} = \sqrt{\text{Offset}^2 - \text{Roll}^2}$$
$$\text{Height} = \sqrt{16 \times 16 - (12 \times 12)} \quad \text{or} \quad \sqrt{256 - 144} \quad \text{or} \quad \sqrt{112}$$
$$\text{Height} = \sqrt{112} \quad \text{or} \quad 10.583 \quad \text{or} \quad 10\tfrac{19}{32} \text{ inches.}$$

Steam Traps

Steam distribution systems compose a major share of the piping in many industrial plants. The function of these systems is to carry heat from a central source to the various points of use. This is done by adding heat to water in a closed vessel or boiler at the central location. The pressure generated in this closed vessel, as the water is converted to steam, provides the transporting force. When the heat is removed from the steam, at the point of use, it reverts to water. To keep steam systems operating efficiently, the water that has given up its heat, called *condensate*, must be removed from the system. To simply open a valve and discharge the condensate would allow steam to escape when the condensate was cleared out. If a valve were open wide enough, it would also result in a lowering of the steam pressure and reduction in temperature. To attempt to open and close the valve as condensate accumulated would require a full-time operator and constant attention. An automatic condensate valve called a *steam trap* is used in steam systems to let out the water and hold back the steam. This valve discharges the water without loss of pressure as might occur with an open valve. Maintenance of pressure in a steam system is very important, as the temperature of the steam varies with the pressure (i.e., the higher the pressure, the higher the temperature).

Several basic types of steam traps are in common use, and although they all operate on the same general principle, each type has advantages for specific applications. The float trap shown in Figure 27-8 is one of the earliest types of steam traps. It consists of a metal vessel or container, which contains a ball that floats on the condensate and controls a valve.

The diagram of a float trap in Figure 27-8 shows the important elements of this type of trap. The float (A) is lifted by the condensate as it accumulates and its level rises. The lever (B) attached to the float moves with it, causing the valve (C) to gradually open. As the condensate flow increases and raises the float, the opening of the valve is increased. Thus, the float opens and closes the valve, adjusting it to suit the rate of condensate flow. Although this style of trap is simple and reliable in operation, its large size is objectionable for many applications, and its ability to discharge air accumulations from the system is limited.

Similar to the float trap is the open-bucket trap. In this type, an open-top bucket is used as a float within the vessel, as illustrated in Figure 27-9.

In operation, the condensate floats the bucket (A), which pushes upward on the rod (B), closing the valve (C). As the condensate continues to accumulate, it spills over the top of the bucket, sinking it, thus opening the valve. When the valve opens, pressure inside the vessel forces the water up through the tube (D) and out of the vessel. When the bucket has emptied sufficiently to float again, the rod is pushed up, the valve closes, and the cycle is repeated. One of the major problems with the open bucket is with the float trap; it cannot discharge air efficiently and therefore becomes air-bound. To overcome this, a vent valve or petcock is usually installed in the top of the body to bleed off the air. This requires a manual operation, and too frequently vents are left open, resulting in considerable steam loss and possible pressure drop.

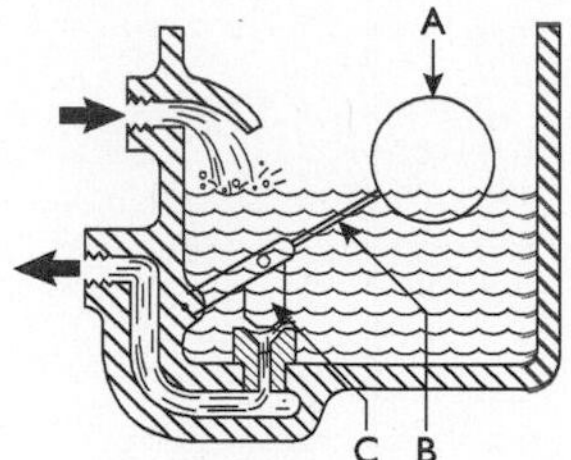

Figure 27-8 Early float trap.

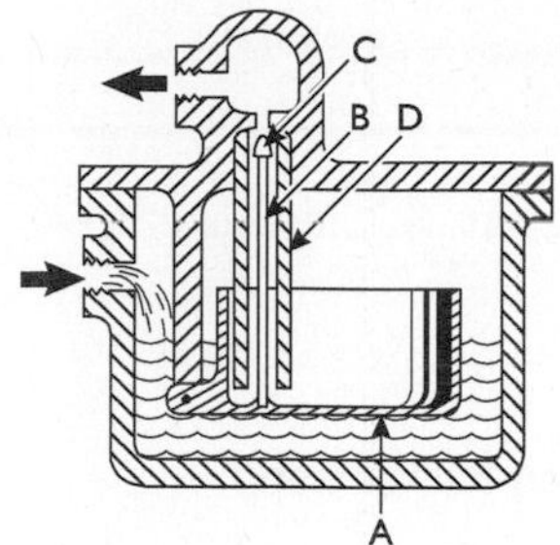

Figure 27-9 A later trap called the open-bucket trap.

A widely used style of trap, similar to the open-bucket trap, is the inverted-bucket trap. In this style, illustrated in Figure 27-10, the bucket is open at the bottom rather than the top.

Condensate enters the inverted-bucket trap either at the side or bottom, as shown in Figure 27-10, and flows up through the tube (A). It spills over the tube, filling the vessel and flowing out through the valve (B). As long as only condensate is flowing in the trap, the bucket (E) stays down and the condensate is discharged. When steam flows up through the tube, the bucket is lifted or floats. The upward movement causes the linkage mechanism (C) at the top of the bucket to close the valve and interrupt the discharge. The steam in the bucket slowly condenses into water; also, a small amount bleeds off through the vent hole (D) in the top of the bucket. The bucket gradually loses its buoyancy and sinks, opening the valve again to discharge more condensate. Another important function of the vent (D) in the top of the bucket is to permit the venting of air. Without the vent, air would be trapped in the bucket and would accumulate, keeping the valve closed or air-bound. Thus, the vent hole automatically bleeds off the air as it accumulates and prevents the trap from becoming air-bound.

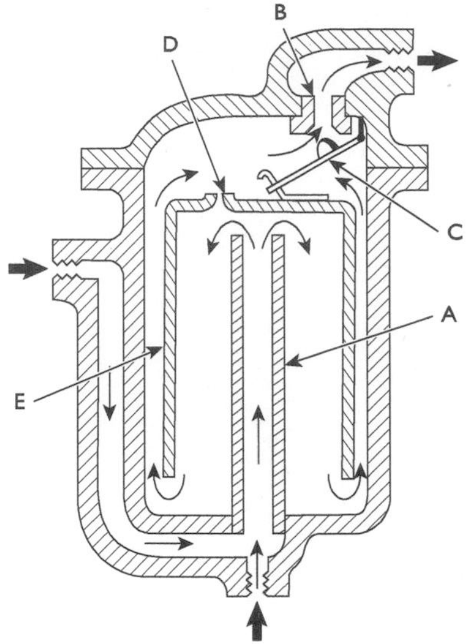

Figure 27-10 An improved version of the open-bucket trap.

The float, bucket, and inverted-bucket traps are all controlled by the buoyancy of a float mechanism in condensate. The thermostatic trap's operation differs from the float types in that it responds to temperature changes, hot steam causing it to open and the cooler condensate causing it to close. A commonly used style of thermostatic trap is the bellows type, shown in Figure 27-11.

The operating element (A) is a corrugated bellows filled with a liquid, such as alcohol and water, that has a boiling point below that of water. Attached to the bottom of the bellows is a valve (B), which closes when the bellows expands. When cool condensate is flowing, the bellows are contracted and the valve is open, allowing the condensate to be discharged. As the condensate temperature approaches steam temperature, the liquid inside the bellows is vaporized, causing a pressure increase inside the bellows, expanding and lengthening the bellows, and closing the valve. The valve remains closed until there is cooling of the bellows, causing it to contract and open the valve. The cooling of the bellows is largely dependent on heat radiation from the body of the trap. Therefore, the cooler the location of the trap, the faster it will respond to line temperature changes. This style of trap performs well where quick start-ups are required, as it is wide open when cool; being wide open, it can discharge large volumes of air. It should not be used in applications that have water-hammer problems because the bellows are sensitive to damage from slugs of water.

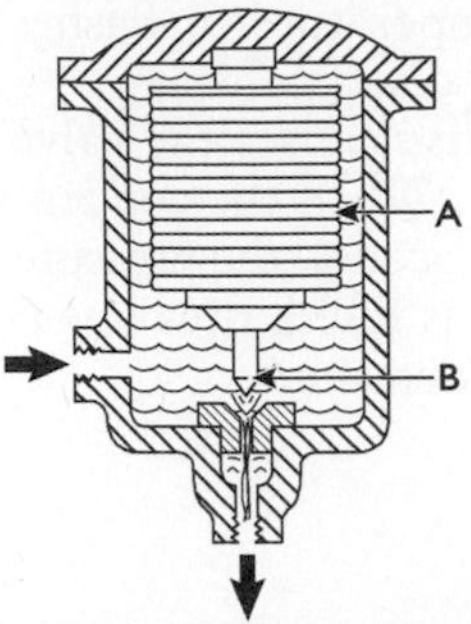

Figure 27-11 A commonly used bellows-type thermostatic trap.

The float and thermostatic trap is a combination of float-trap and thermostatic-trap elements in a single trap. By adding a thermostatic element to a float trap, air-handling ability, particularly on start-ups, is provided. The float mechanism provides the variable flow control necessary to handle varying condensate loads. Figure 27-12 shows a diagram of a typical float-thermostatic trap.

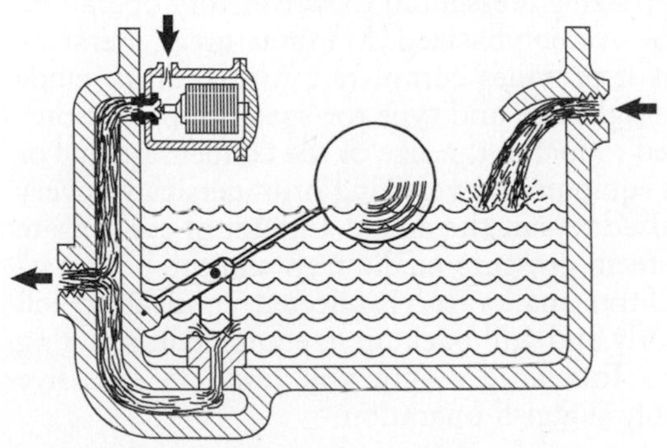

Figure 27-12 The float and thermostatic trap.

A third type of steam trap is the thermodynamic trap, which utilizes the heat energy in hot condensate and steam to control its operation. One of the most widely used types of thermodynamic trap is the piston style shown in Figure 27-13.

This style of trap is operated by the flashing action of hot condensate discharged into a chamber that is at lower pressure than the discharging condensate. As cool condensate flows into the trap, pressure is exerted on the bottom side of the piston disc (A), thus lifting the piston and holding open the valve (B). When the condensate flowing through the trap nears steam temperature, it flashes into vapor in the chamber (C) above the piston disc.

This increase in pressure above the piston disc causes the valve to close, preventing the discharge of live steam. When the condensate cools to the point where flashing no longer occurs, the pressure above the piston disc is reduced and the piston is lifted, thus opening the valve to allow discharge of the cool condensate.

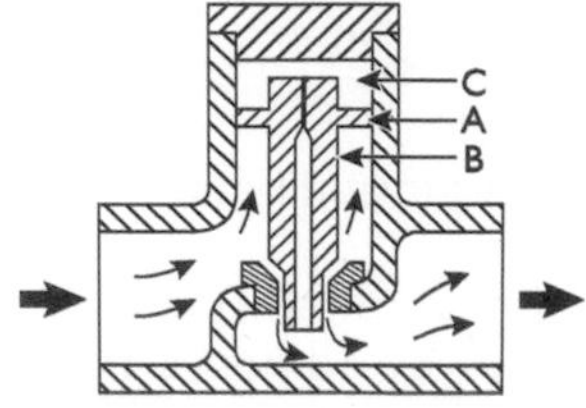

Figure 27-13 A thermodynamic trap.

In discussing the various types and styles of steam traps and their principle of operation, no reference has been made to size or capacity. As correct steam trap sizing is essential to satisfactory operation, its importance cannot be overemphasized. Manufacturers' literature should be consulted, as it provides complete information to guide the user in selecting the right size and type for specific applications. If the trap is simply sized to match the size of the condensate line or the pipe-thread sizes in equipment, oversizing or undersizing is very likely. A trap must be sized to suit the actual amount of condensate formed as well as the steam pressure at the trap and the backpressure in the return line. If too small a trap is selected, condensate will not be discharged properly and will back up in equipment, resulting in inefficient operation. Too large a trap will result in excessive steam waste and probably sluggish operation.

Steam Trap Installation

To efficiently drain condensate from steam systems, steam traps must be correctly located and the piping properly arranged. Most manufacturers will furnish specific instructions for installation of their products in connection with various types of equipment. The following basic rules, however, will go a long way toward providing trouble-free operation, as they apply to practically all steam trap installations. The numbers are keyed to the installation diagram in Figure 27-14.

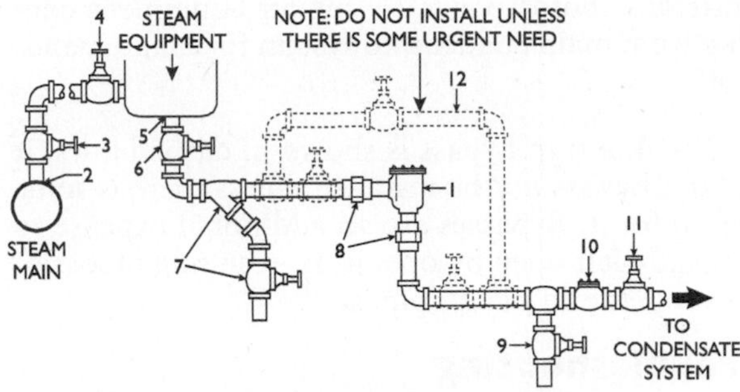

Figure 27-14 Typical steam trap installation.

1. Provide a separate trap for each piece of equipment or apparatus. Short-circuiting (steam follows path of least resistance to trap) may occur if more than one piece of apparatus, coil, etc., is connected to a single trap.
2. Tap steam supply off the top of the steam main to obtain dry steam and avoid steam line condensate.
3. Install a supply line valve close to the steam main to allow maintenance and/or revisions without steam main shutdown.
4. Install a steam supply valve close to the equipment entrance to allow equipment maintenance work without supply line shutdown.
5. Connect condensate discharge line to lowest point in equipment to avoid water pockets and water hammer.
6. Install shutoff valve upstream of condensate-removal piping to cut off discharge of condensate from equipment and allow service work to be performed.
7. Install strainer and strainer flush valve ahead of trap to keep rust, dirt, and scale out of working parts and to allow blowdown removal of foreign material from strainer basket.
8. Provide unions on both sides of trap for its removal and/or replacement.
9. Install test valve downstream from trap to allow observance of discharge when testing.
10. Install check valve downstream from trap to prevent condensate flow-back during shutdown or in the event of unusual conditions.

11. Install downstream shutoff valve to cut off equipment condensate piping from main condensate system for maintenance or service work.

Note: In Figure 27-14, a trap bypass is shown in dashed lines. It is recommended that a bypass *not* be installed unless there is some unusually urgent need for it. Bypasses are an additional expense to install and are frequently left open by operators, resulting in loss of steam and inefficient operation of equipment.

Steam Trap Troubleshooting

In cases of improper functioning of steam equipment, a few simple checks of the steam system should be made before looking for trap malfunction. The following three checks are preliminary steps that should precede checking the operation of the steam trap:

1. Check the steam supply pressure. It should be at or above the minimum required.
2. Check to be sure all valves required to be open are in the full-open position (supply valve, upstream shutoff valve, downstream shutoff valve).
3. Check to be sure all valves required to be closed are in tight closed position (bypass valve, strainer valve, test valve).

The initial step in checking the operation of a steam trap is to check its temperature. Because a properly functioning steam trap is an automatic valve that allows condensate to be discharged but closes to prevent the escape of steam, it should operate very close to the steam temperature. For exact checks, a surface pyrometer should be used. A convenient and dependable operating test is simply to sprinkle water on the trap. If the water spatters, rapidly boils, and vaporizes, the trap is hot and probably very close to steam temperature.

If it is found that the trap is hot, probably very close to steam temperature, the next check is to determine if condensate or steam is being discharged. When a test valve is provided for this purpose, the check is made by closing the downstream shutoff valve and opening the test valve. The discharge from the test valve should be carefully observed to determine if condensate or live steam is escaping. If the trap being tested is a type that has an opening and closing cycle, condensate should flow from the test valve and then should stop as the trap shuts off. The flow should resume when the trap opens again. Steam should not discharge from the test valve if the trap is operating properly. If the trap is a continuous discharge type, such as the float-thermostatic, there should be a continuous discharge of condensate but no steam.

In the event that the installation does not have a test valve, the trap may be checked by listening to its operation. The ideal instrument for this is of course an industrial stethoscope; however, such an instrument is seldom available to the mechanic. A suitable device for this purpose is a screwdriver or metal rod. By holding one end against the trap and the other end against the ear, the sound of the trap's operation may be heard. If the trap is operating properly, the flow of condensate should be heard for a few seconds, a click as the valve closes, and then silence, indicating that the valve has closed tight. This cycle of sounds should repeat in a regular pattern. The listening procedure, however, is not suitable for checking continuous discharge traps because these automatically regulate to an open position in balance with the condensate flow.

Another check on the trap's operation is to open the strainer valve and observe the discharge at this point. There should be an initial gush of condensate and steam as the valve is opened, followed by a continuous flow of live steam. If condensate flows for a prolonged period before steam is observed, the condensate is not being properly discharged from the system.

Unsatisfactory performance of a steam unit may result from numerous conditions other than improper steam trap operation. Some of the common faults that cause troubles are the following:

1. *Inadequate steam supply.* This may be due to the steam supply line being too small for the load, a stoppage in the line, a partially closed valve, a loss of steam supply, etc.
2. *Incorrectly sized trap.* A steam trap must be sized for the load, steam pressure, and backpressure.
3. *Improper connection.* The condensate connection to the steam equipment must be at the low point or condensate will accumulate in the equipment.
4. *Improper pitch or slope of condensate lines.* All condensate lines must pitch or slope toward the trap to avoid pocketing of condensate. Dips or pockets that hold condensate will cause sluggish trap operation and water hammer.
5. *Inadequate condensate lines.* The condensate line must be large enough to carry off the condensate discharge from the trap without building up excessive backpressure.

In addition, Table 27-4 allows detailed troubleshooting of complex steam system problems.

Table 27-4 Steam Systems Troubleshooting Guide

Condition	*Reason*	*Corrective Action*
Trap Blows Live Steam	1. NO PRIME (bucket traps)	1.
	a. Trap not primed when originally installed	a. Prime the trap.
	b. Trap not primed after cleanout	b. Prime the trap.
	c. Open or leaking bypass valve	c. Remove or repair bypass valve.
	d. Sudden pressure drops	d. Install check valve ahead of trap.
	2. VALVE MECHANISM DOES NOT CLOSE	2.
	a. Scale or dirt lodged in orifice	a. Clean out the trap.
	b. Worn or defective valve or disc mechanism	b. Repair or replace defective parts.
	3. RUPTURED BELLOWS (thermostatic traps)	3. Replace bellows.
	4. BACKPRESSURE TOO HIGH (thermodynamic trap)	4.
	a. Worn or defective parts	a. Repair or replace defective parts.
	b. Trap stuck open	b. Clean out the trap.
	c. Condensate return line or pig tank undersize	c. Increase line or pig tank size.
	5. BLOWING FLASH STEAM	5. Normal condition
	Forms when condensate released to lower or atmospheric pressure	No corrective action necessary
Trap Does Not Discharge	1. PRESSURE TOO HIGH	1.
	a. Trap pressure rating too low	a. Install correct trap.

	b. Orifice enlarged by normal wear	b. Replace worn orifice.
	c. Pressure-reducing valve set too high or broken	c. Readjust or replace pressure-reducing valve.
	d. System pressure raised	d. Install correct pressure-change assembly.
	2. CONDENSATE NOT REACHING TRAP	2. Clean out and install strainer.
	a. Strainer clogged	
	b. Obstruction in line to trap inlet	
	c. Bypass opening or leaking	
	d. Steam supply shut off	
	3. TRAP HELD CLOSED BY DEFECTIVE MECHANISM	3. Repair or replace mechanism.
	4. HIGH VACUUM IN CONDENSATE RETURN LINE	4. Install correct pressure-change assembly.
	5. NO PRESSURE DIFFERENTIAL ACROSS TRAP	5.
	a. Blocked or restricted condensate return line	a. Remove restriction.
	b. Incorrect pressure-change assembly	b. Install correct pressure-change assembly.
Continuous Discharge from Trap	1. TRAP TOO SMALL	1.
	a. Capacity undersized	a. Install properly sized larger trap.
	b. Pressure rating of trap too high	b. Install correct pressure-change assembly.
	2. TRAP CLOGGED WITH FOREIGN MATTER	2.
	a. Dirt or foreign matter in trap internals	a. Clean out and install strainer.
	b. Strainer plugged	b. Clean out strainer.

	3. BELOWS OVERSTRESSED (Thermostatic traps)	3. Replace bellows.
	4. LOSS OF PRIME	4. Install check valve on inlet side.
	5. FAILURE OF VALVE TO SEAT	5.
	a. Worn valve and seat	a. Replace worn parts.
	b. Scale or dirt under valve and in orifice	b. Clean out the trap.
	c. Worn guide pins and lever	c. Replace worn parts.
Sluggish or Uneven Heating	1. TRAP HAS NO CAPACITY MARGIN FOR HEAVY STARTING LOADS	1. Install properly sized larger trap.
	2. INSUFFICIENT AIR-HANDLING CAPACITY (bucket traps)	2. Use thermic buckets or increase vent size.
	3. SHORT-CIRCUITING (group traps)	3. Trap each unit individually.
	4. INADEQUATE STEAM SUPPLY	4.
	a. Steam supply pressure valve has changed	a. Restore normal steam pressure.
	b. Pressure reducing valve setting off	b. Readjust or replace reducing valve.
Backpressure Troubles	1. CONDENSATE RETURN LINE TOO SMALL	1. Install larger condensate return line.
	2. OTHER TRAPS BLOWING STEAM INTO HEADER	2. Locate and repair other faulty traps.
	3. PIG TANK VENT LINE PLUGGED	3. Clean out pig tank vent line.
	4. OBSTRUCTION IN CONDENSATE RETURN LINE	4. Remove obstruction.
	5. EXCESS VACUUM IN CONDENSATE RETURN LINE	5. Install correct pressure-change assembly.

Main Supply Valve Opening

The opening of valves controlling steam flow in steam supply lines, called steam mains, requires care and correct procedure. The expansion or growth of the piping system as the temperature increases when steam is introduced must be carefully controlled. Also, the air in the line and large volumes of condensate formed as the line heats up must be removed. To facilitate removal of condensate during normal operation, as well as at start-up, steam lines incorporate *drip pockets*, *drip legs*, and *drip valves*, as illustrated in Figure 27-15.

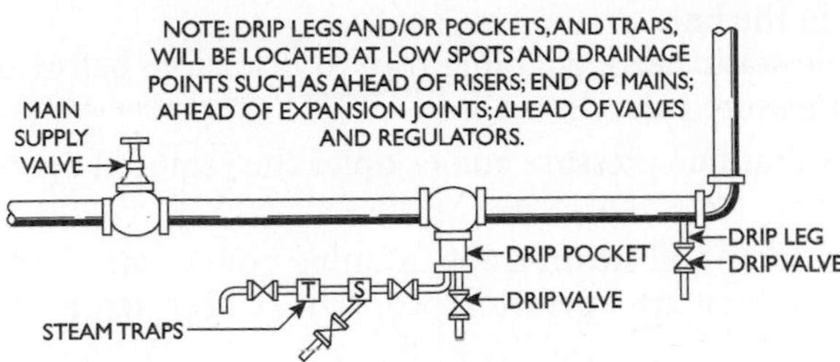

Figure 27-15 A drip pocket, drip leg, and drip valve in a main supply line.

A drip pocket, shown in Figure 27-15, is a line extending down from the main, usually the same diameter as the main, that acts as a condensate reservoir. The condensate that accumulates in the drip pocket is usually drained through a steam trap.

A drip leg is also a line extending down from the main, usually of a smaller diameter than the main, with a valve at the bottom or end of the pipe. The valve is the only opening in a drip leg; there is no steam trap. The valve is used to blow down, or exhaust, the steam in the line. It is also used as an air vent and condensate discharge valve at start-up.

A drip valve is the valve at the bottom of drip pockets and drip legs for free blow of the steam to atmosphere and to provide a vent and condensate discharge opening at start-up.

The following procedure should be followed when opening a steam main supply valve:

1. Open all drip valves, full open, to act as air vents and condensate discharge openings. Check setting of distribution valves to be sure steam goes only to those branch lines ready to receive it.
2. Open main supply valve slowly and in stages to control steam-flow volume and provide gradual heat-up of the line.

3. Watch discharge at drip valves. Do not close drip valves until warm-up condensate has been discharged (except for next item).
4. Condensate should not be drained from drip pockets. An accumulation of condensate is necessary in drip pockets, as they are in the line to do the following:
   a. To let condensate escape by gravity from the fast-moving steam
   b. To store condensate until the pressure differential is great enough for the steam trap to discharge
   c. To provide storage of condensate until there is positive pressure in the line
   d. To provide static head, enabling trap to discharge before a positive pressure exists in the line
5. Check to see that line pressure comes up to the required operating pressure.
6. Check operation of all steam traps draining condensate from line to be sure they are operating properly. (Check temperature discharge, etc.)

Flanged Pipe Connections

Flanged pipe connections are widely used in industrial plants where the pipelines or equipment must be disassembled for cleaning or maintenance work. In many cases, the flanged connection is also more economical, particularly where valves, fittings, and equipment are installed. Flanges are commonly attached to the pipe by screw threads or by welding. Several types of flange facings are in common use, the simplest of which are the plain flat face and the raised face.

The plain flat-faced flange, which is faced straight across, may be used with either a full-face or a ring gasket for pressures less than 125 lb on steam and water. A fairly thick gasket is required to make a tight joint. Generally, the full-faced gasket is preferred because it may be installed more readily with greater assurance of concentricity than the ring gasket.

Higher-pressure cast-iron flanges and steel flanges are made with a raised face. In this style of flange, the face of the flange between the bore and the inside of the bolt holes is 1⁄32 to 1⁄16 in. above the surface of the remainder of the flange. Ring gaskets are employed with this style of face, and a greater pressure can be exerted on the gasket when the bolts are tightened than is possible with a full-face gasket.

Pipe-flange gaskets are made from a wide variety of materials in many construction styles. The function of a gasket is to provide a loose, compressible substance between the flange faces with sufficient body resiliency and strength to make the flange connection leak-proof. Gaskets are made from such sheet materials as rubber, rubber composites, asbestos, paper, cork, lead, copper, and aluminum. They are also fabricated from a variety of materials by weaving, winding, corrugating, etc., to give internal springiness. Whatever the shape, size, or material from which a gasket is made, it must be soft enough to conform to the flange facing when tightened. It must have the capacity to flow when compressed to fill the irregularities of the flange surfaces. In addition, it must be nonporous and sturdy enough to resist the heat and pressure to which it is subjected.

The assembly and tightening of a pipe-flange connection is a relatively simple operation; however, certain practices must be followed to obtain a leak-proof connection. The gasket must line up evenly with the inside bore of the flange face with no portion of it extending into the bore. To locate the gasket in this position, bolts should be inserted in one mating flange. With a full-face gasket, the bolts will go through the gasket bolt holes to hold it in position. With a ring gasket, a few bolts spread within a half circle will position the gasket concentric with the flange. The bare mating flange should then be brought up to the gasketed flange. If it is a tight fit and the mating flange must be sprung into place, bring the flanges into alignment using a pinwrench. *Do not* drive bolts into misaligned flanges and tighten to align. On tight-fitting flanges, it may be easier to position the flanges and then spread them to place the gasket in position, taking care that the gasket is positioned concentric with the flange.

To aid in assembling pipe flanges, some of the following practices are helpful. On vertical flange faces that are deeply tooled, small gobs of grease will hold the gasket in position. On smooth finish faces, a thin coating of cement or shellac may be used. On connections that may be frequently opened, cement should be used on one face and grease on the other so that the joint can be easily opened and the gasket will remain in position on one face. All flange bolts should be generously lubricated for easy tightening and removal if required.

When tightening pipe-flange bolts, a planned bolt-tightening sequence should be followed to ensure that the flange faces are kept parallel and bolts are tightened uniformly. Starting at a convenient point, lightly tighten the first bolt; then, move directly across for the second bolt, approximately ¼ of the way around the circle for the third, and directly across for the fourth. Continue this pattern, even numbers directly across the circle and odd numbers about ¼ of the way around the circle, until all bolts are tight. It is especially important that this sequence be followed on the initial tightening go-around. The final tightening go-around need not follow this pattern; however, it is very important that the final tightening be uniform, finishing up with each bolt pulling the same load. Figure 27-16 illustrates a 16-bolt flange with the bolts numbered to indicate the tightening sequence.

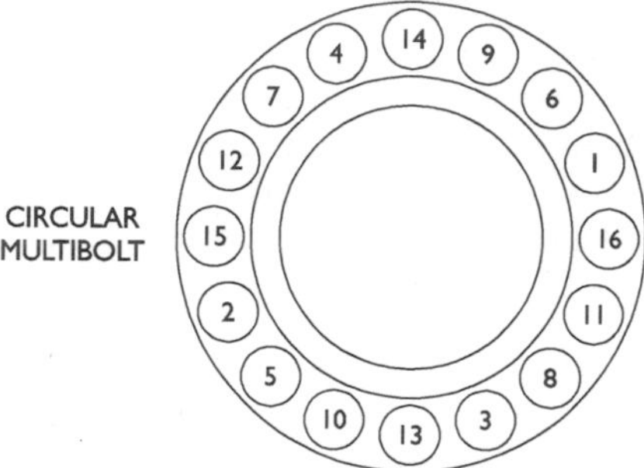

Figure 27-16 Bolt-tightening sequence for round pipe flange.

The noncircular, or oval, flange sequence of bolt tightening differs from the circular flange bolt-tightening sequence just described. The bolts are tightened across the short centerline first and then alternately from side to side moving away from the short centerline. The reason for this order of tightening is to apply the tightening force in a manner that will "iron out" the gasket, starting at the short centerline and working along the long sides of the flange toward the ends. Figure 27-17 illustrates a typical oval flange with the bolts numbered to indicate the correct tightening sequence.

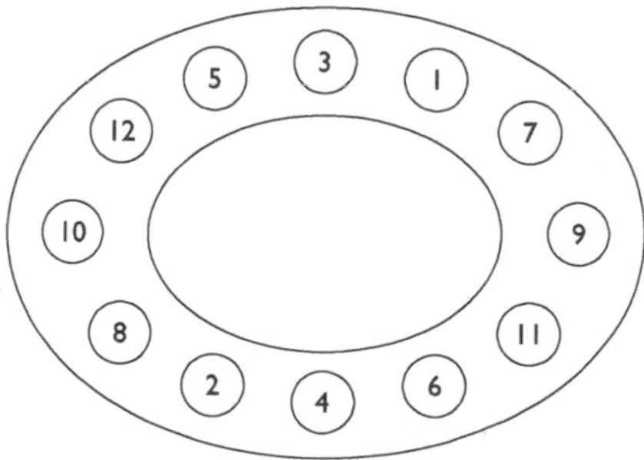

Figure 27-17 Bolt-tightening sequence for oval pipe flange.

Pipe Flange Bolt-Hole Layout

Mating pipe flanges to other flanges or circular parts may require the layout for drilling or punching of matching holes. To make such a bolt-hole layout, the bolt circle diameter and the number of holes required must be known. In some cases, it may also be necessary to know whether the bolt holes are to be "on the centerline" or are to "straddle the centerline" with respect to a specific point or location. The usual practice is to specify "on" or "off" the centerline. Figure 27-18 illustrates an "off the centerline" layout.

Although bolt holes may be laid out with a protractor using angular measurement to obtain uniform spacing, this method is most satisfactory when there are six or fewer holes. Layouts that have a greater number of holes may be more easily accomplished using a system of multipliers or constants to calculate the chordal distance between bolt-hole centers. Table 27-5 lists these constants for bolt circles with 4 through 40 holes.

After the distance between bolt holes for a given bolt circle diameter and the number of bolt holes have been determined, the hole centers can be located around the bolt circle by measurements. Refer to Figure 27-18 for illustration of the dimensions used in the following bolt-hole layout procedure:

A—Bolt circle diameter

B—Bolt hole center-to-center distance (chordal length)

C—One-half center-to-center distance (for "straddle centerline" layout)

Layout Procedure

1. Lay out horizontal and vertical centerlines.
2. Lay out bolt circle.
3. Find value of B (multiply the bolt circle diameter by constant).
4. For "straddle centerline" layout, divide B by 2 to find value of C.
5. Measure distance C off the centerline and mark the center of the first bolt hole.
6. Set dividers to dimension B. From first hole center point, lay out other hole center points by swinging arcs from point to point on bolt circle.

Table 27-5 Flange Hole Constants

Number of Bolt Holes	*Constant*
4	.707
6	.500
8	.383
10	.309
12	.259
16	.195
20	.156
24	.131
28	.112
32	.098
36	.087
40	.079

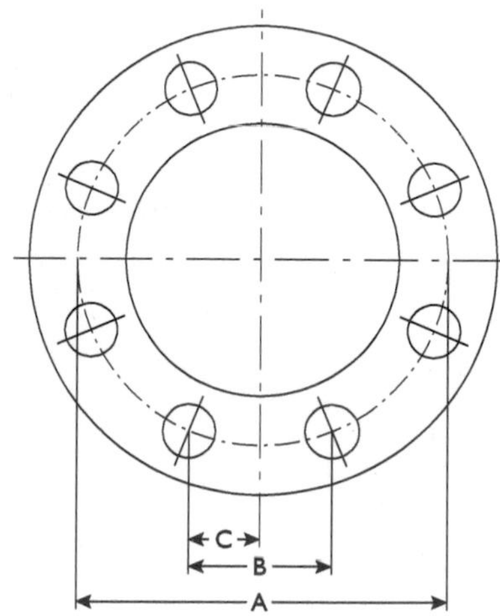

Figure 27-18 An "off the centerline" bolt-hole layout.

Automatic Sprinkler Systems

All modern industrial buildings are equipped with fire-protection piping in the form of automatic sprinkler systems. Although millwrights and industrial mechanics are seldom involved in the installation of such systems, they are frequently responsible for maintenance, service, resetting, etc. The following explanation of some of the systems and equipment in common use is intended to aid mechanics in understanding their operation so that they can provide adequate maintenance service as required. Because the National Board of Fire Underwriters and allied organizations have established elaborate rules for all forms of protective apparatus, the appropriate insurance carrier should be consulted before any changes, adjustment, or revisions are made to the present sprinkler systems.

Wet Sprinkler Systems

A wet sprinkler system is described by the National Board of Fire Underwriters as "a system employing automatic sprinklers attached to a piping system containing water and connected to a water supply so that water discharges immediately from sprinklers opened by a fire." A vital component of a wet sprinkler system is the alarm check valve. The prime function of the alarm check valve, also called the wet sprinkler valve, is to direct water to alarm devices and sound the sprinkler alarm. It does *not* control the flow of water into the system.

The alarm check valve is located in the pipe riser at the point where the water line enters the building. An underground valve with an indicator post is usually located a safe distance outside the building. In design and operation, the alarm check valve is a globe-type check valve. A groove cut in the valve seat is connected by passage to a threaded outlet on the side of the valve body. When the valve lifts from the seat, water can enter the groove and flow to the outlet. Refer to Figure 27-19, which shows a wet sprinkler alarm system. There is also a large drain port above the seat to which is connected the drain valve. Two additional ports, one above and one below the seat, will allow for the attachment of pressure gauges.

A plug or stopcock-type valve called the alarm control cock is connected to the outlet from the seat groove. This cock controls the water flow to the alarm devices, allowing their silencing. This cock *must* be in the alarm position when the system is set or the alarm check valve will be unable to perform its function of sounding an alarm.

Operation of Wet Sprinkler Alarm Systems

When a sprinkler head opens, or when for any reason water escapes from a wet sprinkler system, the flow through the alarm check valve causes the internal valve or clapper to lift. Water entering the seat groove flows through the alarm control cock to the retarding

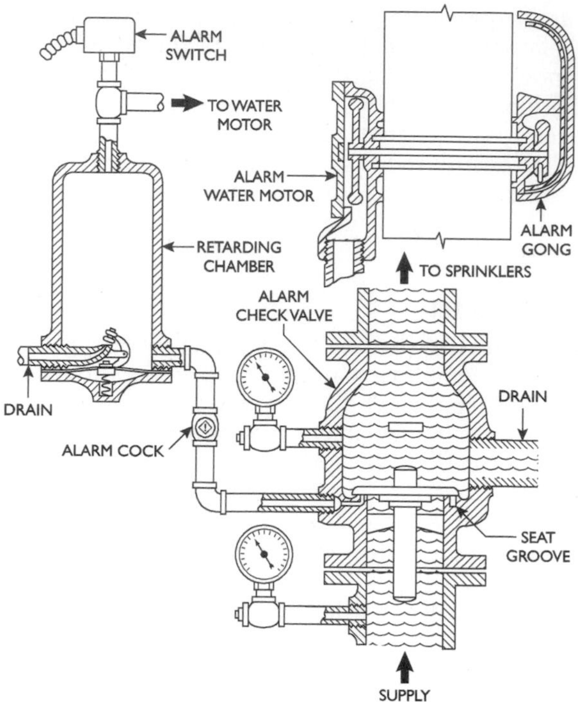

Figure 27-19 A wet sprinkler alarm system.

chamber. The function of the retarding chamber is to avoid unnecessary alarms that might be caused by slight leakage. It will allow a small volume of water to flow without actuating the alarm. When there is a large flow, as occurs when a sprinkler opens, the chamber is quickly filled and pressure closes the diaphragm-actuated drain valve in the bottom of the retarding chamber. The electrical alarm is then actuated by the water pressure, and flowing water is directed to the water motor, causing it to be rotated and to whirl hammers inside the alarm gong, sounding an alarm mechanically.

Placing a Wet Sprinkler System in Service

The placing of a wet sprinkler fire protection system in service is commonly referred to as "setting" the system. Although this operation is not complex or difficult, it does require that correct procedure be followed to guarantee dependable protection. The following procedure is general and may require minor modifications to suit special features found in some systems:

1. Check system to be sure it is ready to be filled with water. In cases where the system has been shut down because a head has opened, be sure the head has been replaced with a new one of proper rating.
2. Open vent valves; —these are usually located at high points in the system.
3. Place alarm cock in "closed" position. This will prevent sounding of the alarm while the water is flowing during the filling operation.
4. Place drain valve in nearly closed position. In this slightly open position, a small amount of water will flow from the valve while the system is filling.
5. Open indicator post valve slowly. When the system has filled, there will be a quieting of the sound of rushing water. When this occurs, turn the valve to full-open position, and then back it off about a ¼ turn to be sure it is not jammed wide open.
6. Observe the water flow at the vent valves. When a steady flow of water occurs, all air has been expelled and the vent valves should be closed.
7. Check the water flow by quickly opening the drain valve and then closing it. The water pressure should drop approximately 10 lb when the valve is opened and immediately return to full line pressure when the drain valve is closed. (Excessive pressure drop indicates insufficient water flow.)
8. Open alarm cock. The system is now operative or in "set" condition.
9. Test by opening drain valve several turns. Water should flow through the water motor, sounding the mechanical gong alarm, and to the pressure-actuated electrical switch, causing the electrical alarm to be energized.
10. Close drain valve. If the alarms have functioned properly, the wet sprinkler alarm system is operational and may be reported as in service or "set."

Dry Sprinkler Systems

A dry-pipe sprinkler system is described by the National Board of Fire Underwriters as

A system employing automatic sprinklers attached to a piping system containing air under pressure, the release of which as from the opening of sprinklers permits the water pressure to open a valve known as a "dry-pipe valve." The water then flows into the piping system and out the open sprinklers.

The dry-pipe valve is located in the pipe riser at the point where the water line enters the building. An underground valve with indicator post is usually located a safe distance outside the building.

The dry-pipe sprinkler valve is a dual valve, as both an air valve and a water valve are contained inside its body. These two internal valves may be separate units, one positioned above the other, or they may be combined in a single unit, one within the other, as illustrated in Figure 27-20. The function of the air valve is to retain the air in the piping system and hold the water valve closed, thus restraining the flow of water. As long as sufficient air pressure is maintained in the system, the air valve can do this because it has a larger surface area than the water valve. The air in the system acts on this large surface and provides enough force to hold the water valve closed against the pressure of the water.

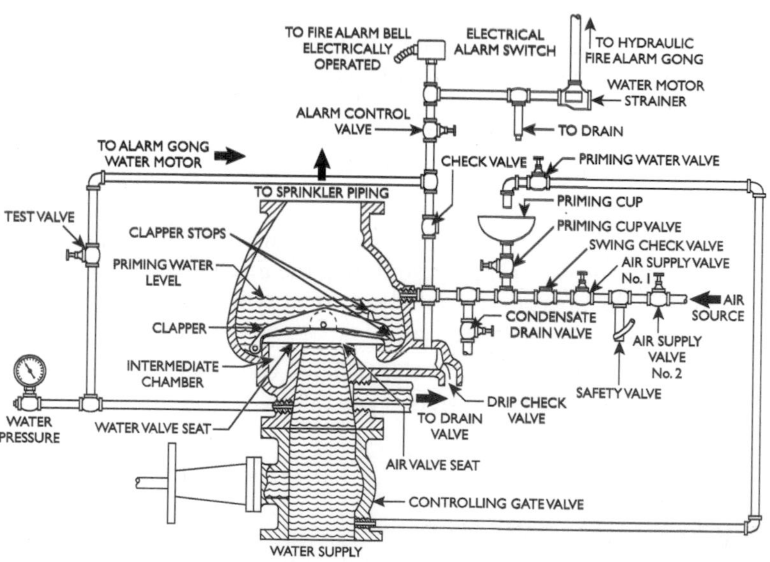

Figure 27-20 Diagram of a dry-pipe sprinkler system.

Dry-pipe valves incorporate an intermediate chamber, either between the internal air and water valves or under them. This chamber is normally dry and open to atmospheric pressure through the open drip check valve. Two gauges are connected, one above the internal valves to indicate air pressure and one below to indicate water pressure. A drain line with a shutoff valve is located below the internal valves.

A three-position test cock is connected to the supply water below the water seat, to the intermediate chamber, and to the alarm pipeline. The cock is used to test the alarms and silence the alarms when necessary. It *must* be set in the "alarm" position when the dry-pipe valve is set, or the alarms will not be actuated if the system is tripped.

Operation of a Dry-Pipe Valve

When a sprinkler head opens, or when for any reason air escapes from a dry-pipe system, the air pressure above the internal air valve is reduced. When the air pressure falls to the point where its force is exceeded by the force of the water below the internal water valve, both valves are thrown open. This allows an unobstructed flow of water through the dry-pipe valve into the sprinkler system piping to the open sprinklers. As the valves are thrown open, water fills the intermediate chamber. To avoid unnecessary alarms, the chamber is equipped with a drip check valve. This drip check valve is normally open to the atmosphere and allows drainage of any slight water leakage past the internal water valve seat. When a sprinkler head opens and the internal valves are thrown open, the chamber instantly fills with water and the drip check valve is forced closed. The water now under pressure in the chamber flows to the electrical alarm switch, closing the contacts and actuating the electrical alarm. Water also flows through the alarm water motor, causing the gong alarm hammers to whirl and hydraulically sound the alarm on the gong attached to the outside of the building.

Placing a Dry-Pipe Sprinkling System in Service

The "setting" of a dry sprinkler system requires a general understanding of the system, its components, and their functions. The efficiency and reliability of the protection afforded by an automatic sprinkler system is dependent on the system being properly set and maintained. Although the following procedure is general and may require modifications to suit individual situations, it covers the essential steps or operations that must be performed.

1. Close the valve controlling water flow to the system. This may be located under the dry-pipe valve, or it may be an underground valve with an indicator post located a safe distance outside of the building. If a fire has occurred and water is flowing from opened sprinklers, the approval of the person in authority must be obtained before closing the valve.
2. Open the drain valve and allow the water to drain from the sprinkler system piping.
3. Open all vent valves and drains throughout the system. Vents will be located at the high points and drains at all trapped and low portions of the piping system.
4. Manually push open the drip check valve. Also, open the drain valve for the dry-pipe valve body if one is provided.
5. Remove the cover plate from the dry-pipe valve body and carefully clean the rubber facings and seat-ring surfaces of the internal air and water valves. *Do not* use rags or abrasive wiping materials. Wipe the seats clean with the bare fingers.
6. Unlatch the clapper and carefully close the internal water and air valves.
7. Replace the dry-pipe valve cover, and close the body drain valve if one is provided.
8. Open the priming-cup and the priming-water valve to admit priming seal water into the dry-pipe valve body to the level of the pipe connection. The priming water provides a more positive seal to prevent air from escaping past the air valve seat into the intermediate chamber.
9. Drain excess priming water by opening the condensate drain valve. Close tightly when water no longer drains from valve.
10. Open the air-supply valve and admit air to build up a few pounds of pressure in the system.
11. Check all open vents and drains throughout the system to be sure all water has been forced from the low points. As soon as dry air exhausts at the various open points, the openings should be closed. Close the air-supply valve.
12. Replace any open sprinklers with new sprinkler heads of the proper rating.
13. Open the air-supply valve and allow system air pressure to build up to the required pressure. The amount of air pressure required to keep the clapper closed varies directly with water-supply

pressure. Consult manufacturers' tables for specific pressure settings. Figure 27-21 shows the operation of the clapper valve.

14. Open the system water-supply valve slightly to obtain a small flow of water to the dry-pipe valve.
15. When water is flowing clear at the drain valve, slowly close it, allowing water pressure to gradually build up below the clapper, as observed on the water-pressure gauge.
16. When water pressure has reached maximum below the clapper, close the drain valve tight and open the supply valve to full-open position. Back the supply valve off about ¼ turn from full open to be sure it is not jammed.
17. Test the alarms by placing the three-position test cock in the "test" position. Water should flow to the electrical alarm switch, actuating it and energizing the electrical alarm. It should also flow through the water motor, sounding the mechanical gong alarm on the outside of the building.
18. If the alarms have functioned properly, the three-position test cock should be placed in "alarm" position. The system is now operative or in "set" condition.

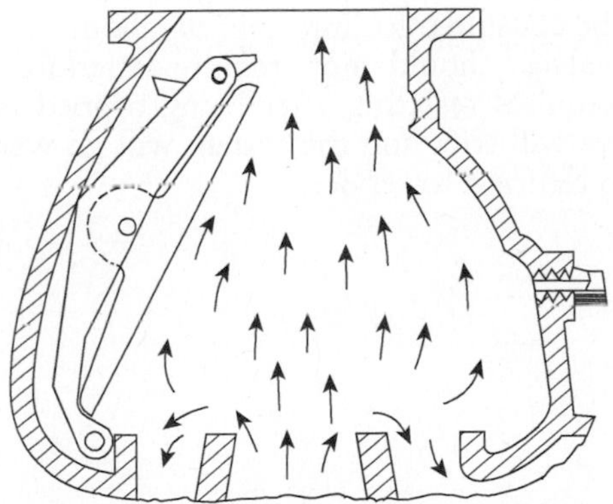

Figure 27-21 Clapper valve in operating position with water flowing.

Note: The test cock alarm test is a test of the functioning of the alarm system only and does not indicate the condition of the dry-pipe valve. The operation of the dry-pipe valve may be tested by opening a vent valve to allow the air in the piping system to escape, causing the dry-pipe valve to trip. It must then be reset by going through the steps indicated above.

Large Dry-Pipe Sprinkler Systems

Water flow is not immediate in dry-pipe sprinkler systems because of the time delay while air pressure falls to the trip point and the air in the system is exhausted and water reaches the open sprinklers. Large dry-pipe systems are equipped with an auxiliary valve to speed up the action. Two devices are commonly used for this purpose; one is called an exhauster and the other an accelerator.

Exhausters are valves with relatively large ports that trip on very slight pressure drop and speed up the escape of the air in the sprinkler system piping. Thus, the tripping delay is shortened, and there is less resistance to the entrance and flow of water.

Accelerator valves are also actuated by a relatively small pressure drop, but speed-up in this case is accomplished by accelerating the tripping of the dry-pipe valve. This is done by directing the air from the system into the intermediate chamber. The force of this air pressure under the internal valve seat is added to the force of the water pressure to rapidly trip the valve and admit water.

When resetting a dry-pipe sprinkler system equipped with either an accelerator or an exhauster valve, a careful check should be made to determine if the device is automatic in operation or if it too must be reset. As there are numerous styles and designs, manufacturers' instructions should be consulted for information about specific valves. Automatic valves should not require attention; however, if a device that requires resetting after being tripped is neglected, the dry-pipe valve will trip, and the system will go wet when an attempt is made to return it to service.

Chapter 28

Pipe Valves

Pipe valves may be described as devices that control the flow of liquids, gases, etc., through pipes by opening and closing the passage. To accomplish this function, a great variety of valve designs have been developed, and these in turn are made in many specific design variations as well as a variety of materials. Pipe valves require a good understanding by mechanics of the fundamentals of sound practices in valve and piping installations and maintenance.

Valve Styles

Although industrial mechanics will probably be most concerned with smaller-sized valves such as the globe valve shown in Figure 28-1 or the angle valve shown in Figure 28-2, they should have an understanding of how to choose valves for specific service.

Figure 28-1 A typical globe valve.

Figure 28-2 A typical angle valve.

For example, an installation may require a valve incorporating an outside screw and yoke, such as the valve shown in Figure 28-3.

Figure 28-3 An outside screw and yoke gate valve.

Or there may be a question as to which style of check valve to use, the horizontal lift shown in Figure 28-4 or the swing shown in Figure 28-5.

Figure 28-4 A horizontal lift-check valve.

Figure 28-5 A swing-check valve.

Other considerations might be the provision of a safety device, such as the safety pop valve shown in Figure 28-6, or a quick-closing quarter-turn valve, such as the square-head cock shown in Figure 28-7.

When in doubt as to which valves or fittings are best suited, do not guess. It is a risky way to equip any piping system, and it is completely unnecessary. Manufacturers and distributors will gladly furnish information and assistance, or much of the exact information may be found in the manufacturer's catalog.

First, you should know where the piping materials are going to be used. Are they to be used on high- or low-pressure and/or high- or low-temperature installations? What kind of fluid will be sent

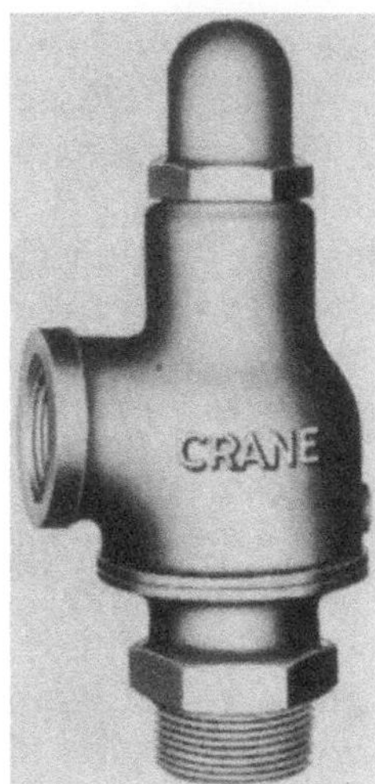

Figure 28-6 A typical safety pop valve.

Figure 28-7 A quick-closing quarter-turn valve.

through them? Will the conditions of operation be moderate or severe? How much headroom must you allow for valve stems? What size will your pipelines be? Will the valve have to be dismantled frequently for inspection and servicing? Is the installation to be relatively permanent, or must the piping be broken into frequently?

When you can answer these questions, you will know which valves to choose. You will know their operating characteristics, which materials you will need, their relative strength in service, and what kind of end connections are best suited to the installation. In short, you will know which of the many types and sizes of valves are needed to do the job most efficiently.

Valve Materials

Piping materials commonly used in industry fall into three basic material groups: bronze, iron, and steel. There are also variants in each of these groups, and each variant has individual service characteristics. It pays to have knowledge of the materials from which valves, fittings, and pipe are usually made and to understand the pressure, temperature, and structural limitations of each material.

Bronze used in the common varieties of valves is an alloy of copper, tin, lead, and zinc. It should not be used for temperatures exceeding 450°F. A higher-grade bronze, one that is a special high-grade alloy, will allow the use of bronze in piping equipment for higher pressures and temperatures up to 550°F.

Iron valves and fittings are made in three general grades: cast iron, high-tensile iron (limited to 450°F), and malleable iron suitable for slightly higher pressures and temperatures. Cast iron is commonly used for small valves and fittings having light metal sections. High-tensile iron is a high-strength alloy cast iron and is principally used for castings of large valves.

Malleable iron is particularly suited for use in screwed fittings, unions, etc., and also is used to some extent for valves and flanges. It is characterized by pressure tightness, stiffness, and toughness and is especially valuable for piping materials subject to expansion and contraction stresses and shock.

Steel is recommended for high pressures and temperatures and for service where working conditions, either internal or external, may be too severe for bronze or iron. Its superior strength and toughness, and its resistance to piping strains, vibration shock, low temperature, and damage by fire, afford reliable protection when safety and utility are desired. Many different types of steel are both necessary and available because of the widely diversified services steel valves and fittings perform.

Service Rating Marks

In addition to the maker's brand and size mark on most valves and on many fittings, a basic service rating also is shown. Pressure and temperature ratings are expressed in terms of steam unless otherwise indicated. For example, the rating of the globe valve shown in Figure 28-1 is 150 lb steam. For lower temperatures than that of steam at the pressures shown, the safe working pressure of a material is usually greater than the steam rating. Many valves have two service ratings, as shown on the swing-check valve in Figure 28-5. In addition to the steam rating, explained above, cold service ratings are usually designated by the mark WOG, which stands for cold water, oil, or gas, nonshock.

Cast- and forged-steel valves and fittings bear a mark such as 150, 300, or 600. The figures denote the maximum pressure at a certain temperature for which an item is suited. A certain 300-lb steel valve, such as shown in Figure 28-8, may be suited for 300-lb pressure at temperatures up to 850°F. If the temperature exceeds that point up to (let's say) 1000°F, however, the valve is not recommended for pressures over 80 lb. This important effect of temperature makes it imperative that both the pressure and temperature conditions of a service be known and that rating not be exceeded.

Temperature and pressure are not always the only factors to be considered, however. Frequently, steel materials are used for the range for which brass or iron is recommended, such as shock, vibration, line stresses, or fire hazard.

Valve Functions

The function of valves was described earlier as the control of the flow of liquids, gas, etc., through pipes by opening and closing the passage. This may be further defined by dividing this function into five principal groupings: starting and stopping flow, regulating or throttling flow, preventing backflow, regulating pressure, and relieving pressure.

Figure 28-8 A steel screw and yoke valve rated up to 300 lb of pressure.

Starting and stopping flow, illustrated in Figure 28-9, is the service for which valves are most generally used. Gate valves are excellently suited for such service. Their seating design, when open, permits fluid to move through the valve in a straight line with minimum restriction of flow and loss of pressure at the valve. The gate principle, which will be explained later, is not practical for throttling.

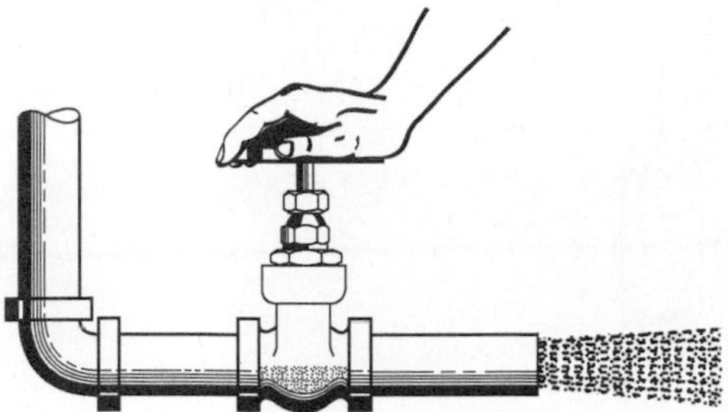

Figure 28-9 The function of a valve to start or stop flow of liquid.

Regulating or throttling flow is done most efficiently with globe or angle valves. Their seating causes a change in direction of flow through the valve body, thereby increasing resistance to flow at the valve. This in itself tends to throttle as shown in Figure 28-10. Globe and angle valve disc construction permits closer regulation of flow. These valves are seldom used in sizes above 12 inches, owing to the difficulty of opening and closing the large valves against pressure.

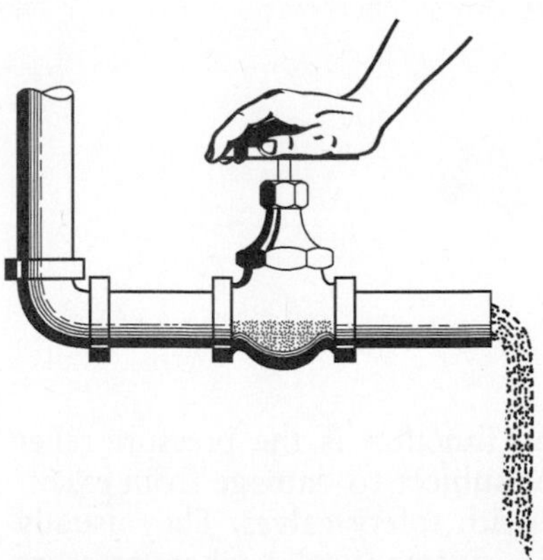

Figure 28-10 A valve regulating or throttling flow of liquids.

Check valves perform the single function illustrated in Figure 28-11: checking or preventing reversal of flow in piping. They come in two basic types: swing and lift checks. Flow keeps these valves open, and gravity and reversal of flow close them automatically. As a general rule, swing checks are used with gate valves, and lift checks are used with globe valves.

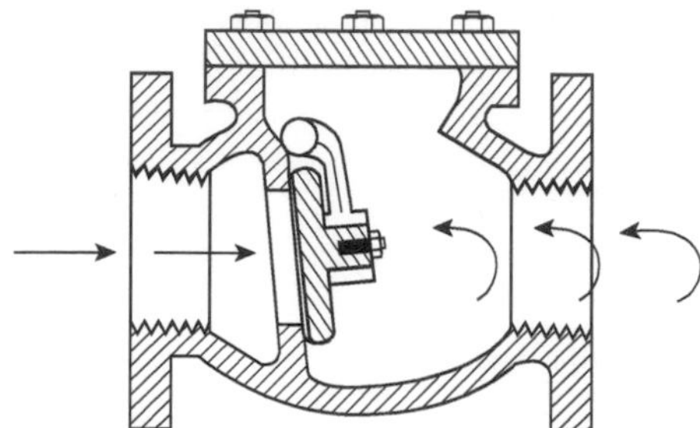

Figure 28-11 The basic function of a check valve.

The basic function of regulating valves is illustrated in Figure 28-12. The illustration shows the incoming pressure at 100 lb and the reduced pressure leaving the valve at 10 lb. Reducing valves not only reduce pressure but also maintain it at the point desired. Reasonable fluctuations of inlet pressure to a regulator valve do not affect outlet pressure for which it is set.

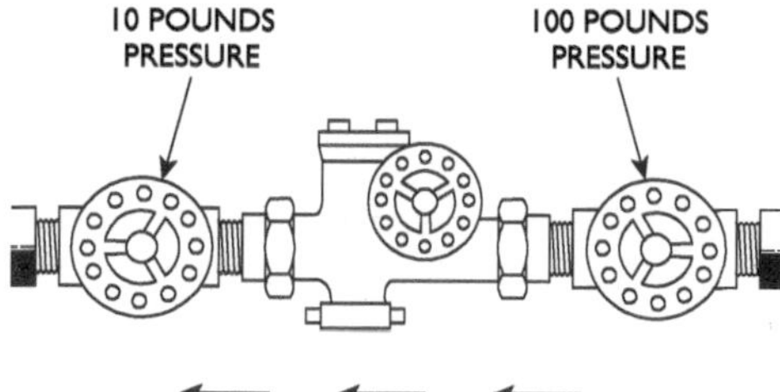

Figure 28-12 The basic function of a regulating valve.

The fifth grouping of valves by function is the pressure-relief valve. Boilers and other equipment subject to damage from excessive pressures should be equipped with safety valves. They usually are spring-loaded valves that open automatically when pressure exceeds limits for which the valve is set. These valves are of two general designs that externally appear very similar, as illustrated in Figure 28-13. The two types of relief valves are known as "safety valves" and "relief valves," and their internal construction will be discussed later. Generally, however, safety valves are used for steam, air, or other gases. Relief valves are usually used for liquid.

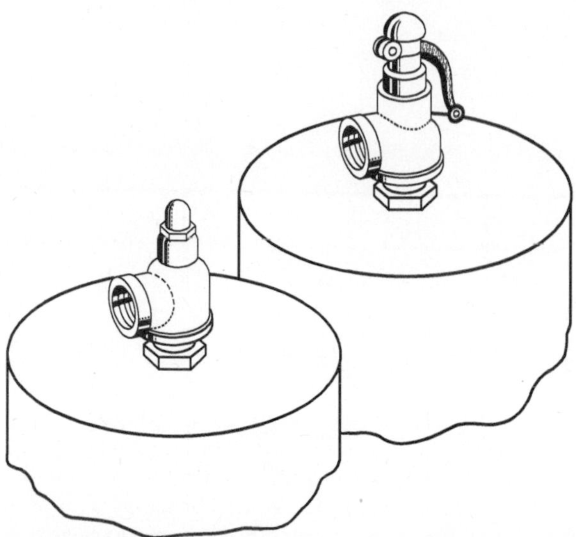

Figure 28-13 A pressure-relief valve.

Valve Construction and Operation

The cross-sectional illustration in Figure 28-14 shows a full-open gate valve. When the gate is in the raised position (as shown), fluids flow through in a straight line. This construction offers little resistance to flow and reduces pressure drop to a minimum. A gate-like disc, actuated by a stem screw and handwheel, moves up and down at right angles to the path of flow and seats against two seat faces to shut off flow.

Gate valves are best for services that require infrequent valve operation and where the disc is kept either fully open or fully closed. They are not practical for throttling. With the usual type of gate valve, close regulation is practically impossible. Velocity of flow against a partly open disc may cause vibration and chattering and result in damage to the seating surfaces. Also, when throttled, the disc is subjected to severe erosive effects called wire drawing.

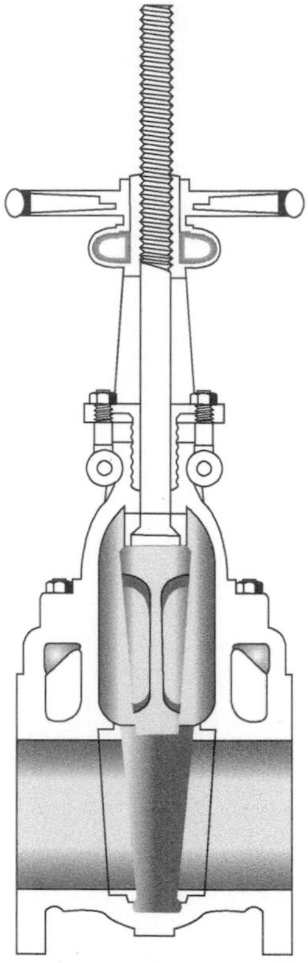

Figure 28-14 A cross-sectional view of the gate valve shown in Figure 28-3. The valve is in open position.

The internal construction of globe and angle valves, shown in Figures 28-15 and 28-16, causes a change in direction as fluids flow through. This seating construction increases resistance and permits close regulation of fluid flow.

Disc and seat can be quickly and conveniently reseated or replaced. This feature makes them ideal for services that require frequent valve maintenance. Shorter disc travel makes it quicker for opening and closing when valves must be operated frequently. Angle valves, used when making a 90° turn in a line, use these operating characteristics to change flow direction. An angle valve reduces the number of joints and saves makeup time. It also gives less restriction to flow than the elbow and globe valve it displaces.

Check valves may be termed *nonreturn* valves in view of the fact that they are used to prevent backflow in lines. In operating principle,

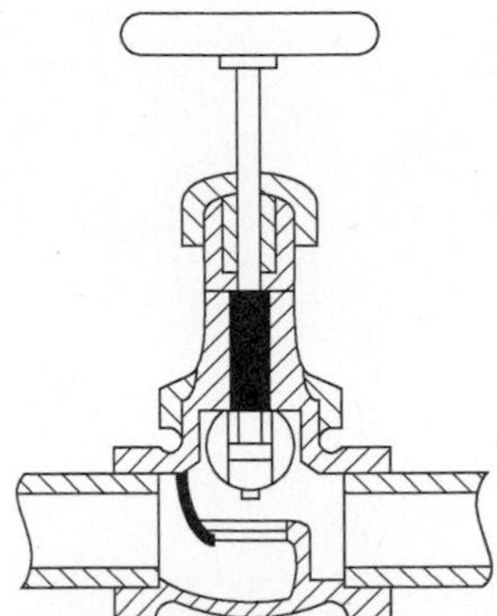

Figure 28-15 A cross-sectional view of the globe valve shown in Figure 28-1.

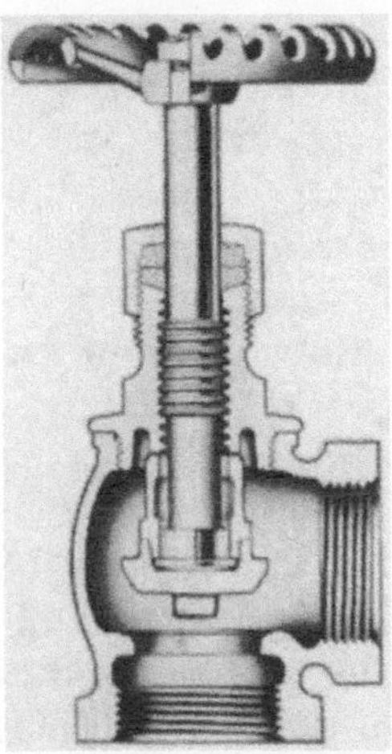

Figure 28-16 A cross-sectional view of the angle valve shown in Figure 28-2.

all check valves conform to one of two basic patterns. The illustration in Figure 28-17 shows the internal construction of the swing-check type.

The flow through these valves moves in approximately a straight line comparable to that in a gate valve. In lift-check valves (Figure 28-18), flow moves through the body in a changing course as in globe or angle valves. In both swing and lift types, flow keeps the valve open while gravity and reversal of flow cause it to close automatically.

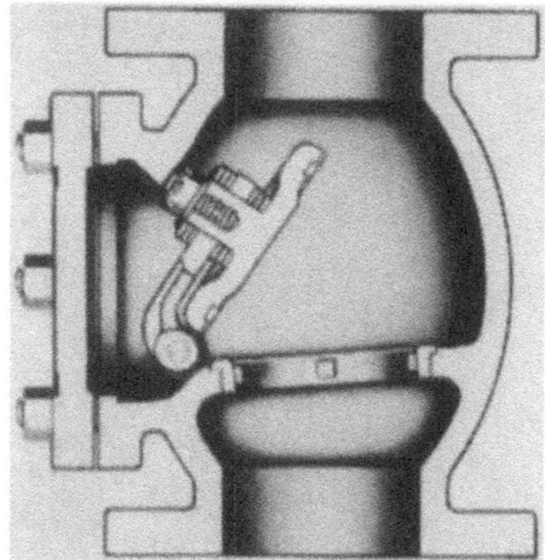

Figure 28-17 Internal view of a swing-check valve similar to valve shown in Figure 28-5.

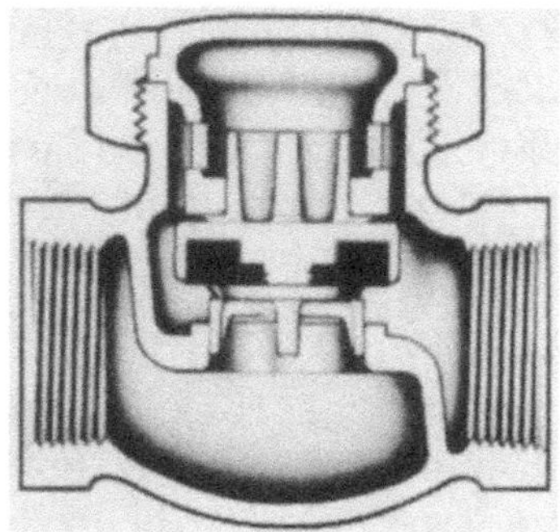

Figure 28-18 Internal view of a lift-check valve.

With the increasing use in industrial processing of volatile, corrosive, and hard-to-hold fluids, the need for absolute tightness around the stem has led to the development of the packless diaphragm valve. This style of valve, shown in Figure 28-19, eliminates the need for any stem packing.

The entire bonnet area is sealed off from the body, thereby preventing exposure of vital operating parts to the effects of line fluids, such as corrosion or erosion. Conversely, the diaphragm valve also prevents contamination of line fluids through contact with the working parts of the valve. The diaphragm serves one function only—sealing the bonnet. It is not subject to crushing and wear. The seating member is a separate circular flat-face disc, securely attached to the stem and joined to the diaphragm with a special

Figure 28-19 A diaphragm valve.

leakproof connection. The internal construction of the valve is shown in open position in Figure 28-20A and in closed position in Figure 28-20B. This independent seating feature permits positive shutoff with minimum loss of fluid, even if the diaphragm should fail.

The disc and/or gate of globe, angle, and gate valves is lifted from the seat or moved into contact with the seat by a threaded stem. Several variations of stem operation are used, each type having its own features. Two styles of rising stems with outside screw and yoke are in common use. In the style shown in Figure 28-21, the stem rises but the handwheel does not, whereas the handwheel shown in Figure 28-22 rises with the stem.

The stem screw remains outside the valve body in the rising-stem construction, whether the valve is opened or closed. Stem threads are not subjected to the effects of fluids in the line such as corrosion, erosion, or sediment. This construction also permits convenient lubrication of stem threads. The rising stem shows at a glance the position of the disc. Adequate headroom must be provided for the rising stem when the valve is opened, and the stem should be protected against damage when raised.

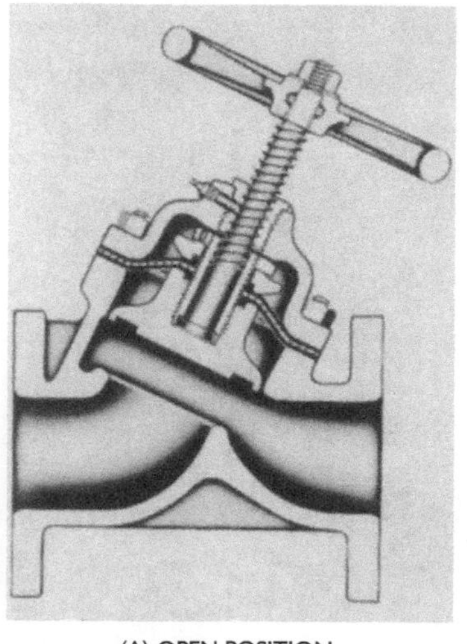

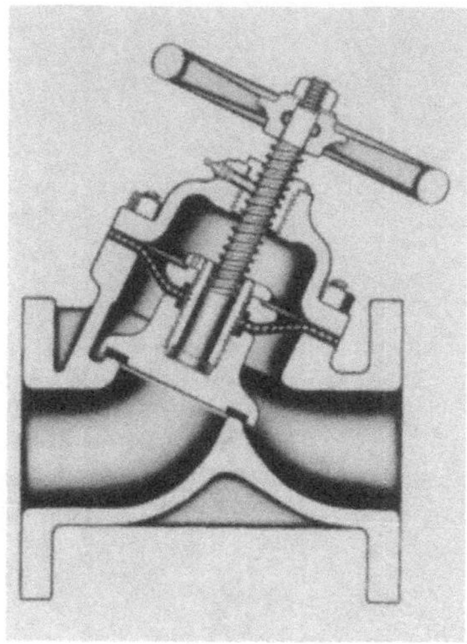

Figure 28-20 A cross-sectional view of the diaphragm valve shown in Figure 28-19.

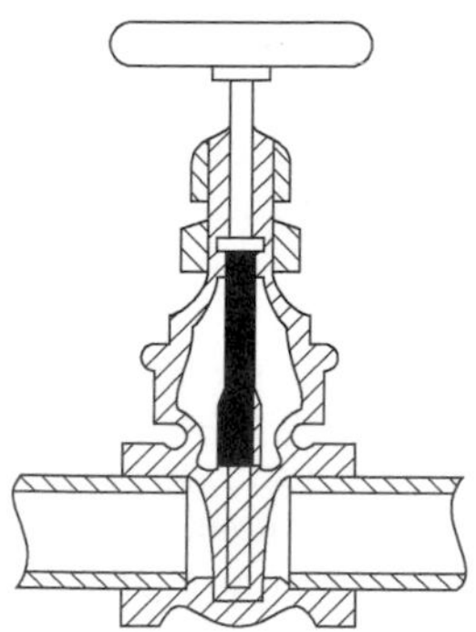

Figure 28-21 A cross-sectional view of the gate valve shown in Figure 28-3. The valve is in closed position.

The simplest and most common style of stem construction is the rising stem with inside screw, used for most gate, globe, and angle valves in the smaller sizes. In this style of construction, shown in Figure 28-23, the position of the stem indicates the position of the disc. The stem should be protected against damage when the valve is open.

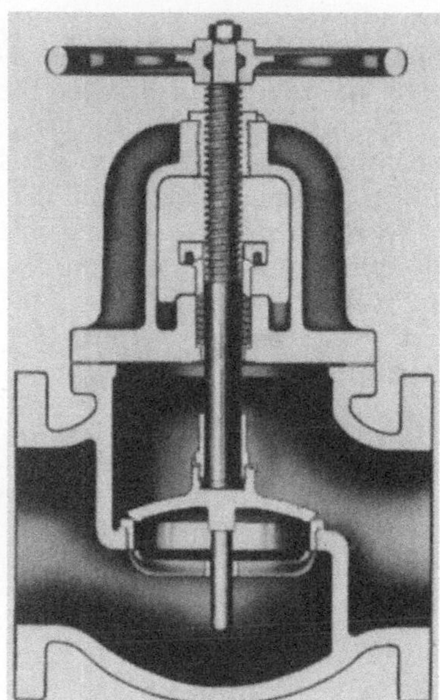

Figure 28-22 A gate valve handwheel rises with the body stem. The handwheel in some styles does not move up and down with the stem.

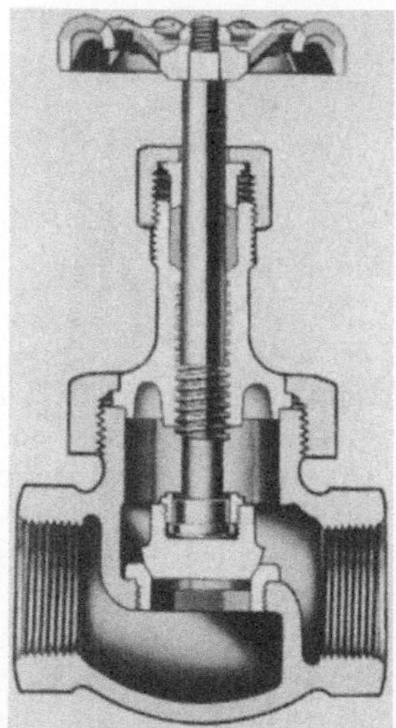

Figure 28-23 The smaller-size globe valve has a rising stem with inside screw.

The nonrising stem with inside screw, shown in Figure 28-24, does not rise when the disc is raised but merely turns with the handwheel. This style is ideal where headroom is limited. Because the stem merely turns when operated, wear on packing is minimized.

In Figure 28-25 the internal construction of a safety valve is shown. They are designed to open fully when they relieve, and they remain in this full-open position until pressure drops, at which time they snap closed. Because of this full-opening and tight-closing feature, they are commonly called "pop" valves.

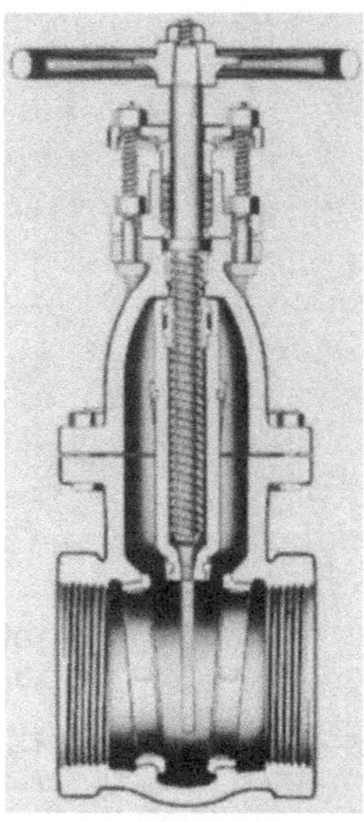

Figure 28-24 The stem does not move in and out of the valve in this unit.

Pop safety valves for steam service should be installed in a vertical position. For air or gas service, they may be installed inverted to allow moisture to collect and seal the mating surfaces. They should be mounted directly to tank or vessel without any intervening pipe or fitting. If discharge piping is used, it should be as short as possible and preferably larger in size than the outlet of the valve. The height of discharge piping should be adequately supported and never imposed on the valve. A loose slip joint in the outlet piping is recommended, as it permits expansion and contraction without unduly affecting the valve. Wrenches should be used with care so as not to abuse or distort the valve. Periodic testing of the valve by pulling the lever is advised.

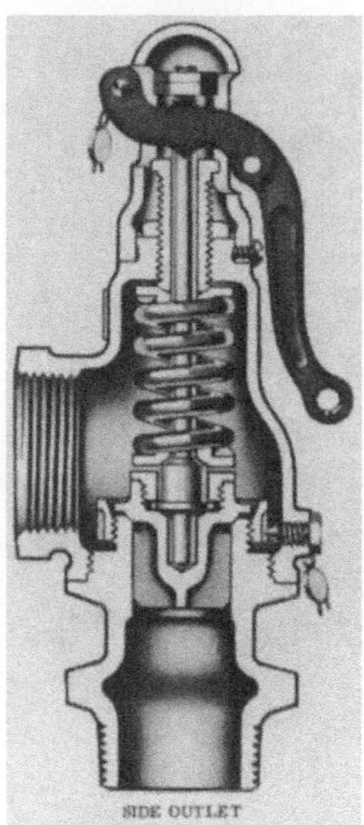

Figure 28-25 A cross-sectional view of a safety valve.

The relief valve, shown in Figure 28-26, is usually used for relief of pressure in a liquid system. Because most liquids can be compressed little if at all, the wide-opening feature of the pop valve is not required. A relief valve in operation adjusts the amount of opening to the volume of liquid it is necessary to discharge to prevent the pressure from exceeding the maximum setting.

Spring-loaded relief valves should be installed with the stem vertical. Whenever piping is installed in the inlet or outlet of these valves, it must be at least as large as the valve connections. Under no circumstances should it be reduced. This piping should be adequately supported to prevent line strains from causing the valve to leak at the seat.

Globe Valves

The most widely used of all pipe valves, in the smaller sizes, is the globe valve. It is made in three principal styles, identified by the design of the disc. These are the plug-type disc, the composition disc, and the conventional disc. The wide bearing surfaces of the

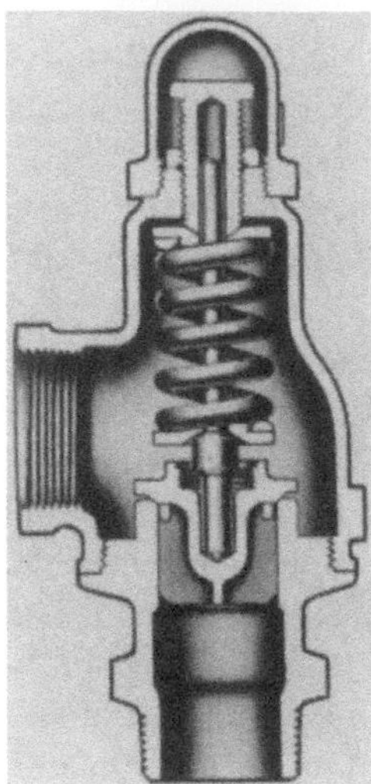

Figure 28-26 A cross-sectional view of the relief valve shown in Figure 28-13.

long, tapered-plug disc and matching seat, shown in Figure 28-27A, offer highest resistance to the cutting effects of dirt, scale, and other foreign matter.

This feature makes plug-type disc seating the best choice for the toughest flow-control services, such as throttling, drip and drain lines, soot blowers, blow-off, and boiler feed lines. Valves with plug-type discs and seats are available in a wide variety of materials either in globe or angle pattern.

The flat face of the composition disc, shown in Figure 28-27B, seats like a cap against the seat opening. The disc unit consists of a metal disc holder, composition disc, and retaining nut.

Discs are available in different compositions suitable to a variety of services such as air, steam, hot and cold water, oil, gas, and gasoline. By changing the type of disc, the valve can be changed over from one service to another. Discs can be renewed quickly in case of leakage. Composition discs are suitable to all moderate-pressure services, except throttling, at temperatures as recommended. They often will withstand embedding of dirt without leaking. They are available usually in brass and iron valves only.

In contrast to the plug-type disc, note in Figure 28-27C the thin line contact achieved by the taper of the conventional seat with the face of the disc. Because this narrow line bearing serves to break

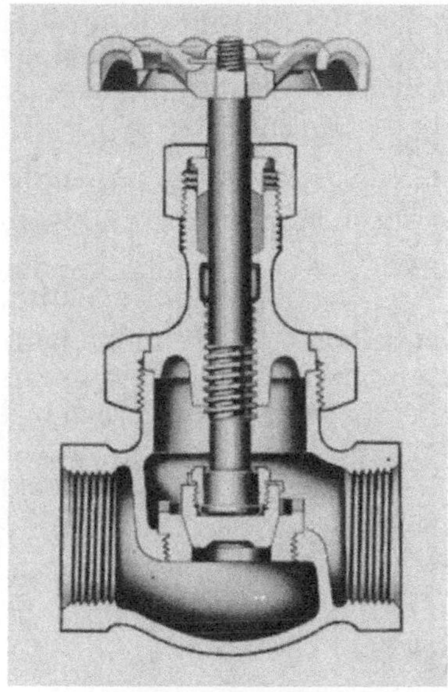
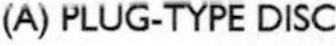

(A) PLUG-TYPE DISC.

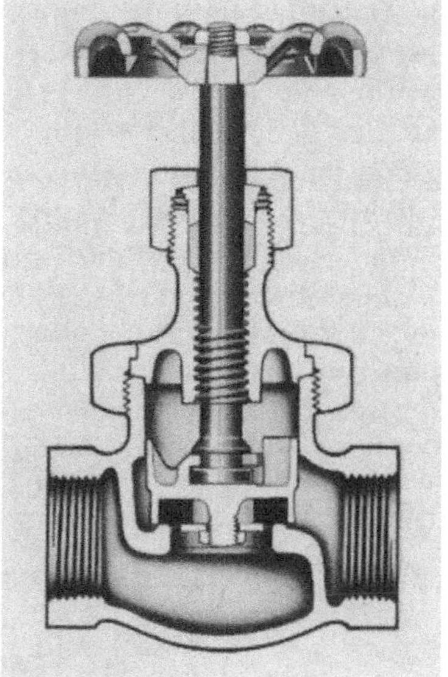

(B) COMPOSITION DISC.

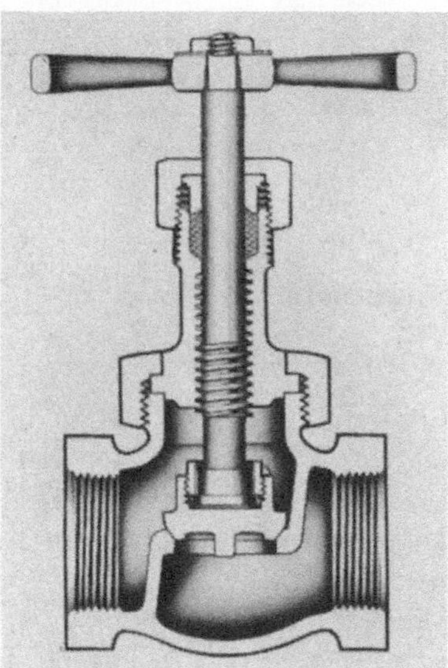

(C) CONVENTIONAL DISC.

Figure 28-27 Three basic designs of globe valves.

down hard deposits that form on the seats in some services, pressure-tight closure is ensured. Conventional disc valves are widely used for many cold and hot services. Valves with conventional seating are available in brass, iron, steel, and alloy materials.

Installation of Valves

A frequently asked question is, "What is the best working position for any valve?" Generally, valves work best when standing upright, with the stem pointing straight up. Any position from straight-up to horizontal is satisfactory but still a compromise. Installing a valve with its stem down is not good practice. With a valve in this inverted position, the bonnet acts as a trap for sediment, which will cut and damage the stem. Upside-down position for valves on liquid lines subjected to freezing temperatures is bad because liquid trapped in the bonnet may freeze and rupture it. See Figure 28-28.

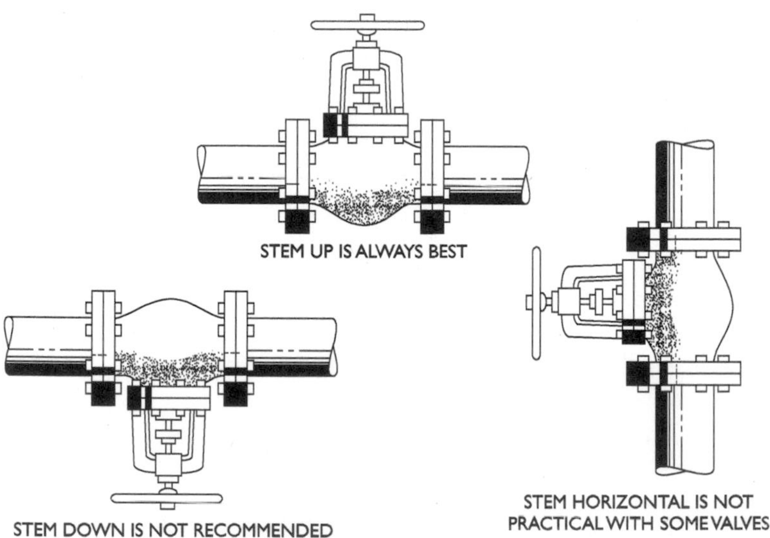

Figure 28-28 Installation position of valves.

Another common question in respect to installation of valves is, "Should the pressure be above or below the disc?" This question, which applies to both globe and angle valves, has several answers because service conditions vary. On lines where continuous flow is desired, it is safer to have pressure below the disc. For example, a disc may become separated from its stem and automatically shut off flow if pressure is above the disc. If this is dangerous for certain installations, then the pressure should be below the disc.

In general, unless pressure under the disc is definitely required, a globe valve or an angle valve will give more satisfactory service when installed with pressure above the disc.

Note: An exception is a valve with a renewable composition disc, which preferably should have pressure below the disc to ensure longer disc life.

The primary caution to be observed when installing any check valve, whether swing or lift check, is to make sure that flow enters at the proper end, i.e., that the disc opens with flow. The arrows in Figure 28-29 indicate the direction of flow through correctly installed swing- and lift-check valves.

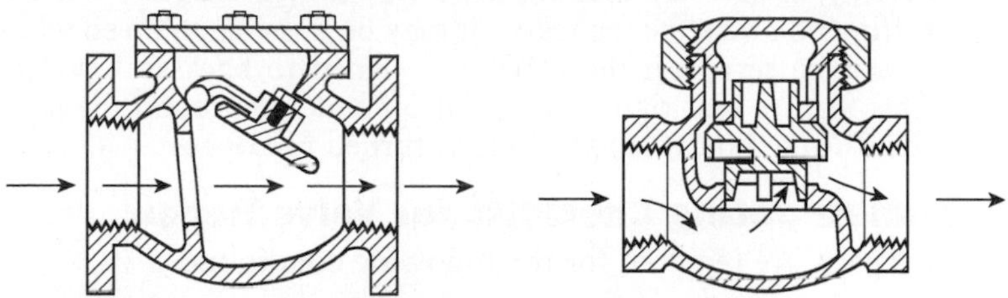

Figure 28-29 Direction of flow through a check valve.

Most manufacturers plainly mark check valves for direction of flow with an arrow similar to that in Figure 28-30. If you follow this marking on the body of the valve, you can be sure the valve is properly installed and that the disc will be properly seated by backflow, or by gravity when there is no flow.

Figure 28-30 A check valve with arrow to show direction of flow.

Maintenance of Valves

Routine maintenance ensures satisfactory valve performance; therefore, do not overlook leaks—big or small. A leak in a valve stem often can be remedied simply and in a hurry, if caught in time. Stem leaks normally can be fixed by slightly tightening the packing nut or gland. Bonnet and flange leaks may be caused by bolts loosening under service strain. If tightening the joint does not stop the leak, then inserting a new gasket probably will.

Wear on stem packing is due mainly to the rising and turning motion of the valve stem, combined with deteriorating effects of service conditions. A few drops of oil on the stem help to reduce friction and wear. Do not forget to lubricate exposed stems.

Stuffing-box leaks on larger valves can be stopped by pulling up the gland follower. On bolted glands, care must be taken to tighten bolts evenly, as severely cocking the follower will bind the stem. If the stuffing box must be repacked, it may be possible to do so while the valve is in service if the valve is designed to back-seat. Valves with this feature seat tightly and seal off leakage around the stem into the stuffing box when the stem is turned full open.

Troubleshooting Checklist for Valve Repair

Valve repairs are required for the following conditions:

1. When external leakage at the stem, body, or shaft cannot be stopped by tightening packing nuts, gland flanges, or body bolting. Check for failure of packing, gaskets, or seals.
2. When closing the valve does not stop fluid flow. Check for wear, corrosion, or erosion of seating surfaces. For check valves, check for sticking or binding of flow-control element or failure of the springs.
3. When opening the valve does not allow fluid to start. Look for a failure of the connection between the valve stem and the flow-control element.
4. When there is a fluid leakage through the valve shell. Investigate possible erosion or corrosion of the inner surface of the casting.

Chapter 29

Structural Steel

Millwrights and mechanics may on occasion be required to fabricate and install simple structural steel members. They may be furnished sketches or drawings or be given instructions only. If no drawings are provided, they must have knowledge of basic standards, practices, terms, etc., to be able to perform this type of work efficiently and safely. The usual practice is to fabricate members in the shop that are to be erected in the field. This requires that holes in the fabricated member match the hole locations in the field, that proper clearances be provided, and that interferences be avoided. The first requirement is to know and understand commonly used structural steel terms.

Structural steel is produced at rolling mills in a wide variety of standard shapes and sizes. In this form, it is referred to as *plain material.* In addition to plates and bar stock, the four shapes most widely used are standard angles, standard beams, standard channels, and standard wide-flange beams and columns.

American Standard Angles

Standard angles consist of two legs set at right angles as shown in Figure 29-1. The symbol used to indicate an angle shape on drawings

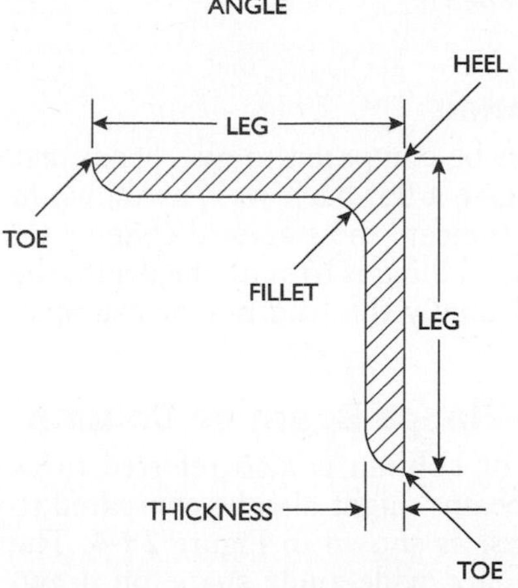

Figure 29-1 End view of standard angle-iron.

is (∠). The usual method of billing is to state the symbol, followed by the long leg, the short leg, the thickness, and finally the length—for example, ∠6 × 4 × 3⁄8 × 12'4". When the legs are equal, both leg lengths are stated—for example, ∠4 × 4 × 3⁄8 × 12'0".

American Standard Beams

Standard beams are generally called "I" beams because of their resemblance to the capital letter "I." See Figure 29-2. The symbol used to indicate the standard beam shape on drawings is the letter (I). The usual method of billing is to state the depth, the symbol, the weight per foot, and finally the length—for example, 15 I 42.9 × 18'4½".

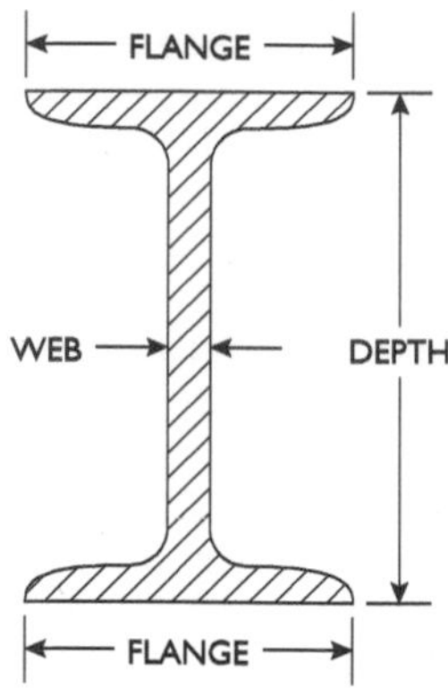

Figure 29-2 End view of standard I beam.

American Standard Channels

The standard channel shape might be compared to an I beam that has been trimmed on one side to give a flat back web, as shown in Figure 29-3. The symbol used to indicate the standard channel on drawings is ([). The usual method of billing is to state the depth, the symbol, the weight per foot, and finally the length—for example, 10[15.3 × 16'5¾".

American Standard Wide-Flange Beam or Column

The standard wide-flange beam or column is also referred to as "B," "C," or "H" shape. These beams might also be compared to an I beam with extra-wide flanges, as shown in Figure 29-4. The symbol used to indicate the standard wide-angle shape on drawings is (WF). The usual method of billing is to state the depth, the symbol, the weight per foot, and finally the length—for example, 12 WF 45 × 24'8".

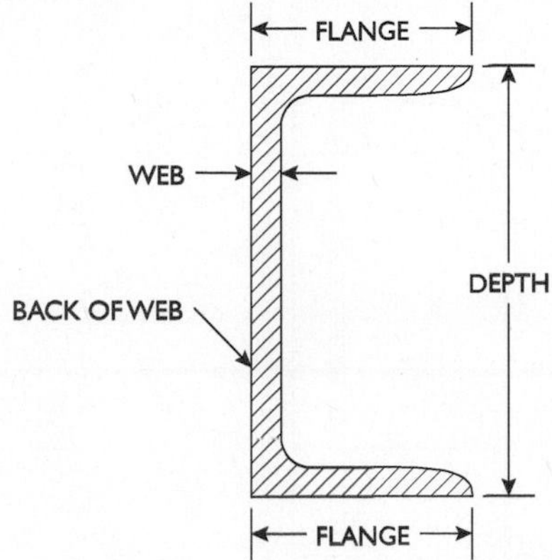

Figure 29-3 End view of standard channel beam.

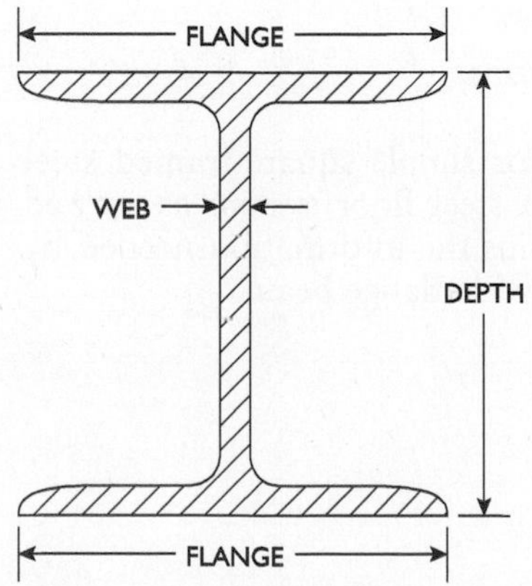

Figure 29-4 End view of standard wide-flange beam.

Simple Square-Framed Beams

Square-framed beams are, as the name implies, beams that intersect or connect at right angles. This is the most common type of steel construction and the style a millwright or mechanic should be familiar with. Two types of connections may be used in square-framed construction. They are the "framed" and "seated" types. In the framed type, shown in Figure 29-5A, the beam is connected by means of fittings (generally a pair of angles) attached to its web. With the seated connection, shown in Figure 29-5B, the end of the beam rests on a ledge or seat.

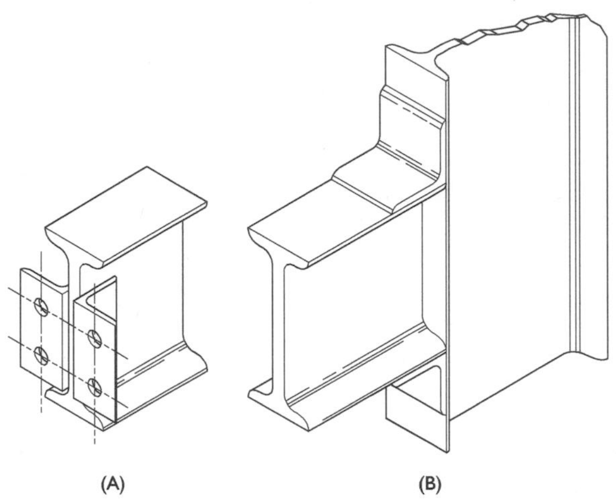

Figure 29-5 Simple square-framed beams.

Figure 29-6 shows a part plan for simple square-framed steel construction. It represents part of a steel floor system as viewed from above. With its notes, it contains the information needed by the mechanic to fabricate the 12-in. wide-flange beam.

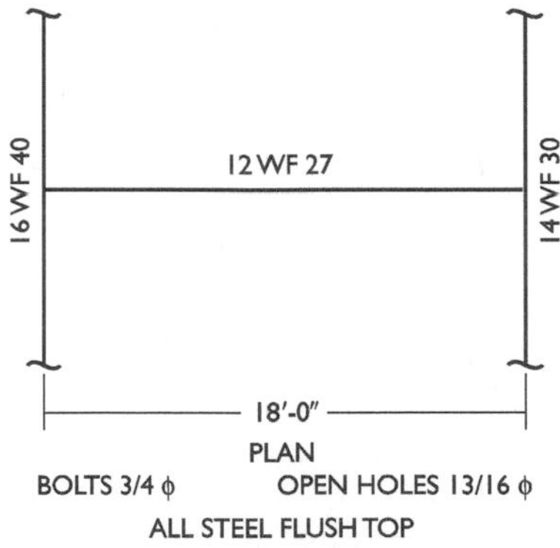

Figure 29-6 Plans showing construction of a simple square-framed beam.

The note indicates that all steel members are to be flush on the top. To accomplish this, it will be necessary to cut away the top angle of the 12-in. wide-flange beam at both ends as shown in Figure 29-7. This is a section through the 16 WF and the 14 WF.

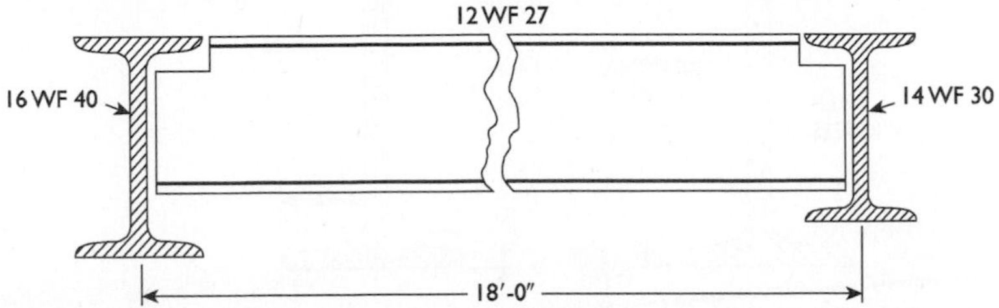

Figure 29-7 The typical simple square-framed beam with flush top.

Clearance Cuts

When connecting one member to another, it is often necessary to notch or cut away one or both flanges of the entering member to avoid flange interference. Such a notch is called a "cope," "block," or a "cut." In some shops, the term "cope" is used if the cut follows closely the shape of the member into which it will fit. When the cut is rectangular in shape with generous clearance, it is usually called a "block-out." Unless there is some reason for a close matching fit, the "block-out" is recommended, as it is the easiest and most economical notch to make.

When making block-out cuts, the dimensions of the rectangular notch may be obtained directly from tables of structural steel dimensions. These dimensions are standardized, and tables may be obtained from steel manufacturers or suppliers.

Included in the appendix are structural steel tables for the commonly used sizes of standard beams, channels, and wide-flange beams. The tables list the "K" and "a" dimensions for members of various sizes and weights. These are standard dimensions for the various measurements illustrated in Figure 29-8.

The "K" and "a" values determine the dimensions of the block-out cut, as they indicate the maximum points of interference. Although the notch is made in the entering member, as shown in Figure 29-9, the values from the table for the supporting member determine the notch dimensions. The steel must be cut to length before the block-out cut is made. The clearance in the notch will then be the same amount as the end clearance of the member.

STANDARD DIMENSION

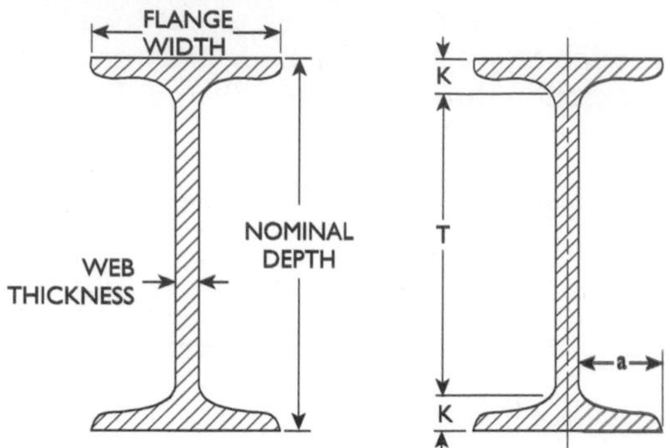

K - DISTANCE FROM FLANGE SURFACE TO START OF FLAT WEB SURFACE.
T - LENGTH OF FLAT WEB SURFACE, FROM START OF FILLET TO START OF FILLET.
a - ONE-HALF FLANGE WIDTH MINUS ONE-HALF WEB THICKNESS, OR FROM WEB SURFACE TO FLANGE TIP.

Figure 29-8 Standard dimensions for various sizes and weights.

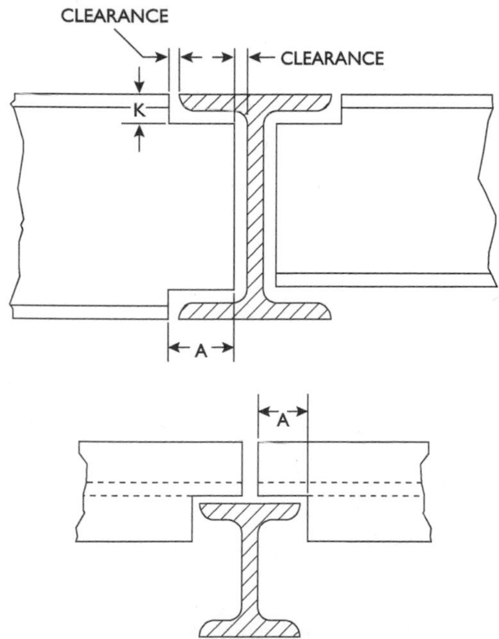

Figure 29-9 A typical square-framed beam showing various cutouts and dimensions.

Determination of end clearance and steel length will be discussed later.

Square-Framed Connections (Two-Angle Type)

So-called "standard" connections should be used when fabricating and installing structural steel members. This is essential if shop-fabricated members are to properly assemble with supporting members at field installation. Standard connections are a part of steel standardization, which makes assembly possible. An understanding of the terms, dimensions, etc., of steel standardization is basic to the use of standard connections.

Spread

The distance between hole centers in the web of the supporting member and the holes in the *attached* connection angles is called the "spread" of the holes. The spread dimension is standardized at 5½ in. as shown in Figure 29-10.

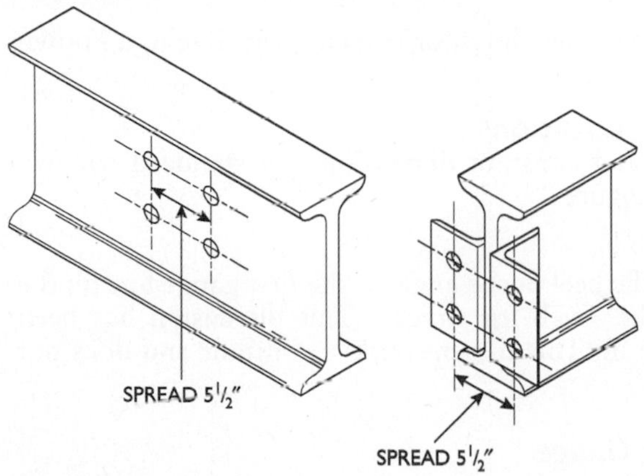

Figure 29-10 Illustrating the spread, which is the distance between attaching holes.

Angle Connection Legs

The legs of the angles used as connections are specified according to the surface to which they are connected, as shown in Figure 29-11. The legs that attach to the entering steel to make the connections are termed "web legs." The legs of the angles that attach to the supporting member are termed "outstanding legs." The lines on which holes in the angle legs are placed are called "gauge lines."

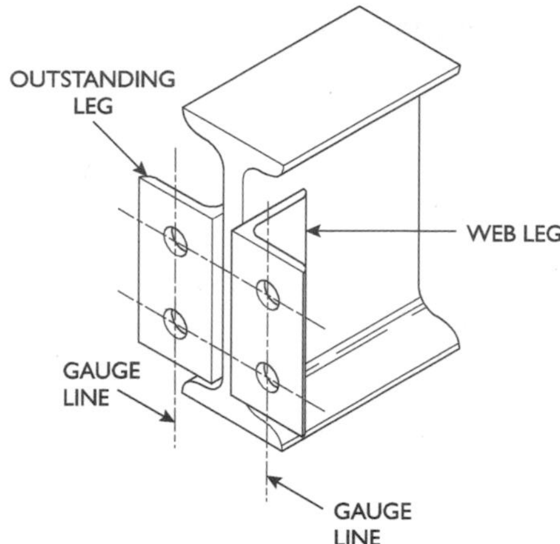

Figure 29-11 Connecting legs used to connect supporting members.

The distance between gauge lines, or from a gauge line to a known edge, is called "gauge."

Connection Hole Locations

The various terms and constant dimensions for standard connections are shown in Figure 29-12.

Web Leg Gauge

The distance from the heel of the angle to the first gauge line on the web leg is called the "web leg gauge." This dimension has been standardized at 2¼ in. This dimension is a constant and does not vary.

Outstanding Leg Gauge

The distance from the heel of the angle to the first gauge line on the outstanding leg is called the "outstanding leg gauge." This dimension varies as the thickness of the web of the member varies. This variation is necessary to maintain a constant 5½-in. spread dimension.

The outstanding-leg-gauge dimension is determined by subtracting the web thickness from 5½ in. and dividing by 2. Another, possibly simpler, way to state outstanding web gauge is the following: 2¾ in. minus ½ the web thickness.

The distance between gauge lines, or from a gauge line to a known edge, is called a gauge. When more than one row of holes is

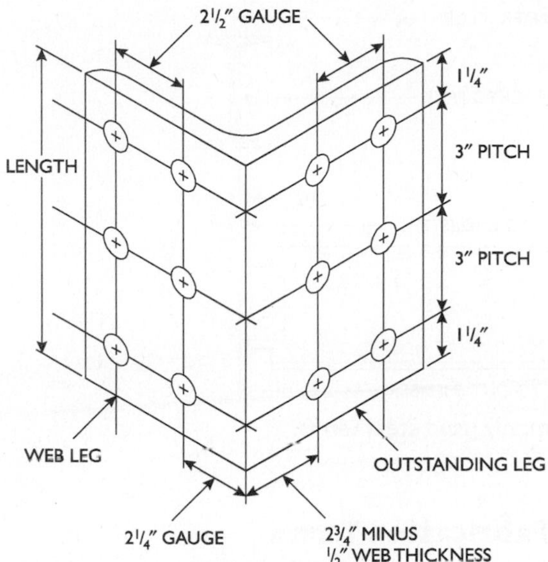

Figure 29-12 Connection hole locations in an angle. These dimensions are standard and should never vary.

used, the gauge is 2½ in. This is a constant dimension and does not vary.

Pitch

The distance between holes on any gauge line is termed "pitch." This dimension is also a constant, being standardized at 3 in.

End Distance

The "end distance" is equal to one-half of the remainder left after subtracting the sum of all pitch spaces from the length of the angle. By common practice, the angle length is selected to give a 1¼-in. end distance.

Fabricating Terms

Illustrated in Figure 29-13 are commonly used structural-steel fabrication terms. Use of these terms, because they are descriptive, aids in the understanding of steel fabrication and reduces the probability of errors.

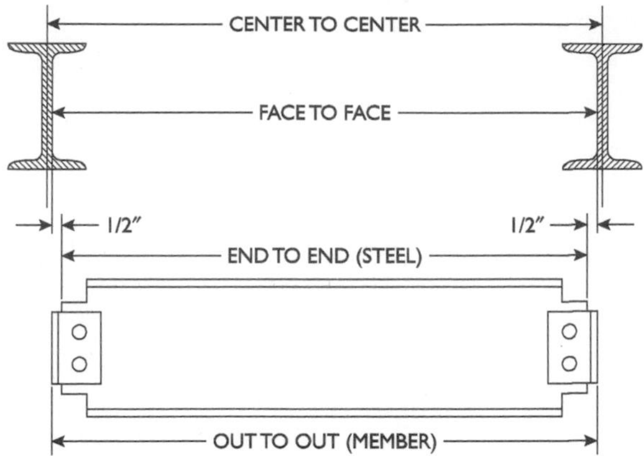

Figure 29-13 Commonly used steel terms.

Definitions of Fabrication Terms

Steel—Refers to various structural steel shapes and forms.

Member—Refers to an assembly of a length of steel and connection fittings.

Center-to-Center—Refers to the distance from the centerline of one member to the centerline of another member.

Face-to-Face—Refers to the distance between the facing web surfaces of two members. It is the center-to-center distance minus the ½-web thickness of each member.

End-to-End—Refers to the overall length of the steel. The end-to-end length should be 1 in. shorter than the face-to-face dimension. Making the steel *1 in. shorter* than the opening into which it will be placed provides clearance for assembly as well as allowance for inaccuracies.

Out-to-Out—Refers to the overall length of the member. To provide assembly clearance, the connection angles are positioned on the member so that the out-to-out distance is slightly less than the face-to-face distance.

Dimensions of Two-Angle Connections

Detailed drawings for fabrication of structural steel members specify dimensions, hole locations, etc., for the connection angles.

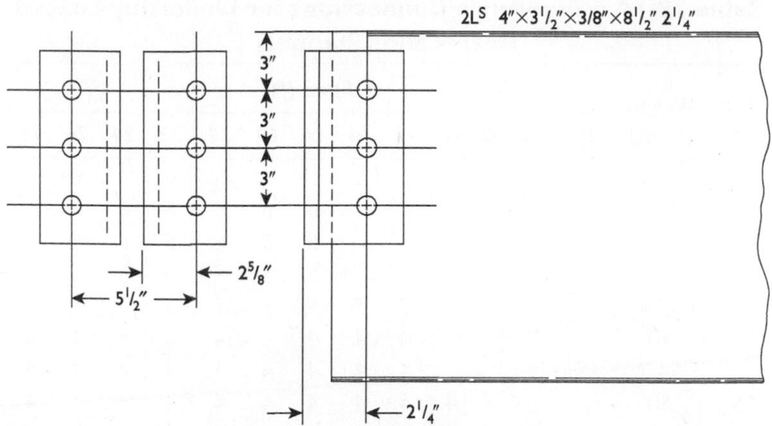

Figure 29-14 Typical two-angle connection.

Figure 29-14 illustrates a typical drawing of a two-angle connection. Note that the connection angles are slightly above center. The center of the uppermost hole is 3 in. below the top of the beam. Whenever practicable, the uppermost holes are set at 3 in. below the top of the beam, as this makes for standardization and tends to reduce errors in matching connections.

Mechanics are not usually expected to select the size of steel to be used. Selection is usually made by engineers or draftsmen who have knowledge of anticipated loadings and who are competent to make the necessary strength calculations. Connection details, however, may not be provided, and the selecting of connections may be left to the mechanic. To aid in the selection of connections, the drawings in Figure 29-15 and Table 29-1 have been prepared. For purposes of simplification, the connections have been given numbers 1 through 6. Although these are not standard connection numbers, all dimensions are standard. Note also that only connections using ¾-in. diameter bolts are shown. Again, this is for purposes of simplification, as ¾-in. bolts are adequate for use with the steel sizes that a mechanic might fabricate and install.

To select angle connections, first choose from Table 29-1 the shape of steel being used. Find the size in the first column on the left and the weight per foot in the next column. Follow along the weight line to the column that corresponds to the span length. The number in this column indicates the connecting angle number to

Table 29-1A Two-Angle Connections for Uniformly Loaded Beams and Channels

Size (in.)	*Weight (lbs/ft)*	*Span (ft)*													
		2	***4***	***6***	***8***	***10***	***12***	***14***	***16***	***18***	***20***	***22***	***24***	***26***	***28***
24	120										6	6	6	6	6
	100								6	6	6	6	6	6	6
	79.9								6	6	6	6	6	6	6
20	95								5	5	5	5	5	5	5
	65.4						5	5	5	5	5	5	5	5	5
18	70						4	4	4	4	4	4	4	4	4
	54.7						4	4	4	4	4	4	4	4	4
15	50				4	4	4	4	4	4	4	4	4	4	4
	42.9				4	4	4	4	4	4	4	4	4	4	4
12	50				3	3	3	3	3	3	3	3	3	3	3
	31.8				3	3	3	3	3	3	3	3	3	3	3
10	35				2	2	2	2	2	2	2	2	2	2	2
	25.4			2	2	2	2	2	2	2	2	2	2	2	2
8	23		2	2	2	2	2	2	2	2	2	2	2	2	2
	18.4		2	2	2	2	2	2	2	2	2	2	2	2	2
7	20		1	1	1	1	1	1	1	1	1	1	1	1	1
	15.3		1	1	1	1	1	1	1	1	1	1	1	1	1
6	17.2		1	1	1	1	1	1	1	1	1	1	1	1	1
	12.5		1	1	1	1	1	1	1	1	1	1	1	1	1
5	14.7		1	1	1	1	1	1	1	1	1	1	1	1	1
	10	1	1	1	1	1	1	1	1	1	1	1	1	1	1

use. In Figure 29-15 will be found the drawings of the various connections 1 through 6. All the necessary dimensions to fabricate angles that conform to established standards are given. Use Table 29-1 in conjunction with Figure 29-15.

Steel Elevation

Unless otherwise stated on the drawing, it is presumed that all square-framed beam members are parallel or at right angles to one another, that their webs are in a vertical plane, and that they are in a level position from end to end. Because dimensions giving the vertical position of beams cannot be shown in a plan viewed from above, this information is usually furnished by a note. This may be

Table 29-1B Two-Angle Connections for Uniformly Loaded Wide-Flange Beams

Size	*Weight*	*Span (ft)*												
(in.)	*(lbs/ft)*	*4*	*6*	*8*	*10*	*12*	*14*	*16*	*18*	*20*	*22*	*24*	*25*	*28*
24	120												6	6
	76							6	6	6	6	6	6	6
21	96									5	5	5	5	5
	62						5	5	5	5	5	5	5	5
18	70											4	4	4
	50						4	4	4	4	4	4	4	4
16	96									4	4	4	4	4
	64						4	4	4	4	4	4	4	4
	36				4	4	4	4	4	4	4	4	4	4
14	38						3	3	3	3	3	3	3	3
	34					3	3	3	3	3	3	3	3	3
	30					3	3	3	3	3	3	3	3	3
12	36				3	3	3	3	3	3	3	3	3	3
	31					3	3	3	3	3	3	3	3	3
	27				3	3	3	3	3	3	3	3	3	3
10	29			2	2	2	2	2	2	2	2	2	2	2
	25			2	2	2	2	2	2	2	2	2	2	2
	21		2	2	2	2	2	2	2	2	2	2	2	2
8	20	2	2	2	2	2	2	2	2	2	2	2	2	2
	17	2	2	2	2	2	2	2	2	2	2	2	2	2

Note: Connections must not be used for shorter spans than indicated, as their capacity does not equal that of the steel in shorter spans.

done by giving the vertical distance, or elevation, above some established horizontal plane in feet and inches.

Elevation of steel might also be given by a note reading, "All steel flush at elevation 64'8" except as noted."

The elevation relationship affects directly any clearance cutting or notching that may be necessary to connect one member to another. For example, the connection shown in Figures 29-6 and 29-7 required that the 12 WF beam be notched at the top on both ends. Figure 29-16 shows the same size steel; however, the 16 WF beam is 4 in. above the other two members rather than all flush as in Figure 29-7. In this case, the end connecting to the 14 WF beam

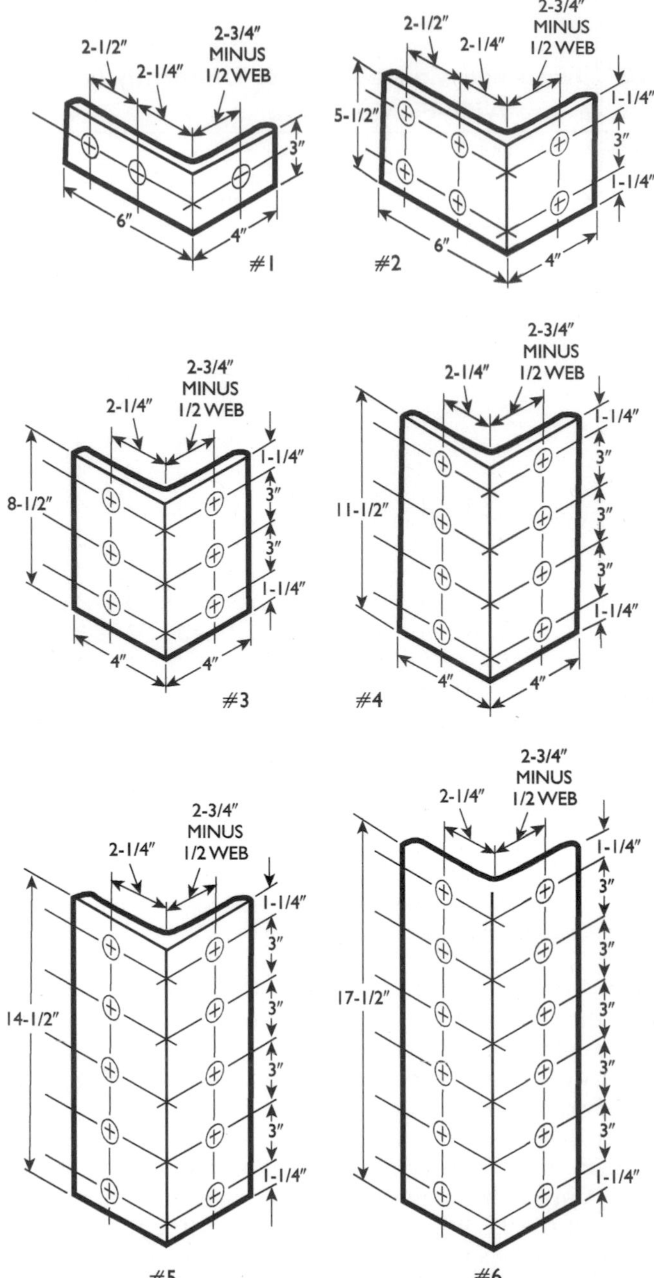

Figure 29-15 Various angle-connection dimensions.

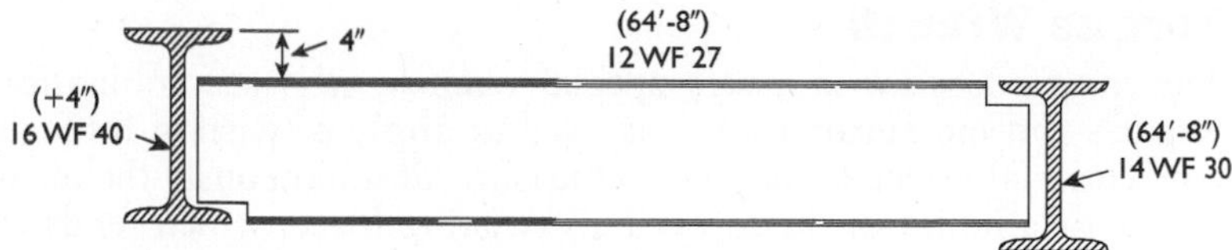

Figure 29-16 Same example as shown in Figure 29-7 except for notch cutting.

will require notching at the top, as in the previous example, but the end connecting to the 16 WF must be notched at the bottom.

To find out whether it is necessary to notch out steel to provide clearance for flanges, the fillet or "K" distance of the member to be entered must be checked. When a member enters at a level below that of the supporting steel, and the amount is greater than the "K" dimension, no notch at the top is necessary. Members entering flush or at a higher level must be notched at the top to provide clearance. Bottom notching depends on the size of the steel as well as the elevation. Equal size members connecting with flush tops require notching at both top and bottom. Checking the "K" dimension in relation to any differences in the elevation of the bottom surfaces of connecting steel members will indicate if notching is necessary.

Structural Steel Bolts

In most cases, ordinary commercial machine bolts are adequate for the minor steel installation work the millwright or mechanic might do. If, however, extensive work is to be done or near-capacity loading of the steel is anticipated, special structural steel bolts should be used. These are high-strength steel bolts made especially for structural steel joints and are supplied with a suitable nut and a plain hardened washer. They are made of a medium carbon steel, quenched and tempered to ASTM specification A325. These bolts may be identified by three grade marks and the number A325 on the head.

When using structural steel joint bolts, they should be tightened with a torque wrench to ensure proper tension. It is important that a high enough torque be applied to develop the bolts' maximum clamping power. Because the A325 bolt will be used only once, more torque can be applied to this type of bolt than to one that might be reused. The tightening torque for a ¾-in. diameter A325 bolt will be in the area of 300 ft-lbs; the specific value should be obtained from the manufacturer's recommendations.

Torque Wrench

The *torque wrench* is a very special wrench; it is a combination wrench and measuring tool. It is used to apply a twisting force as conventional wrenches do and to measure simultaneously the magnitude or amount of the force. This twisting force, which tends to rotate an object about its axis, is called "torque."

There are numerous styles of torque wrenches, some that are direct reading and others that have signaling mechanisms to warn when the predetermined torque is reached. The calibration of all is based on the fundamental law of the lever—that is, *forces* times *distances* equals *torque*. This law is illustrated in Figure 29-17.

Torque values, or units of measure, are stated in terms of the fundamental lever law, that is, a force times a distance. The two torque units of measure commonly used by mechanics are the "inch-pound" and the "foot-pound." They are the product of a force in pounds and a lever length in either inches or feet.

An example of inch-pounds is shown in Figure 29-18 where a force of 5 lb is being applied on a wrench at a distance of 10 in. from the center of rotation. The torque in this example is 5 lb times 10 in., or 50 inch-pounds.

A similar example is illustrated in Figure 29-19 where a force of 60 lb is being applied on a wrench at a distance of 2 ft from the center of rotation. The torque in this example is 60 lb times 2 ft, or 120 foot-pounds.

To convert inch-pounds to foot-pounds, simply divide by 12. For example, a torque value of 60 inch-pounds corresponds to 5 foot-pounds (60 divided by 12). Similarly, to convert foot-pounds to inch-pounds, multiply by 12. For example, a torque value of 12 foot-pounds corresponds to 144 inch-pounds (12 multiplied by 12).

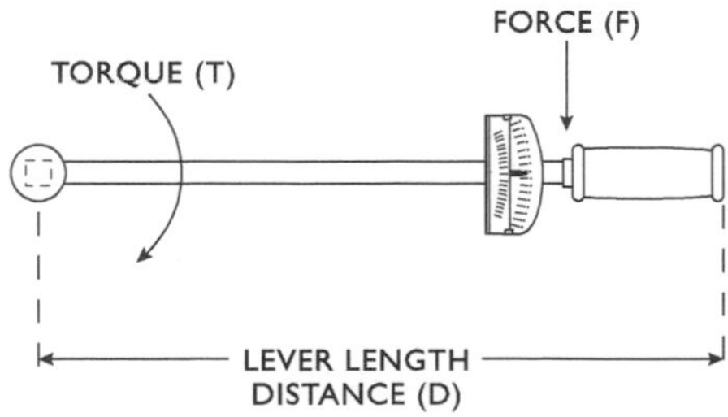

Figure 29-17 Illustrating the fundamentals of a torque wrench.

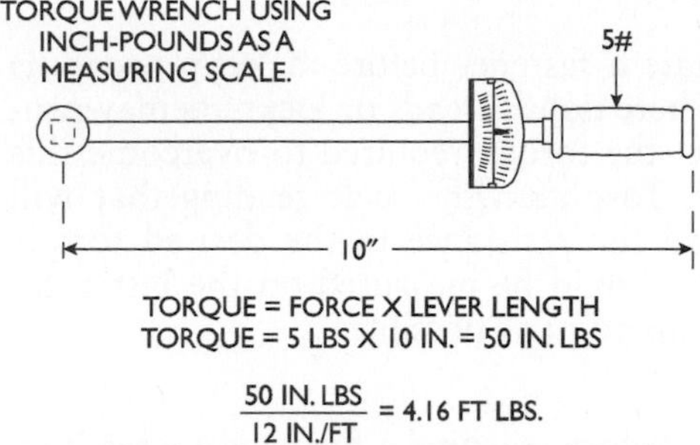

Figure 29-18 Torque wrench using inch-pounds as a measuring scale.

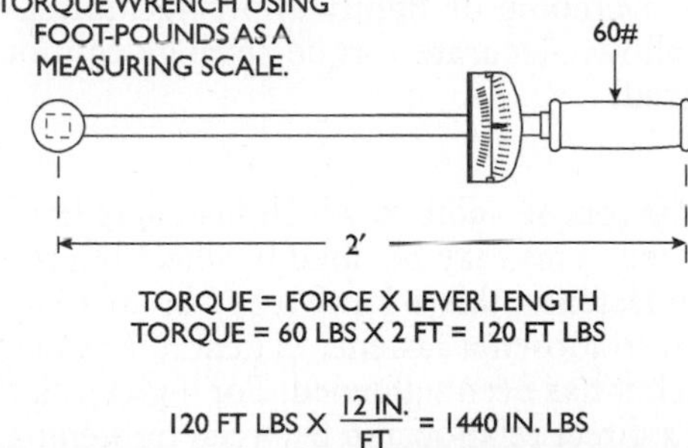

Figure 29-19 Torque wrench using foot-pounds as a measuring scale.

When using a torque wrench, it is important that the correct size wrench be selected. One that has adequate capacity is necessary to provide accuracy. The correct size wrench is one that will read in the 25 to 75 percent range on the scale when the required torque is applied. Avoid using an oversize torque wrench; it is difficult to obtain true readings if the pointer just starts moving up the scale. The other extreme should also be avoided, as extra capacity is needed in the event of seizure or if there is run-down resistance.

Another consideration might be whether a torque wrench should be pushed or pulled. Either way can produce accurate results; however, pulling is the preferred method. This is primarily because there is greater hazard to fingers and knuckles when pushing, should some part fail unexpectedly. The following terms apply to the use of a torque wrench as both a fastener-tightening tool and a measuring tool.

Run-Down Resistance

The torque required to rotate a fastener before makeup occurs is the *run-down resistance*. Where tight threads or locknuts may produce a run-down resistance, the torque required to overcome this resistance should be noted. To obtain the scale reading that will give the desired torque, add the resistance to the desired torque value. Run-down resistance should be measured on the last rotation or as close to the makeup point as possible.

Set or Seizure

In the last stages of rotation before reaching a final torque reading, seizing or set of the fastener may occur. When this happens, there is a noticeable popping effect. To break the set, back off, and again apply the tightening torque. Lack of lubricant contributes to this effect. Frequently it may be overcome by lightly applying lubrication where conditions will allow. Accurate torque settings cannot be made if the fastener is seized.

Breakaway Torque

Occasions may arise where the torque value to which fasteners have been tightened must be checked. This may be done by checking the torque required to loosen the fastener; this is known as "breakaway torque." The torque required to loosen a fastener is generally some value lower than that to which it has been tightened. For a given size and type of fastener, there is a direct relationship between tightening and loosening torque. When this relationship has been determined by actual test, applied tightening torque may be checked by loosening and checking breakaway torque.

Torque wrenches are made with a male drive square at the head end, and attachments are used to connect the wrench to the part to be torqued. Common attachments are socket wrenches, screwdrivers, socket-head wrenches, etc. Although the greatest use of the torque wrench is in conjunction with simple attachments, there are also other adapters and extensions used. Many of them are designed to fit specific fastener styles and to reach applications that would otherwise be impossible to torque. They usually increase the wrench's capacity, given that they lengthen the lever arm. When using adapters and extensions, corrections must be made to the scale readings. The scale correction will be in reverse ratio to the increase in the lever arm length. If the arm length is doubled by adding an adapter or extension, the wrench capacity is doubled, and the scale will show only one-half of the torque applied.

Figure 29-20 shows a sketch of a torque wrench with an adapter attached. Letters have been assigned to the various dimensions and values. The corrected scale reading when using an adapter or extension may be calculated using the following formula:

$$\text{Scale Reading} = \frac{\text{Torque Required} \times \text{Wrench Length}}{\text{Wrench Length} + \text{Extension Length}}$$

Stating the following formula in terms of the letters shown in Figure 29-20 enables easy substitution of values:

$$R = \frac{T \times L}{(L + C)}$$

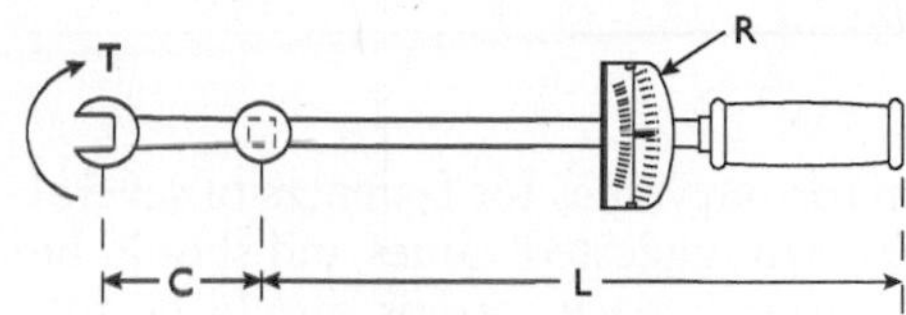

T = TORQUE REQUIRED
R = READING ON SCALE
L = WRENCH LEVER LENGTH
C = EXTENSION LENGTH

$$R = \frac{T \times L}{(L + C)}$$

Figure 29-20 Illustrating the effect of a torque wrench when using an extension.

Example

An adapter is required to reach a fastener that must be tightened to 60 ft-lbs. The torque wrench is 18 in. long and the adapter 6 in. long. What should the scale reading be when a torque of 60 ft-lbs is applied at the end of the adapter?

The known values are as follows:

Torque required — $T = 60$
Wrench length — $L = 18$
Extension length — $C = 6$

$$\text{Scale Reading} = \frac{T \times L}{(L + C)} = \frac{60 \times 18}{18 + 6} = \frac{1080}{24} = 45 \text{ foot-pounds}$$

Fastener Diameter	***Threads per Inch***	***Mild Steel***	***Stainless Steel 18-8***	***Alloy Steel***
1/4	20	4	6	8
5/16	18	8	11	16
3/8	16	12	18	24
7/16	14	20	32	40
1/2	13	30	43	60
5/8	11	60	92	120
3/4	10	100	128	200
7/8	9	160	180	320
1	8	245	285	490

Torque Specifications

The table above lists maximum torque values for fasteners of several materials in common use. These are suggested values and should be used as a guide only. Manufacturers' specifications should be followed on all critical applications.

Chapter 30

Sheet-Metal Work

The term sheet metal is commonly applied to the lighter-gauge metals in rolled sheet form, ranging in thickness from approximately 1⁄64 in. to approximately 1⁄8 in. Although steel sheet is used in the greatest volume, aluminum is also in wide use, as well as copper, brass, and an ever-increasing number of alloys developed to meet the needs of expanding industry and technology. The fabrication of sheet-metal articles by cutting, forming, fastening, etc., called "sheet-metal work," involves certain widely recognized methods and practices. Many of these are relatively simple, but understanding them is a basic requirement to successful accomplishment of sheet-metal work.

Edges

Practically all objects made of sheet metal have some type of formed edge to give a finished appearance, eliminate raw edges, and to add strength. Although some edges provide only a small amount of strength or rigidity, others add substantially to the object's strength. The *single hem edge* shown in Figure 30-1 is probably the most common and also the simplest to make. It is formed by first turning the edge and then squeezing it flat in the brake. The allowance for folding this style of hem is generally about 1⁄4 in. for the lighter gauges and up to 5⁄16 to 3⁄8 in. for heavier-gauge material. Although a wider hem is easier to form, if made too wide, there is a tendency to wrinkle at the edges.

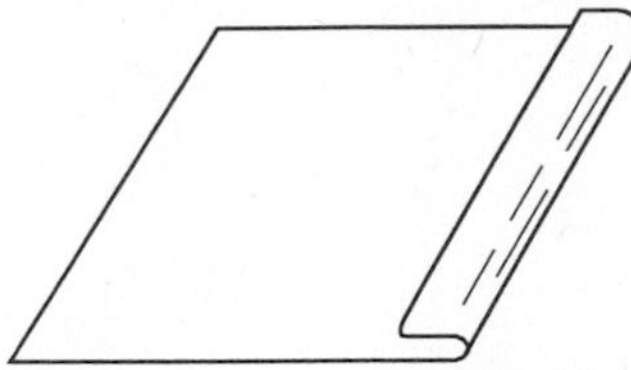

Figure 30-1 The single-hem edge made on sheet metal.

The *double hem* is, as the name implies, a single hem folded over a second time. Because it is double, it provides much greater strength. It is easy to make, as it is simply a single hem done twice. The allowance for a double hem is twice the hem size minus 1⁄16 in. The allowance for the first fold, which becomes the inside hem, shown in Figure 30-2, is the 1⁄16-in. minus dimension. The outside or second fold is made to the hem size—generally 5⁄16 to 3⁄8 in.

Figure 30-2 The double-hem edge.

Wired edges, formed by folding or wrapping the edge of the metal around a wire or rod, are used when greater strength than that provided by the double hem is required. An edge that has the appearance of a wired edge is formed on some manufactured products by rolling the metal into a hollow circle. In sheet-metal shop work, this is extremely difficult, and the wire is almost always used. The allowance for a wired edge depends on the diameter of the wire. A common rule is to allow 2½ times the wire diameter on light-gauge work and to add a slight amount more for the heavier gauges. Several methods of preparing the edge for wiring are in use. The shape of the object largely determines the method used. For example, a straight edge, such as shown in Figure 30-3, can be wired before the object is formed to shape, whereas cylindrically shaped articles are usually wired after they are formed to shape.

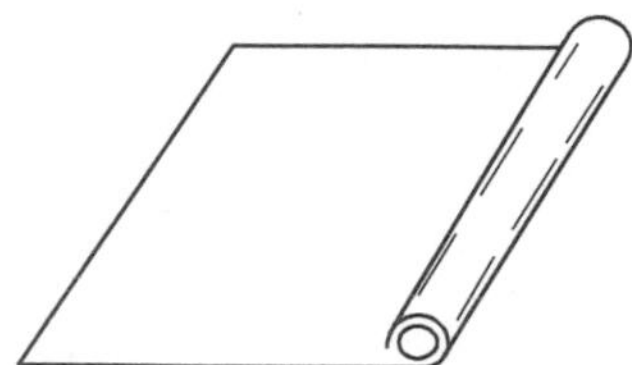

Figure 30-3 The wired edge.

Seams

The joint located along the edges of two pieces of sheet metal, fastening them together, is called a *seam*. There are a variety of methods for joining the edges of sheet metal; however, they may be classified into three general types: mechanical, riveted, and welded.

The *lap seam* is the simplest of all seams and is formed by lapping one edge of metal over the other. This may be either a flat lap (shown in Figure 30-4A) or, if surfaces must be flush, an offset lap (shown in Figure 30-4B). The lap seam is commonly fastened by soldering; however, when a stronger joint is required, it may be fastened with rivets, with or without solder, or it may be welded on the heavier gauges of metals.

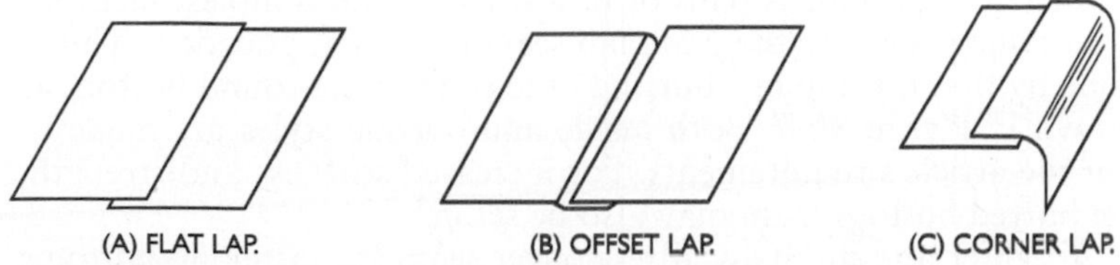

Figure 30-4 Various types of lap seams.

One of the most common types of mechanical seams used for joining sheet metal edges, principally for light- or medium-gauge materials, is the *grooved seam*. To form a grooved seam, the edges of the metal pieces are folded, hooked together, and offset. The seam may be an inside or outside seam, as shown in Figure 30-5, depending on which side is offset. The edges are locked together with a grooving tool called a hand groover or with a grooving machine. Allowance at layout must be made for the amount of material that is to be added for the lock. For light-gauge metal, 3 times the width of the lock is satisfactory. For medium-gauge, the allowance must be increased to compensate for the greater metal thickness.

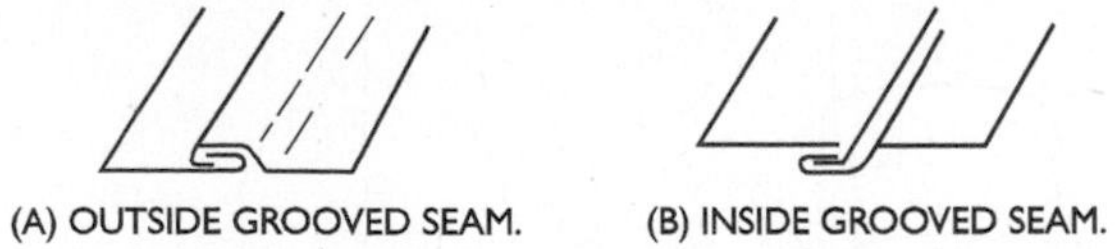

(A) OUTSIDE GROOVED SEAM. (B) INSIDE GROOVED SEAM.

Figure 30-5 The grooved seam.

For joining collars to flanges, or other similar cylindrical shapes to a flat object, the *dovetail seam* is often used. It is made by slitting the end of the cylindrical part and bending every other tab, as shown in Figure 30-6. The bent tabs act as stops, and the remaining straight tabs are bent over the part to be joined. This seam may be soldered for greater rigidity and to make the joint watertight.

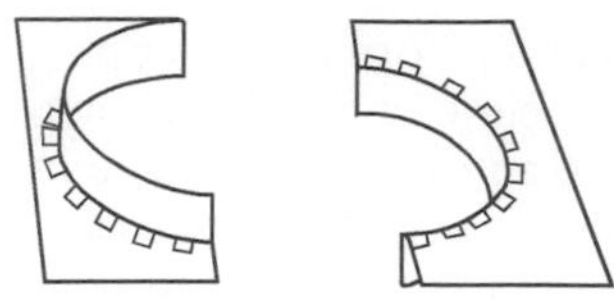

Figure 30-6 The dovetail seam.

Several types of seams are used to fasten bottoms to cylindrically shaped articles such as pails or tanks. One of the simplest methods of fastening is to use the plain lap seam for this application. This is done by first turning or "burring" the edge of the round bottom as shown in Figure 30-7. Both inside and outside styles are made to suit the article's requirements. For increased stiffness and strength, the burred bottom seam may also be set in.

A tighter and mechanically stronger seam for fastening bottoms to cylindrically shaped articles is the bottom double seam. The operations for making this type of seam are shown in Figure 30-8. The single seam is made first by turning an edge on the body of the cylinder. The bottom is turned up or burred and placed on the body and peened down. The seam is completed by bending it over to form a double fold or lock.

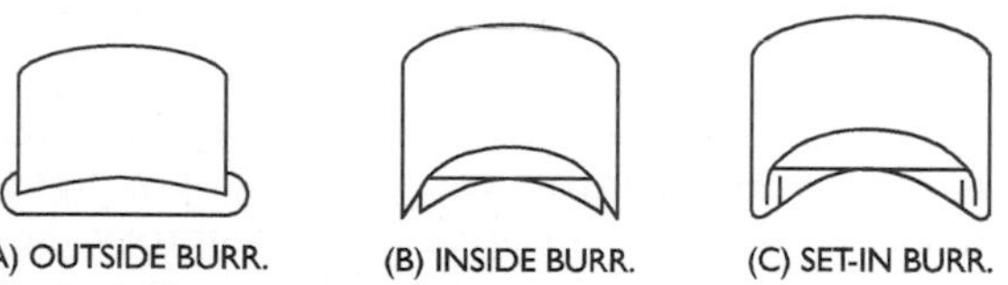

Figure 30-7 Burred bottom seam, which is used to insert a bottom to a cylinder.

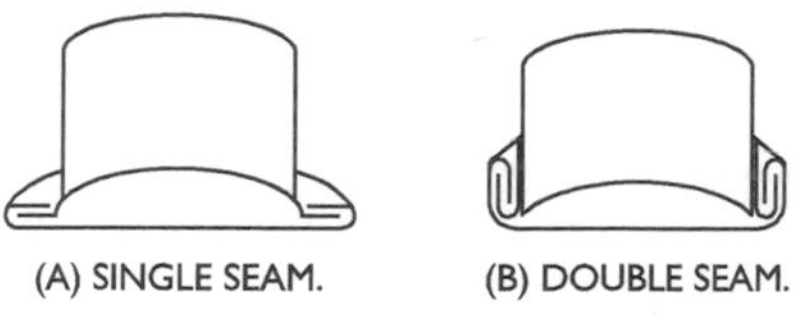

Figure 30-8 The single and double seam.

One of the most versatile and probably the most commonly used seam in sheet-metal work is the *pittsburg lock* seam. Because it is such a satisfactory seam and is so widely used, machines have been developed to form the seam. The machines have a series of rolls, which form the pocket of the seam automatically. The metal is inserted in one end, runs through the rolls, and emerges from the other end with a completely formed pocket. These are great labor savers, as forming a pittsburg lock on the brake requires several operations and considerable handwork. An outstanding advantage of the pittsburg seam is that the single lock section of the seam, shown in Figure 30-9, can be turned on a curved article, and the pocket section can be formed on a flat sheet. The pocket section can then be rolled to fit the curve.

Pattern Development

Sheet-metal patterns are laid out or developed by three principal methods—parallel-line development, triangulation, and radial-line development. In parallel-line development, which is probably the simplest and most widely used method, the sides run parallel to one another as in patterns for ducts, elbows, and tee joints. In radial-line development, the lines radiate from a single point or all sides meet at a common center as with cones or pyramids. Triangulation is used to develop irregularly formed shapes that cannot be developed by either of the other two methods. Surfaces are developed part-by-part with one added to another to develop the entire surface. More complex shapes, such as fittings with a changing shape, off-center fittings, offsets, and transitions, are developed by triangulations.

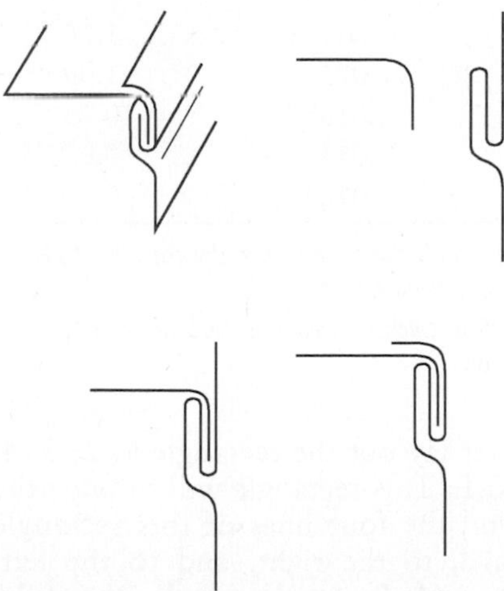

Figure 30-9 The pittsburg lock seam.

The following examples, in the form of problems, are arranged in groups under each method of development, although in some cases more than one method is employed. Table 30-1 gives the weight per square foot for metals of various gauges. This table would be very useful in finding the proper gauge of metal to use when developing any sheet-metal project.

Problem 1: Pattern for Flushing Tank

Draw the plan (front and end views) of the required tank, full size, taking care to make all angles exactly right angles (90°). The end view is not necessary for developing the pattern but is given just for practice in orthographic projection (see Figure 30-10).

Table 30-1 Standard Gauge for Sheet Steel

Number of Gauge	*Approximate Thickness, in Fractions of an Inch*	*Approximate Thickness, in Decimal Parts of an Inch*	*Weight per Square Foot, in Pounds*
11	1/8	.125	5.0
12	7/64	.109	4.375
13	3/32	.094	3.75
14	5/64	.078	3.125
16	1/16	.062	2.5
18	1/20	.05	2.0
20	3/80	.0375	1.5
22	1/32	.03125	1.25
24	1/40	.025	1.00
26	3/160	.01875	0.75
28	1/64	.0156	0.625
30	1/80	.0125	0.5

Note: To find the weight of sheet steel, multiply the thickness in decimals by 40.8. The result will be the weight in pounds per square foot.

Example: If a piece of sheet steel is .005 in. thick, its weight is .005 times 40.8, which equals .204 pound per square foot.

In developing the pattern, first lay out the rectangle 1, 2, 3, 4, exactly equal to the plan I, J, K, L. This rectangle will provide the bottom part of the tank. Extend the four lines of this rectangle indefinitely—upward, downward, to the right, and to the left. Then, set the dividers to a distance (A, B) equal to the height of the tank, and with this distance as radius, describe arcs at the four corners of the rectangle, using the points 1, 2, 3, 4 as centers. These arcs cut off the extended lines of the bottom part, thus giving the rectangles 6, 1, 4, 5 for the front part; 2, 7, 3, 8 for the rear part; 1, 10, 2, 9 for the left; and 3, 4, 11, 12 for the right end of the tank.

Additional strips of stock should be added along the edges 1, 6; 2, 7; 3, 8; and 4, 5 for laps of proper width. The point of intersection of the two diagonals in the bottom part furnishes the center of the hole for the flushing pipe. Because there is no cover to this tank, its pattern is made up of only five rectangles. When a pattern is to be made for a closed square container, six rectangles are required.

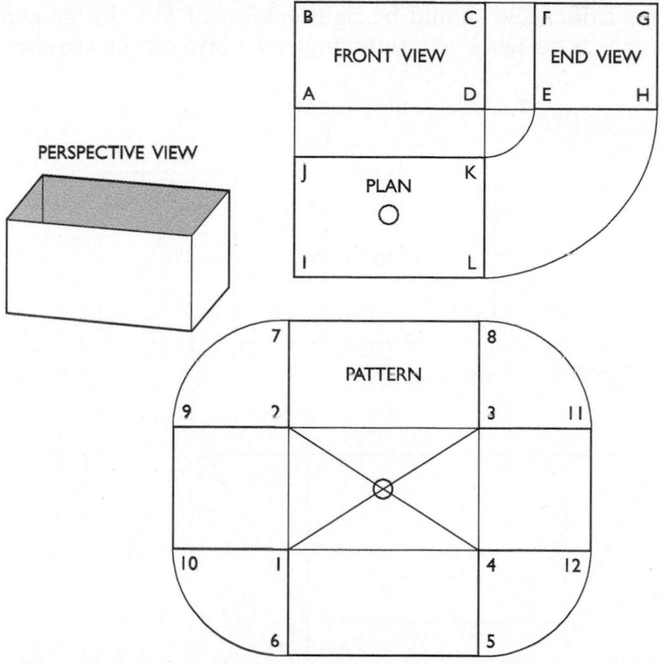

Figure 30-10 Flushing tank and the development of the pattern.

Problem 2: Pattern for a Cube

This is virtually the same as problem 1 except for the additional rectangle 1, 2, 3, 4 forming the top or cover, considering the cube to be a container with a cover (see Figure 30-11).

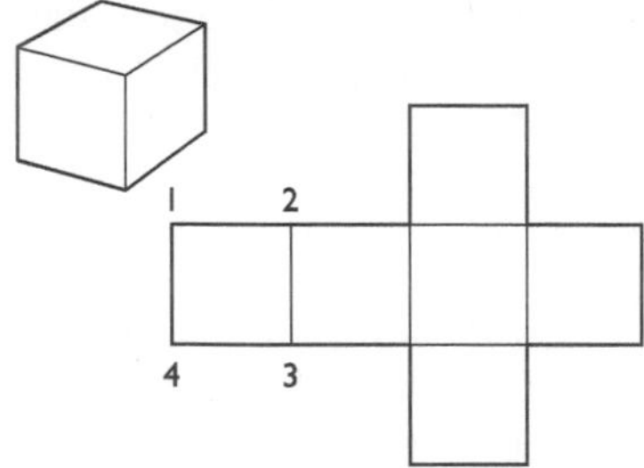

Figure 30-11 Pattern development of a cube.

Problem 3: Pattern for a Wedge-Shaped Drainer

Draw the front and end views as shown in Figure 30-12. All of the angles in the front view should be right angles, or 90°. In the end view, the angle at corner F is a right angle. To lay out the required

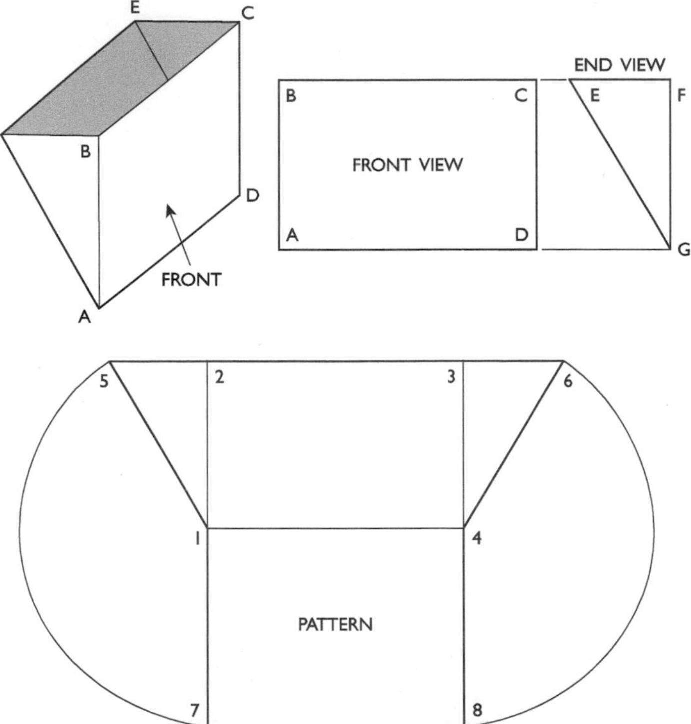

Figure 30-12 Wedge-shaped drainer and the pattern development.

pattern, draw the rectangle 1, 2, 3, 4 exactly equal to the front view A, B, C, D. Extend the line 2, 3 to the right and to the left; also, extend downward the lines 1, 2 and 3, 4.

Set dividers to a distance equal to the oblique line G, E. With this distance as radius, describe the arc 7, 5, with point 1 as center. Then, describe the arc 8, 6, with point 4 as center. These arcs determine the rectangle 7, 1, 4, 8 for the front part of the drain as well as the triangular parts 1, 5, 2 and 4, 6, 3 for the two ends of the drain. Additional strips of stock should be added along the edges 1, 5 and 4, 6 for laps of a suitable width.

Problem 4: Pattern for a Tee Pipe

The problem involves the development of two intersecting cylinders of equal diameters. The solution given here is important because it applies to a large class of elbows and other objects composed of intersecting cylinders. The method used here may also be used where the intersecting cylinders are not of the same diameter (see Figure 30-13).

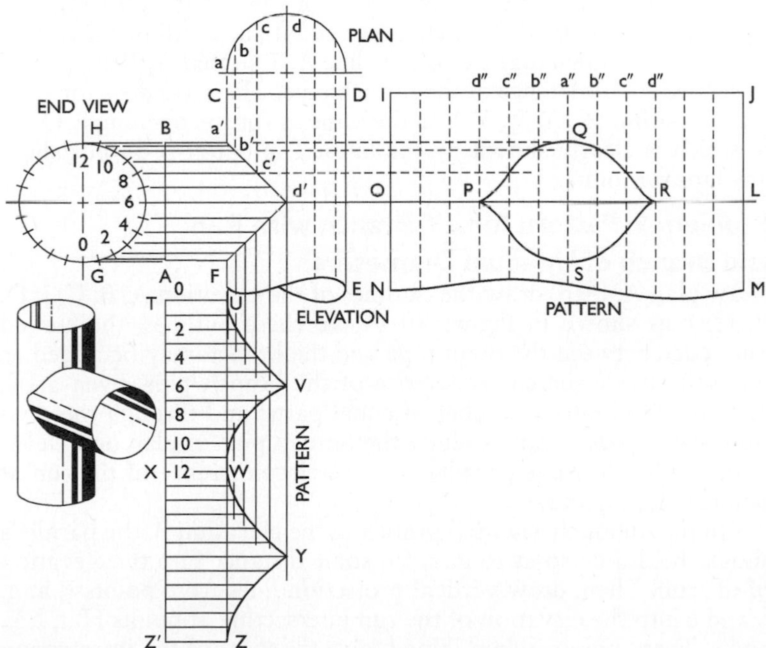

Figure 30-13 Pattern for a right-angle branch.

A hole must be cut in the vertical or *run* pipe to provide an opening of suitable shape for the branch pipe, and the end of the *branch* pipe must be shaped to fit the opening in the run.

N, I, J, M shows the development of the run pipe. The length I, J is that of the circumference of the pipe, half of which is shown in the plan. See Figure 30-13. The equally measured divisions on this circumference are put together to make up the line I, J. At right angles to this line, the lengthwise edges of the run pipe are shown by the lines I, N and J, M. Only a part of the length of this pipe is developed—the part around the required opening P, Q, R, S.

The longitudinally drawn projecting lines from the points of division in the plan furnish the points a', b', c', and d' on the joint line a', d'. From these, another series of projecting lines is drawn to the development intersecting the longitudinal lines on the development that starts from points d", c", b", a", b", c", and d", thus giving points defining the outline of the opening P, Q, R, S, as clearly shown.

In a like manner, the other pattern Z, T, U, Z' is obtained from A, B, a', b', c', d'. The circumference of the branch pipe G, H is divided at points 0, 1, 2, 3, etc., into equal parts. All of these divisions are laid down together on the line A, T so that A, T is equal to the entire circumference of the branch pipe. The procedure for tracing the outline Z', Y, W, V, U is the same as for the opening P, Q, R, S, which is clearly shown by the lines. The patterns do not show any laps for joints.

Problem 5: Pattern for a Y Branch with Run and Branch of Unequal Diameters

Draw plan A; then, draw the outlines of the elevation A, B, C, j, D, E, H, F as shown in Figure 30-14. To these outlines, the curved joint edge between the main pipe and the branch may be added as follows: Divide the circumference of the branch pipe given at K, and also at L, into a number of equal parts, and through the division points pass parallels along the branch pipe in plan and elevation. In plan A, these parallels cut the circumference of the run at points i, h, g, f, and e.

On the obliquely situated branch in the elevation B, the parallels should be drawn so as to pass for some distance into the elevation of the run. Then, draw vertical projection lines from points i, h, g, f, and e into the elevation of the run intersecting at points H, r, l, k, and j. These points, when joined, give the view of the intersection between the branch and run pipes.

Now, proceed to lay out the pattern V, U, T, S, W for the branch. Extend the end line D, E of the branch indefinitely toward the desired pattern, and upon this extended line, make W, S equal in length to the circumference of the branch. At every division point, erect a perpendicular to W, S, thus dividing the space for the pattern into a number of elementary parts. Now, from points H, r, l, k, and j (on the elevation), project a number of lines upon the pattern, parallel to the stretched-out line S, W. These projection lines cut the elementary lines on the pattern at the points T, 6, 7, 8, U, 9, 10, 11, and V and the points in the curved outline of the pattern of the branch.

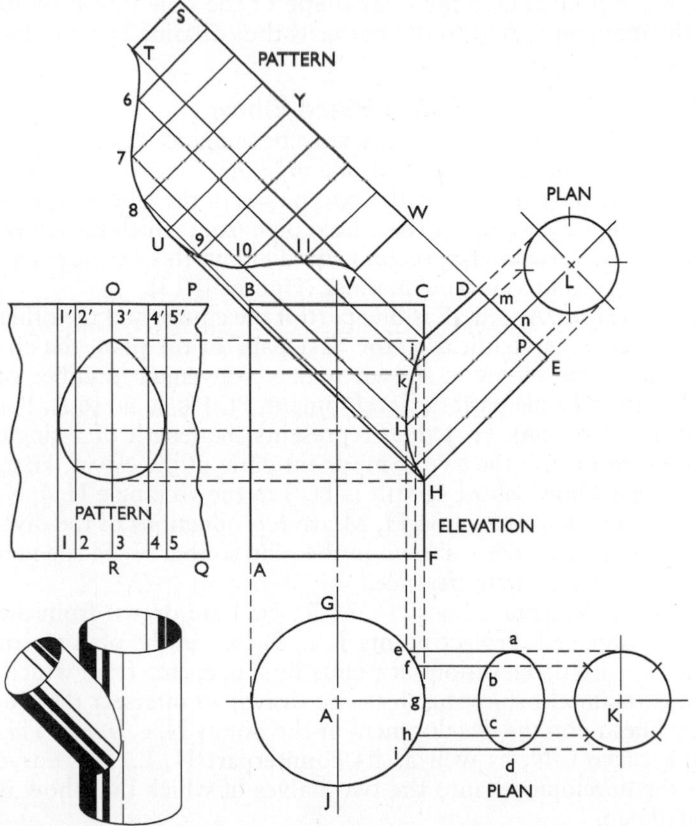

Figure 30-14 Pattern for an oblique branch.

The run pipe has to be cut out on an oval line to receive the branch. For the plotting of this oval on the pattern, draw the short part M, N, P, Q of the development of the run. In the middle of this, erect the perpendicular centerline P, O. Starting from this line, set off to the left the distances 3, 2 and 2, 1, respectively equal to g, f and f, e (in plan), and do the same in opposite order to the right of the centerline, so that 3, 4 equals 2, 3 and 4, 5 equals 2, 1.

Through the points 1, 2, 3, 4, 5, draw the vertical elements 1, 1'; 2, 2'; 3, 3'; etc.

In the elevation, project points j, k, l, r, and H horizontally over to the pattern of the run. These horizontal projection lines cut the vertical elements 1, 1'; 2, 2'; etc., at a series of points, which, when joined, give the oval shape of the hole that is to be cut in the main pipe. Add to the patterns thus obtained proper lap for joints.

Problem 6: Pattern for a Two-Piece Elbow

A two-piece elbow for round pipes may be imagined to have been made up of two adjoining parts of one pipe that was cut at a miter of 45°. Hence, the patterns for the two halves of the elbow may be obtained, first, by developing the whole pipe from which the elbow parts are to be derived and, second, by dividing this development into two portions in a suitable manner (Figure 30-15).

In the elevation, A, B, E, F is one part of the elbow, and the other part, B, C, D, E, is identical to the first part. In the plan, the circumference of the elbow is shown with a convenient number of equal divisions. In the pattern development, H, J is equal to A, B + F, E (in the elevation). H, J then represents the length of a single pipe that is to furnish the two component parts of the elbow. First, develop the cylinder whose length is H, J in the rectangle H, J, K, M, wherein the divisions upon H, M are reproductions of the divisions on the circumference shown in the plan so that H, M is equal to the stretched-out circumference.

Vertical or elementary lines (1', 2', 3', etc.) are drawn from the points just obtained. Project points 1, 2, 3, etc., in the plan to cut the miter line EB (in elevation) at points B, a, b, c, etc., from which, in turn, horizontal projecting lines are drawn to intersect the elementary lines upon the development at the points N, a, b, etc. This gives the curve I, N, as well as its counterpart N, L. This curve divides the development into the two halves of which the elbow is to be made up.

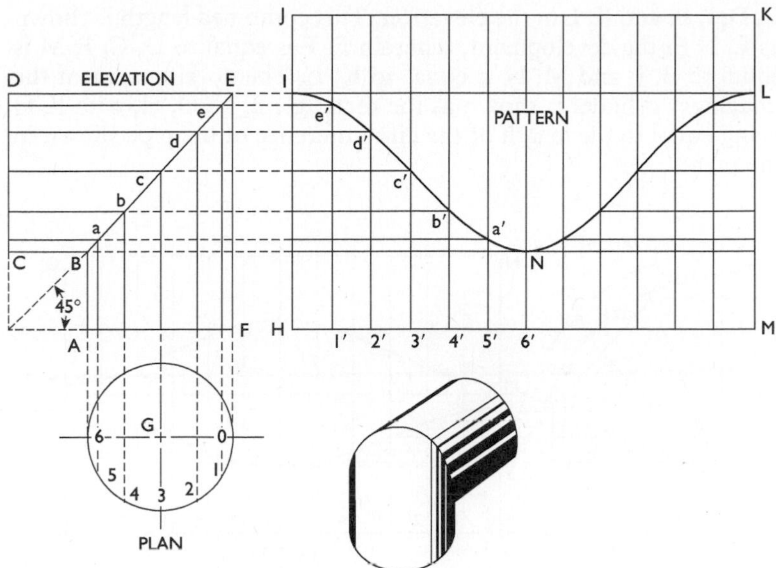

Figure 30-15 Pattern for a two-piece elbow.

F, M is equal to B, I; M, N is equal to L, K. The development of the imaginary cylinder is shown as the rectangle E, N, O, H. E, H is equal to the length of the circumference of the pipe shown in the plan.

The development of the pattern is by elementary lines through points 0, 1', 2', 3', etc., parallel to E, N and spaced equal to the arc distance between similar points, 0, 1, 2, 3, etc., in the plan. By projecting points 1, 2, and 3 in the plan up to miter line B, C in the elevation, they will intersect with corresponding elementary lines 0, 1', 2', 3', etc., which will give points defining the curve F, G. To obtain the curve M, P, draw the centerline X, X' so as to bisect the distances M, F and P, G. Then, set off upon each of the elementary lines 0, 1', 2', 3', etc., above X, X', the amount by which the curve F, G deviates from X, X', thus obtaining similar points which define curve M, P.

Problem 7: Pattern for a Three-Piece Elbow

The general shape of the elbow is shown in the elevation in Figure 30-16. The imaginary cylinder from which the three parts of the desired elbow are to be made will be equal to the combined lengths

C, D; I, B; and K, L in the elevation. This combined length is shown as E, N in the development, wherein E, F is equal to D, C; F, M is equal to B, I; and M, N is equal to K, L. The development of the imaginary cylinder is shown as the rectangle E, N, O, H, with E, H being equal to the length of the circumference of the pipe shown in the plan.

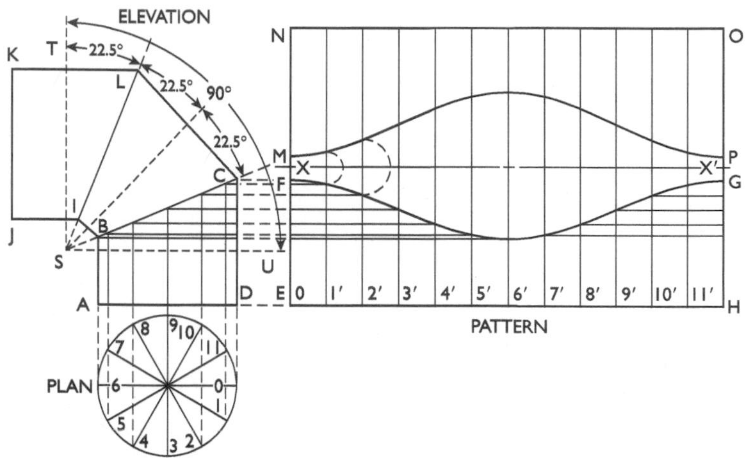

Figure 30-16 Pattern for a three-piece elbow.

The development of the pattern is by elementary lines through points 0, 1', 2', 3', etc., parallel to E, N and spaced equal to the arc distance between similar points 0, 1, 2, 3, etc., in the plan. By projecting points 1, 2, and 3 in the plan up to miter line B, C in the elevation, they will intersect with corresponding elementary lines 0, 1', 2', 3', etc., which will give points defining the curve F, G. To obtain the curve M, P, draw the centerline X, X' so as to bisect the distances M, F and P, G. Then, set off upon each of the elementary lines 0, 1', 2', 3', etc., above X, X', the amount by which the curve F, G deviates from X, X', thus obtaining curve M, P.

Problem 8: Pattern for a Four-Piece Elbow

In the elevation shown in Figure 30-17, the four pieces forming the elbow are A, K, S, I; K, X, T, S; X, Y, Z, T; and Y, j, d, Z. Of these four parts, the two larger parts—A, K, S, I and Y, j, d, Z—are equal. The same is true of the two remaining smaller parts—K, X, T, S and X, Y, Z, T.

To lay out these parts in the elevation at a right angle, a, b, c is drawn, the sides of which intersect at right angles, which are the two largest branches of the joint. It is evident that the point b must be equidistant from both pipes. The right angle a, b, c is divided first into three equal parts, and then each one of these parts is divided in turn into two equal parts. The right angle is thus divided into six equal parts, of which K, b, a is one part; K, b, X equals two parts; X, b, Y equals two parts; and Y, b, c is one part. It will be noticed that this construction does not depend on the diameter of the pipe.

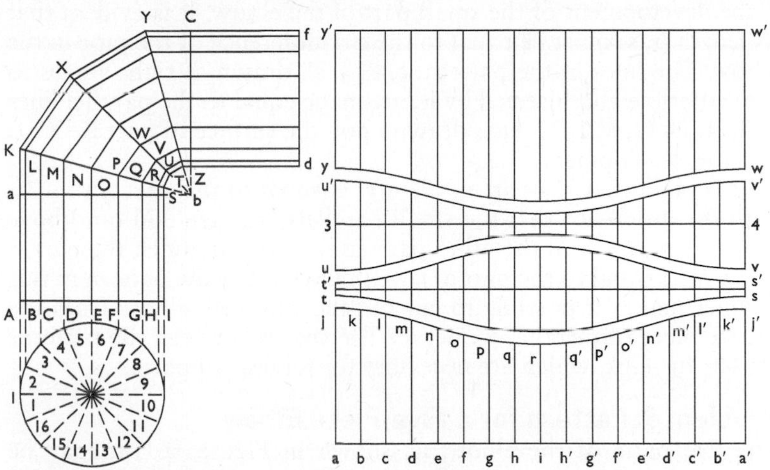

Figure 30-17 Pattern for a four-piece elbow.

The problem of developing the four-part elbow resolves itself into developing only two of its parts, one large branch and one smaller part of the elbow, the remaining parts being correspondingly equal to these. The circumference of the pipe, as seen in the plan, is divided into sixteen equal parts by the points 1, 2, 3, 4, 5, etc. Through these points are drawn lines parallel to the centerline of the pipe that is to be developed.

In the development, the vertical branch of the elbow (A, K, S, I of the elevation) will be used for this purpose. The parallels upon the surface of this branch are A, K; B, L; C, M; D, N; E, O; F, P; G, Q; H, r; and I, s. Through the points K, L, M, N, O, P, Q, R, and S, draw parallels for the part K, X, T, S, which will be developed next. Some of these parallels are S, T, R, U, Q, V, R, w.

To develop the vertical branch of the four-piece elbow, draw upon the straight line a, a' sixteen equal parts, which altogether are equal to the circumference of the cylinder that is to be developed. Let the division points, a, b, c, d, e, f, etc., correspond to the circle in the plan. Through points a, b, c, d, e, etc., draw vertical lines equal to the parallel lines drawn on the surface of the vertical branch of the joint. Thus, a, j is made equal to A, K; b, k equal to B, L; c, l equal to C, M; and so on until r, i is made equal to S, I.

The part laid out so far is a, j, k, m, n, o, p, q, r, i. This is one half of the development; the other half, i, r, j', a', being exactly the same as the first one, may be laid out in the same way. The part t, t', s, s' is the development of the small part of the elbow. It is evident that its length (t, s) must be equal to the circumference of the pipe in the elbow. The lines in the pattern, t, t', s, s', drawn at right angles to the centerline and bisected by it, are made equal to the parallel lines S, T, R, U, Q, V, P, W, etc., drawn upon the surface of part K, X, T, S in the development.

It is plain that the part u, u', v, v' is equal to the part t, t', s, s', with the difference that the small parallels in it are laid out above the large parallels in the other part; in the same manner, the part y, y', w, w' is equal to the part a, j, a', j'. Laying out the pattern in this manner makes it possible to cut out the complete elbow from the square piece of metal a, y', w', a'. The spaces between the patterns are left for laps, which are necessary for joining all parts.

Problem 9: Pattern for a Five-Piece Elbow

The five parts of the elbow, as shown in Figure 30-18, may be thought of as so many parts of one long pipe, cut to the miter angle at proper distances. The length of that cylinder, this time, will be made up of the sum of the alternately consecutive outlines of the five parts of the elevation of the required elbow. Thus, in the development of the whole cylinder shown at E, U, V, H, the vertical edge E, U is equal to the combined lengths of D, C; B, I; M, N; A, Z; and K, L, laid off on the development as the lengths E, F; F, O; O, P; P, T; and T, U.

Along the horizontal edge of the development, on E, H, lay off all the equal parts into which the circumference of the elbow pipe is divided, and from the points of division on the edge E, H, draw vertical elementary lines across the development. From the division points on the circumference, draw projecting lines upward to the miter line B, C, cutting it in a number of points from which, in turn, horizontal projecting lines are drawn meeting the vertical elementary lines on the development in points, forming the curve F, G. The miter line, as is seen in the elevation, has an angle of 11.25°.

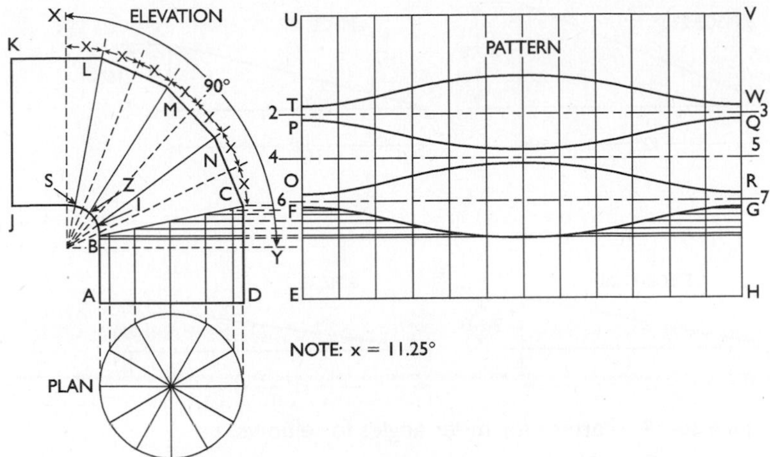

Figure 30-18 Pattern for a five-piece elbow.

The curve O, R is plotted so as to be an exact counterpart of the first curve at the other side of the centerline 6, 7, which bisects O, F and R, G. Curve O, R is the curve that would be obtained by revolving curve F, G 180° on the axis of 6, 7.

Having obtained the second curve, O, R, a second centerline 4, 5 is drawn bisecting O, P and R, Q, and above this centerline the curve P, Q is plotted so as to deviate from the centerline along each vertical elementary line exactly as much as the curve O, R deviates from the centerline. In a like manner, with the aid of the third centerline, 2, 3, the curve T, W is laid out opposite and equal to the curve P, Q. The centerline 2, 3 bisects P, T and Q, W. For simplicity in explaining the development, no laps are provided for joints.

Problem 10: Miter Angles for Round Pipe Elbows

In all elbows shown in Figure 30-19, the miter upon the end piece depends on the number of pieces used in the elbow.

For a two-piece elbow, the miter angle is 45°.
For a three-piece elbow, the miter angle is 22½°.
For a four-piece elbow, the miter angle is 15°.
For a five-piece elbow, the miter angle is 11¼°.
For a six-piece elbow, the miter angle is 9°.

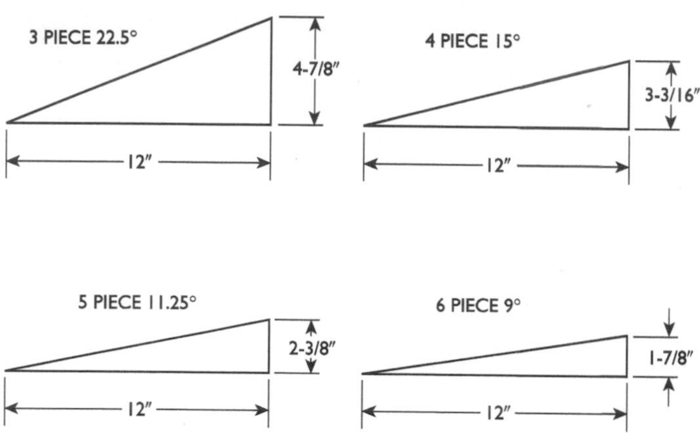

Figure 30-19 Pattern for miter angles for elbows.

The angles can be traced with the aid of a protractor. However, sufficiently accurate angles may be laid out for the different miters by the handy method of right triangles shown in Figure 30-19. The miter angle for a three-piece elbow may be obtained by constructing a triangle, one leg of which is 12 in. long and the other leg 4⅞ in. long. The hypotenuse of this triangle gives the desired miter. The proportions for the various other miters are given in Figure 30-19. The miter for any elbow not given in the illustrations may be found as follows:

Divide by 90 the number of pieces in the elbow less one, multiplied by two.

Thus, for a three-piece elbow,

$$90 \div (3 - 1)2 = 22\tfrac{1}{2}°$$

Problem 11: Pattern for an Offset on a Rectangular Pipe

The pattern layout for an offset is shown in Figure 30-20, and the general appearance of the offset is shown at L in Figure 30-21. In making an offset on any pipe, it is necessary to keep in mind that all parts of it should offer the same area for the even flow of the water within the pipe. Hence, in the elevation, the cross section of the oblique part of the pipe, F, G, J, K, has the same width as the vertical part. To obtain this result, the joint F, K must be drawn so as to bisect the angle E, F, G, and likewise the joint G, J must bisect the angle K, J, I.

For the patterns of the offset, the four sides of the plan A, B, C, D are laid off consecutively upon the line M, N, O, P, Q, and in the points of measurement, elementary lines are drawn at right angles

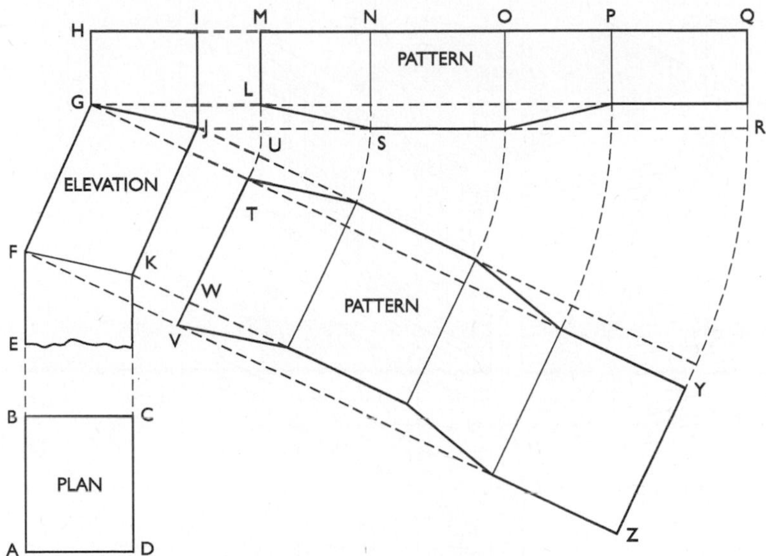

Figure 30-20 Offset pattern for a rectangular pipe.

to the line M, Q. Upon these lines are projected the vertical measurements of the part to be developed. Thus, M, L is equal to G, H, and S, N is equal to J, I. The drawing of the pattern L, M, Q, R plainly shows its derivation from the upright part G, H, I, J by means of the projection lines.For the development of the oblique part of the square pipe, F, G, J, K, project lines from points F, G, J, and K at right angles to the length of the oblique section. These lines, cutting the line U, V at the points V, W, T, U, give starting points for horizontal projectors that determine the lengths of the different portions of the development W, U, Y, Z. Joint laps should be added to the pattern thus obtained.

Problem 12: Pattern for Round Branch from Rectangular Leader

This pattern layout, shown in Figure 30-22, has a general appearance as shown at F in Figure 30-21. In the elevation, the round branch A, E, F, J cuts the rectangular leader along a curved line that is symmetrical lengthwise and crosswise; the curve is an ellipse. In the half plan of the branch, the circumference of the branch is divided into a number of equal arcs (half of the circumference being shown in the dividing points a, b, c, etc.).

Figure 30-21 Pictorial view of an offset and branch elbows.

Through b, c, and d, project the elementary lines G, D; H, C; and L, B. The elements through the point of division intersect the square pipe at the points A, B, C, etc. Now, for the pattern of the round pipe, make the line K, L a continuation of the line F, J and equal to the combined length of the divisions on the circumference of the oblique pipe.

Lay off points 1, 2, 3, etc., on K, L, spaced equal to arc distances a, b; b, c; c, d; etc., in half plan; through points 1, 2, 3, etc., draw elements perpendicular to K, L and project over points A, B, C, etc., parallel to K, L and intersecting the element at R, Q, P, etc.

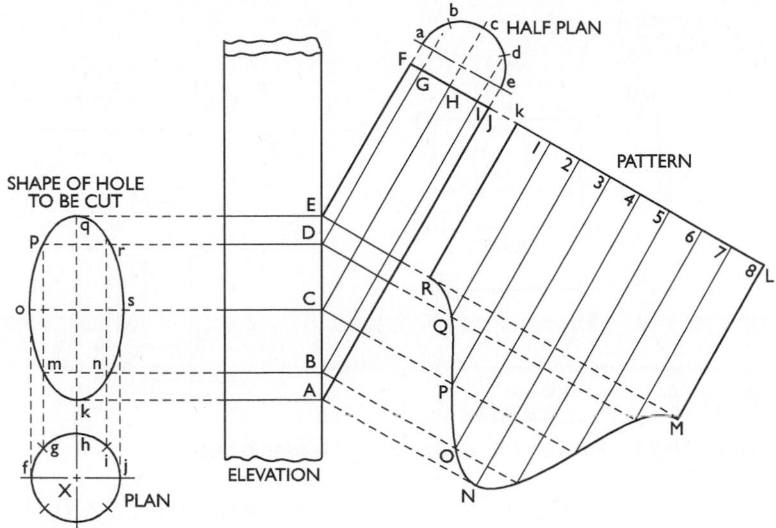

Figure 30-22 Pattern for a Y branch.

Through these points, describe the curve R, N, M and join lines K, R and L, M, thus completing the pattern. Laps for joints must be added.

The opening to be cut in the rectangular leader is an ellipse. It is obtained by projecting points A, B, C, etc., from elevation and points f, g, h, etc., from plan X. The intersections yield points k, m, o, etc., which define the ellipse.

Problem 13: Pattern for a Box-Shaped Leader Trough

Draw the plan, elevation, and end view of the trough full size as shown in Figure 30-23. Note that the plan and elevation are divided into two halves by a centerline V, U. In the development, first draw centerline 21, 22. Because the trough consists of an upright part (B, C, D, E) surmounting the tapering part (A, B, E, F), the pattern will have two corresponding portions. The rectangle 8, 1, 4, 5 provides the pattern for the two end parts 1, 8, 2, 7 and 6, 3, 4, 5, as well as for the rear part 7, 2, 3, 6 of the right band. The front part of the latter is attached at the bottom of the pattern. It is marked 20, 17, 18, 19 and is exactly equal to 7, 2, 3, 6. Note that the centerline 22, 21 passes through the middle of the upper band 8, 1, 4, 5, as well as through the middle of the part 20, 17, 18, 19.

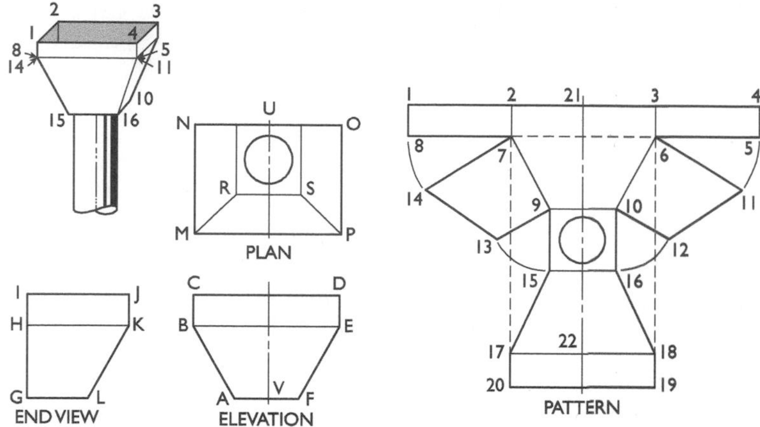

Figure 30-23 Pattern for a box-shaped leader trough.

To the lower edge (9, 10), attach the rectangle 15, 9, 10, 16, which is exactly equal to the projection in a vertical plane of A, B, E, F, the tapering part of the front view. Because the elevation does not show the true length of this tapering part but instead its projection in a vertical plane, lines 9, 7, 6, 10 are made equal to A, B, E, F.

To the lower edge (9, 10), attach the rectangle 15, 9, 10, 16, which is equal to the bottom R, U, S of the trough. At points 7 and 9, erect the lines 7, 14 and 9, 13 at right angles to the line 9, 7. Similarly, at points 10 and 6, erect the lines 10, 12 and 6, 11 at right angles to the line 10, 6. Then, the irregular figure 13, 14, 7, 9 forms the left side of the tapering part, and the figure 12, 11, 10, 6 forms the right side of the tapering trough. Draw lines 2, 20 and 3, 19 parallel to the centerline 21, 22 and with a radius equal to 12, 11. From center 16, describe an arc cutting the line 3, 19 at point 18; with the same radius, from center 15, describe an arc cutting the line 2, 20 at point 17. Connect points 17 and 18 by a straight line and obtain the front face 17, 15, 16, 18 of the trough.

Finally, to the edge 17, 18, attach the band 20, 17, 18, 19, which is equal to 7, 2, 3, 6. At the center of the square 9, 15, 16, 10, describe a circle for the opening. Add to the pattern proper laps for the seams.

Problem 14: Pattern for a Pan

Draw the elevation and end view as shown in Figure 30-24. The flare on all sides being equal in this pan, the oblique edges in both views make the same angle with the bottom lines. Also, the oblique lines J, I; H, G; D, A; and C, B are equal in length. The pattern is to be made of one piece. Furthermore, its joints are not to be soldered but made watertight by turning some of the stock of the sides flat upon the end walls of the pan, in such a manner that the folded metal should reach exactly to the wired edge.

To draw the pattern, project the corner points I, H, D, and C of the end view and elevation so as to form, by intersection, the rectangle O, L, M, N; this is the bottom of the pan. Space off distances I, K; L, P; M, Q; and O, R, each equal to the oblique line H, G. Draw lines through these points, and project from points J, G, A, and B lines intersecting at W, Z, E, etc., giving the outer edges of the pan. Connect points W, Z, E, etc., to the rectangle O, L, M, N, which gives the four sides of the pattern. Thus, K, L; L, P; M, Q; and O, R each equal the oblique line G, H.

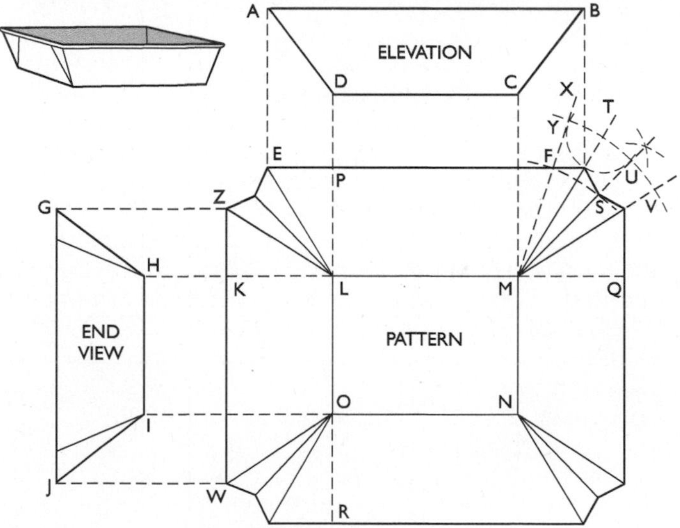

Figure 30-24 Pattern for a flat pan.

Upon these outer edges, project the distances J, G and A, B to the points W, Z, E, and so on; these points, when connected to the corners of the rectangle O, N, M, L, furnish the shapes for the four side parts of the pattern.

Chapter 31

Welding

Because of the development of many new welding processes and also because of the development of many new steels and other metals that can now be welded, welding has become the most important metal-joining process. In addition, welding has risen to the forefront because it allows flexibility and design freedom previously unavailable through conventional joining methods.

Personnel engaged in welding operations must select the one best welding process for the job at hand and also select the proper equipment, the most appropriate filler metal, and the welding procedure for doing the job. To do this properly requires a great deal of know-how, of which this section is a convenient source. Much of the information provided aids in selecting the correct process, as this is directly related to the metal being welded. The processes are explained in detail, and procedure data for welding different metals with the different processes are provided. These data are in concise form and give only the necessary basic information. Some of the newer and less frequently used welding processes are also described in detail with pertinent procedural information.

Shielded Metal-Arc Welding

This is perhaps the most popular welding process in use today. It was first used as a maintenance tool in 1918. In the early 1930s, shielded metal-arc welding became a popular manufacturing method. The high quality of the metal produced by the shielded arc process, plus the high rate of production, made it a tool that rapidly replaced other fastening methods. The process has been developed so that it can be used in all positions and will weld a wide variety of metals. The most popular use, however, is for the welding of mild carbon steels and the low-alloy steels.

The equipment is extremely rugged and simple, and the process is flexible in that the welding operator needs to take only the welding electrode holder and ground cable with him to the point of operation. See Figure 31-1. In operation, an arc is struck between the metal to be welded and the covered electrode or stick. The arc creates sufficient heat to melt the edges of the base material and at the same time melts the electrode, which is transferred across the arc and deposited on the base material. This forms a uniform and metallurgical metal structure equal in strength to the metal being welded. Covered electrodes have been designed to match the properties of most low- and medium-alloy steels and many other metals and alloys. Other electrodes have been developed to provide special surfaces for wear resistance, etc.

Gas Metal-Arc Welding

This is one of the newer arc welding processes. It is an arc welding process in which the heat is generated between the electrode and the metals to be welded. Gas shielding is used to keep the atmosphere away from the molten metal. The difference between gas metal-arc welding and gas tungsten-arc welding is that the electrode does melt in the arc. However, the gas metal-arc electrode is in the form of wire, which is continuously and automatically fed into the arc to maintain a steady arc. This electrode wire melted into the heat of the arc is transferred across the arc and becomes the deposited weld metal. See Figure 31-2. The deposited weld is a metallurgical joint as strong as the base metal.

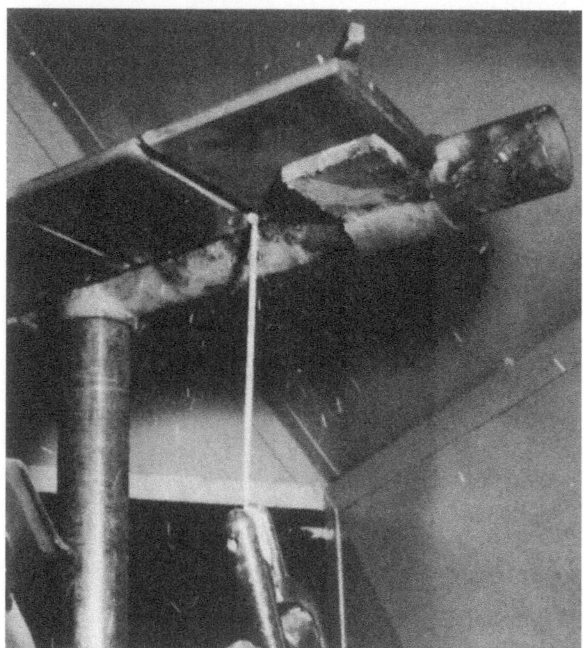

Figure 31-1 The welding electrode used in arc welding.

There are many variations of the gas metal-arc welding process, which is commonly called MIG welding. In its early days, it was used exclusively for the welding of nonferrous metals, primarily aluminum. In this case, an inert gas (argon) was usually employed. Later, however, CO_2 gas was employed for shielding, but in this case, the welding was done on mild steels or low-alloy and stainless steels. Today, there are many variations based on materials to be welded and types of gases used. It is a more complicated process than any of the above and requires more expensive equipment. However, it welds faster and allows the welding of many different metals. In time, it may become the most popular arc welding process.

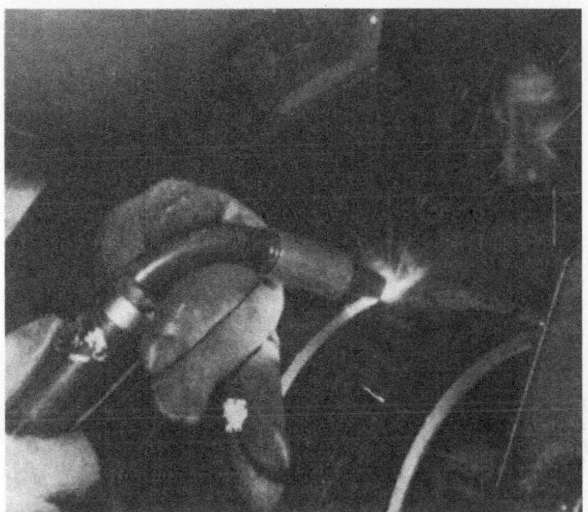

Figure 31-2 Gas metal-arc welding.

Gas Tungsten-Arc Welding

Gas tungsten-arc welding was invented by the aircraft industry and used extensively to weld "hard-to-weld" metals, primarily magnesium, aluminum, and stainless steels. It is an arc welding process in that an arc is struck between a nonconsumed tungsten electrode and the base metal. The entire area is shielded by an envelope of inert gas.

The heat produced by the arc melts the edges of the metal to be welded but does not melt the tungsten electrode. See Figure 31-3. Metal to fill the groove or to make the deposit is added independently into the arc. It then becomes the weld metal, producing a metallurgical joint as strong as the base material. The shielding gas prevents the oxygen and nitrogen of the air from coming in contact with the molten metal. This ensures the high quality of the deposit. The gas tungsten-arc welding process, which is commonly known as TIG welding, produces extremely high-quality welds. It is used primarily for the nonferrous metals, such as aluminum, magnesium, nickel alloys, copper alloys, stainless steels, and refractory metals. Argon is usually used as the shielding gas, although helium or argon-helium mixtures are sometimes employed.

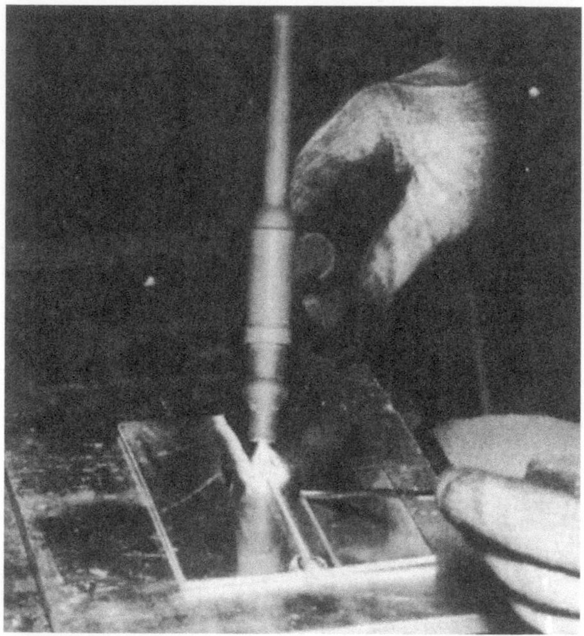

Figure 31-3 The gas tungsten-arc welding process.

Oxyacetylene Welding

Oxyacetylene gas welding is perhaps the oldest of the gas welding processes. It came into use about 125 years ago and is still being used in much the same manner. The process is extremely flexible and one of the most inexpensive as far as equipment is concerned. Today, its most popular application is in maintenance welding, small-pipe welding, auto body repairs, welding of thin materials, and sculpture work. The high temperature generated by the equipment is used for soldering, hard soldering, or brazing. Also, using a special torch, a variation of the process allows for flame cutting. This is accomplished by bringing the metal to a high temperature and then introducing a jet of oxygen that burns the metal apart. It is a primary cutting tool for steel.

In the oxyacetylene gas welding process, coalescence is produced by heating with a gas flame obtained from the combustion of acetylene with oxygen, with or without the use of filler metal. An oxyacetylene flame is one of the hottest of flames—6300°F. This hot flame melts the two edges of the pieces to be welded and the filler metal (added to fill the gaps or grooves) so that the molten metal can mix rapidly and smoothly. The acetylene and the oxygen gases flow from separate cylinders to the welding torch, where they are mixed and burned at the torch tip. Figure 31-4 shows the oxyacetylene gas welding process.

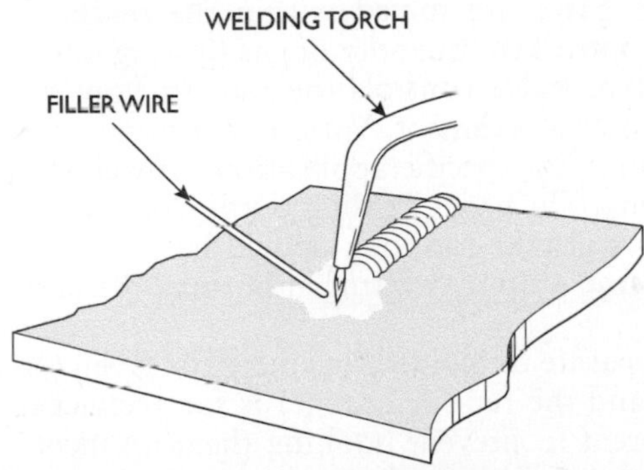

Figure 31-4 Oxyacetylene gas welding process.

The proportions of oxygen and acetylene determine the type of flame. The three basic types are *neutral*, *carburizing*, and *oxidizing*. The neutral flame is generally preferred for welding. It has a clear, well-defined white cone, indicating the best mixture of gases and no gas wasted. The carburizing flame has an excess of acetylene, a white cone with a feathery edge, and adds carbon to the weld. The oxidizing flame, with an excess of oxygen, has a shorter envelope and a small, pointed white cone. This flame oxidizes the weld metal and is used only for specific metals. Flame cutting is accomplished by adding an extra oxygen jet to burn the metal being cut. The equipment required for oxyacetylene welding is shown in Figure 31-5.

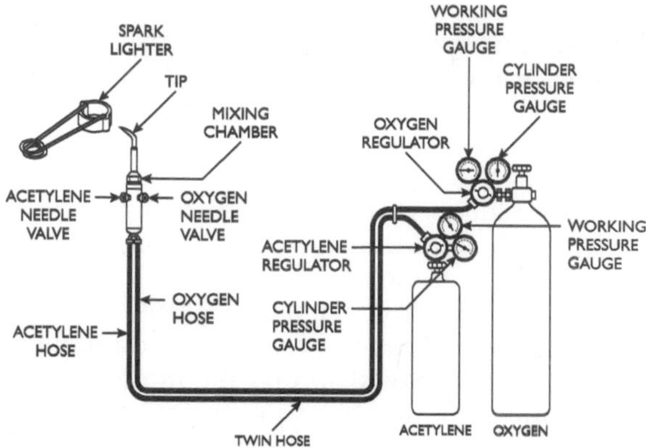

Figure 31-5 Oxyacetylene welding equipment.

The standard torch can be a combination type used for welding, cutting, and brazing. The gases are mixed within the torch. A thumbscrew needle valve controls the quantity of gas flowing into a mixing chamber. A lever-type valve controls the oxygen flow for cutting with a cutting torch or attachment. Various types and sizes of tips are used with the torch for specific applications of welding, cutting, brazing, or soldering. The usual welding outfit has three or more tips. Too small a tip will take too long or will be unable to melt the base metal. Too large a tip may result in burning the base metal.

The gas hoses may be separate or molded together. The green (or blue) hose is for oxygen, and the red (or orange) is for acetylene. The hose fittings are different to prevent hooking them up incorrectly. Oxygen hose has fittings with right-hand threads, and acetylene hose has fittings with left-hand threads.

Gas regulators keep the gas pressure constant, ensuring steady volume and even flame quality. Most regulators are dual stage and have two gauges; one tells the pressure in the cylinder, and the other shows the pressure entering the hose. Gases for the process are oxygen and, primarily, acetylene. Other gases, including hydrogen, city gas, natural gas, propane, and mapp gas, are used for specific applications. With its higher burning temperature, acetylene is the preferred gas in most instances.

Gas cylinders for acetylene contain porous material saturated with acetone. Because acetylene cannot safely be compressed over 15 psi, it is dissolved in the acetone, which keeps it stable and allows pressure of 250 psi. Because of the acetone in the acetylene cylinders, they should always stand upright. The oxygen cylinder capacities vary from 60 to 300 cu. ft with pressures up to 2400 psi. The maximum charging pressure is always stamped on the cylinder.

Use of the Cutting Torch

Oxyacetylene cutting, often called flame or oxygen cutting, is the most widely used process for thermal cutting of carbon steel. Just as a saw is to a carpenter, the cutting torch is to the steel fabricator.

The process involves preheating the steel to a temperature of 1500°F to establish an ignition temperature for the steel and then introducing a stream of oxygen under pressure from the tip of the torch. This causes the metal to oxidize or burn rapidly and produces a cut or kerf along the direction of travel.

Use of the cutting torch follows a set of steps. The first step is to select a tip or nozzle that is suitable for the job. This tip is screwed into the torch securely. The torch with a cutting tip installed is shown in Figure 31-6. Although manufacturers' charts are available for cutting data, the values shown in Table 31-1 will prove valuable in matching the tip to the job.

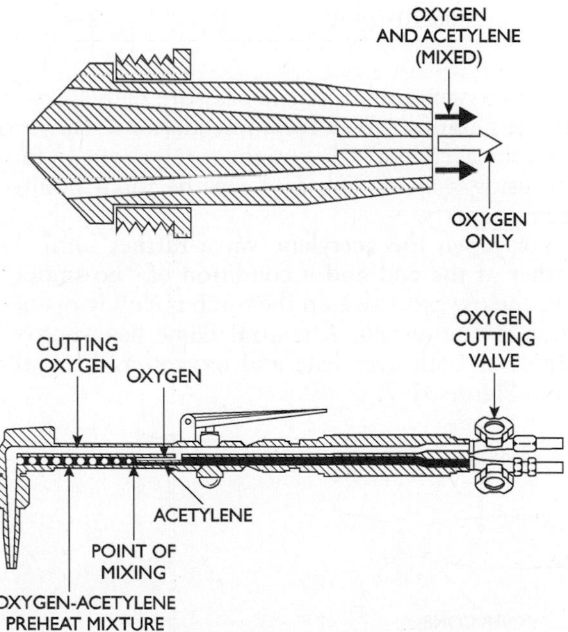

Figure 31-6 Oxyacetylene torch cutting tip.

Table 31-1 Cutting Torch Tip Selections

Thickness of Steel Plate (in.)	*Diameter of Cutting Tip Orifices (in.)*	*Approximate Cutting Speed (in. per min.)*
1/8	0.020–0.040	16–32
1/4	0.030–0.060	16–26
3/8	0.030–0.060	15–24
1/2	0.040–0.060	12–23
3/4	0.045–0.060	12–21
1	0.045–0.060	9–18
1½	0.060–0.080	6–14
2	0.060–0.080	6–13
3	0.065–0.085	4–11
4	0.080–0.090	4–10
5	0.080–0.095	4–8
6	0.095–0.105	3–7
8	0.095–0.110	3–5
10	0.095–0.110	2–4
12	0.110–0.130	2–4

Make sure that the oxygen and acetylene pressure regulators on the tanks are set to the manufacturer's recommendations. The next step is to partly open the acetylene valve on the torch about 1/4 turn and light the torch using a spark lighter. Keep the torch pointed away from people or property.

The next step is to open the acetylene valve further until the flame starts to feather at the end and a condition of "no-smoke" exists. At this point, the oxygen valve on the torch is slowly opened until a neutral flame is established. A neutral flame has approximately equal volumes of both acetylene and oxygen and has the appearance shown in Figure 31-7.

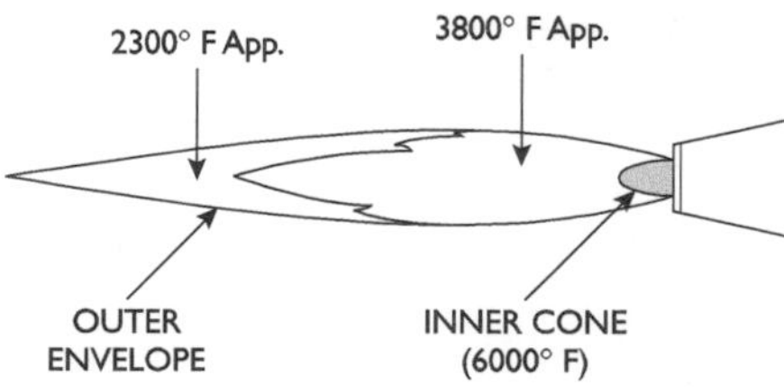

Figure 31-7 Neutral flame.

After the neutral flame is established, the oxygen lever on the torch is depressed. The individual flames that preheat flames around the oxygen orifices at the tip may change from neutral to slightly carburizing, as shown in Figure 31-8.

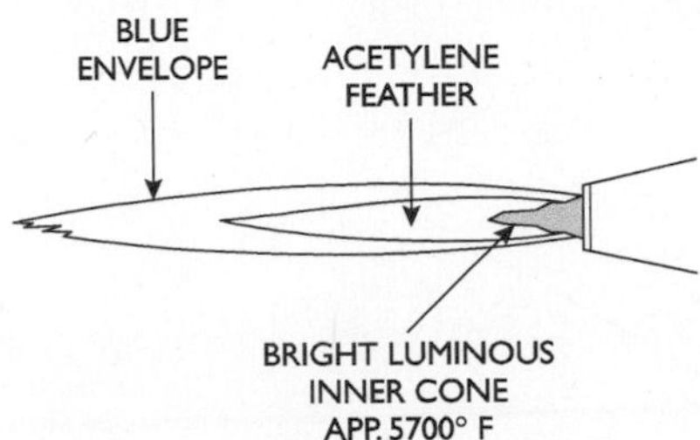

Figure 31-8 Slightly carburizing flame.

Keeping the oxygen lever in the depressed position, the oxygen valve is further adjusted until these preheat flames again appear neutral, as shown previously in Figure 31-7. At this point, after releasing the lever, the torch is set for preheating.

Usually the torch is held at the edge of the carbon steel material and held there until the steel becomes red-hot. Then, the oxygen lever is depressed, and a jet of oxygen is sent onto the red-hot metal, which ignites or oxidizes the metal, as shown in Figure 31-9. This is the point where cutting begins. Hold the torch steady until the cut is all the way through the metal. Then, move the torch slowly along the line to be cut, forming a kerf through the metal.

When the cut is finished, release the oxygen lever, close the acetylene valve, and close the oxygen valve—in that order.

Definitions of Welding Terms

Alternating Current, or AC—Electricity that reverses its direction periodically. For 60-hertz current, the current goes in one direction and then in the other direction 60 times in the same second so that the current changes its direction 120 times in one second.

Ammeter—An instrument for measuring either direct or alternating electric current (depending on its construction). Its scale is usually graduated in amperes, millamperes, microamperes, or kilamperes.

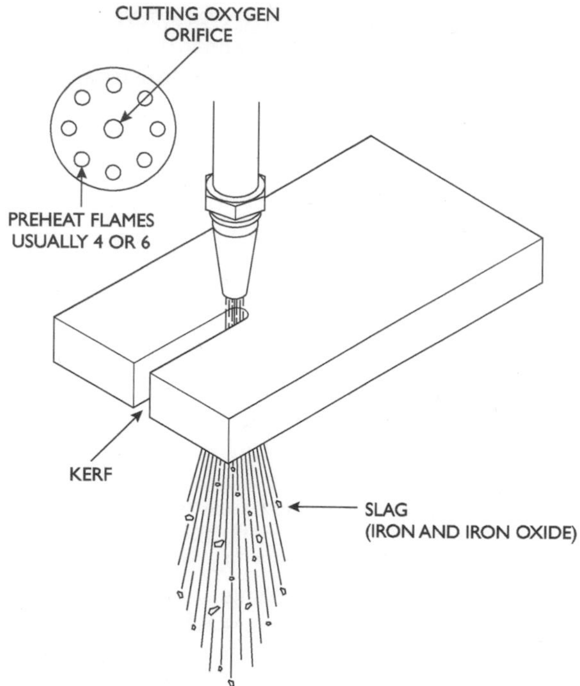

Figure 31-9 Cutting with the torch.

Arc Blow—Magnetic disturbance of the arc that causes it to waver from its intended path.

Arc Length—The distance from the end of the electrode to the point where the arc makes contact with work surface.

Arc Voltage—The voltage across the welding arc. It is measured with a voltmeter.

As-Welded—The condition of weld metal, welded joints, and weldments after welding prior to any subsequent aging, thermal, mechanical, or chemical treatments.

Backing—Material (metal, carbon, granulated flux, etc.) backing up the joint during welding.

Back-Step Welding—A welding technique wherein the increments of welding are deposited opposite the direction of progression.

Base Metal—The metal to be welded, soldered, or cut.

Braze—A weld wherein coalescence is produced by heating to suitable temperature and by using a filler metal that has a liquidus above 800°F (427°C) and below the solidus of the base metals. The filler metal is distributed between the closely fitted surfaces of the joint by capillary attraction.

Butt Weld—A weld made in the joint between two pieces of metal approximately in the same plane. See Figure 31-10.

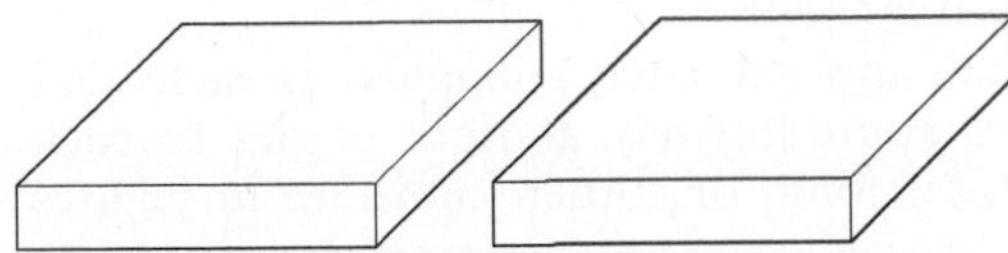

Figure 31-10 A butt joint.

Carbon Steel—Carbon steel is a term applied to a broad range of material containing the following:

Carbon 1.7% max.	Low-Carbon Steels	0.15% C max.
Manganese 1.65% max.	Mild-Carbon Steels	0.15–0.29% C
Silicon 0.60% max.	Medium-Carbon Steels	0.30–0.59% C
	High-Carbon Steels	0.60–1.70% C

Cast Iron—A wide variety of iron-base materials containing 1.7 to 4.5 percent carbon; 0.5 to 3 percent silicon; 0.2 to 1.3 percent manganese; 0.8 percent max. phosphorus; and 0.2 percent max. sulfur. Molybdenum, nickel, chromium, and copper can be added to produce alloyed cast irons.

Covered Electrode—A filler-metal electrode used in arc welding, consisting of a metal-core wire with a relatively thick covering that provides protection for the molten metal from the atmosphere, improves the properties of the weld metal, and stabilizes the arc.

Crater—A depression at the termination of a weld bead or in the weld pool beneath the electrode.

Depth of Fusion—The depth of fusion of a groove weld is the distance from the surface of the base metal to that point within the joints at which fusion ceases.

Direct Current, or DC—Electric current that flows in only one direction. In welding, an arc welding process wherein the power supply at the arc is direct current. It is measured by an ammeter.

Elongation—Extension produced between two gauge marks during a tensile test. Expressed as a percentage of the original gauge length, which should also be given.

FabCo Welding—See Flux-Cored Arc Welding.

Face of Weld—The exposed surface of a weld, on the side from which welding was done.

Fillet Weld—A weld of approximately triangular cross section joining two surfaces approximately at right angles to each other in a lap joint, tee joint, or corner joint. See in Figures 31-11 and 31-12.

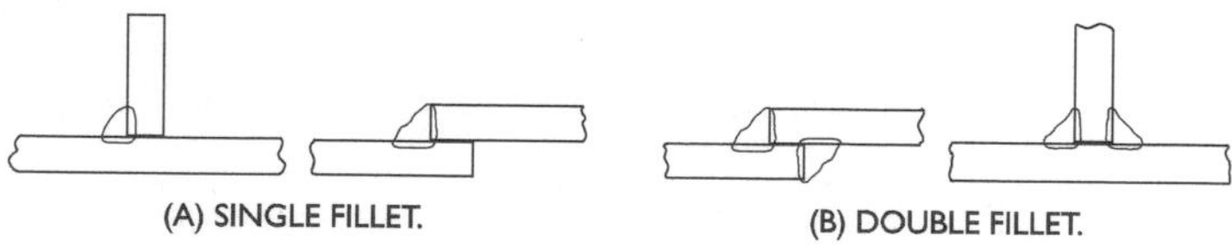

Figure 31-11 Typical fillet weld.

Flat position—The position of welding wherein welding is performed from the upper side of the joint, and the face of the weld is approximately horizontal—sometimes called downhand welding.

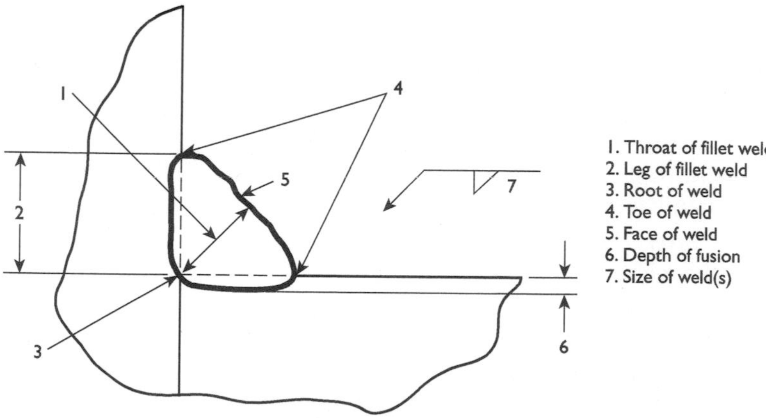

Figure 31-12 Nomenclature of fillet welding.

Flux—A fusible material used to dissolve and/or prevent the formation of oxides, nitrides, or other undesirable inclusions formed in welding.

Flux-Cored Arc Welding (FCAW)—An arc welding process wherein coalescence is produced by heating with an arc between a continuous filler-metal (consumable) electrode and the work. Shielding is obtained from a flux contained within the electrode. Additional shielding may or may not be obtained from an externally supplied gas or gas mixture.

Gas Metal-Arc Welding (GMAW) (MIG)—An arc welding process wherein coalescence is produced by heating with an arc between a continuous filler-metal (consumable) electrode and the work. Shielding is obtained entirely from an externally supplied gas or gas mixture.

Gas Shielded-Arc Welding—See MIG and TIG welding.

Gas Tungsten-Arc Welding (GTAW) (TIG)—An arc welding process wherein coalescence is produced by heating with an arc between a single tungsten (nonconsumable) electrode and the work. Shielding is obtained from a gas—argon or helium, or a mixture of them. Pressure may or may not be used, and filler metal may or may not be used.

Groove Weld—A weld made in the groove between two members to be joined. See Figures 31-13 and 31-14.

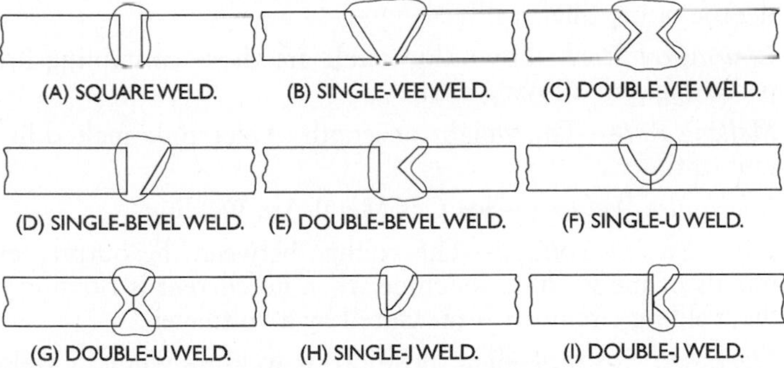

Figure 31-13 Various types of groove welds.

Heat Affected Zone—That portion of the base metal that has not been melted, but the mechanical or microstructure properties of which have been altered by the heat of welding or cutting.

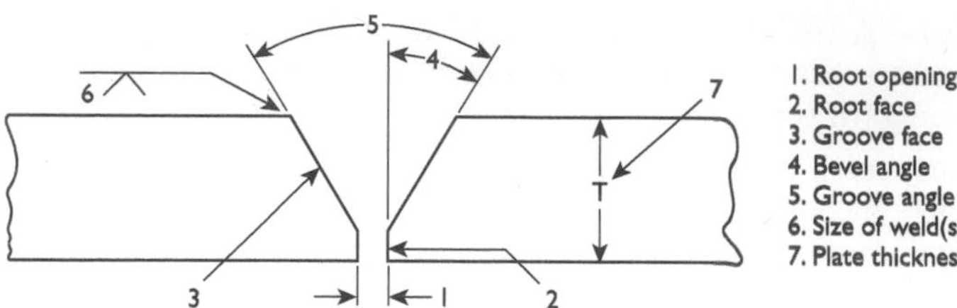

Figure 31-14 Nomenclature of groove welding.

Impact Resistance—Energy absorbed during breakage by impact of specially prepared notched specimen, the result being commonly expressed in foot-pounds.

Lap Joint—A joint between two overlapping members. See Figure 31-15.

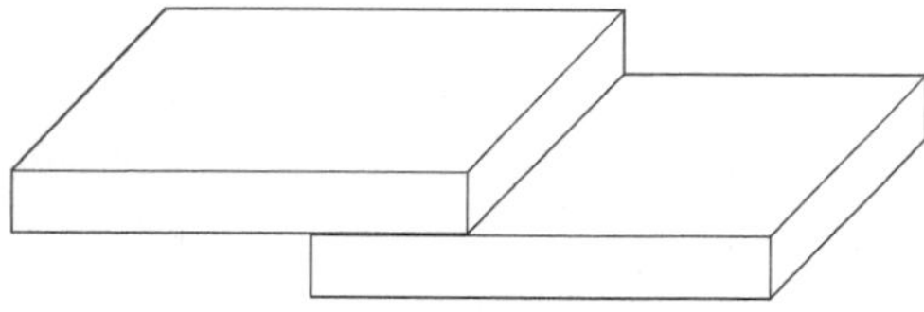

Figure 31-15 A lap joint.

Leg of Fillet Weld—The distance from the root of the joint to the toe of the fillet weld.

Low-Alloy Steel—Low-alloy steels are those containing low percentages of alloying elements.

Melting Rate—The weight or length of electrode melted in a unit of time.

Microwire Welding—See Gas Metal-Arc Welding.

Open-Circuit Voltage—The voltage between the output terminals of the welding machine when no current is flowing in the welding circuit. It is measured by a voltmeter.

Overhead Position—The position of welding wherein welding is performed from the underside of the joint.

Overlap—Protrusion of weld metal beyond the bond at the toe of the weld.

Pass—A single longitudinal progression of a welding operation along a joint of weld deposit. The result of a pass is a weld bead.

Peening—Mechanical working of metal by means of impact blows with a hammer or power tool.

Penetration—The distance the fusion zone extends below the surface of the part(s) being welded.

Porosity—Gas pockets or voids in metal.

Post-heating—Heating applied to the work after welding, brazing, soldering, or cutting operation.

psi—Pounds per square inch.

Preheating—The heat applied to the work prior to the welding, brazing, soldering, or cutting operation.

Puddle—That portion of a weld that is molten at the place the heat is applied.

Radiography—The use of radiant energy in the form of X-rays or gamma rays for the nondestructive examination of metals.

Reduction of Area—The difference between the original cross-sectional area and that of the smallest area at the point of rupture; usually stated as a percentage of the original area.

Reversed Polarity—The arrangement of arc welding leads wherein the work is the negative pole and the electrode is the positive pole in the arc circuit. Abbreviated DCRP.

Rheostat—A variable resistor that has one fixed terminal and a movable contact (often erroneously referred to as a "two-terminal potentiometer"). Potentiometers may be used as rheostats, but a rheostat cannot be used as a potentiometer because connections cannot be made to both ends of the resistance element.

Root of Weld—The points, as shown in cross section (Figures 31-12 and 31-14), at which the bottom of the weld intersects the base metal surfaces.

Root Opening—The separation between the members to be joined at the root of the joint.

Shielded Metal-Arc Welding—An arc welding process wherein coalescence is produced by heating with an electric arc between a covered metal electrode and the work. Shielding is obtained from decomposition of the electrode covering. Pressure is not used, and filler metal is obtained from the electrode.

Silver Solder—A term erroneously used to denote silver-base brazing filler metal.

Size of Weld—

Fillet Weld—For equal fillet welds, the leg length of the largest isosceles right triangle that can be inscribed within the fillet-weld cross section. For unequal fillet welds, the leg lengths of the largest right triangle that can be inscribed within the fillet-weld cross section.

Groove Weld—The joint penetration (depth of chamfering plus the root penetration when specified).

Slag Inclusion—Nonmetallic solid material entrapped in weld metal or between the weld metal and base metal.

Spatter—In arc and gas welding, the metal particles expelled during welding and which do not form a part of the weld.

Straight Polarity—The arrangement of arc welding leads wherein the work is the positive pole and the electrode is the negative pole of the arc circuit. Abbreviated DCSP.

Stress-Relief Heat Treatment—The uniform heating of structures to a sufficient temperature below the critical range to relieve the major portion of the residual stresses followed by uniform cooling.

Stringer Bead—A type of weld bead made without appreciable transverse oscillation.

Tack Weld—A weld (generally short) made to hold parts of a weldment in proper alignment until the final welds are made. Used for assembly purposes only.

Tensile Strength—The maximum load per unit of original cross-sectional area obtained before rupture of a tensile specimen. Measured in pounds per square inch (psi).

Throat of a Fillet Weld—Shortest distance from the root of a fillet weld to the face.

TIG Welding—See Gas Tungsten-Arc Welding.

Toe of Weld—The junction between the face of the weld and the base metal.

Tungsten Electrode—A nonfiller-metal electrode used in arc welding, consisting of a tungsten wire.

Ultimate Tensile Strength—The maximum tensile stress which will cause a material to break (usually expressed in pounds per square inch).

Underbead Crack—A crack in the heat-affected zone not extending to the surface of the base metal.

Undercut—A groove melted into the base metal adjacent to the toe of weld and left unfilled by weld metal.

Uphill Welding—A pipe-welding term indicating that the welds are made from the bottom of the pipe to the top of the pipe. The pipe is not rotated.

Vertical Position—The position of welding wherein the axis of the weld is approximately vertical.

Weaving—A technique of depositing weld metal in which the electrode is oscillated.

Weld—A localized coalescence of metal wherein coalescence is produced by heating to suitable temperatures, with or without the application of pressure and with or without the use of filler metal. The filler metal either has a melting point approximately the same as the base metal or has a melting point below that of the base metals but above 800°F.

Weld Metal—That portion of a weld which has been melted during welding.

Welding Procedure—The detailed methods and practices including joint-welding procedures involved in the production of a weldment using a specific process.

Welding Rod—Filler metal, in wire or rod form, used in gas welding and brazing processes, and those arc welding processes wherein the electrode does not furnish the filler metal.

Weldment—An assembly the component parts of which are joined by welding.

Whipping—A term applied to an inward and upward movement of the electrode employed in a vertical welding to avoid undercut.

Various welding positions are shown in Figures 31-16 and 31-17. The fillet welds are shown in Figure 31-11. Figure 31-12 illustrates various groove welds, which also includes pipe.

Selection of a Welding Power Source

Many different types and sizes of arc welding machines are available today. It is important that users of welding machines have sufficient technical information so that they can wisely select the best machine most suited for their particular work. The following information is provided so that you can be informed concerning the different types of machines available and also so that you can select the one ideally suited for your work.

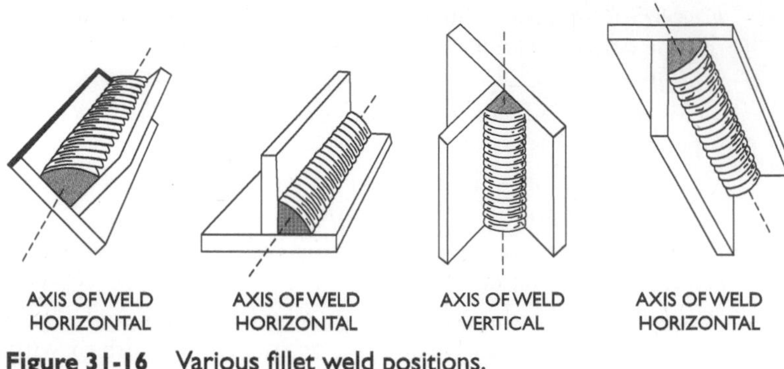

Figure 31-16 Various fillet weld positions.

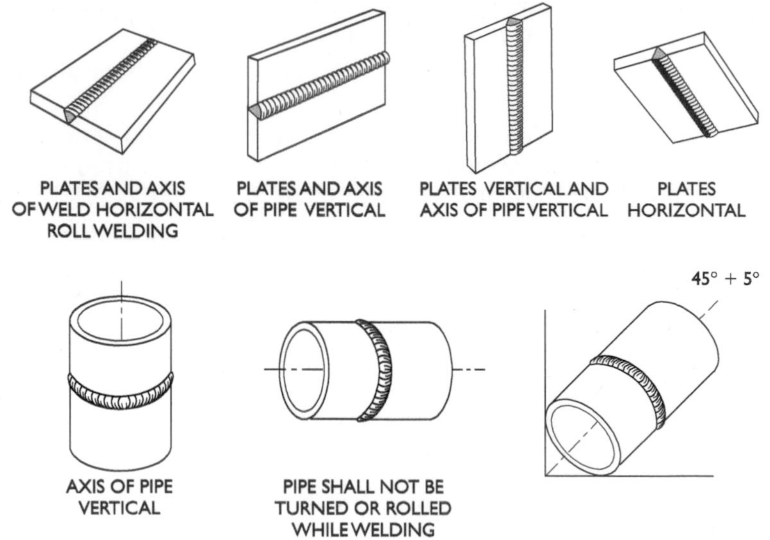

Figure 31-17 Various groove weld positions.

Arc welding machines can be classified in many different ways, such as rotating machines, static machines, electric motor driven machines, internal combustion engine driven machines, transformers/rectifiers, limited-input welding machines, conventional, constant-voltage welding machines, and single- and multiple-operator machines.

There are two basic categories of power sources: the conventional, or *constant-current* (cc), welding machine with the drooping volt-ampere curve; and the *constant-voltage* (cv), or modified-constant-voltage, machine with the fairly flat characteristic curve. The conventional (cc) machine can be used for manual welding and, under some conditions, for automatic welding. The constant-voltage machine is used only for continuous-electrode-wire arc welding processes operated automatically or semiautomatically. These types of machines are best understood by comparing their respective volt-ampere characteristic output curves. This type of curve is obtained by loading the welding machine with variable resistance and plotting the voltage of the electrode and work terminals for each ampere output. Figure 31-18 shows an example of this curve.

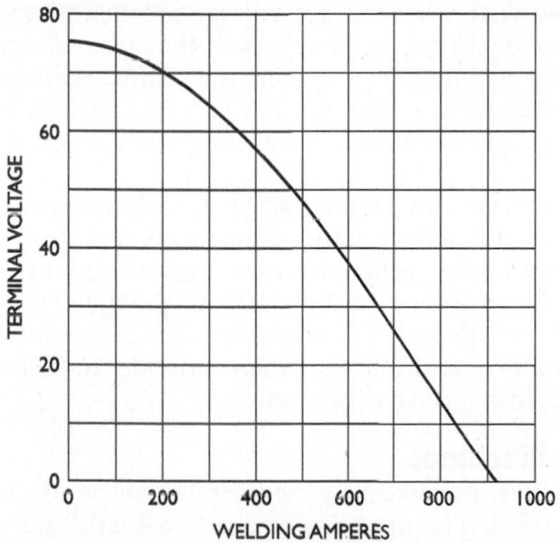

Figure 31-18 Voltage-ampere characteristic curve of the constant-current welder.

Conventional or Constant-Current (CC) Welding Machines

The conventional or constant-current (cc) welder is used for manual covered (stick) electrode arc welding (SMAW), the Gas-Tungsten (TIG) Process (GTAW), Carbon-Arc Welding (CAW), arc gouging, and for Stud Welding (SW). It can be used for automatic welding with larger-sized electrode wire, but only with a *voltage-sensing* wire feeder.

The constant-current (cc) welder produces a volt-ampere output curve as shown in Figure 31-18. A brief study of the curve will reveal that a machine of this type produces maximum output voltage with no load (zero current), and as the load increases, the output voltage decreases. Under normal welding conditions, the output voltage is between 20 and 40 volts. The open-circuit voltage is between 60 and 80 volts. Constant-current machines are available that produce either ac or dc welding power or both ac and dc.

On the constant-current welding machine, when welding with covered electrodes, the actual arc voltage is largely controlled by the welding operator and has a direct relationship to the arc length. As the arc length is increased (a long arc), the arc voltage increases. If the arc length is decreased (a short arc), the arc voltage decreases. The output curve shows that when the arc voltage increases (long arc), the welding current decreases, and when the arc voltage decreases (short arc), the welding current increases. Thus, without changing the machine setting, the operator can vary the current in the arc of "welding heat" a limited amount by lengthening or shortening the arc.

Constant-current machines can produce ac or dc welding power and can be rotating (generators) or static (transformers or transformer/rectifier) machines. The generator can be powered by a motor for shop use or by an internal combustion engine (gasoline, gas, or diesel) for field use. Engine-driven welders can have either water- or air-cooled engines, and many of them provide auxiliary power for emergency lighting, power tools, etc.

Generator Welding Machines

On dual-control machines, normally a generator, the slope of the output curve can be varied. The open circuit, or "no-load," voltage is controlled by the fine adjustment control knob. This control is also the fine welding-current adjustment during welding. The range switch provides coarse adjustment of the welding current. In this way, a soft or harsh arc can be obtained. With the flatter curve and its low open-circuit voltage, a change in arc voltage will produce a greater change in output current. This produces a digging arc preferred for pipe welding. With the steeper curve and its high open-circuit voltage, the same change in arc voltage will produce less of a change in output current. This is a soft, or quiet arc, useful for sheet-metal welding. In other words, the dual-control, conventional or constant-current welder allows the most flexibility for the welding operator. These machines can be driven by an electric motor or internal combustion engine. See Figure 31-19.

Figure 31-19 A typical generator welding machine.

The constant-current, or drooping volt-ampere, characteristic machine can also be used for automatic welding processes. However, to utilize this type of welding machine, the automatic wire-feeding device must compensate for changes in arc length. This requires rather complex control circuits that involve feedback from the arc voltage or in voltage sensing. This type of system is not used for small-diameter electrode wire welding processes.

Transformer Welding Machines

The *transformer* welding machine (Figure 31-20) is the least expensive, lightest, and smallest of any of the different types of welders. It produces alternating current (ac) for welding. The transformer welder takes power directly from the line, transforms it to the power required for welding, and by means of various magnetic circuits, inductors, etc., provides the volt-ampere characteristics proper for welding. The welding current output of a transformer welder may be adjusted in many different ways. The simplest method of adjusting output current is to use a tapped secondary coil on the transformer. This is a popular method employed by many of the small limited-input welding transformers. The leads to the electrode holder and the work are connected to plugs, which may be inserted in sockets on the front of the machine in various locations to provide the required welding current. On some machines, a tap switch is employed instead of the plug-in arrangement. In any case, exact current adjustment is not entirely possible.

Figure 31-20 A typical transformer welder.

On industrial types of transformer welders, a continuous-output current control is usually employed. This can be obtained by mechanical means or electrical means. The mechanical method may involve moving the core of the transformer proper or moving the position of the coils within the transformer. Any of the methods that involve mechanical movement of the transformer parts require considerable movement for full range adjustment. The more advanced method of adjusting current output is by means of electrical circuits. In this method, the core of the transformer or reactor is saturated by an auxiliary electric circuit, which controls the amount of current delivered to the output terminals. By merely adjusting a small knob, it is possible to provide continuous current adjustment from minimum to maximum output.

Although the transformer type of welder has many desirable characteristics, it also has some limitations. The power required for a transformer welder must be supplied by a single-phase system, and this may create an imbalance in the power supply lines, which is objectionable to most power companies. In addition, transformer welders have a rather low power-factor demand unless they are equipped with power-factor-correcting capacitors. The addition of capacitors corrects the power factor under load and produces a reasonable power factor that is not objectionable to electric power companies.

Transformer welders have the lowest initial cost and are the least expensive to operate. Transformer welders require less space and are normally quiet in operation. In addition, alternating-current welding power supplied by transformers reduces arc glow, which can be troublesome on many welding applications. They do not, however, have as much flexibility for the operator as the dual-controlled generator.

Transformer/Rectifier Welding Machines

The previously described transformer welders provide alternating current (ac) at the arc. Some types of electrodes can be operated successfully only with direct-current (dc) power. A method of supplying direct-current power to the arc, other than use of a rotating generator, is by adding a rectifier, an electrical device that changes alternating current into direct current. Rectifier welding machines can be made to utilize a three-phase input, as well as single input. The three-phase input machine overcomes the line imbalance mentioned before.

In this type of machine, the transformers feed into a rectifier bridge, which then produces direct current for the arc. In other cases, where both ac and dc may be required, a single-phase type of ac welder is connected to the rectifier. By means of a switch that can change the output terminals to the transformer or to the rectifier, the operator can select either ac or dc straight- or reverse-polarity current for his welding requirement. In some types of ac-dc welders, a high-frequency oscillator plus water and gas control valves are installed. This then makes the machine ideally suited for TIG tungsten inert gas welding, as well as for manual coated electrode welding.

The transformer/rectifier welders are available in different sizes and, as mentioned, for single-phase or three-phase power supply. They may also be arranged for different voltages from different power supplies. The transformer/rectifier unit is more efficient electrically than the generator and provides quiet operation. See Figure 31-21.

Multiple-Operator Welding System

This system uses a heavy-duty, high-current, and relatively high-voltage power source that feeds a number of individual-operator welding stations. At each welding station, a variable resistance is adjusted to drop the current to the proper welding range. Based on the duty cycle of the welding operators, one welding machine can supply welding power simultaneously to a number of welding operators. The current

Figure 31-21 A typical transformer/rectifier welder.

supplied at the individual station has a drooping characteristic similar to the single-operator welding machines previously described. The welding machine size and the number and size of the individual welding-current control stations must be carefully matched for an efficient multiple-operator system.

Constant Voltage (CV) Welding Machines

A *constant-voltage* or *modified-constant-voltage* power source is a welding machine that provides a nominally constant voltage to the arc regardless of the current in the arc. The characteristic curve of this type of machine is shown by the volt-ampere curve illustrated in Figure 31-22. *This type of machine can be used for only semiautomatic or automatic arc welding* using a continuously fed electrode wire. Furthermore, these machines are made to produce only direct current (dc).

In continuous-wire welding, the burn-off rate of a specific size and type of electrode wire is proportional to the welding current. In other words, as the welding current increases, the amount of wire burned off increases proportionally. This is graphically shown in Figure 31-23. It can be seen that if wire were fed into an arc at a specific rate, it would automatically require, or draw from a constant-voltage power source, a proportionate amount of current. The constant-voltage welder provides the amount of current required from it by the load imposed on it. In this way, a very basic type of automatic welding control can be employed. The wire is fed

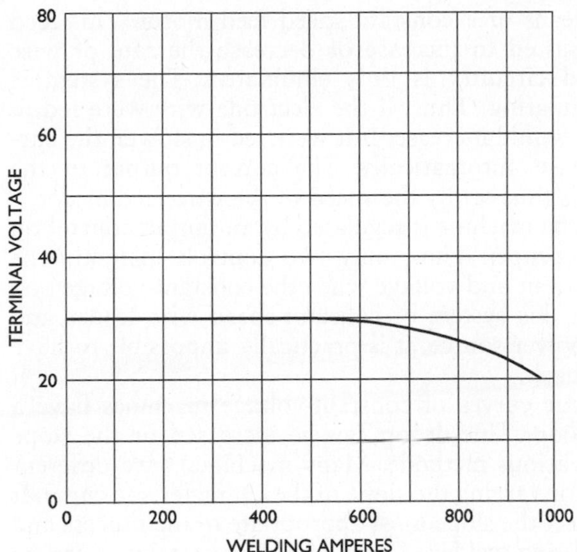

Figure 31-22 Voltage-ampere characteristic curve of the constant-voltage welder.

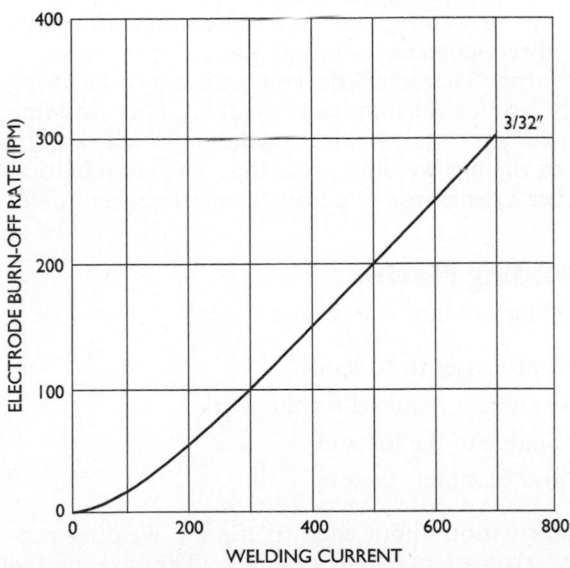

Figure 31-23 Welding current versus electrode burn-off rate.

into the arc by means of a constant-speed feed motor. This feed motor can be adjusted to increase or decrease the rate of wire feed. Complicated circuitry is thus eliminated. The system is inherently self-regulating. Thus, if the electrode wire were fed in faster, the current would increase. If it were fed in slower, the current would decrease automatically. The current output of the welding machine is thus set by the speed of the wire feed motor.

The voltage of the machine is regulated by an output control on the power source proper. Thus, only two controls maintain the proper welding current and voltage when the constant-voltage system is used. With this system of constant-speed wire feeder and constant-voltage power source, it is practically impossible to have stubbing or burn-back.

The characteristic curves of constant-voltage machines have a slight inherent droop. This droop can be increased or the slope made steeper by various methods. Many machines have different taps, or controls, for varying the slope of the characteristic curve. It is important to select the slope most appropriate to the process and the type of work being welded. Constant-voltage machines can be either generator types or transformer/rectifier welders. Generators can be either motor- or engine-driven.

Combination CV-CC Welding Machines

The most flexible type of welding machine is a combination type that can provide direct-current welding power with either a drooping- or flat-output characteristic volt-amp curve by using different terminals and/or changing a switch. This welding machine is the most universal welder available. It allows the welder to use any of the arc welding processes. The combination machine can be either a generator or a transformer/rectifier power source.

Specifying a Welding Machine

Selection of the welding machine is based on

1. The process or processes to be used.
2. The amount of current required for the work.
3. The power available to the job site.
4. Convenience and economic factors.

The previous information about each of the arc welding processes indicates the type of machine required. The size of the machine is based on the welding current, and duty cycle and voltage are determined by analyzing the welding job and considering weld joints, weld sizes, etc., and by consulting welding procedure tables. The incoming power available dictates this fact. Finally, the job situation, personal preference, and economic considerations narrow the field to the final selection. The local welding equipment supplier should be consulted to help make your selection.

To order a welding machine properly, the following data should be given:

1. Manufacturer's designation or catalog number
2. Manufacturer's identification or model number
3. Rated load voltage
4. Rated load amperes (current)
5. Duty cycle
6. Voltage of power supply (incoming)
7. Frequency of power supply (incoming)
8. Number of phases of power supply (incoming)

Welding Machine Duty Cycle

Duty cycle is defined as the ratio of arc time to total time. For a welding machine, a 10-minute time period is used. Thus, for a 60-percent-duty-cycle machine, the welding load would be applied continuously for 6 minutes and would be off for 4 minutes. Most industrial constant-current (drooping) machines are rated at 60 percent duty cycle. Most constant-voltage (flat) machines used for automatic welding are rated at 100 percent duty cycle.

The chart (Figure 31-24), "Percent of worktime versus current load," or percentage duty cycle, represents the ratio of the square of the rated current to the square of the load current multiplied by the rated duty cycle. Instead of working out the formula, use this chart. Draw a line parallel to the sloping lines through the intersection of the subject machine's rated current output and rated duty cycle. For example, a question might arise whether a 400-amp 60-percent-duty-cycle machine could be used for a fully automatic requirement of 300 amps for a 10-minute welding job. It shows that the machine can be used at slightly over 300 amperes at a 100 percent duty cycle. Conversely, there may be a need to draw more than the rated current from a welding machine, but for a short period. For example, it shows that the 200-amp 60-percent-rated machine can be used at 250 amperes providing the duty cycle does not exceed 40 percent (or 4 minutes out of each 10-minute period). Use this chart to compare various machines. Relate all machines to the same duty cycle for a true comparison.

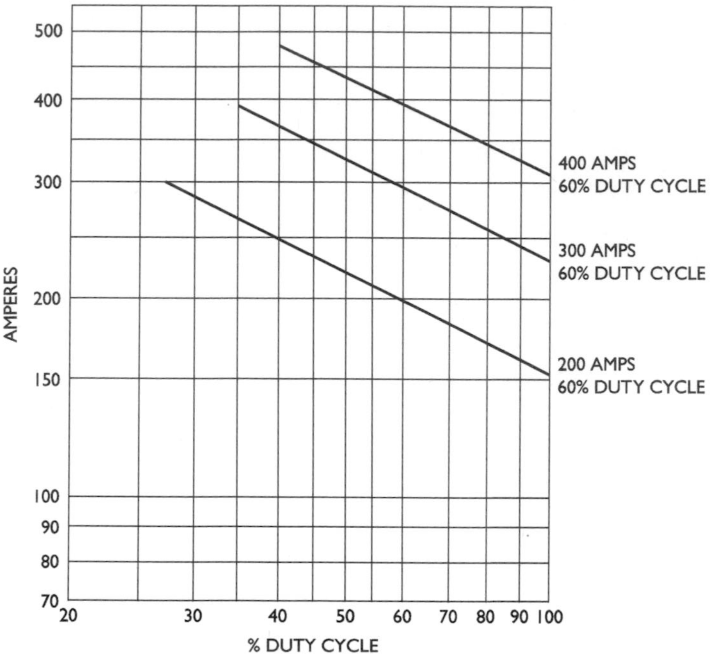

Figure 31-24 Percent of worktime versus current load.

Shielded Metal-Arc (Stick) Welding Process (SMAW)

Shielded metal-arc welding is an arc welding process wherein coalescence is produced by heating with an arc between a covered metal electrode and the work. Shielding is obtained from decomposition of the electrode covering. Filler metal is obtained from the electrode. The process is shown in Figure 31-25, which shows the covered electrode, the core wire, the arc area, the shielding atmosphere, the weld, and solidified slag.

This manually controlled process welds all nonferrous metals ranging in thickness from 18 gauge to maximum encountered. For material thickness over ¼ in., a beveled-edge preparation is used and the multipass welding technique employed. The process allows for all-position welding. The arc is under the control of, and is visible to, the welder. Slag removal is required.

The major components (Figure 31-26) required for shielded metal-arc welding are the following:

1. The welding machine (power source)
2. The covered electrode
3. The electrode holder
4. The welding circuit cables

The *welding machine* (power source) is the most important item of welding equipment involved. Its primary purpose is to provide electric power of the proper current and voltage sufficient to maintain a welding arc. Shielded metal-arc welding can be accomplished by either alternating current (ac) or direct current (dc). Straight (electrode negative) or reverse (electrode positive) polarity can be used with direct current. Welding machines are of many different types designed for specific welding applications. The choice of the proper welding machine for specific applications can be partially based on the following:

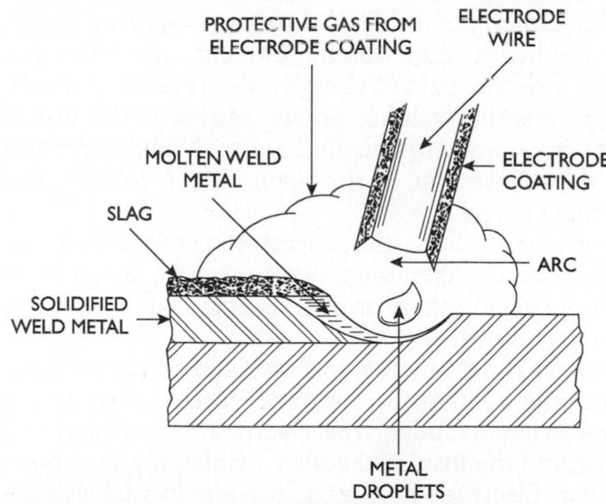

Figure 31-25 Shielded metal-arc welding.

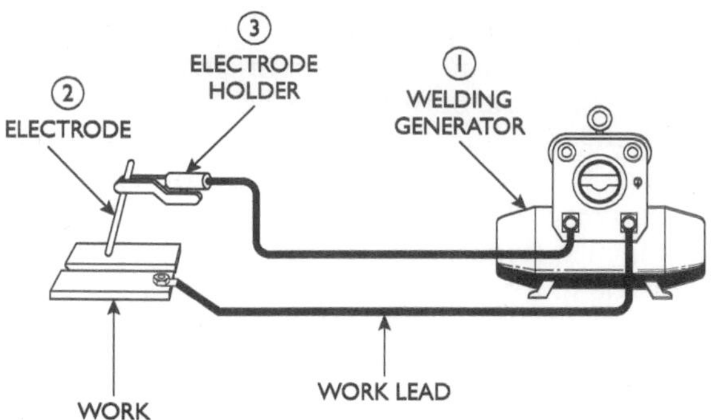

Figure 31-26 Major components required for shielded metal-arc welding.

The ac transformer type is the smallest, lightest, and least expensive of any of the welding power sources. It provides alternating welding current at the arc and is quiet in operation. It is a single-control type of machine usually with one knob for varying the current output. Other types have plug-in connectors or tap switches for the same purpose. The transformer-rectifier type of welding machine converts ac power to dc power and provides direct current at the arc. This type of machine usually has a single control and is quiet in operation. It is more expensive than the transformer welder. There is also the ac-dc transformer-rectifier type of welding machine. This is a specially designed power source that allows either ac or dc welding. A built-in switch allows selection of either type of welding current, and a polarity switch allows the use of either straight or reverse polarity. Probably the most versatile welding power source is the direct-current generator. The conventional dual-control single-operator generator allows the adjustment of the open-circuit voltage (and slope) and the welding current.

When electric power is available, the generator is driven by an electric motor. Away from power lines, the generator can be driven by an internal combustion gasoline engine or a diesel engine. It can also be belt-driven by a power takeoff.

The *electrode holder* is held by the operator and firmly grips the electrode and carries the welding current to it. Electrode holders are supplied in various types including the electrically insulated and noninsulated pincer and the insulated collet. Insulated pincer types are the most popular. Electrode holders also come in various sizes and are designated by their current-carrying capacity.

The *welding circuit* consists of the welding cables and connectors used to provide the electrical circuit for conducting the welding current from the machine to the arc. The electrode cable forms one side of the circuit and runs from the electrode holder to the "electrode" terminal of the welding machine. The work lead is the other side of the circuit and runs from the work clamp to the "work" terminal of the welding machine. Welding cables are normally made of many fine copper wires, but on occasion aluminum is used. The outer sheath or insulation is rubber or neoprene. Cable size is selected based on the maximum welding current used. Sizes range from AWG No. 6 to AWG No. 4/0 with amperage ratings from 75 amps upward. Cable length should be no longer than is required for the particular job.

Covered electrodes, which become the deposited weld metal, are available in sizes from 1⁄16 to 5⁄16 in. diameter of the core wire and length 9 in., 14 in., and 18 in., with the 14-in. length being the most common. The covering on the electrode dictates the usability of the electrode and provides

1. Gas from the decomposition of certain ingredients in the coating, to shield the arc from the atmosphere.
2. Deoxidizers for purifying the deposited weld metal.
3. Slag formers to protect the deposited weld metal from oxidation.
4. Ionizing elements to make the electrode operate more smoothly.
5. Alloying elements to provide higher-strength deposited metal.
6. Iron powder to improve the productivity of the electrode.

The usability of different types of electrodes is standardized and defined by the American Welding Society. The AWS identification system indicates the strength of the deposited weld metal, the welding positions that may be employed, the usability factor of the electrode, and in some cases, the deposited metal analysis. Individual electrodes are identified by the AWS classification number printed on it. Color code marking was used in the past but is no longer employed by major electrode manufacturers.

The selection of electrodes for specific job applications is quite involved but can be based on the following eight factors:

1. Base-metal strength properties
2. Base-metal composition

3. Welding position
4. Welding current
5. Joint design and fit-up
6. Thickness and shape of base metal
7. Service conditions and/or specification
8. Production efficiency and job conditions

Welding Safety

Protective clothing must be worn by the welder to shield skin from exposure to the brilliant light given off by the arc. A helmet is required to protect the face and eyes from the arc. A dark-colored filter glass in the helmet allows the welder to watch the arc while protecting the eyes from the bright light. Various shades of filter glass lenses are available depending on welding current. Ventilation must be provided when welding in confined areas.

Gas Metal-Arc (MIG) Welding Process (GMAW)

Gas metal-arc (MIG) welding is an arc welding process wherein coalescence is produced by heating with an arc between a continuous filler-metal (consumable) electrode and the work. Shielding is obtained entirely from an externally supplied gas mixture. There are four major classifications of this process depending on the type of shielding gas and/or the type of metal transfer:

MIG using pure inert gas shielding on nonferrous metals.

Microwire using short-circuiting transfer and allowing all-position welding.

CO_2 using a shielding gas of CO_2 and larger electrode wire.

Spray using argon/oxygen shielding gas.

Gas metal-arc welding is shown in Figure 31-27. This diagram shows the electrode wire, the gas shielding envelope, the arc, and the weld. The process may be either semiautomatic or automatic, with the semiautomatic method the more widely used.

The outstanding features of GMAW welding are the following:

1. It will make top-quality welds in almost all metals and alloys used in industry.
2. Minimum post-weld cleaning is required.
3. The arc and weld pool are clearly visible to the welder.

4. Welding is possible in all positions depending on electrode wire size and process variation.
5. Relatively high-speed welding process provides economy.
6. There is no slag produced that might be trapped in the weld.

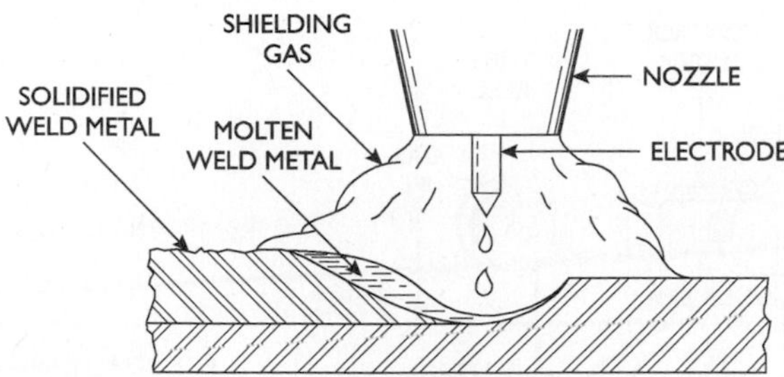

Figure 31-27 Gas metal-arc welding.

Variations of the process offer special advantages. For example, microwire will weld most steels in thinner gauges than previously possible with an arc process and is a low-hydrogen process. CO_2 welding produces high-speed welds at minimum cost due to low-cost gas. The spray transfer using argon/oxygen produces high-speed welds with minimum cleanup, and the true MIG process welds the same nonferrous metals as GTAW, but at a higher rate of speed.

The major equipment components required for the GMAW process are shown in Figure 31-28. These are the following:

1. The welding machine (power source)
2. The wire-feed drive system and controls
3. The welding gun and cable assembly (for semiautomatic welding) or the welding torch (for automatic welding)
4. The shielding-gas supply and controls
5. The electrode wire

The welding machine or power source for consumable-electrode welding is called a "constant-voltage" (CV) type of welder, which means that its characteristic volt-ampere curve is essentially flat. Its output voltage is essentially the same with different welding-current levels. The output voltage is adjusted by a rheostat on the welding machine, which can be either a transformer/rectifier or a motor- or engine-driven generator. Constant voltage (CV) power sources do not have a welding-current control and cannot be used for welding with electrodes. The welding-current output is determined by the load on the machine, which is dependent on the electrode-wire feed speed. Direct-current reverse polarity is normally used for GMAW. Machines are available from 150 to 1000 amps.

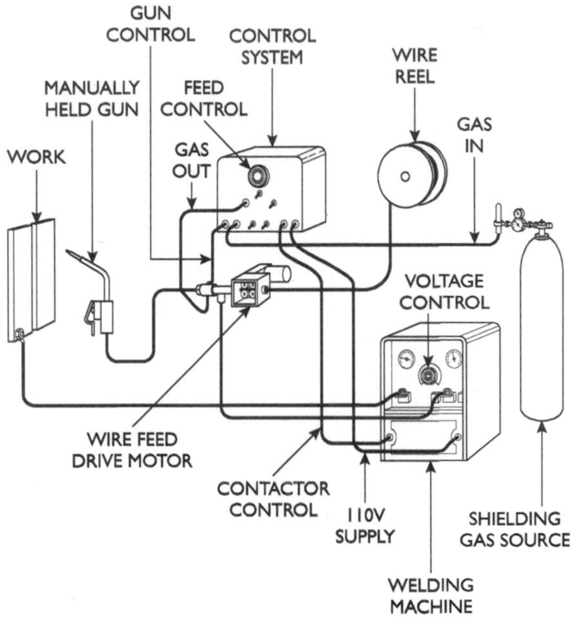

Figure 31-28 Equipment required for gas metal-arc welding.

The wire-feeder system must be matched to the constant-voltage power supply. The CV system of welding relies on the relationship between the electrode wire burn-off rate and the welding current. This relationship is fairly constant for a given electrode wire size, composition, and shielding atmosphere. At a given wire-feed speed rate, the welding machine will supply the proper amount of current to maintain a steady arc. Thus, the electrode-wire feed rate determines the amount of welding current supplied to the arc. The wire-feed speed control adjusts the welding current. The CV welding system is a self-regulating system and is the more practical system when using small-diameter electrode wire. A variation is the miniaturized wire feeder built into the welding gun that is popular for welding with small-diameter aluminum wire.

The welding gun and cable assembly is used to carry the electrode wire, the welding current, and the shielding gas to the welding arc. The electrode-wire guide is centered in the nozzle with the shielding gas supplied concentric to it. The gun is held fairly close to the work to properly control the arc and provide an efficient gas-shielding envelope. Guns for heavy-duty work at high currents and guns using inert gas and medium to high current must be water-cooled. Guns for microwire welding are not water-cooled. Guns may be of two different designs: the pistol-grip or the curved-head (gooseneck). The gooseneck type is most popular for microwire all-position welding, and the pistol-grip type is usually used for welding with larger electrode wires in the flat position. For fully automatic welding, a welding torch is normally attached directly to the wire-feed motor. Automatic torches are either air- or water-cooled depending on the welding application as mentioned above. For CO_2 welding, a side-delivery gas system is often used with automatic torches.

The shielding gas displaces the air around the arc to prevent contamination by the oxygen or nitrogen in the atmosphere. This gas-shielding envelope must efficiently shield the area in order to obtain high-quality weld metal. The shielding gas normally used for gas metal-arc welding is argon, helium, or mixtures for nonferrous metals; CO_2 for steel; CO_2 with argon and sometimes helium for steel and stainless steel; and argon plus small amounts of oxygen for steel and stainless steel. The type of gas for shielding and the flow rates are given by welding procedure tables for welding various metals with the different variations of the process. Shielding gases for welding must be specified as "welding grade." This provides for a specific purity level and moisture content. Gas flow rates depend on the type of gas used, metal being welded, welding position, welding speed, and draftiness. Refer to procedure tables for the information.

The electrode wire composition for gas metal-arc welding must be selected to match the metal being welded, the variation within the GMAW process, and the shielding atmosphere. Refer to procedure tables and selection charts for this information. The electrode wire size depends on the variation of the process and the welding position. All electrode wires are solid and bare except in the case of carbon steel wire, where a very thin protective coating (usually copper) is employed. Electrode wires are available on spools, coils, and reels of various diameters. They are usually packed in special containers to protect them from storage deterioration.

Gas Tungsten-Arc (TIG) Welding Process (GTAW)

Gas tungsten-arc (GTAW) is an arc welding process wherein coalescence is produced by heating with an arc between a single tungsten (nonconsumable) electrode and the work. Shielding is obtained from an inert gas or inert gas mixture. Filler metal may or may not be used. (This process is sometimes called TIG welding.) Figure 31-29 shows the GTAW or TIG process. It shows the arc, the tungsten electrode, and the gas shield envelope all properly positioned above the workpieces. The filler-metal rod is shown being fed manually into the arc and weld pool.

The outstanding features of GTAW welding are the following:

1. It will make top-quality welds in almost all metals and alloys used in industry.
2. Practically no post-weld cleaning is required.
3. The arc and weld pool are clearly visible to the welder.
4. There is no filler metal across the arc stream, so there is no weld spatter.
5. Welding is possible in all positions.
6. There is no slag produced that might be trapped in the weld.

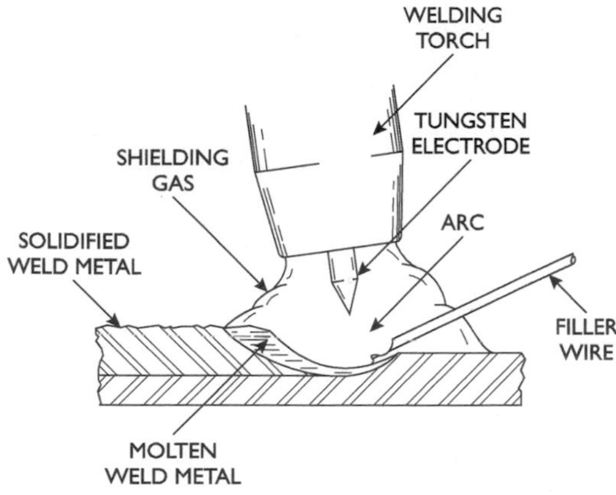

Figure 31-29 Gas tungsten-arc welding process.

GTAW can be used for welding aluminum, magnesium, stainless steel, silicon bronze, silver, copper and alloys, nickel and alloys, cast iron, and mild steel. It will weld a wide range of metal thicknesses.

GTAW welding is also used to make the root pass on carbon-steel pipe weld joints.

The major equipment components required for the GTAW process (see Figure 31-30) are the following:

1. The welding machine (power source)
2. The GTAW welding torch and the tungsten electrodes
3. Filler metal wires
4. The shielding gas and controls

Several optional accessories are available; these include a remote foot rheostat, which permits the welder to control current while welding, thus allowing him to make corrections and fill craters. Also available are water-circulating systems to cool the torch, arc timers, etc.

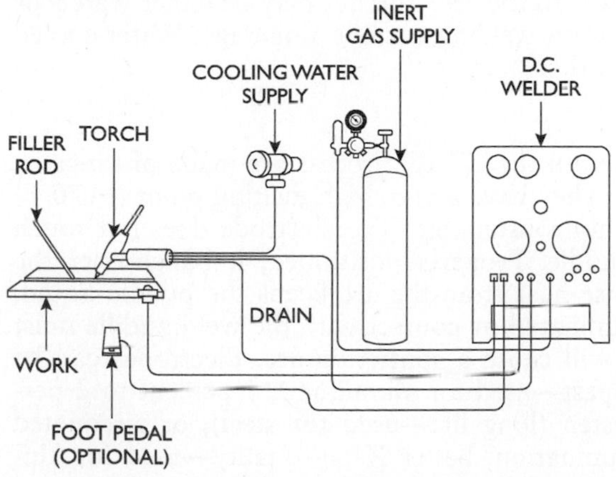

Figure 31-30 Equipment necessary for gas tungsten-arc welding.

Welding Machine

A specially designed welding machine (power source) is used with the GTAW process. It may be an ac/dc rectifier or a direct-current (dc) generator that can be either motor or engine driven. Alternating current (ac) or direct current (dc) may be used; either straight or reverse polarity can be used with direct current. A high-frequency current unit may be added for GTAW welding. High-frequency current is used only in starting the welding arc when using dc current, but it is always used continuously with ac current. Your selection of ac or dc depends on the material being welded. Alternating current is recommended for welding aluminum and magnesium; direct current is recommended for welding stainless steel, cast iron, mild steel, copper, nickel and alloys, and silver. A typical GTAW welding machine operates with a range of 3 to 350 amps, with 10 to 35 V at a 60 percent duty cycle.

It is also possible to use ordinary ac or dc power sources (designed primarily for covered-electrode welding) in conjunction with a high-frequency attachment. However, best results are obtained with a welding machine specifically designed for TIG welding. Welding machines designed for TIG welding can also be used for stick welding.

Torch

The GTAW torch holds the tungsten electrode and directs shielding gas and welding power to the arc. Torches may be either water- or air-cooled, depending on welding current amperage. Water-cooled torches are widely used.

Electrodes

The electrodes used with the GTAW process are made of tungsten and tungsten alloys. They have a very high melting point (6170°F) and are practically nonconsumable. The electrode does not touch the molten weld puddle. Properly positioned, it hangs over the work, and the intense heat from the arc keeps the puddle liquid. Electrode tips contaminated by contact with the weld puddle must be cleaned or they will cause a sputtering arc. Electrodes can be pure tungsten (cheapest—used for aluminum), 1 percent to 2 percent thoriated tungsten (long life—used for steel), or zirconated tungsten (less contamination, better X-ray quality—used for aluminum). The different tungsten types are easily recognized by a color code. They come in either a cleaned or ground finish and in 3- to 24-in. lengths.

Filler Metal

Filler metal may or may not be used. It is normally used, except when very thin metal is to be welded. The composition of the filler metal should be matched to that of the base metal. Filler-metal charts show the recommended types. The size of the filler-metal rod depends on the thickness of the base metal and the welding current. Filler metal is usually added to the puddle manually, but automatic feed is sometimes used.

Shielding Gas

An inert gas, either argon, helium, or a mixture of both, shields the arc from the atmosphere. Argon is more commonly used because it is easily obtainable and, being heavier than helium, provides better shielding at lower rates. For flat and vertical welding, a gas flow of 15 to 30 cfh is usually sufficient. Overhead position welding requires a slightly higher rate.

Welding Safety

Protective clothing must be worn by the welder to shield skin from exposure to the brilliant light given off by the arc. A helmet is required to protect the face and eyes from the bright light. Ventilation must be provided when welding in confined areas.

Electroslag Welding Process (EW)

Electroslag welding (EW) is a welding process that produces coalescence through electrically melted flux, which melts the filler metal and the surfaces of the work to be welded. The weld pool is shielded by a heavy slag, which moves along the full cross section of the joint as welding progresses. The conductive slag is kept molten by its resistance to the flow of electric current passing between the electrode and the work. The cavity formed by the parts to be welded and the molding shoes contains the molten flux pool, the molten weld metal, and the solidified weld metal. Shielding from the atmosphere is provided by the pool of molten flux. The melted base metal, electrode, and guide tube connect at the bottom of the flux pool and form the molten weld metal. As the molten weld metal slowly solidifies, it joins the plate together.

Consumable Guide Electroslag Welding

This is a method of electroslag welding wherein filler metal is supplied by an electrode and its guiding member. In the consumable-guide system, the electrode is directed to the bottom of the joint by a guide. The guide carries the welding current until the electrode passes from it, and it is normally a heavy-wall tube. The guide melts off just above the flux bath and then solidifies as a portion of the solidified weld metal. There is no arc except at the start of the weld before the granulated flux melts from the heat of the arc. Figure 31-31 shows a sketch of the process.

Welding is usually done in the vertical position using water-cooled retaining shoes in contact with the joint sides to contain both the molten weld metal and the molten flux. The surface contour of the weld is determined by the contour of the molding or retaining shoes. The consumable-guide system normally uses fixed or nonsliding

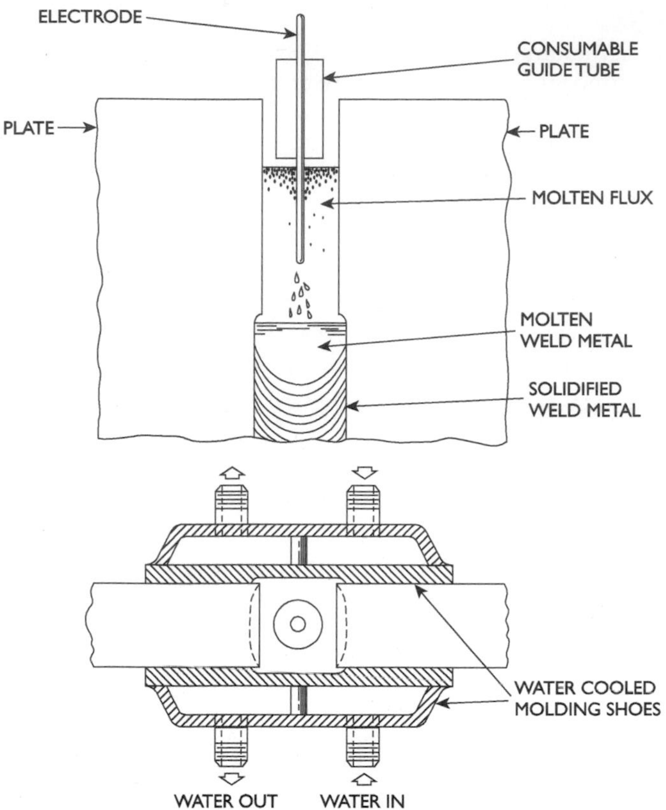

Figure 31-31 The process of electroslag welding.

weld-metal shoes. Butt, corner, and tee joints can be welded in lengths up to 10 ft high. The welding head does not move vertically and is normally mounted on the work at the top of the weld joint. It holds the consumable-guide tube, and for thick plate the guide may be oscillated. Multiple electrodes and guides may also be employed, and they may also be oscillated. Other joint configurations and weld types can also be welded. For these applications, the electrode-guide and retaining shoes are formed to the contour of the joint. The consumable-guide system is an automatic welding process and once started is carried to completion. The surface of the solidified weld metal is covered by a thin layer of slag. This slag is easily removed; however, this flux loss must be compensated for by the operator adding flux during welding. A starting tab is necessary to build up the proper depth of flux and to ensure fusion of the plates. Likewise, a run-off tab is required to run off the molten flux to fully weld the joint. Both tabs are removed flush with the ends of the joint.

Application

The outstanding features of the consumable-guide version of electroslag welding are the following:

1. Extremely high metal deposition rates.
2. Ability to weld very thick materials in one pass.
3. Joint preparation and fit-up requirements are minimum.
4. Little or no distortion.
5. Low flux consumption.

In addition, there is no weld-metal spatter during welding; thus, the deposition of the electrode is 100 percent. Only one setup is made, and there is no plate manipulation during welding.

Materials that can be joined by the consumable-guide electroslag welding process are low-carbon steel, low-alloy high-strength steel, medium-carbon steel, alloy steel, quenched and tempered steel, stainless steel, and high nickel-chromium alloy. Subsequent heat treatment is required for some materials depending on service requirements. The process can be used for welding joints from as short as 4 in. to as high as 10 ft. A single electrode is used on materials ranging from ½ to 2 in. thick. From 2 to 5 in. thick, the electrode and guide tube is oscillated in the joint. From 5 through 12 in., two electrode and guide tubes are used and are oscillated in the joint. Oscillation ensures an even distribution of heat in the joint and maintains uniform penetration into the plates. By using this lateral movement, the number of electrodes required for a joint is reduced.

Equipment

The major equipment components required for a consumable-guide system are the following:

1. The automatic welding head, including the mounting, adjusting, and oscillating mechanism
2. The control box for wire feeding and lateral motion
3. The welding machine or power source (one per electrode)
4. The water-cooled weld-metal retaining shoes

Figure 31-32 shows the major equipment components. The welding head is extremely compact and portable so that it can be taken to work to be welded rather than the reverse, thus saving on material-handling costs. Also, the head mounts on the work to be welded, thus eliminating expensive holding devices and fixtures.

The welding machine normally used is a 750- or 1000-amp, constant-voltage transformer/rectifier power source. The power source must be rated at 100 percent duty cycle and should include contactors and provisions for remote adjustment. Reverse polarity or electrode positive is normally employed. The same welding machine can be used for other continuous-electrode welding processes.

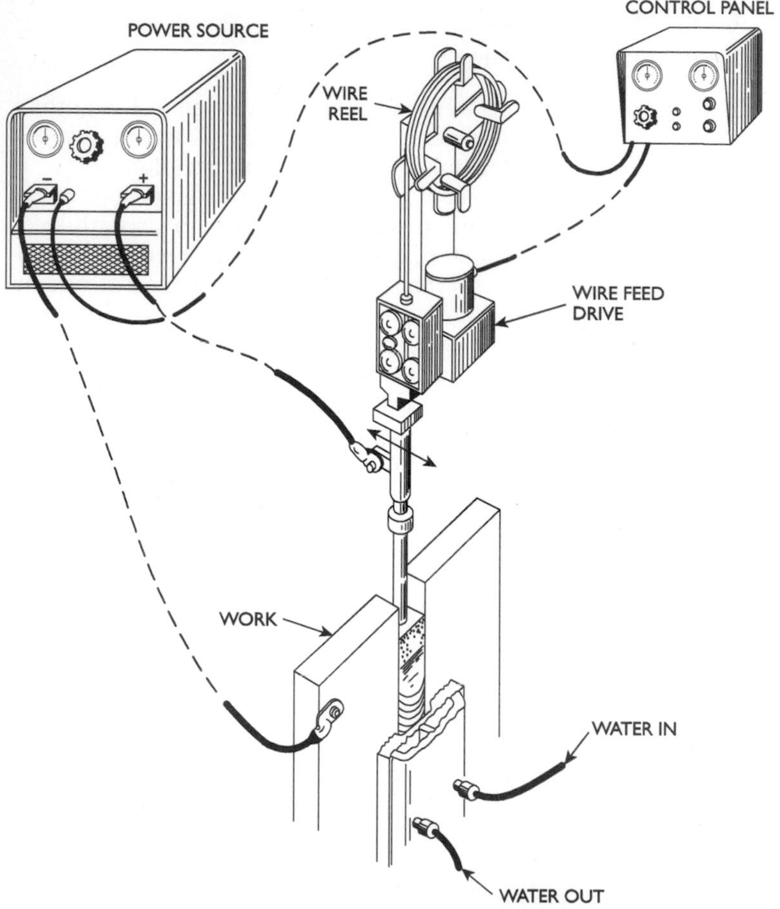

Figure 31-32 Needed equipment for electroslag welding.

Material

The consumable-guide system uses a granular flux with specially balanced composition to provide satisfactory shielding and deoxidizing properties. In addition, the flux must have proper viscosity and electrical conductivity in the molten state. The amount of flux consumed depends on the fit of the molding shoes and is normally constant. The electrode type must be matched to the base metal being welded. The system normally uses a ⅝ in. outside diameter heavy-wall tube for the guide. However, other shapes and types may be employed. For long joints, the consumable-guide tube may be fitted with intermittent insulating material to avoid short-circuiting against the side of the joint. The electrode is usually solid wire of 3/32 in. diameter. The composition for mild steel is similar to that used for CO_2 gas-shielded metal-arc welding. Wire can be supplied on coils or from large reels with appropriate reel dispensing equipment.

Flux-Cored Arc Welding Process (FCAW)

Flux-cored arc welding (FCAW) is an arc welding process wherein coalescence is produced by heating with an arc between a continuous filler-metal (consumable) electrode and the work. Shielding is obtained from a flux contained within the electrode. Additional shielding may or may not be obtained from an externally supplied gas or gas mixture. See Figure 31-33, which shows the fabricated (usually tubular) flux-cored electrode wire, the gas-shielding envelope, either externally supplied or internally generated, the arc, the weld metal, and the slag covering. The process may be either semiautomatic or automatic, with the semiautomatic method being more widely used.

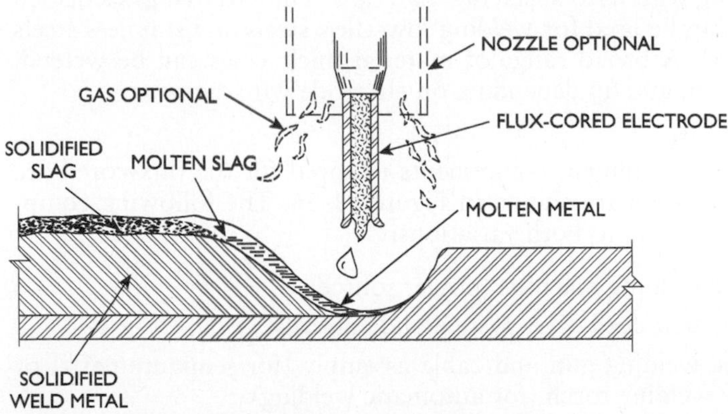

Figure 31-33 The flux-cored arc welding process.

There are two major variations of the process. The original uses the external shielding gas (FabCO). The "gasless" (*Fabshield*) utilizes fluxing elements inside the cored electrode wire to generate shielding gas. In either case, the shielding gas prevents the atmospheric oxygen and nitrogen from reaching the molten weld metal.

The two variations of the process provide different welding features. With "external gas shielding" the features are as follows:

1. Smooth, sound welds
2. Deep penetration
3. Good properties for X-ray
4. High-quality deposited weld metal

The "gasless" variation offers the following features:

1. Elimination of external gas supply, control, and gas nozzle
2. Moderate penetration
3. Ability to weld in drafts or breeze

Both variations have the following features:

1. The arc is visible to the welder.
2. All-position welding is possible, based on size of electrode wire employed.
3. Any weld joint type can be made.

Both processes are restricted to the welding of steels and for overlaying with hard surfacing materials. The external gas shielded version can be used for welding low-alloy steels and stainless steels (*FabLoys*). A broad range of material thicknesses can be welded, from 1⁄16 in. and up depending on electrode wire diameter.

Equipment

The major equipment components required for the flux-cored arc welding process are shown in Figure 31-34. The following equipment is common to both variations:

1. The welding machine (power source)
2. The wire-feed drive system and controls
3. The welding gun and cable assembly (for semiautomatic) or the welding torch (for automatic welding)
4. The flux-cored electrode wire

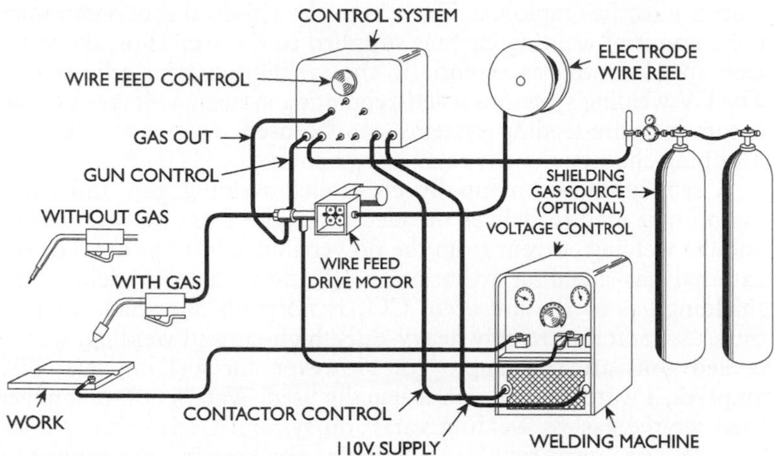

Figure 31-34 Equipment components required for flux-cored arc welding.

The "external gas shielding" version requires the external gas-shielding source, flowmeter/regulator, the gas valve, and the gas nozzle on the gun with accompanying hoses.

The welding machine or power source for flux-cored welding is normally a "constant voltage" (CV) type of welding machine, meaning that its output characteristic volt-amp curve is essentially flat. The output voltage of the machine is adjusted by a control on the welding power source, which can be either a transformer/rectifier or a motor- or engine-driven generator (CV power sources do not have a current control and cannot be used for welding with stock electrodes). The welding current output is determined by the load, which is dependent on the electrode wire feed rate. Direct-current reverse polarity is normally used for flux-cored arc welding. However, some electrodes use dc straight polarity. Alternating current is used abroad. Machines are available from 200 to 1000 amps. It is important that power sources used for flux-cored arc welding be rated from 80 to 100 percent duty cycle. The power source should have a contactor and meters. It should also provide 110-V ac power for the wire feeder.

The wire-feeding mechanism feeds the flux-cored electrode wire automatically from a spool or coil through the cable assembly and welding gun into the arc. The wire-feed system must be matched to the power supply system. For FCAW welding, the CV power source is usually used; therefore, constant-speed but adjustable wire-feeding system must be employed. The electrode-wire feed rate determines the amount of welding current supplied to the arc. Thus, the wire-feed speed control is essentially the welding current adjustment. The CV welding system is a self-regulating system. Voltage-sensing, automatic wire-feeding systems can be used when matched to a drooping-characteristic type of power source.

A semiautomatic manually controlled welding gun and cable assembly is used to deliver the electrode wire from the wire feeder and the welding current from the power source into the arc. For the external gas-shielding variation, it functions also to deliver the shielding gas to the arc area. CO_2 is normally the shielding gas; thus, except for extremely heavy-duty, high-current welding, water-cooled guns are not employed. However, for welding over 600 amperes, a water-cooled gun is usually used. Water-cooling is never used for the gasless welding variation system. Guns of the gooseneck or pistol-grip type are available. For certain applications, a special attachment is made on the gun to provide higher deposition rates. This involves an insulated extension, which, in a sense, adds to the effective stickout (electrically) of the electrode wire. This provides for preheating of the wire prior to its reaching the arc.

The shielding gas displaces the air around the arc and prevents contamination by oxygen and nitrogen of the atmosphere. The shielding gas normally used for flux-cored arc welding is CO_2 for steel. However, for stainless steel and certain alloy steels, argon-CO_2 mixtures or argon-oxygen mixtures are used. This is dependent on the base metal and the flux-cored electrode wire type. Gas flow rates depend on the type of gas used, the metal being welded, the welding position, the welding current, and draftiness. Obviously, simplicity is achieved with the non-gas-shielded system, as the gas supply and controls are not required.

The type of flux-cored tubular arc welding electrode wire must be selected to match the alloy, composition, and strength level of the base metal being welded. Various diameters are available to allow for welding in different positions, etc. Electrode wires are available on spools and coils and are packed in special containers to protect them from moisture.

The American Welding Society classifies electrodes according to the following (as an example)—E70T-1. The prefix "E" indicates an electrode. The number "70" indicates the required minimum as added strength in 1000 psi. The letter "T" indicates a tubular or fabricated flux-cored electrode. The suffix number "1" indicates chemistry of weld metal, shielding gas, and usability factor.

Stud Welding (SW)

Stud welding (SW) is an arc welding process that produces coalescence by heating with an arc drawn between a metal stud, or similar part, and the other work part. When the surfaces to be joined are properly heated, they are brought together under pressure. Partial shielding may be obtained by the use of a ceramic ferrule surrounding the stud. Shielding gas or flux may or may not be used. Figure 31-35 shows one of the more commonly used methods for stud welding. The stud gun holds the stud in contact with the workpiece (step A) until the welder depresses the gun trigger, causing the welding current to flow from the power source through the stud (which acts as an electrode) to the work surface. The welding current activates a solenoid within the gun, which draws the stud away from the work surface (step B) and establishes an arc. The intense heat melts both the work surface and the stud end at the same time. Arc duration is controlled by a timing device built into the control unit. When welding current is shut off, the gun solenoid releases its pull on the stud, and a spring-loaded action pushes the stud down into the molten pool of the workpiece (step C). The molten stud end solidifies with the molten pool on the work surface, and the stud weld is completed (step D).

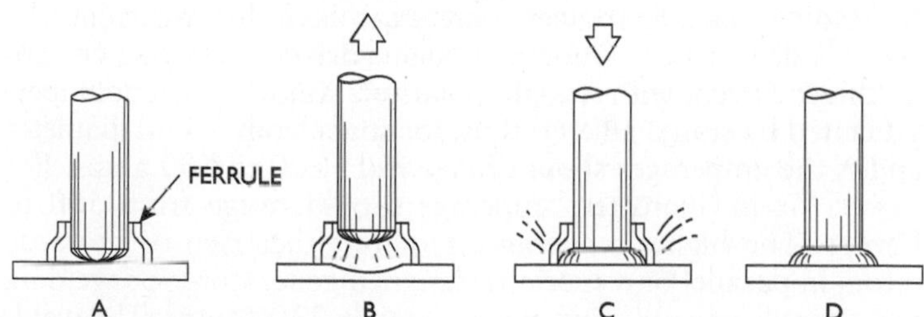

Figure 31-35 Stud welding.

The process can be either semiautomatic or fully automatic, with the semiautomatic method the more widely used. The automatic process can be multihead. Semiautomatic stud welding can be done in all positions and is being successfully used in shipbuilding, bridges and buildings, for application of boiler insulation, for insulating railway refrigerator cars, and for many other applications.

The major equipment components of the process (Figure 31-36) include the following:

1. The welding machine (power source)
2. The control unit

3. The stud gun
4. The studs
5. The ferrules
6. Ground and control cables required to complete the welding circuit

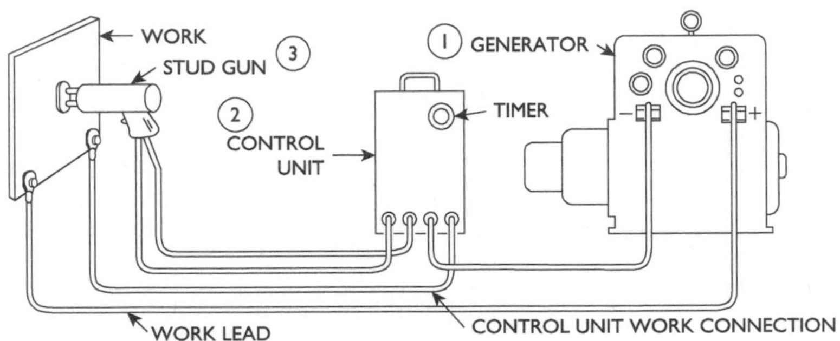

Figure 31-36 Equipment needed for stud welding.

The welding machine (power source) can be a direct-current (dc) rectifier or a dc generator (motor or engine driven). Welding current is dcsp (direct current with straight polarity). Welding current amperage is dictated by stud diameter; thus, for studs from 5⁄16 in. diameter and under, the amperages should range from 200 to 500 amps. For studs over 5⁄16 in. diameter, amperage should range from 500 to 2300 amps. The higher amperages require either two or more dc generators in parallel or a specially designed generator-type welding machine capable of producing approximately 2300 amps. The welding machine should have a high overload capacity and a relatively high open-circuit voltage of 95 to 100 volts. Stud welding uses a constant-current power source.

The control unit consists of a welding current contactor, a timing cycle device, and necessary connections. A recently developed technique controls the speed at which the stud is pushed into the molten base metal. This practically eliminates spatter and provides more control over the weld shape and quality.

The stud gun holds the stud and has a solenoid that provides the withdrawal action to establish the arc. A spring-loaded mechanism within the gun applies the pressure required to push the stud end into the molten pool of the workpiece. The gun should be adjusted to provide proper arc length and must be held so that the stud is perpendicular to the work surface. The gun is normally hand-held; however, the process can be automated.

Steel studs range in diameter from ⅛ to 1 in., varying in size and shape, and can be threaded or plain. Except for the smaller diameters, studs contain a charge of welding flux in the arcing end. These fluxes protect the weld against atmospheric contamination and contain scavengers that purify the weld metal. Studs can be designated for special applications. Companies supplying studs offer many varieties.

A ferrule is used with each stud. It shields the arc, protects the welder, and eliminates the need for a face helmet. The ferrule concentrates heat during welding and confines molten metal to the weld area. It helps prevent oxidation of the molten metal during the arcing cycle and prevents charring of the work piece. It is made of ceramic material and is broken off and discarded after the weld is made.

Submerged Arc Welding Process (SAW)

Submerged arc welding (SAW) is an arc welding process that produces coalescence by heating with an arc or arcs between a bare metal electrode or electrodes and the work. The arc is shielded by a blanket of granular, fusible material on the work. Pressure is not used, and filler metal is obtained from the electrode and sometimes from a supplementary welding rod. Figure 31-37 shows the work to be welded, the consumable-electrode wire, the weld, the protective slag layer, and the flux material. The process may be either automatic or semiautomatic, with the automatic method being more widely used.

The outstanding features of the submerged arc process are the following:

1. High welding speed
2. High metal deposition rates
3. Deep penetration
4. Smooth weld appearance
5. Good X-ray quality welds
6. Easily removed slag covering
7. Wide range of material thickness weldable

Welding is done in the flat and horizontal positions, and the arc is *not* visible to the welder. The process is used to weld low- and medium-carbon steel, low-alloy high-strength steel, quenched and

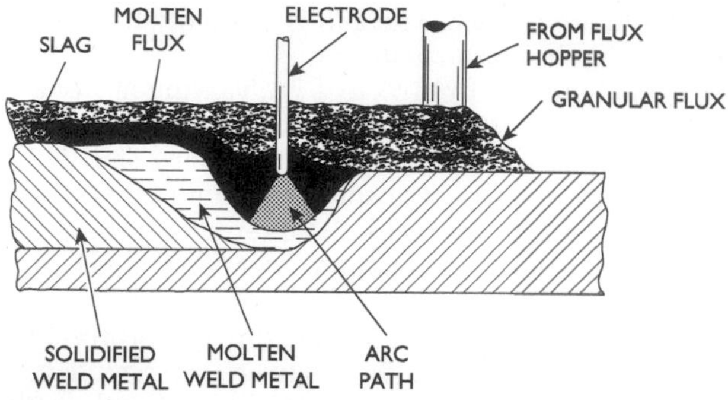

Figure 31-37 Submerged-arc welding.

tempered steel, and many stainless steels. It is also used for hard surfacing and build-up work. Metal thicknesses from 16 gauge to ½ in. are weldable with no edge preparation, and with edge preparation and multiple passes, proper joint design, and weld backup, maximum thickness is practically unlimited.

The major equipment components required for submerged arc welding (Figure 31-38), are as follows:

1. The welding machine (power source)
2. The wire-feeding mechanism and controls
3. The welding torch for automatic welding, or the welding gun and cable assembly for semiautomatic welding
4. The flux hopper and feeding mechanism, and normally a flux recovery system
5. Travel mechanism for automatic welding

The welding flux and electrode wire must be matched to the base metal; this is explained in the following paragraphs.

The welding machine or power source for submerged-arc welding is especially designed for the process. Both ac power and dc power are used. In either case, the power source should be rated at a 100 percent duty cycle. This is because submerged-arc welding operations are continuous, and the length of time in operation will normally exceed the 10-min. base period used for figuring duty cycle. For dc submerged-arc welding, constant-voltage (CV) or the constant-current (CC) power source can be used. The CV type is more common for small-diameter electrode wires, whereas the CC type is used more commonly for large-diameter electrode wires. In either case, the wire feeder must be matched to the power source being used. Welding machines for submerged-arc welding range in size from 200 to 1200 amps. Alternating current is used primarily with the automatic method. It is also used in conjunction with dc for multiple-electrode submerged-arc welding.

The wire-feeding mechanism with controls is used to feed the consumable-electrode wire into the submerged arc. If a drooping characteristic power source is employed, a voltage-sensing wire feeder must be used. This type of wire feeder maintains a specified arc voltage and feeds the wire to maintain this value. If, however, a CV power source is used, the constant-speed wire feeder may be employed. Here, the wire feeder feeds the electrode wire at a rate to draw the prescribed current from the welding machine. The voltage is adjusted by changing the output voltage of the welding machine. In either case, the control system also initiates the arc, controls travel speed and fixtures (when involved), and performs the other necessary functions to make an automatic process operate.

For semiautomatic operations, a welding gun and cable assembly is used to carry the electrode wire to the welding area in accordance to the manual manipulation of the gun. For automatic welding, the torch is usually attached to the wire-feed motor, and the flux hopper is fixtured to the torch. In either case, it is necessary to have the proper layer of flux over the welding arc to protect it from the atmosphere.

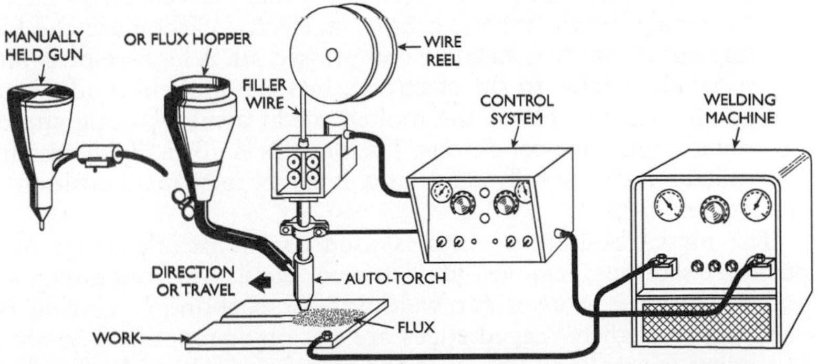

Figure 31-38 Necessary equipment for submerged-arc welding.

Welding is done under a blanket of granular, fusible material commonly called a flux. This flux operates much in the same way as the covering on a coated electrode. It protects the weld metal from contamination by atmospheric oxygen and nitrogen and also acts as a scavenger to clean and purify the weld deposit. Additionally, it may be used to add alloy elements to the deposited weld metal. A portion of the flux is melted by the intense heat of the welding arc and becomes molten. The molten flux cools and solidifies, forming a slag on the surface of the completed weld. The upper or nonmelted portion of the flux can be recovered and reused. Various grades and types of submerged-arc welding flux are available. It is important to select the proper flux for the base metal being welded and to match the chemistry of the welding electrode.

Electrode wire is used for submerged-arc welding. The wires are solid and bare except for a very thin protective coating on the surface, usually copper, to prevent rusting. The electrode wire contains special deoxidizers that help clean and scavenge the weld metal to produce sound, quality welds. Alloying elements may also be included in the electrode wire to provide additional strength of the weld metal. The electrode-wire composition must be matched to that of the base metal but must also be used with the appropriate submerged-arc flux. Consult a flux-wire combination chart for guidance. The electrode wire sizes available are $^{1}/_{16}$, $^{5}/_{64}$, $^{3}/_{32}$, $^{1}/_{8}$, $^{5}/_{32}$, $^{3}/_{16}$, $^{7}/_{32}$, and $^{1}/_{4}$ in. in diameter. Wire is usually available in coils ranging from 50 lbs to as high as 1000 lbs.

Air Carbon-Arc Cutting and Gouging

The carbon-arc cutting process cuts and gouges out metal by melting the metal with the intense heat of an electric carbon arc and by blowing out the molten metal by compressed air. A high-velocity air jet, traveling parallel to the electrode, hits the molten puddle just behind the arc and blows the molten metal away. Cutting speed varies with operating conditions. The process is normally manually controlled but can be fully automatic. It can be used in all positions. See Figure 31-39.

The air carbon-arc process is used to cut metal, gouge out defective sections, remove old or inferior welds, for root gouging, and to prepare grooves for welding. Air carbon-arc cutting is used where slightly ragged edges are not objectionable. The area of the cut is small, and because the metal is made molten and is removed quickly, the surrounding area does not reach high temperatures. This reduces the tendency toward distortion and cracking.

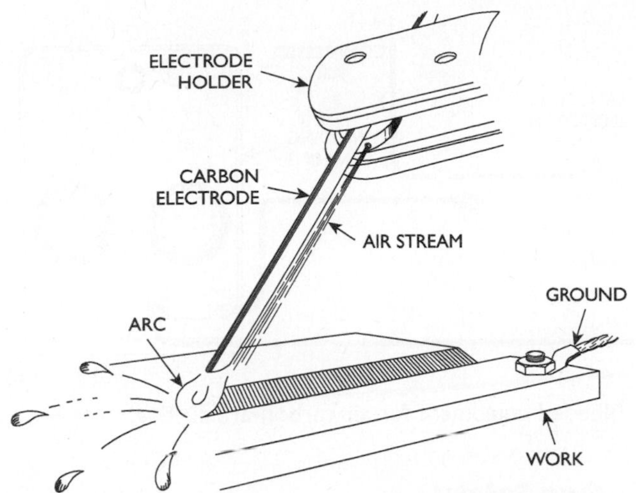

Figure 31-39 Air carbon-arc cutting and gouging.

The torch has a rotating head, which allows the electrode to be set at any angle yet keep the air stream in the puddle. The torch has a concentric cable carrying compressed air and electric current. Air pressure, although not critical, normally ranges from 80 to 100 psi and is obtained from a shop line, or compressor. The power source is usually a dc welding machine with drooping characteristics, either rectifier or generator, with provision for reverse polarity. An ac conventional welding machine can also be used for special applications; however, ac carbon electrodes are required. *Constant-voltage (CV) machines may be used, but precautions must be taken to operate them well within their rated output.* A specially designed welding machine, either rectifier or generator, is used for heavy-duty applications. Current depends on electrode diameter and varies from 200 amps for 3⁄16 in. diameter electrodes to 1300 amps for 5⁄8 in. diameter electrodes. Pure carbon and copper-coated carbon electrodes are used. Electrode diameter sizes are 3⁄16, 1⁄4, 5⁄16, 3⁄8, 1⁄2, and 5⁄8 in. Alternating-current types are required for some work.

The operator starts his airflow. Then, after striking an arc, he pushes the electrode, with a slight cushion, rapidly across the base metal. Speed and electrode angle, as well as electrode diameter, determine groove depth. The needed equipment is shown in Figure 31-40.

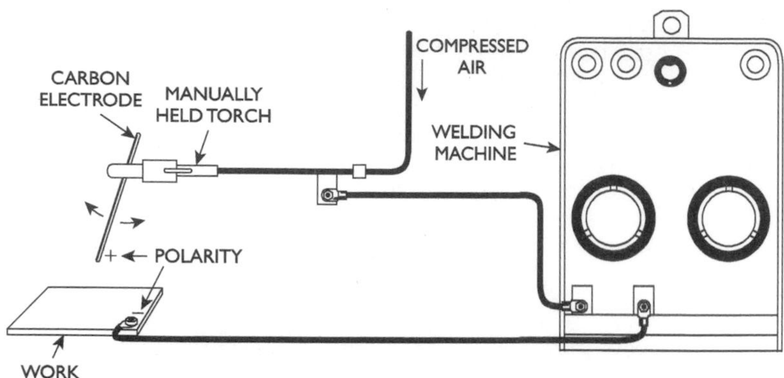

Figure 31-40 Needed equipment for air carbon-arc cutting.

General Welding Safety

After a half century of commercial welding, welding has proved itself not injurious to health. However, as in most trades, if a welder is careless, some features of welding can cause discomfort and actual danger.

Essentially, welding is not a hazardous occupation if proper precautionary measures are always observed. This requires continuous awareness of possibilities of danger and habitual safety precaution by the welder. In addition, it requires that the supervisor be alert, responsible, and tough in enforcing safety regulations.

1. Always wear dry, fire-resistant protective clothing, cuffless trousers that cover shoe tops, leather gloves, jacket, apron, and proper dark lenses.
2. Always keep a safe, clean work area.
3. Make sure there are no flammable materials nearby.
4. Do not weld in the vicinity of explosive materials or near carbon tetrachloride.
5. Always make sure you have enough ventilation to give three or four complete changes of air per hour.
6. Use air exhaust at the weld whenever welding lead, cadmium, chromium, manganese, brass, bronze, zinc, or galvanized metals.
7. Never weld or cut in a confined area without ventilation.
8. Keep all welding equipment in good condition.

9. If it is necessary to couple lengths of cable together, make sure joints are insulated and all electrical connections are tight.
10. When electrode holder is not in use, hang it on welding machine or special holder. *Never let it touch a gas cylinder.*
11. Always have welding machine properly grounded, usually to cold-water pipe.
12. Make sure pedal controls are guarded to prevent accidental starts.
13. If need arises to weld in damp or wet conditions, wear rubber boots and/or stand on dry cardboard or wood.
14. Stand only on solid items, floor, or ground.
15. When welding in high places without railing, use safety belt or lifeline.
16. Always wear proper eye protection, especially when grinding or cutting.
17. Keep your booth curtains closed to protect the eyes of others.
18. Never weld or cut directly on a concrete floor.
19. When using a water-cooled torch, check for water leakage.

Safe Handling of Gas Cylinders

1. Be very careful when you move any gas cylinder—never move it roughly. Always have cap on cylinders when moving. Never roll horizontally.
2. Never use welding gas as compressed air for blowing away dirt or debris.
3. Before attaching a regulator to a cylinder, open and close the valve quickly. This is commonly called "cracking" the cylinder.
4. Open valve on cylinder slowly after regulator is attached.
5. Be sure all connections are clean and gas-tight. Check with saliva or soapy water.
6. When the regulator is not in use, the adjusting screw should be screwed out until diaphragm is free.
7. Always protect the hose from rupture or mechanical damage.
8. Always close the cylinder and release the pressure from the regulators and hose when your work is done.
9. Always leave safety plugs alone.
10. Always keep the cylinders in an upright position.

Chapter 32

Sharpening Saws

The term "sharpen" is used in its broad sense to include all the operations necessary to put a used saw into first-class condition. There are five steps in the sharpening of a saw:

1. Jointing
2. Shaping
3. Setting
4. Filing
5. Dressing

Sharpening Handsaws

Handsaws are of two main types: the crosscut and the ripsaw. The crosscut saw, as the name implies, is used to cut across the grain and to cut wet or soft woods. Ripsaws, on the other hand, are used to cut wood along the grain. The saws are similar in construction, but ripsaws are slightly heavier; they also differ in the rake of the teeth. Other types of handsaws include backsaws, miter saws, dovetail saws, compass saws, keyhole saws, coping saws, etc.

When sharpening handsaws, the first step is to place the saw in a suitable clamp or saw vise, as illustrated in Figure 32-1. In the absence of a good saw vise, a homemade clamp may easily be made in which the saw can be supported. The saw should be held tight in the clamp so that there is no noticeable vibration. The saw is then ready to be jointed.

Jointing

Jointing is done when the teeth are uneven or incorrectly shaped or when the teeth edges are not straight. If the teeth are irregular in size and shape, jointing must precede setting and filing. To joint a saw, place it in a clamp with the handle to the right. Lay a flat file lengthwise on the teeth, and pass it lightly back and forth over the length of the blade on top of the teeth until the file touches the top of every tooth. The teeth will then be of equal height, as shown in Figure 32-2. Hold the file flat; do not allow it to tip to one side or the other. The jointing tool or handsaw jointer will aid in holding the file flat.

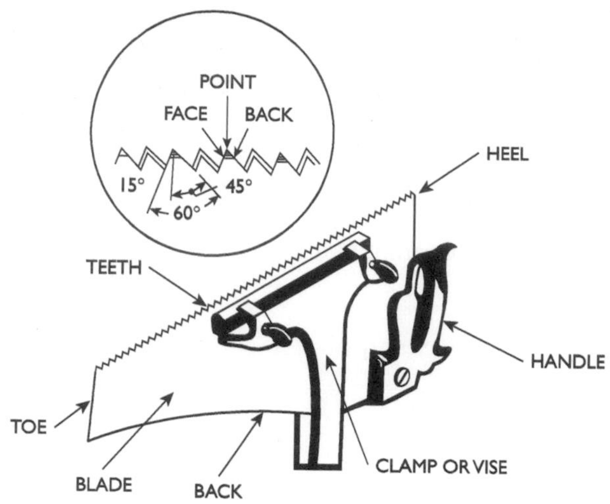

Figure 32-1 The method of fastening a handsaw in a clamp or vise.

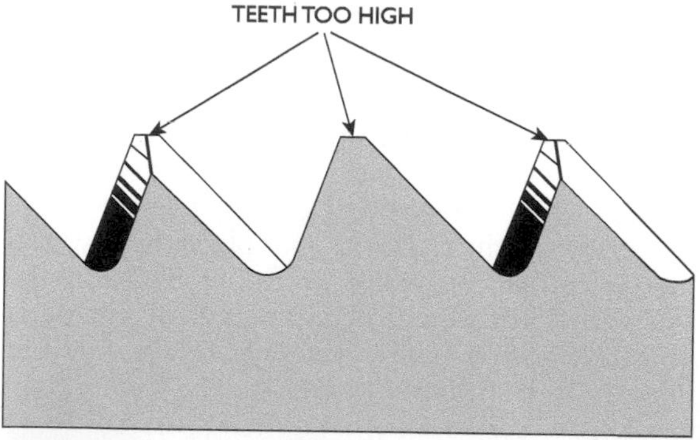

Figure 32-2 The method of jointing saw teeth. Place the saw in a clamp with the handle to the right. Lay a mill file lengthwise flat on the teeth. Pass it lightly back and forth along the length of the teeth until the file touches the top of every tooth. If the teeth are extremely uneven, joint the highest tooth first, and then shape the teeth that have been jointed and joint the teeth a second time. The teeth will then be the same height. Do not allow the file to tip to one side or the other; hold it flat.

Shaping

Shaping consists of making the teeth uniform in width. This is normally done after the saw has been jointed. The teeth are filed with a regular handsaw file to the correct uniform size and shape. The gullets must be of equal depth. For the crosscut saw, the front of the tooth should be filed at an angle of 15° from the vertical, whereas the back slope should be at an angle of 45° from the vertical, as illustrated in Figure 32-3. When filing a ripsaw, the fronts of the teeth are filed at an angle of 8° from the vertical, and the back slope is filed at an angle of 52° from the vertical, as shown in Figure 32-4. Some good workmen, however, prefer to file ripsaws with more of an angle than this, often with the front side of the teeth almost square, or 90°. This produces a faster-cutting saw, but, of course, it pushes harder, and it will grab when cutting at an angle with the grain.

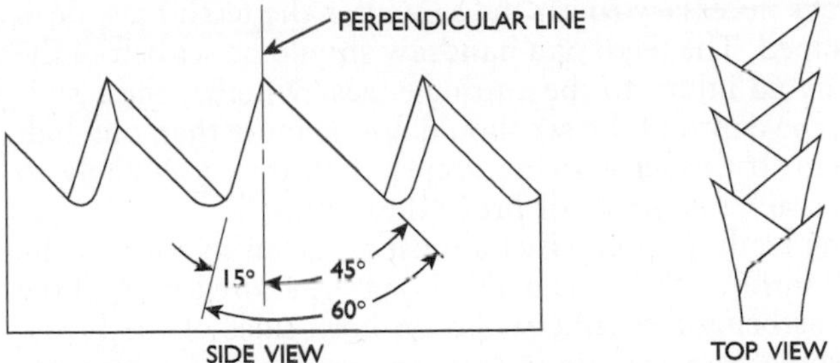

Figure 32-3 Side and tooth-edge views of a crosscut saw. The angle of a crosscut saw tooth is 60°, the same as that of a ripsaw. The angle of the front of the tooth is 15° from the perpendicular, whereas the back angle is 45°.

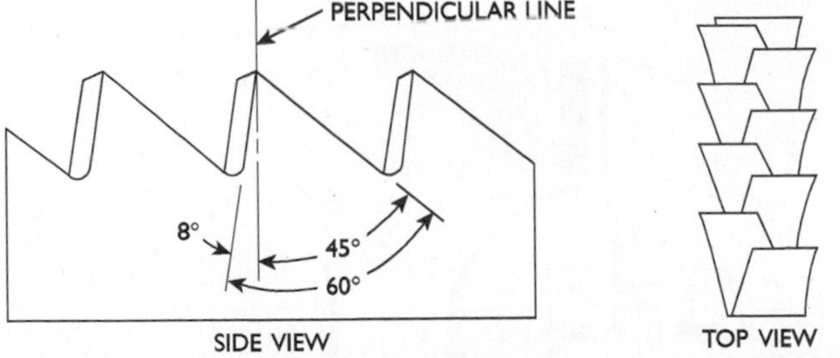

Figure 32-4 Side and tooth-edge views of a typical ripsaw. The tooth of a ripsaw has an angle of 60°, that is, 8° from the perpendicular on the front and 52° on the back of the tooth.

When shaping teeth, disregard the bevel of the teeth, and file straight across at right angles to the blade with the file well down in the gullet. If the teeth are of unequal size, press the file against the teeth level with the largest fault tops until the center of the flat tops made by jointing is reached. Then, move the file to the next gullet, and file until the rest of the flat top disappears and the tooth has been brought to a point. Do not bevel the teeth while shaping. The teeth, now shaped and of even height, are ready to be set.

Setting

After the teeth are made even and of uniform width, they must be set. Setting is a process by which the points of the teeth are bent outward by pressing with a tool known as a saw set. Setting is done only when the set is not sufficient for the saw to clear itself in the kerf. It is always necessary to set the saw after the teeth have been jointed and shaped. The teeth of a handsaw should be set before the final filing to avoid injury to the cutting edges. Whether the saw is fine or coarse, the depth of the set should not be more than one-half that of the teeth. If the set is made deeper than this, it is likely to spring, crimp, crack the blade, or break the teeth.

When setting teeth, particular care must be taken to see that the set is regular. It must be the same width along the entire length of the blade, as well as being the same width on both sides of the blade. The saw set should be placed on the saw so that the guides are positioned over the teeth with the anvil behind the tooth to be set, as shown in Figure 32-5. The anvil should be correctly set in the frame, and the handles should be pressed together. This step causes the plunger to press the tooth against the anvil and bend it to the angle of the anvil bevel. Each tooth is set individually in this manner.

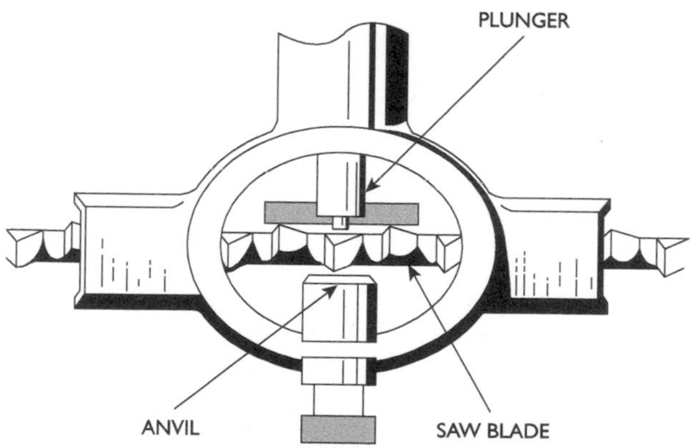

Figure 32-5 The position of the saw set on the saw for setting the teeth.

Filing

Filing a saw consists of simply sharpening the cutting edges. Place the saw in a filing clamp with the handle to the left. The bottom of the gullets should not be more than ½ in. above the jaws of the clamp. If more of the blade projects, the file will chatter or screech. This dulls the file quickly. If the teeth of the saw have been shaped, pass a file over the teeth, as described in jointing, to form a small flat top. This acts as a guide for the file; it also evens the teeth.

To file a crosscutting handsaw, stand at the first position shown in Figure 32-6. Begin at the point of the saw with the first tooth that is set toward you. Place the file in the gullet to the left of this tooth, and hold the handle in the right hand with the thumb and three fingers on the handle and the forefinger on top of the file or handle. Hold the other end of the file with the left hand, the thumb on top and the forefinger underneath. The file may be held in the file-holder guide, as shown in Figure 32-7. The guide holds the file at a fixed angle throughout the filing process while each tooth is sharpened.

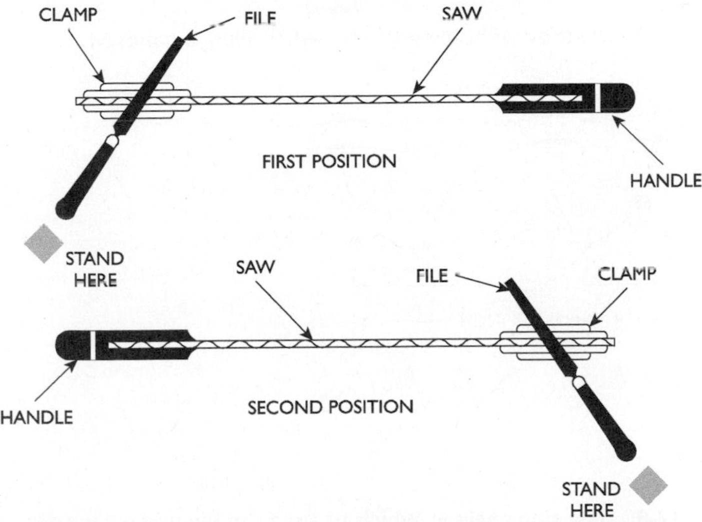

Figure 32-6 Standing positions for filing a crosscut saw. The saw clamp should be moved along the blade as filing progresses.

Hold the file directly across the blade. Then, swing the file left to the desired angle. The correct angle is approximately 65°, as shown in Figure 32-8. Tilt the file so that the breast (the front side of the tooth) may be filed at an angle of approximately 15° from the vertical, as illustrated in Figure 32-8. Keep the file level and at this angle; do not allow it to tip upward or downward. The file should cut on the push stroke and should be raised out of the gullet on the reverse stroke. It cuts the teeth on the right and left on the forward stroke.

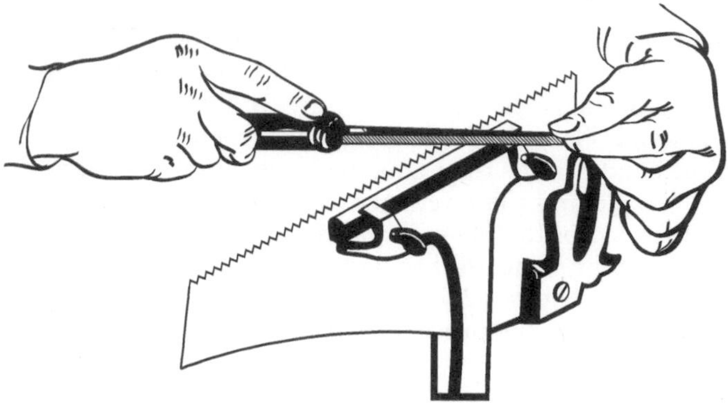

Figure 32-7 Method of holding the file when filing a handsaw.

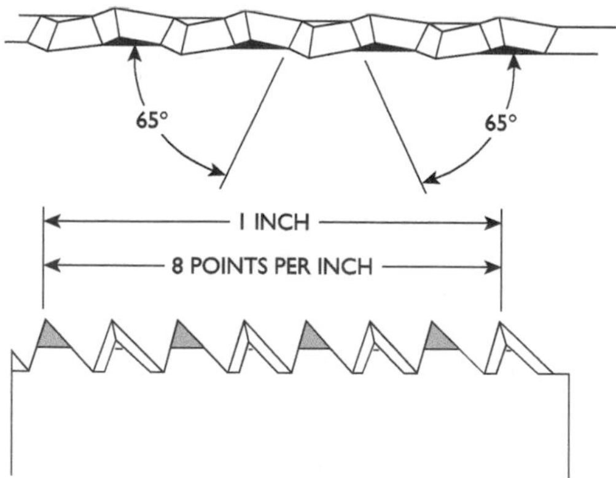

Figure 32-8 The side angle at which to hold the file when filing a crosscut saw that has eight points per inch.

File the teeth until half of the flat top is removed. Then, lift the file, skip the next gullet to the right, and place the file in the second gullet toward the handle. If the flat top on one tooth is larger than the other, press the file harder against the larger tooth so as to cut that tooth faster. Repeat the filing operation on the two teeth that the file now touches, always being careful to keep the file at the same angle. Continue in this manner, placing the file on every second gullet until the handle end of the saw is reached.

Turn the saw around in the clamp with the handle to the left. Stand in the second position, and place the file to the right of the first tooth set toward you, as shown in Figure 32-6. This is the first gullet that was skipped when filing from the other side. Turn the file handle to the right until the proper angle is obtained, and file away the remaining half of the flat top on the tooth. The teeth that the file touches are now sharp. Continue the operation until the handle end of the saw is reached.

When filing a ripsaw, one change is made in the preceding operation: The teeth are filed straight across the saw at right angles to the blade. The file should be placed on the gullet so as to file the breast of the tooth at an angle of 8° from the vertical, as shown in Figure 32-4. Stand in the positions shown in Figure 32-9. When sharpening a ripsaw, file every other tooth from the side. Compare Figure 32-10 with Figure 32-8 to see the difference in the teeth angles. Then, turn the saw around, and sharpen the remaining teeth

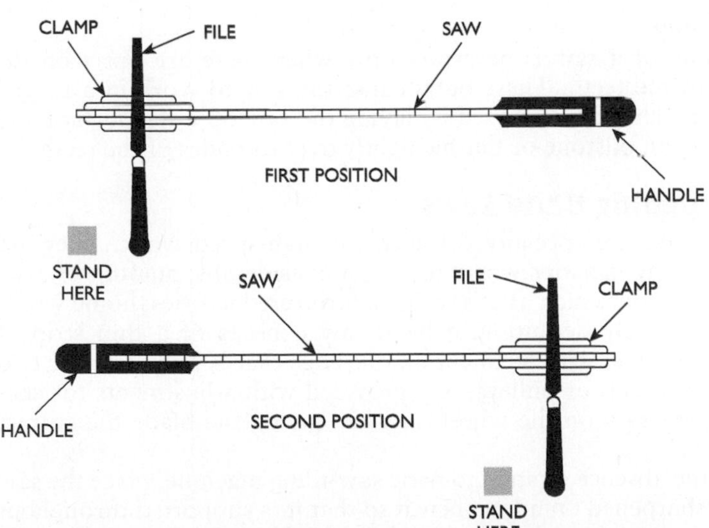

Figure 32-9 Standard positions for filing a typical ripsaw. Again, the clamp must be moved along the blade as filing progresses.

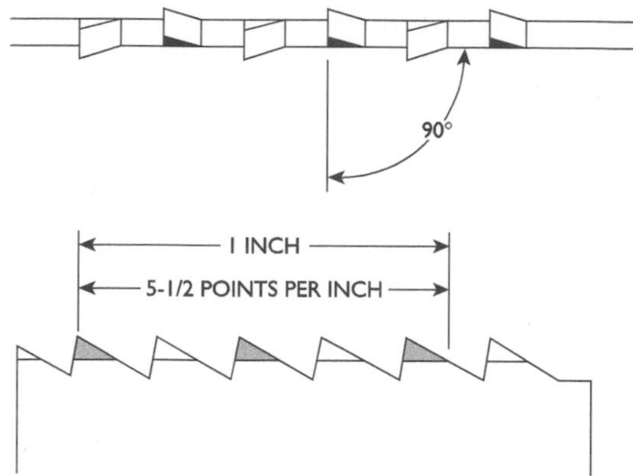

Figure 32-10 The side angle at which to hold the file when filing a ripsaw.

as described in the preceding paragraphs. When filing teeth, care must be taken in the final sharpening process to file all the teeth to the same size and height; otherwise the saw will not cut satisfactorily. Many good saw filers file ripsaws from only one side, taking care that the file is held perfectly horizontal. For the beginning, however, turning the saw is probably the most satisfactory method.

Dressing

Dressing of a saw is necessary only when there are burrs on the sides of the teeth. These burrs cause the saw to work in a ragged fashion. They are removed by laying the saw on a flat surface and running an oilstone or flat file lightly over the sides of the teeth.

Sharpening Band Saws

Band saws are specially designed for high-speed work. They are used chiefly in shops where there is a considerable amount of sawing to be done, such as in sawmills, furniture factories, home workshops, etc. By definition, a band saw consists of a thin strip of tempered steel with teeth cut on one edge that is strained over two vertical wheels or pulleys. It is provided with adjustments for centering the saw on the wheel rims and giving the blade the correct tension.

In the absence of an automatic saw-filing machine, place the saw to be sharpened on a long bench so that it is supported throughout its entire length at the same level during filing. Make sure that the teeth point to the left. Use a suitable clamp that will hold approximately 50 teeth or more at one setting, as shown in Figure 32-11.

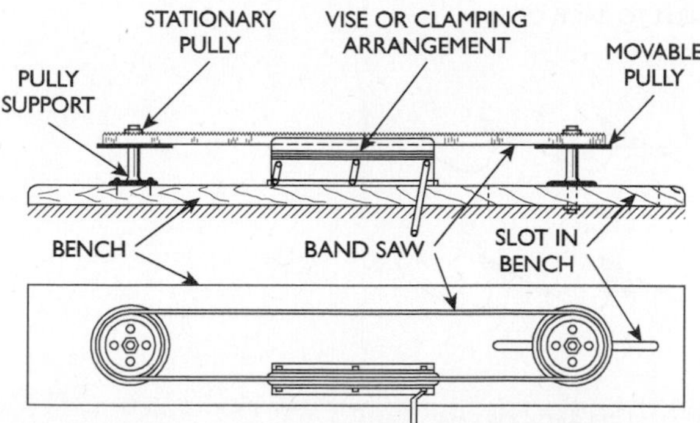

Figure 32-11 A homemade band-saw filing frame. This arrangement consists essentially of two equal-size pulleys or wheels mounted horizontally to receive the band saw. The right-hand pulley is movable along the center in a slot to fit various saw sizes.

The saw is then moved so that one section after another is worked on until the entire length of the saw has been sharpened.

An adequate emergency vise for occasional use is made by simply clamping two boards about 10 inches wide in the woodworker's vise; the boards are high enough to bring the saw to a comfortable height for filing. The bottom of the teeth gullets should be about ⅛ in. above the jaws of the clamp to prevent chattering.

It is customary to slightly joint the section before commencing to file the teeth. This is done with a saw jointer or by running a mill file lightly over the teeth to make them all of uniform height. The band saw can be jointed satisfactorily by holding a piece of an emery wheel or an oilstone against the toothed edge while the saw is in operation. The best guide when shaping the teeth is a new saw, or you can follow the shape of the teeth in the saw being sharpened, provided they are not too worn.

When filing, hold the file in a horizontal position. File each tooth straight across the saw at right angles to the blade, raising the file on the back stroke. If the point of any tooth is not brought up sharp after the stroke of the file, do not do any extra filing to sharpen this particular tooth. Instead, continue until the section being worked on is filed. Each section may require two or three filings before the filing job is completed.

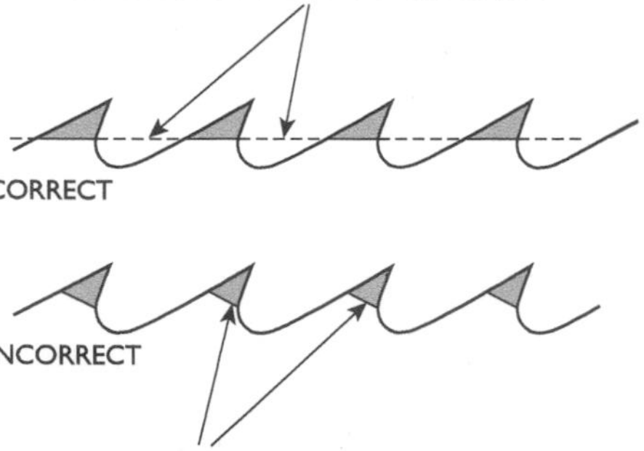

LINE OF SET PARALLEL TO BACK OF SAW

CORRECT

INCORRECT

LINE OF SET NOT PARALLEL TO BACK OF SAW

Figure 32-12 The outlines of correctly and incorrectly set band-saw teeth. The line of set should be kept parallel with the back of the blade and not at an angle with the blade.

If setting is necessary, it should be done before the teeth are filed. If the saw is to do only straight-line cutting, satisfactory results are obtained only when the saw is correctly set. Sufficient set is necessary to clear the blade in the cut, particularly when cutting on curved lines.

When setting, do not set the teeth more than halfway down, because if they are set too deeply, the body of the teeth will become distorted. Also, remember to keep the line of the set parallel to the back of the blade, as shown in Figure 32-12, and not at an angle to the back.

Brazing

When the band saw breaks, regardless of the cause, it may be joined together again either by brazing or by silver soldering. Blades obtained directly from the manufacturer are often welded electrically, but this process requires special equipment that is not readily available in the average saw filing or refitting shop. The brazing and silver-soldering methods also require certain special equipment, and because of the complexity of the processes, the average user is advised to return the broken blade to the manufacturer or take it to a reliable saw-filing shop that is properly equipped for brazing or welding.

Sharpening Circular Saws

Jointing

The first step in sharpening circular saws is that of jointing. This operation consists in making all points of the saw exactly the same distance from the center. This may be done by running the saw slowly backward by hand on the mandrel while holding a piece of emery stone or a mill file lightly against the top of the teeth, or it may be done while the saw is running if extreme care is taken. Continue this operation until the tops of all the teeth show that they have been touched by the emery stone or file.

Shaping

Large-diameter saws may be filed while in place if their blades are sufficiently rigid to prevent chattering. Small circular saws should be removed from their arbors for filing and may be held in a clamp in the same manner as handsaws. After jointing, place the saw in a filing clamp, and shape the teeth as near to the original shape as possible. All the teeth should be the same shape, with gullets of even depth and width. A slim taper file is satisfactory for circular-crosscut and circular-combination saws. To file the teeth of circular ripsaws, the mill bastard file with one round edge is quite satisfactory; the round edge permits the filer to shape the gullets and teeth with the same file.

Setting

After the teeth have been shaped, they should be set. The setting operation is most commonly accomplished with a saw set, which makes the setting of any saw comparatively easy. The purpose of setting saw teeth—that is, springing over the upper part of each tooth (not more than half of the tooth nearest the point), one to the right, the next to the left, and so on alternately throughout the entire tooth edge—is to make the saw cut a kerf slightly wider than the thickness of the blade. This provides clearance and prevents friction, which would cause the saw to bind and push hard in the cut.

Whether the saw is fine or coarse, the depth of the set should not go lower than half the tooth. This is important. If the depth of the set goes deeper than this, it is likely to spring, crimp, or crack the blade, or break a tooth.

A taper-ground saw requires little setting because the blade, being of uniform thickness along the entire tooth edge, tapers thinner to the back and also tapers from butt to point along the back, which provides the measure of clearance necessary for easy running. Soft, wet woods require more set and coarser teeth than dry, hard woods. For fine work on either hard or soft dry woods, it is best to use a saw that has fine teeth and only a slight set.

To set the teeth of a circular saw, the saw blade should project well above the clamp jaws. Place the die and anvil of the saw set on the tooth to be set, taking care not to carry the set too far down on the tooth. If this were done, the body of the blade (below the gullets) would be distorted. Be sure that every other tooth is set in the same direction as it was when the saw was new. After setting, any teeth that are not in alignment with the others should be corrected.

The raker teeth of flat-ground combination saws should not be set. The teeth and rakers of hollow-ground combination saws also should not be set. Electric handsaws should have more set than bench saws—approximately 0.018 to 0.025 in. on each side.

The teeth of circular saws may be set in two ways, swage-set and spring-set. Some factories use one method of setting, and some use the other. There is an advantage in the swage-set method, however, in that all teeth cut entirely across the work, whereas the spring-set teeth cut alternately.

Swage-Set

Because swage-set teeth cut entirely across the face, this method of setting will produce smoother work than spring-set teeth. Many plants do not use swaged saws, and many mill men are under the impression that it is harder to keep them in proper cutting order. With a few inexpensive tools and a little care, any ordinary filer who can keep up a spring-set blade properly can soon master the technique of swaging. To swage-set a saw blade,

1. Make it round by holding a piece of grindstone or soft emery against the rim as the saw revolves.
2. File all teeth to a keen point.
3. Swage from the under or top side, according to the swage used, 1⁄16 in. on each side of the tooth.
4. Joint again and file square across, or grind to a keen edge without changing the hook.
5. Side file to bring all teeth to a uniform width. Sharpen saws from two to four times during a full day's sawing.

Spring-Set

In handling spring-set blades, it is important that the saw blade is always put on the spindle in the same position. A good method for accomplishing this is to have a mark on the spindle; turn the mark up, and put the saw blade on with the etched trademark directly over the spindle mark. When jointed in this position and put back the same way, the blade will run true.

For spring-setting, follow the same steps as outlined under swage setting, but, instead of swaging, bend the teeth alternately, first one to the right, then one to the left. File the front of the teeth square with the side of the saw, beveling each tooth slightly on the back. Cutoff teeth are given more bevel at the points where the cutting takes place. All saws are left stiffer when spring-set than when swage-set.

Filing

After setting, file the teeth as close as possible to the same shape they were when the saw was new. In filing, do not reduce the length of the teeth; simply bring them up to a sharp point. If the teeth are uneven, the saw cannot cut properly. Make all the same shape, with gullets of even depth. Do not file sharp corners or nicks in the bottom of the gullets. This usually results in cracks in the gullets. Bevel the teeth on crosscut saws on both the face and back edges; more bevel, however, is filed on the face than on the back of the teeth.

File ripsaw teeth straight across to a chisel-like edge. Then, give the teeth a slight bevel on the back of the teeth. When filing any saw, take care that the bevel does not run down into the gullets. The bevel on both the face and back should be about one-third the length of the teeth.

When filing a flat-ground combination saw, which crosscuts, rips, and miters, use the same method for beveling and scoring teeth that is used in sharpening a cutoff saw. Some combination saws have rakers, or cleaner teeth, to remove the material left in the cut by the beveled cutting teeth. The points of these rakers, or cleaner teeth, should be filed approximately 1/64 in. shorter for hardwood and 1/32 in. shorter for softwood than the points of the beveled cutting teeth. After filing these teeth shorter, square the face of each raker tooth, and bring it to a chisel-like edge by filing on the back of the tooth only.

When sharpening a hollow-ground combination saw, follow the method used with a flat-ground combination saw, but do not set the teeth because the hollow-grinding operation provides ample clearance.

Gumming

Saws running at present-day speeds must have plenty of throat room to take care of the sawdust. Insufficient throat room is an invitation to cracks. After repeated filings, it may be necessary to lengthen the teeth by gumming. Use a thin, free-cutting, roundnose wheel. Grind lightly, going around several times. Grinding heavily on one tooth will heat the rim and destroy the tension. If this occurs, the saw will not run true until it is rehammered. Fast grinding also causes case hardening, which results in cracks. Do not use a thick wheel, which would make wide gullets at the base of the tooth. A narrow tooth does not have the proper strength and is liable to vibrate in the cut.

Small Circular Ripsaws

These saws should be kept perfectly round and true on the edge, and the gullets should be round at the bottom and of equal depth and width. They should never be filed to sharp corners at the bottom of the teeth, as this will cause them to crack. The best results can be obtained only by keeping the points of the teeth sharp and in proper shape to cut. They should be set or swaged for clearance, and this work should be carefully done. If swaged, the corners should be of uniform depth and should be sufficiently stout so that they will not crumble off in the cut.

Saws are frequently accused of being either too hard or too soft, when, in reality, the trouble is entirely due to the manner in which they are filed. For instance, if the teeth are lacking in hook and are extremely stout at the points, they will cut hard even when sharp. When they become slightly dull, which they will in a short time because of the blunt shape of the points, they will not cut at all and are liable to crack in this condition. An emery wheel or round file is indispensable to the proper care of these saws, as it is impossible to maintain the desired shape of the blade with the use of only a common flat file. Machines are now made to keep these teeth in perfect shape.

In order to maintain the original pitch and back line of each tooth, it is necessary that approximately the same amount of filing be done on the back as on the front of the tooth. If more filing is done on the face of the tooth than on the back, the original shape will soon be destroyed, and it will be almost impossible to restore the blade to the proper shape without retoothing the entire saw. Sharp corners at the bottom of the teeth will also cause cracking at the rim.

No matter whether ripsaws are swage- or spring-set, they should all be filed straight across in front and in back of the teeth. It is a mistake to think that ripsaws will cut better if beveled than if dressed square across. A beveled tooth has a tendency to split the fiber instead of cutting it off squarely across. The bevel also produces a lateral motion, which causes the teeth to chatter and vibrate in the cut; many saws become cracked from this cause.

Small Circular Cutoff Saws

When filing small circular cutoff saws, as in the filing of small ripsaws, it is essential in all cases that they be kept perfectly round and true on the edge. The teeth should be uniform in width and shape, and the gullets should be equal in width and depth. Every tooth should have the proper amount of bevel, and this bevel should be similar on both sides of the tooth when a "V" tooth is used. The amount of set should be the least that will clear the plate sufficiently to prevent friction. The set should never extend too far into the body of the tooth, and the tooth should never be set too close to the point.

Fine-Toothed Cutoff Saw

For smooth work, these saws require no pitch or hook to the teeth. If rapid work is desired, a pitch to the center will provide speed, but the work will not be done quite so smoothly and will require an extra finishing operation.

Large Circular Saws

The points of teeth on large circular ripsaws, as on small saws, are the only portion of the saw that should come in contact with the timber. They must be kept sharp by the use of a file or emery wheel and should be set by springing or spread by swaging. They should be swaged and side-dressed so that the extreme point of the tooth is the widest, diminishing in thickness back from the point.

A saw that is fitted full-swage will stand up better under a fast feed than if it were fitted spring-set, but it takes more power to drive the saw because more friction is produced on the edge by the wider points of the teeth. However, for log sawing, this style is most reliable. Because the swage wears faster on the log side, and thus produces an unequal strain on the saw, it is a mistake to run a saw without first swaging it almost every time it is filed. Where the timber is clean and free from grit, a saw may sometimes be run two or three times after being swaged before needing to be swaged again, and, if it is carefully filed, it will do rather good work.

The sharpening of crosscut saws differs from that of ripsaws only in the shape of the teeth and the manner of filing them. Large crosscut saws for cutting off large logs where power feed and rapid work are required should have the pitch line from 4 to 8 in. in front of the center of the saw for softwood. For hardwood, a trifle more hook is preferable.

With extremely stout teeth, the strain is transmitted to the bottom of the gullets, usually resulting in cracks at the rim. To remedy this situation, gum out the teeth deeper. When a crack starts, drill a hole at the bottom of the crack to prevent it from extending farther into the plate. Bevel only the point of the tooth for ordinary work; in special cases, such as cutting cedar logs into shingle bolts, a larger, wider bevel is necessary. For heavy work, where a smooth cut is not necessary, a crosscut saw should be filed with the front of the tooth slightly beveled.

Inserted-Tooth Saws

The teeth of these saws are drop-forgings or are sometimes topped with one of the ultrahard alloys. They are made separate from the disc and are arranged to be inserted and locked firmly in place on the rim of the disc. Inserted-tooth circular saws have the following advantages over solid-tooth saws:

1. The teeth, being drop-forged from bar steel, are regular in size and shape and are generally of better material than it is possible to use for the entire saw blade.
2. The teeth are capable of having a better-shaped throat—a special advantage for coarse feeds and for soft, wet, or fibrous woods.
3. They effect a great savings in time, files, and blades over gumming and sharpening.
4. The diameter of the saw is not reduced, as is the case with constant filing of solid-tooth saws. One file will go as far toward keeping a good inserted-tooth saw in order as ten files would with a solid-tooth saw.

Before inserting new points, the grooves in the plate and shanks should be wiped perfectly clean and should be well oiled, so that the points will draw easily into the plate. When inserting a point, pick it up with the left hand. After dipping the grooved part in oil, place it in position, holding it even with the sides of the shank. Great care must be taken to have the point seat clean and free from particles of fine dust or gum that may have collected there during the use of the saw; this is often the cause of saws being out-of-round.

Chapter 33
Wood Fastenings

Up to the end of the colonial period, all nails used in the United States were handmade. They were forged on an anvil from nail rods, which were sold in bundles. These nail rods were prepared either by rolling iron into small bars of the required thickness or by the much more common practice of cutting plate iron into strips by means of rolling shears.

Just before the Revolutionary War, the making of nails from these rods was a household industry among the New England farmers. The struggle of the colonies for independence intensified an inventive search for shortcuts to the mass production of material entering directly or indirectly into the prosecution of the war; thus came about the innovation of cut nails made by machinery. With its introduction, the household industry of nail making rapidly declined. At the close of the 18th century, 23 patents for nail-making machines had been granted in the United States, and their use had been generally introduced into England, where they were received with enthusiasm.

In France, lightweight nails for carpenters' use were made of wire as early as the days of Napoleon I, but these nails were made by hand with a hammer. The handmade nail was pinched in a vise with a portion projecting. A few blows of a hammer flattened one end into a head. The head was beaten into a countersunk depression in the vise, thus regulating its size and shape. In the United States, wire nails were first made in 1851 or 1852 by William Hersel of New York.

In 1875, Father Goebel, a Catholic priest, arrived from Germany and settled in Covington, Kentucky; there he began the manufacture of wire nails that he had learned in his native land. In 1876, the American Wire and Screw Nail Company was formed under Father Goebel's leadership. As the production and consumption of wire nails increased, the vogue of cut nails, which dominated the market until 1886, declined.

The approved process in the earlier days of the cut-nail industry was as follows: Iron bars, rolled from hematite or magnetic pig were fagotted, reheated to a white heat, drawn, rolled into sheets of the required width and thickness, and then allowed to cool. The sheet was then cut across its length (its width being usually about a foot) into strips a little wider than the length of the required nail. These plates, heated by being set on their edge on hot coals, were seized in a clamp and fed to the machine, end first. The cut-out pieces, slightly tapering, were squeezed and headed up by the machine before going to the trough.

The manufacture of tacks, frequently combined with that of nails, is a distinct branch of the nail industry, affording much room for specialties. Originally it was also a household industry carried on in New England well into the 18th century. The wire, pointed on a small anvil, was placed in a pedal-operated vise, which clutched it between jaws furnished with a gauge to regulate the length. A certain portion was left projecting; this portion was beaten with a hammer into a flat head.

Antique pieces of furniture are frequently held together with iron nails that are driven in and countersunk, which hold quite firmly. These old-time nails were made of four-square wrought iron and tapered somewhat like a brad but with a head that, when driven in, held with great firmness.

The raw material of the modern wire nail factory is drawn wire, just as it comes from the wire-drawing block. The stock is low-carbon Bessemer or basic open-hearth steel. The wire, fed from a loose reel, passes between straightening rolls into the gripping dies, where it is gripped a short distance from its end, and the nail head is formed by an upsetting blow from a heading tool. As the header withdraws, the gripping dies loosen, and the straightener carriage pushes the wire forward by an amount equal to the length of the nail. The cutting dies advance from the sides of the frame and clip off the nail, at the same time forming its characteristic chisel point. The gripping dies seize the wire again, and an ejector flips the nail out of the way just as the header comes forward and heads the next nail. All of these motions are induced by cams and eccentrics on the main shaft of the machine, and the speed of production is at a rate of 150 to 500 or more complete cycles per minute. At this stage, the nails are covered with a film of drawing lubricant and oil from the nail machine, and their points are frequently adorned with whiskers—a name applied to the small diamond-shaped pieces stamped out when the point is formed and which are occasionally found on the finished nail by the customer.

These oily nails, in lots of 500 to 5000 lb, are shaken with sawdust in tumbling barrels, from which they emerge bright and clean and free of their whiskers, ready for weighing, packing, and shipping.

The "Penny" System

This method of designating nails originated in England. Two explanations are offered as to how this curious designation came about. One is that the six-penny, four-penny, ten-penny, etc. nails derived their names from the fact that one hundred nails cost six pence, four pence, etc. The other explanation, which is the more probable of the two, is that one thousand ten-penny nails, for instance, weighed ten pounds. The ancient, as well as the modern, abbreviation for penny is *d,* being the first letter of the Roman coin denarius; the same abbreviation in early history was used for the English pound in weight. The word *penny* has persisted as a term in the nail industry.

Kinds of Nails

Nails arc the carpenter's most useful fastener, and a great variety of types and sizes are available to meet the demands of the industry. One manufacturer claims to produce more than 10,000 types and sizes. Some common types of nails are illustrated in Figure 33-1.

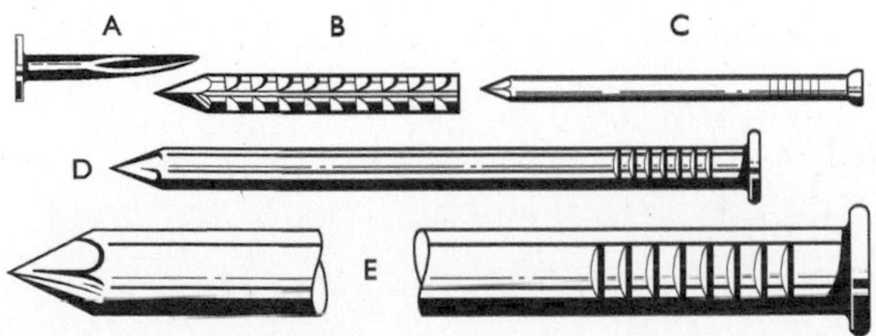

Figure 33-1 Various nails grouped by general size: A. tack; B. sprig or dowel pin; C. brad; D. nail; E. spike.

The following shapes of points are available:

- Common blunt pyramidal
- Long sharp
- Chisel-shaped
- Blunt, or shooker
- Side-sloped
- Duck-bill, or clincher

The shanks, as shown in Figure 33-2, may be as follows:

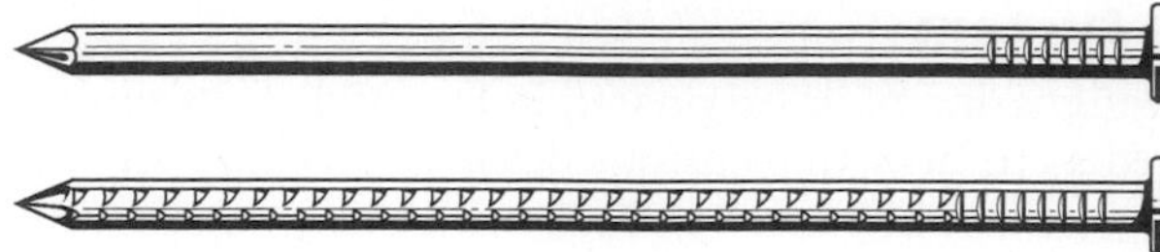

Figure 33-2 Smooth and barbed box nails. Note the sharp point and thin, flat head.

- Barbed
- Longitudinally grooved
- Round smooth
- Spiral grooved
- Annular grooved

Nails may be finished:

- Bright
- Galvanized, usually hot-dipped
- Cadmium-plated
- Blued
- Painted
- Cement-coated

Nails may be made of:

- Mild steel
- Copper
- Brass
- Aluminum
- Case-hardened steel

The heads may be:

- Flat
- Oval or oval countersunk
- Round
- Countersunk
- Double-headed
- Cupped
- Lead-headed

Besides these, there are many nails with other types of heads adapted for special uses.

Tacks

Tacks are small, sharp, pointed nails that usually have tapering sides and thin, flat heads. The regular lengths of tack range from ⅛ to 1⅛ in. The regular sizes are designated in ounces, shown in Table 33-1.

Table 33-1 Wire Tacks

Size oz.	Length in.	No. per lb	Size oz.	Length in.	No. per lb	Size oz.	Length in.	No. per lb
1	⅛	16,000	4	7/16	4000	14	13/16	1143
1½	3/16	10,666	6	9/16	2666	16	⅞	1000
2	¼	8000	8	⅝	2000	18	15/16	888
2½	5/16	6400	10	11/16	1600	20	1	800
3	⅜	5333	12	¾	1333	22	1 1/16	727
						24	1⅛	666

Sprigs

The name *sprig* is sometimes given to a small headless nail that is usually called a barbed dowel pin. Sprigs are regularly made in sizes ½ to 2 in.; No. 8 steel wire gauge or 0.162-in. diameter.

Brads

Brads are small slender nails with small, deep heads; sometimes, instead of a head, they have a projection on one side. There are several varieties adapted to many different requirements. Although brads are generally thought of as being very small, the common variety is made in sizes from ¼ in.

Nails

The term "nails" is popularly applied to all kinds of nails except extreme sizes such as tacks, brads, and spikes. Broadly speaking, however, the term includes all of these. The most generally used are called common nails, and are regularly made in sizes from 1 in. (2*d*) to 6 in. (60*d*), as shown in Table 33-2 and Figures 33-3, 33-4, and 33-5. Some special types of nails are illustrated in Figures 33-6 through 33-10.

Spikes

By definition, an ordinary spike is a stout piece of metal from 3 to 12 in. in length and thicker in proportion than a common nail. It is provided with a head and a point and is frequently curved, serrated, or cleft to render extraction difficult. It is used to a great extent to attach railroad rails to ties and is also used in the construction of docks, piers, and other work requiring large timbers.

Table 33-2 Common Nails

	Plain			Coated			
Size	*Length in.*	*Gauge No.*	*No. per lb*	*Length in.*	*Gauge No.*	*No. per Keg*	*Net Wgt. lb*
2d	1	15	876	1	16	85,700	79
3d	1¼	14	568	1⅛	15½	54,300	64
4d	1½	12½	316	1⅜	14	29,800	61
5d	1¾	12½	271	1⅝	13½	25,500	70
6d	2	11½	181	1⅞	13	17,900	65
7d	2¼	11½	161	2⅛	12½	15,300	72
8d	2½	10¼	106	2⅜	11½	10,100	71
9d	2¾	10¼	96	2⅝	11½	8900	68
10d	3	9	69	2⅞	11	6600	63
12d	3¼	9	63	3⅛	10	6200	80
16d	3½	8	49	3¼	9	4900	80
20d	4	6	31	3¾	7	3100	83
30d	4½	5	24	4¼	6	2400	84
40d	5	4	18	4¾	5	1800	82
50d	5½	3	14	5¼	4	1300	79
60d	6	2	11	5¾	3	1100	82

It should be noted that spike and common-nail sizes overlap; sizes common to both are from 3 to 6 in., the spike being thicker for equal sizes. There are two kinds of ordinary or round wire spikes classed with respect to the shape of the ends as flat head, diamond point and oval head, chisel point. The sizes and other proportions for ordinary spikes are given in Table 33-3.

Holding Power of Nails

Numerous tests have been made at various times to determine the holding power of nails. Tests at the Watertown Arsenal on different sizes of nails from *8d* to *60d* gave average results in pounds, as shown in Table 33-4.

A. M. Wellington found the force required to withdraw 9/16 × 9/16-in. spikes, driven 4¼ in. into seasoned oak, to be 4281 lb; the same spikes driven into unseasoned oak, to be 6523 lb.

Professor W. R. Johnson found that a plain ⅜-in. square spike, driven 3⅜ in. into seasoned yellow pine or unseasoned chestnut required approximately 2000 lb of force to extract it; from seasoned white oak, it required approximately 4000 lb; and from well-seasoned locust, it required 6000 lb.

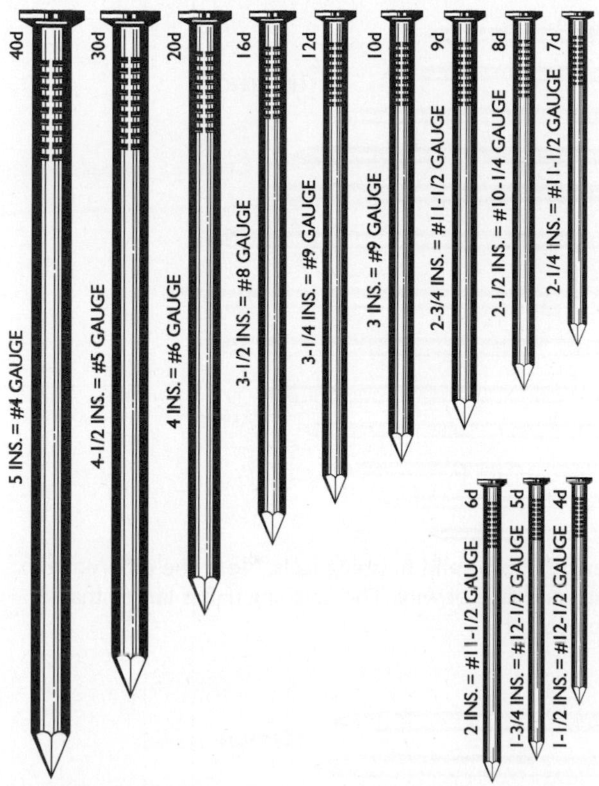

Figure 33-3 Common wire nails. The standard nail for general use is regularly made in sizes from 1–in. (2d) to 6 in. (60d).

Experiments in Germany, by Funk, give from 2465 to 3940 lb (the mean of many experiments was 3000 lb) as the force necessary to extract a plain ½-in. square iron spike 6 inches long, wedge-pointed for 1 inch and driven 4½ in. into white or yellow pine. When driven 5 in., the force required was approximately $\frac{1}{10}$ part greater. Similar $\frac{9}{16}$-in. square spikes, 7 in. long, driven 6 in. deep, required from 3700 to 6745 lb of force to extract them from pine;

the mean of the results was 4873 lb. In all cases, about twice as much force was required to extract them from oak. The spikes were all driven across the grain of the wood. When driven with the grain, spikes or nails hold with less than half as much force.

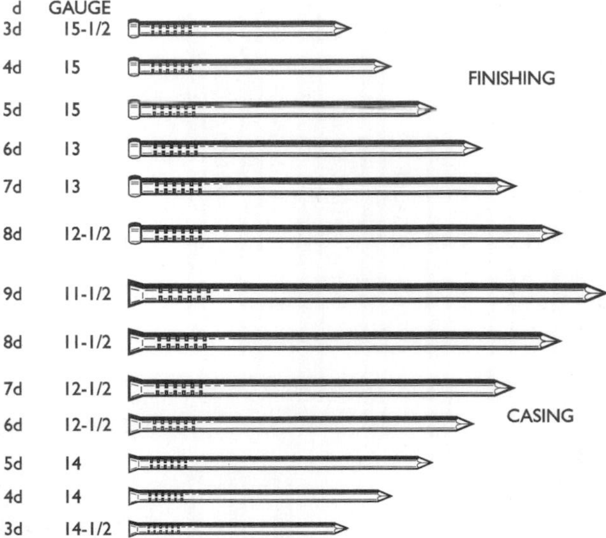

Figure 33-4 Various casing and finishing nails. Note the difference in the shape of heads and size of wire. The finishing nail is larger than a casing nail of equal length.

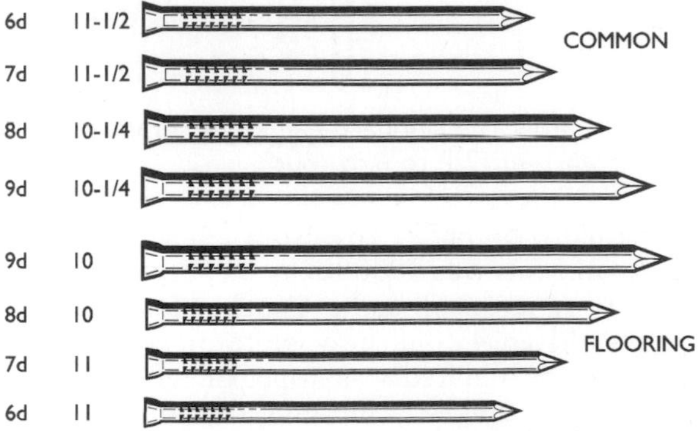

Figure 33-5 Flooring and common nails. Note the variation in head shape and gauge number.

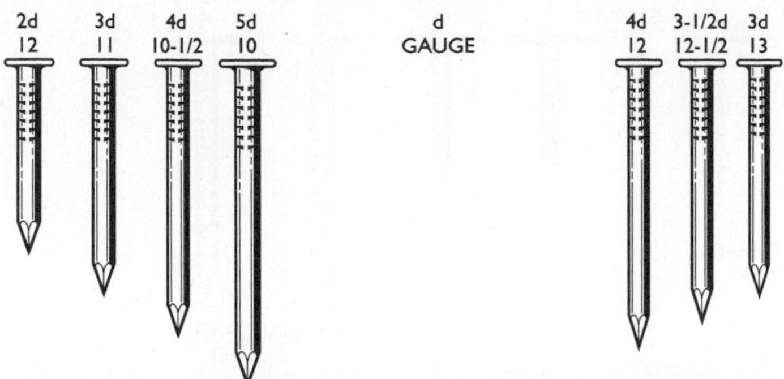

Figure 33-6 A few sizes of slating and shingle nails. Note the difference in wire gauge.

Table 33-3 Ordinary Spikes

Size	*Length in.*	*Gauge No.*	*Degree of Countersink*	*Head Diam.*	*Head Rad.*	*No. per lb*
10d	3	6	123	13⁄32	7⁄16	41
12d	3¼	6				38
16d	3½	5	123	7⁄16	7⁄16	30
20d	4	4	123	15⁄32	7⁄16	23
30d	4½	3	123	½	7⁄16	17
40d	5	2	123	17⁄32	7⁄16	13
50d	5½	1				10
60d	6	1	123	9⁄16	7⁄16	9
7 inch	7	5⁄16 inch	123	⅝	⅝	7
8 inch	8	⅜ inch	123	¾	¾	4
9 inch	9	⅜ inch				3½
10 inch	10	⅜ inch				3
12 inch	12	⅜ inch				2½

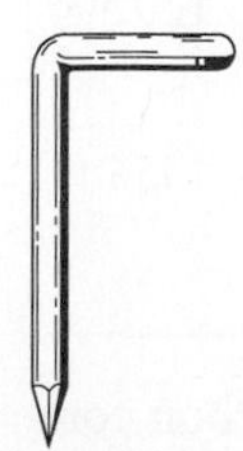

Figure 33-7 Hook-head, metal-lath nail. This is a bright, smooth nail with a long, thin, flat head. It is also made blued or galvanized.

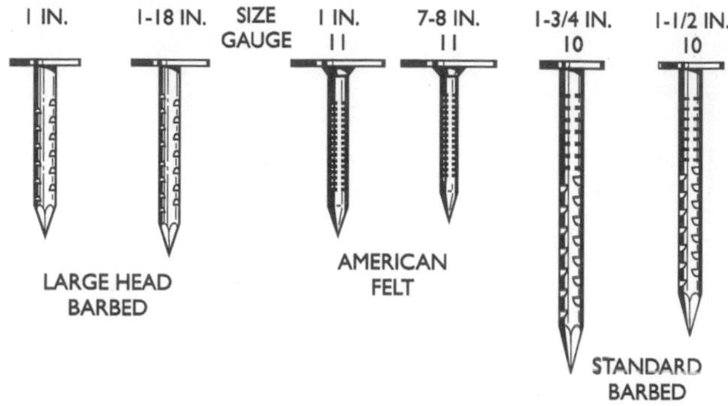

Figure 33-8 Various roofing nails.

Table 33-4 Withdrawal Force (lbs. per sq. in. of Surface)

Wood	*Wire Nail*	*Cut Nail*
White Pine	167	405
Yellow Pine	318	662
White Oak	940	1216
Chestnut		683
Laurel	651	1200

Boards of oak or pine nailed together by from 4 to 16 ten-penny common, cut nails and then pulled apart in a direction lengthwise to the boards and across the nails (tending to break the latter in two by a shearing action) averaged 300 to 400 lb per nail to separate them. Chestnut offers about the same resistance as yellow pine.

A. W. Wright of the Western Society of Engineers obtained the following results with spikes driven into dry cedar (cut 18 months):

Table 33-5 Holding Power of Spikes

Size of spikes	5 × ¼ in. sq.	6 × ¼	6 × ½	5 × ⅜
Length driven in	4 × ¼ in.	5 in.	5 in.	4¼ in.
Pounds resistance to drawing, average lbs.	857	857	1691	1202
Max. lbs	1159	923	2129	1556
From 6 to 9 tests each min. lbs.	766	766	1120	687

With respect to types of shanks, the plain-shank, low-carbon common nails have sufficient holding power for most work. Barbed nails do not have quite as much holding power as plain-shank nails when

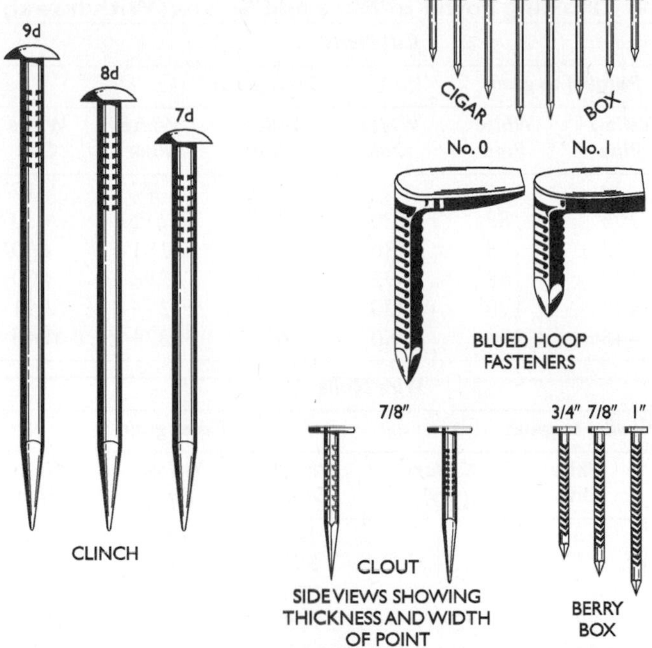

Figure 33-9 Miscellaneous nails.

driven into dry wood, but driven into wet or green wood, they do not lose their grip when the wood dries out as do nails with plain shanks. Coated nails are often used for short-time holding power, such as for boxing and crating. In the smaller sizes, the holding power of cement-coated nails may be as much as 150 percent greater than that of plain uncoated nails, especially when driven into soft wood, although in very hard woods, the holding power may be little, if any, greater. In any case, the extra holding power is lost in a relatively short time.

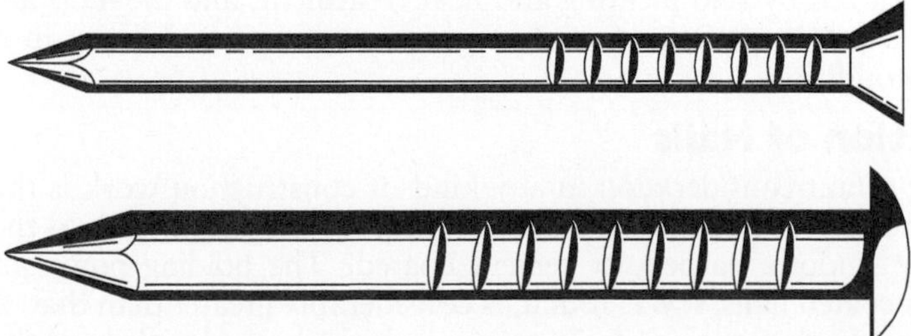

Figure 33-10 Hinge nails.

Table 33-6 Holding Power of Nails and Spikes (Withdrawal)

Cut Nails

	Parallel to grain			*Cross grain*		
Size	**Yellow Pine**	**White Pine**	**White Oak**	**Yellow Pine**	**White Pine**	**White Oak**
6*d*	89			154	77	317
8*d*	206	89	520	327	211	630
10*d*	222	108	580	324	181	650
20*d*	320	148	692	407	298	800
50*d*	439	170	820	570	316	991
60*d*	445	200	950	639	324	1040

Wire Nails

	Parallel to grain			*Cross grain*	
Size	**White Pine**	**Cedar (dry)**	**White Oak**	**Yellow Pine**	**White Pine**
6*d*	30		129	108	60
10*d*	50		390	132	70
60*d*			731	465	
⅜ in.	370	283	1188	590	450
1⁄16 in.	344				436
½ in.	113	338	744	700	364

The increase in holding power of deformed-shank and cement-coated nails is for withdrawal resistance only. There is little or no difference in the shearing resistance of all nails with the same sizes of shank. In buildings, nails are always placed in shear if at all possible, and the extra cost of deformed or coated nails may or may not be justified.

Lathers have the habit of putting nails in their mouths when at work. For sanitation, the different types of nails they use are blued and sterilized by acid pickling and heat treatment, and the nails are then packed in moisture-proof containers to ensure delivery in a sterile condition.

Selection of Nails

An important consideration in any kind of construction work is the type and size of nail to use. The first factor is the finish. Should the nails be smooth, barbed, or cement coated? The holding power of cement-coated nails, it was found, is considerably greater than that of the same-sized smooth nails. In most cases, the barbed nails have the least holding power. Thus, nails can be graded with regard to holding power as follows: first, cement-coated; second, smooth; third, barbed.

Next to be considered is the diameter size of the nail. Short, thick nails work loose quickly. Long, thin nails are apt to break at the joints of the lumber. The simple rule to follow is to use as long and as thin a nail as will drive easily.

Definite rules have been formulated by which to determine the size of nail to be used in proportion to the thickness of the board into which it is to be nailed:

1. When using box nails in timber of medium hardness, the penny of the nail should not be greater than the thickness, in eighths of an inch, of the board into which the nail is being driven.
2. In very soft woods, the nails may be one penny larger, or even in some cases, two pennies larger.
3. In hard woods, nails should be one penny smaller.

The kind of wood is, of course, a big factor in determining the size of nail to use. The dry weight of the wood is the best basis for the determination of its grain substance or strength. The greater its dry weight, the greater its power to hold nails. However, the splitting tendency of hard wood tends to offset its additional holding power. Smaller nails can be used in hard timber than in soft timber, as shown in Figure 33-11. Positive rules governing the size of nails to be used as related to the density of the wood cannot be laid down. Experience is the best guide.

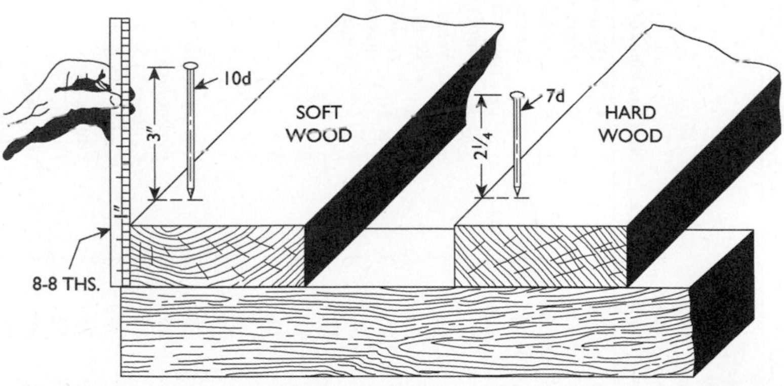

Figure 33-11 Application of Rules 2 and 3 in determining the proper size nail to use.

Wire Nails—Kinds and Quantities Required

The following example illustrates the usefulness of Table 33-7.

Table 33-7 Wire Nails—Kinds and Quantities Required (American Steel and Wire Co.)

Length, in.	Am. Steel & Wire Co.'s Steel Wire Gauge	Approx. No. to lb.	Nailings	Sizes and Kinds of Material		Trade Names	Pounds per 1000 feet B. M. on center as follows: 12 in.	16 in.	20 in.	36 in.	48 in.
							Pounds				
2½	10¼	106	2	1 × 4	I. Used square	8d common	60	48	37	23	20
2½	10¼	106	2	1 × 6	edge, as	8d common	40	32	25	16	18
2½	10¼	106	2	1 × 8	platforms,	8d common	31	27	20	12	10
2½	10¼	106	2	1 × 10	floors,	8d common	25	20	16	10	8
2½	10¼	106	3	1 × 12	sheathing, or	8d common	31	24	20	12	10
4	6	31	2	2 × 4	shiplap.	20d common	105	80	65	60	32
4	6	31	2	2 × 6	II. When used D.	20d common	70	54	43	27	22
4	6	31	2	2 × 8	& M., blind	20d common	53	40	53	21	17
4	6	31	3	2 × 10	nailed, only ½	20d common	60	50	40	25	20
4	6	31	3	2 × 12	quantity named required.	20d common	52	41	33	21	17
6	2	11	3	2 × 4		60d common	197	150	122	76	61
6	2	11	2	3 × 6		60d common	131	97	82	52	42
6	2	11	2	3 × 8		60d common	100	76	61	38	34
6	2	11	3	3 × 10		60d common	178	137	110	70	55
6	2	11	3	3 × 12		60d common	145	115	92	53	46
2½	12½	189	2	Base, per 100 ft. lin		8d finish		1			
2½	10¼	106	2	Byrket lath		8d common		48			
2½	12½	189	1	Ceiling, ¾ × 4		8d finish	18	14			
2	13	309	1	Ceiling, ½ and ⅝		6d finish	11	8			
2½	12½	189	2	Finish, ⅞		8d finish	25	12			
3	11½	121	2	Finish, 1⅛		10d finish	12	10			
2½	10	99	1	Flooring, 1 × 3		8d flooring	42	32			
2½	10	99	1	Flooring, 1 × 4		8d flooring	32	26			
2½	10	99	1	Flooring, 1 × 6		8d flooring	22	18			

4	6	31		Framing, 2 × 4 to 2 × 6 requires 9 or more sizes and vary greatly	20d common	20	16	14
3½	8	49			16d common	10	10	8
3	9	69			10d common	8	6	5
6	2	11		Framing, 3 × 4 to 3 × 14	60d common	30	25	20
2½	11½	145	2	Siding, drop, 1 × 4	8d casing	45	35	
2½	11½	145	2	Siding, drop, 1 × 6	8d casing	30	25	
F2½	11½	145	2	Siding, drop, 1 × 8	8d casing	23	18	
2	13	309	1	Siding, bevel, ½ × 4	6d finish	23	18	
2	13	309	1	Siding, bevel, ½ × 6	6d finish	15	13	
2	13	309	1	Siding, bevel, ½ × 8	6d finish	12	10	
				Casing, per opening	6d and 8d casing	About ½ lb per side.		
1¼	14	568	12"o.c.	Flooring, ⅜ × 2	3d brads	About 10 lb per 1000 sq. ft.		
1½	15	778	16"o.c.	Lath, 48"	3d fine	6 lb per 1000 pieces.		
⅞	12	469	2"o.c.	Ready roofing	Barbed roofing	¾ lb to the square.		
⅞	12	469	1"o.c.	Ready roofing	Barbed roofing	1½ lb to the square.		
⅞	12	180	2"o.c.	Ready roofing (⅝ heads)	American felt roofing	1½ lb to the square.		
⅞	12	180	1"o.c.	Ready roofing (⅝ heads)	American felt roofing	3 lb to the square.		
1¼	13	420		Shingles[a]	3d shingle	4½ lb; about 2 nails to each 4 in.		
1½	12	274		Shingles	4d shingle	7½ lb; about 2 nails to each 4 in.		
⅞	12	180	4	Shingles	Ameican felt roofing	12 lb, ⅝ in. heads; 4 nails to shingle.		
⅞	12	469	4	Shingles	Barbed roofing	4½ lb, ⅝ in. heads; 4 nails to shingle.		
1	16	1150	2"o.c.	Wall board, around	2d Barbed Berry, flat head entire edge	5 lb, ⅝ in. heads; per 1000 sq. ft.		
1	15½	1010	3"o.c.	Wall board, intermediate	2d casing or flooring nailings	2½ lb; ⅝ in. heads; per 1000 sq. ft.		

[a]Wood shingles vary in width; asphalt shingles are usually 8 in. wide. Regardless of width, 1000 shingles are the equivalent of 1000 pieces, 4 in. wide.

Example What size, type, and quantity of nails are required to lay 1 in. × 3 in. flooring for a hall 50 ft × 100 ft, with joists spaced 16 in. on centers?

Look in the fifth column, headed "Sizes and Kinds of Materials," and find "Flooring, 1 × 3." Follow the line to the right; the size and type specified in the "Trade Names" column is 8*d* flooring.

$$\text{B.M. (board measure) for flooring 1 inch thick} = 50 \times 100 = 5000$$

$$\text{quantity of nails} = 32 \times \frac{5000}{1000} = 160 \text{ pounds}$$

Continue on the same line. In the column for 16-in. centers under "Pounds per 1000 feet B.M.," it is found that 32 lb of nails are required per 1000 feet B.M.

Driving Nails

One advantage of wire nails is that it is not necessary to hold them in a certain position when driving them to prevent splitting. However, in some instances it is advisable to first drill holes nearly the size of the nail before driving, to guard against splitting. Also, in fine work, where a large number of nails must be driven, such as in boat building, holes should be driven. This step prevents crushing the wood and possible splitting because of the large number of nails driven through each plank. The size of drill for a given size nail should be found by experiment.

In cheap boat construction, steel clinch nails should be used; however, in the best construction, only copper nails should be used, and these should be riveted over small copper washers.

The right and wrong ways to drive a nail are shown in Figure 33-12. Figure 33-13 illustrates the necessity of using a good hammer to drive a nail. The force that drives the nail is due to the inertia of the hammer. The inertia depends on the suddenness with which its motion is brought to rest on striking the mail. With hardened steel, there is practically no give, and all the energy possessed by the hammer is transferred to the nail. With soft and/or inferior metal, all the energy is not transferred to the nail; therefore, the drive per blow is less than with hardened steel.

Nails for Hardwood Boxes

3/8	in. thickness use 4*d* cement coated nails
7/16 or 1/2	in. thickness use 5*d* cement coated nails
9/16 or 5/8	in. thickness use 6*d* cement coated nails
7/8	in. thickness use 7*d* cement coated nails

Nails for Softwood Boxes

1/4	in. thickness use 4*d* cement coated nails
3/8	in. thickness use 5*d* cement coated nails
7/16 or 1/2	in. thickness use 6*d* cement coated nails
9/16 or 5/8	in. thickness use 7*d* cement coated nails
7/8	in. thickness use 8*d* cement coated nails
1/4	in. thickness: use special large 3d or regular 4*d* cement-coated nails.

Screws

Wood screws are not often used in structural carpentry because their advantage over nails is only in their greater withdrawal resistance, and structural loadings that place the fastenings in withdrawal are always avoided when possible. Their lateral, or shearing, resistance is not appreciably greater than that of driven nails of the same diameter. In some types of cabinet work, screws are used to allow ready dismounting, or disassembly, and their appearance is good. They are used in installing all kinds of builder's hardware because of their great withdrawal resistance and because they are more or less readily removed in case repairs or alterations are necessary.

By definition, a wood screw is a fastening implement consisting of a shank with a rather coarse, sharp, right-hand thread, a sharp gimlet point that enters the wood readily, and a head that may be any one of various contours and is provided with a slot, or crossed slots, to receive the tip of a driving tool.

Table 33-8 Approximate Number of Wire Nails per Pound

Amer. Steel & Wire Co.'s Steel Wire Gauge	Length																			
	3/16	1/4	3/8	1/2	5/8	3/4	7/8	1	1 1/8	1 1/4	1 1/2	1 3/4	2	2 1/4	2 1/2	2 3/4	3	3 1/2	4	4 1/2	5
3/8								29	26	23	20	17	15	15	12	11	11	8.9	7.9	7.1	6.4
5/16								43	38	34	29	25	22	20	18	16	15	13	11	10	9.0
1								47	44	40	34	29	26	23	21	20	18	16	14	12	11
2								60	54	48	41	35	31	28	25	23	21	18	16	14	13
3								67	60	55	47	41	36	32	29	27	25	21	18	16	15
4								81	74	66	55	48	41	37	34	31	29	25	22	20	18
5								90	81	74	61	52	45	41	38	35	32	28	24	22	21
6				213	174	149	128	113	101	91	76	65	58	52	47	43	39	34	29	26	24
7				250	205	174	148	132	120	110	92	78	70	61	55	51	47	40	35	31	28
8				272	238	198	174	153	139	126	106	93	82	74	66	61	56	48	42	38	34
9				348	286	238	213	185	170	152	128	112	99	87	79	71	67	58	50	45	41
10				469	373	320	277	242	616	196	165	142	124	111	100	91	84	71	62	55	49
11				510	117	366	323	285	254	233	200	171	149	136	122	111	103	87	77	69	61
12				740	603	511	442	397	351	327	268	229	204	182	161	149	137	118	103	95	87
13			1356	1017	802	688	590	508	458	412	348	297	260	232	209	190	175	153	138	123	110
14		2293	1664	1290	1037	863	765	667	586	536	459	398	350	312	278	256	233	201	176	157	1400

15		2899	2213	1619	1316	1132	971	869	787	694	578	501	437	390	351	317	290	256	220	196	177
16		3932	2770	2142	1708	1414	1229	1099	973	872	739	635	553	496	452	410	370	318	277	248	226
17		5316	3890	2700	2306	1904	1581	1409	1253	1139	956	831	746	666	590	532	486	418	360	322	295
18		7520	5072	3824	3130	2608	2248	1976	1760	1590	1338	1150	996	890	820	740	680	585	507	448	412
19		9920	6860	5075	4132	3508	2816	2556	2284	2096	1772	1590	1390	1205	1060	970	895	800			
20	18,620	14,050	9432	7164	5686	4795	4230	3596	3225	2893	2412	2070	1810	1620	1450	1315	1215	1035			
21	23,260	17,252	12,000	8920	7232	6052	5272	4576	4020	3640	3040	2665	2310	2020	1830						
22	28,528	21,508	14,676	11,776	9276	7672															
23	35,864	27,039	18,026	13,519	10,815	9013															
24	44,936	34,018	22,678	17,008	13,607	11,339															
25	57,357	43,243	28,828	21,622	17,297	14,414															

These approximate numbers are an average only, and the figures given may be varied either way, by changes in the dimensions of the heads or points. Brads and on-head nails will run more to the pound than the table shows, and large or thick-headed nails will run less.

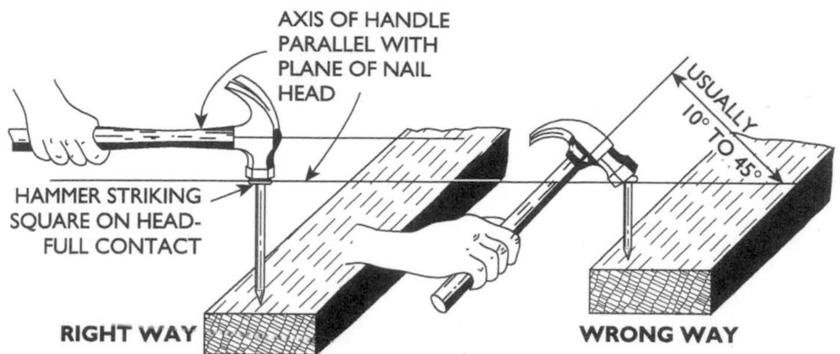

Figure 33-12 Right and wrong ways to drive a nail. Hit the nail squarely on the head. The handle should be horizontal when the hammer head hits a vertical nail.

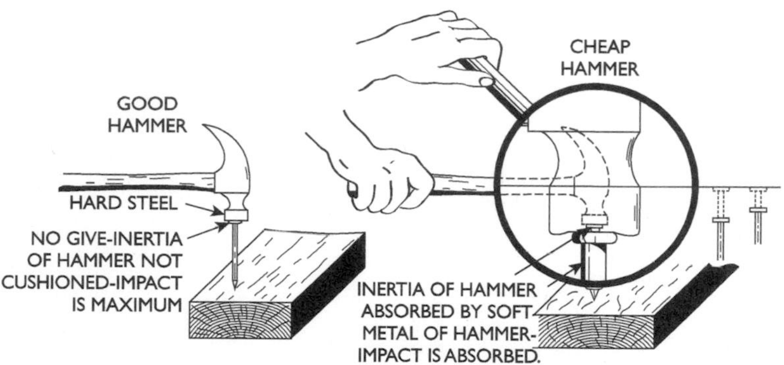

Figure 33-13 Why an inferior hammer should not be used.

Screws of many types are made for specialized purposes, but stock wood screws are usually obtainable in either steel or brass, and, more rarely, of high strength bronze. Three types of heads are standard: the flat countersunk head, with the included angle of the sloping sides standardized at 82°; the round head, whose height is also standardized, but whose contour seems to vary slightly among the products of different manufacturers; and the oval head, which combines the contours of the flat head and the round head. All of these screws are available with the Phillips slot, or crossed slots, instead of the usual single straight slot.

The Phillips slot allows a much greater driving force to be exerted without damaging the head. By far the greater part of all wood screws used, probably 75 percent or more, are of the flat-head type.

Material

For ordinary purposes, steel screws, with or without protective coatings, are commonly used. In boat building or other such work where corrosion will probably be a problem if screws are used, the screws should be of the same metal or at least the same *type* of metal as the parts they contact. While it is possible and indeed probable that a single brass screw that is driven through an aluminum plate, if it is kept dry, will show no signs of corrosion, many brass screws driven through the aluminum plate in the presence of water or dampness will almost certainly show signs, perhaps serious, of galvanic corrosion.

Dimensions of Screws

When ordering screws, it is important to know what constitutes the length of a screw. The overall length of a 2-in. flat-head screw is not the same as that of a 2-in. round-head screw. To avoid confusion and mistakes, the lengths for various types of screws are shown in Figure 33-14 and should be carefully noted.

It should also be noted that, unlike the ordinary wire gauges, the 0 in the screw gauge, shown in Figure 33-15, indicates the diameter of the smallest screw, and the diameter of the screws *increase* with the number of the gauge.

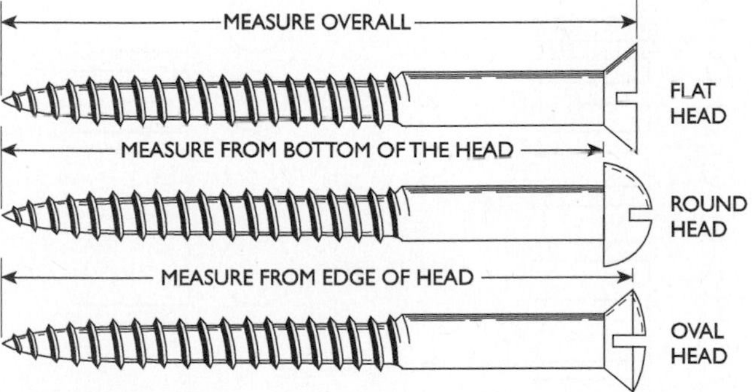

Figure 33-14 Various wood screws and how their lengths are measured.

Shape of the Head

The buyer will find a multiplicity of head shapes to select from; the variety of heads regularly carried in a well-stocked hardware store is usually great enough to meet every possible requirement. However, in order to avoid possible disappointment in remote supply bases (small dealers), it is better to select from these three forms of heads, which may be regarded as standard:

- Flat
- Round
- Oval

All of these heads are available in either the straight-slotted or Phillips type.

The other forms may be regarded as special or semispecial, that is, carried by large dealers only or obtainable only on special order.

Flat heads are necessary in some cases, such as on door hinges, where any projection would interfere with the proper working of the hinge; flat-head screws are also employed on finish work where flush surfaces are desirable. The round and oval heads are normally ornamental when exposed.

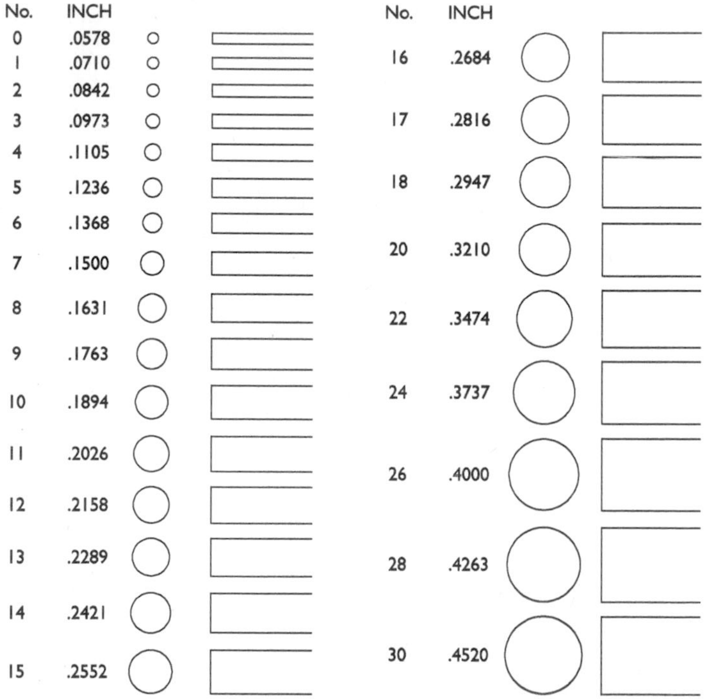

No.	INCH	No.	INCH
0	.0578	16	.2684
1	.0710	17	.2816
2	.0842	18	.2947
3	.0973	20	.3210
4	.1105	22	.3474
5	.1236	24	.3737
6	.1368	26	.4000
7	.1500	28	.4263
8	.1631	30	.4520
9	.1763		
10	.1894		
11	.2026		
12	.2158		
13	.2289		
14	.2421		
15	.2552		

Figure 33-15 Wood screw gauge numbers.

The diameter of the head in relation to the gauge number of the screw is shown in Table 33-9. Some of the many special screw heads available are shown in Figure 33-16.

How to Drive a Wood Screw

Consult Table 33-10 to determine the size of drill to use in drilling the shank-clearance hole. This hole (Figure 33-17) should be slightly smaller than the shank diameter of the screw and about ¾ the shank length for soft and medium-hard woods. For extremely hard woods, the length of the hole should equal the shank length.

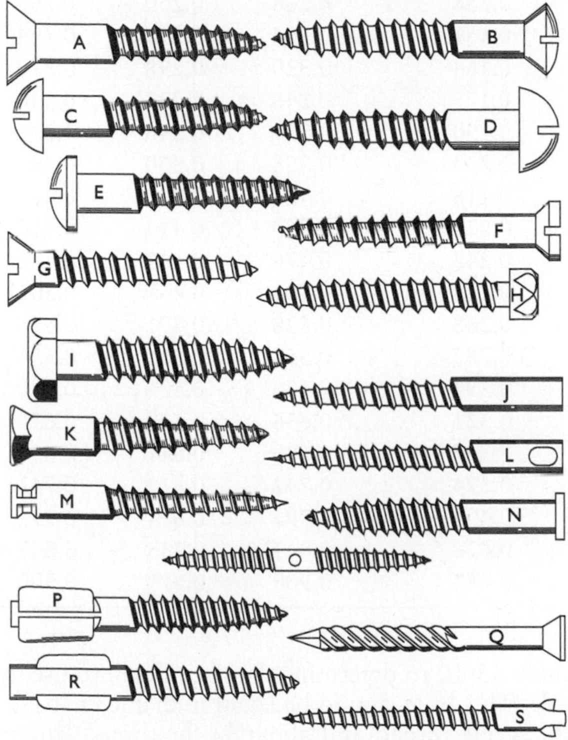

Figure 33-16 Various wood screws showing the variety of head shapes available: A. flat head; B. oval head; C. round head; D. piano head; E. oval filister head; F. countersunk filister head; G. felloe; H. close head; I. hexagon head; J. headless; K. square bung head; L. grooved; M. pinched head; N. round bung head; O. dowel; P. winged; Q. drive; R. winged; S. winged head. Heads A through G may be obtained with Phillips-type heads.

Table 33-9 Head Diameters

Screw Gauge	Screw Diameter	Head Diameter		
		Flat	Round	Oval
0	0.060	0.112	0.106	0.112
1	0.073	0.138	0.130	0.138
2	0.086	0.164	0.154	0.164
3	0.099	0.190	0.178	0.190
4	0.112	0.216	0.202	0.216
5	0.125	0.242	0.228	0.242
6	0.138	0.268	0.250	0.268
7	0.151	0.294	0.274	0.294
8	0.164	0.320	0.298	0.320
9	0.177	0.346	0.322	0.346
10	0.190	0.371	0.346	0.371
11	0.203	0.398	0.370	0.398
12	0.216	0.424	0.395	0.424
13	0.229	0.450	0.414	0.450
14	0.242	0.476	0.443	0.476
15	0.255	0.502	0.467	0.502
16	0.268	0.528	0.491	0.528
17	0.282	0.554	0.515	0.554
18	0.394	0.580	0.524	0.580
20	0.321	0.636	0.569	0.636
22	0.347	0.689	0.611	0.689
24	0.374	0.742	0.652	0.742
26	0.400	0.795	0.694	0.795
28	0.426	0.847	0.735	0.847
30	0.453	0.900	0.777	0.900

Again consult Table 33-10 to determine the size of drill to use in drilling the pilot hole. This hole should be equal in diameter to the root diameter of the screw thread and about ¾ the thread length for soft and medium-hard woods. For extremely hard woods, the pilot hole depth should equal the thread length.

If the screw being inserted is the flat-head type, the hole should be countersunk. A typical countersink is shown in Figure 33-18.

The foregoing process involves three separate steps. All of them can be performed at once by using a device of the type shown in

Figure 33-19. This tool will drill the pilot hole, the shank-clearance hole, and the countersink all in one operation.

These tools are made in many sizes, one for each screw size, and they are available in complete sets or separately. The screw size is marked on the tool.

Table 33-10 Safe Loads for Wood Screws

Kind of Wood	*Gauge Number* 4	8	12	16	20	24	28	30
White oak	80	100	130	150	170	180	190	200
Yellow pine	70	90	120	140	150	160	180	190
White pine	50	70	90	100	120	140	150	160

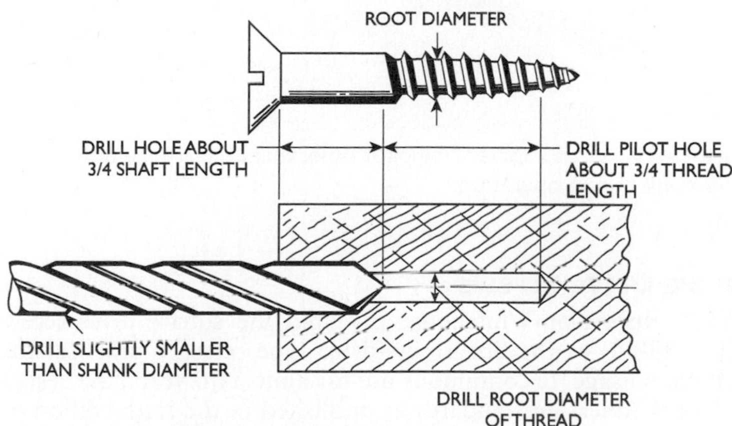

Figure 33-17 Drilling shank-clearance and pilot holes.

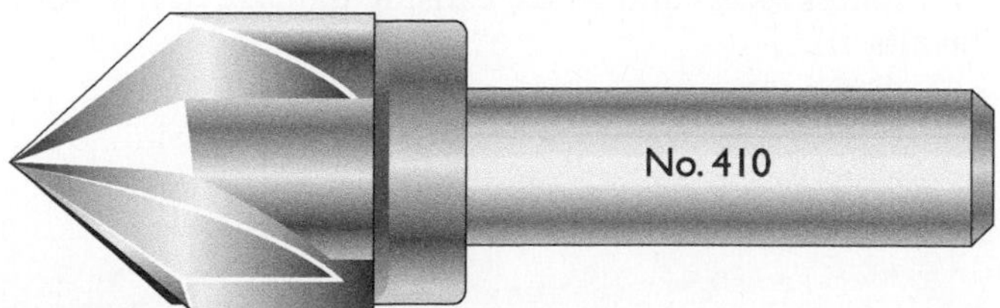

Figure 33-18 A typical countersink.

Strength of Wood Screws

Table 33-10 gives the safe resistance, or safe load (against pulling out), in pounds per linear inch of wood screws when inserted across the grain. For screws inserted with the grain, use 60 percent of these values.

The lateral load at right angles to the screw is much greater than that of nails. For conservative designing, assume a safe resistance of a No. 20 gauge screw at double that given for nails of the same length, when the full length of the screw thread penetrates the supporting piece of the two connected pieces.

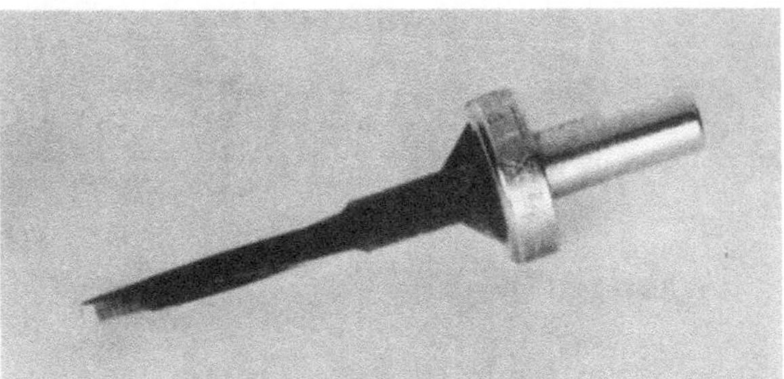

Figure 33-19 A tool for drilling pilot hole, shank-clearance hole, and countersink in one operation.

Square Drive Screws

In 1908 Robertson-Whitehouse invented the square drive screw (Figure 33-20) as the first recess-drive type fastener practical for production usage. In continuous use for almost 90 years, the design is a North American standard, as published in the sixth edition of Industrial Fasteners Institute Metric and Inch Standards. Square drive screws offer markedly more resistance to "Cam-Out" (that disaster waiting to happen whereby the screw driver slips out of the screw slot or Phillips recess and skids, causing damage to the wood and the screw itself).

Figure 33-20 Square drive screw.
Courtesy of McFeely's Square Drive Screws

The screws are available in flat head, pan head, trim head, round washer head, large round washer head, and oval head. Standard thread styles include the following:

Figure 33-21 Deep thread.
Courtesy of McFeely's Square Drive Screws

Deep thread. Characterized by a reduced-diameter shank, which results in a deep thread profile. A single lead thread, meaning that one revolution advances the screw one pitch. The deep thread form provides superior resistance to pull out.

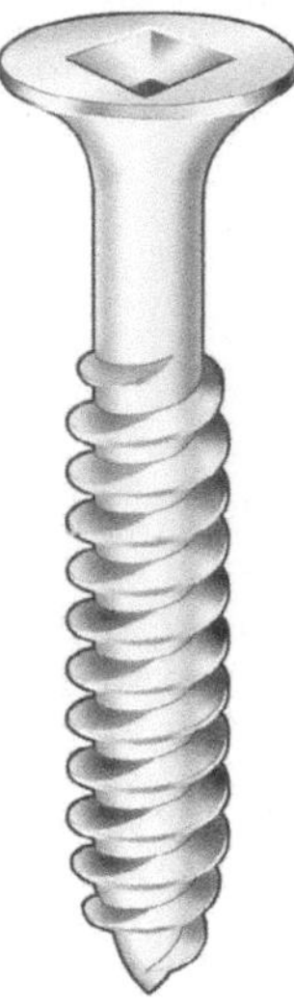

Figure 33-22 Wood screw style thread.
Courtesy of McFeely's Square Drive Screws

Wood screw style thread. A single lead, extra-thick thread used primarily on solid brass or silicon bronze screws to accommodate the limitations of these soft materials. Unlike traditional wood screw thread, the shank is of uniform diameter throughout most of its length.

Double lead thread. Two threads are wrapped around the shank. One revolution advances the screw two pitch lengths. This provides a great advantage when used on long screws or in situations requiring rapid assembly. Drywall screws typically use this thread since pull-out strength isn't as important as speed.

Tapping style thread. Also known sometimes as Type "A" or "Wood tapping" but is basically a sheet metal type of thread. It is a single lead, fine-thread design (meaning more threads per inch). The entire screw shank is threaded.

Screw Dimensions

Screw dimensions are listed in Table 33-11 in two different ways. The maximum and minimum decimal dimensions are shown as well as the maximum and minimum fractional dimensions.

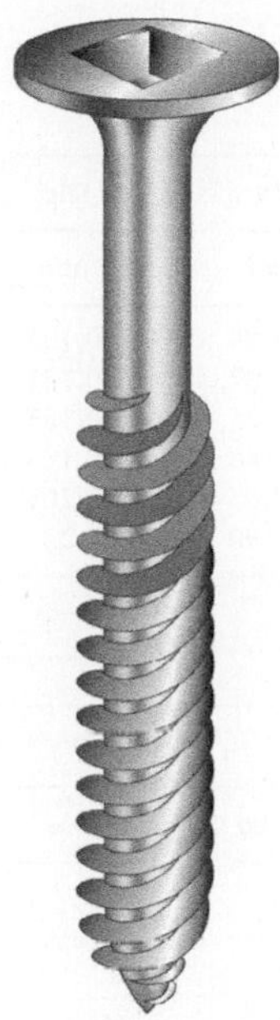

Figure 33-23 Double lead thread.
Courtesy of McFeely's Square Drive Screws

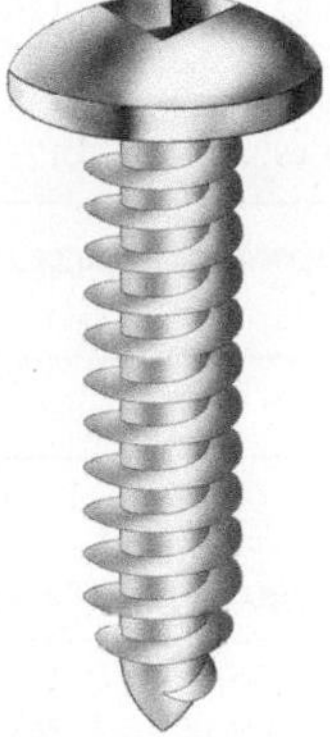

Figure 33-24 Tapping style thread.
Courtesy of McFeely's Square Drive Screws

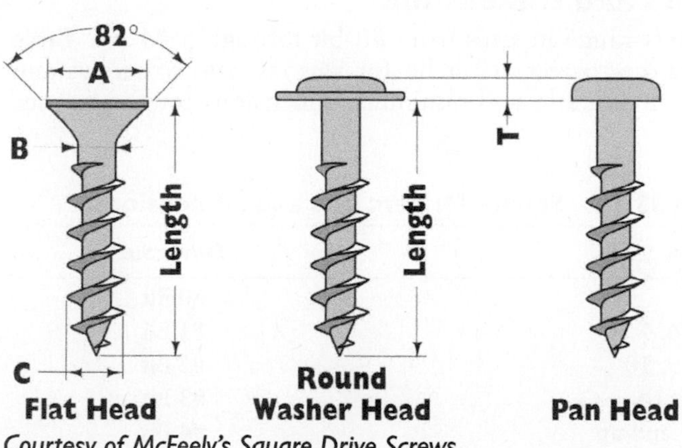

Courtesy of McFeely's Square Drive Screws

Table 33-11 Square Drive Screw Dimensions

	"A"		"A"			"A"			"B"		"C"	
	Flat Head		Pan Head			Round Washter Head			Body Dia		Thread Dia	
Size	max	min	max	min	T max	max	min	T max	max	min	max	min
4	0.225	0.195	0.219	0.205	0.086	N/A	N/A	N/A	0.095	0.084	0.116	0.105
6	0.279	0.244	0.270	0.256	0.103	N/A	N/A	N/A	0.118	0.107	0.142	0.131
8	0.332	0.292	0.322	0.306	0.120	0.376	0.352	0.110	0.136	0.125	0.168	0.157
10	0.385	0.340	0.373	0.357	0.137	0.443	0.411	0.125	0.157	0.146	0.194	0.183
12	0.438	0.389	0.425	0.407	0.153	N/A	N/A	N/A	0.176	0.165	0.220	0.209
14	0.507	0.452	0.492	0.473	0.175	N/A	N/A	N/A	0.201	0.190	0.246	0.235

Note: IFI Standards specify a tolerance on screw length of +0", – 1/16".

	"A"	"A"		"A"		"B"	"C"
	Flat Head	Pan Head		Round Washter Head		Body Dia	Thread Dia
Size	Max Dia	Max Dia	T max	Max Dia	T max	Max	Max
4	7/32	7/32	3/32	N/A	N/A	3/32	1/8
6	9/32	17/64	7/64	N/A	N/A	1/8	9/64
8	21/64	21/64	1/8	3/8	7/64	9/64	11/64
10	25/64	3/8	9/64	7/16	1/8	5/32	3/16
12	7/16	27/64	5/32	N/A	N/A	11/64	7/32
14	1/2	31/64	11/64	N/A	N/A	13/64	1/4

Driver Bits and Dimensions

The driver bits range in sizes from #0 bit through a #4 bit. Table 33-12 shows the correct driver bit for various sizes of screws and also includes maximum and minimum dimensions for the bit tips themselves.

Table 33-12 Square Drivers Bits and Dimensions

Screw Size	Driver Size
#4	#0 Bit
#5, 6, 7	#1 Bit
#8, 9, 10	#2 Bit
#12, 14	#3 Bit
5/16" and up	#4 Bit

Bit Size	Max	Min	Fraction
#0	.071	.0696	1/16+
#1	.091	.0090	3/32-
#2	.1126	.111	7/64+
#3	.133	.1315	1/8+
#4	.191	.1895	3/16+

Lag Screws

By definition, a lag screw, as shown in Figure 33-25, is a heavy-duty wood screw with a square or hexagonal head so that it may be turned by a wrench. These are large, heavy screws that are used where great strength is required, such as for heavy timber work. Table 33-13 gives the dimensions of ordinary lag screws.

How to Put in Lag Screws

First, bore a hole slightly larger than the diameter of the shank to a depth that is equal to the length that the shank will penetrate (see Figure 33-26). Then bore a second hole at the bottom of the first hole equal to the root diameter of the threaded shank and to a depth of approximately one-half the length of the threaded portion. The exact size of this hole and its depth will, of course, depend on the kind of wood; the harder the wood, the larger the hole.

The resistance of a lag screw to turning is enormous when the hole is a little small, but this can be considerably decreased by smearing the threaded portion of the screw with soap or beeswax.

Figure 33-25 Ordinary lag screw.

Table 33-13 Lag Screws

Length 3	3½	4	4½	5	5½	6	6½	7	7½	8	9	10	11	12
Dia. 5/16 to 7/8	5/16 to 1	5/16 to 1	5/16 to 1	5/16 to 1	5/16 to 1	5/16 to 1	7/16 to 1	7/16 to 1	7/16 to 1	7/16 to 1	7/16 to 1	½ to 1	½ to 1	½ to 1

Strength of Lag Screws

Table 33-14 gives the safe resistance, or load to pulling out, in pounds per linear inch of thread for lag screws inserted across the grain.

Table 33-14 Safe Loads for Lag Screws (Inserted Across the Grain)

Kind of Wood	Diameter of Screw in Inches				
	1/2	5/8	3/4	7/8	1
White pine	590	620	730	790	900
Douglas fir	310	330	390	450	570
Yellow pine	310	330	390	450	570

Bolts

By definition, a bolt is a pin or rod that is used for holding anything in place, and often has a permanent head on one end. A bolt is generally regarded as a rod that has a head at one end and is threaded at the other to receive a nut; the nut is usually considered part of the bolt, and prices quoted on bolts by dealers normally include the nuts.

Kinds of Bolts

There is a multiplicity of bolt forms to meet various requirements.

The *common machine bolt* has a square head at one end and a short length of thread at the other end. Two of these are shown in Figure 33-27.

When a loop, or "eye," is provided instead of a head, it is called an *eye bolt.*

A *countersunk bolt* has a beveled head, which fits into a countersunk hole.

A *key-head bolt* has a head shaped so that, when inserted into a suitable groove or slot, it will not turn when the nut is screwed onto the other end.

Another method of preventing turning consists of forming a short portion of the bolt body square at the head end, the head itself being spherical in shape; such a type is known as a *carriage bolt.*

A headless bolt threaded for a certain distance at both ends is called a *stud bolt.* See Figure 33-28.

In addition to the types just mentioned, there are numerous others, such as *milled coupling, railroad track, stove,* and *expansion.* Several different types of bolts are illustrated in Figure 33-29.

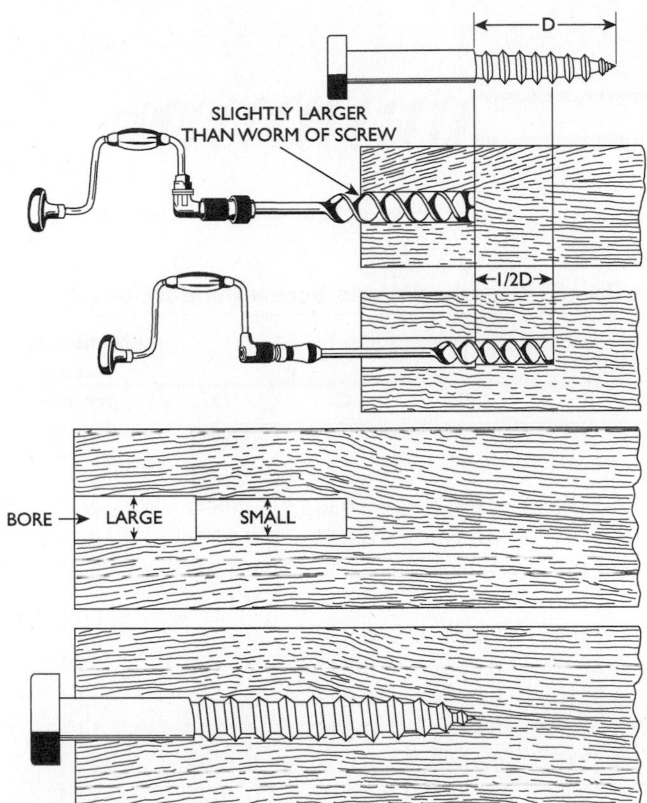

Figure 33-26 Drilling holes for lag screws.

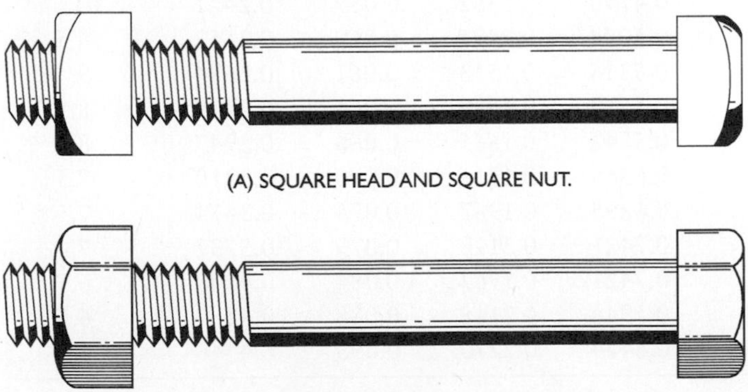

Figure 33-27 Machine bolts.

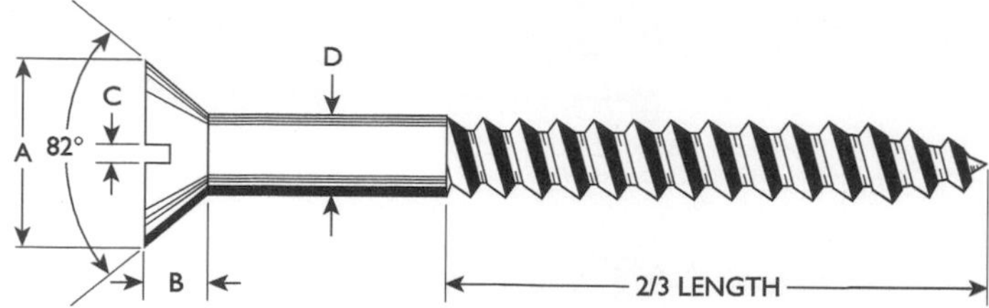

Table 33.15 Standard Wood Screw Proportions

Screw Numbers	*A*	*B*	*C*	*D*	*Number of Threads per Inch*
0				0.0578	30
1				0.0710	28
2	0.1631	0.0454	0.030	0.0841	26
3	0.1894	0.0530	0.032	0.0973	24
4	0.2158	0.0605	0.034	0.1105	22
5	0.2421	0.0681	0.036	0.1236	20
6	0.2684	0.0757	0.039	0.1368	18
7	0.2947	0.0832	0.041	0.1500	17
8	0.3210	0.0809	0.043	0.1631	15
9	0.3474	0.0984	0.045	0.1763	14
10	0.3737	0.1059	0.048	0.1894	13
11	0.4000	0.1134	0.050	0.2026	12.5
12	0.4263	0.1210	0.052	0.2158	12
13	0.4427	0.1286	0.055	0.2289	11
14	0.4790	0.1362	0.057	0.2421	10
15	0.5053	0.1437	0.059	0.2552	9.5
16	0.5316	0.1513	0.061	0.2684	9
17	0.5579	0.1589	0.064	0.2815	8.5
18	0.5842	0.1665	0.066	0.2947	8
20	0.6368	0.1816	0.070	0.3210	7.5
22	0.6895	0.1967	0.075	0.3474	7.5
24	0.7421	0.2118	0.079	0.3737	7
26	0.7421	0.1967	0.084	0.4000	6.5
28	0.7948	0.2118	0.088	0.4263	6.5
30	0.8474	0.2270	0.093	0.4546	6

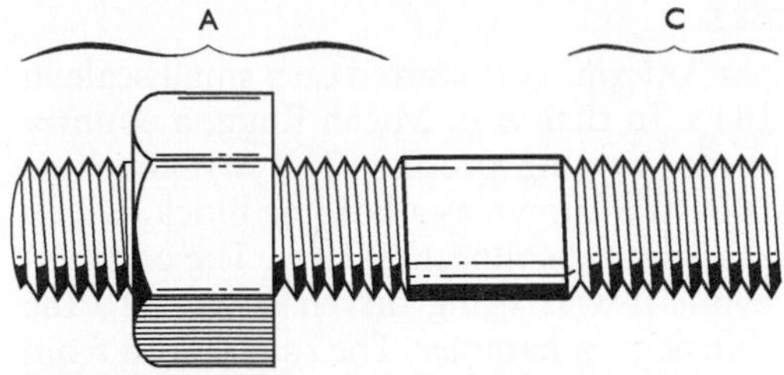

Figure 33-28 Stud bolt with hexagon nut. A is the nut and C is the attachment end.

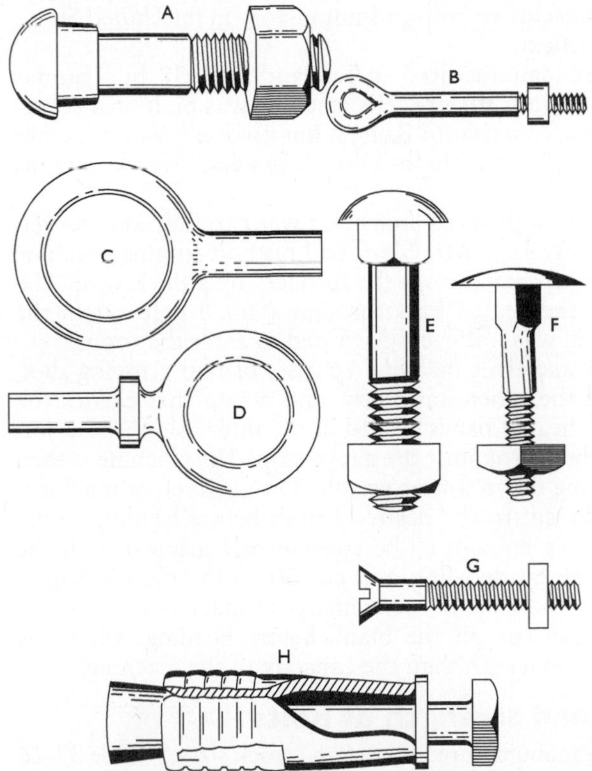

Figure 33-29 Various bolt styles: A. railroad track bolt; B. welded eye bolt; C. plain forged eye bolt; D. shouldered eye bolt; E. carriage bolt; F. step bolt; G. stove bolt; H. expansion bolt.

Manufacture of Bolts

The bolt-and-nut industry in America was started on a small scale in Marion, Connecticut, in 1818. In that year, Micah Rugg, a country blacksmith, made bolts by the forging process. The first machine used for this purpose was a device known as a heading block, which was operated by a foot treadle and a connecting lever. The connecting lever held the blank while it was being driven down into the impression in the heading block by a hammer. The square iron from which the bolt was made was first rounded so that it could be admitted into the block.

At first, Rugg only made bolts to order, and charged at the rate of 16 cents apiece. This industry developed quite slowly until 1839, when Rugg went into partnership with Martin Barnes. Together they built the first exclusive bolt-and-nut factory in the United States at Marion, Connecticut.

Bolts were first manufactured in England in 1838 by Thomas Oliver of Darlston, Staffordshire. His machine was built on a somewhat different plan from that of Rugg's, but no doubt was a further development of the first machine. Oliver's machine was known as the "English Oliver."

The construction of the early machines was carefully kept secret. It is related that in 1842, a Mr. Clark had his bolt-forging machine located in a room separated from the furnaces by a thick wall. The machine received the heated bars through a small hole cut in the wall; the forge man was not even permitted to enter the room.

A modern bolt-and-rivet machine consists of two gripping dies, one movable and the other stationary, and a ram that carries the heading tool. The heated bar is placed in the impression in the stationary gripping die and against the gauge stop. The machine is then operated by pressing down a foot treadle. On this type of machine, the bar is generally cut to the desired length before heading, especially when it is long enough to be conveniently gripped with the tongs, but it can be headed first and cut off to the desired length afterward. It is also possible in some makes of machines to insert a cutting tool that can cut off the blank before heading, when the work is not greater in length than the capacity of the machine.

Proportions and Strength of Bolts

Ordinary bolts are manufactured in certain "stock sizes." Table 33-16 gives these sizes for bolts from ¼ in. up to 1¼ in., with the length of thread.

Table 33-16 Properties of U.S. Standard Bolts—Unified National Coarse Threads

Diameter	*Number of Threads per inch (National Coarse Thread)*	*Head*	*Head*	*Head*
1/4	20	3/8	13/32	1/2
5/16	18	1/2	35/64	43/64
3/8	16	9/16	5/8	3/4
7/16	14	5/8	11/16	53/64
1/2	13	3/4	53/64	1
9/16	12	7/8	31/32	1 5/32
5/8	11	15/16	1 1/32	1 1/4
3/4	10	1 1/8	1 15/64	1 1/2
7/8	9	1 5/16	1 29/64	1 47/64
1	8	1 1/2	1 21/32	1 63/64
1 1/8	7	1 11/16	1 55/64	2 15/64
1 1/4	7	1 7/8	2 1/16	2 31/64
1 3/8	6	2 1/16	2 17/64	2 47/64
1 1/2	6	2 1/4	2 31/64	2 63/64
1 5/8	5 1/2	2 7/16	2 11/16	3 15/64
1 3/4	5	2 5/8	2 57/64	3 31/64
1 7/8	5	2 13/16	3 3/32	3 47/64
2	4 1/2	3	3 5/16	3 63/64

For many years, the coarse-thread bolt was the only type available. In recent years, bolts with a much finer thread, called the National Fine thread, have become easily available. These have hex heads and hex nuts. They are much better finished than the stock coarse-thread bolts and consequently are more expensive. Cheap rolled-thread bolts, with the threaded portions slightly upset, should not be used by the carpenter. When they are driven into a hole, either the hole is too large for the body of the bolt or the threaded portion reams it out too large for a snug fit. Good bolts have cut threads that have a maximum diameter no larger than the body of the bolt. See Table 33-17 for thread specifications.

Table 33-17 Unified National Fine Threads

Diameter	*Threads per inch*
1/4	28
5/16	24
3/8	24
7/16	20
1/2	20
9/16	18
5/8	18
3/4	16
7/8	14
1	14

When a bolt is to be selected for a specific application, Table 33-16 should be consulted.

Example

How much of a load may be applied to a 1-in. bolt for a tensile strength of 10,000 lb per square inch?

Referring to Table 33-18, we find on the line of a 1-in. bolt a value of 5510 lb corresponding to a stress on the bolt of 10,000 lb per square inch.

Example

What size bolt is required to support a load of 4000 lb for a stress of 10,000 lb per square inch?

$$\text{Area at root of thread} = \text{given load} \div 10{,}000$$
$$= 4000 \div 10{,}000 = 0.400 \text{ sq. in.}$$

Referring to Table 33-18, in the column headed "Area at Bottom of Thread," we find 0.419 square inch to be the nearest area; this corresponds to a 7/8-in. bolt.

Of course, for the several given values of pounds stress per square inch, the result could be found directly from the table, but the calculation above illustrates the method that would be employed for other stresses per square inch not given in the table.

Table 33-18 Proportions and Strength of U.S. Standard Bolts

Bolt Diameter	Area at Bottom of Threads	Tensile Strength		
		10,000 lbs/in²	*12,500 lbs/in²*	*17,500 lbs/in²*
1/4	0.027	270	340	470
5/16	0.045	450	570	790
3/8	0.068	680	850	1190
7/16	0.093	930	1170	1630
1/2	0.126	1260	1570	2200
9/16	0.162	1620	2030	2840
5/8	0.202	2020	2520	3530
3/4	0.302	3020	3770	5290
7/8	0.419	4190	5240	7340
1	0.551	5510	6890	9640
1 1/8	0.693	6930	8660	12,130
1 1/4	0.890	8890	11,120	15,570
1 3/8	1.054	10,540	13,180	18,450
1 1/2	1.294	12,940	16,170	22,640
1 5/8	1.515	15,150	18,940	26,510
1 3/4	1.745	17,450	21,800	30,250
1 7/8	2.049	20,490	25,610	35,860
2	2.300	23,000	28,750	40,250

Example

A butt joint with fish plates is fastened by six bolts through each timber. What size bolts should be used, allowing a shearing stress of 5000 lb per square inch in the bolts, when the joint is subjected to a tensile load of 20,000 lb?

$$\text{Load carried per bolt} = 20{,}000 \div \text{number of bolts}$$
$$= 20{,}000 \div 6 = 3333 \text{ lbs.}$$

Each bolt is double shear, hence,

$$\text{Equivalent single shear load} = \tfrac{1}{2} \text{ of } 3333 = 1667 \text{ lbs.}$$

and

$$\text{area per bolt} = \frac{1667}{5000} = 0.333 \text{ sq. in.}$$

Referring to Table 33-18, the nearest area is 0.302, which corresponds to a ¾-in. bolt. In the case of a dead, or "quiescent," load, ¾-in. bolts would be ample; however, for a live load, take the next larger size, or ⅞-in. bolts.

The example does not give the size of the timbers, but the assumption is that they are large enough to carry the load safely. In practice, all parts should be calculated as described in the chapter on the strength of timbers. The ideal joint is one so proportioned that the total shearing stress of the bolts equals the tensile strength of the timbers.

Chapter 34

Carpentry

The carpenter and millwright trades are closely related in many areas, and in some cases they overlap. Carpenters may be responsible for machinery and equipment installation as well as building construction. On the other hand, it is common practice for some portion of the millwright's duties to be carpenter work. Although the journeyman carpenter trade includes everything from rough construction through trim and finish work, the millwright is usually concerned with construction-type carpentry only. This chapter therefore is limited to those carpentry operations that may be performed by millwrights.

Commercial Lumber Sizes

Two words are used to describe the wood that is the principal material used by carpenters: *timber* and *lumber*. Timber is usually applied to any wood suitable for structural use while in its natural state. Lumber is the timber after cutting and sawing into standard commercial-size pieces.

Lumber is carried in stock by dealers in various sizes and is supposed to be seasoned. Seasoning, either naturally or by exposure to heat, is the process of removing about 85 percent of the moisture contained in freshly cut timber. Lumber is usually classified according to the three types into which it is rough-sawed: "dimension," stock that is 2 in. thick and from 4 to 12 in. wide; "timbers," 4 to 8 in. thick and 6 to 10 in. wide; and "common boards," 1 in. thick and from 4 to 12 in. wide.

Rough lumber is *dressed*, or *surfaced*, by removing about ⅛ in. from each side and from ⅜ to ½ in. from the edges. This planing operation is commonly referred to as "surfacing," and the letter "S" is used to indicate the operation. The term "dressed," with the letter "D" to indicate the operation, was at one time used interchangeably, but present practice is to use only the term "surfaced" and the letter "S." The commonly used abbreviated descriptions of the planing operation using the letter "S" are as follows:

S1S—Surfaced One Side
S2S—Surfaced Two Sides
S4S—Surfaced Four Sides
S1S1E—Surfaced One Side, One Edge
S1S2E—Surfaced One Side, Two Edges

Sawmill practice for many years was to rough-saw lumber to the dimensions indicated by the nominal size; i.e., a 2 × 4 was sawed 2 in. thick by 4 in. wide. Unseasoned lumber, when dried, would reduce in dimensions, and when the surfaces were planed the finished dimensions became 3⁄8 to 1⁄2 in. under the nominal dimensions. The standards used conformed to this practice; for example, 2-in. dimension stock, when surfaced, became 1 5⁄8 -in. thick. This practice of sawing to nominal dimensions was followed for many years, and older buildings are constructed of lumber milled to old standards. This can cause problems when revisions are made using the present sizes of lumber and millwork to revise old structures. Adjustments must be made to adapt the present sizes of lumber and millwork to the old standard dimensions. The *old* standard dimensions are listed in Table 34-1.

Table 34-1 Old Standard Lumber Sizes

Lumber	*Nominal Size*		*Actual S4S Size*	
Classification	*Thickness, in.*	*Width, in.*	*Thickness, in.*	*Width, in.*
	2	4	1 5⁄8	3 5⁄8
	2	6	1 5⁄8	5 5⁄8
Dimension	2	8	1 5⁄8	7 1⁄2
	2	10	1 5⁄8	9 1⁄2
	2	12	1 5⁄8	11 1⁄2
	4	6	3 5⁄8	5 1⁄2
	4	8	3 5⁄8	7 1⁄2
	4	10	3 5⁄8	9 1⁄2
Timbers	6	6	5 1⁄2	5 1⁄2
	6	8	5 1⁄2	7 1⁄2
	6	10	5 1⁄2	9 1⁄2
	8	8	7 1⁄2	7 1⁄2
	10	10	9 1⁄2	9 1⁄2
	1	4	25⁄32	3 5⁄8
	1	6	25⁄32	5 5⁄8
Common	1	8	25⁄32	7 1⁄2
boards	1	10	25⁄32	9 1⁄2
	1	12	25⁄32	11 1⁄2

The practice of sawing to nominal dimensions is no longer followed. Instead, finished standard dimensions have been established, and these are used in the lumber industry to determine sawmill dimensions. This allows more efficient use of logs, as lumber can be rough-sawed to lesser dimensions, providing only enough extra material to surface-plane to the new, reduced standard finish dimensions. Table 34-2 lists the standard lumber sizes *now* in use in the lumber industry.

Guiding and Testing Tools

In good carpentry, much depends on accuracy in measurement and in fitting parts together at the required angle. In order to ensure this accuracy, various tools of guidance and direction are used so that joints, etc., can be made with precision.

Straightedge

This tool is used to guide the pencil or scriber when marking a straight line and when testing a faced surface, such as the edge of a board, to determine if it is straight. Anything having an edge known to be straight, such as the edge of a steel square, may be used; however, a regular straightedge is preferable.

Table 34-2 Standard Lumber Sizes

Lumber Classification	Nominal Size		Actual S4S Size	
	Thickness, in.	Width, in.	Thickness, in.	Width, in.
	2	4	1½	3½
	2	6	1½	5½
Dimension	2	8	1½	7¼
	2	10	1½	9¼
	2	12	1½	11¼
	4	6	3½	5½
	4	8	3½	7½
	4	10	3½	9½
Timbers	6	6	5½	5½
	6	8	5½	7½
	6	10	5½	9½
	8	8	7½	7½
	10	10	9½	9½
	1	4	¾	3½
	1	6	¾	5½
Common	1	8	¾	7¼
boards	1	10	¾	9¼
	1	12	¾	11¼

The straightedge may be made of either wood or steel, and its length may be from a few inches to several feet. For ordinary work, a carpenter can make a sufficiently accurate straightedge from a strip of good straight-grained wood, as shown in Figure 34-1, but for accurate work, a steel straightedge, such as the three shown in Figure 34-2, should be used. Wood is objectionable for precision work because of its tendency to warp or spring out of shape.

Figure 34-3 shows the correct and incorrect methods of holding a straightedge as a guiding tool, and Figure 34-4 shows how and how not to hold the pencil when marking stock.

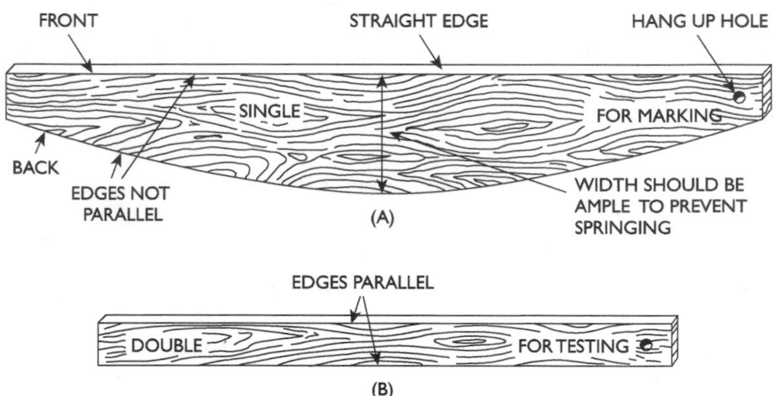

Figure 34-1 Wooden straightedges: (A) single, (B) double.

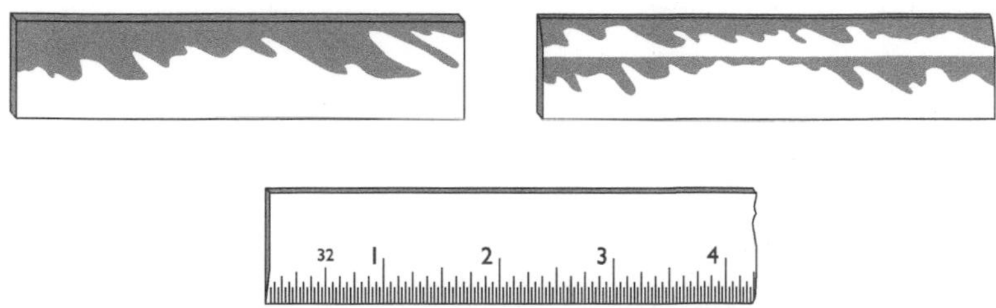

Figure 34-2 Typical steel straightedges. These tools are used where straight lines are scribed or where surfaces must be tested for flatness. Depending on their use, straightedges are made in lengths of 12 to 72 in., widths of $1\frac{3}{8}$ to $2\frac{1}{2}$ in., and thicknesses of $\frac{1}{16}$ to $\frac{1}{2}$ in.

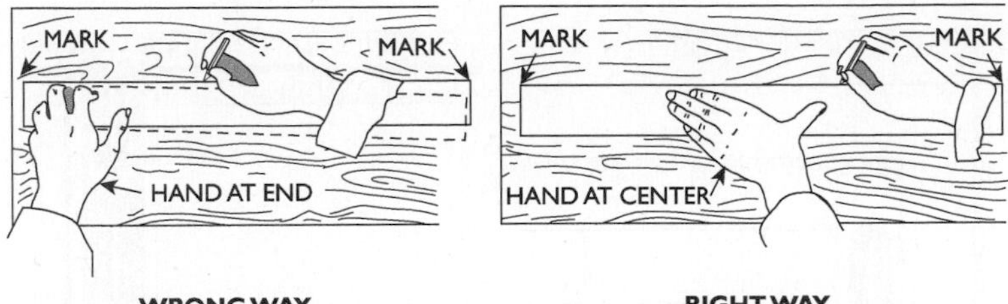

Figure 34-3 The incorrect and correct methods of using the straightedge as a guiding tool. To properly secure a straightedge, the hand should press firmly on the tool at its center, with the thumb and other fingers stretched wide apart.

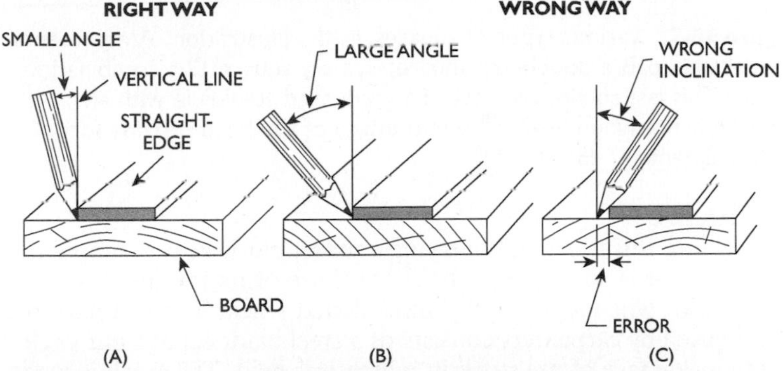

Figure 34-4 Right and wrong inclination of the pencil in marking with the straightedge. The pencil should not be inclined from the vertical more than is necessary to bring the pencil lead in contact with the guiding surface of the straightedge (A). When the pencil is inclined more, and pressed firmly, considerable pressure is brought against the straightedge, tending to push it out of position (B). If the inclination is in the opposite direction, the lead recedes from the guiding surface, thus introducing an error, which is magnified when a wooden straightedge is used because of the greater thickness of the straightedge (C).

Square

The square is a 90°, or right-angle, standard and is used for marking or testing work. There are several common types of squares, as shown in Figure 34-5. They include the following:

- Try square
- Miter square
- Combined try-and-miter square
- Framing or so-called "steel" square
- Combination square

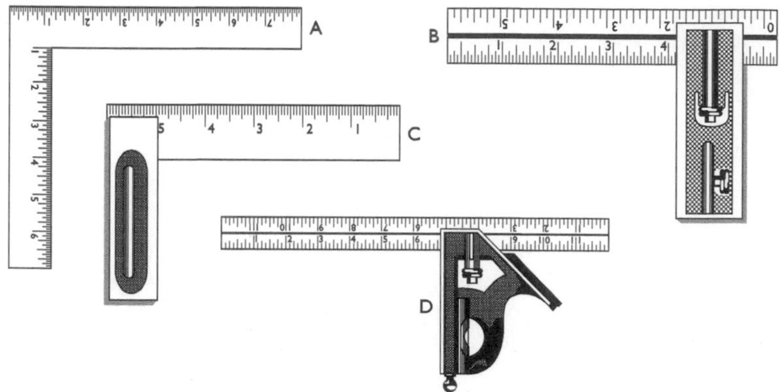

Figure 34-5 Various types of squares. In the illustration, A represents a steel square; B, a double try square; C, a try square; D, a combination square. This last square consists of a graduated steel rule with an accurately machined head. The two edges of the head provide for measurements of 45° and 90°.

Try Square

In England, this is called the trying square, but here it is simply the try square. It is so called probably because of its frequent use as a testing tool when squaring up mill-planed stock. The ordinary try square used by carpenters consists of a steel blade set at right angles to the inside face of the stock in which it is held. The stock is made of some type of hardwood and is always faced with brass in order to preserve the wood from injury.

The usual sizes of try squares have blades ranging from 3 to 15 in. long. The stock is approximately ½ in. thick, with the blade inserted midway between the sides of the stock. The stock is made thicker than the blade so that its face may be applied to the edge of the wood and the steel blade may be laid on the surface to be marked. Usually the blade is provided with a scale of inches divided into eighths.

Miter and Combined Try-and-Miter Squares

The term "miter," strictly speaking, signifies any angle except a right angle, but as applied to squares, it means an angle of 45°.

In the miter square, the blade (as in the try square) is permanently set, but at an angle of 45° with the stock, as shown in Figure 34-6.

A try square may be made into a combined try-and-miter square when the end of the stock to which the blade is fastened is faced off at 45°, as along the line MS in Figure 34-7. When the 45° face (MS) of the stock is placed against the edge of a board, the blade will be at an angle of 45° with the edge of the board, as in Figure 34-8.

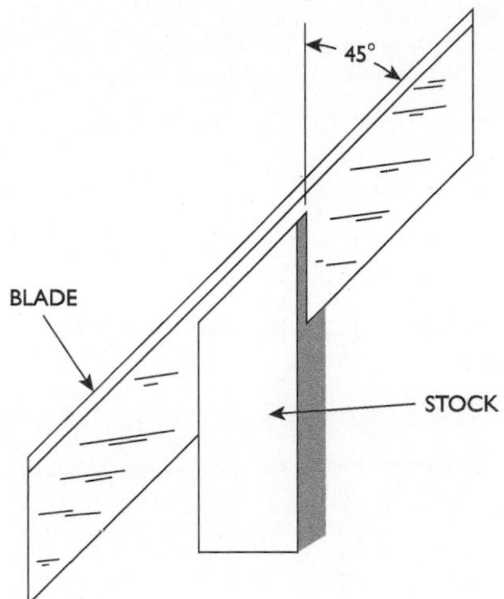

Figure 34-6 A typical miter square; it differs from the ordinary try square in that the blade is set at an angle of 45° with the stock, and the stock is attached to the blade midway between its ends.

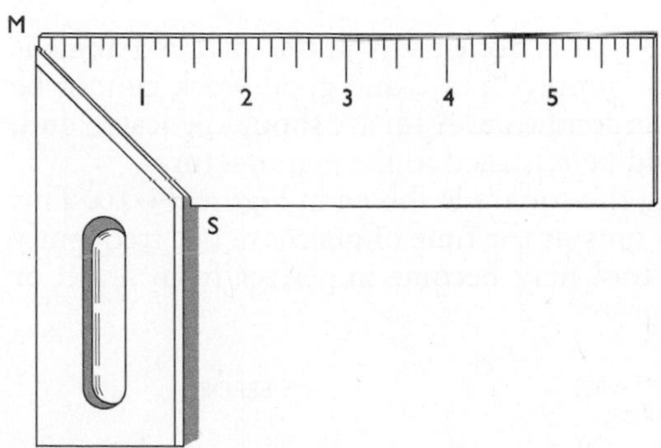

Figure 34-7 A combined try square and miter square. Because of its short 45° face (MS), it is not as accurate as the miter square, but it serves the purpose for ordinary marking and eliminates the need for extra tools.

An improved form of the combined try-and-miter square is shown in Figure 34-9. Because of the longer face (LF) compared with the short face (MS) in Figure 34-7, the blade describes an angle of 45° with greater precision. Its disadvantage is that it is awkward to carry because of its irregular shape. However, its precision greatly outweighs its disadvantage.

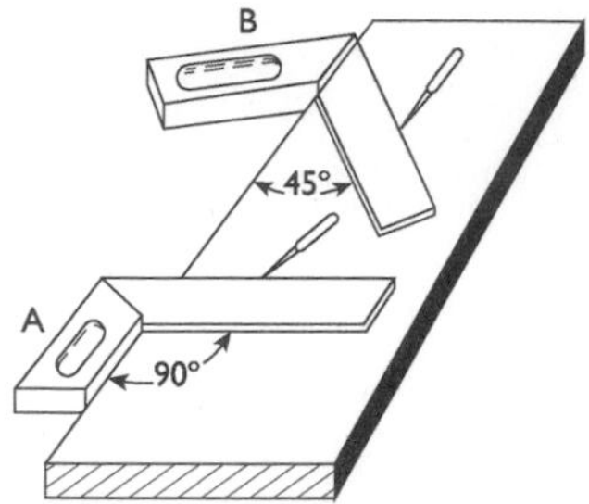

Figure 34-8 The combined try-and-miter square as used for a 90° marking at A and a 45° marking at B.

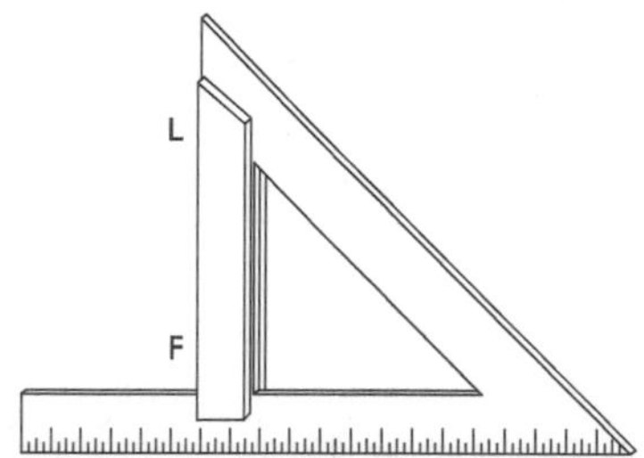

Figure 34-9 An improved form of the combined try-and-miter square.

A square with a blade that is not exactly at the intended angle is said to be out of true, or simply "out," and good work cannot be done with a square in this condition. A square should be tested and, if found to be out, should be returned to the manufacturer.

The method of testing the square is shown in Figure 34-10. This test should be made not only at the time of purchase but frequently afterward, because the tool may become imperfect from a fall or rough handling.

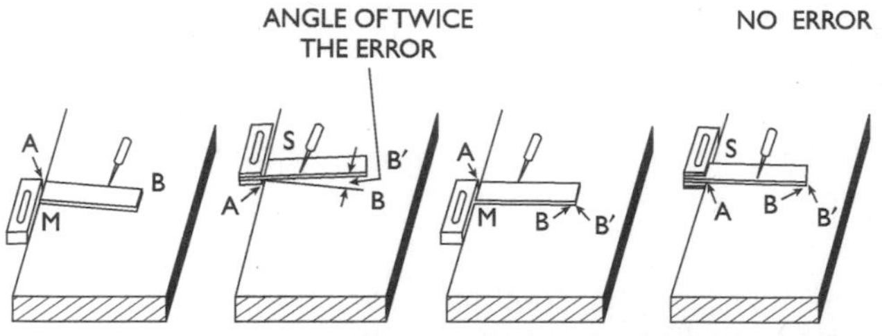

Figure 34-10 The method of testing a try square. If the square is "out" (angle not 90°), scribed lines AB and AB' for positions M and S of the square (left side) will not coincide. Angle BAB' is twice the angle of error. If the square is perfect, lines AB and AB' for positions M and S will coincide (right side).

Under no circumstances should initials or other markings be stamped on the brass face of the ordinary try square, because the burrs that project from bending the brass face will throw the square out of true; for this reason, manufacturers will not take back a square with any marks stamped on the brass face.

Framing or "Steel" Square

The ridiculousness of calling a framing square a steel square is evident from the fact that all types of squares may be obtained that are made entirely of steel. It is properly called a framing square because with its framing table and various other scales, it is adapted especially for use in house framing, although its range of usefulness makes it valuable to any woodworker. Its general appearance is shown in Figure 34-11.

The framing square consists of two essential parts—the tongue and the body, or blade. The tongue is the shorter, narrower part, and the body is the longer, wider part. The point at which the tongue and the body meet on the outside edge is called the heel.

There are several grades of squares known as polished, nickeled, blued, and royal copper. The blued square with figures and scales in white is perhaps the most desirable. A size that is widely used has an 18-in. body and a 12-in. tongue, but there are many uses which require the largest size, whose body measures 24 by 2 in. and whose tongue measures 16 or 18 by 1½ in.

The feature that makes this square such a valuable tool is its numerous scales and tables, which include the following:

- Rafter or framing table
- Essex table
- Brace table
- Octagon scale
- Hundredths scale
- Inch scale
- Diagonal scale

Rafter or Framing Table

This is always found on the body of the square. It is used for determining the length of common valley, hip, and jack rafters and the angles at which they must be cut to fit at the ridge and plate. This table appears as a column six lines deep under each inch graduation from 2 to 18 in., as seen in Figure 34-12A, which shows only the

12-in. section of this table; at the left of the table are letters indicating the application of the figures given. Multiplication and angle symbols are applied to this table to prevent errors in laying out angles for cuts.

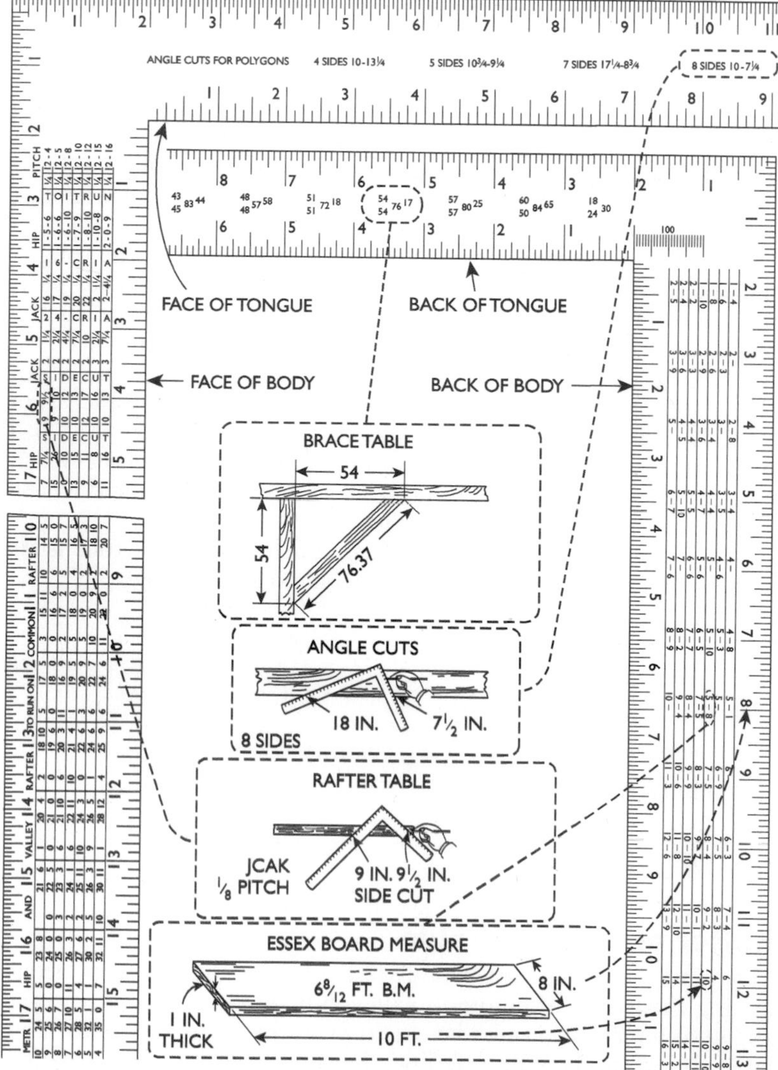

Figure 34-11 The front and back views of a typical framing square.

Essex Table

This is always found on the body of the square, as shown in Figure 34-12B. This table gives the board measure in feet and twelfths of a foot of 1-in. thick boards of usual lengths and widths. On certain squares, it consists of a table eight lines deep under each graduation, as seen in the figures that represent the 12-in. section of this table.

Brace Table

This table is found on the tongue of the square, a section of which is shown in Figure 34-12C. The table gives the length of the brace to be used where the rise and run are from 24 to 60 in. and are equal.

Octagon Scale

This scale is located on the tongue of the square, as shown Figure 34-12D, and is used for laying out a figure with eight sides on a square piece of timber. On this scale, the graduations are represented by 65 dots located 5/24 in. apart.

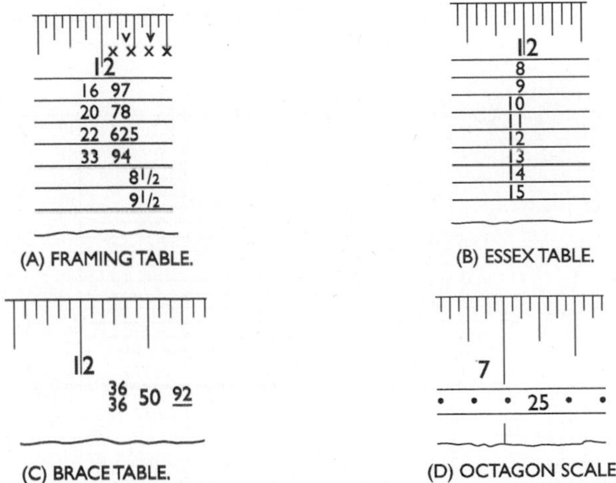

Figure 34-12 Typical framing-square markings.

Hundredths Scale

This scale is found on the tongue of the square; by means of a divider, decimals of an inch may be obtained. It is used particularly in reference to brace measure.

Inch Scales

On both the body and the tongue, there are (along the edges) scales of inches graduated in 1⁄32, 1⁄16, 1⁄12, 1⁄10, 1⁄8, and 1⁄4 in. Various combinations of graduations can be obtained according to the type of square. These scales are used in measuring and laying out work to precise dimensions.

Diagonal Scale

Many framing squares are provided with what is known as a diagonal scale, as shown in Figure 34-13; one division (ABCD) of this scale is shown enlarged for clearness in Figure 34-14. The object of the diagonal scale is to give minute measurements without having the graduations close together where they would be hard to read. In construction of the scale (Figure 34-14), the short distance AB is 1⁄10 in. Obviously, to divide AB into ten equal parts would bring the divisions so close together that the scale would be difficult to read. Therefore, if AB is divided into ten parts, and the diagonal BD is drawn, the intercepts 1a, 2b, 3c, etc., drawn through 1, 2, 3, etc., parallel to AB, will divide AB into 1⁄10, 2⁄10, 3⁄10, etc., in. Thus, if a distance of 3⁄10 AB is required, it may be picked off by placing one leg of the divider at 3 and the other leg at c, thereby producing 3c = 3⁄10 AB.

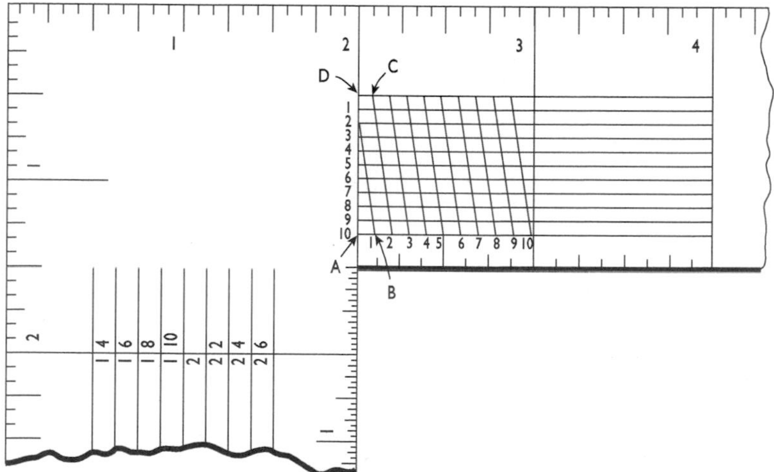

Figure 34-13 The diagonal scale on a framing square is used to mark off hundredths of an inch with dividers.

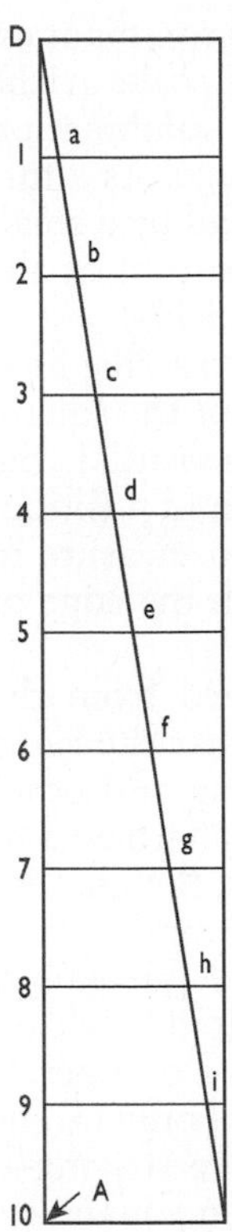

Figure 34-14 Section ABCD of Figure 34-13, enlarged to illustrate the principle of the diagonal scale.

Because of the importance of the framing square and the many problems to be solved with it, applications of the square are given at length in a later chapter.

Combination Square

This tool (Figure 34-15), as its name indicates, can be used for the same purposes as an ordinary try square, but it differs from the try square in that the head can be made to slide along the blade and clamp at any desired place; combined with the square, it is also a level and a miter. The sliding of the head is accomplished by means of a central groove in which a guide travels in the head of the square. This permits the scale to be pulled out and used simply as a rule. It is frequently desired to vary the length of the try-square blade; this is readily accomplished with the combination square. It is also convenient to square a piece of wood with a surface and at the same time tell whether one or the other is level, or plumb. The spirit level in the head of the square permits this to be done without the use of a separate level. The head of the square may also be used as a simple level.

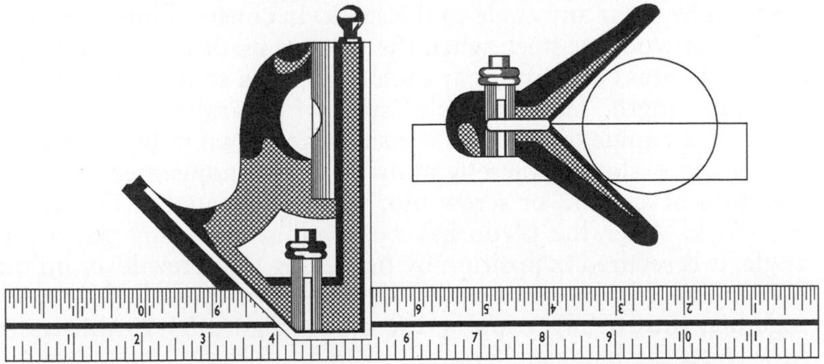

Figure 34-15 A typical combination square with a grooved blade, level, and centering attachments.

Because the scale may be moved in the head, the combination square makes a good marking gauge by setting the scale at the proper position and clamping it there. The entire combination square may then be slid along as with an ordinary gauge. As a further convenience, a scriber is held frictionally in the head by a small brass bushing. The scriber head projects from the bottom of the square stock in a convenient place to be withdrawn quickly.

In laying out, the combination square may be used to scribe lines at miter angles as well as at right angles, as one edge of the square head is at 45°. Where micrometer accuracy is not essential, the blade of the combination square may be set at any desired position, and the square may then be used as a depth gauge to measure in mortises, or the end of the scale may be set flush with the edge of the square and used as a height gauge.

The head may be unclamped and entirely removed from the scale, and a center head can then be substituted so that the same tool can quickly be used to find the centers of shafting and other cylindrical pieces. In the best construction, the blade is hardened to prevent the corner from wearing round and destroying the graduations, thus keeping the scale accurate at all times. This combination square, combining as it does a rule, square, miter, depth gauge, height gauge, level, and center head, permits more rapid work on the part of the carpenter, saves littering the bench with a number of tools, each of which is necessary but which may be used only rarely, and tends toward the goal for which all carpenters are striving—greater efficiency. Some of the uses for the combination square are illustrated in Figure 34-16.

Sliding "T" Bevel

A bevel is virtually a try square with a sliding adjustable blade that can be set at any angle to the stock. In construction, the stock may be of wood or steel; when the stock is made of wood, it normally has brass mountings at each end and is sometimes concave along its length. The blade is of steel with parallel sides, and its end is at an angle of 45° with the sides, as shown in Figure 34-17. The blade is slotted, thereby allowing linear adjustment and the insertion of a pivot, or screw pin, which is located at the end of the stock. After the blade has been adjusted to any particular angle, it is secured in position by tightening the screw lever on the pivot; this action compresses the sides of the slotted stock together, thus firmly gripping the blade. Figure 34-18 illustrates how to set the blade angle.

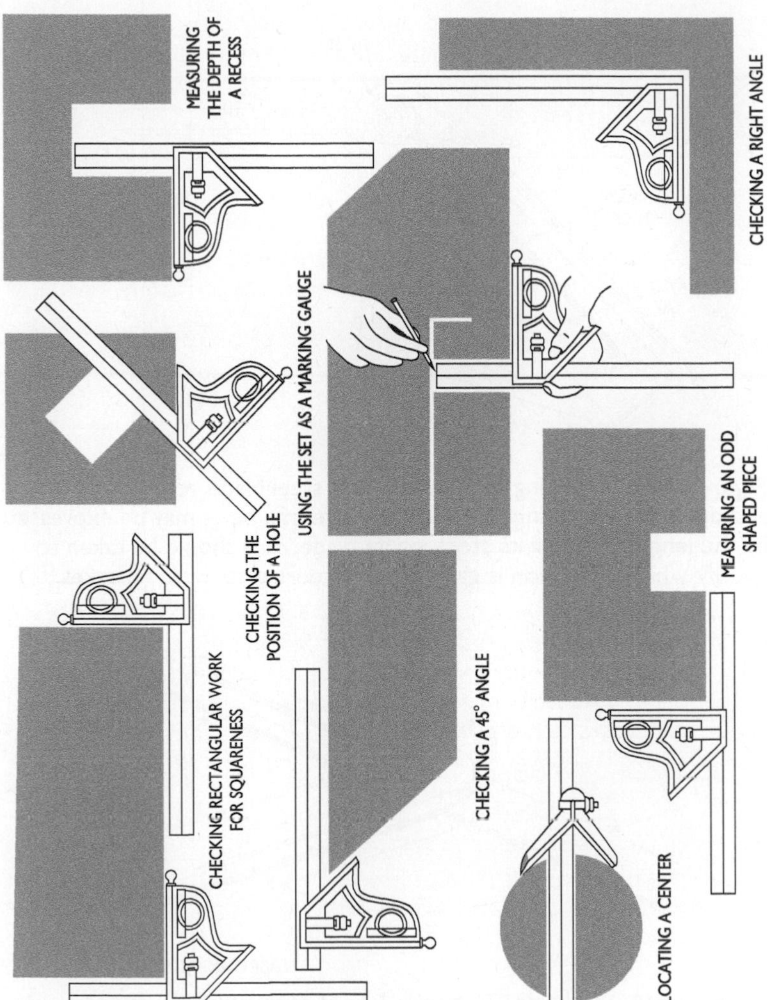

Figure 34-16 Some of the many uses of the combination square.

When selecting a bevel, care should be taken to see that the edges are parallel and that the pivot screw, when tightened, holds the blade firmly without bending it. In the line of special bevels, there are various modifications of the standard or ordinary form of the bevel just described. Two of these are shown in Figures 34-19 and 34-20.

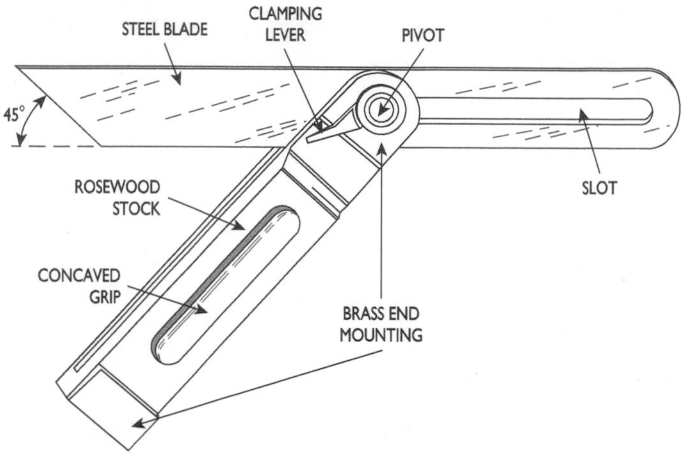

Figure 34-17 A sliding "T" bevel with a steel blade, rosewood stock, and brass end mountings. Because the size of a bevel may be expressed by the length of either its stock or its blade, care should be taken to specify which dimension is given when ordering, to avoid mistakes.

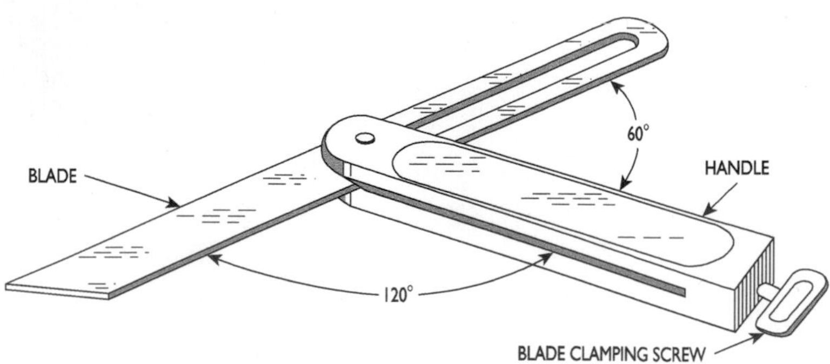

Figure 34-18 A sliding "T" bevel. A tool of this type is used to mark and test cutting angles.

Shooting Board

In its simplest form, a shooting board consists of two dressed boards, one fastened on the face of the other so that the lower board projects a few inches in advance of the upper board; the projection of the lower board forms a bed, or alignment guide, for the shooting plane. The object of this device is to hold the plane at a 90° angle to the side of the board so that the edge of the board after planing will be square, or at right angles, with the side of the board.

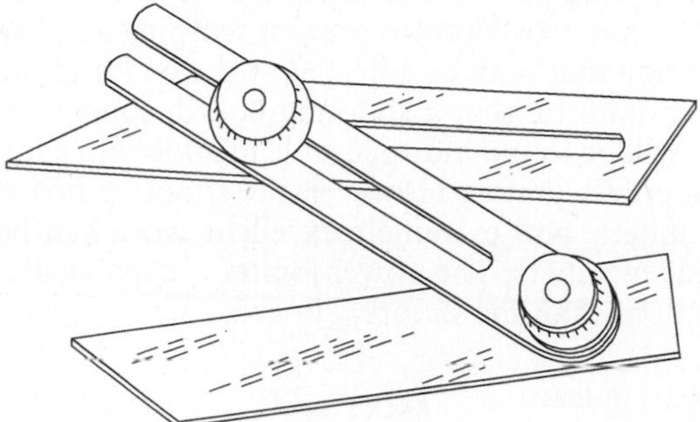

Figure 34-19 A double-slot steel bevel. As shown, both the stock and the blade are slotted, thus permitting adjustments that cannot be obtained with a common bevel.

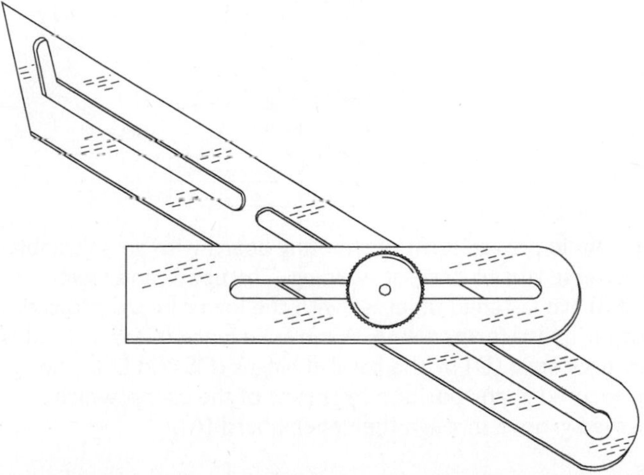

Figure 34-20 A typical combination bevel. One leg is pivoted to a straightedge, as shown, so that it can swing over the stock and be clamped at any angle. The slotted auxiliary blade may be slipped on the split blade and clamped at any desired angle to be used in conjunction with the stock for laying out work.

When "shooting an edge," the piece to be planed is held by the left hand against a stop with its edge projecting slightly over the stop. Then, the shooting plane, laid on its side, is moved steadily along its bed by the right hand, thus obtaining a planed edge at right angles with the side of the board being planed. An improved form of board, as shown in Figure 34-21, has provisions to prevent warping and has a fence with parallel motion that may be adjusted to the width of the board to be planed. It is doubtful that a shooting board, sometimes called a "chute board," will be found indispensable in a modern shop that is equipped with a power jointer. However, the shooting board can be set up to dress miters, and extremely excellent work can be done with relatively wide members. The power jointer is capable of a greater variety of operations than the shooting board.

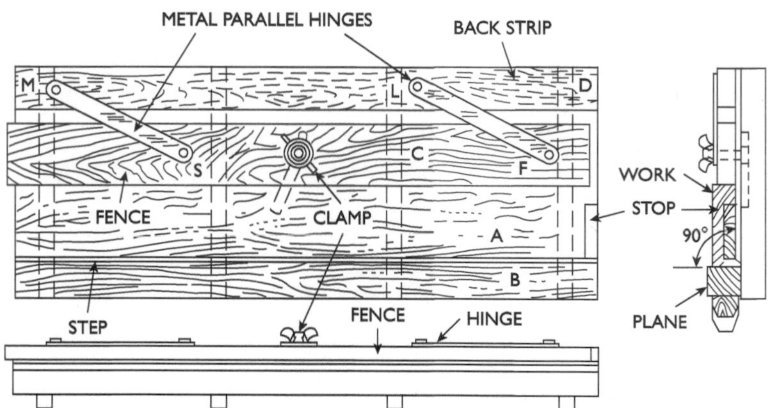

Figure 34-21 An improved form of shooting board with an adjustable fence and transverse ribs to prevent warping. The upper and lower boards (A and B) are fastened together with the lower board projecting so that the upper board forms a step. A movable fence (C) is pivoted to the stationary back strip (D) by the parallel hinges (MS and LF). The fence may be secured in any position by means of the clamp, which works in a curved groove through the upper board (A).

Miter Box

This device is used to guide the saw in cutting work to form miters, and it consists of a trough formed by a bottom and two side pieces of wood screwed together, with a saw cut through the sides at angles of 45° and 90°, as shown in Figure 34-22. Note that there are two 45° cuts; these are for cutting right and left miters.

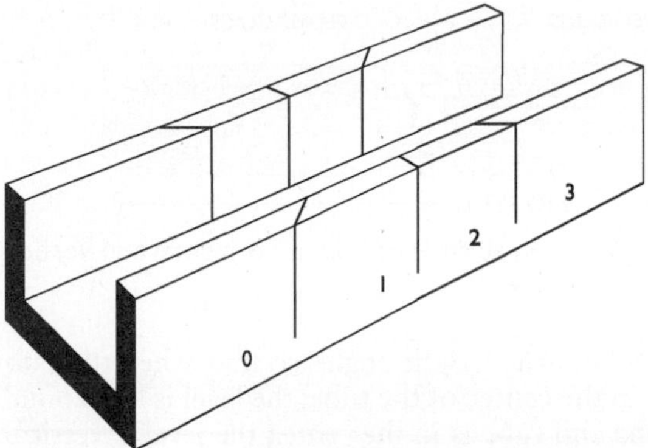

Figure 34-22 A typical wooden miter box. A nonadjustable 45° and 90° miter box, as shown, may easily be constructed from three pieces of hardwood screwed together as indicated.

Miter Shooting Board

For fine work, it is necessary to accurately dress the ends of the pieces cut by the saw in a miter box. This is done with the aid of a miter shooting board, as shown in Figure 34-23. It is essentially a simple shooting board with two 45° stops fastened on top of the upper board.

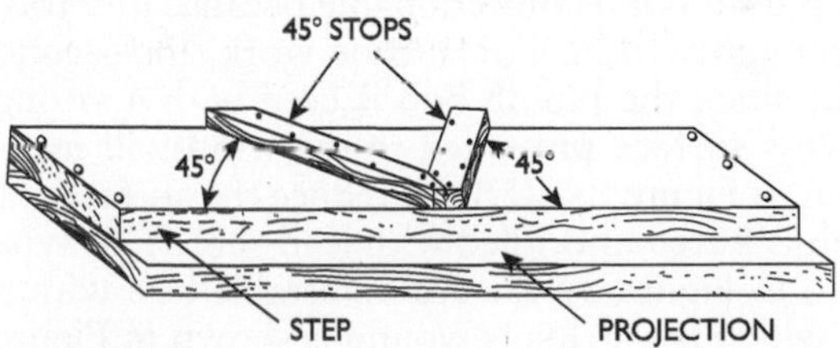

Figure 34-23 A miter shooting board for dressing miters after sawing in a miter box.

Level

This tool is used for both guiding and testing—to guide in bringing the work to a horizontal or vertical position and to test the accuracy of completed construction. It consists of a long rectangular body of wood or metal that is cut away on its side and near the end to receive glass tubes, which are almost entirely filled with a nonfreezing liquid that leaves a small bubble free to move as the level is moved. A typical level is shown in Figure 34-24.

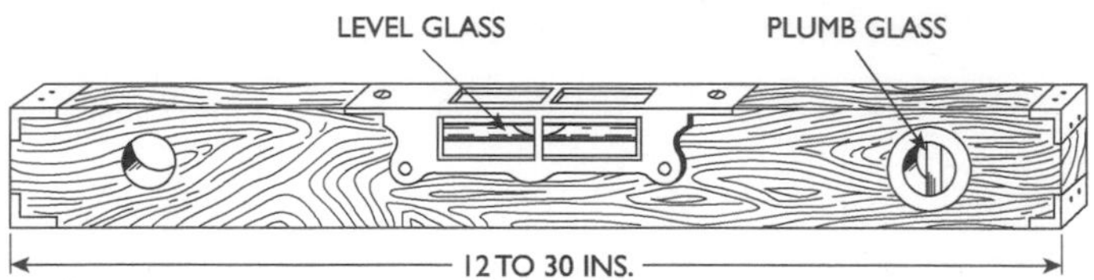

Figure 34-24 A typical wooden spirit level with a horizontal and vertical tube.

The side and end tubes are at right angles so that when the bubble of the side tube is in the center of the tube, the level is horizontal; when the bubble of the end tube is in the center, the level is vertical. By holding the level on a surface supposed to be horizontal or vertical, it may be ascertained whether the assumption is correct or not.

Plumb Bob

The word "plumb" means perpendicular to the plane of the horizon. Because the plane of the horizon is perpendicular to the direction of gravity at any given point, the force due to gravity is utilized to obtain a vertical line in the device known as a plumb bob.

This tool consists of a pointed weight attached to a string. When the weight is suspended by the string and allowed to come to rest, as in Figure 34-25A, the string will be plumb (vertical). The ordinary top-shaped solid plumb bob is objectionable because of a too-blunt point and not enough weight. For outside work, the second objection is important; when the plumb bob is used with a strong wind blowing, the excess surface presented to the wind will magnify the error, as shown in Figure 34-25B. To reduce the surface for a given weight, the bob is bored and filled with lead shot. This type of plumb bob is shown in Figure 34-26. An adjustable bob with a self-contained reel on which the string is wound is shown in Figure 34-27. The convenience of this arrangement is apparent. When using the plumb rule, the relative positions of the suspension string and vertical line, or axis, should be viewed at a point near the bob, because when out of plumb, the distance between the two is greater than at points higher up. When the tool is designed as in Figure 34-28, it should be read from the scale. Of course, on ordinary work, no such refinements are necessary.

Plumb Rule

This tool utilizes the principle of the plumb bob for vertical testing or guiding, instead of the spirit tube. It is essentially a plumb bob working in a rectangular case, as shown in Figure 34-28.

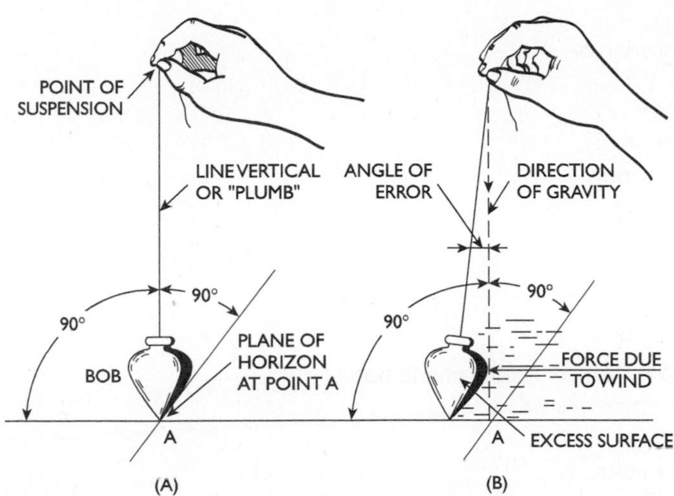

Figure 34-25 The solid plumb bob. A, using the plumb bob to define a vertical, or plumb, line. B, measurement error due to the force of the wind on the excess surface of the plumb bob.

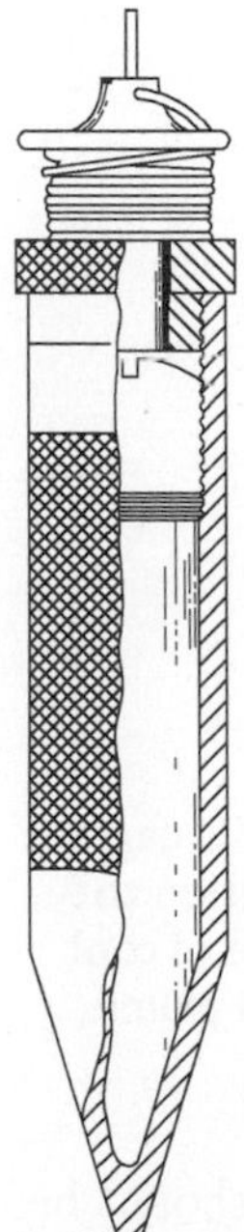

Figure 34-26 A shot-filled plumb bob. This type of plumb bob has the advantage of great weight in proportion to its surface area, and it is considerably better suited than the one shown in Figure 34-25 when working outside in a wind.

Laying Out

The term *laying out* here means the process of locating and fixing reference lines that define the position of the foundation and outside walls of a building to be erected.

Selection of Site

Preliminary to laying out (sometimes called *staking out*), it is important that the exact location of the building on the lot be properly selected. In this examination, it may be wise to dig a number of small, deep holes at various points, extending to a depth a little below the bottom of the basement.

The *ground water*, which is sometimes present near the surface of the earth, will (if the holes extend down to its level) appear in the bottom of the holes. This water nearly always stands at the same level in all the holes.

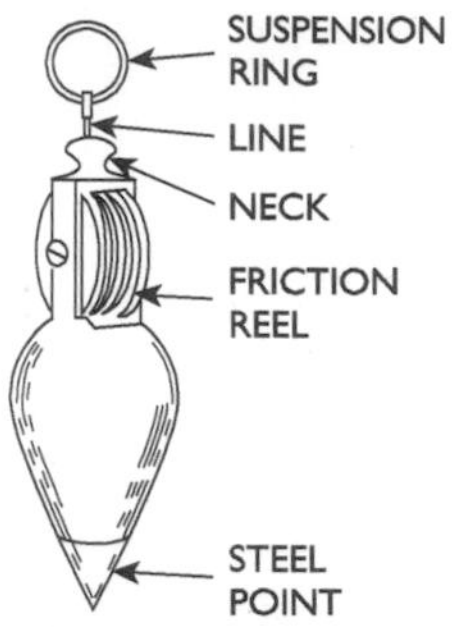

Figure 34-27 An adjustable plumb bob.

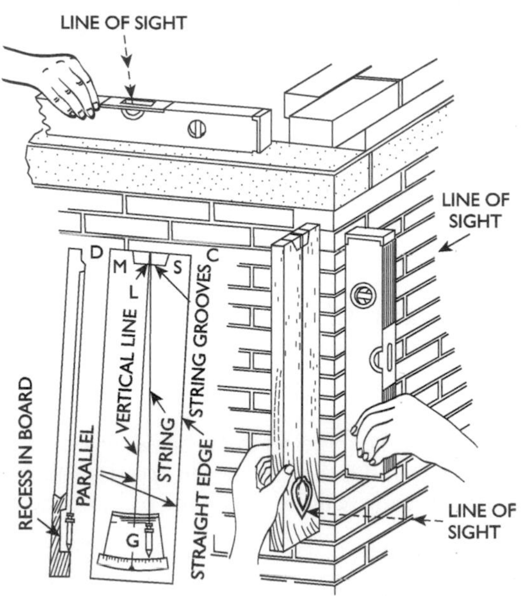

Figure 34-28 A typical plumb rule. To obtain the length of 1 degree on the scale, multiply distance LG (from the point of suspension to the tip of the bob) by a constant factor of 0.01745. This type of tool can be used in conjunction with a level to check the vertical plumb on a wall.

If possible, when selecting the site for the house, it should be located so that the bottom of the basement is above the level of the groundwater. This may mean locating the building at some elevated part of the lot or reducing the depth of excavation. The availability of storm and sanitary sewers, and their depth, should have been previously investigated. The distance of the building from the curb is usually stipulated in city building ordinances.

Staking Out

After the approximate location has been selected, the next step is to *lay out the building lines.* The position of all corners of the building must be marked in some way so that when the excavation is begun, the workmen may know the exact boundaries of the basement walls. There are several methods of laying out these lines:

1. With layout square.
2. With surveyor's instrument.
3. By method of diagonals.

The first method will do for small jobs, but the efficient carpenter or contractor will be equipped with a level or transit with which the lines may be laid out with precision and more convenience than by the first method.

The Lines

Several lines must be located at some time during construction, and they should be carefully distinguished. They are:

1. The *line of excavation,* which is the outside line.
2. The *face line* of the basement wall inside the excavation line, and in the case of masonry building.
3. The *ashlar line,* which indicates the outside of the brick or stone walls.

In the case of a wooden structure, only the two outside lines need to be located, and often only the line of the excavation is determined at the outset.

Laying Out with Layout Square

Start the layout from any point on the ground at which it is desired to place one corner of the building. By driving a stake at this point far enough to be outside the excavation line (about 3 ft), erect a batter board as shown in Figure 34-29. The batter board *must be leveled, and must be at the same elevation.*

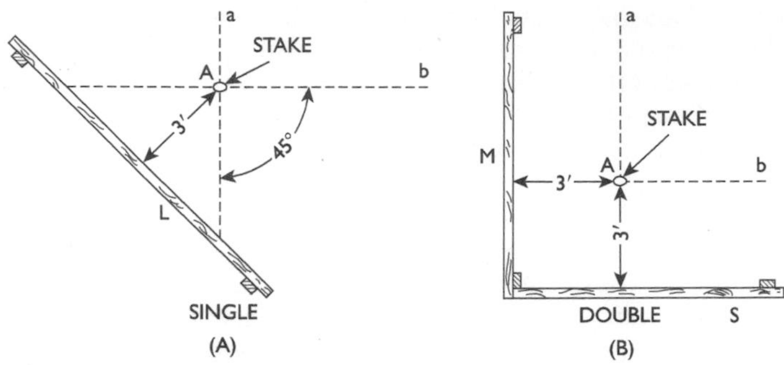

Figure 34-29 Single and double batter boards. After locating a corner of the proposed building, drive down stake A and erect either a single batter board, as shown in Figure 34-29A, or a double batter board, as shown in Figure 34-29B. Note the general direction of the building lines Aa and Ab, and locate the single board L or double board MS three feet back from the stake.

Suppose that the building is of rectangular shape and that the front of the building is to be parallel with the street. Starting at the stake A, Figure 34-29 (using the double batter board), lay out a line parallel with the street as in Figure 34-30. Drive a stake, B, at a distance equal to the length of the front of the building. The exact location of the ends of the line may be indicated by a nail driven into each stake. Because the building is of rectangular shape, lines must be laid out at A and B at 90° or right angles to the line AB.

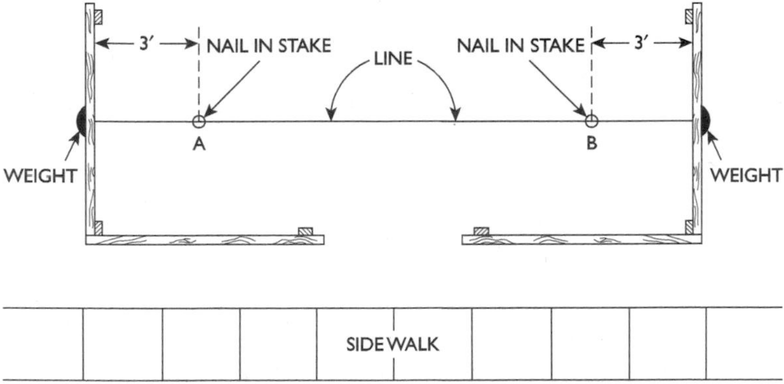

Figure 34-30 Locating the front of the building. Points A and B are two stakes with nails driven in each. The distance between these two nails is the length of the front side of the building.

The right angle is obtained by means of a large square constructed as in Figure 34-31. The figure shows the right way to make the square by having boards A and B be the same length. It must be evident that if A and B are cut off where they are joined to C, making B shorter than A, the extra length of A does not add to the precision, as the latter depends upon the length of the shortest side. This square is shown in Figure 34-32 at corner A.

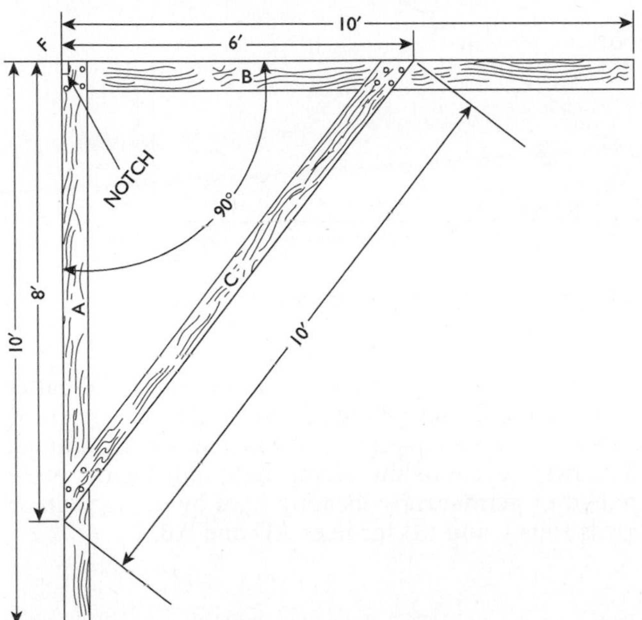

Figure 34-31 To make a large layout square, use 1 × 6 boards (A, B, and C, 10 feet long) and square off ends of A and B with precision. Mark off with care eight feet on board A, six feet on board B, and 10 feet on board C. Place board A on top of board B, and board C on top of both A and B, and fasten with nails. If this work is done with precision, an accurate right angle will be obtained at point F. The square should be notched at point F to permit it to rest under the layout lines.

In using the square, the legs and lines are brought into alignment by means of a plumb bob. Having placed one leg under line AB, adjust line AD on the batter boards until it is directly over the nail in stake A and the other leg of the square. When the four lines AB, BC, CD, and DA are located and the work checked by measuring diagonals AC and BD (which must be equal), the lines are located

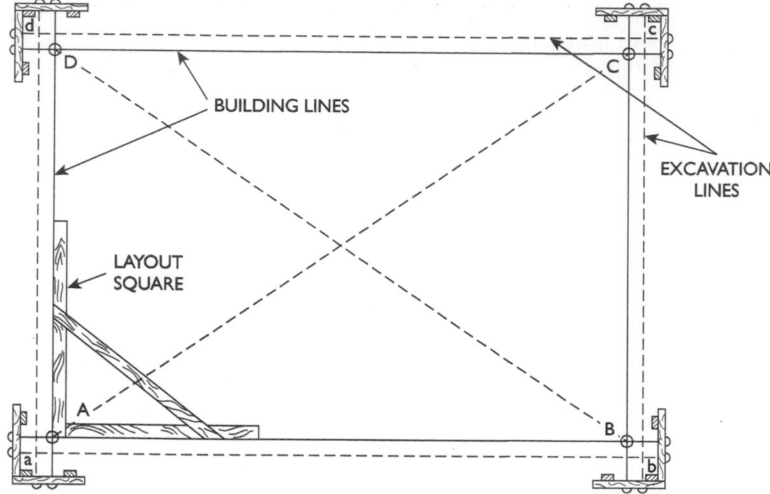

Figure 34-32 Layout for a building using batter boards.

permanently by placing them in vertical slits sawed in the batter boards. Stakes B, C, and D may now be driven at the corners, using a plumb bob to locate the intersections of the lines on the ground. Figure 34-33 shows the use of the plumb bob, and Figure 34-34 shows the method of permanently locating lines by sawing slits in the batter boards (slits L and M) for lines AD and AB.

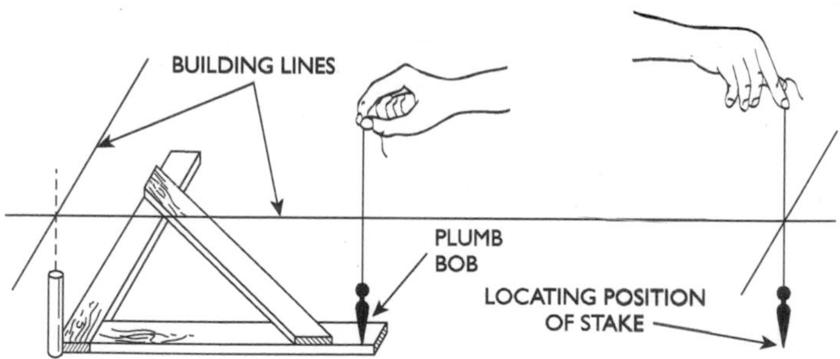

Figure 34-33 Method of bringing lines and layout square into alignment, and location of points for corner stakes by means of a plumb bob.

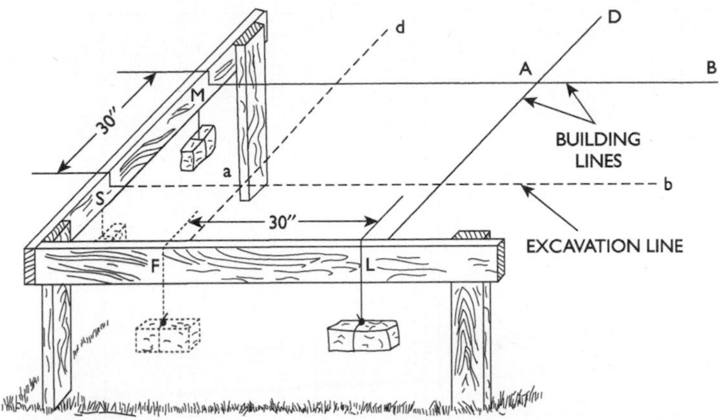

Figure 34-34 Permanent location of layout lines is made by cutting in the batter board. Slits L and M locate the building lines. Approximately 30 inches away are lines F and S, which are excavation lines.

After permanently locating the four building lines, mark off on the batter boards the distance the excavation lines are from the building lines, and cut slits at these points, as in Figure 34-34. In excavating, the lines are placed in the outer or excavation slits and may be later moved into the other slits as the work progresses. These lines are held taut by means of weights, as shown.

Laying Out with Transit Instruments

A transit may be used, and as this is an instrument of precision, the work of laying out is more accurate than when the layout square is employed. In Figure 34-35, let ABCD be a building already erected, and at a distance from this (at right angle), building GHJK will be erected. Level up the instrument at point E, making AE the distance the new building will be from points A and B. Make points B and F the same distance apart as points A and E. At this point, drive a stake in the ground at point G, making FG the required distance between the two buildings. Place point H on the same line as point G, making the distance between the two points as required.

Place the transit over point G, and level it up. Focus the transit telescope on point E or F and lock into position. Turn the horizontal circle on the transit until one of the zeros exactly coincides with the vernier zero. Loosen the clamp screw and turn the telescope and vernier 90°. This will locate point K, which will be at the desired distance from point G. For detailed operation of the transit, see the

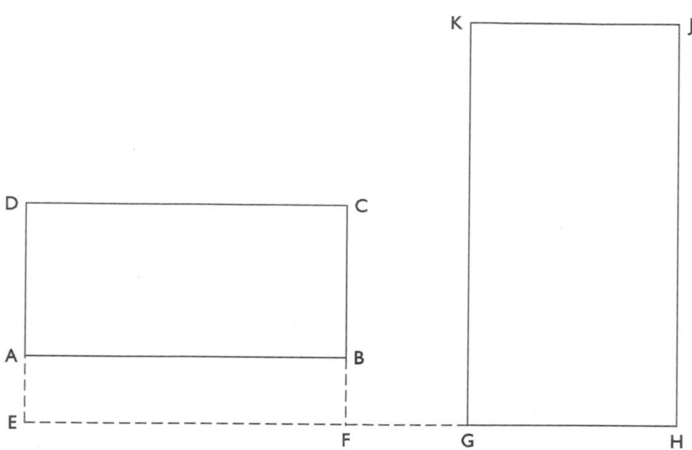

Figure 34-35 Illustrating method of laying out with transit instrument.

manufacturer's information. The level may be used in setting floor timbers, in aligning shafting, and locating drains. Figure 34-36 illustrates a builder's transit, which is used in various earth excavations.

Method of Diagonals

All that is needed in this method is a line, stakes, and a steel tape measure. Here, the right angle between the lines at the corners of a rectangular building is found by calculating the length of the diagonal that forms the hypotenuse of a right-angle triangle. By applying the following rule, the length of the diagonal (hypotenuse) is found.

Rule

The length of the hypotenuse of a right-angle triangle is equal to the square root of the sum of the squares of each leg.

Thus, in a right-angle triangle ABC, of which AC is the hypotenuse,

$$AC = \sqrt{AB^2 + BC^2}. \tag{1}$$

Suppose, in Figure 34-37, ABCD represents the sides of a building to be constructed, and it is required to lay out these lines to the dimensions given. Substitute the values given in equation (1); thus,

$$AC = \sqrt{30^2 + 40^2} = \sqrt{900 + 1600} = \sqrt{2500} = 50$$

To lay out the rectangle of Figure 34-37, first locate the 40-ft line AB with stake pins. Attach the line for the second side to B, and measure off on this line the distance BC (30 ft), point C being indicated

Figure 34-36 A builders transit, which is used in various earth excavations.

by a knot. This distance must be accurately measured with the line at the *same tension* as in A and B.

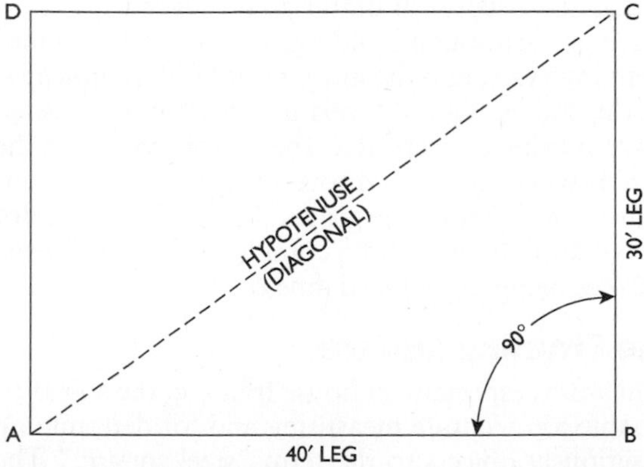

Figure 34-37 Illustrating how to find the length of the diagonal in laying out lines of a rectangular building by the method of diagonals.

With the end of a steel tape fastened to stake pin A, adjust the position of the tape and line BC until the 50-foot division on the tape coincides with point C on the line. ABC will then be a right angle, and point C will be properly located.

The lines for the other two sides of the rectangle are laid out in a similar manner. After thus obtaining the positions for the corner stake pins, erect the batter boards and permanent lines as shown in Figure 34-34. A simple procedure may be used in laying out the foundations for a small rectangular building. Be sure that the opposite sides are equal, and then measure *both* diagonals. No matter what this distance may be, they will be equal if the building is square. No calculations are necessary, and the method is precise.

Points on Laying Out

In most localities, it is customary for the carpenter to be present and to assist the mason in laying out the foundations. For ordinary residence work, a surveyor or the city engineer is employed to locate the lot lines. Once these lines are established, the builder is able to locate the building lines by measurement.

A properly prepared set of plans will show both the present contour of the ground upon which the building is to be erected and the new grade line that is to be established after the building is completed. The most convenient method of determining old grade lines and of establishing new ones is by means of a transit, or with a Y level and a rod. Both instruments work on the same principle in grade work. As a rule, a masonry contractor has his own Y level and uses it freely as the wall is constructed, especially where levels are to be maintained as the courses of material are placed.

In locating the earth grade about a building, stakes are driven into the ground at frequent intervals, and the amount of "fill" is indicated by the heights of these stakes. Grade levels are usually established after the builders have finished, except that the mason will have the grade indicated for him where the wall above the grade is to be finished differently from the wall below the grade. When a Y level is not available, a 12- or 14-ft straightedge and a common carpenter's level may be used, with stakes being driven to define the level.

How to Use the Framing Square

On most construction work, especially in house framing, the so-called steel square is invaluable for accurate measuring and for determining angles. The author seriously objects to the term "steel square." The proper name is *framing square*, because the square with its markings was designed especially for marking timber framing. However, the wrong name has become so firmly rooted that it will have to be put up with.

The square with its various scales and tables has already been explained. The purpose at this point is to explain these markings in more detail and also to explain their application by examples showing actual uses of the square. The following names are commonly used to identify the different portions of the square and should be noted and remembered.

Body. The long, wide member.

Face. The sides visible (both body and tongue) when the square is held by the tongue in the right hand with the body pointing to the left (see Figure 34-38).

Tongue. The short, narrow member.

Back. The sides visible (both body and tongue) when the square is held by the tongue in the left hand with the body pointing to the right (see Figure 34-38).

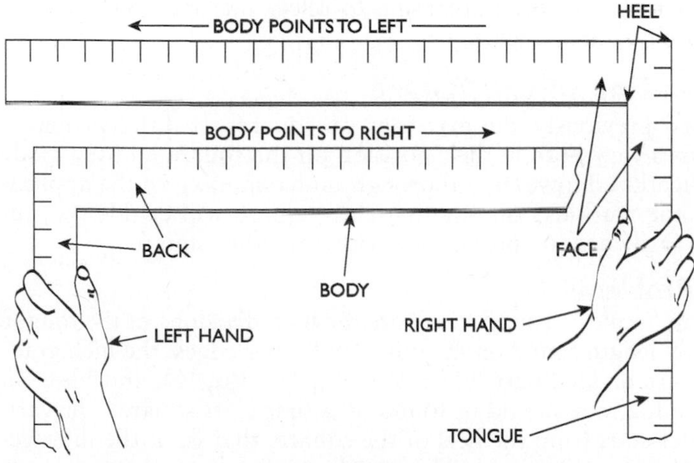

Figure 34-38 The face and back sides of a framing square. The body of the square is sometimes called the blade.

The square most generally employed has an 18-in. tongue and a 24-in. body. The body is 2 in. wide, and the tongue is 1½ in. wide, $\frac{3}{16}$ in. thick at the heel or corner for strength, diminishing, for lightness, to the two extremities to approximately $\frac{3}{32}$ in. The various markings on squares are of two kinds:

1. Scales, or graduations
2. Tables

When buying a square, it is advisable to get one with all the markings rather than a cheap square on which the manufacturer has omitted some of the scales and tables. The following comparison illustrates the difference between a cheap and a complete square.

	Tables	*Graduations*
Cheap square	Rafter, Essex, Brace	1/16, 1/12, 1/8, 1/4
Complete markings	Rafter, Essex, Brace	1/100, 1/64, 1/32, 1/16
	Octagon, Polygon cuts	1/12, 1/10, 1/8, 1/4

The square with the complete markings will cost more, but in the purchase of tools, you should make it a rule to purchase only the finest made. The general arrangement of the markings on squares differs somewhat with different makes; it is advisable to examine the different makes before purchasing to select the one best suited to your specific requirements.

Application of the Square

As stated previously, the markings on squares of different makes sometimes vary both in their position on the square and the mode of application. However, a thorough understanding of the application of the markings on any first-class square will enable the student to easily acquire proficiency with any other square.

Scale Problems

The term "scales" is used to denote the inch divisions of the tongue and body length found on the outer and inner edges; the inch graduations are divided into 1/4, 1/8, 1/10, 1/12, 1/16, 1/32, 1/64, and 1/100. All these graduations should be found on a first-class square. The various scales start from the heel of the square, that is, at the intersection of the two outer, or two inner, edges.

A square with only the scale markings is adequate to solve many problems that arise when laying out carpentry work. An idea of its range of usefulness is shown in the following problems.

Problem 1

To describe a semicircle given the diameter.

Drive brads at the ends of the diameter LF, as shown in Figure 34-39. Place the outer edges of the square against the nails, and hold a lead pencil at the outer heel M; any semicircle can then be described, as indicated. This is the outer-heel method, but a better guide for the pencil is obtained by using the inner-heel method, which is also shown in the figure.

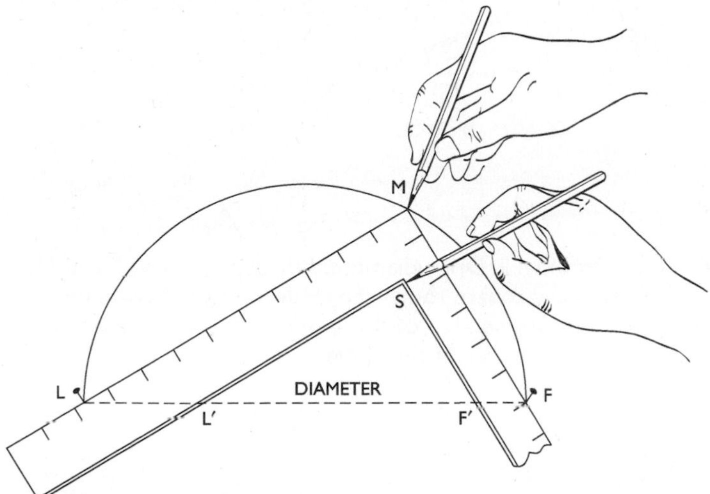

Figure 34-39 Problem 1. The outer-heel method is described in the text. For the inner-heel method, the pencil is held at S, and the distance L' F' should be taken to equal the diameter, with the inner edges of the square sliding on the brads.

Problem 2

To find the center of a circle.

Lay the square on the circle so that its outer heel lies in the circumference. Mark the intersections of the body and tongue with the circumference. The line that connects these two points is a diameter. Draw another diameter (obtained in the same way); the intersection of the diameters is the center of the circle, as shown in Figure 34-40.

Problem 3

To describe a circle through three points which are not in a straight line.

Join the three points with straight lines; bisect these lines, and, at the points of bisection, erect perpendiculars with the square. The intersection of these perpendiculars is the center from which a circle may be described through the three points, as shown in Figure 34-41.

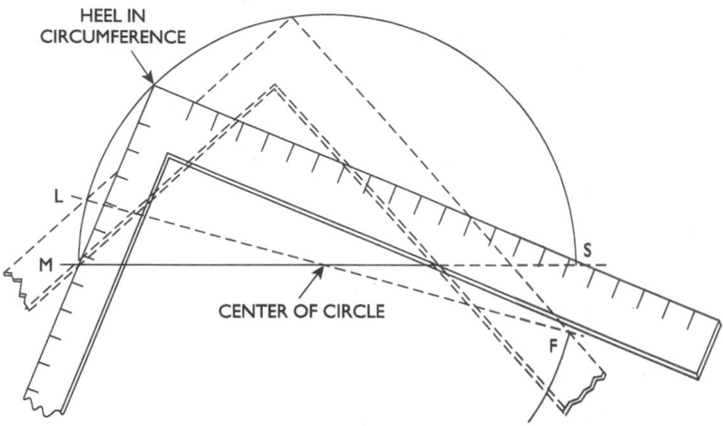

Figure 34-40 Problem 2. Draw diameters through points LF and MS, where the sides of the square touch the circle with the heel in the circumference. The intersection of these two lines is the center of these two lines in the center of the circle.

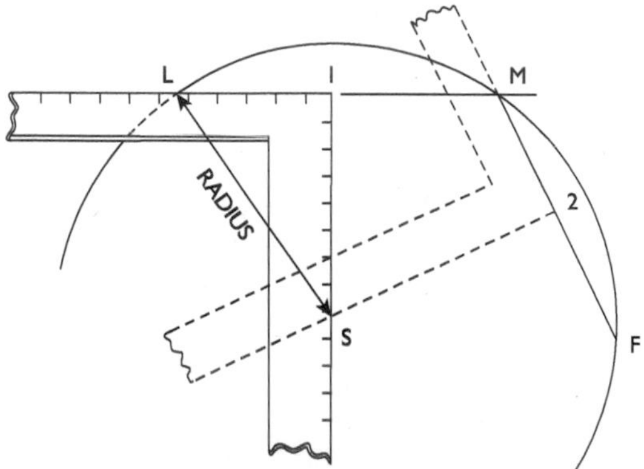

Figure 34-41 Problem 3. Let points L, M, and F be three points that are not in a straight line. Draw lines LM and MF, and bisect them at points 1 and 2, respectively. Apply the square with the heel at points 1 and 2, as shown; the intersection of the perpendicular lines thus obtained, point S, is the center of the circle. Lines LS, MS, and FS represent the radius of the circle, which may now be described through points L, M, and F.

Problem 4

To find the diameter of a circle whose area is equal to the sum of the areas of two given circles.

Lay off on the tongue of the square the diameter of one of the given circles, and on the body the diameter of the other circle. The

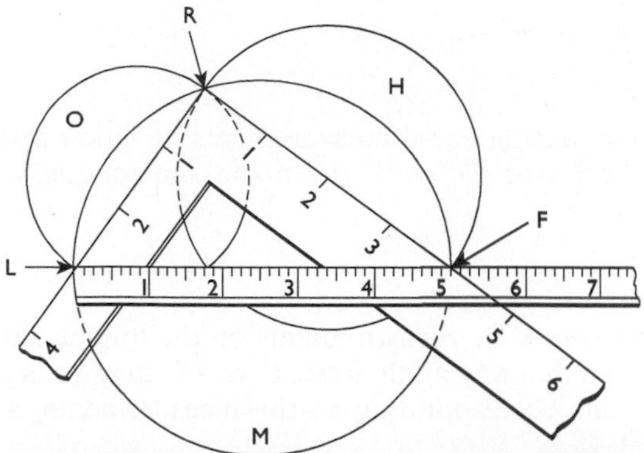

Figure 34-42 Problem 4. Let O and H be the two given circles, with their diameters LF and RF at right angles. Suppose the diameter of O is 3 in. and the diameter of H is 4 in. Points L and F, at these distances from the heel of the square, will be 5 in. apart, as measured with a 2-ft rule. This distance LF, or 5 in., is the diameter of the required circle. Proof: $(LF)^2 = (LR)^2 + (RF)^2$, or $25 = 9 + 16$.

distance between these points (measure across with a 2-ft rule) will be the diameter of the required circle, as shown in Figure 34-42.

Problem 5

To lay off angles of 30° and 60°.

Mark off 15 in. on a straight line, and lay the square so that the body touches one end of the line and the 7½-in. mark on the tongue is against the other end of the line, as shown in Figure 34-43. The tongue will then form an angle of 60° with the line, and the body will form an angle of 30° with the line.

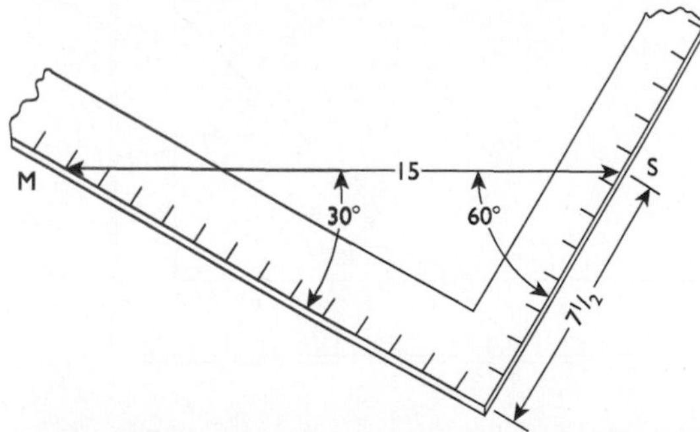

Figure 34-43 Problem 5. Draw line MS, 15 in. long. Place the square so that S touches the tongue 7½ in. from the heel and point M touches the body. The triangle thus formed will have an angle of 30° at M and an angle of 60° at S.

Problem 6

To lay off an angle of 45°.

The diagonal line connecting equal measurements on either arm of the square forms angles of 45° with the blade and tongue, as shown in Figure 34-44.

Problem 7

To lay off any angle.

Table 34-3 gives the values for measurements on the tongue and the body of the square so that any angle from 1° to 45° may be laid out by joining the points corresponding to the measurements, as explained in Figure 34-45.

Problem 8

To find the octagon of any size timber.

Place the body of a 24-in. square diagonally across the timber so that both extremities (ends) of the body touch opposite edges. Make marks at 7 in. and 17 in., as shown in Figure 34-46. Repeat this process at the other end, and draw lines through the pairs of marks. These lines show the portion of material that must be taken off the corners.

The side of an inscribed octagon can be obtained from the side of a given square by multiplying the side of the square by 5 and dividing the product by 12. The quotient will be the side of the octagon. This method is illustrated in Figure 34-47.

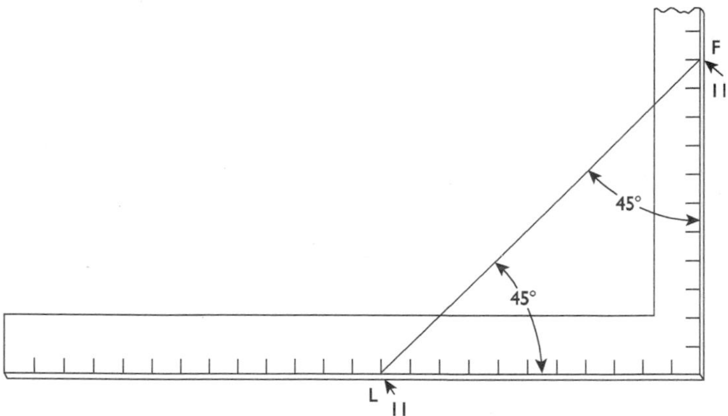

Figure 34-44 Problem 6. Take equal measurements L and F on the body and tongue of the square. The triangle thus formed will have an angle of 45° at L and at F.

Table 34-3 Angle Table for the Square

Angle	*Tongue*	*Body*
1	0.35	20.00
2	0.70	19.99
3	1.05	19.97
4	1.40	19.95
5	1.74	19.92
6	2.09	19.89
7	2.44	19.85
8	2.78	19.81
9	3.13	19.75
10	3.47	19.70
11	3.82	19.63
12	4.16	19.56
13	4.50	19.49
14	4.84	19.41
15	5.18	19.32
16	5.51	19.23
17	5.85	19.13
18	6.18	19.02
19	6.51	18.91
20	6.84	18.79
21	7.17	18.67
22	7.49	18.54
23	7.80	18.40
24	8.13	18.27
25	8.45	18.13
26	8.77	17.98
27	9.08	17.82
28	9.39	17.66
29	9.70	17.49
30	10.00	17.32
31	10.28	17.14
32	10.60	16.96
33	10.89	16.77
34	11.18	16.58
35	11.47	16.38
36	11.76	16.18

(Continued)

Angle	Tongue	Body
37	12.04	15.98
38	12.31	15.76
39	12.59	15.54
40	12.87	15.32
41	13.12	15.09
42	13.38	14.89
43	13.64	14.63
44	13.89	14.39
45	14.14	14.14

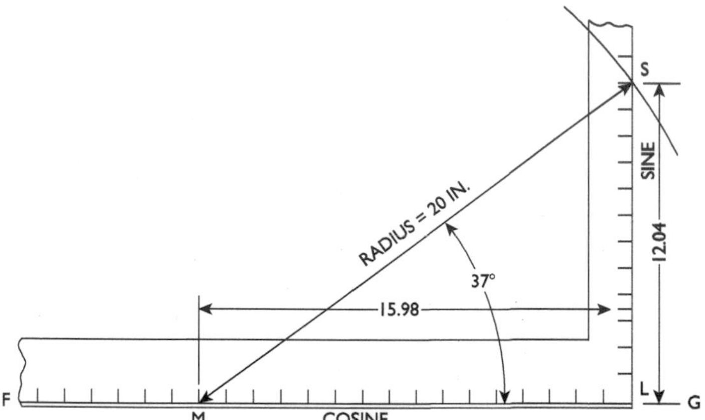

Figure 34-45 Problem 7. Let 37° be the required angle. Place the body of the square on line FG, and from Table 34-1 lay off LS (12.04) on the tongue and LM (15.98) on the body. Draw line MS; then, angle LMS will equal 37°. Line MS will be found to be equal to 20 in. for any angle, because the values given in Table 34-1 for LS and MS are natural sines and cosines multiplied by 20.

The side of a hexagon is equal to the radius of the circumscribing circle. If the side of a desired hexagon is given, arcs should be struck from each extremity at a radius equal to its length. The point where these arcs intersect is the center of the circumscribing circle, and having described it, it is sufficient to lay off chords on its circumference equal to the given side to complete the hexagon.

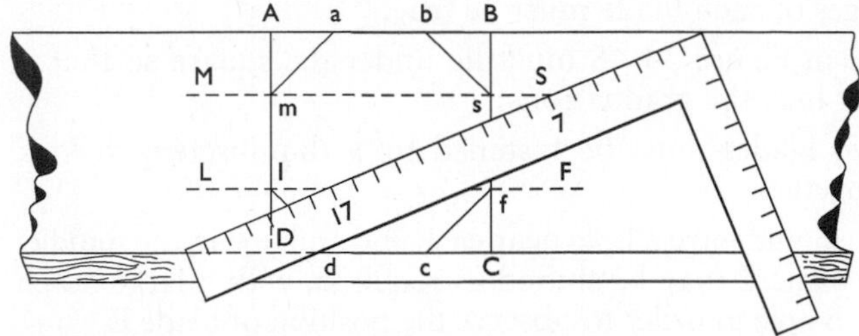

Figure 34-46 Problem 8. Lay out square ABCD. Place the body of the 24-in. square as shown, and draw parallel lines MS and LF through points 7 and 17. These lines intercept sides ml and sf of the octagon. To lay off side sb, place the square so that the tongue touches point s and the body touches I, with the heel touching line AB. The remaining sides are obtained in a similar manner.

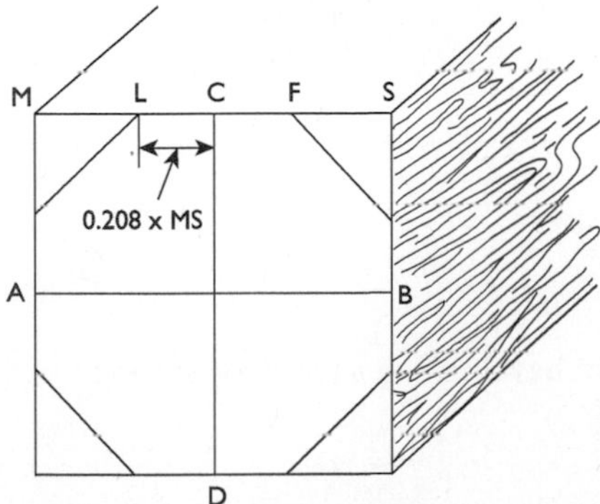

Figure 34-47 Problem 8 (second method). Let lines AB and CD be centerlines, and let line MS be one side of the square timber. Multiply the length of the side by 0.208; the product is half the side of the inscribed octagon. Therefore, lay off CF and CL, each 0.208 times side MS; LF is then one side of the octagon. Set dividers to distance CL, and lay off the other sides of the octagon from the centerlines to complete the octagon.

Square-and-Bevel Problems

By the application of a large bevel to the framing square, the combined tool becomes a calculating machine, and by its use, arithmetical processes are greatly simplified. The bevel is preferably made of steel blades. The following points should be observed in its construction:

- The edges of each blade must be true.
- Blade E in Figure 34-48 must lie under the square so that it does not hide the graduations.
- The two blades must be fastened by a thumbscrew to lock them together.
- Blade L should have a hole near each end and one in the middle, so that blade E may be shifted as required, with a large notch near each hole in order to observe the position of blade E.

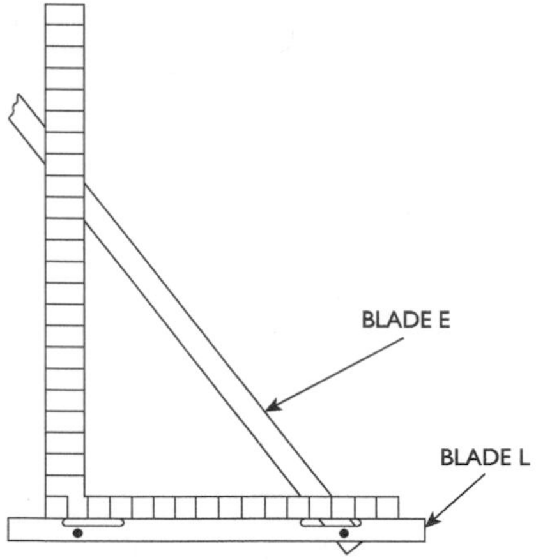

Figure 34-48 The application of a bevel to the square for solving square-and-bevel problems.

Problem 9

To find the diagonal of a square.

Set blade E to 10⅜ on the tongue and to 15 on the body. Assume an 8-in. square. Slide the bevel sideways along the tongue until blade E is against point 8. The other edge will touch 11 5/16 on the body; this is the required diagonal.

Problem 10

To find the circumference of a circle from its diameter.

Set the bevel blade to 7 on the tongue of the square and to 22 on the body. The reading on the body will be the circumference corresponding to the diameter at which E is set on the tongue. To reverse the process, use the same bevel, and read the required diameter from the tongue, the circumference being set on the body.

Problem 11

Given the diameter of a circle, find the side of a square of equal area.

Set the bevel blade to 10⅝ on the tongue and to 12 on the body. The diameter of the circle, on the body, will give the side of the equal square on the tongue. If the circumference is given instead of the diameter, set the bevel to 5½ on the tongue and to 19½ on the body, thereby finding the side of the square on the tongue.

Problem 12

Given the side of a square, find the diameter of a circle of equal area.

Using the same bevel as in Problem 11, blade E is set to the given side on the tongue of the square, and the required diameter is read off the body.

Problem 13

Given the diameter of the pitch circle of a gear wheel and the number of teeth, find the pitch.

Take the number of teeth, or a proportional part, on the body of the square and the diameter, or a similar proportional part, on the tongue, and set the bevel blade to those marks. Slide the bevel to 3.14 on the body, and the number given on the tongue multiplied by the proportional divisor will be the required pitch.

Problem 14

Given the pitch of the teeth and the diameter of the pitch circle in a gear wheel, find the number of teeth.

Set the bevel blade to the pitch on the tongue and to 3.14 on the body of the square. Move the bevel until it marks the diameter on the tongue. The number of teeth can then be read from the blade. If the diameter is too large for the tongue, divide it and the pitch into proportional parts, and multiply the number found by the same figure.

Problem 15

Given the side of a polygon, find the radius of the circumscribing circle.

Set the bevel to the pairs of numbers in Table 34-4, taking one-eighth or one-tenth inch as a unit. The bevel, when locked, is slid to the given length of the side, and the required length of the radius is read on the other leg of the square. For example, if a pentagon (five sides) must be laid out with a side of 6 inches, the bevel is set to the figures in column 5 with the lesser number set on the tongue—in this case, 74⁄8 = 9¼ on the tongue and 87⁄8 = 10⅞ on the body of the square. Slide the bevel to 6 on the body. The length of the radius, 5³⁄₃₂, will be read on the tongue.

Table 34-4 Inscribed Polygons

No. of Sides	*3*	*4*	*5*	*6*	*7*	*8*	*9*	*10*	*11*	*12*
Radius	56	70	74	60	60	98	22	89	80	85
Side	97	99	87	60	52	75	15	95	45	44

Problem 16

To divide the circumference of a circle into a given number of equal parts.

From the column marked Y in Table 34-5, take the number opposite the given number of parts. Multiply this number by the radius of the circle. The product will be the length of the cord to lay off on the circumference.

Problem 17

Given the length of a cord, find the radius of the circle.

This is the same as Problem 16, but the present form may be more expeditious for calculations. The method is useful for determining the diameter of gear wheels when the pitch and number of teeth have been given. Multiply the length of the cord, width of the side, or pitch of the tooth by the figures found corresponding to the number of parts in column Z of Table 34-5. The result is the radius of the desired circle.

Table Problems

The term "table" is used here to denote the various markings on the framing square with the exception of the scales already described. Because these tables relate mostly to problems encountered in cutting lumber for roof-frame work, it is first necessary to know something about roof construction so as to be familiar with the names of the various rafters and other parts. Figure 34-49 is a view of a roof frame showing the various members. In the figure, it will be noted that there is a plate at the bottom and a ridge timber at the top; these are the main members to which the rafters are fastened.

Main or Common Rafters

The following definitions relating to rafters should be carefully noted:

> The **rise** of a roof is the distance found by following a plumb line from a point on the central line of the top of the ridge to the level of the top of the plate.

Table 34-5 Chords or Equal Parts

No. of Parts		Y	Z
3	Triangle	1.732	.5773
4	Square	1.414	.7071
5	Pentagon	1.175	.8006
6	Hexagon	1.000	1.0000
7	Heptagon	.8677	1.1520
8	Octagon	.7653	1.3065
9	Nonagon	.6840	1.4619
10	Decagon	.6180	1.6184
11	Undecagon	.5634	1.7747
12	Duodecagon	.5176	1.9319
13	Tridecagon	.4782	2.0911
14	Tetradecagon	.4451	2.2242
15	Triangle	.4158	2.4050
16	Square	.3902	2.5628
17	Pentagon	.3675	2.7210
18	Hexagon	.3473	2.8793
19	Heptagon	.3292	3.0376
20	Octagon	.3129	3.1962
22	Nonagon	.2846	3.5137
24	Decagon	.2610	3.8307
25	Undecagon	.2506	3.9904
27	Duodecagon	.2322	4.3066
30	Tridecagon	.2090	4.7834
36	Tetradecagon	.1743	5.7368
40	Triangle	.1569	6.3728
45	Square	.1395	7.1678
50	Pentagon	.1256	7.9618
54	Hexagon	.1163	8.5984
60	Heptagon	.1047	9.5530
72	Octagon	.0872	11.462
80	Nonagon	.0785	12.733
90	Decagon	.0698	14.327
100	Undecagon	.0628	15.923
108	Duodecagon	.0582	17.182
120	Tridecagon	.0523	19.101
150	Tetradecagon	.0419	23.866

The **run** of a common rafter is the shortest horizontal distance from a plumb line through the center of the ridge to the outer edge of the plate.

The **rise per foot run** is the basis on which rafter tables on some squares are made. The term is self-defining. Other roof components are illustrated in Figure 34-50.

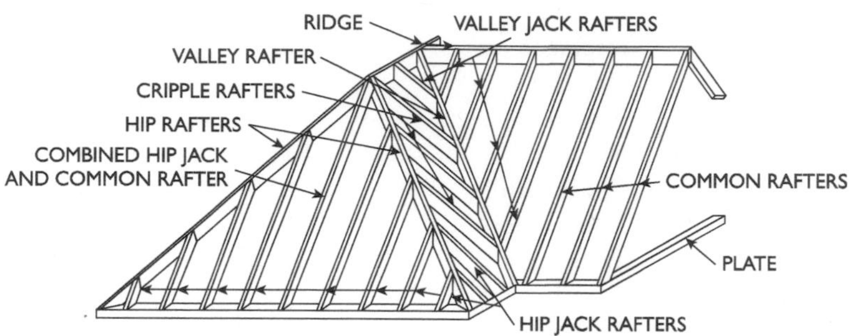

Figure 34-49 A typical roof frame, showing the ridge, the plate, and various types of rafters.

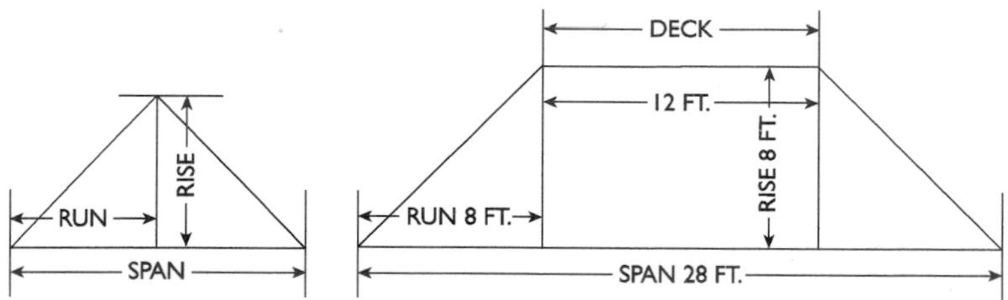

Figure 34-50 The terms rise, run, span, and deck are illustrated in two types of roofs. If the rafters rise to a deck instead of a ridge, subtract the width of the deck from the span. For example, assume the span is 28 feet and the deck is 12 ft; the difference is 16 feet, and the pitch is $8/(28 - 12) = \frac{1}{2}$.

To obtain the rise per foot run, multiply the rise by 12 and divide by the run; thus,

$$\text{rise per foot run} = \frac{\text{rise} \times 12}{\text{run}}$$

The factor 12 is used to obtain a value in inches, as the rise and run are normally given in feet.

Example

If the rise is 8 ft and the run is 8 ft, what is the rise per foot run?

$$\text{rise per foot run} = \frac{8 \times 12}{8} = 12 \text{ inches}$$

The rise per foot run is always the same for a given pitch and can be readily remembered for all ordinary pitches. Thus,

Pitch	½	⅓	¼	⅙
Rise per foot run (in.)	12	8	6	4

The pitch can be obtained if the rise and run are known, as shown in Figure 34-51, by dividing the rise by twice the run, or

$$\text{pitch} = \frac{\text{rise}}{2 \times \text{run}}$$

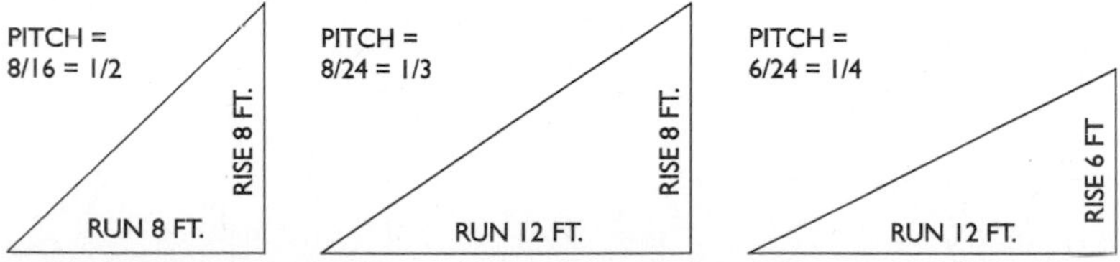

Figure 34-51 To obtain the pitch of any roof, divide the rise of the rafter by twice the run.

In roof construction, the rafter ends are cut with slants that rest against the ridge and the plate, as shown in Figure 34-52A. The slanting cut that rests against the ridge board is called the *plumb*, or *top*, cut, and the cut that rests on the plate is called the *seat*, or *heel*, cut.

The length of the common rafter is the length of a line from the outer edge of the plate to the top corner of the ridge board or, if there is no ridge board, from the outer edge of the plate to the vertical centerline of the building, as shown in Figure 34-52B. The run of the rafter, then, in the first case is one-half the width of the building less one-half the thickness of the ridge, if any; if there is no ridge board, the run is one-half the width of the building. Where there is a deck, the run of the rafters is one-half the width of the building less one-half the width of the deck. The question is sometimes asked, What constitutes half pitch, or full pitch? This nomenclature originated in England and is still used there, but American framers prefer to use the less confusing notation of so many inches of rise per foot of run. For those who prefer to use fractional designations, however, the pitch table (Table 34-6) is given. The notation, however, is obsolete for the most part and is rarely used by the modern framer.

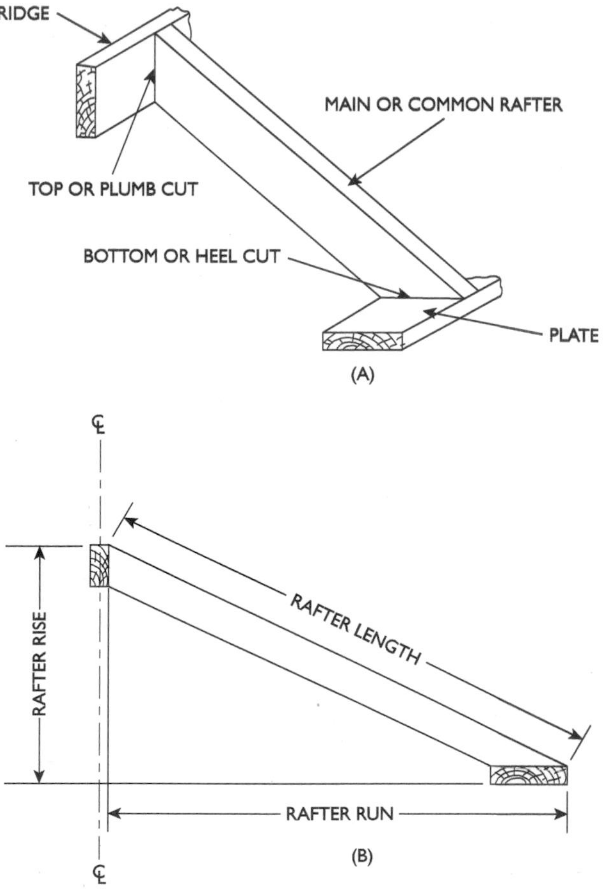

Figure 34-52 A portion of the roof frame, showing the top, or plumb, cut and the bottom, or heel, cut is illustrated in A. The length of a common rafter is shown in B.

Table 34-6 Pitch Table

Pitch	1	11/12	5/6	3/4	2/3	7/12	1/2	5/12	1/3	1/4	1/6	1/12
Run	12	12	12	12	12	12	12	12	12	12	12	12
Rise	24	22	20	18	16	14	12	10	8	6	4	2

Now, with a 24-in. square, draw diagonals connecting 12 on the tongue (corresponding to the run) to the value from Table 34-6 on the body (corresponding to the rise) to obtain the pitch angle for any combination of run and rise. This procedure is further illustrated in Figure 34-53.

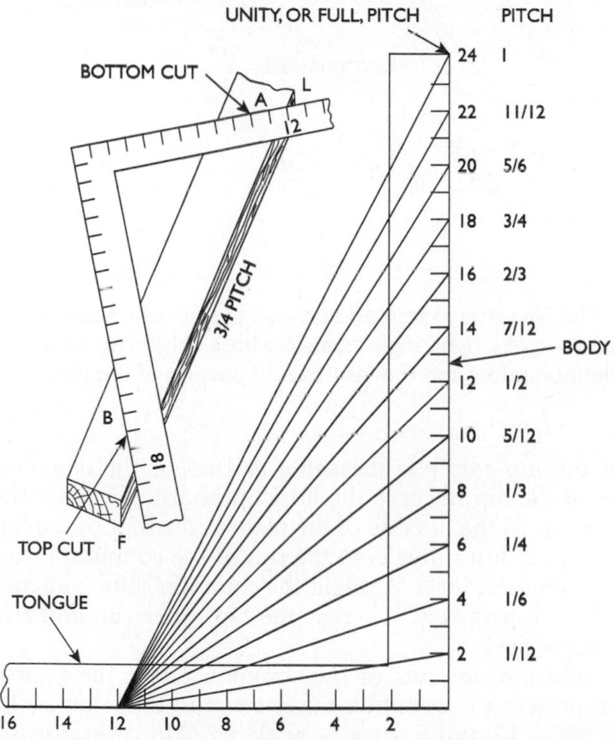

Figure 34-53 The application of the framing square for obtaining the various pitches given in Table 34-6.

Hip Rafters

The hip rafter represents the hypotenuse, or diagonal, of a right-angle triangle; one side is the common rafter, and the other side is the plate, or that part of the plate lying between the foot of the hip rafter and the foot of the adjoining common rafter, as shown in Figure 34-54.

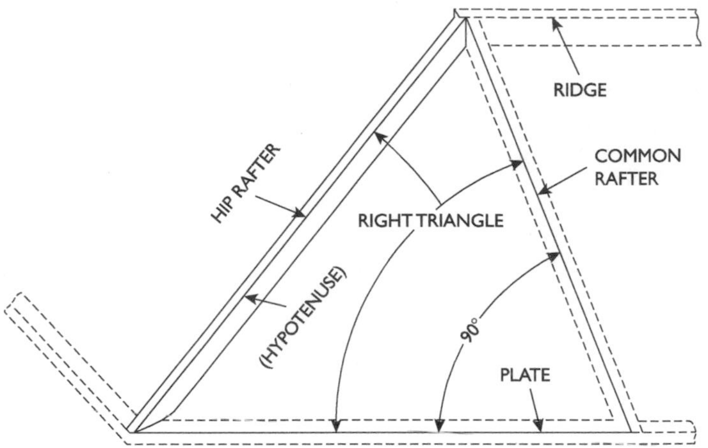

Figure 34-54 The hip rafter is framed between the plate and the ridge and is the hypotenuse of a right-angle triangle whose other two sides are the adjacent common rafter and the intercepted portion of the plate.

The rise of the hip rafter is the same as that of the common rafter. The run of the hip rafter is the horizontal distance from the plumb line of its rise to the outside of the plate at the foot of the hip rafter. This run of the hip rafter is to the run of the common rafter as 17 is to 12. Therefore, for a 1⁄6 pitch, the common rafter run and rise are 12 and 4, respectively, whereas the hip rafter run and rise are 17 and 4, respectively.

For the top and bottom cuts of the common rafter, the figures that are used represent the common rafter run and rise, that is, 12 and 4 for a 1⁄6 pitch, 12 and 6 for a 1⁄4 pitch, etc. However, for the top and bottom cuts of the hip rafter, use the figures 17 and 4, 17 and 6, etc., as the run and rise of the hip rafter. It must be remembered, however, that these figures will not be correct if the pitches on the two sides of the hip (or valley) are not the same.

Valley Rafters

The valley rafter is the hypotenuse of a right-angle triangle made by the common rafter with the ridge, as shown in Figure 34-55. This corresponds to the right-angle triangle made by the hip rafter with the common rafter and plate. Therefore, the rules for the lengths and cuts of valley rafters are the same as for hip rafters.

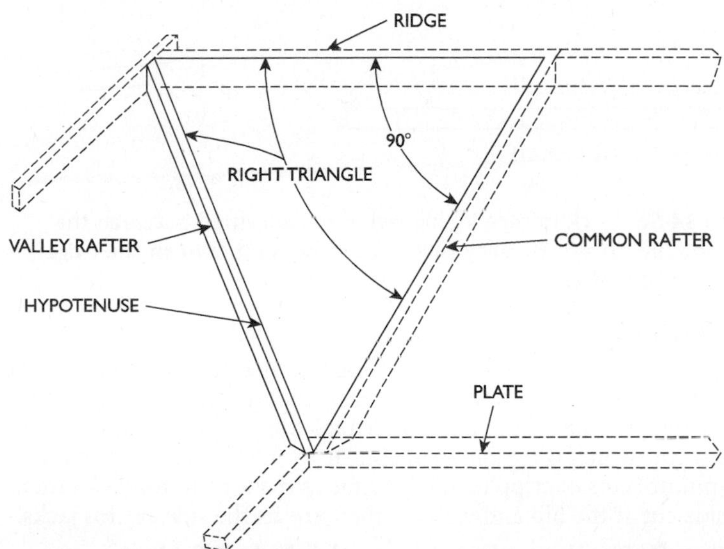

Figure 34-55 The valley rafter is framed between the plate and the ridge and is the hypotenuse of a right-angle triangle whose other two sides are the adjacent common rafter and the intercepted portion of the ridge portion.

Jack Rafters

Jack rafters are usually spaced either 16 or 24 in. apart, and, as they lie equally spaced against the hip or valley, the second jack rafter must be twice as long as the first, the third three times as long as the first, and so on, as shown in Figure 34-56. One reason for the 16- and 25-in. spacings on jack rafters is that laths are 48 in. long; therefore, the rafters must be 16 or 24 in. apart so that the lath may be conveniently nailed to it.

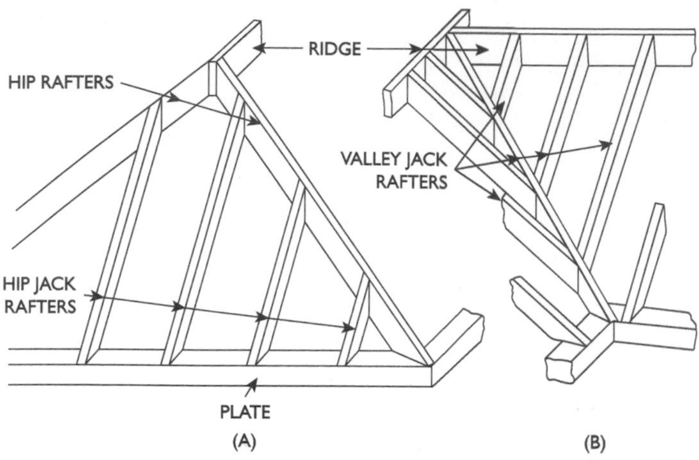

Figure 34-56 Jack rafters. A, hip jack rafters, framed between the plate and hip rafters; B, valley jack rafters, framed between the ridge and the valley rafter.

Cripple Rafters

A cripple rafter is a jack rafter that touches neither the plate nor the ridge; it extends from the valley rafter to the hip rafters. The cripple-rafter length is that of the jack rafter plus the length necessary for its bottom cut, which is a plumb cut similar to the top cut. Top and bottom (plumb) cuts of cripples are the same as the top cut for jack rafters. The side cut at the hip and valley is the same as the side cut for jacks.

Finding Rafter Lengths without the Aid of Tables

In the directions accompanying framing squares and in some books, frequent mention is made of the figures 12, 13, and 17. The reader is told for common rafters to "use figure 12 on the body and the rise of the roof on the tongue" and for hip or valley rafters to "use figure 17 on the body and the rise of the roof on the tongue," and no explanation of how these fixed numbers are obtained is provided. The intelligent workman should not be satisfied with knowing which number to use, but he should want to know *why* each particular number is used. This can be readily understood by referring to Figure 34-57. In this illustration, let ABCD be a square whose sides are 24 in. long, and let abcdefgL be an inscribed octagon. Each side of the octagon (ab, bc, etc.) measures 10 in.—that is, LF = one-half side = 5 in. and, by construction, FM = 12 in. Now, let FM represent the run of a common rafter. Then, LM will be the run of an octagon rafter, and DM will be the run of a hip or valley rafter. The values for the run of octagon and hip or valley rafters (LM and DM, respectively) are obtained as follows:

$$LM = \sqrt{(FM)^2 + (LF)^2} = \sqrt{(12)^2 + (5)^2} = 13$$
$$DM = \sqrt{(FM)^2 + (DF)^2} = \sqrt{(12)^2 + (12)^2} = 16.97$$

Example

What is the length of a common rafter that has a 20-ft run and a ⅜ pitch?

For a 10-ft run,

$$\text{the span} = 2 \times 10 = 20 \text{ feet}$$

with ⅜ pitch,

$$\text{rise} = \frac{3}{8} \times 20 = 7.5 \text{ feet}$$

$$\text{rise per foot run} = \frac{\text{rise} \times 12}{\text{run}} = \frac{7.5 \times 12}{10} = 9 \text{ inches}$$

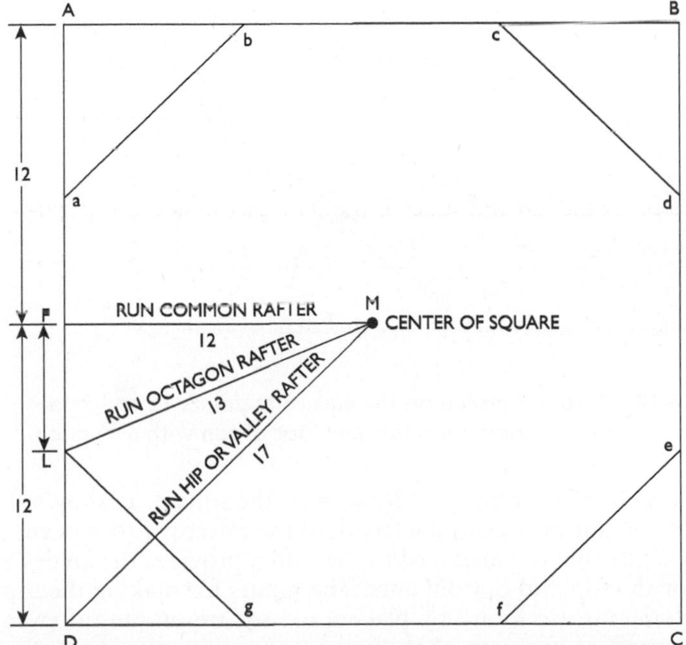

Figure 34-57 A square and an inscribed octagon are used to illustrate the method of obtaining and using points 12, 13, and 17 in the application of a framing square to determine the length of rafters without the aid of rafter tables.

On the body of the square shown in Figure 34-58, take 12 in. for 1 ft of run, and on the tongue, take 9 in. for the rise per foot of run. The diagonal, or distance between the points thus obtained, will be the length of the common rafter per foot of run with a ⅜ pitch. The distance FM measures 15 in., or by calculation,

$$FM = \sqrt{(12)^2 + (9)^2} = 15 \text{ inches}$$

Because the length of run is 10 ft,

$$\begin{aligned}\text{length of rafter} &= \text{length of run} \times \text{length per foot} \\ &= 10 \times 15/12 \\ &= \frac{150}{12} \\ &= 12.5 \text{ feet}\end{aligned}$$

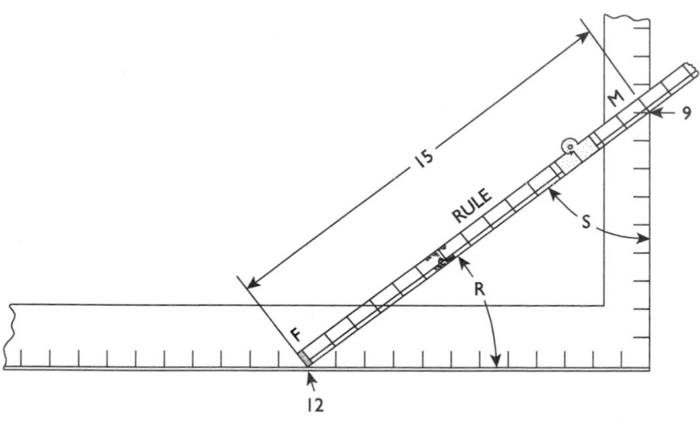

Figure 34-58 A rule is placed on the square at points 12 and 9 to obtain the length of a common rafter per foot of run with a ⅜ pitch.

The combination of figures 12 and 9 on the square, as shown in Figure 34-58, not only gives the length of the rafter per foot of run, but also, if the rule is considered as the rafter, provides the angles S and R for the top and bottom cuts. The points for making the top and bottom cuts are found by placing the square on the rafter so that a portion of one arm of the square represents the run and a portion of the other arm represents the rise. For the common rafter with a ⅜ pitch, these points are 12 and 9; the square is placed on the rafter as shown in Figure 34-59.

Example

What length must an octagon rafter be to join a common rafter that has a 10-ft run (as rafters MF and ML in Figure 34-57)?

From Figure 34-57, it is seen that the run per foot of an octagon rafter, as compared with a common rafter, is as 13 is to 12 and that the rise for a 13-in. run of an octagon rafter is the same as for the run of a 12-in. common rafter. Therefore, measure across from points 13 and 9 on the square, as MS in Figure 34-60, which gives the length (15¾ in.) of an octagon rafter per foot of run of a common rafter. The length multiplied by the run of a common rafter gives the length of an octagon rafter. Thus,

$$15\tfrac{3}{4} \times 10 = 157\tfrac{1}{2} \text{ inches} = 13 \text{ feet, } 1\tfrac{1}{2} \text{ inches}$$

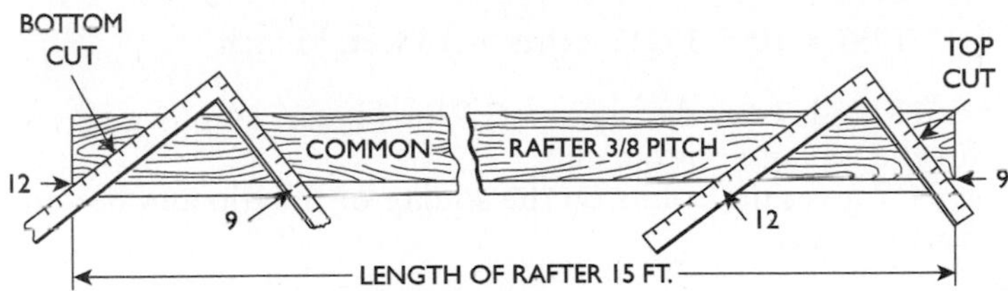

Figure 34-59 The square is placed on the rafter at points 12 and 9, as shown, thereby giving the proper angles for the bottom and top cuts.

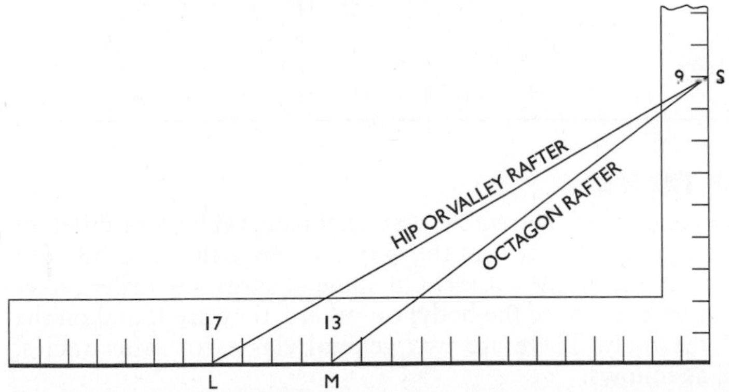

Figure 34-60 Measurements using the square for octagon and hip or valley rafters, illustrating the use of points 13 and 17. Line MS (13,9) is the octagon rafter length per foot of run of a common rafter with a ⅜ pitch; line LS (17,9) is the hip or valley rafter length per foot of run of a common rafter with a ⅜ pitch.

Points 13 and 9 on the square (MS in Figure 34-60) give the angles for the top and bottom cuts.

Example

What length must a hip or valley rafter be to join a common rafter that has a 10-ft run (as rafters MF and MD in Figure 34-57)?

Figure 34-57 shows that the run per foot of a hip or valley rafter, as compared with a common rafter, is as 17 is to 12 and that the rise per 17-in. run of a hip or valley rafter is the same as for a 12-in. run of a common rafter. Therefore, measure across from points 17 and 9 on the square, as LS in Figure 34-60; this gives the length (19¼ in.) of the hip or valley rafter per foot of common rafter. This length, multiplied by the run of a common rafter, gives the length of the hip or valley rafter; thus,

$$19\tfrac{1}{4} \times 10 = 192\tfrac{1}{2} \text{ inches} = 16 \text{ feet}, \tfrac{1}{2} \text{ inch}$$

Points 17 and 9 on the square (LS in Figure 34-60) give the angles for the top and bottom cuts.

Table 34-7 gives the points on the square of the top and bottom cuts of various rafters.

Table 34-7 Square Points for Top and Bottom Cuts

	Pitch	1	11/12	5/6	3/4	2/3	7/12	1/2	5/12	1/3	1/4	1/6	1/12
Tongue	Common							12					
	Octagon							13					
	Hip or valley							17					
	Body	24	22	20	18	16	14	12	10	8	6	4	2

Rafter Tables

The arrangement of these tables varies considerably with different makes of squares, not only in the way they are calculated but also in their positions on the square. On some squares, the rafter tables are found on the face of the body; on others, they are found on the back of the body. There are two general classes of rafter tables, grouped as follows:

1. Length of rafter per foot of run
2. Total length of rafter

Obviously, where the total length is given, there is no figuring to be done, but when the length is given per foot of run, the reading must be multiplied by the length of run to obtain the total length of the rafter. To illustrate these differences, directions for using several types of squares are given in the following sections. These differences relate to the common and hip or valley rafter tables.

Reading the Total Length of the Rafter

One popular type of square is selected as an example to show how rafter lengths may be read directly without any figuring. The rafter tables on this particular square occupy both sides of the body instead of being combined in one table; the common rafter table is found on the back, and the hip, valley, and jack rafter tables are located on the face.

Common Rafter Table

The common rafter table, Figure 34-61, includes the outside-edge graduations of the back of the square on both the body and the tongue; these graduations are in twelfths. The inch marks may represent inches or feet, and the twelfths marks may represent twelfths of an inch or twelfths of a foot (inches). The edge-graduation figures above the table represent the run of the rafter; under the proper figure on the line representing the pitch is found the rafter length required in the table. The pitch is represented by the figures at the left of the table under the word *pitch*. Thus,

12 Ft of Run							
Feet of rise	4	6	8	10	12	15	18
Pitch	1/6	1/4	1/3	5/12	1/2	5/8	3/4

The length of a common rafter given in the common rafter table is from the top center of the ridge board to the outer edge of the plate. In actual practice, deduct one-half the thickness of the ridge board, and add for any eave projection beyond the plate.

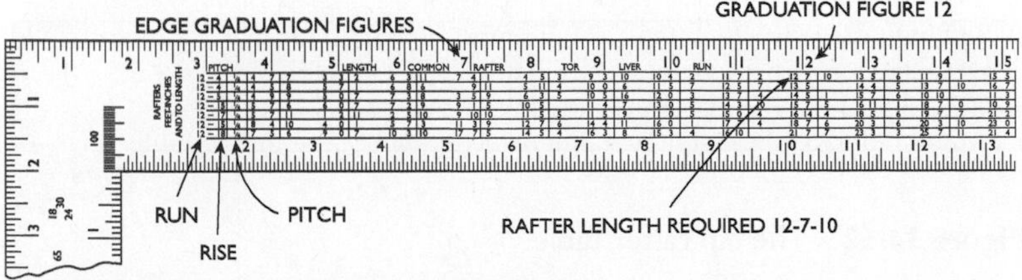

Figure 34-61 The common rafter table.

Example

Find the length of a common rafter for a roof with a $\frac{1}{6}$ pitch (rise = $\frac{1}{6}$ the width of the building) and a run of 12 ft (found in the common rafter table, Figure 34-61, the upper or $\frac{1}{16}$-pitch ruling).

Find the rafter length required under the graduation figure 12. This is found to be 12, 7, 10, which means 12 ft, 7$\frac{10}{12}$ in. If the run is 11 ft and the pitch is $\frac{1}{2}$ (the rise = $\frac{1}{2}$ the width of the building), then the rafter length will be 15, 6, 8, which means 15 ft 6$\frac{8}{12}$ in. If the run is 25 ft, add the rafter length for a run of 20 ft to the rafter length for a run of 5 ft. When the run is in inches, then in the rafter table read inches and twelfths instead of feet and inches. For instance, if, with a $\frac{1}{2}$ pitch, the run is 12 ft, 4 in., add the rafter length of 4 in. to that of 12 ft as follows:

For a run of 12 feet, the rafter length is	16 feet, 11$\frac{8}{12}$ inches.
For a run of 4 inches, the rafter length is	5$\frac{8}{12}$ inches.
	Total—17 feet, 5$\frac{4}{12}$ inches.

The run of 4 in. is found under the graduation 4 and is 5, 7, 11, which is approximately 5$\frac{8}{12}$ in. If the run was 4 ft, it would be read as 5 ft, 7$\frac{11}{12}$ in.

Hip Rafter Table

This table, as shown in Figure 34-62, is located on the face of the body and is used in the same manner as the table for common rafters explained above. In the hip rafter table, the outside-edge graduation figures represent the run of common rafters. The length of a rafter given in the table is from the top center of the ridge board to the outer edge of the plate. In actual practice, deduct one-half the thickness of the ridge board, and add for any eave projection beyond the plate. When using this table, find the figures on the line with the required pitch of the roof.

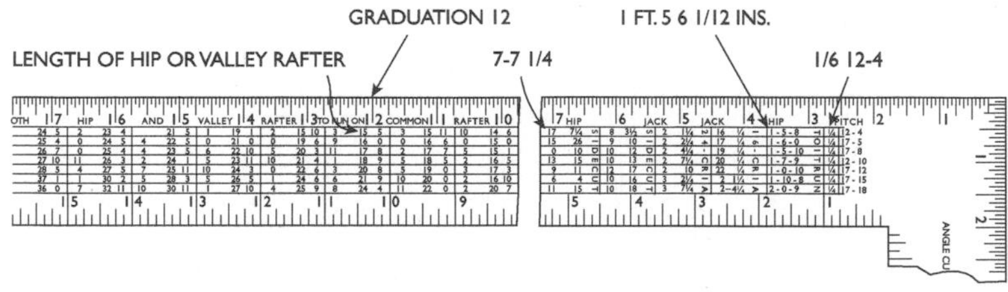

Figure 34-62 The hip rafter table.

Under **PITCH**, the set of three columns of figures gives the pitch. The seven pitches in common use are given as, for example, 1/6-12-4; this means that for a 1/6-pitch, there is a 12-in. run per 4-in. rise.

Under **HIP**, the set of figures gives the length of the hip and valley rafter per foot of run of common rafter for each pitch, as 1 ft, 5 6/12 in. for a 1/6 pitch.

Under **JACK** (16 in. on center), the set of figures gives the length of the shortest jack rafter, spaced 16 in. on center, which is also the difference in length of succeeding jack rafters.

Example

If the jack rafters are spaced 16 in. on center for a 1/6-pitch roof, find the lengths of the jacks and cut bevels.

The jack top and bottom cuts (or plumb and heel cuts) are the same as for the common rafter. Take 12 on the tongue of the square; that is, mark on the 9½ sides, as shown in the illustration that represents the rise per foot of the roof, or, if the pitch is given, take the figures in Table 34-7 that correspond to the given pitch. Thus, for a 1/6 pitch, these points are 12 and 4. Figure 34-63 shows the square on the jack in this position for marking top and bottom cuts.

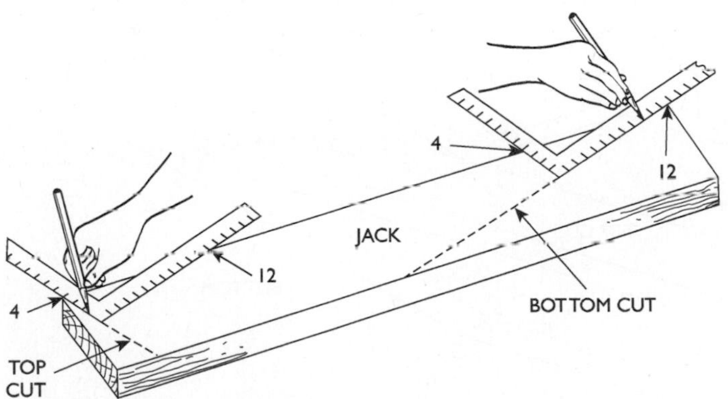

Figure 34-63 The square is applied to a jack rafter for marking top and bottom cuts. The vertical and horizontal cuts for jack rafters are the same as for common rafters.

Look along the line of 1/6 pitch, in Figure 34-64, under **JACK** (16-in. center), and find 16 7/8, which is the length in inches of the shortest jack and is also the amount to be added for the second jack. Deduct one-half the thickness of the hip rafter, because the jack rafter lengths given in this table are to centers. Also, add for any projection beyond the outer edge of the plate.

Look along the line of 1/6 pitch, in Figure 34-64, under **JACK**, and find 9-9½ for a 1/6 pitch. These figures refer to the graduated scale on the edge of the arm of the square. To obtain the required bevel, take 9 on one arm and 9½ on the other, as shown in Figure 34-65. It should be carefully noted that the last figure, or figure to the right, gives the point on the marking side of the square; that is, mark on the 9½ sides, as shown in the illustration.

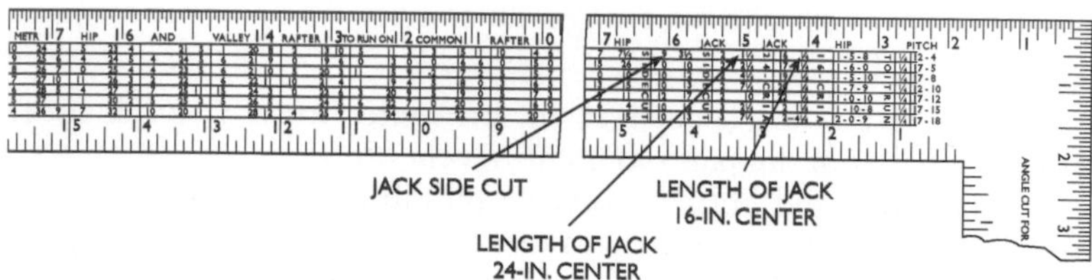

Figure 34-64 A rafter table.

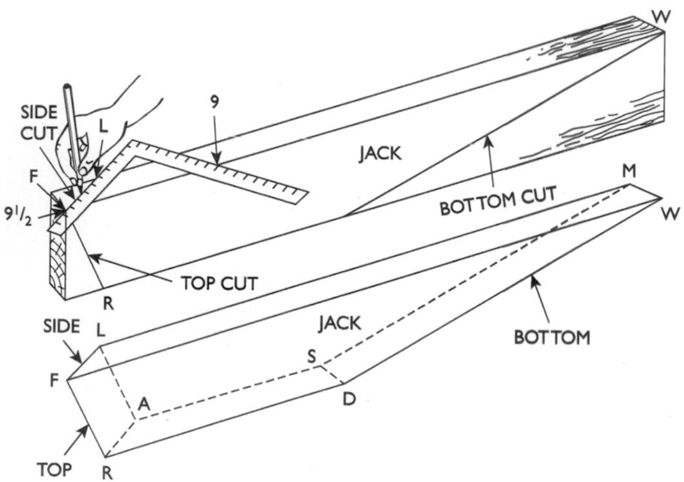

Figure 34-65 Marking and cutting a jack rafter with the aid of the square. FR and DW are the marks for the top and bottom cuts, respectively. With the jack rafter cut as marked, LARF represents the section cut at the top, and MSDW represents the section cut at the bottom.

Under **JACK** (24 in. on center), the set of figures gives the length of the shortest jack rafter spaced 24 inches on center, which is also the difference in length of succeeding jack rafters. Deduct one-half the thickness of the hip or valley rafter, because the jack rafter lengths given in the table are to centers. Also, add for any projection beyond the plate.

Under **HIP**, the set of figures gives the side cut of the hip and valley rafters against the ridge board or deck, as 7–7¼ for a ⅙ pitch (mark on the 7¼ side).

To get the cut of the sheathing and shingles (whether hip or valley), reverse the figures under **HIP**, as 7¼–7 instead of 7–7¼. For the hip top and bottom cuts, take 17 on the body of the square, and on the tongue, take the figure that represents the rise per foot of the roof.

Figure 34-66 shows the marking and cut of the hip rafter, and Figure 34-67 shows the rafter in position resting on the cap and the ridge. The section L'A'R'F' resting on ridge is the same as L'A'R'F' in Figure 34-66.

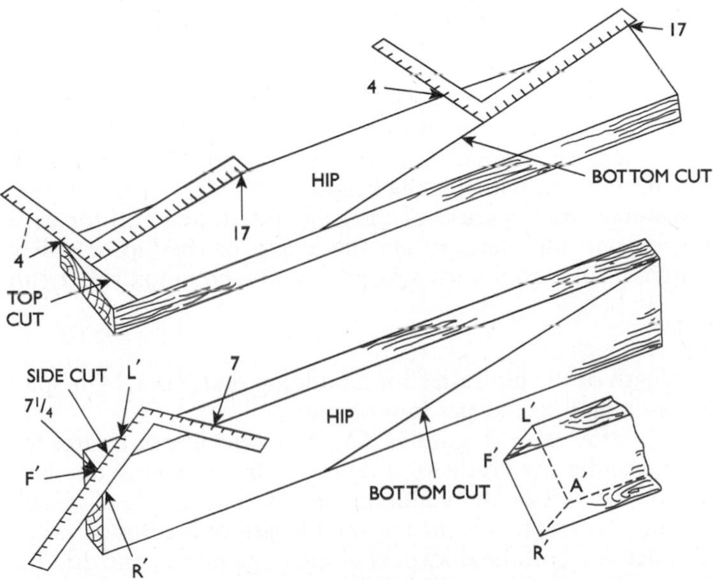

Figure 34-66 The square, as applied to hip rafters, for marking top, bottom, and side cuts. Note that the number 17 on the body is used for hip rafters. Section L'A'R'F' shows the bevel required for the ridge.

Under **HIP AND VALLEY**, the set of figures gives the length of run of the hip or valley rafter for each pitch of the common rafter.

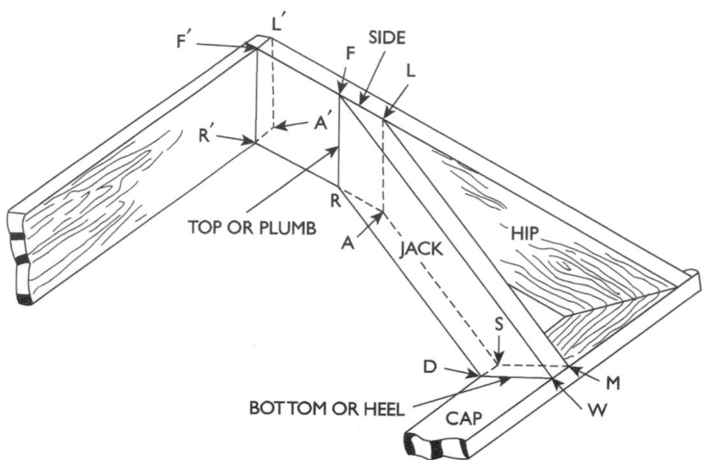

Figure 34-67 The jack rafter in position on the roof between the hip rafter and the cap.

For instance, for a roof with a 1⁄6 pitch under the figure 12 (representing the run of the common rafter, or one-half the width of the building), along the 1⁄6-pitch line of figures find 17, 5, 3, which means 17 ft, 5 3⁄12 in., which is the length of the hip or valley rafter. Deduct one-half the thickness of the ridge board and add for eave overhang beyond the plate, to get the length of the hip or valley rafter required for a roof with a 1⁄6 pitch and a common rafter run of 12 ft.

Example

Find the length of the hip rafter for a building that has a 24-ft span and a 1⁄6 pitch (a 4-in. rise per foot of run).

In the hip rafter table (Figure 34-62), along the line of figures for 1⁄6 pitch and under the graduation figure 12 (representing one-half the span, or the run of the common rafter), find 17, 5, 3, which means 17 ft, 5 3⁄12 in.; this is the required length of the hip or valley rafter. Deduct one-half the thickness of the ridge board, and add for any overhang required beyond the plate.

For the top and bottom cuts of the hip or valley rafter, take 17 on the body of the square and 4 (the rise of the roof per foot) on the tongue. The mark on the 17 side gives the bottom cut; the mark on the 4 side gives the top cut.

For the side cut of the hip or valley rafter against the ridge board, look in the set of figures for the side cut in the table (Figure 34-62) under **HIP** along the line for 1/6 pitch, and find the figure 7–7¼. Use 7 on one arm of the square and 7¼ on the other; mark on the 7¼ arm for the side cut.

Reading Length of Rafter per Foot of Run

There are many methods used by carpenters for determining the lengths of rafters, but probably the most dependably accurate method is the "length-per-foot-of-run" method. Because many, perhaps most, of the better rafter-framing squares now have tables on their blades giving the necessary figures, they may almost be considered as standard. The tables may not be arranged in the same manner on all these squares, and on some they may be more complete than on others. The use of these tables is taught in all Carpenters' Union Apprenticeship schools.

Under the heading **LENGTH COMMON RAFTERS PER FOOT RUN** in Figure 34-68 will be found numbers, usually from 3 to 20. With each number is a figure in inches and decimal hundredths. The integers represent the rises of the rafters per foot of run, and the inches and decimals represent the lengths of the rafters per foot of run. As an example of the use of these tables, take a building 28 ft, 2 in. wide, thereby making the run of the rafters 14 ft (allowing for a 2-in. ridge). Let the desired pitch be 4 in. per foot. Under the number 4 on the square will be found the length per foot of run—12.64 in. The calculation for the length of the rafter is as follows:

$$12.64 \times 14 = 176.96 \text{ inches}$$

$$\frac{176.96}{12} = 14.75 \text{ feet, or 14 feet, 9 inches}$$

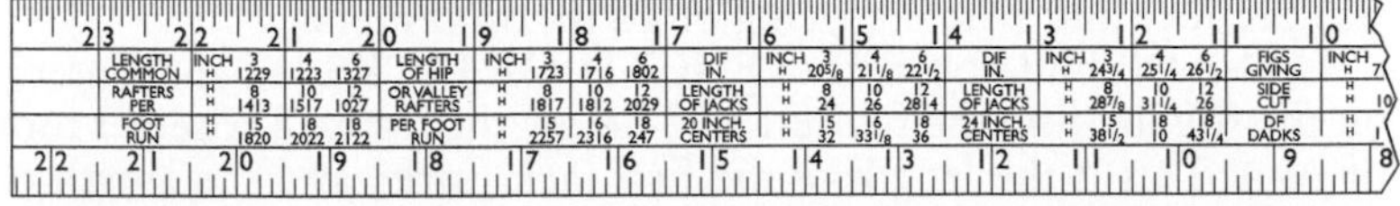

Figure 34-68 The length-per-foot-run tables on one type of rafter-framing square.

If the run is in feet and inches, it is most convenient to reduce the inches to the decimal parts of a foot, according to the following table:

1 in.	=	0.083	ft
2 in.	=	0.167	ft
3 in.	=	0.250	ft
4 in.	=	0.333	ft
5 in.	=	0.417	ft
6 in.	=	0.500	ft
7 in.	=	0.583	ft
8 in.	=	0.667	ft
9 in.	=	0.750	ft
10 in.	=	0.833	ft
11 in.	=	0.917	ft

Example

Find the length of a hip or valley rafter with an 8-in. rise per foot on a 20-ft building with the run of the common rafters measuring 10 ft.

Look for **LENGTH OF HIP OR VALLEY RAFTERS PER FOOT RUN** (Figure 34-68), and read under the 8-in. rise the figure 18.76. This is the calculation:

$$18.76 \times 10 = 187.6 \text{ inches}$$

$$\frac{187.6}{12} = 15 \text{ feet, } 7.6 \text{ inches}$$

One edge of all good steel squares is divided into tenths of inches, so this length may be measured off directly on the rafter pattern with the steel square.

Example

Find the difference in lengths of jack rafters on a roof with an 8-in. rise per foot and with a spacing of 16 in. on centers.

Under **DIFFERENCE IN LENGTH OF JACKS** (16-in. centers) on the square, find the figure 19.23 below the figure 8 (rise per foot of run). This is the length of the first jack rafter, and the length of each succeeding jack will be 19.23 in. greater—38.46 in., 57.69 in., 76.92 in., etc.

Example

Find the side cuts of jacks on a square similar to the one shown in Figure 34-69.

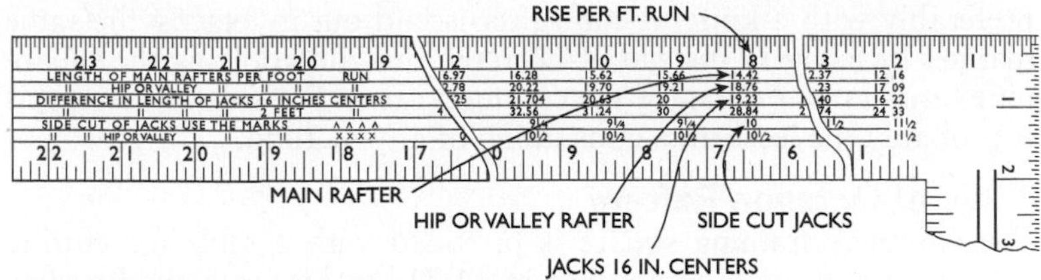

Figure 34-69 Typical rafter tables.

The fifth line is marked **SIDE CUT OF JACKS USE THE MARKS.** If the rise is 8 in. per foot, find the figure 10 under figure 8 in the upper line. The proper side cut will then be 10 × 12, cut on 12. The side cuts for hip or valley rafters are found in the sixth line; for the 8 × 12 roof, it is 10⅞ × 12, cut on 12.

No treatise on rafter framing is complete without an explanation of one of the oldest and most useful, though probably not the most accurate, methods of laying out a rafter with a steel square. Any square may be used if it has legible inch marks representing the desired pitch. It is the same method used for the layout of stairs. Figure 34-70 shows the layout for a rafter with a 9-ft run that has a pitch of 7 in. × 12 in., making the rise of the rafter 5 ft, 3 in. The steel square is applied nine times; carefully mark each application,

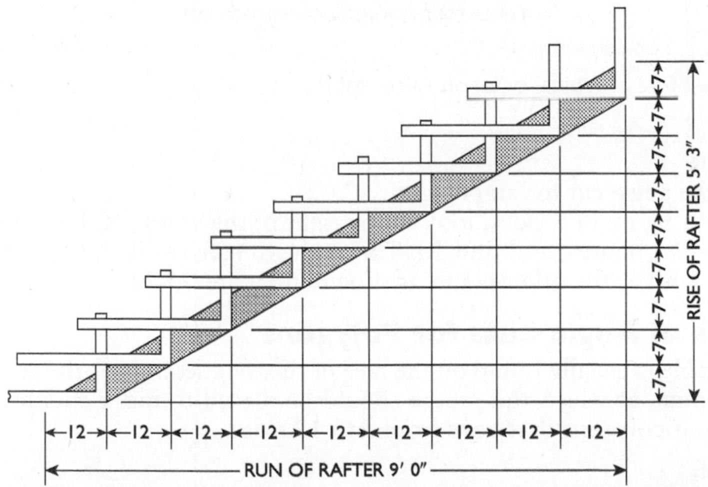

Figure 34-70 The method of stepping off a rafter with a square; the square is applied in consecutive steps, hence the name of the method.

preferably with a knife. A hip rafter is laid out in exactly the same manner by using 17 instead of 12 in the run and applying the square nine times as was done for the common rafter. For short rafters, this is probably the least time-consuming of any method.

Table of Octagon Rafters

The complete framing square is provided with a table for cutting octagon rafters, as shown in Figure 34-71. In this table, the first line of figures from the top gives the length of octagon hip rafters per foot of run. The second line of figures gives the length of jack rafters spaced 1 foot from the octagon hip. The third line of figures gives the reference to the graduated edge that will give the side cut for octagon hip rafters. The fourth line of figures gives the reference to the graduated edge that will give the side cuts for jack rafters. The tables are used in a manner similar to that used for the regular rafter tables just described and therefore need no further explanation. The last line, or bottom row of figures, gives the bevel of intersecting lines of various regular polygons. At the right end of the body on the bottom line can be read MITER CUTS FOR POLYGONS—USE END OF BODY.

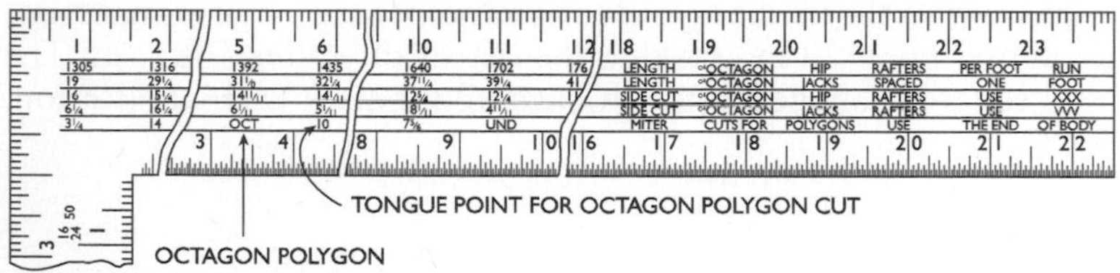

Figure 34-71 Typical octagon rafter tables.

Example

Find the angle cut for an octagon.

For a figure of 8 sides, look to the right of the word OCT in the last line of figures, and find 10. This is the tongue reading; the end of the body is the other point, as shown in Figure 34-72.

Table of Angle Cuts for Polygons

This table is usually found on the face of the tongue. It gives the setting points at which the square should be placed to mark cuts for common polygons that have from 5 to 12 sides.

Example

Find the bevel cuts for an octagon.

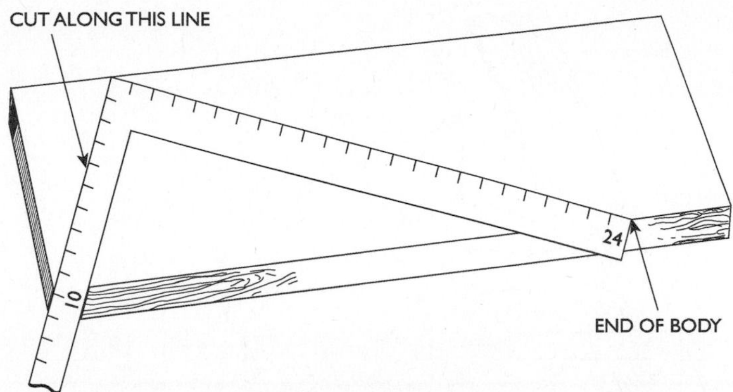

Figure 34-72 The square in position for marking an octagon cut; it is set to point 10 on the tongue and to point 24 on the body.

On the face of the tongue (Figure 34-73), look along the line marked **ANGLE CUTS FOR POLYGONS,** and find the reading "8 sides 18—7½." This means that the square must be placed at 18 on one arm and at 7½ on the other to obtain the octagon cut, as shown in Figure 34-74.

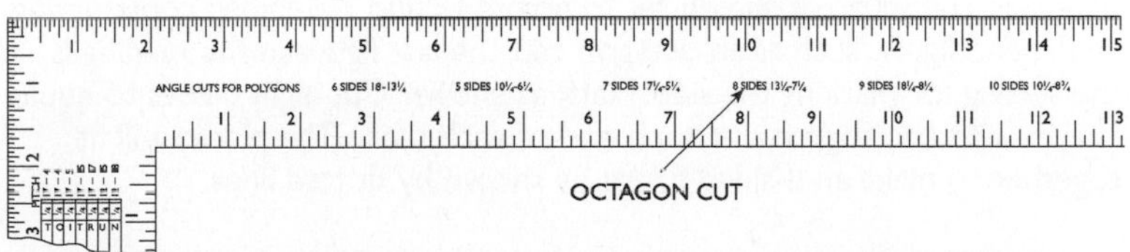

Figure 34-73 Table of angle cuts for polygons on the face of the square.

Table of Brace Measure

This table on the square, shown in Figure 34-75, is located along the center of the back of the tongue and gives the length of common braces.

Example

If the run is 36 in. on the post and 36 in. on the beam, what is the length of the brace?

In the brace table along the central portion of the back of the tongue (Figures 34-75 and 34-76), look at L for

36
50.91
36

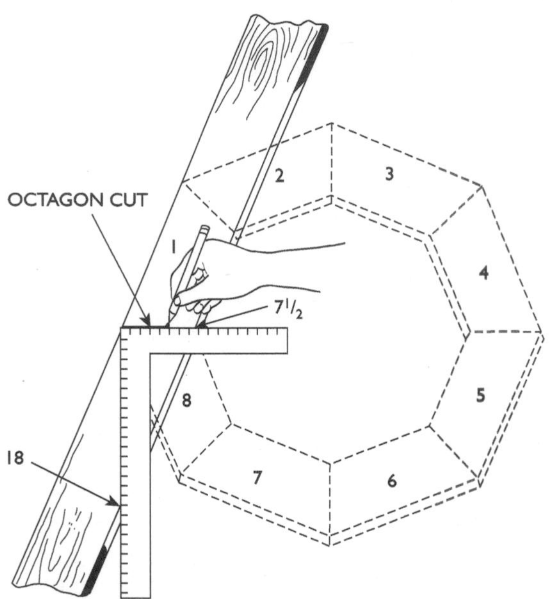

Figure 34-74 The application of the square for marking angle cuts of polygons. The square is shown set to points 18 and 7½. When constructing an 8-sided figure, such as an octagon cap, the last figure in the reading is the setting for marking the side. Mark as shown. Cut eight pieces to equal length, with this angle cut at each end of each piece. The pieces will fit together to make an 8-sided figure, as shown by dotted lines.

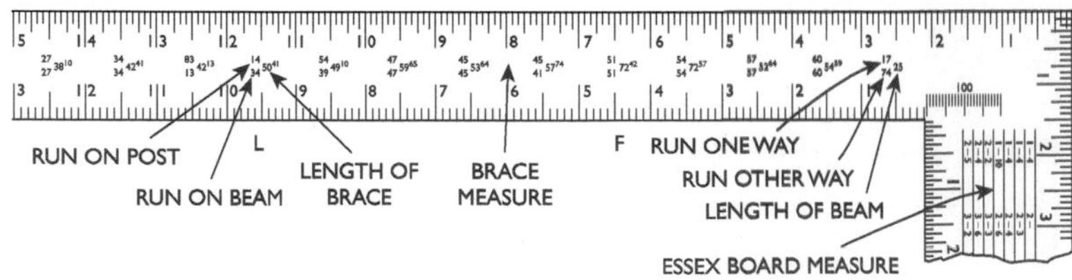

Figure 34-75 Table of brace measure on the back of the square.

This reading means that for a run of 36 in. on the post and 36 in. on the beam, the length of the beam is 50.91 in.

At the end of the table (at F near the body) will be found the reading

18
      30
24

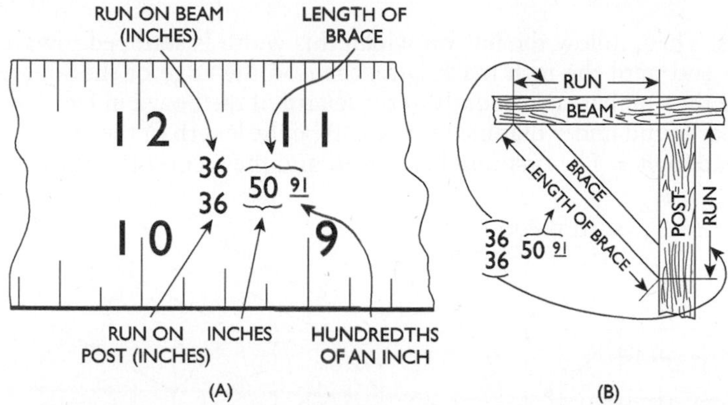

Figure 34-76 A portion of the brace-measure table, with an explanation of the various figures, is shown in A. The brace in position, illustrating the measurements of the brace-measure table, is shown in B.

This means that where the run is 18 in. one way and 24 in. the other, the length of the brace is 30 in.

The best way to find the length of the brace for the runs not given on the square is to multiply the length of the run by 1.4142 ft (when the run is given in feet) or by 16.97 in. (when the run is given in inches). This rule applies only when both runs are the same.

Octagon Table or Eight-Square Scale

This table on the square is usually located along the middle of the tongue face and is used for laying off lines to cut an eight-square or octagon-shaped piece of timber from a square timber.

In Figure 34-77, let ABCD represent the end section, or butt, of a square piece of 6 in. × 6 in. timber. Through the center, draw the lines AB and CD parallel with the sides and at right angles to each other. With dividers, take as many squares (6) from the scale as there are inches in width of the piece of timber, and lay off this square on either side of the point A, such as Aa and Ah; lay off in the same way the same spaces from the point B, such as Bd and Be; also lay off Cb, Cc, Df, and Dg. Then, draw the lines ab, cd, ef, and gh. Cut off at the edges to lines ab, cd,ef, and gh, thus obtaining the octagon , or 8-sided, piece.

Essex Board Measure Table

This table is shown in Figure 34-78 and normally appears on the back of the tongue on the square. To employ this table, the inch graduations on the outer edge of the square are used in combination with the values along the five parallel lines. After measuring the length and width of the board, look under the 12-in. mark for the width in inches. Then, follow the line on which this width is stamped toward either end until the inch mark is reached on the edge of the square where the number corresponds to the length of the board in feet. The number found under that inch mark will be the length of the board in feet and inches. The first number is feet, and the second is inches.

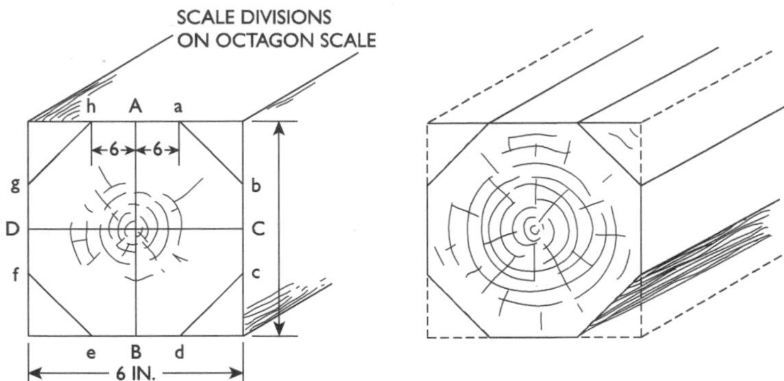

Figure 34-77 A square timber and its appearance after it has been cut to an octagon shape, thus illustrating the application of the octagon scale.

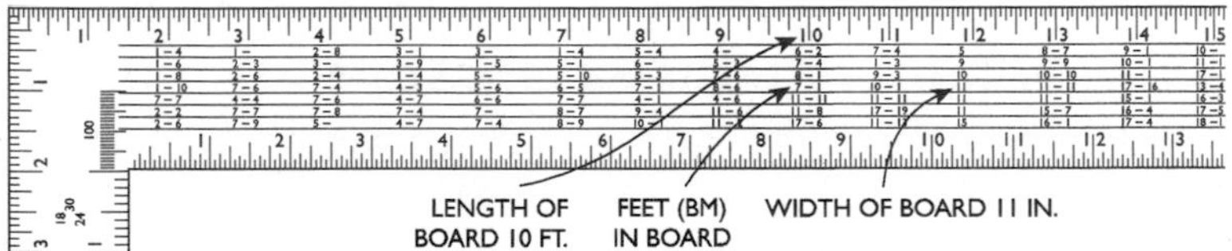

Figure 34-78 Table of Essex board measure on the back of the square.

Instead of a dash between the foot and inch numbers, some squares have the inch division continued across the several parallel lines of the scale appearing on one side of the vertical inch division lines and inches on the other.

Example

How many feet Essex board measure are there in a board 11 in. wide, 10 ft long, and 1 in. thick? In a board 3 in. thick?

Under the 12-in. mark on the outer edge of the square (Figure 34-78) find 11, which represents the width of the board in inches. Then, follow on that line to the 10-in. mark (representing the length of the board in feet), and find on a line 9—2, which means that the board contains 9 ft 2 in. board measure for a thickness of 1 in. If the thickness were 3 in., then the board would contain 9 ft 2 in. × 3, or 27 ft 6 in. BM.

Floor Framing

The frame of a wooden building comprises the sills, corner posts, braces, studding, girts or ledgers, girders, floor beams, caps or plates, rafters, etc.; it is the network of timbers to which the outside covering, floors, and partition walls are attached. The floor frame of a wood-frame building consists specifically of the posts, beams, sill plates, joists, and subfloor. When these are assembled properly on a foundation, they form a level anchored platform for the rest of the building. The posts and center beams of wood or steel, which support the inside ends of the joists, are sometimes replaced with a wood frame or masonry wall when the basement area is divided into rooms.

An important factor in the design of a wood floor system is to equalize shrinkage and expansion of the wood framing at the outside walls and at the center beam. This is usually accomplished by using approximately the same total depth of wood at the center beam as the outside framing. Thus, as beams and joists approach moisture equilibrium, or the moisture content they reach in service, there are only small differences in the amount of shrinkage.

Both wood girders and steel beams are used in wooden building construction. The standard I beam and the wide-flange beam are the most commonly used beam shapes. Wood girders are of two types—solid and built-up. The built-up is preferred because it can be made from drier dimension material and is more stable. The built-up girder, Figure 34-79, is usually made up of two or more pieces of 2-inch dimension lumber spiked together, the ends of the pieces

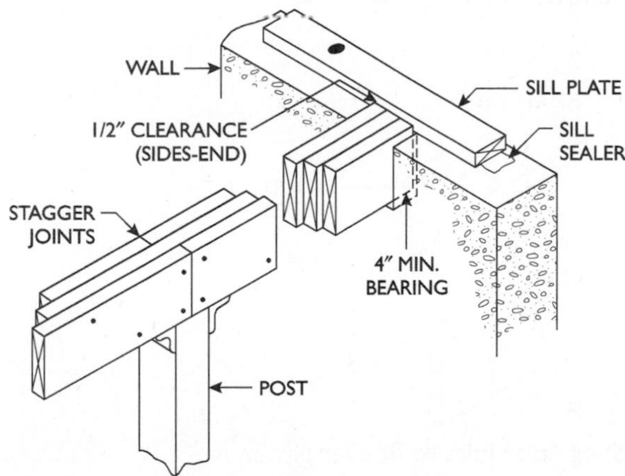

Figure 34-79 Built-up wooden girders.

joining over a supporting post. Ends of wood girders should bear at least 4 inches on the masonry walls or pilasters. When wood is untreated, a ½-in. air space should be provided at each end and at each side of wood girders framing into masonry walls. The top of the girder should be level with the top of the sill plates on the foundation walls, unless ledger strips are used.

Perhaps the simplest method of floor-joist framing is one where the joist bears directly on the girder, in which case the top of the beam coincides with the top of the anchored sill as shown in Figure 34-79. To provide for more uniform shrinkage at the inner beams and the outer wall, and to provide greater basement headroom, joist hangers or a supporting ledger strip are commonly used. Several ways of supporting joists on ledger strips are shown in Figures 34-80, 34-81, and 34-82. In Figure 34-80, a continuous horizontal tie between exterior walls is obtained by nailing notched joists together. In Figure 34-81, the connecting scab at each pair of joists provides this tie and also a nailing area for the subfloor. A steel strap is used to tie the joists together when the tops of the beam and the joists are level as shown in Figure 34-82.

Joists may be arranged with a steel beam generally the same way as illustrated for wood beam. Perhaps the most common methods, in addition to resting directly on top of the beam, are shown in Figures 34-83 and 34-84. In Figure 34-83, a wood ledger or angle iron is bolted to the web of the beam upon which the joist rests. In Figure 34-84, the joists bear directly on the flange of the beam; however, wood blocking is required between the joists near the beam flange to prevent overturning.

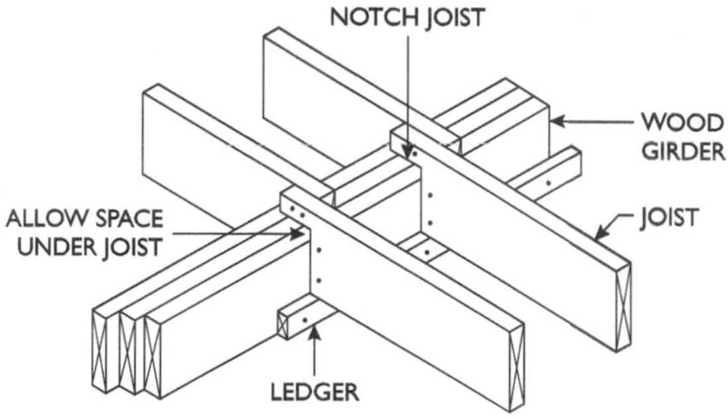

Figure 34-80 Notching floor joist to fit over girder.

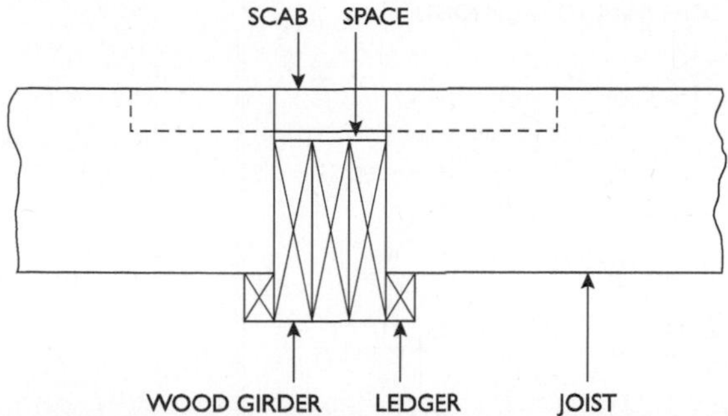

Figure 34-81 Using a connecting scab to tie joist together.

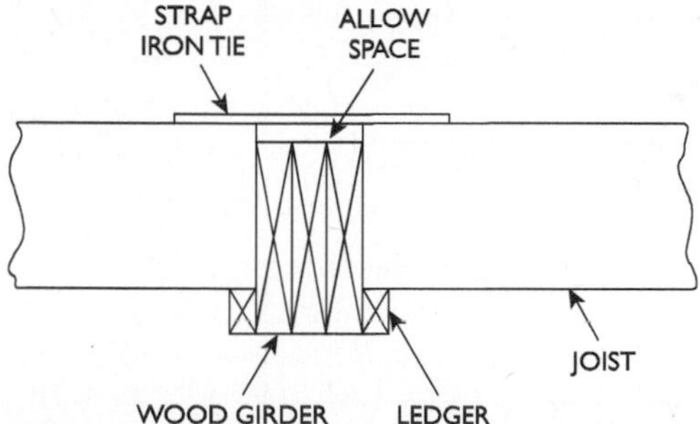

Figure 34-82 A steel strap is used to connect floor joist together.

Two general types of sill construction are used which conform to the two widely recognized styles of wood wall framing, i.e., platform construction and balloon construction. In platform construction, the box sill is commonly used. It consists of a 2-in. or thicker plate anchored to the foundation wall over a sill sealer which provides support and fastening for the joists and header at the ends of the joist as shown in Figure 34-85.

Balloon-frame construction uses a nominal 2-in. or thicker wood sill on which the joists rest. The studs also bear on this member and are nailed both to the floor joists and the sill. The subfloor is laid diagonally or at right angles to the joists, and a fire stop is added between the studs at the floor line, as shown in Figure 34-86. When diagonal subfloor is used, a nailing member is normally required between joists and studs at the wall lines. Because there is less potential shrinkage in exterior walls with balloon framing, it is usually preferred over the platform type for full two-story brick or stone veneer buildings.

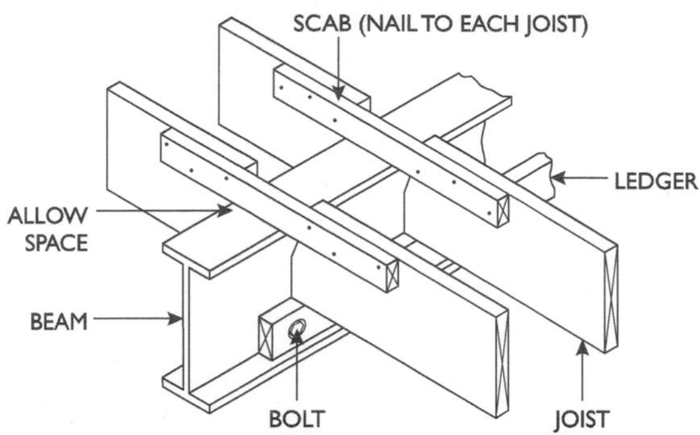

Figure 34-83 A wooden ledger is fastened to the I beam to support the joist.

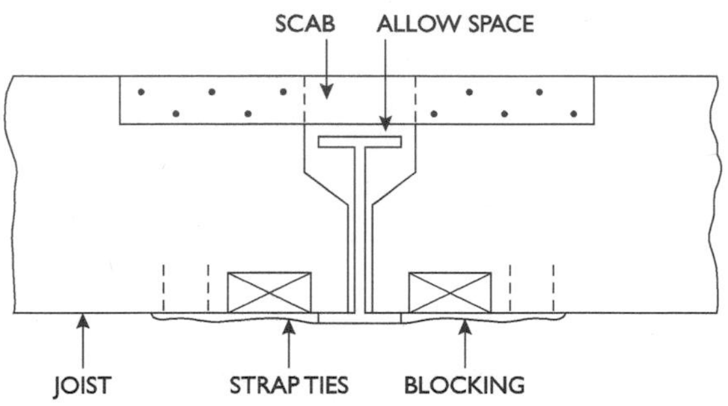

Figure 34-84 Joists are resting directly on I beam and connected at top with a scab.

After the sill plates have been anchored to the foundation walls or piers, the joists are located according to the building design. Sixteen-inch center-to-center spacing is most commonly used. Any joists having a slight bow edgewise should be so placed that the crown is on top. A crowned joist will tend to straighten out when subfloor and normal floor loads are applied. The largest edge knots should be placed on top, as knots on the upper side of a joist are on the compression side of the member and will have less effect on strength.

The header joist is fastened by nailing into the end of each joist. In addition, the header joist and the stringer joists parallel to the exterior walls in platform construction are toenailed to the sill 16 inches on center. Each joist should be toenailed to the sill and center beam. Details of typical platform construction are shown in Figure 34-87.

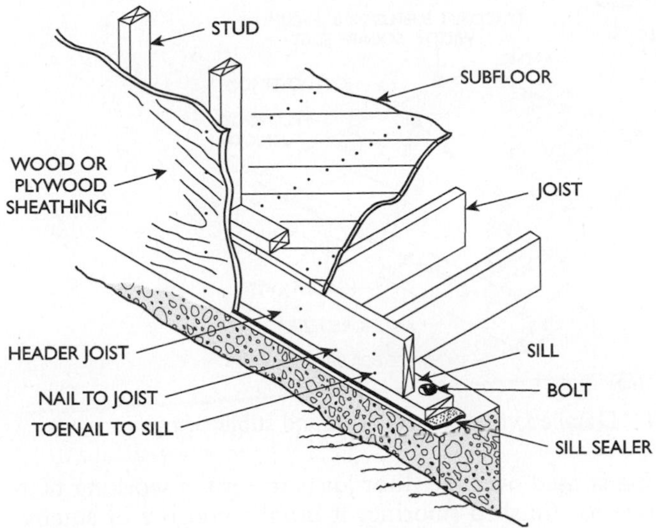

Figure 34-85 Platform construction showing sill, floor joist, and header.

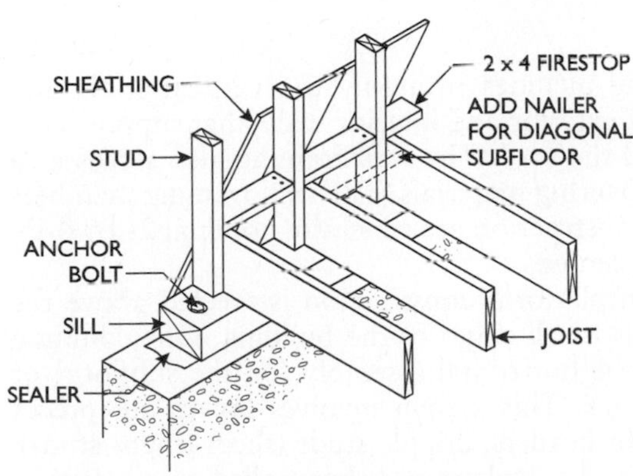

Figure 34-86 Balloon framing showing joist, plate, and wall stud.

Cross-bridging between wood joists has long been used in wooden building construction; however, recently questions of its benefit in relation to cost have been raised. Solid bridging between joists provides a more rigid base for partitions located above joist spaces. Load-bearing partitions should be supported by double joists.

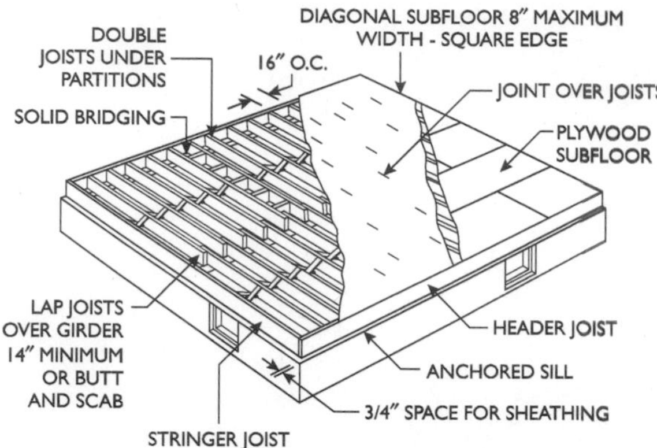

Figure 34-87 Detailed view of floor joist and subflooring.

Subflooring is used over the floor joists to form a working platform and base for finished flooring. It usually consists of square-edge or tongued-and-grooved boards no wider than 8 in. and not less than ¾ in. thick, or plywood ½ to ¾ in. thick, depending on the type of finished floor and spacing of joists.

Wall Framing

The term *wall framing* includes primarily the vertical studs and horizontal members of exterior and interior walls that support ceilings, upper floors, and the roof. The wall framing also serves as a nailing base for wall covering materials. The wall framing members used in conventional construction are generally nominal 2- by 4-in. studs spaced 16 in. on center.

The wall framing in platform construction is erected above the subfloor, which extends to all edges of the building. One common method of framing is the horizontal assembly (on the subfloor) or "tilt up" of wall sections. This system involves laying out precut studs, window and door headers, cripple studs (short length studs), and windowsills. Top and soleplates are then nailed to all vertical members and adjoining studs to headers and sills. Set-in corner bracing should be provided when required. The entire section is then erected, plumbed, and braced as shown in Figure 34-88. After all walls are erected, a second top plate is added that laps the first at corners and wall intersections. This gives an additional tie to the framed walls. These top plates can also be partly fastened in place when the wall is in a horizontal position. Walls are normally plumbed and aligned before the top plate is secured. Temporary braces are nailed to the studs at the top of the wall and to a 2-by-4 block fastened to the subfloor or joist. The temporary bracing is left in place until ceiling and the roof framing are completed and sheathing is applied to the outside walls.

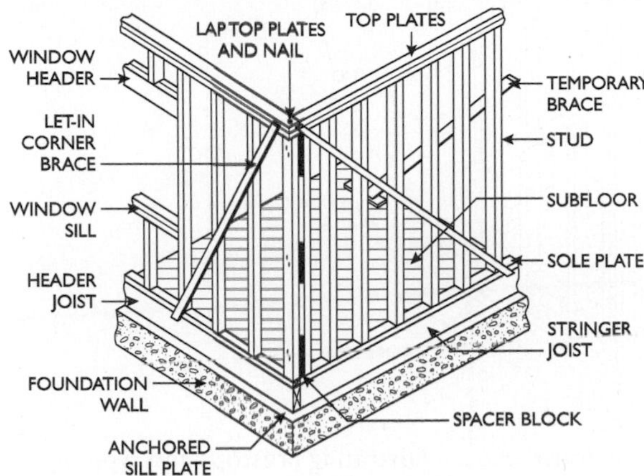

Figure 34-88 Wall-stud construction showing plank-type bracing and top plate.

In balloon construction, the wall studs extend from the sill of the first floor to the top plate or end rafter of the second floor, whereas the platform-framed wall is complete for each floor. In balloon frame construction, both the wall studs and the floor joists rest on the anchored sill, as shown in Figure 34-89. The ends of the second-floor joists bear on a 1- by 4-in. ribbon that has been let into the studs. In addition, the joists are nailed to the studs at these connections. The end joists parallel to the exterior on both the first and second floors are also nailed to each stud.

Firestops are used in balloon framing to prevent the spread of fire through the open wall passages. Firestops are ordinarily of 2- by 4-in. blocking placed between the studs at joist level.

Interior Walls

The interior walls of a wooden frame building with conventional joist and rafter roof construction are normally located to serve as bearing walls for the ceiling joists as well as room dividers. Walls located parallel to the direction of the joists are commonly non-load-bearing. Studs are nominal 2 by 4 in. in size for load-bearing walls but can be 2 by 3 in. in size for non-load-bearing walls. Spacing of the studs is usually controlled by the thickness of the covering material, the most common spacing being 16 in. on center.

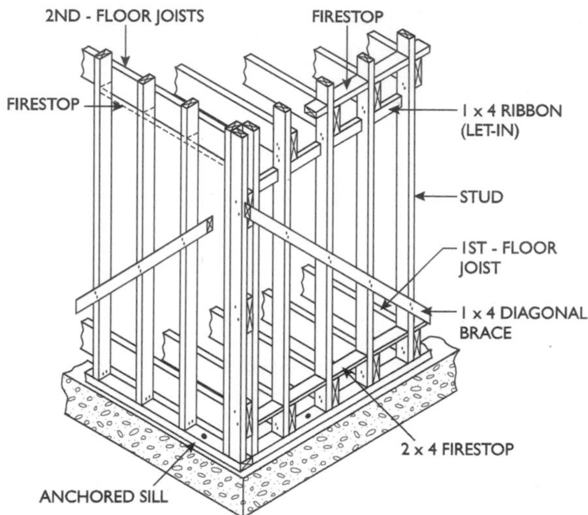

Figure 34-89 Balloon construction illustrating firestops and type of installation for ceiling joists.

The interior walls are assembled and erected in the same manner as the exterior walls, with a single bottom (sole) plate and a double top plate. The upper top plate is used to tie intersecting and crossing walls to each other.

Ceiling Joists

After exterior and interior walls are plumbed and braced and top plates are added, ceiling joists can be positioned and nailed in place. They are normally placed across the width of the building, as are the rafters. The partitions of the building are usually located so that ceiling joists of even lengths (10, 12, 14, and 16 ft or longer) can be used without waste to span from exterior walls to load-bearing interior walls. The sizes of the joists depend on the span, wood species, spacing between joists, and the load on the floor. When preassembled trussed rafters (roof trusses) are used, the lower chord acts as the ceiling joist. The truss also eliminates the need for load-bearing partitions.

Because ceiling joists, in addition to being ties between exterior walls and interior partitions, also serve as tension members to resist thrust of the rafters or pitched roofs, they must be nailed to the plate at outer and inner walls. They are also nailed together, directly or with wood or metal cleats, where they cross or join at a load-bearing partition. Figure 34-90 shows ceiling joist connections with roof rafters at outside wall.

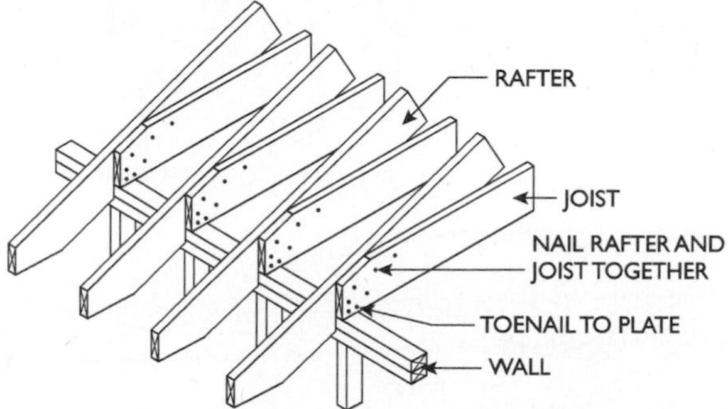

Figure 34-90 Ceiling joist and rafter construction.

Roof Framing

Two basic types of roof style are used on wood frame buildings: flat or slightly pitched roofs in which the roof and ceiling supports are furnished by the same member, and pitched roofs where ceiling joists are used for roof support. Roof joists for flat or low-pitched roofs, sometimes known as shed roofs, are commonly laid level or with a slight pitch, with roof sheathing and roofing on top and with the underside utilized to support the ceiling. Perhaps the simplest form of pitched roof, where both rafters and ceiling joists are required, is the gable roof shown in Figure 34-91A. A variation of the gable roof, used for Cape Cod or similar styles, includes the use of shed and gable dormers as illustrated in Figure 34-91B.

A third common roof style is the hip roof shown in Figure 34-91C. Although these roof types are the most common, others may include such forms as the mansard and the A-frame.

In normal pitched-roof construction, the ceiling joists are nailed in place after the interior and the exterior wall framing are complete. Rafters should not be erected until ceiling joists are fastened in place, as the thrust of the rafters will otherwise tend to push out the exterior walls. Rafters are usually precut to length with proper angle cut at the ridge and eave and with notches provided for the top plates. Fitting of rafters to the top plate is shown in Figure 34-92. Studs for the gable end walls are cut to fit and nailed to the end rafter and the top plate of the end wall.

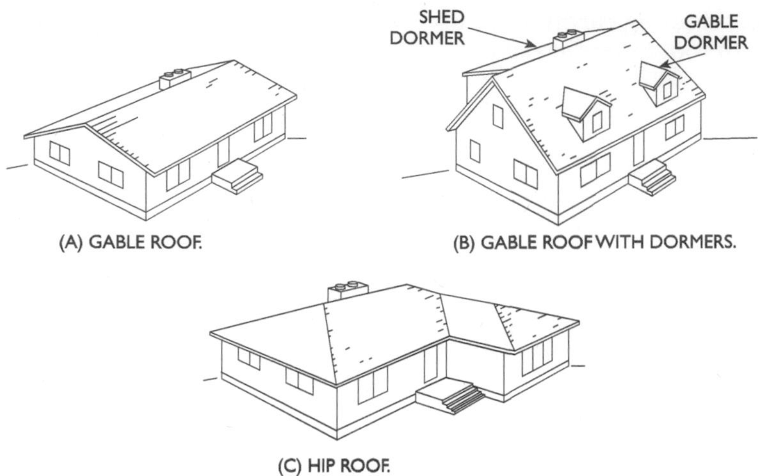

Figure 34-91 Various styles of frame roofs.

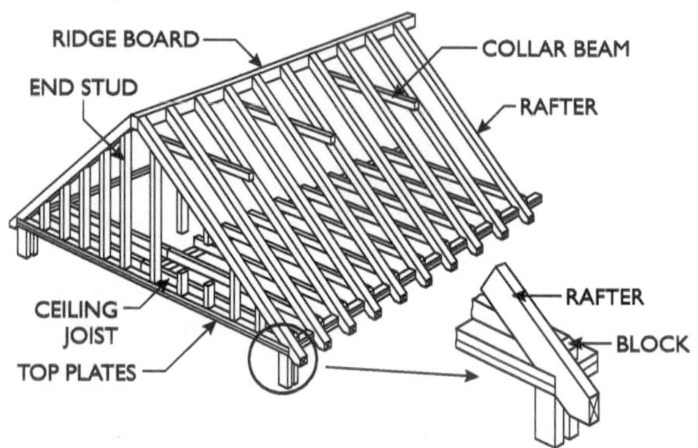

Figure 34-92 Roof construction showing ridge board, end studs, and collar beams.

In construction of small gable dormers, the rafters at each side are doubled. The side studs and the short valley rafters rest on these members in the manner shown in Figure 34-93. Side studs may also be carried past the rafter and bear on a soleplate nailed to the floor framing and the subfloor. This same type of framing may be used for the sidewalls of shed dormers. The valley rafter is also tied to the roof framing at the roof by a header.

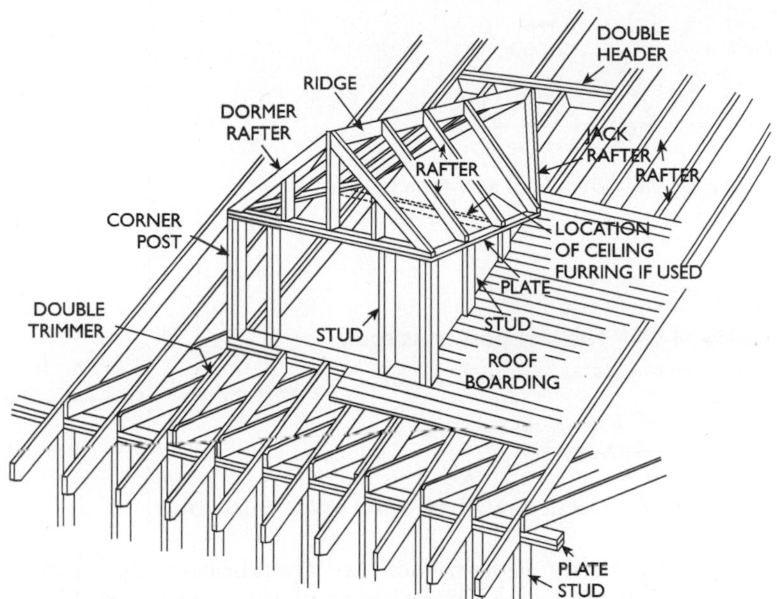

Figure 34-93 Detail view of a hip-roof dormer.

The simple truss or trussed rafter is an assembly of members forming a rigid framework of triangular shapes capable of supporting loads over long spans without intermediate support. The truss has been greatly refined during its development, and the gusset and other preassembled types of wood trusses are being used extensively in the building field. They save money, can be erected quickly, and allow rapid covering of the building.

Trusses are usually designed to span from one exterior wall to the other with length from 20 to 32 ft or more. Because no interior bearing walls are required, the entire building becomes one large room. This allows increased flexibility for interior planning, as partitions can be placed without regard to structural requirements. Wood trusses most commonly used are the W-type shown in Figure 34-94, the king-post in Figure 34-95, and the scissors type in Figure 34-96. These and similar trusses are most adaptable to buildings with rectangular plans so that the constant width requires only one type of truss. Trusses are commonly designed for 2-ft spacing, which requires somewhat thicker sheathing material than is needed for conventional rafter construction using 16-in. spacing.

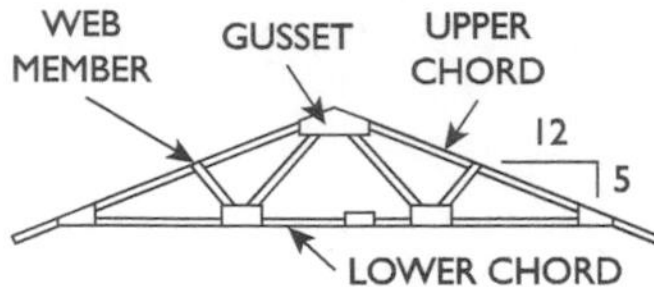

Figure 34-94 Details of a W-type truss roof.

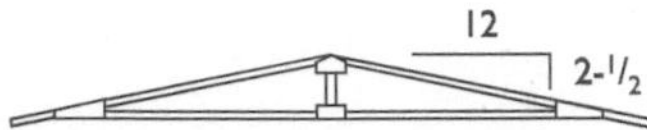

Figure 34-95 The king-post truss roof.

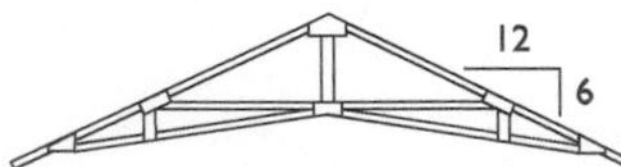

Figure 34-96 Detail view of the scissor-type truss roof.

A great majority of the trusses used are fabricated with gussets of plywood (nailed, glued, or bolted in place) or with metal gusset plates. Designs for standard W-type and king-post trusses with plywood gussets are available from lumber dealers. Many lumber dealers can supply completed trusses ready for erection.

Wall Sheathing

Wall sheathing is the outside covering used over the wall framework of studs, plates, and window or door headers. It forms a flat base upon which the exterior finish can be applied. Certain types of sheathing can provide great rigidity to a building. Perhaps the most common types used in construction are boards, plywood, structural insulating board, and gypsum sheathing.

Wood sheathing is usually composed of nominal 1-inch boards in a shiplap, a tongued-and-grooved, or a square-edge pattern. Widths commonly used are 6, 8, and 10 in. It may be applied horizontally or diagonally. When diagonal sheathing is carried to the foundation, in the manner illustrated in Figure 34-97, great strength and rigidity result.

Plywood is used extensively for sheathing walls, applied vertically, normally in 4- by 8-ft and longer sheets. When well nailed, this method of sheathing can eliminate the need for diagonal corner bracing. For 16-inch stud spacing, 3/8-in. or thicker plywood is recommended, especially when the exterior finish must be nailed to the sheathing.

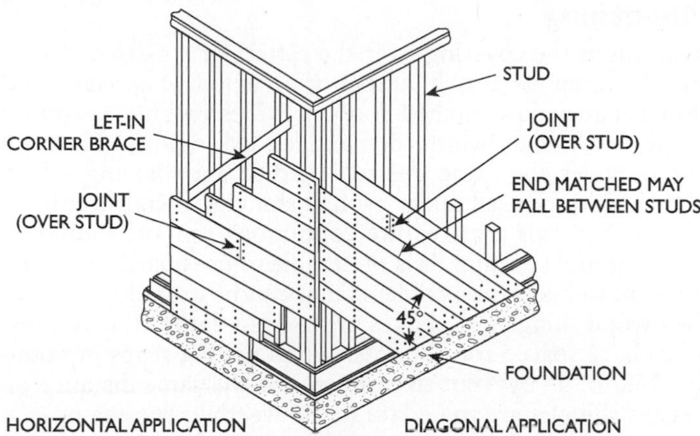

Figure 34-97 Sheathing applied to the outside wall in a horizontal pattern and in a diagonal pattern.

When applying insulating board sheathing, corner bracing is required to provide strength and rigidity. Fastening must be adequate around the perimeter and at intermediate studs and must be adequately secured (nails, staples, or other fastening systems). Nail-base sheathing permits shingles to be applied directly to it as siding if fastened with special annular-grooved nails. Examples of plywood and insulating board sheathing application are illustrated in Figure 34-98.

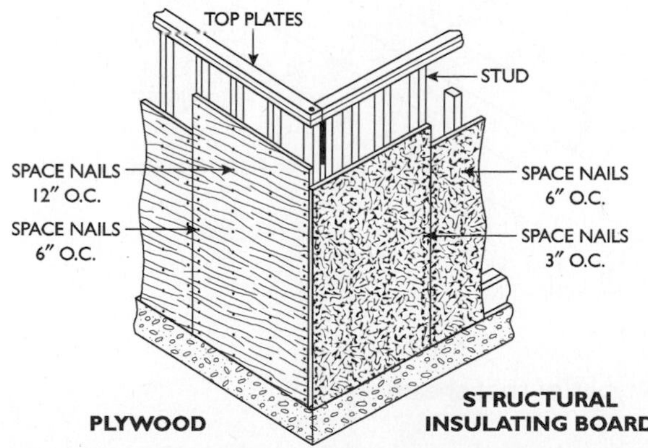

Figure 34-98 Plywood or insulating board applied to outside walls as sheathing.

Roof Sheathing

Roof sheathing is the covering over the rafters or trusses and usually consists of nominal 1-inch lumber or plywood. Diagonal wood sheathing on flat or low-pitched roofs provides racking resistance where areas with high winds demand added rigidity. Plywood sheathing provides the same desired rigidity and bracing effect. Board sheathing to be used under asphalt shingles, metalsheet roofing, or other materials that require continuous support should be laid closed (without spacing). It is preferable to use boards no wider than 6 or 8 in. to minimize problems that can be caused by shrinkage. When wood shingles or shakes are used in damp climates, it is common to have spaced roof boards. Wood nailing strips in nominal 1- by 3-in. or 1- by 4-in. size are spaced the same distance on centers as the shingles are to be laid to the weather. For example, if shingles are to be laid 5 in. to the weather and 1- by 4-in. strips are used, there would be spaces about 1½ in. between each board to provide the needed ventilation spaces. Examples of close laid and spaced roof sheathing are illustrated in Figure 34-99.

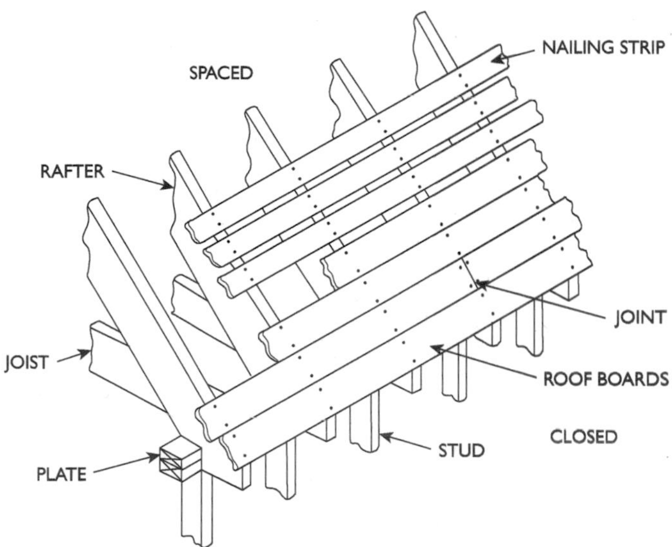

Figure 34-99 Two types of roof sheathing.

When plywood roof sheathing is used, it should be laid with the face grain perpendicular to the rafters as shown in Figure 34-100. End joints are made over the center of the rafters and should be staggered

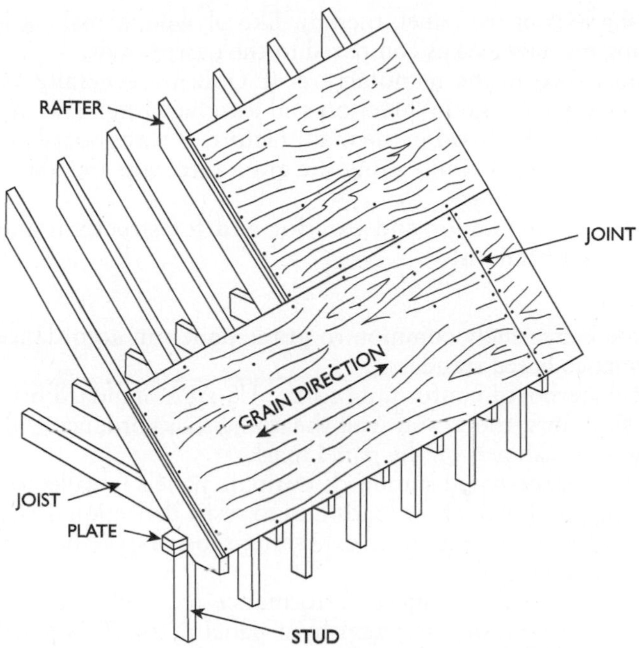

Figure 34-100 When using plywood as sheathing, special attention should be given to direction of wood grain with reference to rafter.

by at least one rafter or more. Plywood should be nailed at each bearing, 6 in. on center along all edges and 12 in. on center along intermediate members. A ⅛-in. edge spacing and a $\frac{1}{16}$-in. end spacing should be allowed between sheets when installing.

Oriented Strandboard (OSB) and Waferboard

Oriented strandboard (OSB) and waferboard are panel products that illustrate a trend toward more efficient use of forest resources while employing less valuable, fast-growing species. At the same time, they provide economy of construction and substantial insulation and structural advantages.

OSB and waferboard are made of aspen or poplar (as well as southern yellow pine). Wafers or strands are bonded together under heat and pressure using a waterproof phenolic resin adhesive or equivalent waterproof binder.

Oriented strandboard has been developed in recent years. Like waferboard, OSB is made of aspen-poplar strands or southern yellow pine. However, the strands in the outer faces of OSB are oriented along the long axis of the panel, thereby, like plywood, making it stronger along the long axis as compared to the narrow axis.

The strands used in the manufacture of OSB are generally 80 mm (3⅛ in.) long in the grain direction and less than 1 mm (1/32 in.) in thickness. The wafers used in the manufacture of waferboard are generally 30 mm long (1¼ in.) along the grain direction and about 1 mm (1/32 in.) in thickness.

Waferboard is for interior and exterior and for structural and non-structural applications.

OSB Grades

It has become increasingly common to grade panels in accordance with performance-based standards.

OSB and waferboard conforming to the US APA Standard may be used for subfloors, roofs, and wall sheathing in accordance with end uses and spans shown on the panel mark.

These performance-based standards evaluate panels installed on framing for their ability to carry loads and to resist deflection under loads and conditions similar to or exceeding those experienced in construction or service.

The ability of a panel to meet performance requirements of a given end use is shown on the panel by a panel mark. This panel mark consists of the following:

End-use mark -1 F, 2F, 1R, 2R, and W

Span mark -16, 20, 24, 32, 40, 48.

In the end-use mark, the "F," "R," and "W" indicate, respectively, floor, roof, and wall sheathing, and the "1" indicates that the panel may be used alone to meet structural requirements. The "2" means that the panel must have an additional support element such as underlay for floors or H clips or blocking for roofs.

The two digit span mark indicates the span between supports in inches. For example, 16 indicates a maximum span of 16 in.

The panel mark 1 R24//2F16/W24 means that the panel may be used without blocking on a roof with trusses at 24 in. on center, on a floor with underlay on joists with centers at 16 in., or on a wall with studs at 24 in. on center.

OSB Sizes

OSB and waferboard are available in a variety of thicknesses as shown in Table 34-8.

Table 34-8 OSB and Waferboard Thickness and Weights

OSB (O-1) and Waferboard (R-1)	
4 ft × 8ft	
in.	*lb*
1/4	27
5/16	33
3/8	40
7/16	47
1/2	53
5/8	67
3/4	80
OSB (O-2)	
4ft × 8ft	
in.	*lb*
1/4	26
5/16	32
3/8	39
7/16	46
1/2	52
5/8	66
3/4	78
1-1/8	120

Chapter 35

Blacksmithing

While the blacksmith trade has practically gone out of existence in modern industrial plants, mechanics on occasion perform some of the common blacksmith operations. While in many cases these operations may be performed in a different manner, such as using an acetylene torch for heating, the basic principles of blacksmithing still apply. The blacksmithing chapter is included as reference material for those mechanics who may be called upon to forge, form, or bend metal by heating and striking.

By definition, a blacksmith is a smith who works with or welds wrought iron, as by beating upon an anvil, and makes or shapes small utensils or parts of machines, horse shoes, etc.; one who forges or welds iron on an anvil. Formerly, a blacksmith was a smith who worked in black metal or iron, as distinguished from a whitesmith, who worked on white metal or tin.

Blacksmith's Tools

The sections of metal worked by the blacksmith are few and simple. The sections of iron and steel dealt with by the smith are round, square, or rectangular; hence, no complicated tools are necessary. The only cutting tools used by the blacksmith are those actually employed to sever the forging from the stock or to remove extraneous metal. The chief art of smithing consists in reducing the sectional bar to the desired form, or drawing it out.

Tongs of various shapes are used to hold the iron bar that is being worked. The anvil and its attachments are important tools that are used in nearly all operations.

Tongs

Next to the anvil and hammer in importance are the tongs, of which there is a great variety. Tongs are used for holding metal that is at too high a temperature to be held in the bare hands, for placing iron in and removing it from the fire, and work of a similar character. Since the hand should be free to manipulate and turn the metal, the tongs are held in position by a link driven over the handles. The elasticity of the handles serves to hold the work securely.

Figure 35-1 represents the simplest form of tongs. These are known from the shape of the jaws as flat, single, or double pick-up

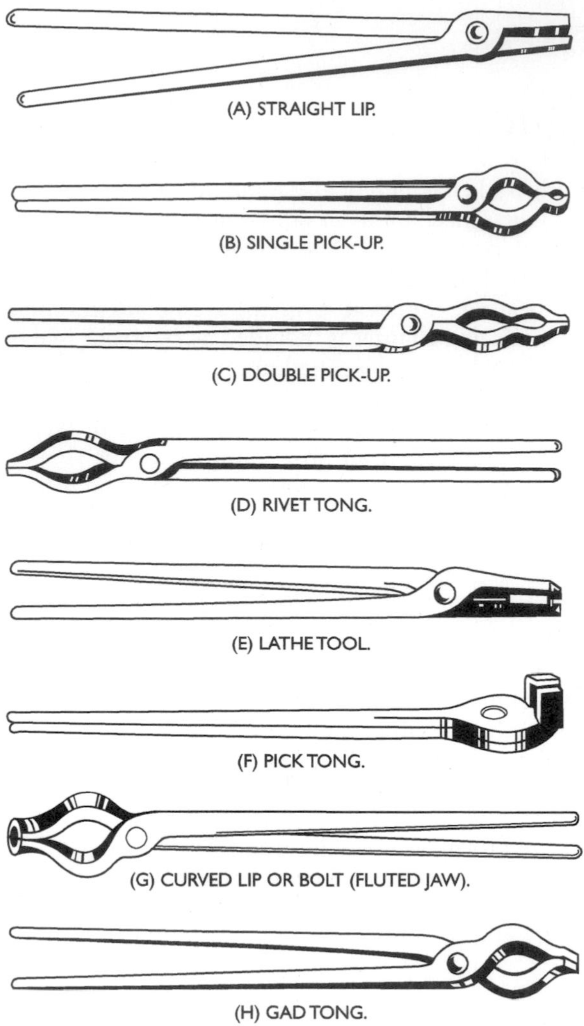

Figure 35-1 Various tongs used to handle hot metal in the blacksmith trade.

tongs. They are used for holding flat pieces of metal and vary in size according to the work in hand. They are usually light and have long handles so that objects on the ground may be reached without stooping.

Round-bit tongs are illustrated in Figure 35-1G. The jaws are concave and are suited for holding cylindrical objects. Tongs of this kind are frequently made of great strength and weight and are adapted to the handling of heavy shafts and pins. Similar tongs are used for holding square pieces—the jaws have a rectangular recess instead of a circular one. A square piece can thus be held more securely by gripping it on the corners than when it is seized on the flat surface. Other types and shapes of tongs are shown in Figure 35-2.

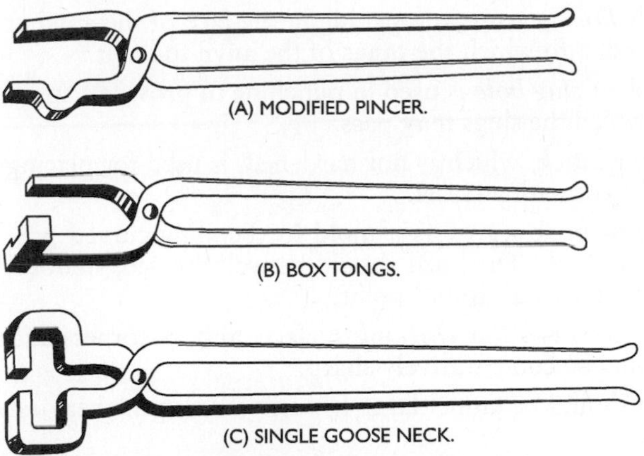

Figure 35-2 Other shapes of tongs.

Anvil

By definition, an anvil is a heavy block of iron and steel upon the surfaces of which the smith beats heated plastic metal to the desired shape. Its construction is shown in Figure 35-3.

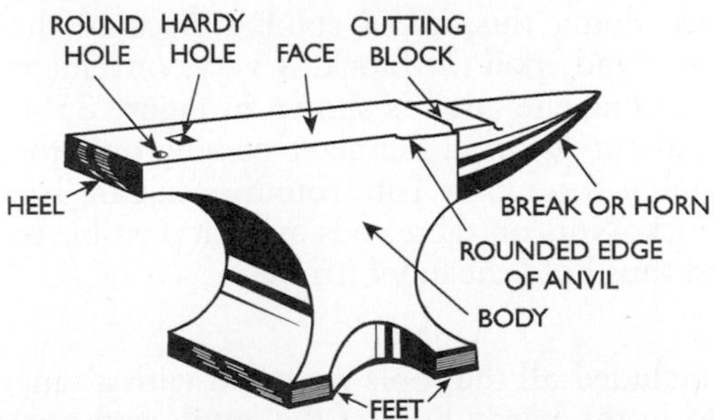

Figure 35-3 Ordinary blacksmith anvil.

The face of a good anvil is of hardened and tempered steel. The smith is very particular about its condition. Care must be exercised to prevent cutting into the face of the anvil or marring it with the edge of the sledge or hammer.

The *edges* of the anvil should not be chipped from careless operations.

The *rounded corner* provides a working surface of a very short radius.

The *hardy-hole* is a square opening in the face of the anvil at the heel end, into which the tangs of the anvil tools fit.

The *punch* or *slug hole* is used in punching to provide a space through which the slugs may pass.

The *cutting block*, which is not hardened, is used for placing stock to be cut with a chisel.

The *horn* or *beak* serves as a mold for bending curved portions of the work. The horn should be well dressed, smooth, and drawn to a small round point.

The *heel* presents a flat working surface, and its corners and edges should be comparatively sharp.

The *body* should be amply large to easily absorb the heaviest blows.

The *feet* (four in number) serve to increase the base upon which the anvil rests as well as to afford the means for clamping it down into position.

The anvil should be placed on the end of a heavy block of wood sunk into the ground to a depth of at least two feet so that it may rest upon a firm but elastic foundation. As the anvil is subjected to constant vibrations, by nature of the work, it must be firmly fastened to the block. When doing this, avoid spiking, because the spikes will soon work loose and spoil the block. A very convenient and reliable method of holding the anvil is shown in Figure 35-4. There are two iron rods about ⅜ in. in diameter passing over the feet of the anvil and running through a 1-in. round or square bar extending through the block. Nuts on these rods make it possible to draw them very tight and thus hold the anvil firmly.

Anvil Tools

Under this heading are included all the tools provided with a tang so that they may be held in the hardy-hole of the anvil, as distinguished from the tools erroneously called "anvil tools," which are

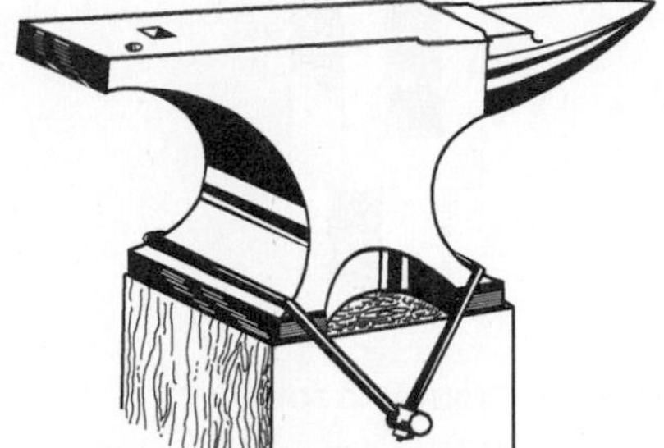

Figure 35-4 Method of anchoring an anvil to a wooden base.

held in the smith's hand and used in combination with the anvil tools; they are properly called *hand forming tools*. The anvil tools ordinarily used for general work, and which should always be included in any blacksmith's shop equipment, are as follows:

1. Bottom hardies
   a. Cold cut
   b. Hot cut
2. Bending fork
3. Bottom fuller
4. Bottom swage
5. Cutting block
6. Punching block

These are illustrated in Figure 35-5.

Hardies. These are bottom cutting chisels used to cut off lengths from bars, or crop ends from forgings. They are called *cold* or *hot* respectively since they are shaped and tempered for cutting cold or hot iron. See Figures 35-5A and 35-5B.

Bending Fork. This tool is made with square, flat, or round fingers, as shown in Figure 35-5C, and is used extensively in a variety of bending operations.

Bottom Fuller. This tool, shown in Figure 35-5D, is simply an inverted wedge with a blunt nose or working edge. It is used for spreading or notching the work.

Bottom Swage. The most common form of swage, shown in Figure 35-5E, has a concave face and is accordingly used to smooth off a round bar. It is also used for drawing metal down to a required diameter.

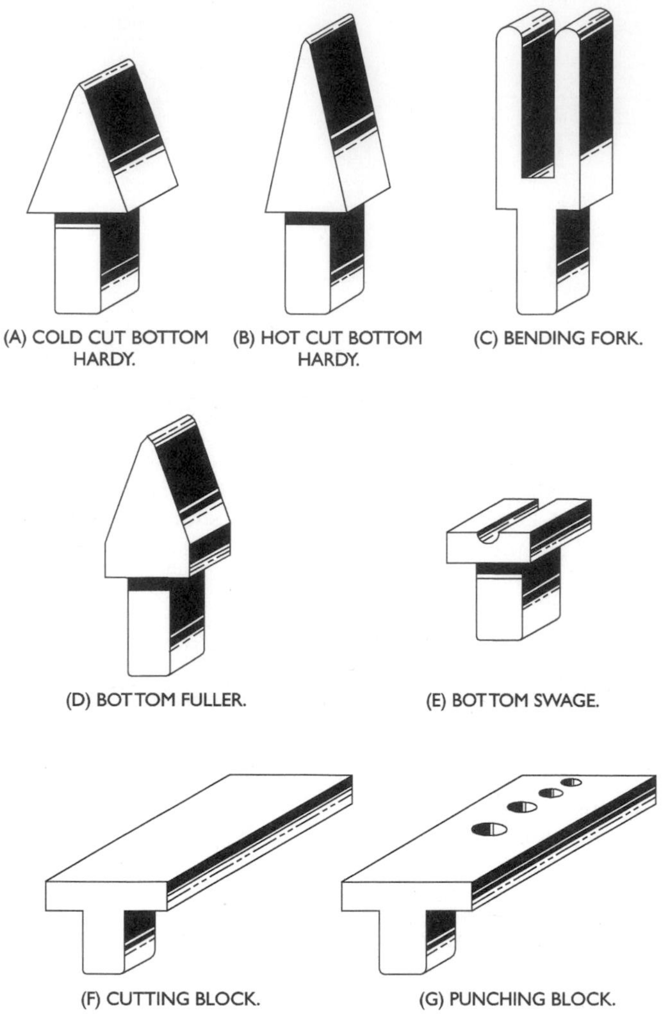

Figure 35-5 Various anvil tools.

Cutting Block. A flat plate of mild steel, shown in Figure 35-5F, to be used for cutting operations, which because of the shape of the work, cannot be conveniently performed on the cutting face of the anvil.

Punching Block. A block similar to the cutting block, but provided with a series of holes of various sizes as shown in Figure 35-5G. The holes provide a space through which slugs punched from the work may pass.

It should be understood that in addition to the standard anvil tools just described, there are numerous special tools that are used occasionally.

Hand Forming Tools

These are virtual counterparts of the anvil tools and are used in combination with other anvil tools. These are called hand forming tools because they are provided with a handle so that they may be held by the smith's hand in *forming* or *shaping* the work. The corresponding anvil tools are used at the same time, but are held by the anvil by inserting the tang into the hardy-hole of the anvil. The hand forming tools ordinarily used are as follows:

1. Top hardies
   a. Cold cut
   b. Hot cut
2. Top fuller
3. Top swage
4. Flatters
   a. Plain
   b. Offset
5. Punches
6. Cupping tool
7. Bucking bar
8. Heading tool
9. Shearing tool

The smith, in forming his metals, holds these tools in position by the handle, while the helper strikes the head of the placed tool with a sledge. Figure 35-6 illustrates these various tools.

Cold-Cut Top Hardy. A stout-bladed tool, shown in Figure 35-6A, used to cut cold metals.

Hot-Cut Top Hardy. A thin-bladed tool, shown in Figure 35-6B. This tool should never be used to cut cold metal.

Top Fuller. A tool that has a rounded nose, shown in Figure 35-6C, and is used for spreading and notching metals.

Top Swage. A tool that has a grooved face, shown in Figure 35-6D, and is used to "swage" or form metals to the shape of the groove, or for drawing metal down to a required diameter.

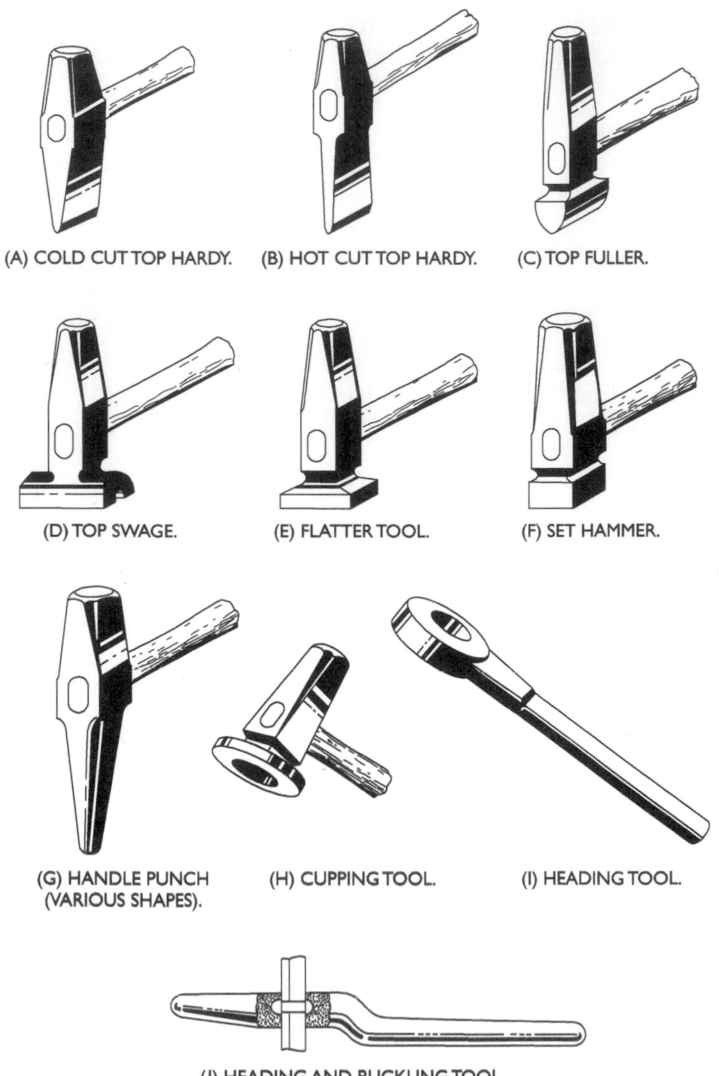

Figure 35-6 Various hand forming tools.

Flatters. The set hammer form of flatter, shown in Figure 35-6F, may be properly called a hand forming tool when used to smooth off and finish small flat surfaces. Its principal use is for striking blows in a definite spot or inaccessible place. A regular flatter is a flat-faced tool, shown in Figure 35-6E, used to smooth and finish the surface of forgings. To get the best smooth finish, do not let the temperature of the forging get too high, dip the face of the flatter in water, have water on the face of the anvil, and do not strike the flatter too heavily.

Punches. These are used for making large holes in hot metal. The general appearance of a punch is shown in Figure 35-6G. The working end is shaped according to the kind of hole desired: round, square, oblong, etc. The size depends upon the hole to be punched. They are invariably used for making holes through hot metal. The ordinary method is to punch part way through from one side and then turn the piece and drive through from the opposite direction. This avoids tearing the metal on the surface.

Cupping Tool. A tool with a rounded cavity, shown in Figure 35-6H, used for finishing the heads of rivets.

Heading Tool. A tool used for making up bolt heads, shown in Figure 35-6I. It has a hole about 1⁄32 in. larger than the diameter of the stock being used; the face end of the hole is beveled.

Bucking Bar. This may be called an *inertia* tool, and to possess that property in sufficiently marked degree, it is made heavy. The bar, shown in Figure 35-6J, has a suitable cavity in its working face and is used to "buck" or back up a rivet while it is being headed.

Shearing Tool. The cutting end of this tool is ground to the proper angle to adapt it to the work for which it is intended.

Swage Block

In connection with the use of the swage, which is used for drawing metal down to a required diameter, a swage block is very convenient. It takes the place of both the anvil and the bottom swage and is illustrated in Figure 35-7. It is usually made of cast iron in an approximately square shape, with a number of grooves of different dimensions cut on the face. These grooves are used according to the diameter and shape of the piece being worked. The grooves are semielliptical, which should also be the shape of the curve of the top swage. The angular grooves are right angled and are adapted to receive different sizes of square iron. The holes through the casting are available for punching, drifting, etc.

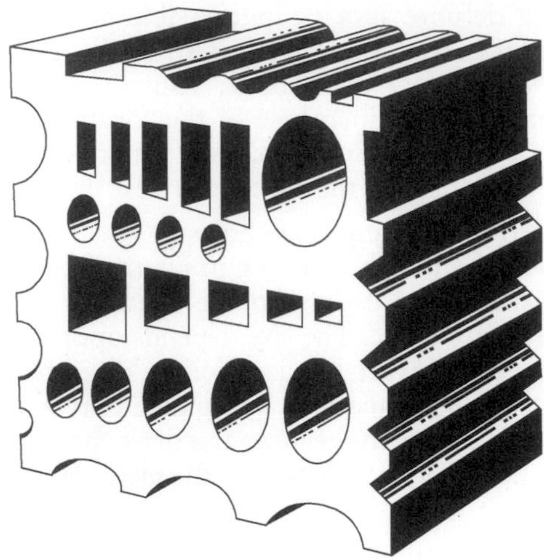

Figure 35-7 A swage block.

Hammers

Ordinarily, five kinds of hammers are used in a blacksmith's shop, and they are known as

- Ball-peen
- Cross-peen
- Straight-peen
- Riveting
- Sledge

By definition the word *peen* (also spelled *pein*) means the end of a hammer head opposite the face when adapted for striking; it is usually shaped for indenting when pointed, conical, hemispherical, or wedge-shaped.

Ball-, cross-, and straight-peen hammers are shown in Figure 35-8, and some of their uses are shown in Figure 35-9. All hammers used directly on hot iron or steel should have the centers of their faces slightly crowning or convex, and the edges well rounded off to prevent their leaving sharp and unsightly marks on the work. Sledge hammers are used by the blacksmith's assistant or helper for making forgings heavier than could be successfully made by hand hammering alone. The sledge hammer is used both directly on the hot metal in roughly blocking it to shape and in finishing it by means of other tools placed on the work and struck with the sledge. The most common patterns of the sledge are shown in Figure 35-10.

(A) BALL PEEN.
(B) CROSS PEEN.
(C) STRAIGHT PEEN.

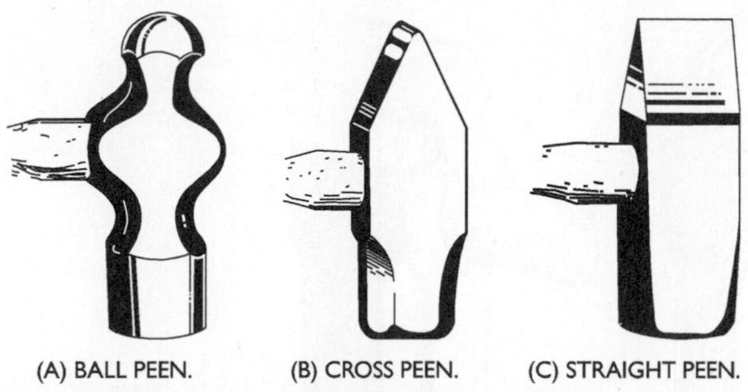

Figure 35-8 Various blacksmith peen hammers.

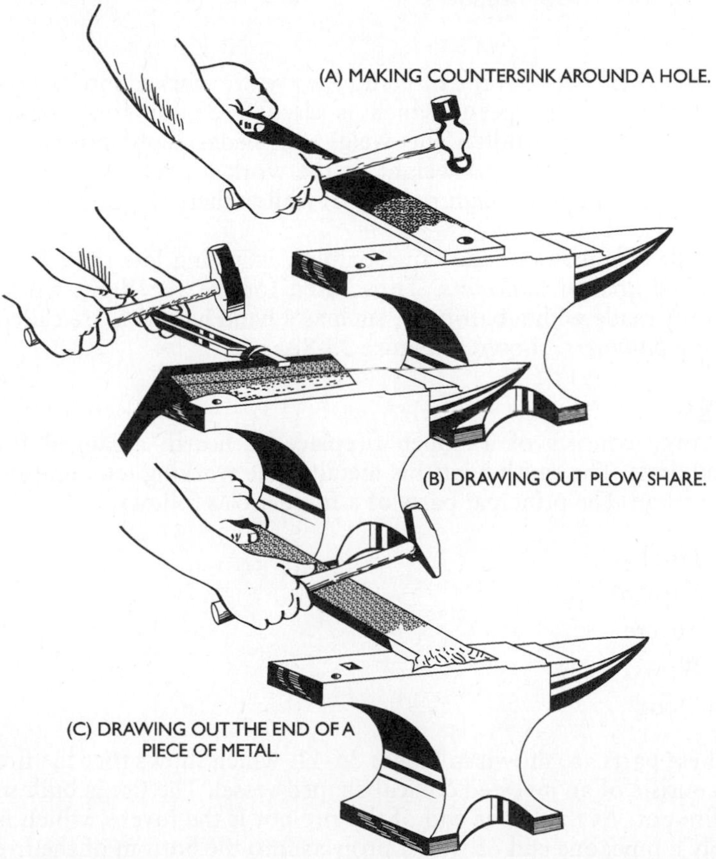

(A) MAKING COUNTERSINK AROUND A HOLE.
(B) DRAWING OUT PLOW SHARE.
(C) DRAWING OUT THE END OF A PIECE OF METAL.

Figure 35-9 Some of the operations performed with the peen hammer.

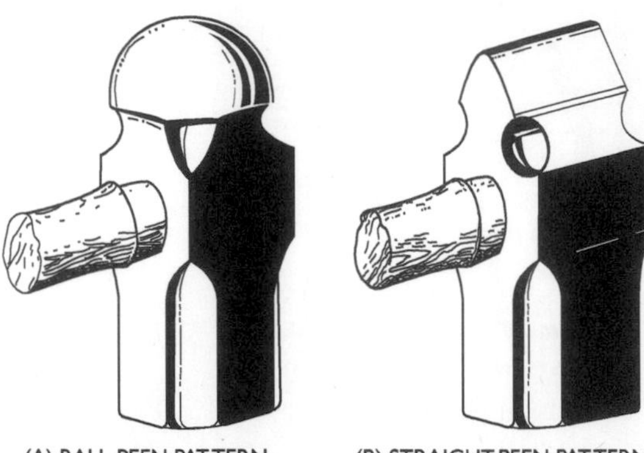

Figure 35-10 Sledge hammers.

The corners between the face and the eye are worked into octagonal shape and the peen, which is circular on the top, stands straight with the handle. The weight of sledge hammers varies according to the size and weight of the work for which they are used; some hammers weigh only 8 lb, while others weigh 20 lb or over.

Smaller hammers of the same pattern, weighing less than 8 lb, are called *quarter hammers*. Those used for the very light work, generally made with a ball-peen, such as a hand hammer, are called *backing hammers*, shown in Figure 35-8A.

Forge

The forge consists of an open fireplace or hearth arranged for forced draft. The smith heats his metal to the working temperature in the forge. The principal parts of a forge are as follows:

- Fire pot
- Hearth
- Tuyere
- Blower
- Hood

These parts are shown in Figure 35-11, which shows that the fire pot consists of an inverted conical-shaped vessel. The fire is built in the fire pot. At the lower end of the fire pot is the tuyere, which is simply a pipe, one end of which projects into the bottom of the fire pot and through which a blast of air obtained from a blower (or bellows) is used for forced draft. Surrounding the fire pot is a large box-like casing or hearth filled with cinders. Coal is tamped around the fire, on which the metal to be heated is placed.

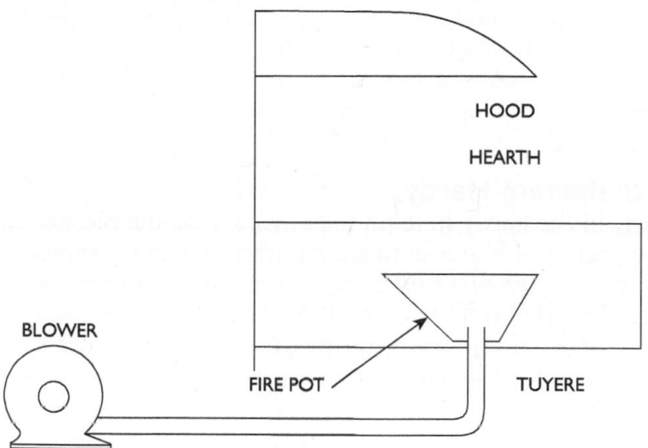

Figure 35-11 The elementary blacksmith forge showing essential parts.

The small circular-type forge, owing to its portability, can be carried into the field for such purposes as tool dressing and small on-site repairs. The best type of portable forge may be taken apart into three pieces by the disconnecting of a few wing nuts. Fan, pan, and legs can be separately packed for transportation. Whenever more than rivet heating or chisel dressing is intended, it is advisable to use a large, square pan forge of a more substantial type.

Forge Operation

The fuel used on the forge is bituminous coal. It should contain very little sulfur and earthly matter. The best-quality coal is called smithing coal, although charcoal or coke may be used. When building a fire, place a block or brick over the tuyere opening and bank the coal in the fire pot; then remove the block or brick and insert shavings in the opening. When the shavings are well ignited, place some coke over the flame and accelerate the fire with the blower. Add a quantity of smithing coal well damped with water and partially burn out the gases.

The depth of the fire should always be liberal, because with a shallow fire, the blast of the blower will blow through the fire, and the excess air will rapidly oxidize the metal being heated. The fire should be limited to as small a space as is necessary to heat the metal. It is regulated by quenching around its exterior portion. Use the blast only when heating the metal, and if it is desired to keep the fire for any length of time, it should be well banked.

Various forging operations are shown in the following illustrations. These operations are cutting, drawing, upsetting, bending, straightening, and punching.

Cutting with Bottom Hardy

Place the hardy in the hardy-hole on the anvil. Grasp the piece to be cut and strike a series of blows with the hammer, striking first on one side and then turning the piece over and striking it on the other side. When turning the piece over, be careful that the edge of the hardy is placed directly under the indentation just made on the other side. The illustration in Figure 35-12 shows the cutting of a cold piece.

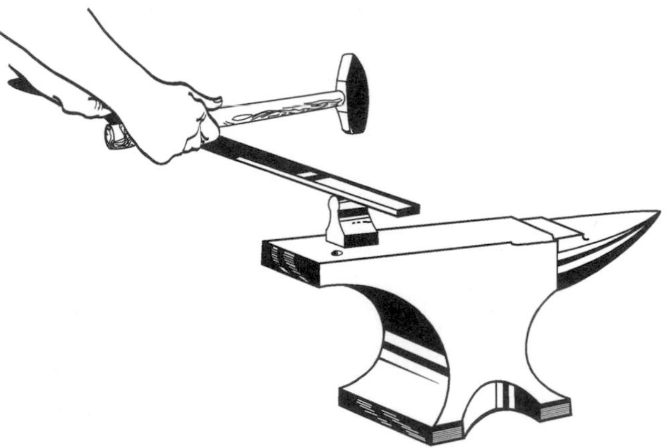

Figure 35-12 Cutting flat iron with a bottom hardy.

Cutting with a Top Hardy

Place the piece to be cut on the cutting face of the anvil. Hold the piece with one hand and the top hardy with the other hand. Place the hardy squarely across the piece and nick or deeply cut by blows delivered by a helper with a sledge. This operation, shown in Figure 35-13, is performed on the cutting face rather than the hardened face to protect the tool and anvil.

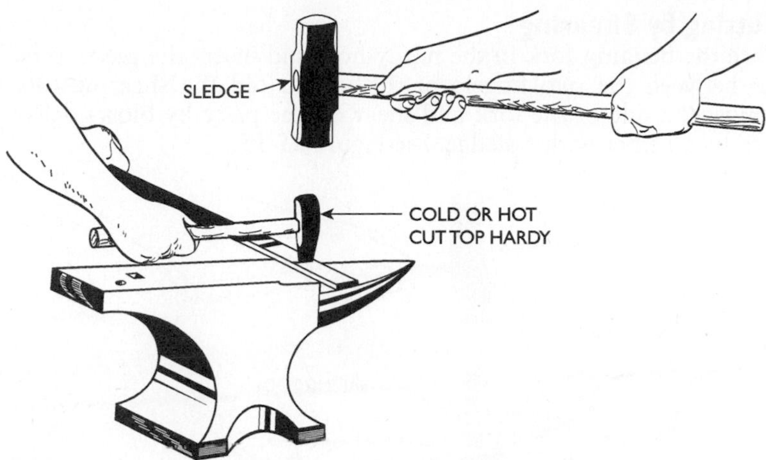

Figure 35-13 Cutting flatiron with the use of a top hardy.

Cutting by Nicking

After the piece to be cut has been nicked, it is placed over an edge of the anvil and broken apart by hammering. Because the breaking operation is a bending action, the hammer blows should not be delivered too near the nick. This type of operation is shown in Figure 35-14.

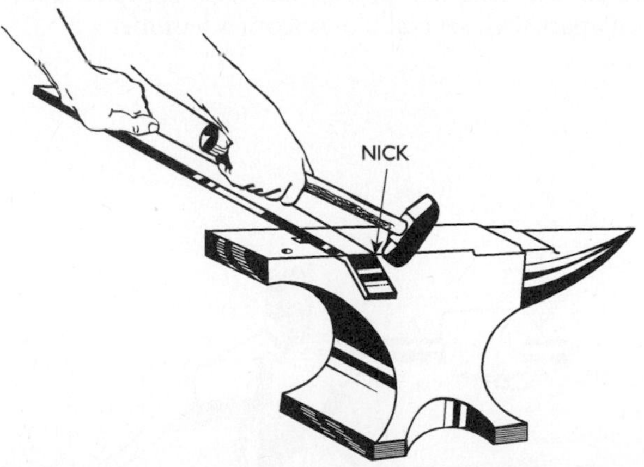

Figure 35-14 Cutting flatiron by first nicking the material.

Cutting by Shearing

Place the bending fork in the hardy-hole and insert the piece to be cut between the two fingers of the fork. Hold the shearing tool against the side of the fork and shear off the piece by blows delivered by a helper with a sledge. See Figure 35-15.

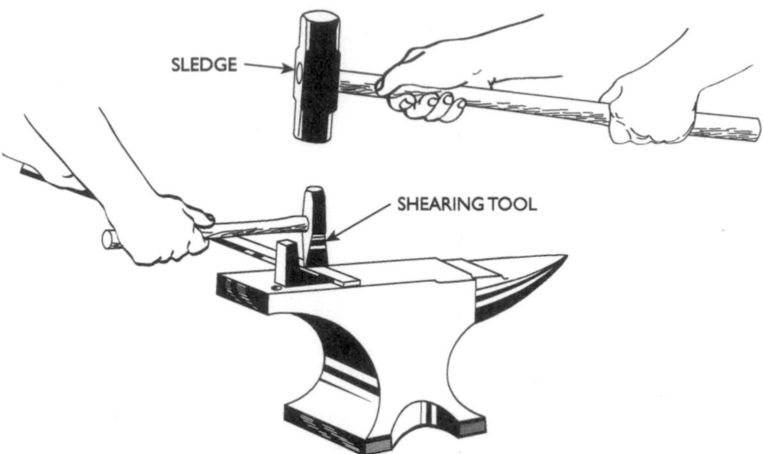

Figure 35-15 Cutting flatiron with a shearing tool.

Cutting with Cold Chisel

Place the piece to be cut in a vise and nick both sides with a cold chisel, as shown in Figure 35-16. After this process has been completed the piece can be broken apart with several blows from a hammer.

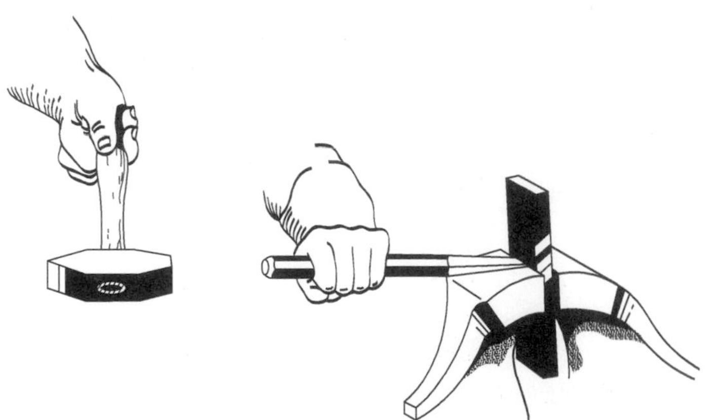

Figure 35-16 Cutting flatiron by using a cold chisel.

Drawing

In order to reduce the diameter of round stock, it is advisable to first draw the piece to a square shape, as shown in Figure 35-17A, then to an octagonal shape (Figure 35-17B), after which it can be hammered to a rough round shape, as shown in Figure 35-17C. After the stock has been reduced in size, it is placed in the swage (Figure 35-17D) where all uneven spots are removed.

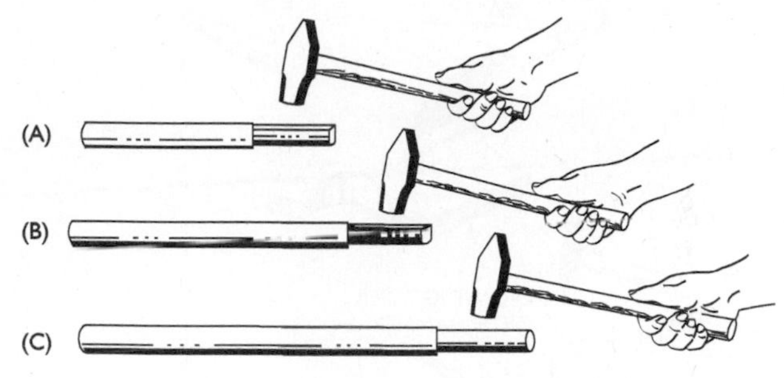

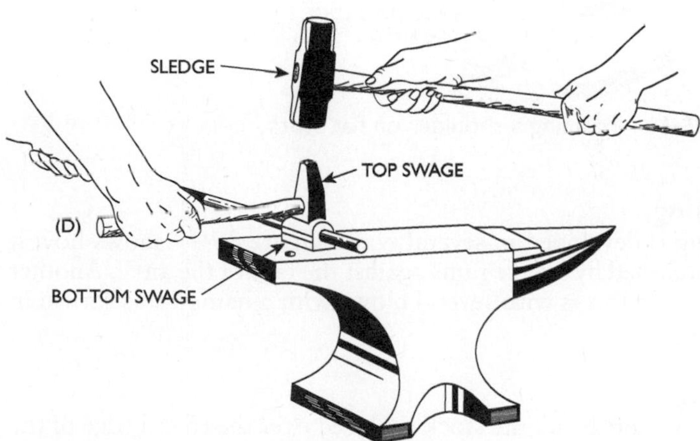

Figure 35-17 Drawing round stock.

When forming a shoulder on a piece of flatiron, it is first placed between the top and bottom fullers, as shown in Figure 35-18A. The taper is formed by using the anvil and hammer as shown in Figure 35-18B.

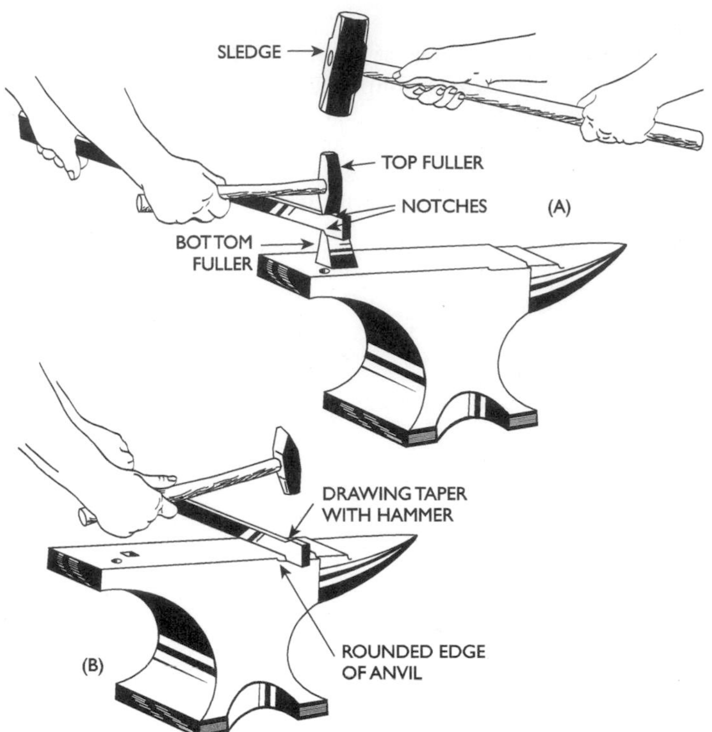

Figure 35-18 Forming a shoulder on flat stock.

Upsetting

Upsetting is developed in several ways. Figure 35-19 shows how it is accomplished by simple rams against the face of the anvil. Another way of doing this is with several blows with a hammer, as shown in Figure 35-20.

Bending

For a very short bend, the stock is placed over the round edge of the anvil and hammered to shape as shown in Figure 35-21. For anvil-horn bends, place stock over the horn at the point where its curvature corresponds with the curve required and hammer to shape, as shown in Figure 35-22.

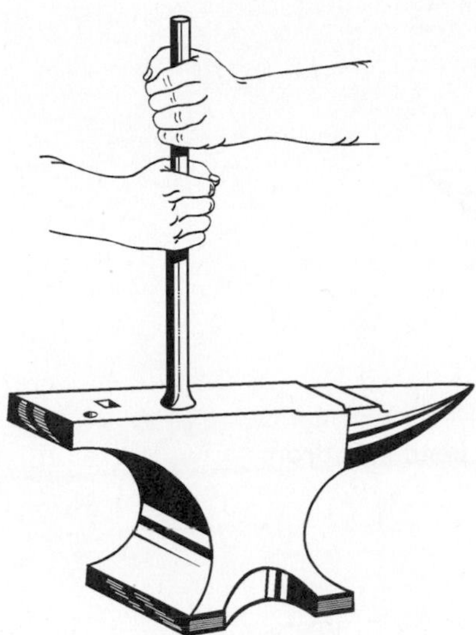

Figure 35-19 Developing a striking head on a rod by ramming the end of the rod against the anvil face.

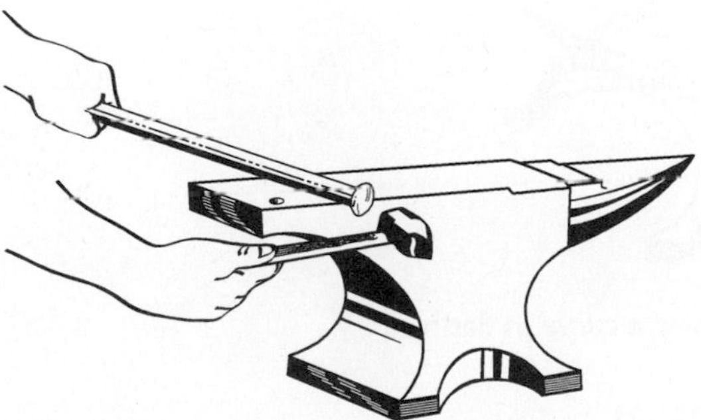

Figure 35-20 Developing a striking head with several blows from a hammer.

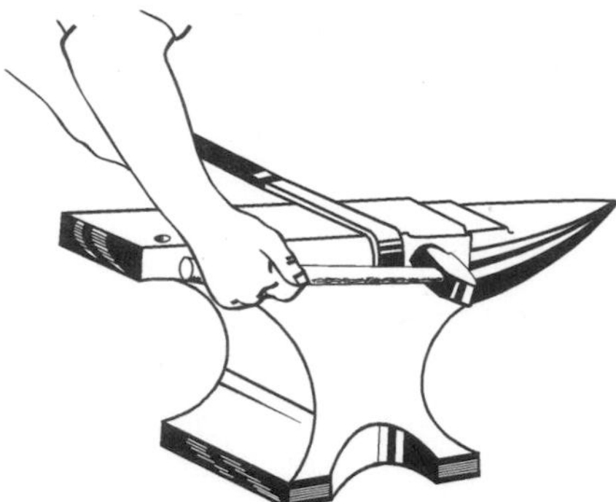

Figure 35-21 Making a very short bend in flatiron.

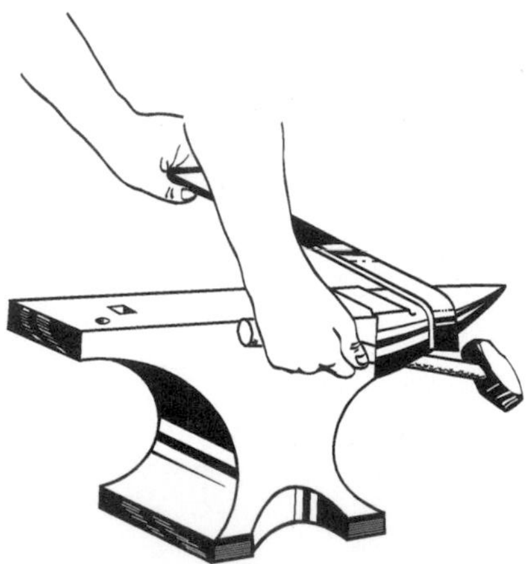

Figure 35-22 Developing a curve in flatiron.

Straightening

Figure 35-23 illustrates a method used to straighten thin flat pieces such as saws and knives. The metal is struck with a hammer along the surface of the bend.

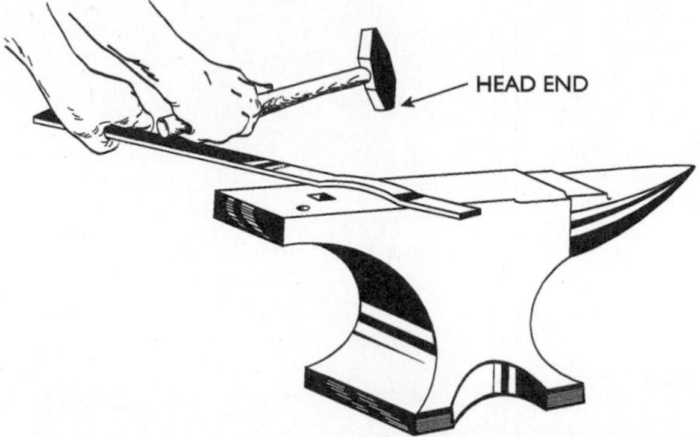

Figure 35-23 One of many methods used to straighten flatiron.

Punching

To punch holes in flatiron, place the punching block in the hardy-hole in the anvil. Select a hole in the punching block that will let the disc or slug punched out of the metal pass through. See Figure 35-24.

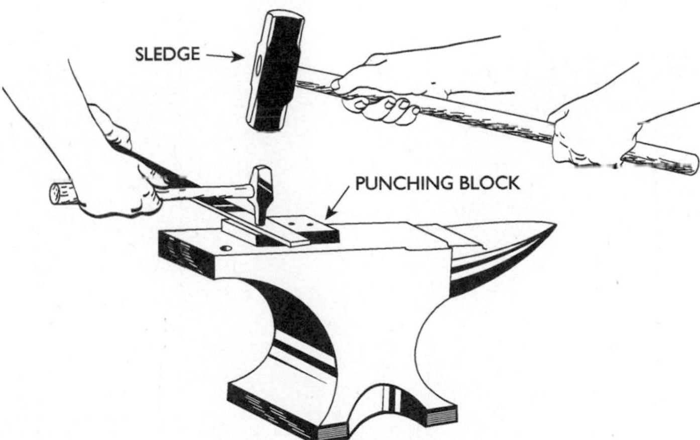

Figure 35-24 Punching a hole in flatiron with the use of a punching block.

Chapter 36

Rigging

The moving of heavy objects such as machinery and equipment, using ropes, cables, slings, rollers, hoists, etc., is called *rigging*. The term is probably derived from the nautical rigging operations, which involved the fitting of ships and the handling of heavy gear primarily with ropes, slings, and pulleys. The rigging operation as practiced today in the construction field and in industry has developed into a much broader operation using all manner of tools and equipment. To perform this rigging operation in a safe manner requires that the tools and equipment be adequate to handle the heavy loads and that safe practices and methods be followed. The greater the weight of an object, the greater the potential for injury and destruction. As the selection of tools and equipment, as well as the method, is usually based on the object's weight, the determination of the weight of objects to be moved is one of the most important functions involved in rigging work.

When the weight of an object is not known, and the information cannot be obtained from catalog data, shipping papers, or other dependable sources, it must be estimated. An estimate is an approximate calculation carefully made—it is not a guess. Safe practice in rigging work requires an accuracy of plus or minus 20 percent for light objects and much closer limits for heavy objects. Estimates within these limits can be easily and quickly made by approximate calculation. Guesses may be outside of safe limits.

The following is a summary and review of the common terms and simple formulas used for weight estimating:

Linear or Length Measure. This is a length or distance measure. The units used for weight-estimating purposes are the inch and the foot.

Square or Area Measure. This is the measure of a surface in terms of squares. The units used for weight-estimating purposes are the square inch and the square foot.

Cubic or Volume Measure. This is the measure of the volume or internal size or capacity of an object in terms of cubes. The units used for weight-estimating purposes are the cubic inch and the cubic foot.

Circumference of a Circle. The length or distance around a circle is called the *circumference* of a circle. The circumference of a circle always has the same relationship to the distance across the circle, called the *diameter.* The circumference is always 3.1416 times longer than the diameter.

Area Calculation

The value 3.1416, which is the ratio of the circumference of a circle to its diameter, is called pi (π). For weight-estimating purposes, the value of $3\frac{1}{7}$ or $\frac{22}{7}$ is often used since it is approximately correct and simplifies calculations. For very rapid calculation of lesser accuracy, the figure may be rounded off to an even 3.

Rectangle

The area of a rectangle is found by multiplying the width by the length. See Figure 36-1.

Area = width × length
Area = 4 × 6 = 24 sq. ft.

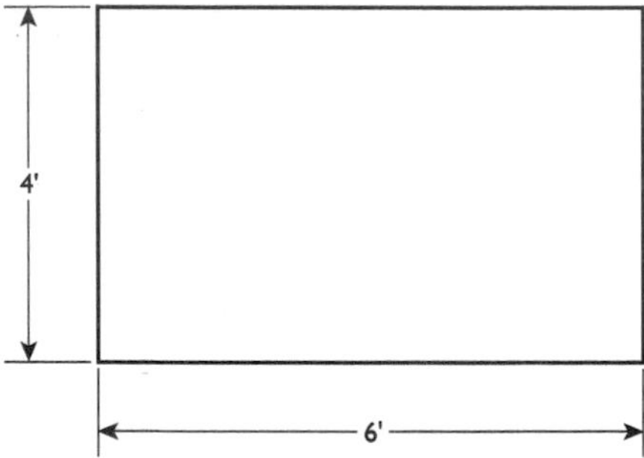

Figure 36-1 A rectangle with 24 sq. ft of area

Circle

The area of a circle can be found by squaring the radius (multiplying the radius by itself) and multiplying the product by π (3.1416 or approximately $\frac{22}{7}$). See Figure 36-2. One-fourth pi (.7854) can also be used to calculate area by multiplying the *diameter* squared times .7845.

Diameter = 8 ft.
Radius = one-half of 8 ft., or 4 ft.
Area = radius × radius × π
Area = $4 \times 4 \times \frac{22}{7}$
Area = $\frac{352}{7}$ or 50.3 sq. ft.

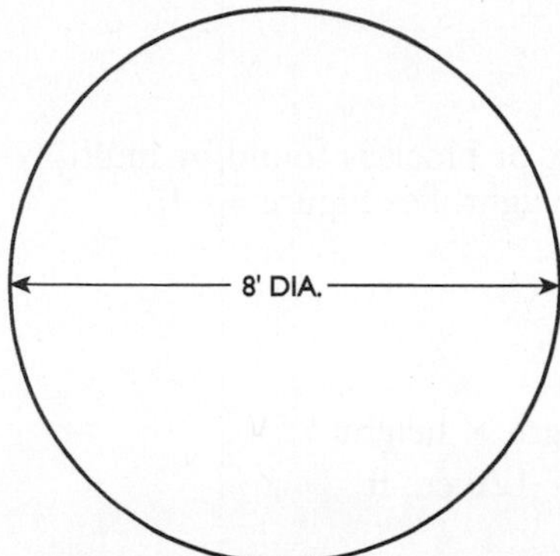

Figure 36-2 A circle with 50.3 sq. ft of area.

Note: For rapid approximate calculations, an even 3 may be used as the value of pi. The calculation may be done mentally, as in $4 \times 4 \times 3$, or 48 sq. ft. While this is easier and more rapid, it results in an error of about 4½ percent on the light side.

Triangle

The area of a triangle is found by multiplying the base length by the height and dividing by 2. See Figure 36-3.

Base length	= 10 ft.
Height	= 8 ft.
Area	= base × height ÷ 2
Area	= 10 × 8 ÷ 2 = 40 sq. ft.

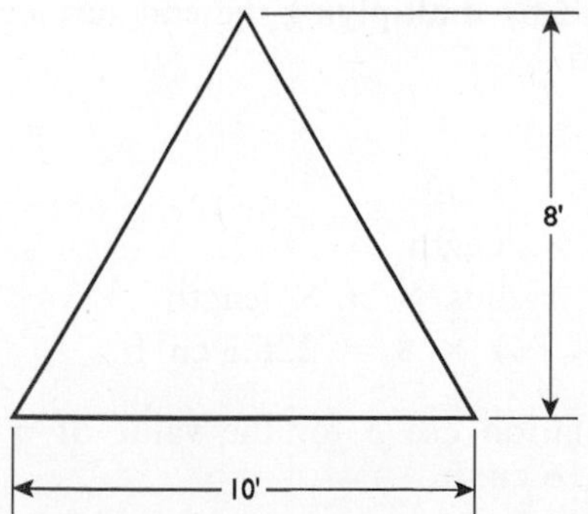

Figure 36-3 A triangle with 40 sq. ft of area.

Volume Calculation

Cube

The volume of a cube such as a box or block is found by multiplying the width by the length by the height. See Figure 36-4.

Width = 4 ft.
Length = 6 ft.
Height = 5 ft.
Volume = width × length × height
Volume = 4 × 6 × 5 = 120 cu. ft.

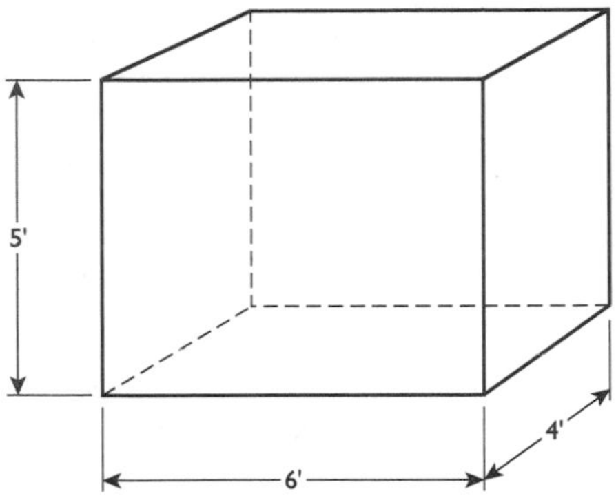

Figure 36-4 Finding the volume of a cube.

Cylinder

The volume of a cylinder is found by multiplying the end surface area by the length. See Figure 36-5.

Diameter = 6 ft.
Length = 8 ft.
Volume = end area × length
Volume = radius × radius × π × length
Volume = 3 × 3 × 22/7 × 8 = 226.3 cu. ft.

Note: For rapid mental calculation use 3 for the value of π. Volume = 3 × 3 × 3 × 8 = 216 cu. ft.

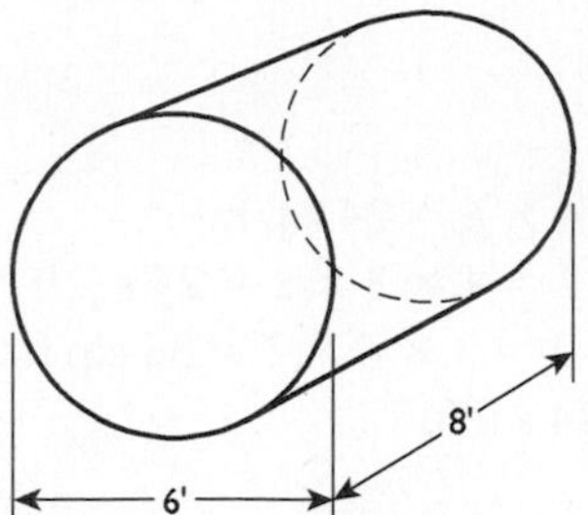

Figure 36-5 Estimating the volume of a cylinder.

Weight Estimation

Most heavy objects handled by riggers are constructed principally of iron or steel. The weight per cubic foot of these two materials ranges from 475 to 490 lb. For ease of calculation, the value of 500 lb per cubic foot is used as the weight of iron and steel for weight-estimating purposes. The resulting error will not exceed 5 percent and will be on the heavy side.

Example 1

Estimating the weight of a steel tank (open top). See Figure 36-6.

Plate thickness = ½ in.
Width = 4 ft
Height = 3 ft
Length = 6 ft

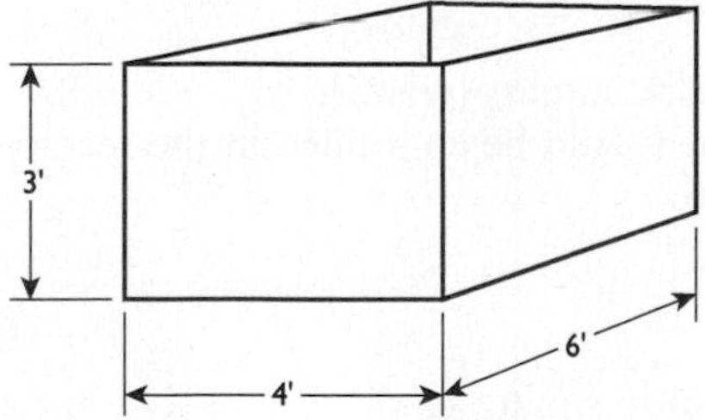

Figure 36-6 Estimating the weight of a steel tank.

Steps

1. Find the area of the bottom and the sides in square feet.
2. Multiply the total area by the plate thickness to obtain the volume of the metal in cubic feet.
3. Multiply the volume of the metal by the weight per cubic foot (500).

1. *Area of Bottom and Sides*

Bottom area = width × length = 4 × 6 = 24 sq. ft.
Front and back area = width = height × 2 = 4 × 3 × 2 = 24 sq. ft.
Sides area = height × length × 2 = 3 × 6 × 2 = 36 sq. ft.
Total Area = 24 + 24 + 36 = 84 sq. ft.

2. *Volume of Metal*

Convert this thickness from inches to feet. ½ ÷ 12 = 1/24 ft.
Volume = Area × thickness = 84 × 1/24 = 3½ cu. ft.

3. *Weight of Metal*

Weight = volume in cu. ft. × weight per cu. ft.
Weight = 3½ × 500 = 1750

Alternate Method

If weight tables are available, the weight may be determined by multiplying the total square-foot area of the tank by the weight per square foot of steel plates as listed in the table.

Weight table lists ½-in. steel plate at 20.4 lb/sq. ft.

Weight = total area × weight/sq. ft.
Weight = 84 × 20.4 = 1713.6 lbs.

Example 2

Estimating the weight of the tank contents (water).

Find the weight of water that would be contained in the rectangular tank shown in Figure 36-6.

Steps

1. Find the area of the bottom in square feet.
2. Multiply the bottom area by the height to get the volume of the tank in cubic feet.
3. Multiply the volume by the weight of water per cubic foot.

1. *Area of Bottom*

Bottom area = width × length = 4 × 6 = 24 sq. ft.

2. *Volume or Contents of Tank*

Volume = Bottom area × height
Volume = 24 × 3 = 72 cu. ft.

3. *Weight of Water*

Note: Weight per cu. ft of water is 62.5 lb.

Weight = volume × weight/cu. ft.
Weight = 72 × 62.5 = 4500 lbs.

Example 3
Estimating the weight of a cylindrical tank. See Figure 36-7.

Plate thickness = ⅜ inch
Diameter = 12 ft.
Length = 16 ft.

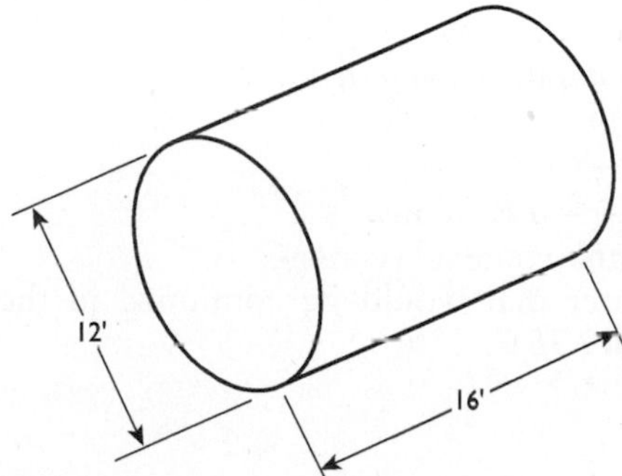

Figure 36-7 Estimating the weight of a cylindrical tank.

Steps

1. Find the area of the tank end in square feet.
2. Find the area of the tank shell in square feet.
3. Find the volume of the metal by multiplying the total area (ends and shell) by the plate thickness.
4. Find the weight by multiplying the volume by the weight per cubic foot.

1. *Area of Ends*

Area = radius × radius × π × 2
Area = 6 × 6 × 22⁄7 × 2 = 226 sq. ft.

2. *Area of Shell* (consider shell as cut and opened out flat)

Area = circumference × length
Area = diameter × π × length
Area = 12 × 22⁄7 × 16 = 624 sq. ft.
Total Area
Area at ends + area of shell: 226 + 624 = 850 sq. ft.
Area = Diameter × π × length

3. *Volume of Metal*

Convert the plate thickness from inches to feet: 3⁄8 ÷ 12 = 1⁄32 ft.
Volume = area × thickness = 850 × 1⁄32 = 26.6 cu. ft.

4. *Weight of Metal*

Weight = volume in cu. ft. × weight/cu. ft.
Weight = 26.6 × 500 = 13,300 lbs.

Example 4
Estimating the weight of the tank contents (water).
Find the weight of the water that would be contained in the cylindrical tank shown in Figure 36-7.

Steps

1. Find the area of the tank end in square feet.
2. Multiply the area of the end by the length to get the tank volume in cubic feet.
3. Multiply the volume by the weight of water per cubic foot (62.5 lb).

1. *Area of tank end*

Area = radius × radius × π
Area = 6 × 6 × 22⁄7 = 792⁄7 = 113 sq. ft.

2. *Volume or cubic content of tank*

Volume = end area × length
Volume = 113 × 16 = 1808 cu. ft.

3. *Weight of water* (water weighs 62.5 lb/cu. ft.)

Weight = volume × weight per cu. ft.
Weight = 1808 × 62.5 = 113,000 pounds.

Example 5
Estimating the weight of a cast-iron roll (solid).
Find the weight of the solid cast-iron roll shown in Figure 36-8.

Roll dia. = 24 inches = 2 ft.
Roll face = 42 inches = 3½ ft.
Shaft dia. = 9 inches = ¾ ft.
Shaft length = 12 inches = 1 ft.

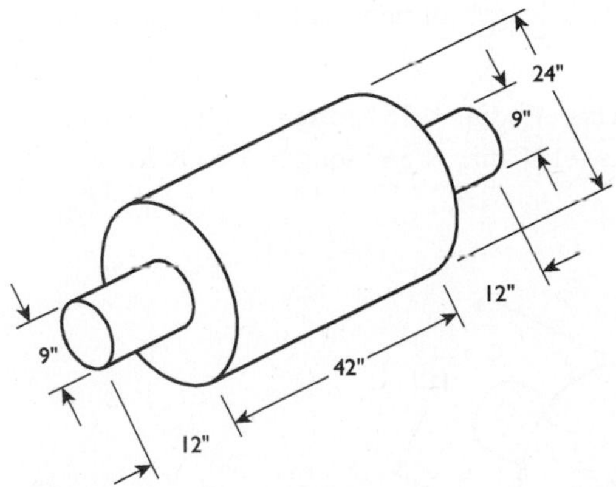

Figure 36-8 Estimating the weight of a cast-iron roll.

Steps

1. Find the volume of the roll shafts in cubic feet.
2. Find the volume of the roll body in cubic feet.
3. Multiply the volume of metal by the weight per cubic foot.

1. *Volume of shafts*

   Volume = end area × length
   Volume = radius × radius × π × length
   Volume = $\frac{3}{8} \times \frac{3}{8} \times \frac{22}{7} \times 1 = \frac{99}{224}$ = .45 cu. ft.
   Volume of 2 shafts = .45 × 2 = .9 cu. ft.

2. *Volume of roll body*

   Volume = radius × radius × π × length
   Volume = $1 \times 1 \times \frac{22}{7} \times 3\frac{1}{2}$ = 11 cu. ft.

3. *Weight of roll* (Cast iron estimating weight is 500 lb/cu. ft.)

   Weight = volume × weight per cu. ft.
   Weight = 11.9 × 500 = 5950 pounds.

Example 6

Estimating the weight of a cast-iron roll (chambered).

Find the weight of the chambered cast-iron roll shown in Figure 36-9. Outside diameter and length dimensions are the same as in Figure 36-8.

Shaft chambers—4" dia. × 15" long = ⅓ ft. × 1¼ ft.
Body chambers—12" dia. × 36" long = 1 ft. × 3 ft.

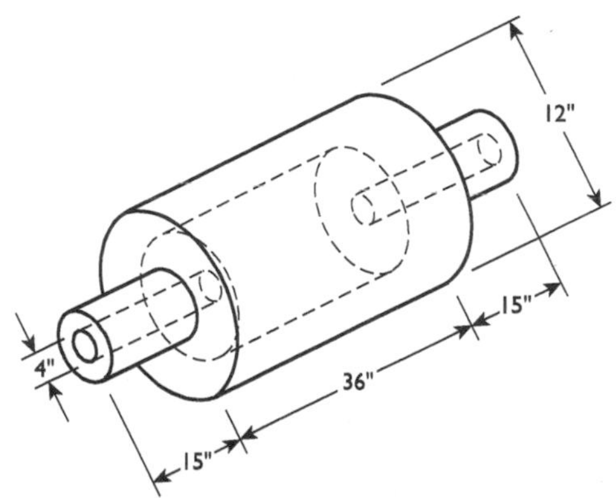

Figure 36-9 Estimating the weight of a chambered cast-iron roll.

Steps

1. Find the volume of the solid shafts.
2. Find the volume of the solid roll body.
3. Find the volume of the chambers in shafts.
4. Find the volume of the chamber in body.
5. Subtract the volume of the chambers from the solid volume to get the metal volume.
6. Multiply the metal volume by the weight per cubic foot.

1. *Volume of solid shafts.* From Example 5—9 cu. ft.
2. *Volume of solid body.* From Example 5—11 cu. ft.
3. *Volume of shaft chambers*

$$\text{Volume} = \text{end area} \times \text{length} \times 2$$
$$\text{Volume} = \text{radius} \times \text{radius} \times \pi \times \text{length} \times 2$$
$$\text{Volume} = \tfrac{1}{6} \times \tfrac{1}{6} \times \tfrac{22}{7} \times \tfrac{5}{4} \times 2 = .2 \text{ cu. ft.}$$

4. *Volume of body chambers*

$$\text{Volume} = \text{end area} \times \text{length}$$
$$\text{Volume} = \text{radius} \times \text{radius} \times \pi \times \text{length}$$
$$\text{Volume} = \tfrac{1}{2} \times \tfrac{1}{2} \times \tfrac{22}{7} \times 3 = 2.3 \text{ cu. ft.}$$

5. *Metal volume*

$$\text{Total solid volume} = .9 + 11 = 11.9 \text{ cu. ft.}$$
$$\text{Total chamber volume} = .2 + 2.3 = 2.5 \text{ cu. ft.}$$
$$\text{Metal volume} = 11.9 \text{ minus } 2.5 = 9.4 \text{ cu. ft.}$$

6. *Weight of roll*

$$\text{Weight} = \text{volume} \times \text{weight/cu. ft.}$$
$$\text{Weight} = 9.4 \times 500 = 4700 \text{ pounds.}$$

Example 7

Estimating the weight of an irregularly shaped object. See Figure 36-10.

The weight of objects having irregular shapes may be estimated with a high degree of accuracy by visualizing the object as a regular shape or a group of regular shapes. For example, the object shown in Figure 36-10A may be visualized as a sold rectangular object with proportionally lesser dimensions as shown in Figure 36-10B. The weight estimation now becomes a simple calculation of the weight of a rectangular object.

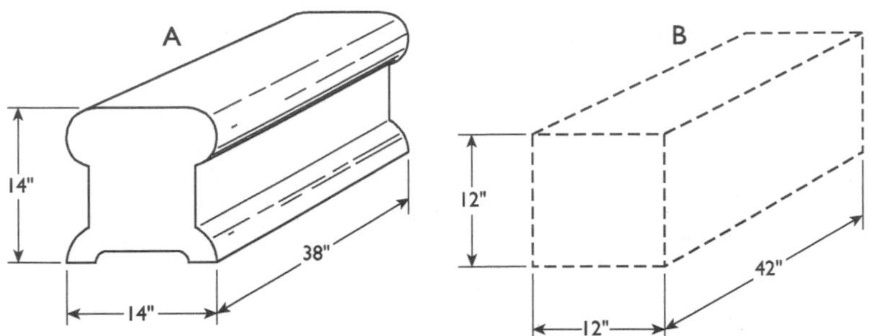

Figure 36-10 Comparing irregularly and regularly shaped objects to estimate weight.

Example 8
Estimating the weight by using weight tables.

In some cases it may be simpler and easier to estimate the weight by using standard weight tables for steel bars, sheets, and structural shapes. When this is done, the arithmetic is simplified, as no volume calculations are necessary. The weights of bars and shapes are listed by their cross-sectional size and weight per linear foot. Steel-plate weights are usually given in weight per square foot, thus an area calculation is necessary to use the plate tables. This method of weight estimating is most applicable for machinery and equipment fabricated from structural members and plate materials.

The tank shown in Figure 36-11 is an example of such a piece of equipment. At first glance it would appear to be rather difficult to estimate the weight of this object, but with a few simple mental calculations and reference to a weight table it becomes an easy task.

Steps

1. Find the area of plate in body section.
2. Multiply the area of plate by weight per square foot.
3. Multiply the total length of 6 in. I-Beam by weight per linear foot.
4. Multiply the total length of 8 in. H-Beam by weight per linear foot.
5. Add all part weights to find total weight.

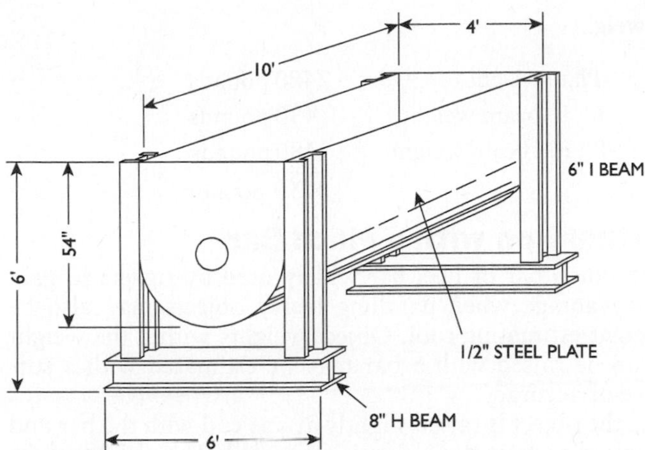

Figure 36-11 Calculating the weight of a tank.

Steel Plate Area. The easiest way to find the plate area would be simply to measure the distance around the body with a tape and multiply this by the length. It can be mentally calculated by multiplying the radius (2 ft) by π (3) to get a distance of 6 ft around the bottom, plus 5 ft for the two 2½-ft sides, or a total of 11 ft. The area of the body sides and bottoms is 11 × 10, or 110 sq. ft. The ends may be visualized as approximately two 4 × 4 plates, or an area of 4 × 4 × 2, or 32 sq. ft. The total area of body plate then is 110 plus 32, or 142 sq. ft.

Weight of steel plate. Weight of ½ in. steel plate from table—20.4 lb/sq. ft.

Weight = area × weight/sq. ft.
Weight = 142 × 20 = 2840 pounds.

Weight of 6 in. I-beam. Weight of 6 in. I-beam from table—17.7 lb/linear ft. Length of 6 in. I-beam—4 pieces 6 ft long = 24 ft.

Weight = length × weight/linear ft.
Weight = 24 × 18 = 430 pounds.

Weight of 8 in. H-beam. Weight of 8 in. H-beam from table—40 lb/linear ft. Length of 8 in. H-beam—2 pieces 6 ft. long = 24 ft.

Weight = length × weight/linear ft.
Weight = 12 × 40 = 480 pounds.

Total weight

Plate weight	2480 pounds
6" I–beam weight	430 pounds
8" H–beam weight	480 pounds
	3750 pounds

Weight Estimation with a Pinch Bar

The common pinch bar or heel bar widely used by riggers to gain mechanical advantage when handling heavy objects may also be used as a weight-estimating tool. Object weights within the weight range that can be raised with a bar may be estimated with a surprising degree of accuracy.

To do this, the object is raised slightly at one end with the bar and the force required to raise the object is estimated. The weight of the end of the object will be as many times greater than the force as the ratio of the leverage of the bar. As objects that are generally uniform in shape and construction will have a total weight of twice the end weight, doubling the estimated weight will give the total weight.

For example, the man shown in Figure 36-12 is using a pinch bar. Notice that the distance to the operator end is 30 times greater than the distance from the heel to the toe point. For each pound of force exerted downward on the end of the bar, there will be 30 lb exerted upward at the toe point. If the man shown in Figure 36-12 can raise the end of the object by pressing downward on the end of the bar with a force estimated at 100 lb, the end weight of the object will be about 3000 lb. The total weight of the object will be twice this amount, or about 6000 lb.

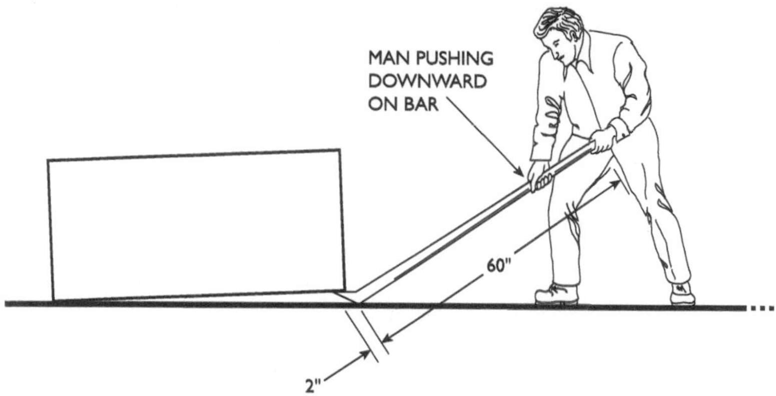

Figure 36-12 Weight estimation with a pry bar.

Ropes

For many years the principal materials from which rope was manufactured were manila (hemp), cotton, and wire. Synthetic materials have in many cases replaced manila and cotton. While the reasons for this are numerous, some of the most important are probably increased resistance to deterioration, greater strength, and greater pliability. Wire rope continues to be used in applications where very great strength is required, as in heavy-duty hoisting, principally by mechanical means.

Most rope is manufactured in three or more strands. Most wire rope used for hoisting has six or more strands. Wire rope is discussed in detail later in this chapter; this section deals with manila, cotton, and synthetic materials. Three-strand construction has been the most widely used style for many years. Since the eight-strand plaited construction has been introduced it has been widely used, particularly in synthetic materials. Increasing the number of strands reduces the size of the strands for a given rope size and increases the rope's pliability.

Table 36-1 Cordage Institute Rope Specifications—Three-Strand Laid and Eight-Strand Plaited (Standard Construction)

Nominal Size		Manila				Polypropylene			
Diameter	Circumference	Linear Density[1] (lb/100 ft)	New Rope Tensile Strength[2] (lb)	Safety Factor	Working Load[3] (lb)	Linear Density[1] (lb/100 ft)	New Rope Tensile Strength[2] (lb)	Safety Factor	Working Load[3] (lb)
3/16	5/8	1.50	406	10	41	.70	720	10	72
1/4	3/4	2.00	540	10	54	1.20	1130	10	113
5/16	1	2.90	900	10	90	1.80	1710	10	171
3/8	1 1/8	4.10	1220	10	122	2.80	2440	10	244
7/16	1 1/4	5.25	1580	9	176	3.80	3160	9	352
1/2	1 1/2	7.50	2380	9	264	4.70	3780	9	420
9/16	1 3/4	10.4	3100	8	388	6.10	4600	8	575
5/8	2	13.3	3960	8	496	7.50	5600	8	700
3/4	2 1/4	16.7	4860	7	695	10.7	7650	7	1090
13/16	2 1/2	19.5	5850	7	835	12.7	8900	7	1270
7/8	2 3/4	22.4	6950	7	995	15.0	10,400	7	1490
1	3	27.0	8100	7	1160	18.0	12,600	7	1800
1 1/16	3 1/4	31.2	9450	7	1350	20.4	14,400	7	2060
1 1/8	3 1/2	36.0	10,800	7	1540	23.8	16,500	7	2360

(continued)

Nominal Size		*Manila*				*Polypropylene*			
Diameter	***Circum-ference***	***Linear Density[1] (lb/100 ft)***	***New Rope Tensile Strength[2] (lb)***	***Safety Factor***	***Working Load[3] (lb)***	***Linear Density[1] (lb/100 ft)***	***New Rope Tensile Strength[2] (lb)***	***Safety Factor***	***Working Load[3] (lb)***
1¼	3¾	41.6	12,200	7	1740	27.0	18,900	7	2700
1⁵⁄₁₆	4	47.8	13,500	7	1930	30.4	21,200	7	3020
1½	4½	60.0	16,700	7	2380	38.4	26,800	7	3820
1⅝	5	74.5	20,200	7	2880	47.6	32,400	7	4620
1¾	5½	89.5	23,800	7	3400	59.0	38,800	7	5550
2	6	108.	28,000	7	4000	69.0	46,800	7	6700
2⅛	6½	125.	32,400	7	4620	80.0	55,000	7	7850
2¼	7	146.	37,000	7	5300	92.0	62,000	7	8850
2½	7½	167.	41,800	7	5950	107.	72,000	7	10,300
2⅝	8	191.	46,800	7	6700	120.	81,000	7	11,600
2⅞	8½	215.	52,000	7	7450	137.	91,000	7	13,000
3	9	242.	57,500	7	8200	153.	103,000	7	14,700
3¼	10	298.	69,500	7	9950	190.	123,000	7	17,600
3½	11	366.	82,000	7	11,700	232.	146,000	7	20,800
4	12	434.	94,500	7	13,500	276.	171,000	7	24,400

[1]**Linear Density.** *Linear Density (pounds per 100 ft) shown is "average." Maximum is 5% higher.*

[2]**New Rope Tensile Strengths.** *New Rope Tensile Strengths are based on tests of new and unused rope of standard construction in accordance with Cordage Institute Standard Test Methods.*

[3]**Working Loads** *are for rope in good condition with appropriate splices, in non-critical applications, and under normal service conditions. Working loads should be exceeded only with expert knowledge of conditions and professional estimates of risk. Working loads should be reduced where life, limb, or valuable property are involved, or for exceptional service conditions such as shock loads, sustained loads, etc.*

Courtesy Cordage Institute.

Table 36-2 Cordage Institute Rope Specifications—Three-Strand Laid and Eight-Strand Plaited (Standard Construction)

Nominal Size		*Polyester*				*Composite[4]*				
Diameter	***Circum-ference***	***Linear Density[1] (lb/100 ft)***	***New Rope Tensile Strength[2] (lb)***	***Safety Factor***	***Working Load[3] (lb)***	***Nominal Diameter (in.)***	***Linear Density[1] (lb/100 ft)***	***New Rope Tensile Strength[2] (lb)***	***Safety Factor***	***Working Load[3] (lb)***
³⁄₁₆	⅝	1.20	900	10	90	³⁄₁₆	.94	720	10	72
¼	¾	2.00	1490	10	149	¼	1.61	1130	10	113

5/16	1	3.10	2300	10	230	5/16	2.48	1710	10	171
3/8	1 1/8	4.50	3340	10	334	3/8	3.60	2430	10	243
7/16	1 1/4	6.20	4500	9	500	7/16	5.00	3150	9	350
1/2	1 1/2	8.00	5750	9	640	1/2	6.50	3960	9	440
9/16	1 3/4	10.2	7200	8	900	9/16	8.00	4860	8	610
5/8	2	13.0	9000	8	1130	5/8	9.50	5760	8	720
3/4	2 1/4	17.5	11,300	7	1610	3/4	12.5	7560	7	1080
13/16	2 1/2	21.0	14,000	7	2000	13/16	15.2	9180	7	1310
7/8	2 3/4	25.0	16,200	7	2320	7/8	18.0	10,800	7	1540
1	3	30.4	19,800	7	2820	1	21.8	13,100	7	1870
1 1/16	3 1/4	34.4	23,000	7	3280	1 1/16	25.6	15,200	7	2170
1 1/8	3 1/2	40.0	26,600	7	3800	1 1/8	29.0	17,400	7	2490
1 1/4	3 3/4	46.2	29,800	7	4260	1 1/4	33.4	19,800	7	2830
1 5/16	4	52.5	33,800	7	4820	1 5/16	35.6	21,200	7	3020
1 1/2	4 1/2	67.0	42,200	7	6050	1 1/2	45.0	26,800	7	3820
1 5/8	5	82.0	51,500	7	7350	1 5/8	55.5	32,400	7	4620
1 3/4	5 1/2	98.0	61,000	7	8700	1 3/4	66.5	38,800	7	5550
2	6	118.	72,000	7	10,300	2	78.0	46,800	7	6700
2 1/8	6 1/2	135.	83,000	7	11,900	2 1/8	92.0	55,000	7	7850
2 1/4	7	157.	96,500	7	13,800	2 1/4	105.	62,000	7	8850
2 1/2	7 1/2	181.	110,000	7	15,700	2 1/2	122.	72,000	7	10,300
2 5/8	8	204.	123,000	7	17,600	2 5/8	138.	81,000	7	11,600
2 7/8	8 1/2	230.	139,000	7	19,900	2 7/8	155.	91,000	7	13,000
3	9	258.	157,000	7	22,400	3	174.	103,000	7	14,700
3 1/4	10	318.	189,000	7	27,000	3 1/4	210.	123,000	7	17,600
3 1/2	11	384.	228,000	7	32,600	3 1/2	256.	146,000	7	20,800
4	12	454.	270,000	7	38,600	4	300.	171,000	7	24,400

[1]*Linear Density. Linear Density (pounds per 100 ft) shown is "average." Maximum is 5% higher.*

[2]*New Rope Tensile Strengths. New Rope Tensile Strengths are based on tests of new and unused rope of standard construction in accordance with Cordage Institute Standard Test Methods.*

[3]*Working Loads are for rope in good condition with appropriate splices, in non-critical applications, and under normal service conditions. Working loads should be exceeded only with expert knowledge of conditions and professional estimates of risk. Working loads should be reduced where life, limb, or valuable property are involved, or for exceptional service conditions such as shock loads, sustained loads, etc.*

[4]*Composite Rope. Materials and construction of this polyester/polypropylene composite rope conform to MIL-R-43942 and MIL-R-43952. For other composite ropes, consult the manufacturer.*

Courtesy Cordage Institute.

Table 36-3 Cordage Institute Rope Specifications—Three-Strand Laid and Eight-Strand Plaited (Standard Construction)

Nominal Size		*Nylon*				*Sisal*			
Diameter	*Circumference*	*Linear Density[1] (lb/100 ft)*	*New Rope Tensile Strength[2] (lb)*	*Safety Factor*	*Working Load[3] (lb)*	*Linear Density[1] (lb/100 ft)*	*New Rope Tensile Strength[2] (lb)*	*Safety Factor*	*Working Load[3] (lb)*
3/16	5/8	1.00	900	12	75	1.50	360	10	36
1/4	3/4	1.50	1490	12	124	2.00	480	10	48
5/16	1	2.50	2300	12	192	2.90	800	10	80
3/8	1 1/8	3.50	3340	12	278	4.10	1080	10	108
7/16	1 1/4	5.00	4500	11	410	5.26	1400	9	156
1/2	1 1/2	6.50	5750	11	525	7.52	2120	9	236
9/16	1 3/4	8.15	7200	10	720	10.4	2760	8	345
5/8	2	10.5	9350	10	935	13.3	3520	8	440
3/4	2 1/4	14.5	12,800	9	1420	16.7	4320	7	617
13/16	2 1/2	17.0	15,300	9	1700	19.5	5200	7	743
7/8	2 3/4	20.0	18,000	9	2000	22.5	6160	7	880
1	3	26.4	22,600	9	2520	27.0	7200	7	1030
1 1/16	3 1/4	29.0	26,000	9	2880	31.3	8400	7	1200
1 1/8	3 1/2	34.0	29,800	9	3320	36.0	9600	7	1370
1 1/4	3 3/4	40.0	33,800	9	3760	41.7	10,800	7	1540
1 5/16	4	45.0	38,800	9	4320	47.8	12,000	7	1710
1 1/2	4 1/2	55.0	47,800	9	5320	59.9	14,800	7	2110
1 5/8	5	66.5	58,500	9	6500	74.6	18,000	7	2570
1 3/4	5 1/2	83.0	70,000	9	7800	89.3	21,200	7	3030
2	6	95.0	83,000	9	9200	108.	24,800	7	3540
2 1/8	6 1/2	109.	95,500	9	10,600			7	
2 1/4	7	129.	113,000	9	12,600	146.	32,800	7	4690
2 1/2	7 1/2	149.	126,000	9	14,000			7	
2 5/8	8	168.	146,000	9	16,200	191.	41,600	7	5940
2 7/8	8 1/2	189.	162,000	9	18,000			7	
3	9	210.	180,000	9	20,000	242.	51,200	7	7300
3 1/4	10	264.	226,000	9	25,200	299.	61,600	7	8800
3 1/2	11	312.	270,000	9	30,000			7	
4	12	380.	324,000	9	36,000	435.	84,000	7	12,000

[1] *Linear Density. Linear Density (pounds per 100 ft) shown is "average." Maximum is 5% higher.*

[2] *New Rope Tensile Strengths. New Rope Tensile Strengths are based on tests of new and unused rope of standard construction in accordance with Cordage Institute Standard Test Methods.*

[3] *Working Loads are for rope in good condition with appropriate splices, in non-critical applications, and under normal service conditions. Working loads should be exceeded only with expert knowledge of conditions and professional estimates of risk. Working loads should be reduced where life, limb, or valuable property are involved, or for exceptional service conditions such as shock loads, sustained loads, etc.*

Courtesy Cordage Institute

Because of the variety of rope materials and constructions now in use, as well as the wide range of factors affecting rope behavior, it is impossible to cover all rope applications in this chapter. Perhaps the single most important consideration in rope use is that of safety. While all aspects cannot be specifically detailed, several general safety considerations are important if the mechanic is to use rope properly and avoid possible rope failure. Almost all rope-failure accidents are caused by improper care and use, rather than poor engineering or original product defect. Some very important safety considerations are as follows:

1. Rope must be adequate for the job. Choosing a rope of the correct size, material, and strength must not be done haphazardly. Consult dealer, distributor, or manufacturer for information and assistance if needed.
2. Do not overload rope. Sudden strains or shock loading can cause failure. Working load specifications may not be applicable when rope is subject to significant dynamic loading. Loads must be handled slowly and smoothly to minimize dynamic effects.
3. Avoid using rope that shows signs of aging and wear. If in doubt, destroy the used rope. If the fibers show wear in any given area, the rope should be respliced, downgraded, or replaced.
4. Avoid chemical exposure. Rope is subject to damage by chemicals. Special attention must be given for applications where exposure to either fumes or actual contact may occur.
5. Avoid overheating. Heat can seriously affect the strength of rope. The frictional heat from slippage on capstan or winch may cause localized heating that can melt synthetic fibers or burn natural fibers.
6. Never stand in line with rope under strain. If a rope or attachment fails, it can recoil with sufficient force to cause physical injury. The snap-back action can propel fittings and rope with possible disastrous result.

The safety factors and working loads given in the rope specification tables are for noncritical applications and normal service conditions. In cases where unusual risks are present, or where there is any question about the loads involved or the conditions of use, the working loads should be substantially reduced and the rope properly inspected.

Rope Selection

Because of the wide range factors affecting the performance of rope, including use, condition, and exposure, it is impossible to make blanket recommendations as to the correct choice of rope. However, Table 36-4, which provides a list of relative values and material characteristics, can be used as a guide.

Table 36-4 Relative Values and Material Characteristics of Rope

Relative Values				
Characteristics	***Manila***	***Nylon***	***Polypropylene***	***Polyester***
Strength	Fair	Excellent	Very Good	Excellent
Shock Load	Fair	Excellent	Very Good	Very Good
Surface Abrasion	Good	Very Good	Good	Excellent
Elasticity	Fair	Excellent	Good	Very Good
Floats	No	No	Yes	No
Resistance				
Characteristics	***Manila***	***Nylon***	***Polypropylene***	***Polyester***
Rot & Mildew	Poor	Excellent	Excellent	Excellent
Sunlight	Excellent	Good	Fair	Excellent
Oil & Gas	Fair	Excellent	Excellent	Excellent
Acids	Poor	Fair	Excellent	Very Good
Alkalis	Poor	Excellent	Good—Except Sodium Hydr.	Good

Rope Characteristics

Rope can be made from many different materials, each differing in characteristics.

Manila. Made from fine Abaca (Hemp) fiber. Excellent resistance to sunlight, low stretch, and easy to knot. Good surface-abrasion resistance.

Nylon. The strongest fiber rope manufactured. High elasticity allows absorption of shock loads. Resistant to rot, oils, gasoline, grease, marine growth, and most chemicals. High abrasion resistance.

Polypropylene. A lightweight fiber with good strength. It floats, is resistant to rot, gasoline, and most chemicals, as well as being waterproof. Some products contain additives to reduce sunlight deterioration.

Polyester. Less strength than nylon fiber but better resistance to sunlight deterioration. Low stretch and excellent surface-abrasion resistance. Other characteristics similar to nylon.

Rope Terms

Ropes and cordage have been so peculiarly a sailor's province that nautical expressions have become the common words and phrases used to describe and discuss them. The following are some in general use:

Belay. To make fast the end of a tackle, fall, etc., at the conclusion of a hoisting operation or the like.

Bend. A fastening of one rope to another or to a ring, thimble, etc.

Bight. The loose part of a rope between two fixed ends.

Haul. To heave or pull on a rope.

Hitch. A fastening of a rope simply by winding it, without knotting, around some object.

Knot. A fastening of one part of a rope to another part of the same, by interlacing them and drawing the loops tight.

Lay. To twist strands together, as in rope making, the fiber or tow receiving a right-handed twist to make yarns, yarns being laid left-handed into strands, and strands right-handed into ropes. Three strands make a hawser, and three hawsers are laid up into a cable.

Make fast. To secure the loose end of a rope to some fixed object.

Marline spike. A long tapered steel instrument used to unlay or separate the strands of rope for splicing, etc., or for working marline around a seizing.

Parcelled. Wrapped with canvas, rags, leather, etc., to resist chafing.

Seize. To lash a rope permanently with a smaller cord.

Serve. To lash with cords, etc., wound tightly and continuously around the object.

Splice. To connect rope ends together by unlaying the strands of each, then plaiting both together so as to make one continuous whole.

Strand. Two or more large yarns twisted together.

Taut. Stretched or drawn tight, strained.

Yarn. Fibers twisted together.

Theory of Knots

The principle of a knot is that "no two parts which would move in the same direction, if the rope were to slip, should lie alongside of and touch each other." Another principle that should be added to the above is that a knot or a hitch must be so devised that the tight part of the rope must bear on the free end in such a manner as to pinch and hold it against another tight part of the rope, or in a hitch, against the object to which the rope is attached.

The elements of a knot or bends that a rope undergoes in the formation of a knot or of a hitch are of three kinds:

1. Bight.
2. Loop or turn.
3. Round turn.

These are shown in Figure 36-13. Knots and hitches are made by combining these elements in different ways, conforming to the principles of a knot given previously. Figures 36-14 through 36-21 show different kinds of knots used in hoisting apparatus.

Figure 36-13 Elements of a knot.

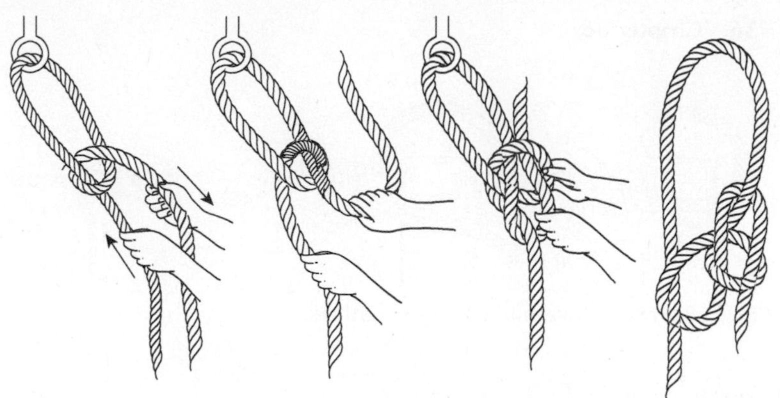

Figure 36-14 Bowline knot made at the end of the rope.

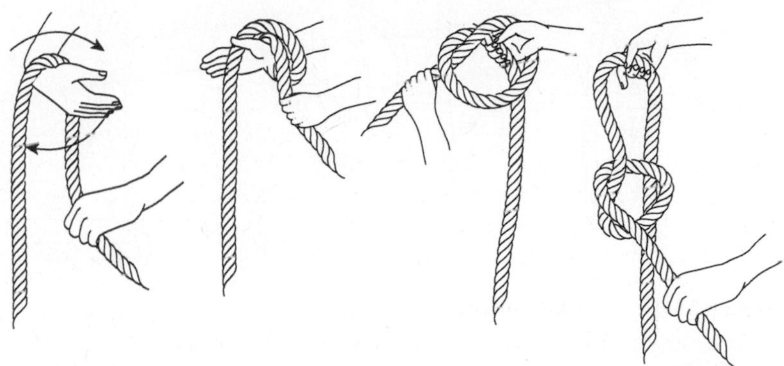

Figure 36-15 Slip knot.

Figure 36-16 Timber-hitch.

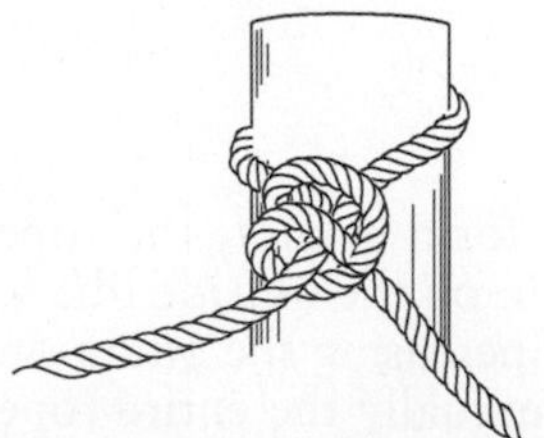

Figure 36-17 Half-hitch.

Figure 36-18 Clove-hitch, sailor's method.

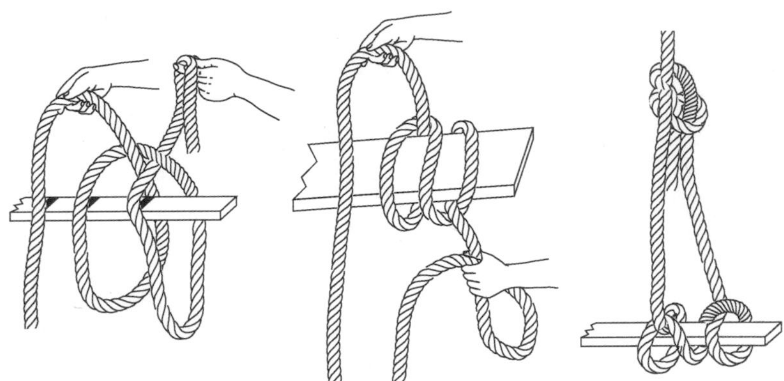

Figure 36-19 Scaffold-hitch.

Figure 36-20 Anchor bend.

Effect of Knots

A rope is weakened by knots. In order to form a knot, the rope must be bent, which brings most of the strain on the outside fibers. The overloading breaks the outside fibers, increasing the strain on the fibers below, which later break, and eventually the entire rope

Figure 36-21 Combined timber and half-hitch.

breaks. From experiments, the approximate efficiency of knots, hitches, and splices varies as follows: straight rope, 100%; eye splice over an iron eye, 90%; short splice, 80%; timber hitch anchor bend, 65%; clove hitch running bowline, 60%; overhand knot, 45%.

Treatment of Rope Ends

The process of building up a rope from strands is called *laying* a rope, and so twisting together strands that have become untwisted is called *relaying*. *Whipping* consists in binding the end of a rope with twine to prevent it untwisting, as in Figure 36-22. Ropes that are to be passed through pulley blocks or through small holes should be finished in this way. A method of doing this so that both ends of the twine are fastened is shown in Figure 36-22.

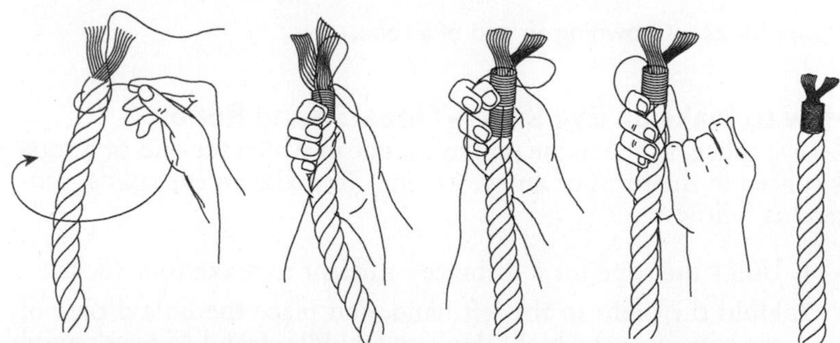

Figure 36-22 Whipping the end of a rope.

Crowning

Crowning is a neat, secure, and permanent method of fastening the strands of a rope when a slight enlargement of the end is not an objection. Figure 36-23 shows how this is done.

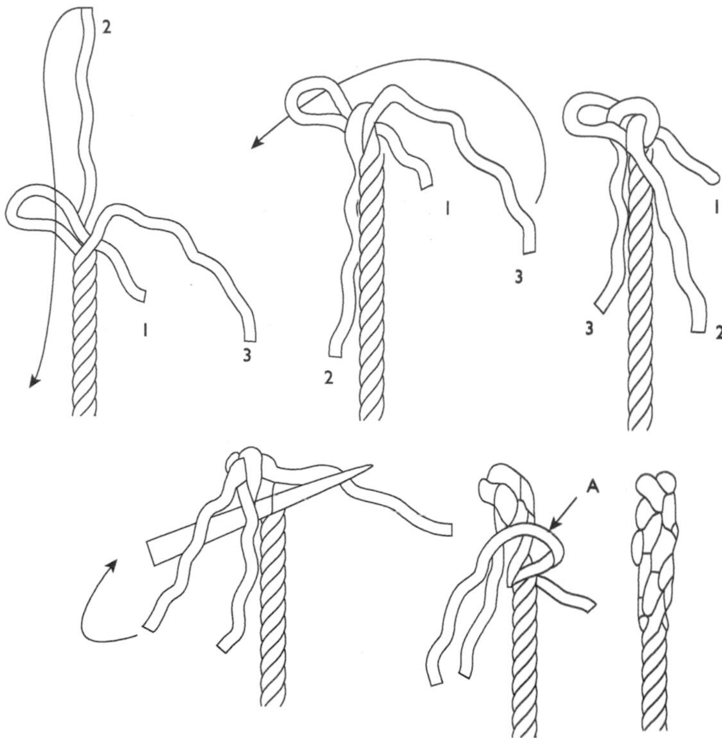

Figure 36-23 Crowning an end of a rope.

How to Make an Eye Splice-Three-Strand Rope

An eye splice, as the name implies, is one in which the end of a rope is spliced in the form of an eye or ring. To make an eye splice, proceed as follows:

1. Unlay the rope for a distance sufficient to make four tucks.
2. Hold the bight in the left hand, and place the unlaid part of the rope over the bight. Hold the middle strand *M* over strand 2 with the thumb and first finger of the left hand.

3. Tuck the middle strand *M* under strand 1, as illustrated in Figure 36-24.
4. Next, lay strand *L* over strand 1 and tuck it under strand 2, as shown in Figure 36-24.
5. Now turn the splice over and tuck strand *R* under strand 3, as shown in Figure 36-24.
6. Continue tucking the strands in this order.

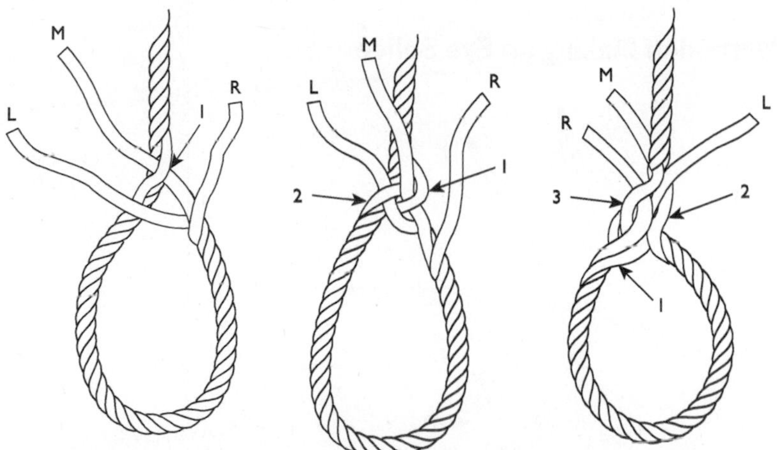

Figure 36-24 Method of making an eye splice.

How to Make an Eye Splice—Eight-Strand Rope

1. Eight-strand plaited rope is no more difficult to splice than three-strand. It is made of eight strands grouped in four pairs. Two of these pairs turn to the left and two to the right. First tie a string tightly around the rope, at a sufficient distance from the end to allow six-to-eight tucks. Next, unlay the pairs of strands back to the string. Making sure not to mix or twist them, tape the ends of the pairs together as seen in Figure 36-25. The two pairs that turn to the left are labeled 1 and 2; the two pairs that turn to the right are labeled 3 and 4.
2. Starting with the right-turning pairs, tuck them under the diametrically opposite left-turning pairs as illustrated in

Figure 36-25. Next, tuck the left-turning pairs under the diametrically opposite right-turning pairs as shown in Figure 36-26.

3. Pull all four ends down firmly. Starting with the right-turning pair, take another full tuck (insert all four pairs). The splice should now look like Figure 36-27.
4. Continue taking full tucks until six tucks have been completed. If a smoother splice is preferred the pairs may be divided (cut off one strand of each pair) and the remaining four strands tucked another two or three times.

Method of Making an Eye Splice

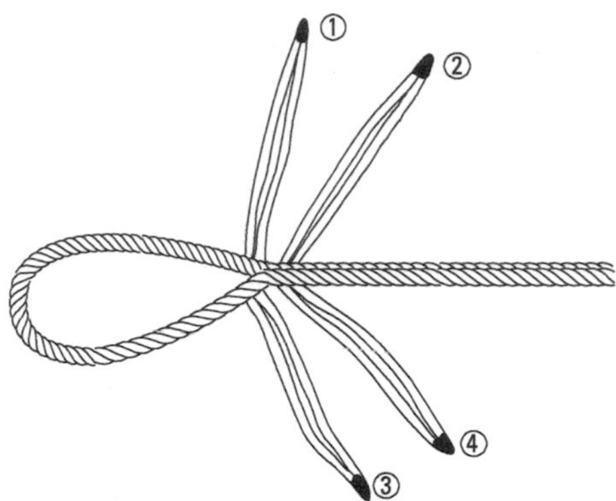

Figure 36-25

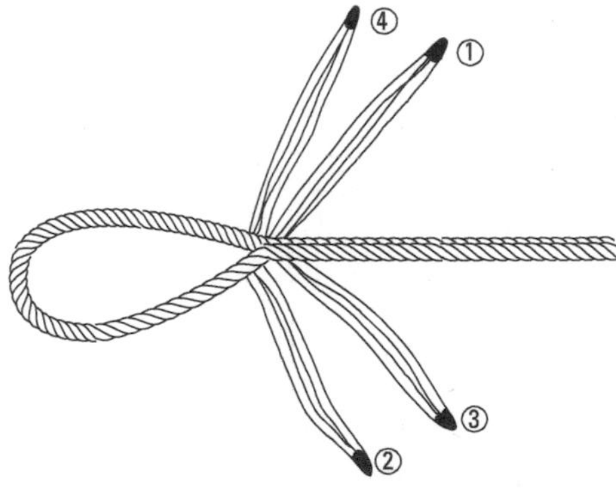

Figure 36-26

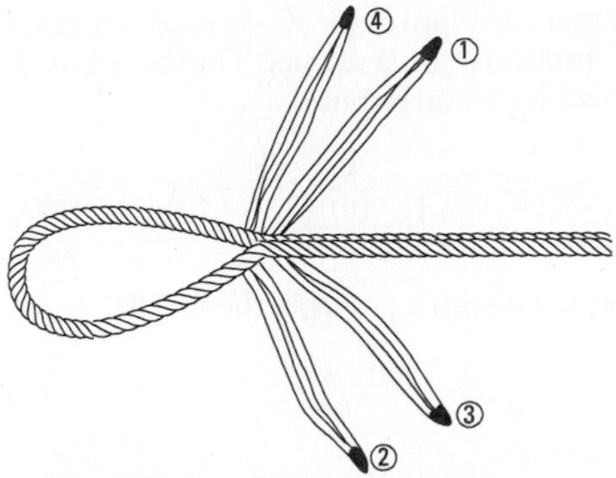

Figure 36-27

Rope Weights

Manila or wire rope for hoisting is usually provided with a rope weight, or *cow-sucker,* shown in Figure 36-28. The object of this weight is to cause the rope to pay out when released.

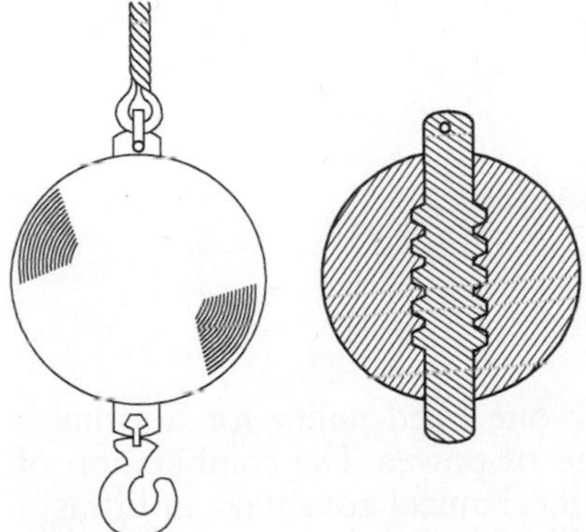

Figure 36-28 Rope weight or cow-sucker.

Chains

Chains are made of round bar iron or steel forged into links by bending to shape and welding. The stud is a distance piece, usually of cast steel, which serves to strengthen the link; it is used on the larger sizes alone and permits a longer link. Most chains for heavy stresses are made this way. A length of stud chain is illustrated in Figure 36-29 and is noteworthy because of the swivel in the middle of its length, permitting rotary motion without fouling the chain. A chain or anchor shackle, as seen in Figure 36-30, is used for connecting lengths (usually 15 fathoms) of a cable. The bolt has a countersunk head and is locked by a cotter pin.

Figure 36-29 A length of stud chain with a swivel in the middle.

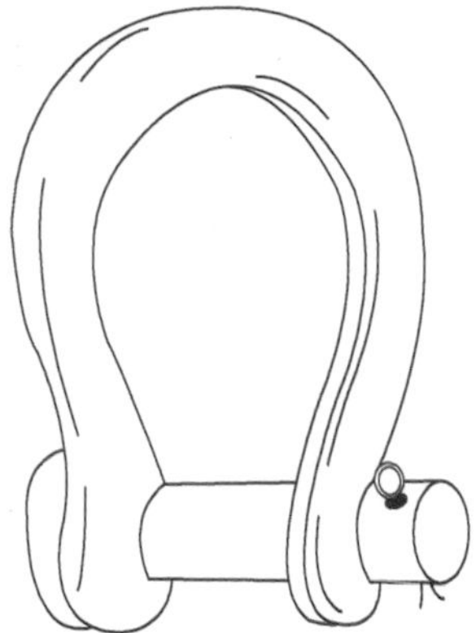

Figure 36-30 A common anchor shackle.

Mechanical Advantage

Combination of Pulleys

The mere passing of a rope over one fixed pulley for hoisting a weight does not give any increase of power. The combination of ropes and pulleys used to gain a mechanical advantage in lifting a load is known as a block and tackle. The block consists of a shell, pulleys, and a hook or an eye. Usually, two blocks are used. A rope or chain is used for connecting the blocks.

The block and tackle gives a mechanical advantage in the application of power. It is easier to haul downward on a pull of 100 lb than it is to lift 100 lb directly from the ground. This advantage should not be confused with mechanical advantage or multiplication of effort. According to the degree of gearing, the 100-lb weight may be lifted by the application of less force, as 50, 25, 10, 5 lb, etc., of pull. For instance, when one end of the rope is fixed and passes under a single pulley to which the load is attached, and the free end is lifted, the travel of the rope or cord is double that of the weight. The power necessary to sustain the latter is half the weight, less friction. It may be stated in the reverse proposition that, with a movable pulley, the weight capable of being lifted is twice the force applied, minus the friction of the apparatus.

Combinations of pulleys are arranged with several sheaves in one case to form a block, to secure this multiplication of power. The upper, or fixed, block gives the advantage of position, and the lower, or movable, block, by multiplying the travel of the rope as compared with that of the weight, increases the power in proportionate ratio. Briefly, the weight capable of being lifted is equal to the force multiplied by the number of ropes supporting the lower, or movable, block.

Examination of the block and tackle shown in Figure 36-31 will make this clear; the arrowheads show the direction of travel of the rope, or fall, as it is termed. The thin cord shown manipulates a patent brake, seen in the upper block, which locks the rope should it be desired to suspend the object lifted. It will be evident, upon consideration, that no two sheaves travel at the same velocity because of the varying speed of the different parts of the rope. It is therefore necessary that the sheaves be independent of each other, revolving loosely upon a spindle fixed in the shell or frame of the block. A basic factor of all mechanical powers is that whatever is lost in time is gained in load lifted, or the reverse. It has been seen that with the weight traveling half as fast as the rope, double the weight could be lifted with the same force, or a force of only half the weight was necessary. In practice, the friction losses must be considered in estimating the power required to lift loads by blocks.

Example

What force (pull) must be applied on a block and tackle having four rope lengths shortened to lift 500 lb?

For four rope lengths (according to Table 36-5), the ratio of load-to-pull is 3.3. Accordingly, the force applied = 500 ÷ 3.3 = 151.5 lb.

Example

If a pull of 100 lb is applied on a block and tackle having a load-to-pull ratio of 4.33, what load can be lifted?

$$100 \times 4.33 = 433 \text{ lb}$$

Differential Blocks

The heavy-duty hoist or chain block invented by Thomas A. Weston in 1854 is comparatively inexpensive and simple in construction. It is based upon the principle of the Chinese windlass

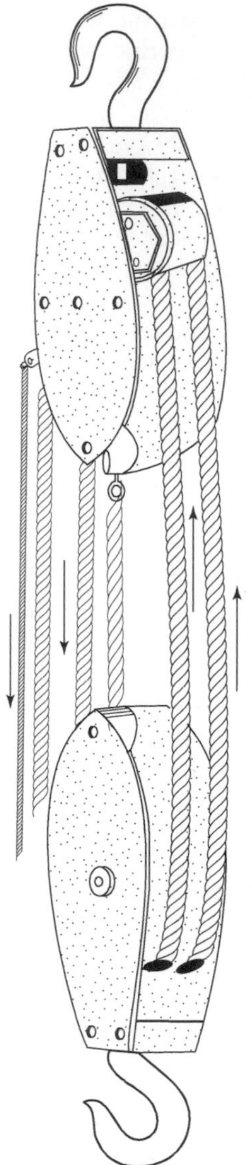

Figure 36-31 Block and tackle.

shown in Figure 36-32. An inspection will show that the differential block depends upon a very slow speed of the weight in comparison with the speed of the haul. This is secured by making the two upper drums nearly the same size, the endless chain being paid out by one while it is reeled in by the other. In other words, the smaller drum tends to lower the weight while the larger one raises it, the total lift equaling the difference of circumference of the two drums.

Table 36-5 Working Loads of Rope Blocks

	Manila Rope		*Wire Rope*	
Number of Rope Lengths Shortened	***Ratio of Load to Pull***	***Efficiency, Percent***	***Ratio of Load to Pull***	***Efficiency, Percent***
2	1.91	96		
3	2.64	88	2.73	91
4	3.30	83	3.47	87
5	3.84	77	4.11	82
6	4.33	72	4.70	78
7	4.72	67	5.20	74
8	5.08	64	5.68	71
9	5.37	60	6.08	68
10			6.46	65
12			7.08	59

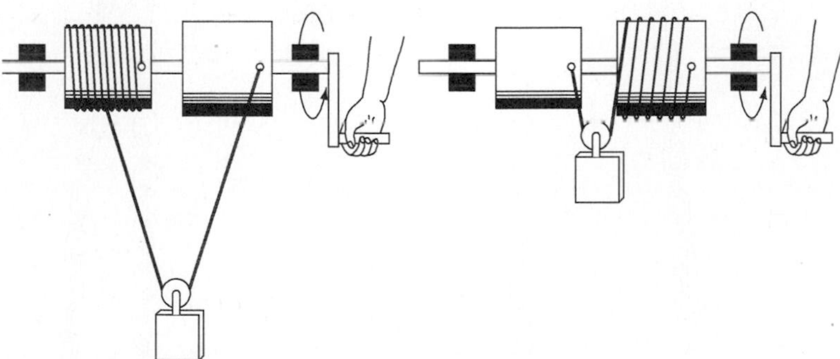

Figure 36-32 The Chinese windlass, illustrating the principle of the differential hoist.

Worm Hoist

In a worm hoist (sometimes called a screw hoist), the power is transmitted from a hand chain to the load chain by worm gearing. The principle of transmission is shown in Figure 36-33.

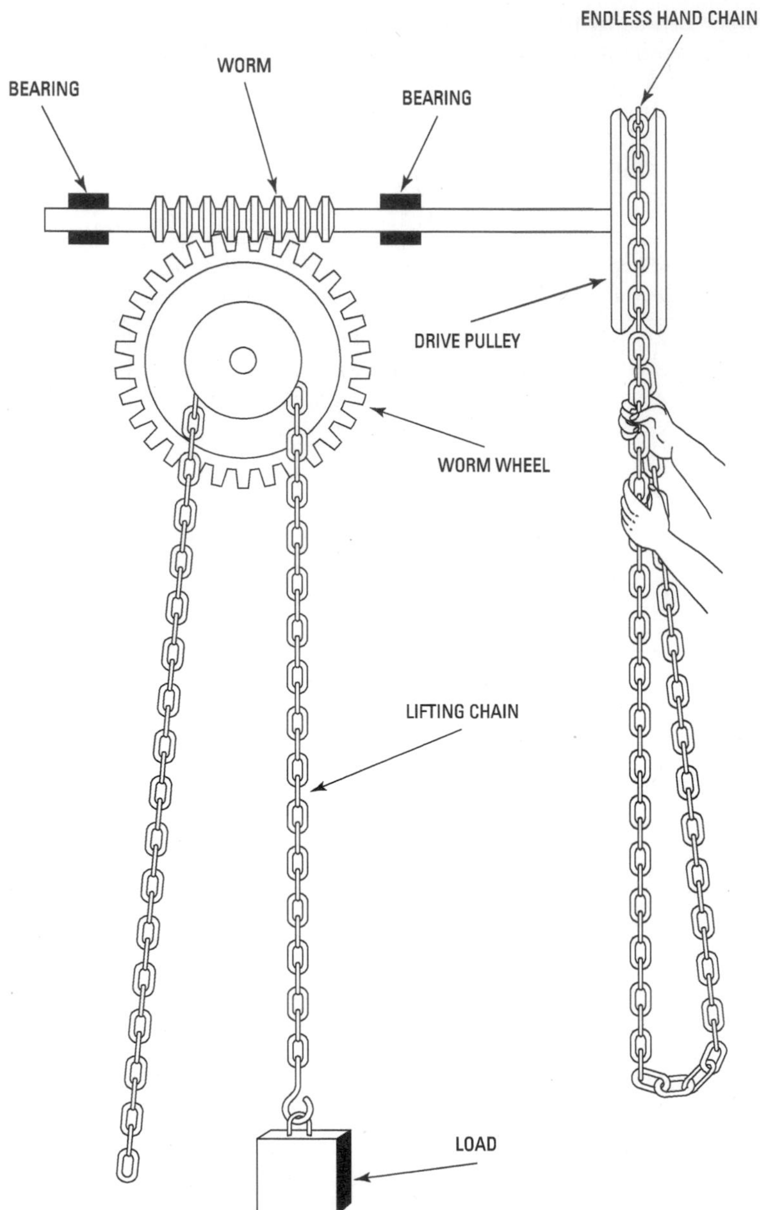

Figure 36-33 Elementary worm hoist.

In the worm hoist, an endless chain passes around a sprocket wheel and rotates a worm, and by multiplying gearing, hauls on the lifting chain. Not only is the gain of power at the expense of time obtained by the multiple gearing of this device, but the worm prevents slipping, and as there is no weight on the hand chain, a jam is not dangerous.

Spur Gear and Drum Hoist

This type of hoist is known as a winch, which by definition is a windlass, particularly one used for hoisting, as on a truck or the mast of a derrick. A winch usually has one or more hand cranks geared to a drum around which the rope or chain pulls a load. Figure 36-34 shows a simple winch with single reduction gearing. It is equipped with a brake for lowering and controlling operations. When using a winch of this description with wire rope, the barrel or drum should be at least twenty times the diameter of the rope.

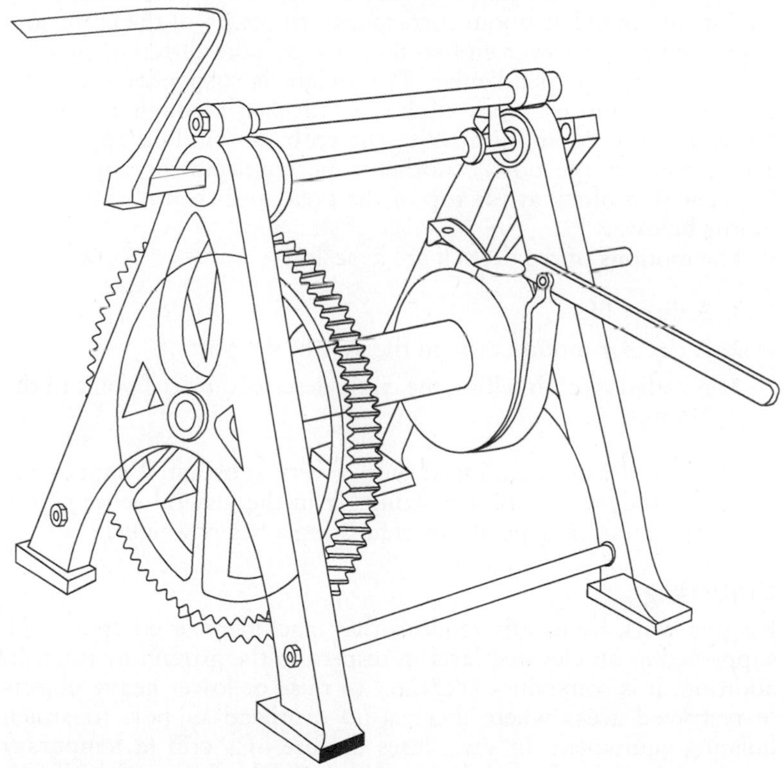

Figure 36-34 Crab or hand winch.

The Supporting Structure

Erectors will find the *gin pole* of great use; this is a stout pole, of suitable length, that serves as a portable derrick. The gin pole and the method of using it in placing a mast in position are shown in Figure 36-35.

The *tripod* or *shear legs* is an old and well-known hoisting device, and is shown in Figure 36-36. It consists of three sticks or legs fastened together at the top, from which point the tackle is suspended. Different methods of fastening the legs together are used.

Another form of supporting structure, consisting of a mast of suitable height and fixed by braces on a triangular base, is shown in Figure 36-37. The term *stiff* leg is applied to distinguish this mast from the mast secured by guy cables, as shown in Figure 36-35.

By definition, a *derrick* is an apparatus for lifting and moving heavy weights. It is similar to the crane, but differs from it by having a boom. The boom corresponds to the jib of the crane and is pivoted at the lower end so that it may take different inclinations from the perpendicular. The weight is suspended from the end of the boom by ropes or chains that pass through a block at the end of the boom, directly to the crab, or winding apparatus, at the foot of the post. Another rope connects the top of the boom with a block at the top of the post, and then passes to the motor below.

The motions of the derrick are as follows:

1. A direct lift.
2. A circular motion around the axis of the post.
3. A radial motion within the circle described by the point of the boom.

The *pile driver* is equipped with a very substantial supporting structure and, while not a hoisting rig in the general sense of the term, its operation depends upon hoisting a heavy weight.

Cribbing

Rigging work frequently requires that machinery or equipment be supported at an elevated level in respect to the ground or floor. In addition, it is sometimes necessary to raise or lower heavy objects in restricted areas where there is no overhead support to attach hoisting equipment. In such cases the use of a crib (a temporary framework of timbers) is often employed. This is done by placing

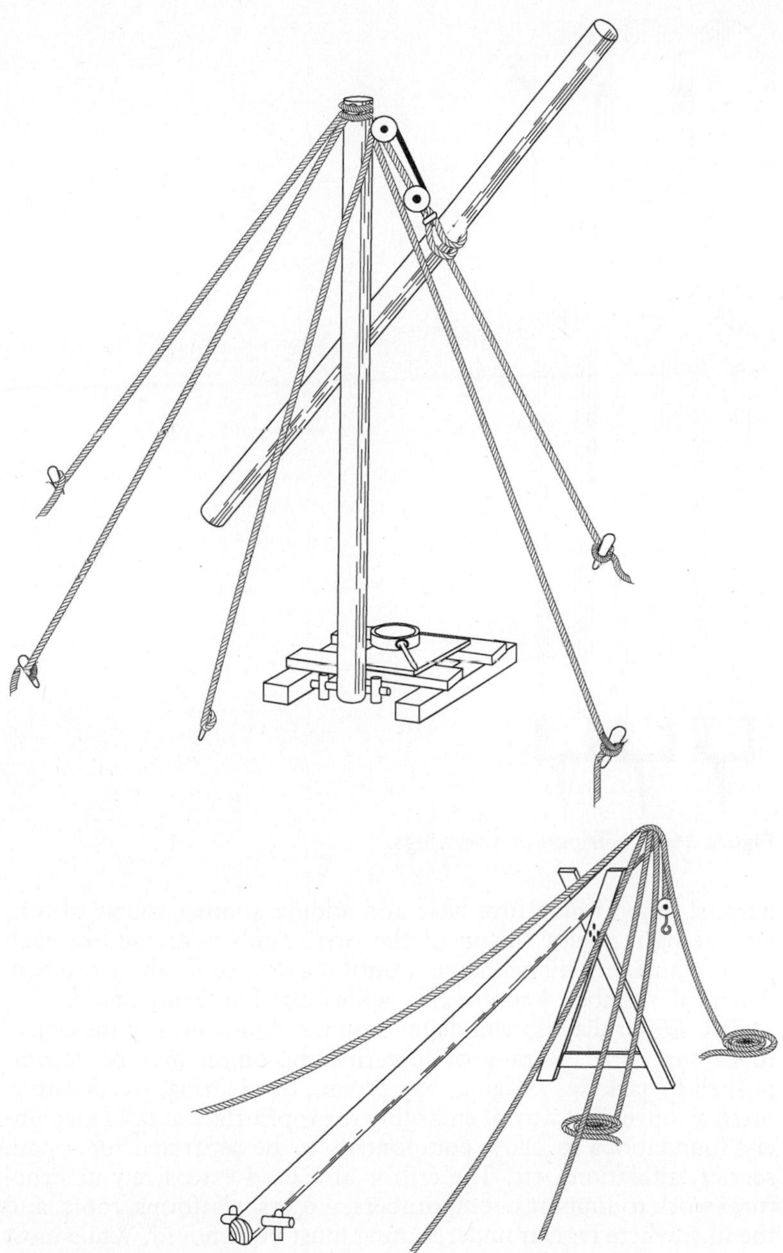

Figure 36-35 Gin pole.

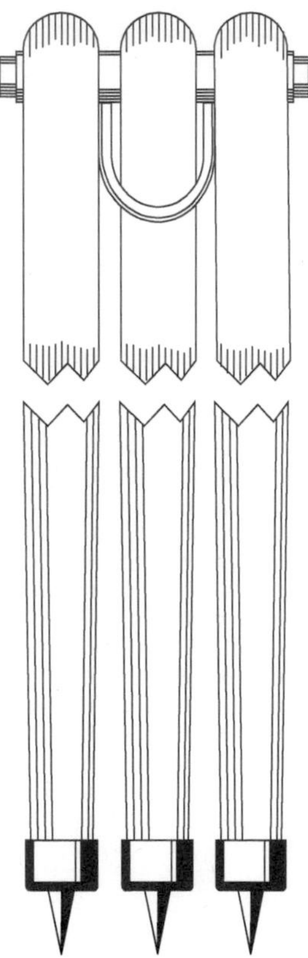

Figure 36-36 Tripod or shear legs.

parallel timbers on a firm base and adding another course of timbers at right angles on top of the first. Timbers are added, each layer in an alternating direction, until the desired height is reached. A typical 3-timber, 4-course crib is illustrated in Figure 36-38.

The size of the crib will depend on the dimensions of the object to be supported. Raising or lowering the object may be accomplished by jacking, wedging, bar prying, lever lifting, etc. A crib is often a convenient way of enlarging the top surface area of machinery foundations to allow components to be separated for repair, service, alterations, etc. The crib is also used extensively in structural work to support floors, timbers, girders, platforms, roofs, and the like, where regular underpinning must be removed. While most any species of wood can be used for temporary cribbing, when the

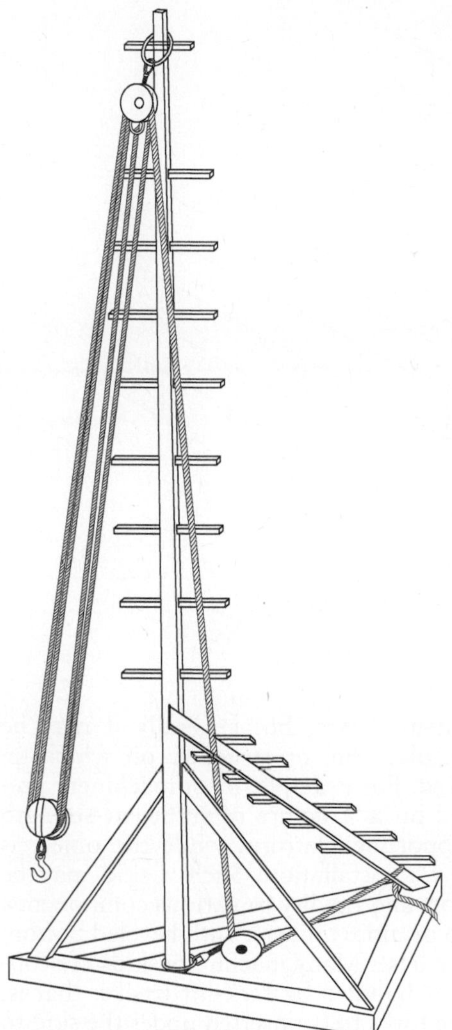

Figure 36-37 Stiff-leg mast.

timbers are to be reused, a relatively dense wood will give the best service. Yellow pine, although rather heavy, gives excellent service; Douglas fir is lighter and easier to handle, but it splinters more and is less resistant to damage from rough handling. A stable crib requires that the timbers be of uniform dimension.

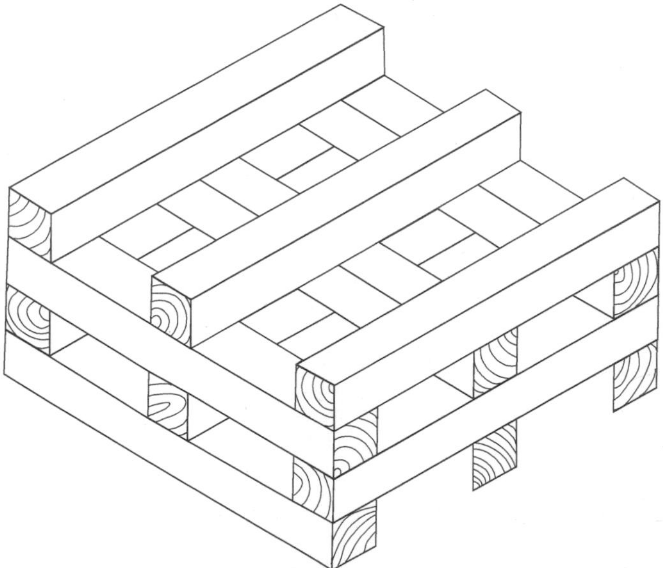

Figure 36-38 Three-timber, 4-course crib.

Skid

A rigging skid may take many forms, but generally it may be described as a frame, base, platform, or structure on which an object rests while being moved. For example, most machinery prepared for shipment is placed on a structure of sufficient strength and rigidity to act as a supporting platform while the object is transported and positioned for installation. Such a skid may be quite elaborate, giving support and rigidity to various components, or it may be as simple as two or more timbers, with beveled bottom ends, attached to the machine base. Many machinery skids are constructed to permit them to be handled by fork-lift trucks; that is, there are provisions to enable forks to be inserted under the skid to lift and move the object.

The moving, relocation, etc., of unskidded machinery and equipment may require that some form of skid be provided to facilitate this activity. An arrangement for this purpose is illustrated in Figure 36-39. Heavy-duty angle iron is cut, bent, and welded at the ends to provide a rounded bottom end. The extensions at the rounded ends are provided to allow placement of a jack for lifting. Cross rods, to suit the dimensions of the object, tie the units together and provide a single, rigid skid assembly. Such a skid facilitates lifting and transporting a heavy object by lift truck, rollers, sliding, or hoisting, and may be used over and over.

Rollers

The simple cylindrical roll is one of the fundamental devices used to facilitate moving heavy objects. It was probably in use for this purpose before recorded history. The principle of operation is quite simple. Rolling motion, with very low friction, is substituted for sliding motion, which has extremely high friction. Objects that would require a prohibitively high degree of force to overcome the friction of skidding or sliding may be moved with relative ease when placed on cylindrical rollers. The roller is a basic rigging tool and is commercially available in various lengths to suit the dimensions of the object to be moved. The best rollers are made from dense, close-grained clear wood. A favorite for this purpose is a wood called "horn beam," which has all the qualities mentioned, as well as being resistant to damage when struck with a sledge. This form of abuse is often necessary, as skidded objects on rollers are steered by striking the rollers with a sledge to turn them in the direction desired.

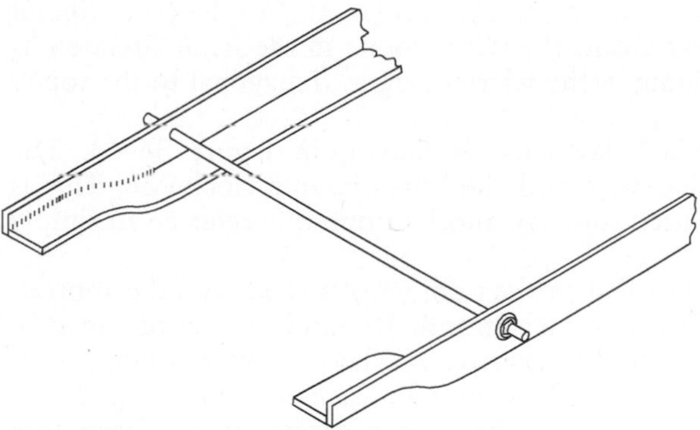

Figure 36-39 Adjustable skid.

Wire Rope

The basic element in the construction of wire rope is a single metallic wire. Several of these wires are laid helically around a center to form a *strand*. Finally a number of strands are laid helically around a *core* to form a wire rope, as illustrated in Figure 36-40.

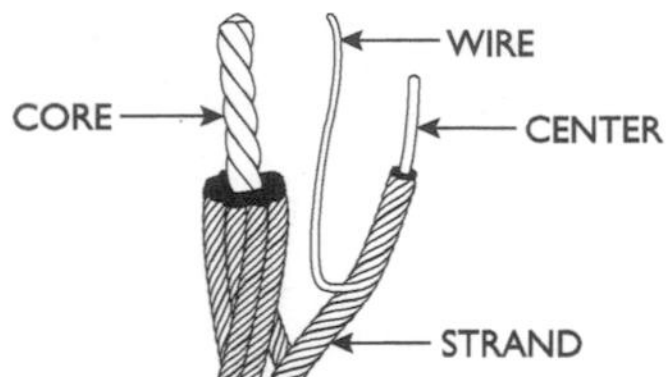

Figure 36-40 Component parts of wire rope.

The primary function of the core is to serve as a foundation for the rope, to keep it round and keep the strands correctly spaced and supported. During construction, the wires that make up the strand may be laid around the center in either a clockwise or counterclockwise direction. The same is true of the strands when they are laid around the core. This direction of rotation is called the *lay* of the rope. In *right* lay, the strands rotate around the core in a clockwise direction, as threads do in a right-hand thread. In *left* lay, the strands rotate counterclockwise, as do left-hand threads.

The terms *regular* and *lang* are used to designate direction of the wires around the center. *Regular* lay means that the wires rotate in a direction opposite to the direction of the strands around the core. This results in the wires being roughly parallel to the centerline of the rope. *Lang* lay means the wires rotate in the same direction as the strands, resulting in the wires being at a diagonal to the rope's centerline.

A "right-regular" lay rope is shown in Figure 36-40. The strands rotate clockwise and the wires counterclockwise. This is the most widely used rope lay and is commonly referred to simply as "regular-lay."

Wire rope is classified by the number of strands and the approximate number of wires in each strand. A strand consists of a specific number of wires of predetermined sizes, laid in layers around a center in a given pattern or construction. Each wire in a strand performs a specific function. The center serves as the base that supports the other wires in the strand. The intermediate layer of wires serves as a supporting arch for the outer layer of wires, which in turn absorbs the wear and tear of contact with sheaves, drums, and other surfaces. Each construction is designed to give each wire freedom of movement in relation to the adjacent wires.

For convenient reference, rope constructions are usually grouped into classifications by the number of wires in their strands as follows:

Classification	*Wires per Strand*
6 × 7	7
6 × 19	16–26
6 × 37	27–49
8 × 19	16–26

Strands made of a large number of small wires are more flexible than strands formed with a small number of wires. But strands with a small number of wires are more resistant to abrasion. Coarse-laid strands are usually composed of 7 wires. Standard flexible or special flexible strands have 19 wires, and sometimes have small, filler wires, which increase the total to 25 wires. Illustrated in Figure 36-41 are "coarse-laid" 6 × 7 and "flexible-laid" 6 × 19 wire rope cross sections.

Applications where greater flexibility than can be obtained from 6 × 19 wire rope is necessary may dictate the use of the 6 × 37 classification. The 6 × 37 wire ropes include all six-strand, round-strand ropes that have 27 to 49 wires to the strand. While the 6 × 19 rope gives emphasis to abrasion resistance in varying degrees, 6 × 37 ropes are designed with greater emphasis on flexibility. As a rule, selection of a wire rope is a compromise. For example, in order to get maximum flexibility it may be necessary to settle for less than optimum resistance to abrasion. The greater flexibility of the 6 × 37 rope is made possible by the greater number of smaller wires per strand, which permits much more movement within the rope to adjust easily to the strain of bends over sheaves or drums.

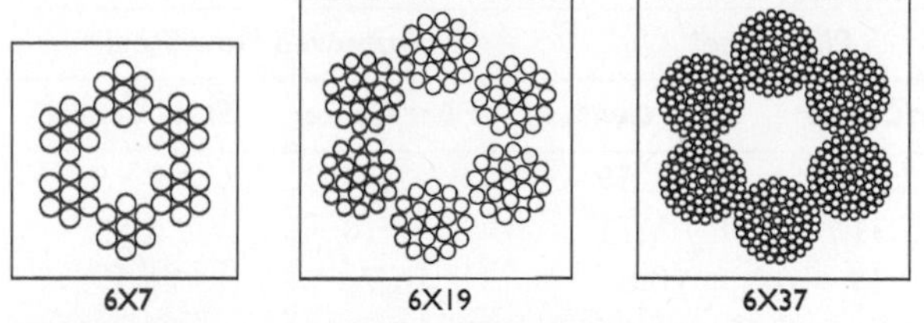

Figure 36-41 Wire-rope classification.

Wire-Rope Strength—Factor of Safety

Wire rope is made in a wide variety of grades to meet the demand for strength, flexibility, toughness, etc. While steel is the most widely used material, wire rope for special applications is also made of iron, bronze, and stainless steel. The two most common grades of wire rope are designated *plow steel* and *improved plow steel.* Plow steel is a tough, strong grade of steel developed to provide the qualities needed in wire rope—principally high strength, flexibility, and toughness. Improved plow steel, as the name implies, is a further improvement in the quality of material, resulting in a superior-quality rope of extra strength.

To use wire rope safely in rigging operations, a knowledge of working loads and safety factors is a basic requirement. The first principle of safe operation is that the load on the rope be limited to a portion of the rope's ultimate strength. The safe load for a wire rope is determined by dividing its breaking strength by a *factor of safety.* It is impossible to set a figure to be used in all cases because operating conditions are never the same. Such a variation exists that safety factors may range from slightly under 5 to 8 or even 10 in extreme situations. Listed in Table 36-6 are the ultimate breaking strengths of 6 × 19 classification plow steel and improved plow steel wire rope.

To determine the safe working load for a wire-rope application, first determine the proper safety factor based on service and conditions. This is often set at a value of 5 for work of a general nature, but must be increased for such considerations as shock loading, speed of acceleration and deceleration, arrangement of sheaves and drums, and fittings used. The number determined will decide the portion of the rope's ultimate strength that may be used when it is divided into the value given in the strength table from the manufacturer of the wire rope.

Table 36-6 Wire Rope Strength (6 × 19 Classification)

	Breaking Strength in Tons of 2000 lb			
	Plow Steel		Improved Plow Steel	
Diameter	Fiber Center	Steel Center	Fiber Center	Steel Center
¼	2.39	2.59	2.74	2.95
⅜	5.31	5.71	6.10	6.56
½	9.35	10.0	10.7	11.5
⅝	14.5	15.6	16.7	17.9
¾	20.7	22.2	23.8	25.6
⅞	28.0	30.1	32.2	34.6
1	36.4	39.1	41.8	44.9

Example

The breaking strength for a ½ in. diameter improved plow steel rope is listed in Table 36-6 at 10.7 tons. If the rope were to be used for a general-purpose application at a safety factor of 5, its maximum safe load would be 1/5 of the breaking strength, or 2.14 tons. If, however, it were to be used for a more severe application, at a safety factor of 8, its maximum safe load would be 1/8 of the breaking strength, or 1.34 tons.

When establishing factors of safety, it must be remembered that the strength of wire rope indicated in Table 36-6 is that of new rope. In actuality, the strength of all rope begins to reduce the moment it is placed in service. Therefore, the more severe the application, the more rapid the reduction in strength, and the greater the weight that must be assigned to these service conditions when establishing a factor of safety.

Wire-Rope Slings

Hoisting heavy objects requires some form of attachment of the load to the hoisting hook. The means of attachment must be of adequate strength to support the load and must be capable of being securely affixed to the hook. Manila rope and iron and steel chains have to a great extent been superseded by wire rope for this purpose, resulting in greater safety and the ability to lift much greater loads. Manila rope is still used widely for light loads and applications where a flexible and/or a relatively soft material are an advantage. Chain has advantages where high temperatures may be encountered and where rough, sharp surfaces or edges may tend to cut or abrade. For special applications, slings may be made of other materials, such as leather, cotton webbing, and roller chain, but by far the greatest number of slings in use today are made of wire rope.

Wire-rope slings are made by providing a fitting at one or both ends of a length of wire rope. Sling end fittings of the more popular styles are shown in Figure 36-42. Any combination of these slings to suit the job requirements is available from most wire-rope manufacturers.

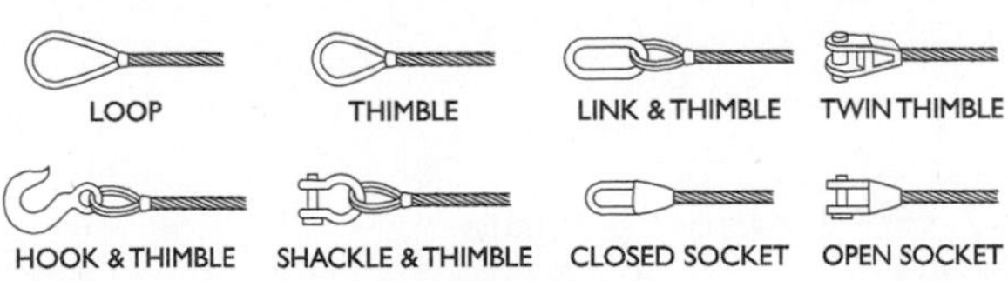

Figure 36-42 Sling end fittings.

Wire-rope slings, when properly designed and fabricated, are the safest slings to use. They do not wear or diminish in strength due to exposure to the elements, as do slings made of hemp rope. Nor are they susceptible to the weakest-link ailment of chains. The appearance of broken wires clearly indicates the fatigue of the metal and signals the termination of the useful life of the sling. The single sling with loop ends, the most widely used of all slings, lends itself readily to use as a single rope sling or in a choker hitch, U-hitch, or basket hitch.

When slings are used in any of these modes, careful consideration must be given to the load imposed on the legs at various angles. It you suspend a load of 2000 lb on two parallel ropes, as occurs with a basket hitch, each rope will be stressed 1000 lb (static load). If slings are attached to a common hook so that their angle to the horizontal is reduced to 60°, the stress is increased to 1415 lb; at 30° it is 2000 lb; and if it approaches close to the horizontal, at 5° the stress is increased over tenfold, to greater than 11,500 lb. If such flat hitches are made, not only is there danger of overloading the sling rope and causing its failure, but the load, crate, or box it is carrying may also be crushed by the force applied at its corners. Loads on ropes at various angles suspending a 2000-lb (1-ton) load are shown in Figure 36-43. The safe load in tons for various sling hitches is shown in Table 36.7.

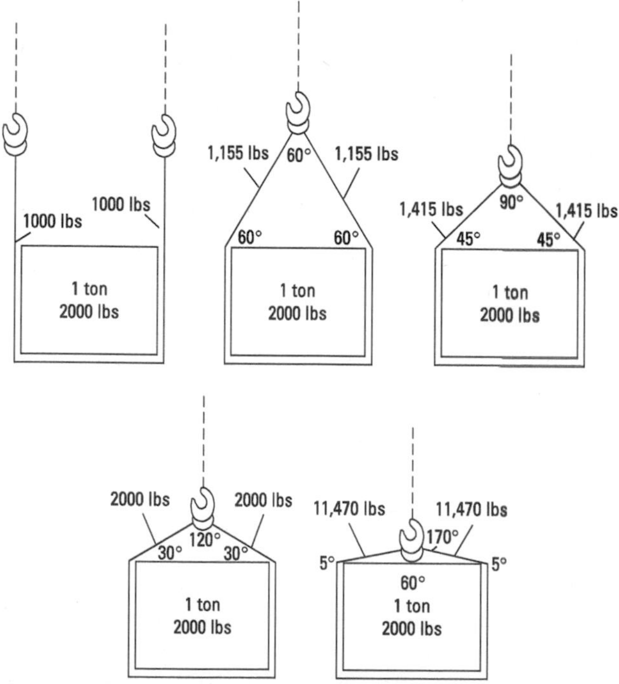

Figure 36-43 Hitch leg loadings.

Table 36-7 Safe Load in Tons

					Total Load on Two-Leg Slings (For three-leg sling multiply by 1½. For four-leg sling multiply by 2.)		
Nominal Size, in	**Single Sling**	**Choker Sling**	**U Sling**	**Basket Sling**	**60° Bridle**	**45° Bridle**	**30° Bridle**
¼	.5	.3			.57	.5	.3
5⁄16	.8	.6	1.1	1.0	.9	.7	.6
⅜	1.1	.8	1.5	1.4	1.3	1.1	.8
½	2.0	1.4	2.7	2.4	2.3	1.9	1.3
⅝	2.9	2.1	4.2	3.8	3.7	3.0	2.1
¾	4.1	3.0	6.0	5.4	5.2	4.2	3.0
⅞	5.6	3.8	7.7	6.8	6.7	5.4	3.8
1	7.2	5.0	10.0	9.3	8.7	7.1	5.0
1⅛	9.0	5.6	11.2	10.5	9.7	7.9	5.6

Wire-Rope Measurement

The correct measurement of a wire rope is that of the circumscribed circle. Stated differently, the correct measurement is that of a true circle enclosing the rope. Therefore, care should be taken when measuring to obtain this diameter, in the manner shown in Figure 36-44.

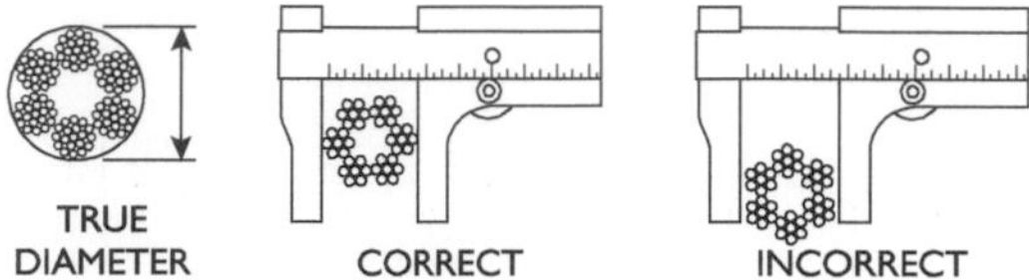

Figure 36-44 The correct and incorrect method of measuring wire-rope diameter.

Wire-Rope Splicing

The tools required for wire-rope splicing are shown in Figure 36-45, and consist of the following:

1. Two T-shaped splicing pins.
2. Two round splicing pins.
3. A tapered spike for removing strands and fiber core.
4. A knife for cutting the fiber core.
5. A pair of wire cutters for cutting off ends of the strands.
6. Two wooden mallets to hammer down any uneven surface.
7. A piece of fiber rope, spliced endless.
8. A hickory stick the size of an ordinary hammer handle, which is used to untwist the strands.

There are four types of splices generally used for wire rope:

1. Standard splice.
2. Chicago, or tied, splice.
3. Splice for ropes with an independent wire-rope core.
4. Thimble splice.

General Splicing Practice

The making of rope splices in the field is usually performed by people who possess a certain amount of mechanical skill and a facility

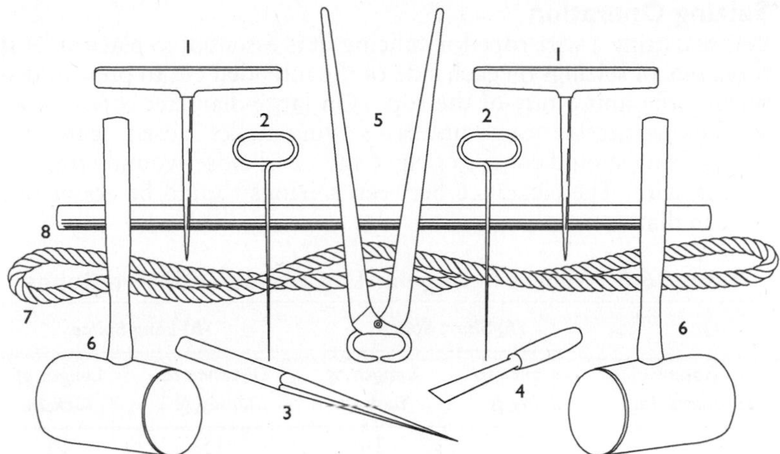

Figure 36-45 Typical wire-rope splicing tools.

in handling tools. It would therefore be well for those who are entirely lacking in experience to make several practice splices prior to attempting to splice a rope that will be subject to severe conditions in actual use. If the first effort to produce a satisfactory result fails, a review of the work will reveal where the mistake was made, and will indicate what must be avoided in the future. It is extremely important in making a splice to use great care in laying the various rope strands firmly into position. If, during any of the various operations, some of the strands are not pulled tightly into their respective places in the finished splice, it is doubtful that satisfactory results will be obtained.

When such a poorly made splice is put into service, the strands that are relatively slack will not receive their full share of the load, thus causing the remaining strands to be excessively stressed. The unbalanced condition will result in a distorted relative position of the rope strands, so that some of these will project above the surface of the rope and be subjected to excessive abrasion and abuse. It is strongly recommended that during each of the splicing steps, particular attention be paid to maintaining, as nearly as possible, the same degree of tightness in all of the strands in the splice. When ropes are to be used in places where failure would endanger human life, the splicing should be done only by men experienced in such work. It is a good practice to test such splices to at least twice their maximum working load before placing them in service.

Seizing Operation

Before cutting a wire rope for splicing, it is essential to place at least three sets of seizings on each side of the intended cut to prevent disturbing the uniformity of the rope. On large-diameter ropes, more seizings are necessary, and unless a serving mallet is used, each standard seizing should consist of eight snug and close-wound wraps of seizing wire. The clearance between seizings should be about one rope in diameter.

Table 36-8 Splices in Regular Lay, Six-Strand Wire Ropes

	(A) Short Splice		*(B) Long Splice*	
Rope Diameter, in.	**Distance to Unlay, ft**	**Length of Tuck, in.**	**Distance to Unlay, ft**	**Length of Tuck, in.**
1/4	7	10	15	15
3/8	8	12	18	18
1/2	9	14	21	21
5/8	10	16	24	24
3/4	11	18	27	27
7/8	12	20	30	30
1	13	22	33	33
1 1/8	14	24	36	36
1 1/4	15	26	39	39
1 3/8	16	28	42	42
1 1/2	17	30	45	45

Making a Standard Splice for Six-Strand Fiber Core Rope

When making a splice, a length of rope equal to half the length of the finished splice is consumed on each of the two ends. This length is indicated as "Distance to Unlay," in Table 36-8. It is necessary to allow twice this amount of extra rope for making the splice. There are two types of splices commonly used for six-strand ropes:

1. The standard "short" splice for ordinary conditions.
2. The "long" splice for rope haulages or inclines where the duty is particularly severe.

Splices should never be used in vertical shafts and are not recommended on inclines where the steepest grade is greater than 45°. Table 36-8 gives the distance to unlay and length of tuck recommended for both "short" and "long" splices in *regular lay ropes;* both of these values should be increased 20 percent over those specified for "long" splices.

In making the splice, measure from each of the two ends to be spliced a distance 8 to 10 inches more than that indicated in Table 36-8. At these points (marked D in Figure 36-46), place three seizings of wire firmly around the rope to prevent the strands from unlaying farther back.

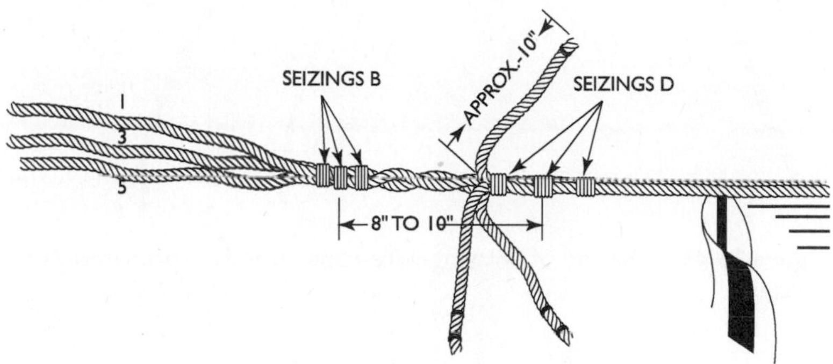

Figure 36-46 Application of seizings to wire rope whose strands have been unlaid.

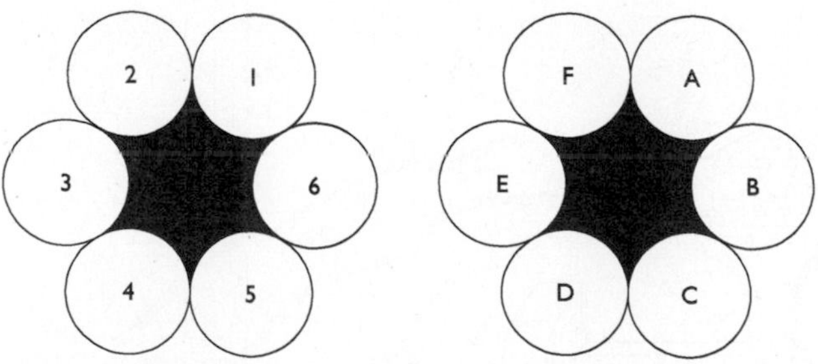

Figure 36-47 Convenient method of marking strands for proper identification.

Now, unlay three alternate strands at each end to these binding wires. It is important that the strands should be alternate; that is, if we assume them to be numbered for one end of the rope and lettered for the other end, as illustrated in Figure 36-47, either strands 1, 3, and 5 or strands 2, 4, and 6 (and also strands A, C, and E or B, D, and F) should be unlaid. Figure 36-48 shows one end of the rope after three strands have been unlaid. The other end of the rope should be similarly prepared.

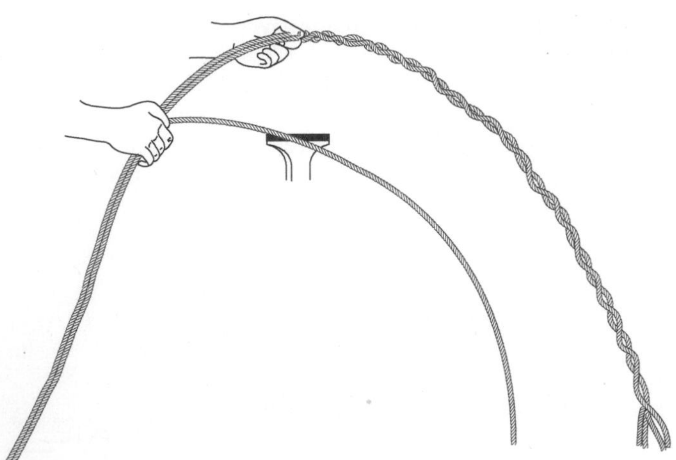

Figure 36-48 Method of unlaying wire-rope strands in preparation for splicing.

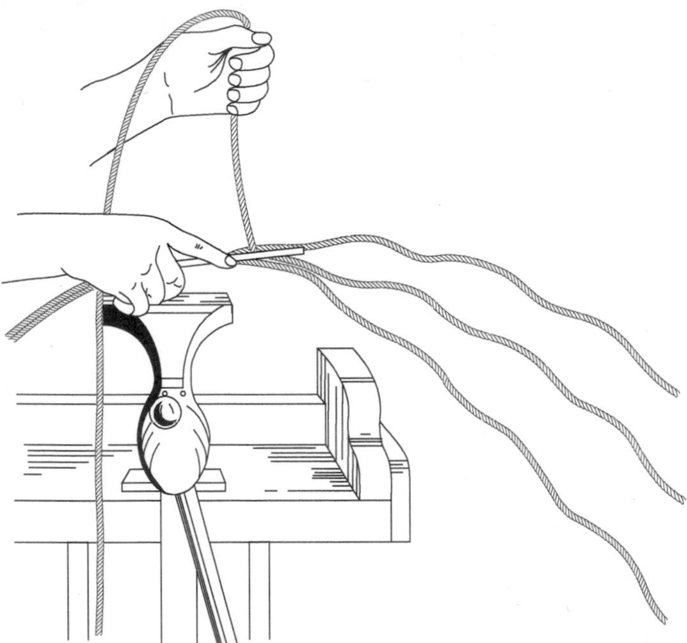

Figure 36-49 The cutting of fiber core close to first seizing of wire rope.

Next, cut off the three strands that have just been unlaid at each end of the rope, leaving the ends projecting about ten inches. Apply three more seizings of wire at B, in Figure 36-47, in order to hold firmly in place the three strands that have not been unlaid. Then separate these three strands as far as seizing B, and cut off the fiber core close to the first seizing of B, as illustrated in Figure 36-49.

Place one end of the rope securely in a vise, as shown in Figure 36-51, and assemble the other end so that corresponding strands from each end interlock regularly with each other in a manner similar to that in which the fingers will interlock when those of one hand are pushed between those of the other. It is extremely important that the two ends of rope are forced firmly against one another and are held in this position until the splice has been sufficiently formed to prevent slippage. It is advisable, therefore, for one person to hold the ropes tightly together, while two other people do the actual splicing. Next, apply seizing F (Figure 36-50) so as to bind the two ropes firmly together.

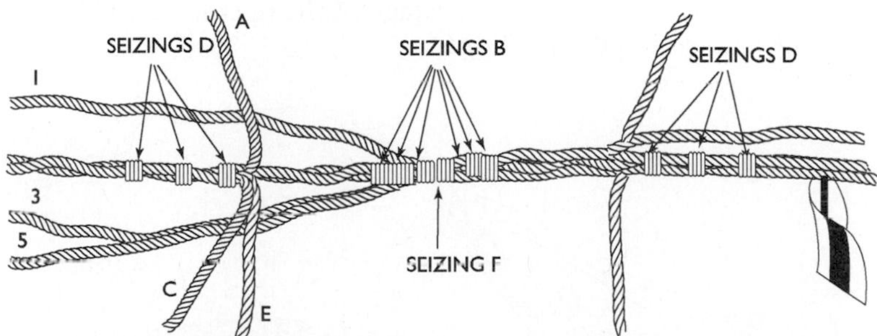

Figure 36-50 Application of additional seizing to hold the two wire-rope ends firmly together.

Now remove seizings B and D, which are to the left of seizing F (Figure 36-50). Unlay one strand, A (Figure 36-51), and follow with strand 1 from the other end, laying it tightly in the open groove left by the unwinding of A and making the twist of the strand agree exactly with the lay of the open groove. One person should rotate the strand being laid as he passes it around the rope in a direction that will tighten the wires. If the proper twist is kept by the person holding the end, little effort is required by the person actually laying the strand in the rope. Forcing the strands in place may result in a poor splice.

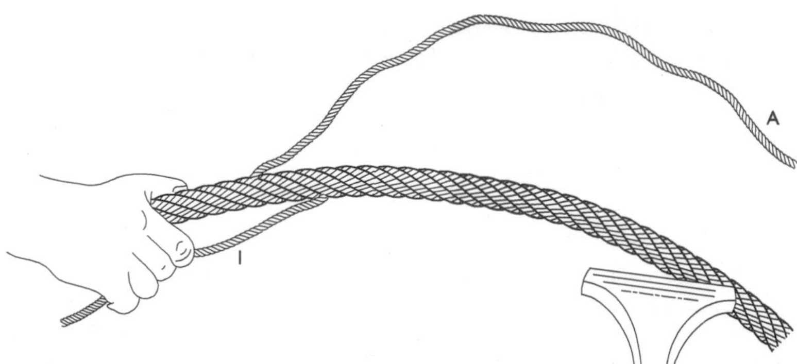

Figure 36-51 Method of laying in strands in wire rope when splicing.

When all but a short end of strand 1 has been laid in, strand A should be cut off leaving an end equal in length to strand 1. These lengths should be as given under "Length of Tuck" in Table 36-8. After this stage has been reached, as shown in Figure 36-52, it is no longer necessary for one person to hold the two ends of the rope together because the splice at this stage of the operation will be sufficiently formed to prevent slippage.

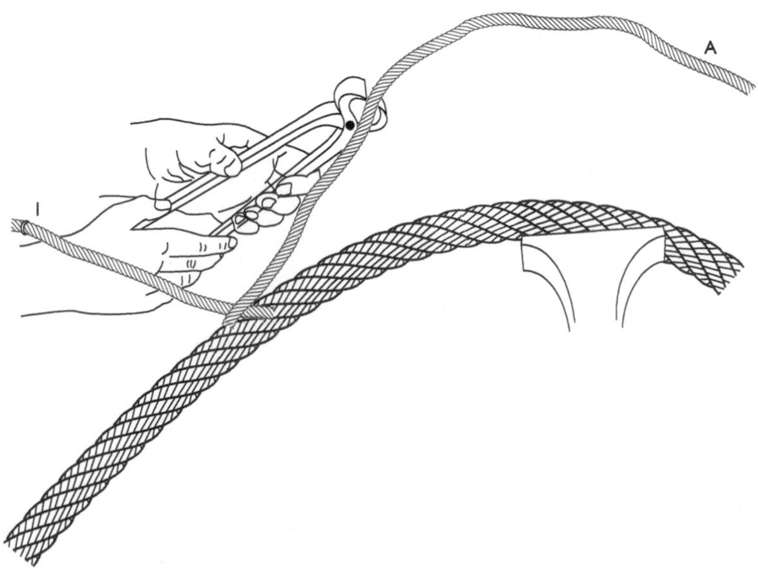

Figure 36-52 The center strand in wire rope during splicing.

At this point, unlay another strand, C, in the same manner that A was unlaid and follow with strand 3, stopping, however, before reaching the ends of strands A and 1. The unlaid strand C should be cut off as A was cut, leaving two short ends, C and 3, equal in length to those of A and 1. Now proceed in a similar manner with the third set of strands, E and 5. Figure 36-53 shows the relative position of strands C and 3 with respect to strands E and 5, and Figure 36-54 shows the relative position of strands A and 1, as well.

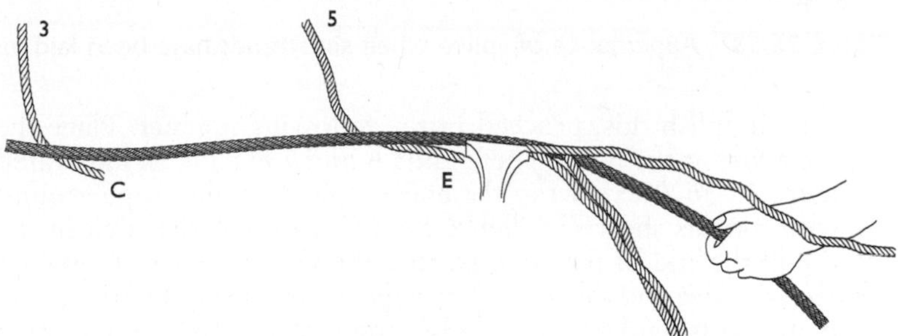

Figure 36-53 Relative position of four end strands when splicing.

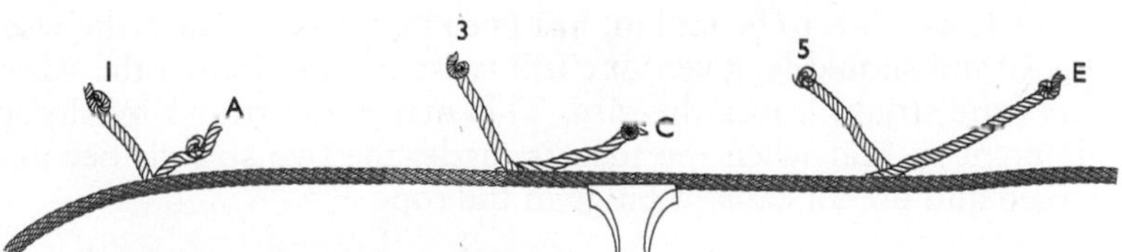

Figure 36-54 Appearance of six end strands and their uniformity in spacing when making a splice.

The points where the ends project would be spaced uniformly over the length of the splice, as shown in Figure 36-54. There now remain the three strands on the other side, which must be laid in the same way. When all six strands have been laid in, as directed, the splice will take the form indicated in Figure 36-55. There will now be six places at which the ends of the strands extend. These ends must be secured without increasing the rope diameter.

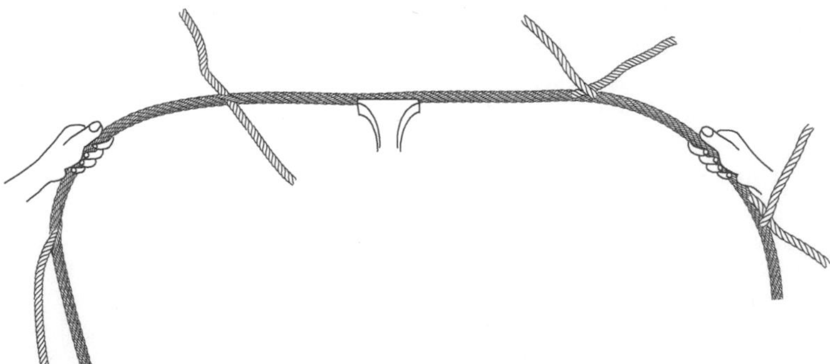

Figure 36-55 Appearance of splice when six strands have been laid in.

To accomplish this, proceed in the following manner: Place the rope in a vise at the point where ends A and 1 extend, as illustrated in Figure 36-56. Next, wrap the endless piece of fiber rope around the wire rope as shown in Figure 36-57 and insert the stick in the loop. Pull the end of the stick so that the wire rope will be unlaid between the vise and the stick. The rope, by means of the stick, may be unlaid sufficiently to insert the point of the spike under two strands, as illustrated in Figure 36-58. Use the pin to force the fiber core into such a position that it may be reached by the knife to cut. Again, referring to Figure 36-58, it will be observed that the end of strand 1, which is to be laid in, has been bent back toward the vise. This strand should be given one full twist so as to loosen the wires where the strand leaves the vise. This makes the strand mushy at this point so that when the tuck is made, the two strands become merged and do not cause a bulge in the rope.

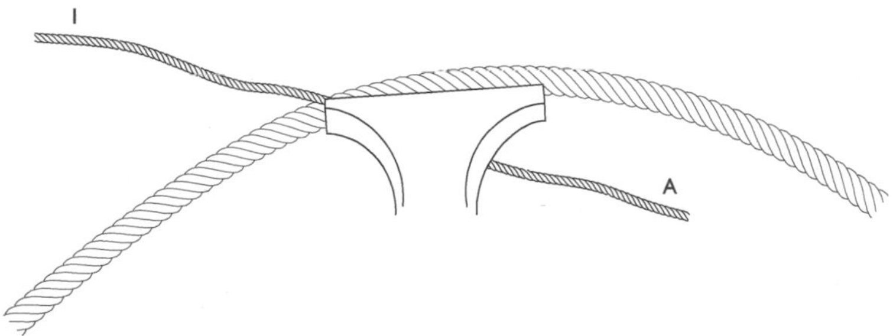

Figure 36-56 The fastening of wire rope in vise preparatory to cutting the fiber core.

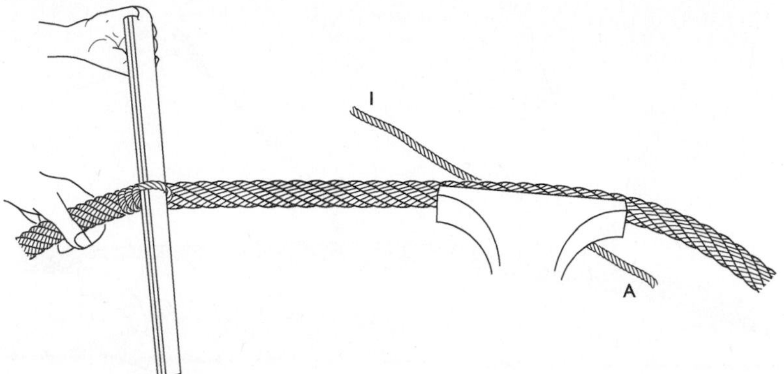

Figure 36-57 How wire rope may be unlaid for a short distance by means of a suitable stick.

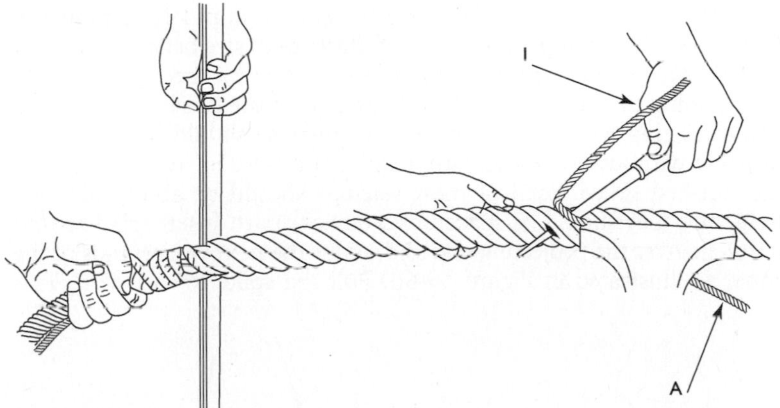

Figure 36-58 Method used to remove fiber core from wire rope.

In addition to removing one complete twist, the strand should be straightened and its curvature removed. The strands of a preformed rope have more curvature than those of a standard rope, which makes it more difficult to straighten preformed strands. Special care should be taken to remove the curvature from the strands when splicing preformed rope. After the fiber core has been cut, it should be removed for a distance equal to the length of the projecting end of the strand. This may be done by moving the spike along the rope as illustrated in Figure 36-59, while the other hand removes the fiber core. The spike should be under two strands of the rope, as indicated.

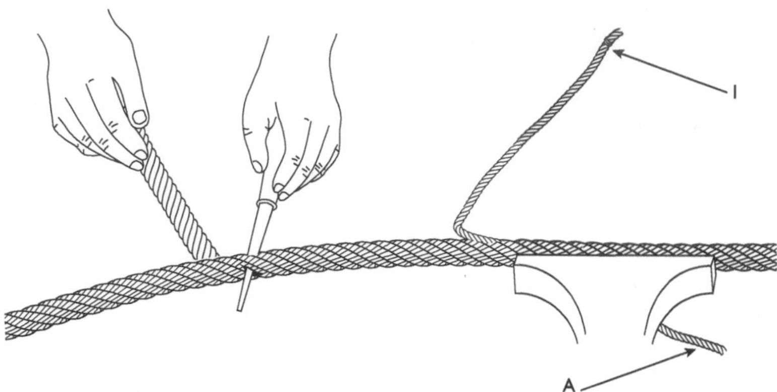

Figure 36-59 Removing the fiber core from wire rope by the use of a splicing stick.

Next, two seizings are placed on the end of strand 1, as shown in Figure 36-60, to compensate for the difference in size between the fiber core that has been taken out and the strand that has been laid in, thus maintaining the same rope diameter as that of the finished splice. The size of seizing wire to be used for this purpose should be as large as possible without causing any appreciable increase in rope diameter in the finished splice. Each of these seizings should be about one inch long, and they should be spaced two inches apart. Insert spike so that it will be over the projecting end and under the next two strands of the rope, as illustrated in Figure 36-60. Pull the spike toward you. This

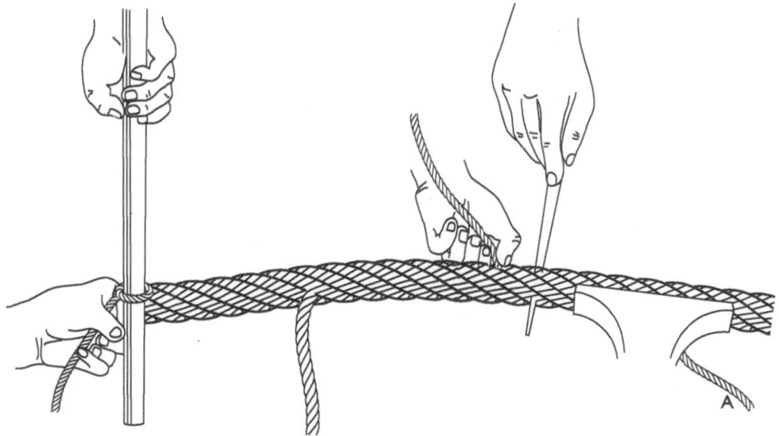

Figure 36-60 Method of laying in strand into splice.

will cause it to travel along the rope, leaving an opening in front. While one hand is employed in moving the spike, the other hand holding the end of the strand should lay this end in the opening as indicated.

Figure 36-61 shows the rope after the end of strand 1 has been laid in place. The splice is now ready for the tucks. Perhaps the success of the splice depends more on this operation than any other in the whole splice. In a *regular lay rope* the tucked strands should lie side by side where they disappear into the core of the rope. They should not be crossed. The opposite is the case with a *lang lay rope*. Here the strands should be crossed before tucking. This difference is made between the *lang lay* and *regular lay* ropes so as to permit the wires to mesh together where the twist was removed. Maximum holding power is obtained in this manner and the tendency of the rope to bulge at the tucks is reduced. If the strands are not laid together in this manner, then they will not mesh into each other, and a rough splice will result with greatly reduced strength.

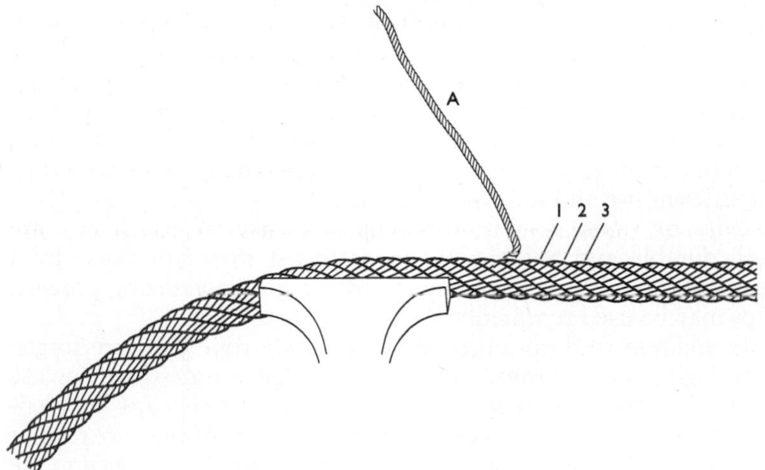

Figure 36-61 Tucking operation in nearly finished splice.

Finally, after the ends have been tucked, the projecting ends of fiber and core should be cut off. Hammer down any inequalities with the wooden mallet as shown in Figure 36-62. When all the strands have been laid in the rope as described, the splice is complete. With practice and careful work, a splice can be made in such a way that it will be impossible to detect after the rope has been running for a short time.

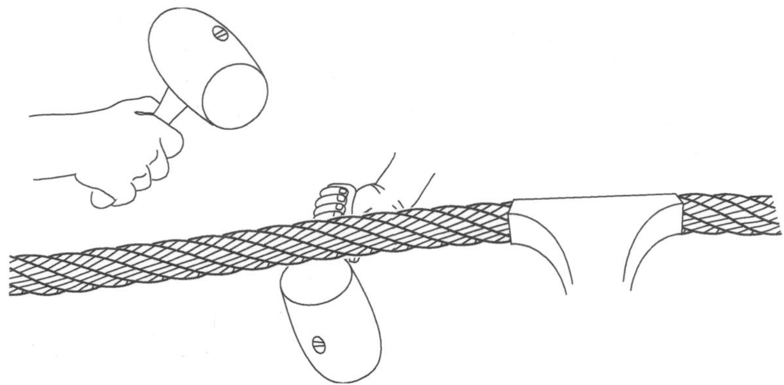

Figure 36-62 Hammering operation with wooden mallet to remove any inequalities in the splice.

Wire-Rope Fittings and Method of Attachment

The safe and efficient operation of wire rope requires that wire-rope fittings be selected with the same care as the rope itself. Although there is a multiplicity of wire-rope fittings in use, serving various classes of work, the most commonly encountered wire-rope fittings are clamps, thimbles, hooks, shackles, and turnbuckles. Each one requires particular care when attaching it to the wire rope for efficient use and safety in operation.

Clips or Clamps—Wire-rope clips are easy to attach and are both durable and reliable. When protected from corrosion by a heavy coating of pure zinc applied by the hot galvanizing process, clips may be used repeatedly.

In addition to drop-forged steel clips, wire-rope clips are forged from high-strength bronze. These bronze clips are designed for use where electrolysis or corrosion makes the use of steel clips impractical. The number of clips to use depends on the size of the wire rope; it may be from two to six or more. The clips should be attached to the rope ends as shown in Figure 36-63, with the base of the clip against the live or long end, and the U-bolt bearing against the dead, or short, end of the rope. This is the only correct method of attaching wire-rope clips.

The *clips* should be spaced *at least six rope diameters apart to ensure maximum holding power.* After the rope has been placed in service and is under tension, the nuts should be tightened again to compensate for any decrease in rope diameter caused by the load.

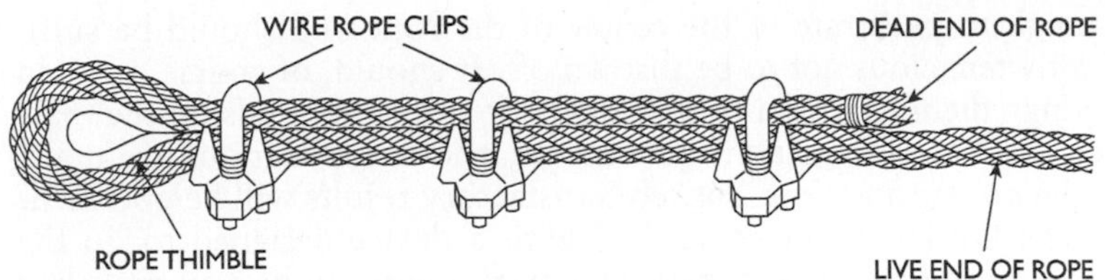

Figure 36-63 Proper application of wire-rope clips.

Wire-Rope Lubrication

When a wire rope is considered a "machine," in that its parts have a motion relative to one another, the need for *lubrication* becomes apparent. Proper lubrication will reduce wear between the individual wires and strands in normal operation and prevent or retard corrosion.

Used ropes should be cleaned before they are lubricated. The cleaning may be accomplished by means of wire brushes or scrapers, by compressed air, or by superheated steam. The object of the cleaning process is to remove all foreign material and old lubricant from the valleys between the strands and from spaces between the outer wires. The lubricant may be applied in any manner suitable to field conditions. It may be brushed onto the rope with a stiff brush, applied by passing the rope through a box of lubricant, or the lubricant may be made to drip onto the rope, preferably at a point where the rope opens slightly from bending. The main purpose of the foregoing methods of lubrication is to apply a uniform coating to the entire length of rope.

When a wire rope is taken out of service for an appreciable length of timc, it should be cleaned and lubricated. It should be stored in a dry place protected from the elements. During the manufacture of the wire rope, the hemp center and wire strands are thoroughly impregnated with a type of lubricant that will satisfactorily protect and lubricate it in service for a period of time. Because of constant motion and friction between the component parts of the wire rope, this lubricant will be pressed to the surface where it combines with dirt and other foreign material and thus loses its lubricating value.

Lubricant lost during operation can be satisfactorily replenished by external application, and if the proper type and quantity is put on at each application, no difficulty should be experienced with the material running down the ropes or being thrown off in the machine room and causing slippage troubles. Such a lubricant should be thin enough to penetrate to the center of the rope and should be sufficiently tenacious not to be thrown off. It should, of course, contain no ingredients harmful to the rope. Several compounds on the market fulfill the foregoing requirements, and if applications are made in the proper amounts, entirely satisfactory results will be obtained.

The lubricant can be applied with a device designed to do the work more quickly and more effectively than can be accomplished by hand. One such device consists merely of a hinged box with two containers, into which holes are drilled near the bottom for dispensing the lubricant to the ropes. Brushes are placed below the containers to spread the lubricant and work it into the valleys between two strands. As the ropes run through this box, lubricant is poured into the containers and a complete application can be made in very little time. The size of the holes in the container, and the speed at which the ropes are operated, should be such that approximately one quart of lubricant is applied to each 600 ft of ⅝-in. diameter rope. For ½-in. diameter ropes, this quantity should be applied to each 900 ft of rope.

If the lubricant is applied to the wire ropes with a brush, application at the point where they operate over the drive sheave is most effective. Sometimes when this work is done by hand, there is a tendency to spread the material *too thin*. It should be borne in mind that if the applied lubricant does not penetrate to the rope core or hemp center, much of the value of the application will be lost. For this reason, it is important that a sufficient quantity be used.

Lubrication Compound—There are several wire-rope lubricants on the market suitable for various operating conditions. One such lubricant, used successfully by shipyard riggers, consists of the following:

For summer use:

3¾ gallons gear compound

5 gallons winter crankcase oil

⅜ gallons apexior (this is an asphalt-base water-resisting paint, flammable, slow-drying)

For winter use:

3½ gallons gear compound

15 gallons winter crankcase oil

½ gallon apexior

In compounding the lubrication in suitable amounts, it will be necessary to observe that the same proportion of the foregoing ingredients should be used in each instance.

Wire-Rope Ends

In the normal use of wire ropes, the most severe deterioration frequently occurs near one end due to the diminished radii close to the drum with accompanied additional bending. The life of a length of rope so damaged may be materially increased by exchanging the drum end with the load end, thus reversing the rope in use. At the same time, worn sections of the remainder of the length will also be relocated and perfect sections will be exposed to the positions of greatest wear. Turning wire-rope ends in this manner will frequently result in service increases as high as 25 percent, possibly more.

Wire-Rope Pointers

Prior to the use of a wire rope, it is good practice to make a thorough examination of the equipment on which it is going to be used. It has been found that in the majority of instances the rapid deterioration of wire rope is caused by the machinery on which it is operating. The principal items to be checked are

1. Sheaves or drums.
2. Reeving.
3. General operating conditions.

Points to be checked on drums are the diameter of the sheave or drum from bearing surface to bearing surface, not from flange edge to flange edge, because it is the radius of the bearing surface that governs the degree of bending to which the rope is subjected as it travels over the sheave or drum. Table 36-9 gives the recommended and minimum sheave or drum diameters for various sizes of standard hoisting rope.

Because the rate of fatigue of wire rope is governed very largely by the radii of the bends it must make in normal operation, it is extremely important that the recommended diameters be equaled or exceeded. A large amount of internal wear is caused by improper sheave and drum diameters, because the radial pressure of a wire rope increases in inverse proportion to the radius of the bend. As the radial pressure increases, there takes place in the rope a crushing action that increases internal friction and causes nicking of one inside wire by another.

Table 36-9 Recommended Sheave and Drum Diameter

Rope	*Recommended Diameter No. Times Rope Diameter*	*Minimum Diameter No. Times Rope Diameter*
6 × 7	72	42
6 × 19	45	30
6 × 30	45	30
6 × 37	27	18
8 × 19	31	21
18 × 7	51	34

Sheave and Drum Grooves

The diameter of the grooves must bear a certain definite relation to the diameter of the wire rope used in it, as illustrated in Figure 36-64. As a rope passes constantly over the sheave, the groove is worn deeper, and the rope diameter becomes constantly smaller until it is less than that of a new rope of listed diameter. When a new rope is placed in such a groove, it will be forced down into the groove and pinched out of shape. Hence, it will be subjected to extreme conditions of abrasion against the side of the groove. In addition, it will be forced out-of-round and therefore will become unbalanced.

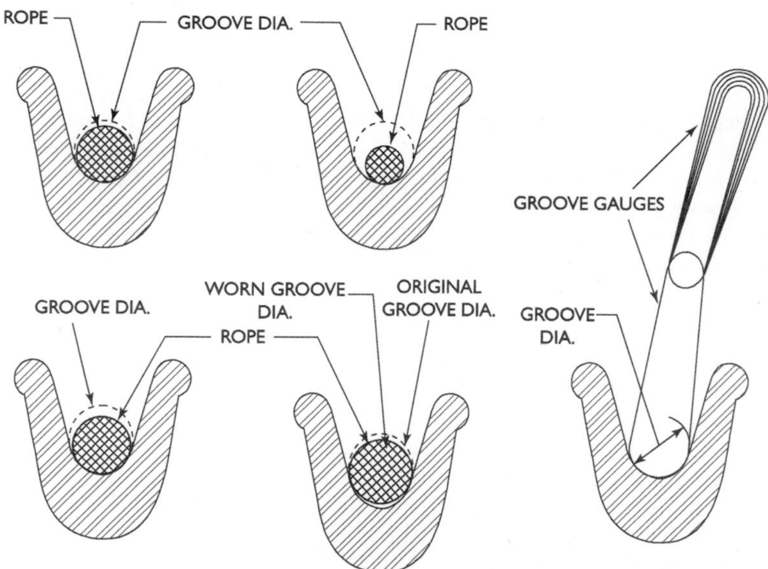

Figure 36-64 Relationship between groove diameter and wire-rope size of typical sheave.

A rope placed in a sheave groove of too large a diameter, on the other hand, will not be afforded proper support. Such a condition tends to allow the rope to flatten out as radial pressure is applied to it. The ideal condition is that in which the rope receives support from the sheave groove around just less than one-half of the circumference. The rope is thus allowed freedom of action with a maximum of support. Gauges for measuring sheave grooves can be obtained from any wire-rope manufacturer.

Sheave and Drum Material

If the material in the sheave or drum is too soft, it will be cut by the wire rope, and the bearing surfaces will become corrugated to fit the contour of the rope used over them. This contour will not conform exactly to that of the next rope used; therefore, this rope will be cut and worn by the corrugations. Where such a condition prevails, the sheaves should be remachined or, preferably, replaced with sheaves or drums of harder material. Such corrugation will occur more readily if the sheave diameter is too small.

Sheave or Drum Operation

Defective bearings will cause sheaves and drums to wobble or to revolve eccentrically and will thus set up a whipping action in the wire rope. Whipping will, of course, greatly increase the fatigue and other deterioration of the rope. Such a condition should be corrected immediately.

General Sheave and Drum Conditions

A sheave or drum that is out-of-round or that has a flat spot in the bearing surface will also cause a whipping action in the wire rope. Such a condition should be corrected immediately. Sheave and drum grooves should be either machined or ground until both surface and contour are smooth and true.

When a sheave flange is broken off, the wire rope may jump this flange and cause serious damage both to the rope itself and to the machinery. If the rope does not jump the flange, it is at least likely to come in contact with the sharp edges of the broken flange and become badly gouged. Such sheaves should be replaced without delay.

Rollers

On some types of installations, rollers are used to support the rope and keep it from scrubbing against objects such as the ground, structural properties, or railroad ties. Rollers should not be used where the rope is forced to make an appreciable angle of departure over them. This often occurs where a knuckle condition or a decided change of grade exists. Knuckle or deflection sheaves of suitable size and material should be installed at these points.

When field conditions require such installations, it is important that the following precautions be observed:

1. The size of rollers should not be less than eight times the rope diameter.
2. The roller bearings should be kept free and well lubricated at all times. If the roller is too heavy, it will have considerable inertia, with the result that it will take an appreciable time to bring its peripheral speed up to the speed of the rope working over it. This will cause excessive abrasive wear.
3. The rollers should be spaced close enough to prevent the rope from touching the ground or other objects. They should be spaced at irregular intervals. This will prevent rhythmic vibration from being set up in the rope with consequent fatiguing action at points where these vibrations are dampened or arrested.

Common Causes of Wire—Rope Failures

In order to obtain maximum safety and service from a wire rope, it is of paramount importance that the rope be of the correct type and size for the service to which it is being assigned. Therefore the rope manufacturer's recommendation as to the proper rope to employ for a particular duty should always be adhered to. Of the many forms of abuse of wire ropes, the most commonly encountered are as follows:

1. Ropes of incorrect size, construction, or grade.
2. Ropes allowed to drag over obstacles.
3. Ropes not properly lubricated.
4. Ropes operated over sheaves and drums of inadequate size.
5. Ropes allowed to overwind or crosswind on drums.
6. Ropes operated over sheaves and drums with improperly fitting grooves or broken flanges.
7. Ropes permitted to jump sheaves.
8. Ropes subjected to moisture or acid fumes.
9. Ropes with improperly attached fittings.
10. Ropes permitted to untwist.

11. Ropes subjected to excessive heat.
12. Ropes permitted to become kinked.
13. Ropes subjected to severe overloads.
14. Ropes destroyed by internal wear caused by grit penetrating between strands and wires.

Hand Signals for Cranes and Hoists

Hand signals are important to learn because background noises can make communication difficult. Two-way FM portable radios are replacing hand signals, but if radio communication is distorted from static or other interference, hand signals are a reliable fallback. If the designated signal person is not in line-of-sight with the operator, then a third person needs to relay the signals. Following are some simple rules for using signals:

1. Only one person should be designated to give signals.
2. Keep your signal in one place. Try to make eye contact, and then make your signals in front of your face or body.
3. If your gloves and clothing are of similar colors, make your signals away from your body where the operator can see them.
4. A set of signals should be *agreed upon* and adopted at each operation where hoisting equipment is used. Only the agreed-upon signals should be used by a designated person, except in an emergency, when *anyone* may give a "STOP" signal.

Common Hand Signals

With forearm vertical and index finger pointing up, make a small horizontal circular motion with hand and forearm.

Figure 36-65 Raise load.

With whole arm extended down and palm downward, point index finger down and move hand and arm in small horizontal circles.

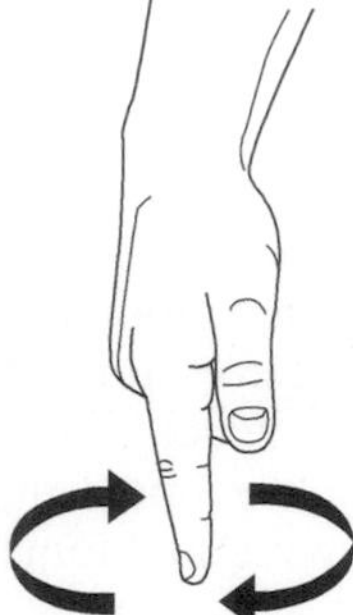

Figure 36-66 Lower load.

Using two arms, one hand with index finger pointing up into downturned palm of other hand, make circular motion with index finger.

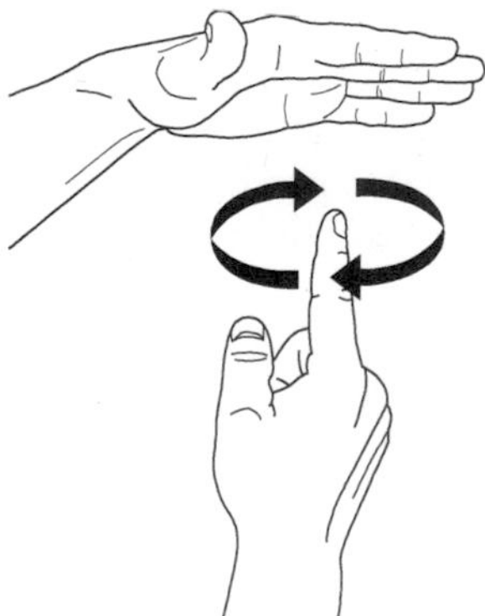

Figure 36-67 Raise load slowly.

Using two arms, one hand with index finger pointing down into upturned palm of other hand, make circular motion with index finger.

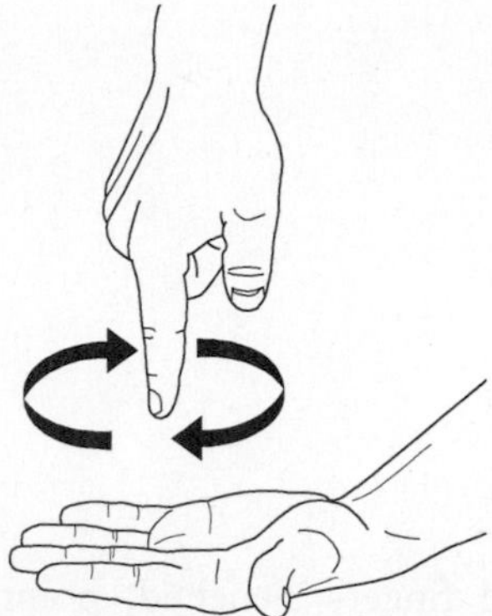

Figure 36-68 Lower load slowly.

With arm extended and finger clenched, point thumb up.

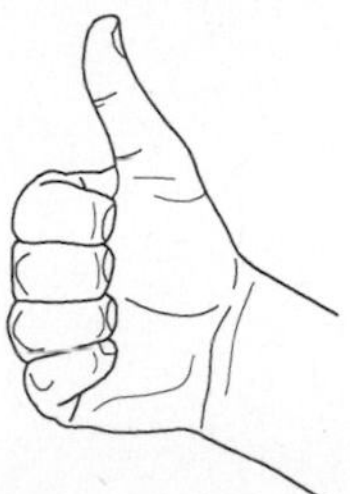

Figure 36-69 Boom up.

With arm extended and finger clenched, point thumb down.

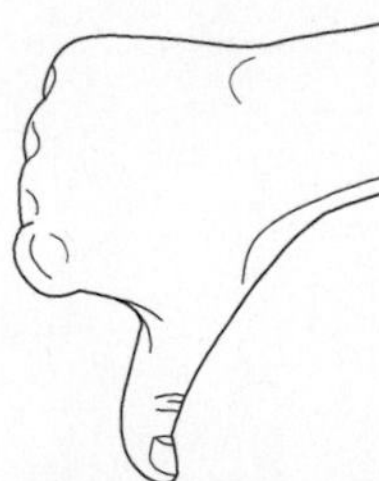

Figure 36-70 Boom down.

With one arm partially extended and fingers clenched, point thumb up into palm of other hand.

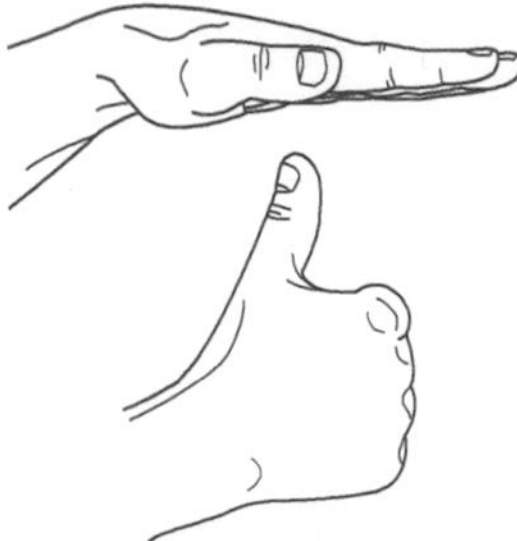

Figure 36-71 Raise boom slowly.

With one arm partially extended and fingers clenched, point thumb down into palm of other hand.

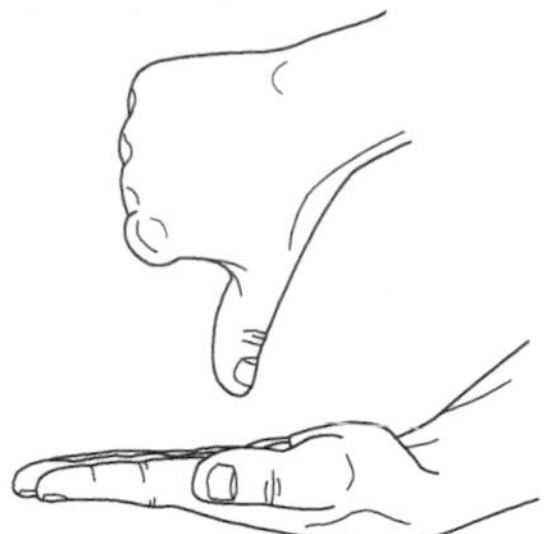

Figure 36-72 Lower boom slowly.

With arm extended and thumb pointing upward, flex the fingers in and out.

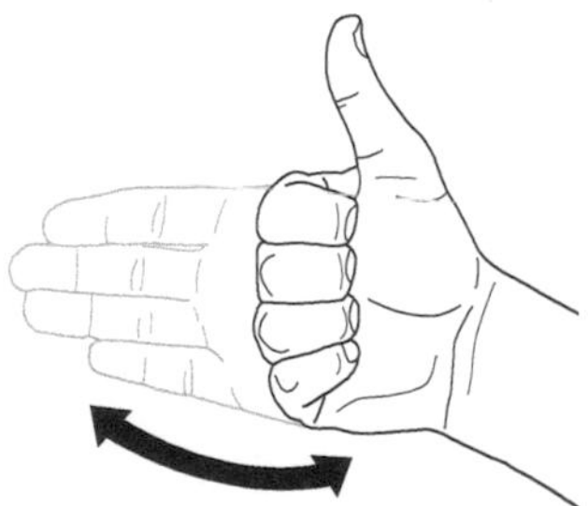

Figure 36-73 Raise boom and lower load.

With arm extended and thumb pointing downward, flex the fingers in and out.

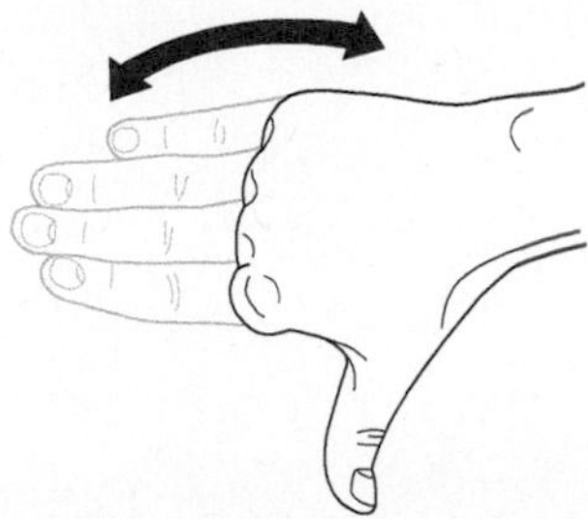

Figure 36-74 Lower boom and raise load.

With both fists clenched in front of body, point the thumbs outward away from each other.

Figure 36-75 Extend boom.

With both fists clenched in front of body, point the thumbs pointing inward toward each other.

Figure 36-76 Retract boom.

With one arm raised, forearm vertical, and palm pointing toward crane operator, use the index finger of the other hand to trace a small circle in the palm.

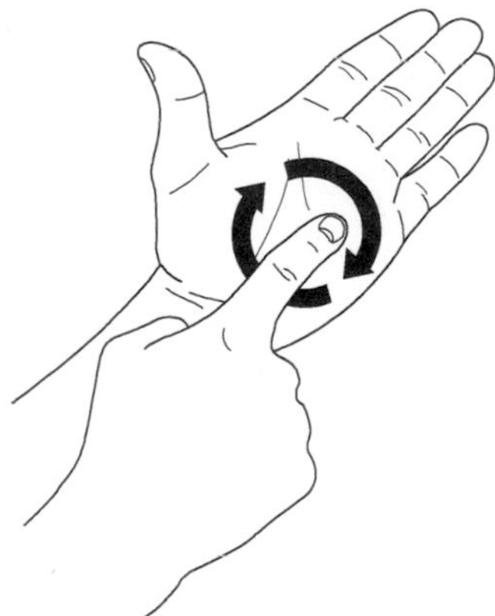

Figure 36-77 Go-slow all movements.

With arm extended and palm down, move arm back and forth horizontally.

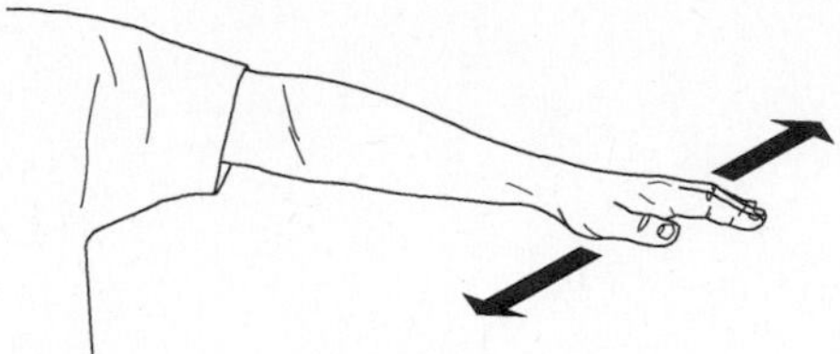

Figure 36-78 Stop.

With *both* arms extended and palms down, move both arms back and forth horizontally. Also one arm can be used with arm extended and palm down, moving hard back and forth *very* rapidly.

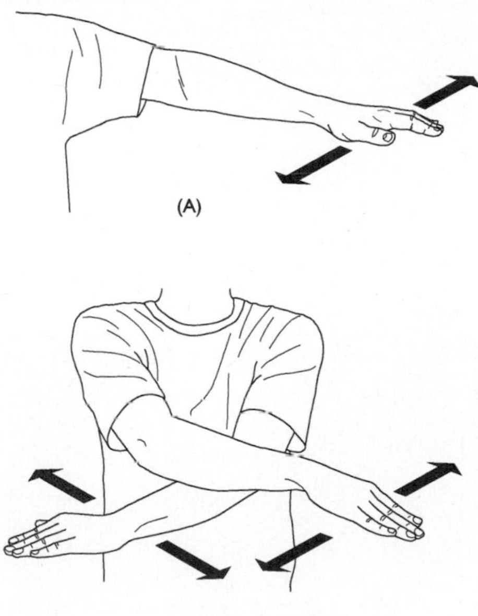

Figure 36-79 Emergency stop.

Chapter 37

Hydraulics and Pneumatics

Hydraulic and pneumatic power transmission equipment is an essential part of modern industrial operations. The term "fluid power" is becoming more and more widely used to describe the use of fluids for power transmission. Any material that is capable of flowing, either liquid or gas, may be classed as a fluid. However, the most widely used forms of fluid power are oil and water hydraulics, compressed air, and vacuum. While hydraulic and pneumatic machinery and equipment have been in use since the beginning of industrialization, their use has expanded rapidly in recent years. The impetus behind this expansion has been the development of synthetic packing materials, improved metal alloys, and advanced techniques for fine metal finishes. These, combined with new engineering know-how, have resulted in fluid-power equipment with a very high level of dependability.

Almost invariably, fluid-power systems and devices include mechanical and electrical machinery and equipment. Installation and maintenance of hydraulic and pneumatic components, therefore, have become important functions of the industrial mechanic. While the range and variety of fluid-power systems is extremely broad, all are based on certain basic principles. The first step to understanding these systems is knowledge and understanding of the principles as they apply to actual machinery and equipment. For example, almost all hydraulic systems work on the hydrostatic principle, but in reality there is nothing really static about them, because in most hydraulic systems there is a fluid flow. While a strict interpretation of physical laws is required for scientific research and experimentation, practical application often involves taking some liberties with these laws.

A basic physical law that has broad application to hydraulics and pneumatics is known as *Pascal's Law,* or the law of *Transmissibility of Pressure.* This states that increases of pressure are transmitted equally throughout a fluid. For example, pressure set up in a liquid confined in a pipe, as shown in Figure 37-1, will act equally in all directions and at right angles to the inside surface of the pipe. Or to apply this to the hydraulic cylinder in Figure 37-2, applying a force F on a piston will cause pressure to be transmitted equally in all directions within the fluid-filled cylinder.

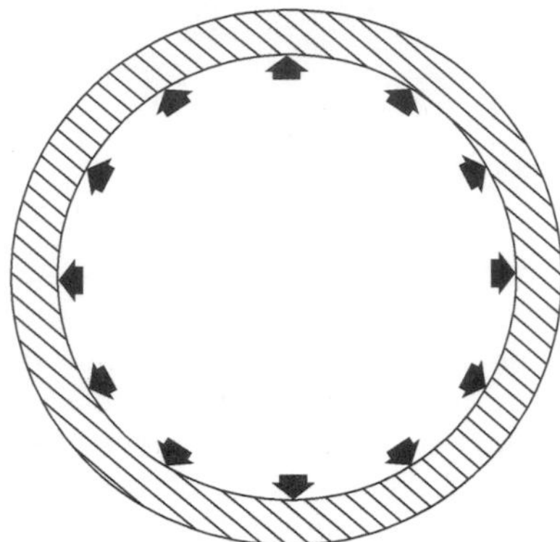

Figure 37-1 The reaction of fluid in a pipe.

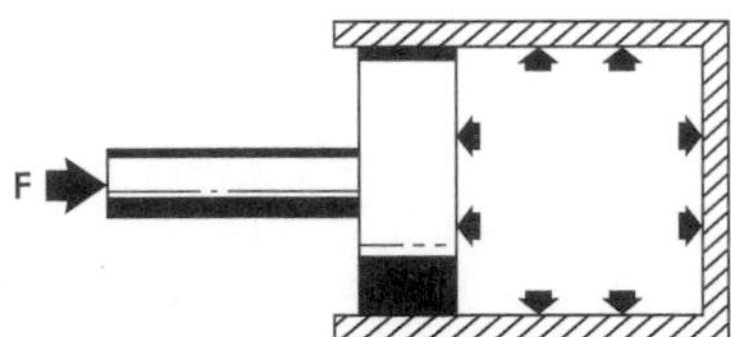

Figure 37-2 The reaction of fluid in a cylinder when pressure is applied.

This basic law explains the principle upon which the transmission of fluid power is based. To illustrate this, the system in Figure 37-3 shows two cylinders of equal diameter connected by a pipeline

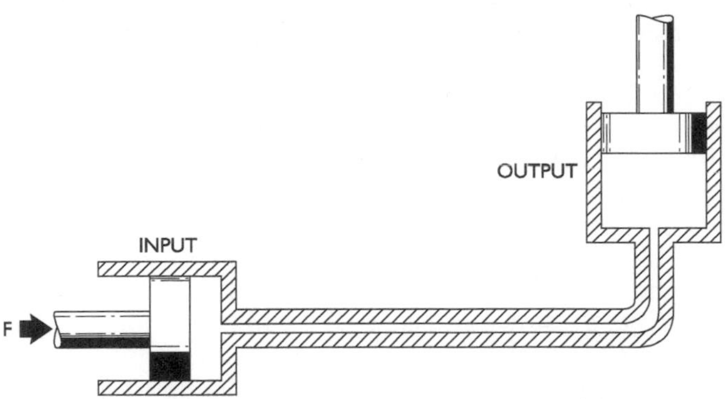

Figure 37-3 Two cylinders connected together. If force is applied to one cylinder, the pressure will be equal in both cylinders.

of smaller diameters. When force *F* is applied to the input piston, the pressure developed within the fluid will be equal at every point in the system. The force developed on the output piston therefore will be equal to the force applied by the input piston. The practical limitation in such a system is that the inside diameter of the pipe be large enough to carry the volume of flow between the pistons. Long distances and small-diameter pipe would introduce friction loss and velocity considerations.

Force may be transmitted in any direction within pipes or hoses with quite small losses. However, hydraulic power is usually limited to relatively short distances, because the pump or other source of power is seldom far from the device it supplies. Frictional losses, therefore, need not be considered unless distances and travel speeds are high in relation to the size of the pipe or hose lines.

The *force-to-pressure* relationship is fundamental to any calculation of hydraulic or pneumatic forces or pressures. The pressure within the fluid is measured as *unit pressure,* the common unit being pounds per square inch (psi). Force, of course, is the pounds portion, and the square inch is the most commonly used area measure.

The total force applied to the piston rod in Figure 37-4 will be evenly distributed over the entire face of the piston. Because the unit pressure will be in terms of the force exerted on one square inch of surface, the force must be divided by the number of square inches over which it is distributed. Thus, if the force is 1500 lb and the area of piston surface is 20 sq. in., the unit pressure will be 75 psi on the gauge.

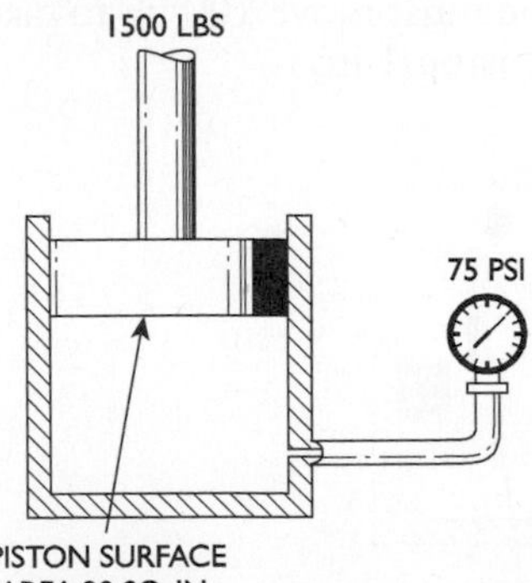

Figure 37-4 The relationship between force and pressure.

In actual hydraulic applications the reverse situation would often exist, in that the generated hydraulic pressure would provide the power to exert a force on the piston. In Figure 37-5 a motor-driven pump is supplying hydraulic pressure at 150 psi. Because there are 15 sq. in. of piston surface for the fluid to act on, the total output force of the piston rod is 150 multiplied by 15, or 2250 lb.

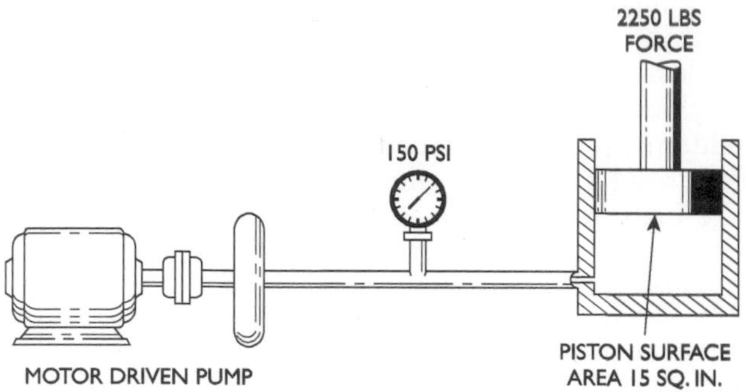

Figure 37-5 The relationship between pressure and force.

A common application of fluid power is the hydraulic multiplication of force, or hydraulic leverage. This principle is in everyday use in the hydraulic jack. The diagram in Figure 37-6 illustrates a jack system with an input force of 80 lb. Because the surface of the input piston is one-half square inch, the hydraulic pressure generated by the 80-lb force is 160 psi. The 160-psi pressure acting on the 50 sq. in. of output piston surface develops an output force of 8000 lb; the input force is multiplied 100 times. This is, of course, balanced by the fact that the input piston must move 100 in. to displace enough fluid to move the output piston 1 in.

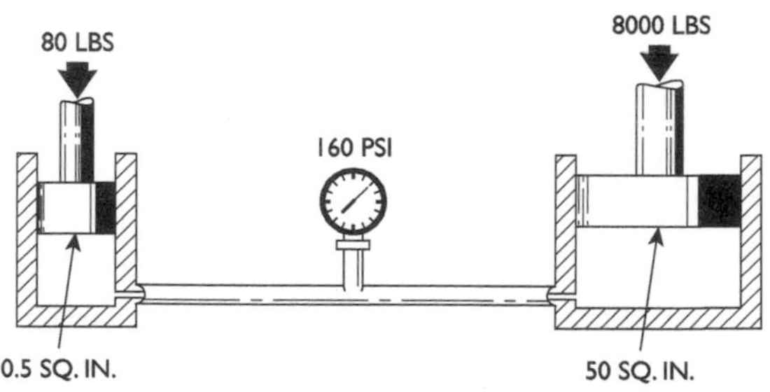

Figure 37-6 The basic principle of a hydraulic jack.

Pressure within a fluid may be developed by the weight of the fluid above. Water hydraulic power supplied by service water systems is usually developed in this way. That is, the weight of the water stored at some elevation, such as a reservoir or a water tower, is the force that develops the pressure. The pressure due to the weight of the fluid above is called *static head* pressure and is stated in terms of *feet*. For example, if an open-top tank 20 ft high were filled with water, there would be a 20-ft head pressure on the bottom of the tank. In terms of gauge pressure of psi, there are 0.433 psi for each foot of head pressure. This may be calculated by dividing the weight of a cubic foot of water (62.4 lb) by 144, the number of square inches in a square foot. To convert gauge pressure to head pressure, multiply the gauge-pressure value by 2.304. This value (2.304) is the height of water in feet required to exert 1 psi of pressure.

Static head pressure is generated by the height of the fluid and is not affected by the volume. The two vessels shown in Figure 37-7 will have the same gauge pressure due to the weight of the water. While the total weight of water in the large tank will be many times greater, the weight of water on a single square inch in either tank will be exactly the same.

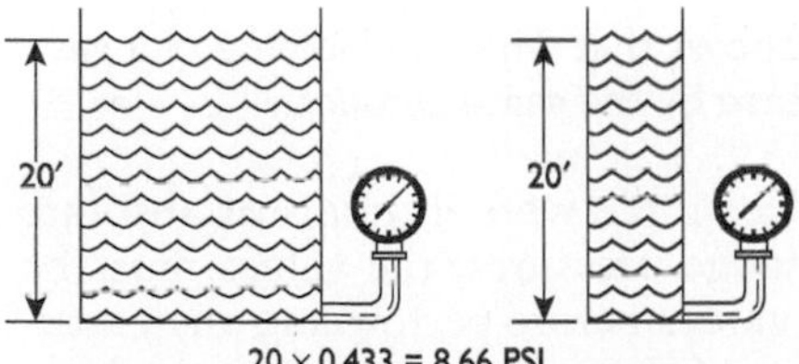

Figure 37-7 The gauge pressure of the fluid will be the same because of the height, or head, pressure.

A basic physical law governing pneumatic fluid-power equipment, known as *Boyle's Law,* states: If the temperature of a confined body of gas is maintained constantly, the absolute pressure is inversely proportional to the volume. This simply means that when the volume of a confined gas is reduced by some amount, its absolute pressure will increase by the same amount, and vice versa. It is important to note that this law is in terms of *absolute* pressure, not gauge pressure. Absolute pressure is 14.7 psi more than the reading of a pressure gauge. The reason for this is that a pressure gauge shows zero pounds pressure when it is open to the atmosphere, although 14.7 lb is actually being exerted upon it. The weight of the earth's atmosphere at sea level is 14.7 psi. The three diagrams in Figure 37-8 illustrate this law in action. In Figure 37-8A a cylinder containing 10 cu. ft of gas is compressed to a pressure of 20 psia (pounds per square inch absolute). In Figure 37-8B additional force has been applied to the plunger, the volume has been reduced to 5 cu. ft, and the pressure has increased to 40 psia. In Figure 37-8C, with additional force applied to the plunger, the volume has again been cut in half to 2.5 cu. ft, and the pressure has again doubled to 80 psia.

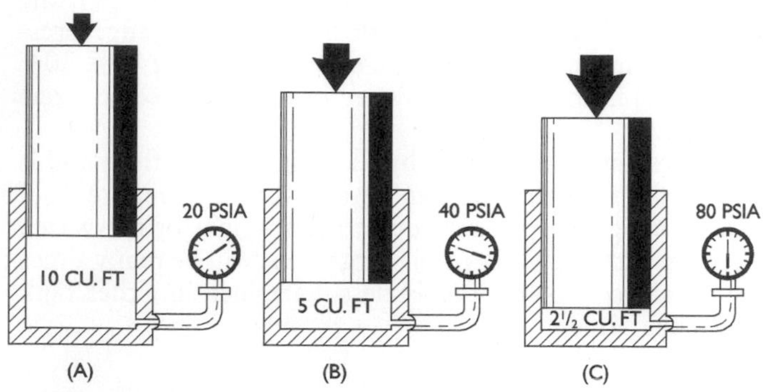

Figure 37-8 Boyle's Law, which shows that when the volume of a gas is reduced, the pressure will increase by the same amount.

The values in the preceding example were in terms of absolute pressure. When dealing with gauge pressures, the values must be converted to absolute pressure values before performing any calculations. This is done by adding 14.7 to the gauge reading. After completing the calculations, the terms are changed back to gauge reading by subtracting 14.7. In Figure 37-9 are illustrated similar cylinders with the pressures in terms of psig (pounds per square inch gauge). Note that while the volume in each case has been reduced by exactly one-half, the resulting pressures have been more than doubled in terms of gauge pressure. However, when the psig pressure values are changed to psia values by adding 14.7, the pressure readings are in exact inverse ratio to the changed volumes. In Figure 37-9A, when 14.7 is added to the 10 psig reading, the absolute pressure is determined to be 24.7 psia. Doubling this gives 49.4 psia for an absolute reading in Figure 37-9B, which when reduced by 14.7, gives 34.7 psig. The same procedure is followed to determine the psia and psig readings in Figure 37-9C.

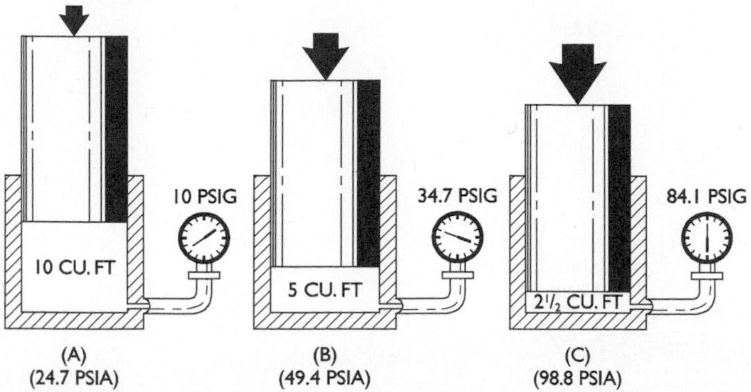

Figure 37-9 The difference between absolute pressure and gauge pressure.

Vacuum fluid power may be described as deriving its force from the weight of the atmosphere. Since the maximum normal atmospheric pressure at sea level is 14.7 lb per sq. in., this is the maximum pressure available in a vacuum system. While this is relatively low pressure compared to compressed air systems, the manner in which this force acts makes it ideal for many applications. Vacuum systems require some form of pump to remove or reduce the atmospheric pressure within the system. The resulting low-atmospheric-pressure condition is called a vacuum. In static situations when this unbalanced condition exists, the pressure of the atmosphere exerts force on all outside surfaces of the system under vacuum. Also, because of the lower pressure within the system, the atmosphere will rush into any area of the system that may be open. Vacuum, therefore, is handled and controlled in pipelines and systems very much like those used for compressed air.

The action of atmospheric pressure is illustrated in Figure 37-10, which shows the basic principle of the barometer, an instrument for measuring atmospheric pressure. A glass tube with one closed end is immersed horizontally in mercury, and all air is allowed to escape so that the tube becomes completely filled with mercury. The closed end of the tube is then raised, keeping the open end submerged. The tube is placed in a vertical position with the open end held off the bottom but still immersed in the mercury. Atmospheric pressure acting on the surface of the pool of mercury will cause the mercury to rise approximately 30 in. in the tube. Vacuum is measured in *inches of mercury* (abbreviated " Hg) based on the measuring instrument.

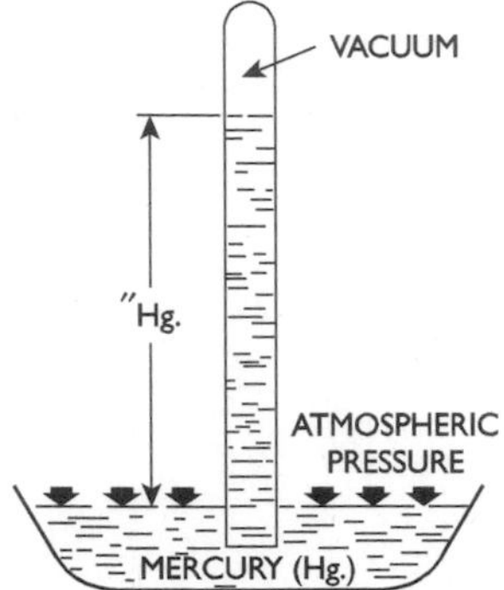

Figure 37-10 Atmospheric pressure.

Theoretically a perfect vacuum is 30" Hg.; however, vacuum pumps capable of pulling 26" to 28" Hg. are considered very efficient. The same diagram may be used to illustrate the action of a vacuum pump. In Figure 37-11 a pump is connected to the top of the tube and a vacuum is pulled in the tube. Depending on the efficiency of the pump, the mercury column will rise some height up to a maximum of 30 in. This also illustrates the action called *suction* in lift pumps, the common siphon, and other fluid devices. Since the column is pushed up by atmospheric pressure, not "sucked" up as we commonly refer to it, no matter how efficient the pump, the column of mercury can rise no higher than it is pushed by the atmosphere.

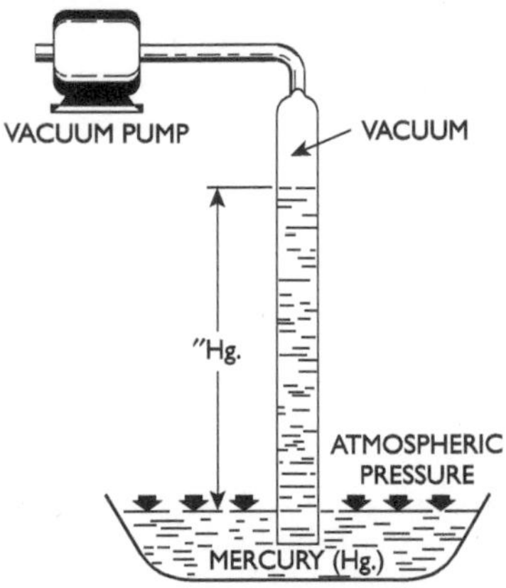

Figure 37-11 A pump is used to pull a vacuum in the tube.

Practically all hydraulic fluid-power systems use some form of pump or fluid pressurizing device. The purpose of the pump is to develop pressure in the hydraulic fluid so that work can be performed. The two most common devices used to convert the fluid power to mechanical force, work, or power are the cylinder and the hydraulic motor. Three types of pumps are used in fluid-power systems: rotary, reciprocating, and centrifugal. The rotary pump is the most commonly used because it provides smooth, continuous flow at pressures suitable for most systems. The reciprocating and centrifugal pumps are used for specialized applications.

The design and principle of operation of the various styles of pumps commonly used in industry are covered in Chapter 25. However, the radial-piston pump is not mentioned, and because of its wide use in hydraulic systems, its principle of operation should be discussed. A cross-sectional diagram of the essential elements of a radial-piston pump is shown in Figure 37-12. It is a positive-displacement pump that takes its name from the pistons, which reciprocate in a radial direction from the shaft.

The center cylinder-block section rotates with the shaft. The pistons rotate with the block, their heads in contact with the inside surface of the stationary reactor ring. The reactor ring may be moved sideways by turning the adjustment wheel connected to the adjusting screw. The pistons are usually held against the reactor-ring surface by springs.

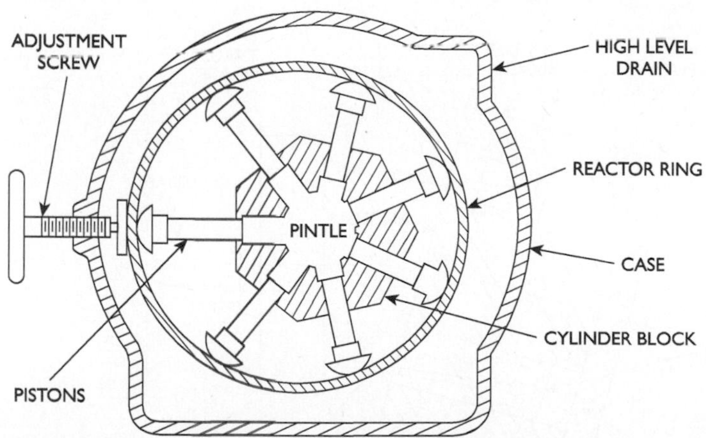

Figure 37-12 Essential parts of a radial-piston pump.

The pistons move in and out as they slide around inside the reactor ring when it is in an eccentric position with respect to the shaft centerline. The more eccentric the action of the reactor ring is, the longer the stroke of the pistons will be. A stationary pintle in the center of the cylinder block does the valving. Each cylinder, as it rotates, is connected alternately to the fluid inlet and then to the pressure outlet. If the reactor ring is positioned concentric with the shaft, the piston stroke becomes zero. The positioning of the reactor ring may also change the direction of pumping, because moving it across the centerline reverses the direction of flow.

Another specialized type of piston pump frequently used as a fluid-power pump is the *axial-piston pump*. In this design, the pistons reciprocate axially or parallel to the shaft centerline. The principle of operation of the axial-piston pump is illustrated in Figure 37-13. All parts illustrated, except the valve plate, rotate as a single unit. The stationary valve plate and the mating surface of the cylinder block have a high-quality lapped finish to prevent leakage of fluid. The piston reciprocation results from the cylinder block and the thrust plate being at an angle to each other. Varying this angle increases or decreases the piston stroke. An adjustment screw (not shown) varies the angle to change the piston stroke and vary pumping volume. As the cylinder block rotates in contact with the stationary valve plate, the cylinders are ported alternately to inlet and outlet connections. The axial-piston pump principle is commonly used in the design of hydraulic motors used in fluid-power systems.

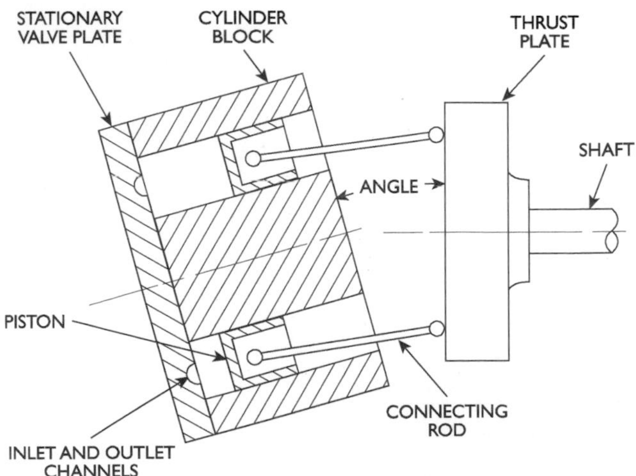

Figure 37-13 Principal parts of an axial-piston pump.

A characteristic of the positive-displacement pump used in most fluid-power systems is that it has a positive output. This means that when fluid flow is not required, the pump must be stopped or its output must be discharged or recirculated. Discharge or sewering is sometimes practiced with water systems; however, oil hydraulic systems are usually of closed design and the oil must be recirculated. To start and stop the pump between cycles is seldom practical unless there are long shutdown periods or the system is very low in horsepower.

Recirculation may be accomplished at low pressure by porting the fluid through a valve to a reservoir or at high pressure by discharge across a relief valve. Discharge across a relief valve should be allowed only for short periods, because power is converted to heat as the oil passes through the relief valve. In most cases, oil is diverted back to the reservoir at little or no pressure by the positioning of a control valve. This is termed *unloading* the pump and is the usual method of handling the oil flow while no work is being done.

Direction and Flow-Control Valves

Valves are a vital element in the control of hydraulic systems. They are the means of starting, stopping, and reversing flow within the system, as well as regulating the flow volume and system pressure. Direction valves are used to start, stop, and reverse flow. They may be either manually operated or controlled by automatic devices. They are usually classed as 2-way, 3-way, 4-way, or 5-way operation. The common gate, globe, and needle valves are examples of 2-way valves. Their primary function is to open and close a line. An equally important function of the 2-way valve in hydraulic systems is to provide a variable restriction for controlling the amount of flow. Because the purpose of a needle valve is to restrict flow, it is commonly used to control flow in hydraulic systems, whereas globe and gate valves are usually used for on-off service.

The spool-type valve is in wide use in hydraulic systems for direction control, as it is quick-acting and easily operated. A typical 2-way spool valve in the open and closed position is shown in Figure 37-14.

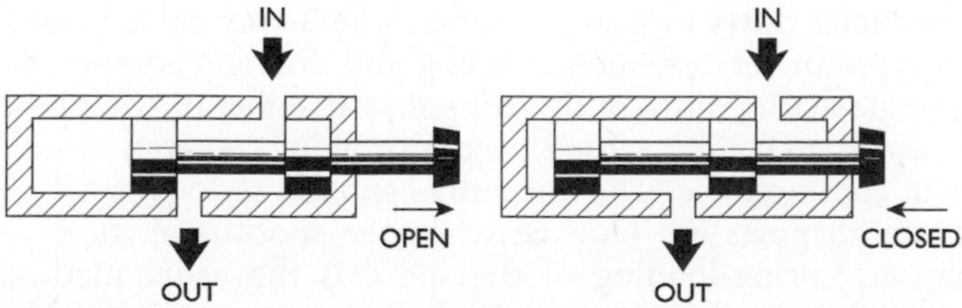

Figure 37-14 A spool-type, 2-way valve shown in open and closed positions.

A condition commonly required in hydraulic systems is that the valve remain in a given position until actuating force is applied. Also, when this force is removed, the valve must return to the same given position. The given position may be open or closed as the situation requires. When the given position is open, the valve is termed a *normally open* valve; when closed, a *normally closed* valve. Figure 37-15 illustrates a normally open 2-way valve and Figure 37-16 a normally closed style.

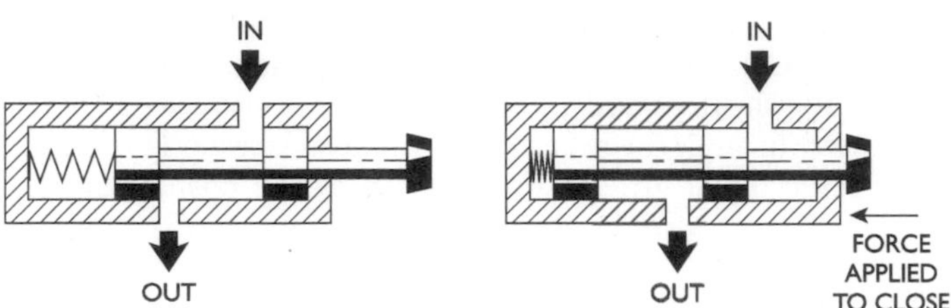

Figure 37-15 Normally open, 2-way spool valve.

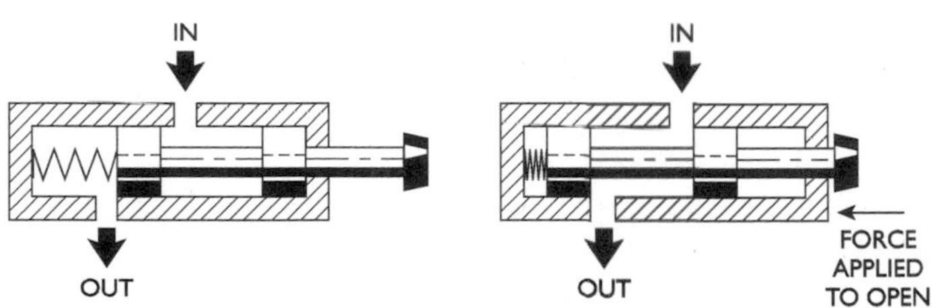

Figure 37-16 Normally closed, 2-way spool valve.

The 3-way valve is used for more sophisticated control than is possible with the 2-way valve. The 3-way valve has three working connections or ports, and its general function is to connect one line to either of two other lines. This may be where the valve accepts an inlet flow and diverts it to either one or two outlets, or it connects either of two inlet flows to a single outlet. The 3-way valve is used for several types of service, such as filling and draining equipment, operating single-acting rams, or diverting flow to either of two units. Normally open, normally closed, and normally centered 3-way valves are in common use. The normally centered style cuts off all flow, because all ports are blocked when the spool is in the centered position. Spring-loading of the spool is the usual method used to place the spool in its normal position automatically. The three positions are illustrated in Figure 37-17. Because flow may be in either direction, the ports are numbered. There may be a single inlet and double outlet, or vice versa, as the service requires. In illustration A, ports 1 and 2 are connected, and in illustration C, ports 1 and 3 are connected. The centered no-flow position is shown in illustration B. When the inlet is connected to port 1, ports 2 and 3 are outlets; and when the outlet is connected to 1, ports 2 and 3 are inlets.

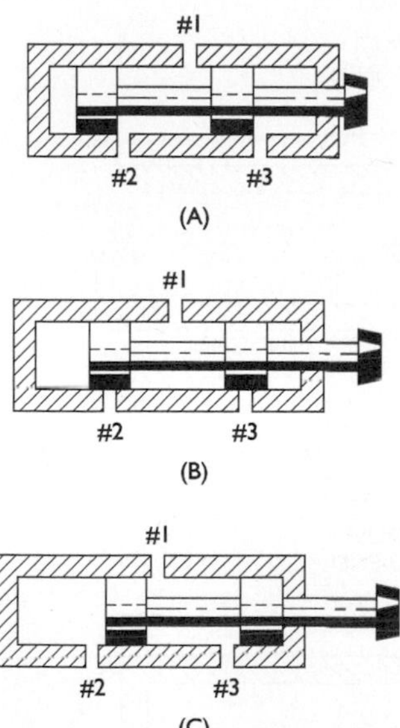

Figure 37-17 The three positions of a typical 3-way valve.

The 4-way valve is used for control of double-acting cylinders. It has three control positions: forward, stop, and reverse. To accomplish this, four working ports are required: inlet port, cylinder rod-end port, cylinder blind-end port, and exhaust or sewer port. In actual construction there are five internal ports, as shown in Figure 37-18. However, two of these ports perform the same function (exhaust or sewer) and are considered a single connection. The two exhaust ports of many 4-way hydraulic valves are connected internally, resulting in four external connections. Common practice with 4-way pneumatic valves is to have separate exhausts, resulting in five external connections, to allow throttling or speed adjustment for each direction of motion.

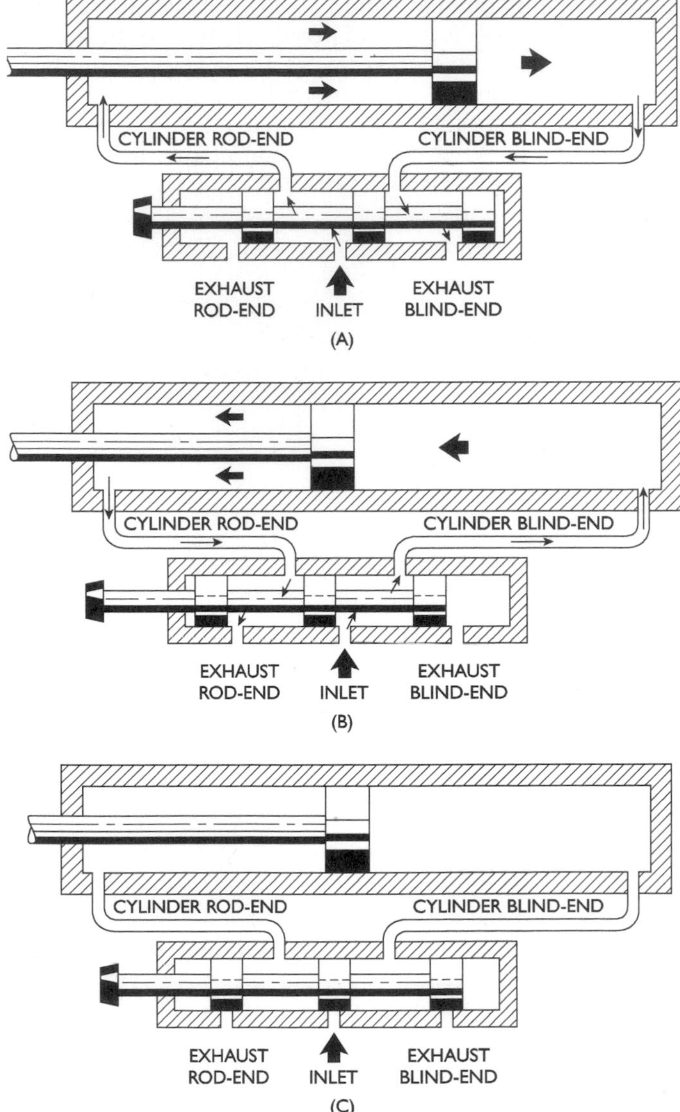

Figure 37-18 The connection of a 4-way valve.

Figure 37-18A shows the 4-way valve positioned to direct inlet flow to the rod-end of the cylinder and the outlet flow from the blind end of the cylinder to exhaust or sewer. The flow direction is reversed in Figure 37-18B by reversing the 4-way valve position. In Figure 37-18C the cylinder motion is stopped by centering the valve, which closes off the inlet and exhaust ports.

The 5-way valve is similar in construction to the 4-way valve; however, all five ports are working ports. The operational diagram, shown in Figure 37-19, is identical to the 4-way valve. Its principal difference is that it has two inlets and a single exhaust. This feature allows it to furnish fluid at two different pressure levels. The 5-way valve is used for directional control of cylinders and fluid motors, providing high force in one direction and low force in the opposite direction. The diagram in Figure 37-19 illustrates the use of a 5-way valve to control a double-acting cylinder. High force is directed to the forward stroke to perform work, while the return stroke receives low-pressure, high-volume flow for fast return.

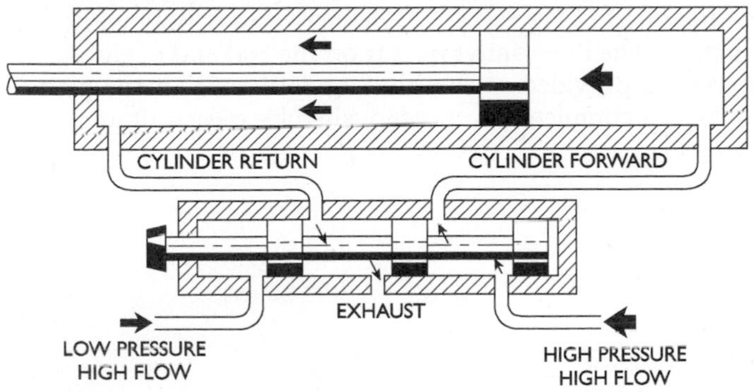

Figure 37-19 Typical connections of a 5-way valve.

Cylinders

The hydraulic or pneumatic cylinder is the device in a fluid-power system that transforms the flow of pressurized fluid into linear motion. Cylinders are sometimes called linear motors or reciprocating motors because they deliver a push-or-pull motion. They usually consist of a circular tube, sealed at both ends, in which a piston and rod move. The rod may be single and project from one end of the cylinder, or the cylinder may be a double-rod, which has a rod projecting from both ends. Double-rod cylinders, in addition to having two working rod ends, also have the same displacement or thrust in both directions.

A second basic design feature of cylinders is their capacity to move in one direction (single-acting) or in both directions (double-acting). One of the most widely used designs of fluid-power cylinder, the single-rod, double-acting style, is shown in Figure 37-20.

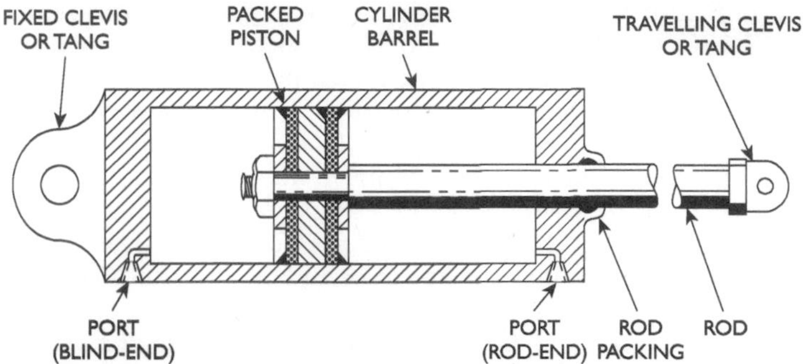

Figure 37-20 A single-rod, double-acting cylinder.

The single-acting cylinder may be either push or pull style. The pull style, illustrated in Figure 37-21, delivers power only while the rod is retracting. The inlet-outlet port is on the rod end of the cylinder and a vent is provided at the blind end. Return of the piston after the stroke is completed is accomplished by means of an internal spring, by an external spring, by gravity, or by mechanical action.

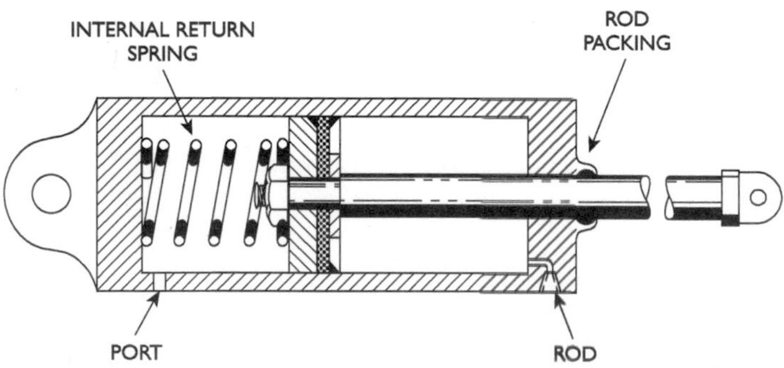

Figure 37-21 A pull-style, single-acting cylinder.

The single-acting, push-style cylinder is illustrated in Figure 37-22. The inlet-outlet port is at the blind end of the cylinder and a vent is provided at the rod end. No rod-packing seal is required since the fluid is sealed by the piston packing. Return of the piston after the stroke is completed by spring, gravity, or mechanical action.

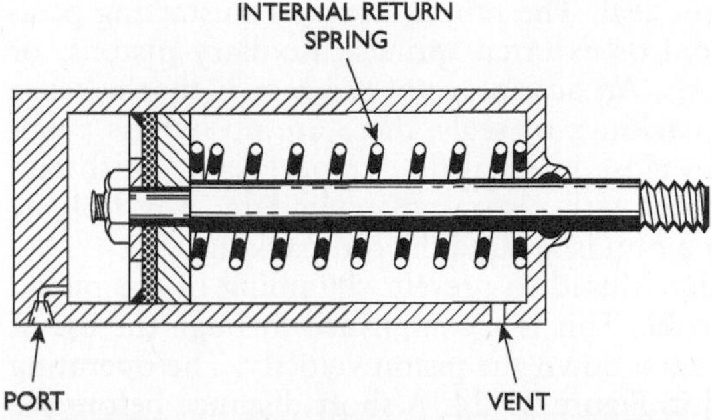

Figure 37-22 A push-style, single-acting cylinder.

The single-acting *ram* is a cylinder with a large-diameter rod in respect to the barrel bore. The rod cross-sectional area is usually greater than one-half the barrel bore cross-sectional area. Another distinctive feature of the ram is that it works on the displacement principle, having no internal moving packing or seals, such as are used with pistons. Some designs incorporate a shoulder on the rod, as shown in Figure 37-23, to prevent it from leaving the barrel. In this design, the net thrust area that the fluid acts upon is that of the rod where it goes through the packing.

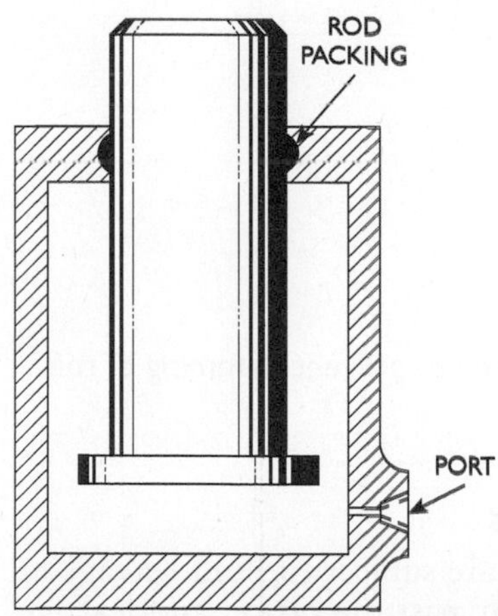

Figure 37-23 A single-acting ram cylinder.

In operation, the pressurized fluid, acting on the rod, causes it to be forced out of the barrel. Fluid leakage out of the barrel is prevented by the packed gland, or seal. The ram is returned to starting position by gravity, internal or external springs, auxiliary pistons, or other mechanical means. An advantage of the ram is that because there are no internal packings or seals, there are no bypass problems. Also, because no close internal fits are necessary, finish and size are not as critical as with piston-type cylinders. Any leakage that might occur with a ram is around the rod packing.

A *cushioned* cylinder is used to prevent slamming of the piston at either end of the stroke. This is accomplished through the use of secondary pockets to slow down the piston velocity. The operating principle is illustrated in Figure 37-24. A short distance before the piston reaches the end of the stroke, the main flow path of the outgoing fluid is cut off. This occurs as the nose of the cushion ram enters the cushion chamber. The fluid remaining between the piston and the cylinder end is then forced through the needle-valve restriction. Adjustment of the needle valve varies the deceleration rate. In some cushion cylinder designs a check valve is also used to allow the entering fluid to bypass the cushion chamber and act on the full area of the piston.

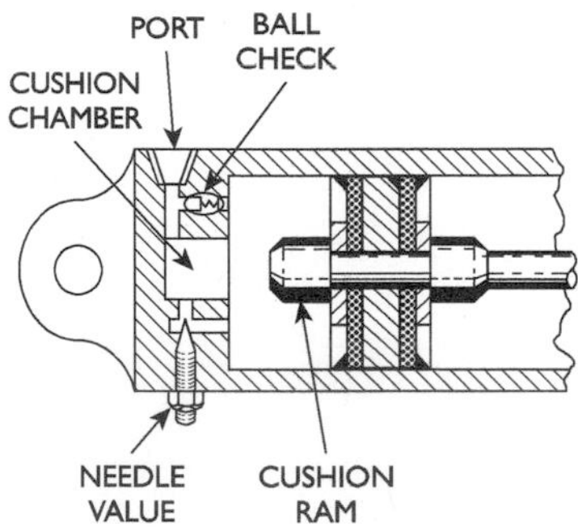

Figure 37-24 A cushioned cylinder used to prevent slamming of the piston.

Cylinder Packings and Seals

Hydraulic and pneumatic cylinders are subject to fluid leakage at two points: bypass leakage internally past the piston and leakage out of the cylinder at the point where the rod leaves the cylinder (packing box). A variety of packings and seals are used to control this leakage. Most commonly used piston seals are cup packings, U-packings, V-packings, O-rings, and metal piston rings. Widely used packing box or rod packings are compression packing, U-packings, V-packings, flange packings, and O-rings. The following is a brief description of each type; more detail can be found in Chapter 18.

Cup packings made from leather or synthetic materials are most satisfactory for low air pressures. The wide lips, which provide good sealing for low pressure, when subjected to high pressures, have excessive drag. Their break-away friction at low operating pressures is also quite low. Double, back-to-back, flat-cup packings used in double-acting cylinders are shown in Figure 37-25.

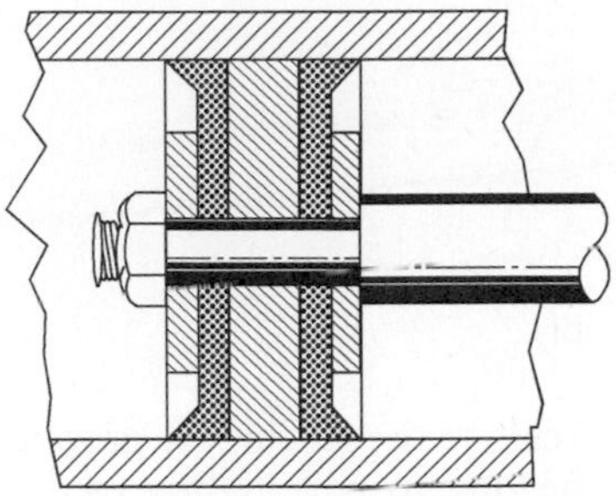

Figure 37-25 The use of double, back-to-back, flat-cup packing.

Double U-cup packings are shown in Figure 37-26. When used in a double-acting cylinder, one cup must be facing in each direction. This packing is used for both air and oil, usually at pressure below 500 psi. When made from homogenous synthetic rubber, U-cup packings are easy to replace because they may be stretched over special low retaining ridges, as shown in Figure 37-26.

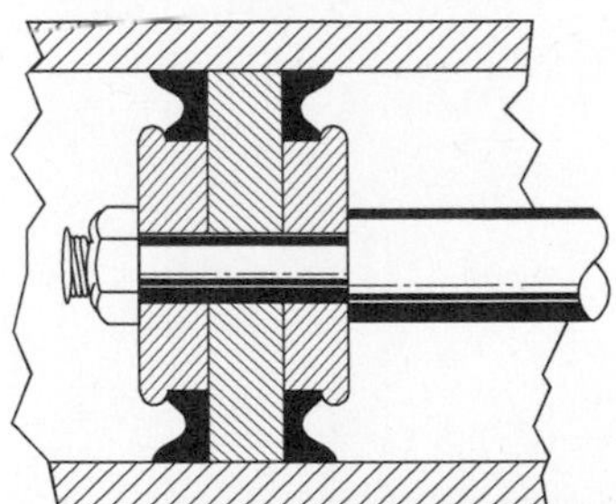

Figure 37-26 The double U-cup packing seal.

The multiple-V packing, shown in Figure 37-27, is used on high-pressure hydraulic systems in which a nonleaking packing combined with long life is required. Double sets, facing toward the pressure, are used on double-acting cylinders. Only one set is required for single-acting cylinders. Adapters or "support" rings are used at each end of the set. V-packings are usually installed in minimum sets of three, with a maximum of six rings per set.

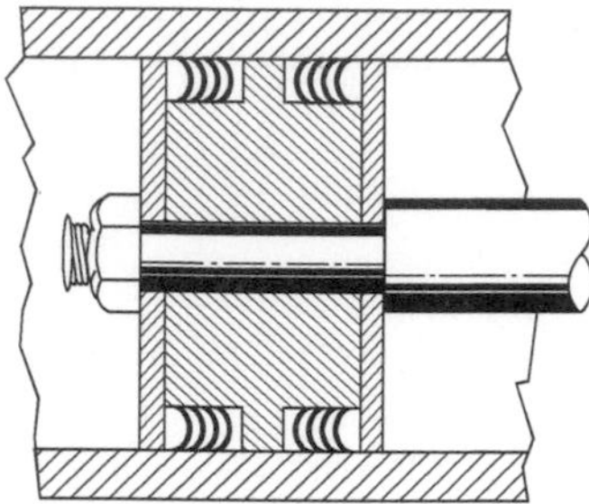

Figure 37-27 Multiple-V packing assembly.

Probably the simplest of all piston seals is the O-ring. A groove in the piston, wide enough to allow the ring to roll slightly as the piston moves back and forth, extends the life of the ring by changing the exposed wearing surface. In spite of this feature, the O-ring has a relatively short life compared with some other seals. Other disadvantages of the O-ring are its high breakaway friction and its tendency to extrude at higher pressures. To control extrusion, backup rings, shown in Figure 37-28, may be used.

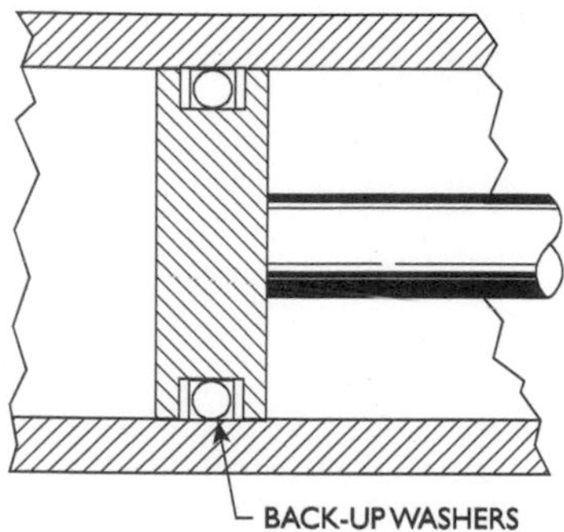

Figure 37-28 Backup washers used in conjunction with O-ring.

The automotive-type piston rings are used when small bypass leakage is not objectionable. While a small amount of bypass leakage always occurs, they give long life and excellent service on high-speed, fast-cycling hydraulic cylinders. They are seldom used on air cylinders. The piston rings are usually made of cast iron, and the cylinder is either made of steel or lined with a steel sleeve. Clearance must be provided at the ring joints, which may be step design or bevel style.

The function of rod packing, also called packing-box or stuffing-box packing, is to control leakage out of the cylinder. It is installed at the point where the rod leaves the cylinder, in a chamber called a packing box or a stuffing box. It seals the rod but does not support or guide it. Support and guidance are provided by a bearing, usually of a metal dissimilar to the rod, frequently bronze. Plastic materials, such as nylon and Teflon, are also used for rod bearings.

The rod packing is manufactured from a variety of materials to meet certain requirements of pressure, temperature, fluid compatibility, wearing qualities, etc. Some of the materials commonly used in rod packings are synthetic rubber, natural rubber, leather, plastics, and treated fabrics.

One of the oldest types of rod packing is the simple compression packing. Figure 37-29 shows a typical installation using a gland follower to compress the packing material. Recommended practice is to install this packing in individual rings, carefully tamping each ring into place. This design of rod packing is used mainly for hydraulics, on those applications where the rod must be wiped very dry and the extra friction is not objectionable.

The multiple V- and U-cup packings used for rod sealing are identical with those used for piston sealing. The U-cup is generally used as rod packing for air or low-pressure hydraulic applications. An adapter to prevent mechanical damage to the lips is recommended.

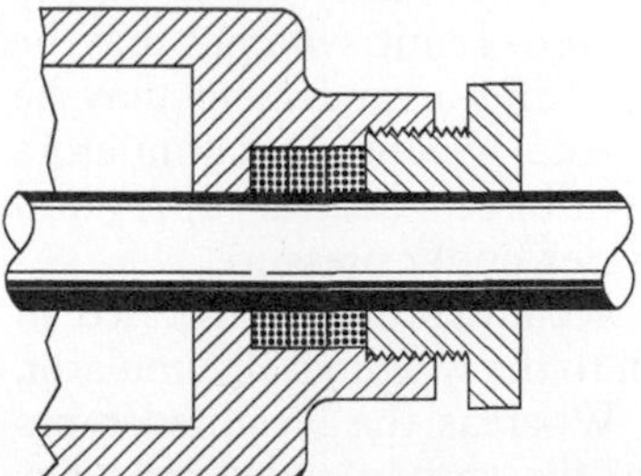

Figure 37-29 Compression packing with gland follower.

Multiple V-packings are supported with adapters on each end and installed with the lips toward the pressure. An important advantage of V-packings in high-pressure applications is that they seldom fail suddenly but gradually become leaky.

The O-ring, when used for rod sealing, is retained in an external groove. A wide variety of synthetic rubbers are available to meet the requirements of pressure temperature and fluid type. Backup washers are used for higher pressures to prevent extrusion.

Another type of rod-packing is the flange, or hat-section, packing. Generally used for air or hydraulic, the wide lip makes a tight seal at relatively low pressures. Flange packings are usually formed or molded from leather or rubber packing materials.

Pressure Accumulators

Many fluid-power systems use an accumulator to store fluid under pressure. By incorporating an accumulator in the system, a relatively small pump can accumulate a quantity of fluid under pressure, which can be used for a high-volume, short-duration discharge. Another important function of accumulators is to cushion shock waves in the circuit, such as might be caused by a pulsating pump or the sudden closing of a valve.

Three types of accumulators are in general use: the *weighted,* which stores energy by lifting a weight; the *spring-loaded,* which uses the energy stored in a spring; and the *pneumatic,* or *gas-charged,* in which the incoming pressurized fluid compresses the gas.

The weighted accumulator, shown in Figure 37-30 has been in use for many years on water hydraulic systems. It is very simple in operation. The pressurized fluid forces the ram up as long as the demand is less than the supply. When the demand exceeds the supply, the weight forces the ram down to provide constant flow and pressure. The function of the check valve is to prevent loss of fluid if the accumulator is to remain pressurized for an extended period. As the accumulator approaches the extreme loaded position, the supply of fluid must be cut off or diverted. Limit switches may be used to stop the supply pump, or a porting arrangement may be used to divert the fluid back to a reservoir. Weighted accumulators are frequently used on large systems that have a central supply and on equipment that has long holding cycles under pressure.

The principle of the spring-loaded accumulator is illustrated in Figure 37-31. While similar in operation to the weighted accumulator, its characteristics of operation differ. Whereas the weighted types provide a relatively constant pressure, the spring-loaded type has

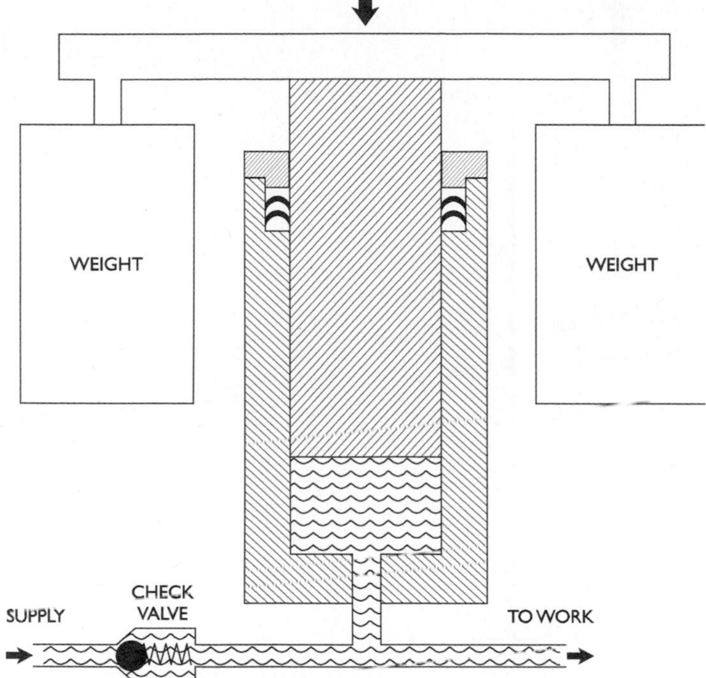

Figure 37-30 A weighted accumulator.

varying energy storage depending on the extent of spring compression. When the spring is at its maximum length, the spring accumulator exerts its minimum force on the compressed liquid. As fluid under pressure enters the accumulator, causing the spring to compress, the pressure on the fluid will rise because of the increased loading required to compress the spring.

Several designs of pneumatic or gas-charged accumulators are used in fluid-power systems. The basic operating principle in all cases is the same, in that the incoming fluid compresses air or gas to store energy. Several methods are used to separate the fluid from the gas and prevent loss of the charging gas into the system. Some small-capacity accumulators employ a diaphragm, as shown in Figure 37-32, to separate the fluid from the gas. The bag-style accumulator, shown in Figure 37-33, is similar in design and operation to the diaphragm style, usually having greater capacity. Several styles of accumulators

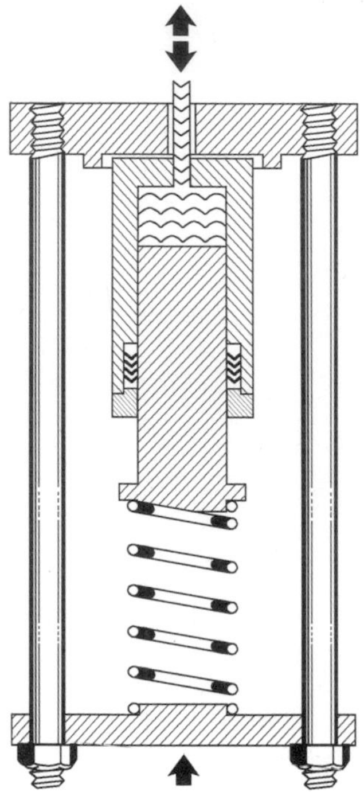

Figure 37-31 The spring-loaded accumulator.

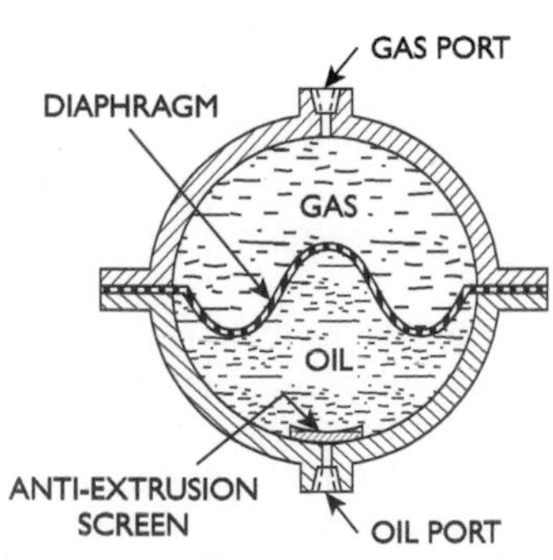

Figure 37-32 One type of diaphragm accumulator.

that use pistons or other barrier devices are also in general use. The operating principle of the mechanical-barrier accumulator is illustrated in Figure 37-34.

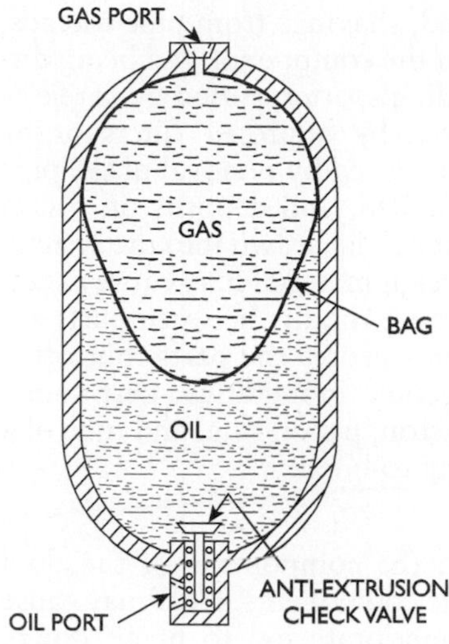

Figure 37-33 A bag-type diaphragm accumulator.

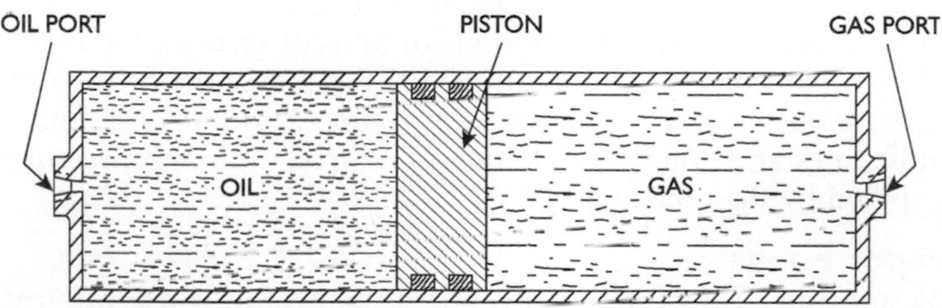

Figure 37-34 A mechanical-barrier accumulator.

Fluid Circuit Failures

The causes of component failures have been studied. These reasons should always be kept in mind when servicing fluid power devices, and it might be well to review some of them.

Dirt

Without a doubt, dirt and foreign substances cause more components to fail than any other single factor.

In a pneumatic system, dirt and foreign substances score the honed cylinder tubes, precision-finished valve liners and valve seats, ground and polished piston rods, valve stems and other precision parts. In pneumatics the foreign material may be in the form of pipe scale, lime deposits, thread compound, shavings from pipe threads, corrosive fumes entering the intake of the compressor and being distributed throughout the system, welding spatter caused by carelessness during construction, rust caused by improper filters or by excessive condensation, sand and dirt caused by removal of the pipe plugs before the components are installed, and deposits on piston rods caused by particles in the air that can be drawn into the system.

In a hydraulic system, dirt and foreign material may cause excessive damage to the components, because the fits between parts are held to very close limits. Dirt not only scores the parts, but often causes valve spools to stick and become inoperative. Dirt sometimes becomes lodged between the piston, piston ring, and tube of a hydraulic cylinder, causing piston ring to break.

Heat

Heat causes considerable trouble to the components of the fluid power system, especially the hydraulic components. Heat may cause valve spools to stick, packing to deteriorate, oil to break down, deposits to cling to the finished surfaces, excessive external and internal leakage, and inaccurate feeds in hydraulic systems. Fluid power systems should be protected from hot blasts. If heat in a hydraulic system is caused by internal conditions, install aftercoolers and, if possible, correct the condition that is causing the heat. Some of the causes of heat are high ambient temperature, restrictions in hydraulic lines and components, high pressures, and high pressures being spilled by the relief valve.

Improper Fluids

Use of improper fluids in a hydraulic system may cause failures. Care should be used in selecting the fluid to be used in the hydraulic system. Check with the pump manufacturer for his recommendations. If the oil is satisfactory for the pump, it is likely that it is satisfactory for the other components of the system.

Troubleshooting a Faulty Installation

Faulty installation may contribute to many system failures. Some things to check for include the following:

- Flow controls reversed.
- Wrong connections made to directional controls.
- Installation of a hydraulic power device so that backpressure is created in the return line to the fluid reservoir. This causes directional and pressure control valves to malfunction.

- Installation of hydraulic power devices in a closed pit or confined space. This causes heat problems.
- Failure to make drain connections to hydraulic valves. Manufacturers of control valves hang a tag on a valve port marked "connect to drain." It means what it says.
- Installation of piping of inadequate size in the pressure or exhaust lines. This slows down the action of the system, creates heat, causes malfunction of the valves, and creates backpressure. In a pneumatic system, it causes sluggishness in the action of the components.
- Improperly mounted cylinders.
- If control valves with mounting feet are not on a flat surface, they often cause trouble, as distortion occurs when the feet are securely bolted down.
- Loose pipelines can impose serious problems, especially on high-pressure systems. Piping needs to be anchored.
- Lack of protection to the piston rods of cylinders that are installed in a dirty atmosphere is another source of trouble. The piston rod may be scored and dirt may cut the rod packing.
- Misalignment of piston rods. Misalignment causes bent piston rods, loss of power, broken covers, worn bearings, scored cylinder walls, and packing leaks.
- Leaks around pipe port and connections due to improper installation. Pneumatic system leaks are expensive, as they can go unnoticed for a long period of time. In hydraulics, oil leaks are messy and they can present a real fire hazard.

Chapter 38

Mensuration and Mechanical Calculations

Mensuration is *the process of measuring*. It is the branch of mathematics that has to do with finding the length of lines, the area of surfaces, and the volume of solids. Accordingly, the problems that follow will be divided into three groups:

1. Measurement of lines
   - One dimension, *length*
2. Measurement of surfaces *(areas)*
   - Two dimensions, *length and breadth*
3. Measurement of solids *(volumes)*
   - Three dimensions, *length*, *breadth*, and *thickness*

Measurement of Lines

Problem 1

To find the length of any side of a right triangle, given the other two sides.

> *Rule—Length of hypotenuse equals square root of the sum of the squares of the two legs; length of either leg equals the square root of the difference of the square of the hypotenuse and the square of the other leg.*

Example

The two legs of a right triangle measure 3 and 4 ft; find the length of the hypotenuse. If the lengths of the hypotenuse and one leg are 5 and 3 ft, respectively, what is the length of the other leg? See Figure 38-1.

$$\text{Fig. 1A.} \quad AB = \sqrt{3^2 + 4^2} = \sqrt{25} = 5$$

$$\text{Fig. 1B.} \quad BC = \sqrt{5^2 - 3^2} = \sqrt{25 - 9} = \sqrt{16} = 4$$

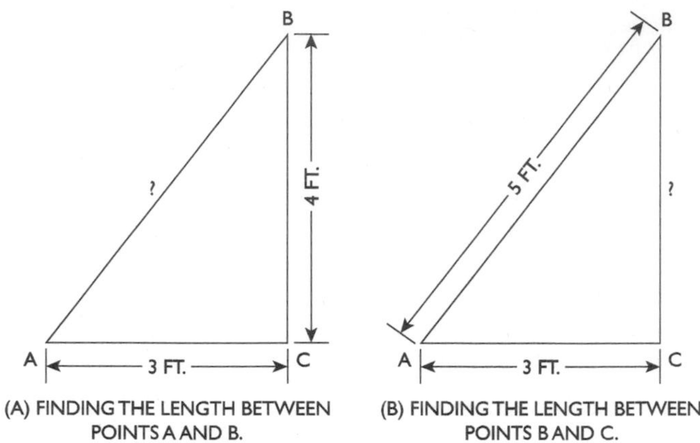

Figure 38-1 Finding the length of a right triangle.

Problem 2
To find circumference of a circle.

Rule—Multiply the diameter by 3.1416 (pi).

Example
What length of molding strip is required for a circular window 5 ft in diameter?

$$5 \times 3.1416 = 15.7 \text{ ft}$$

Because the mechanic does not ordinarily measure feet in tenths, the .7 should be reduced to inches; it corresponds to 8⅜ in. from Table 38-1. The length of molding is 15 ft 8⅜ in. (approx.).

Problem 3
To find the length of an arc of a circle.

Rule—As 360° is to the number of degrees of the arc, so is the length of the circumference to the length of the arc.

Example
If the circumference of a circle is 6 ft, what is the length of a 60° arc?

Let X equal the length of the arc, and solve for X.

$$360 : 60 = 6 : X = \frac{60 \times 6}{360} = \frac{360}{360} = 1 \text{ ft.}$$

Table 38-1 Decimals of a Foot and Inches

Inch	0 in.	1 in.	2 in.	3 in.	4 in.	5 in.	6 in.	7 in.	8 in.	9 in.	10 in.	11 in.
0	.0	.0833	.1677	.2500	.3333	.4167	.5000	.5833	.6667	.7500	.8333	.9167
1/16	.0052	.0885	.1719	.2552	.3385	.4219	.5052	.5885	.6719	.7552	.8385	.9219
1/8	.0104	.0937	.1771	.2604	.3437	.4271	.5104	.5937	.6771	.7604	.8437	.9271
3/16	.0156	.0990	.1823	.2656	.3490	.4323	.5156	.5990	.6823	.7656	.8490	.9323
1/4	.0208	.1042	.1875	.2708	.3542	.4375	.5208	.6042	.6875	.7708	.8542	.9375
5/16	.0260	.1094	.1927	.2760	.3594	.4427	.5260	.6094	.6927	.7760	.8594	.9427
3/8	.0312	.1146	.1979	.2812	.3646	.4479	.5312	.6146	.6979	.7812	.8646	.9479
7/16	.0365	.1198	.2031	.2865	.3698	.4531	.5365	.6198	.7031	.7865	.8698	.9531
1/2	.0417	.1250	.2083	.2917	.3750	.4583	.5417	.6250	.7083	.7917	.8750	.9583
9/16	.0469	.1302	.2135	.2969	.3802	.4635	.5469	.6302	.7135	.7969	.8802	.9635
5/8	.0521	.1354	.2188	.3021	.3854	.4688	.5521	.6354	.7188	.8021	.8854	.9688
11/16	.0573	.1406	.2240	.3073	.3906	.4740	.5573	.6406	.7240	.8073	.8906	.9740
3/4	.0625	.1458	.2292	.3125	.3958	.4792	.5625	.6458	.7292	.8125	.8958	.9792
13/16	.0677	.1510	.2344	.3177	.4010	.4844	.5677	.6510	.7344	.8177	.9010	.9844
7/8	.0729	.1562	.2396	.3229	.4062	.4896	.5729	.6562	.7396	.8229	.9062	.9896
15/16	.0781	.1615	.2448	.3281	.4115	.4948	.5781	.6615	.7448	.8281	.9115	.9968

Problem 4
To find the rise of an arc.

> *Rule 1—The rise of an arc is equal to the square of the chord of half the arc divided by the diameter.*
>
> *Rule 2—Length of chord subtending an angle at the center is equal to twice the radius times the sine of half the angle.*

Example
A circular pattern 10 ft in diameter has six plate forms. Find the width of board required for these forms allowing 3 in. margin for joints. See Figure 38-2.

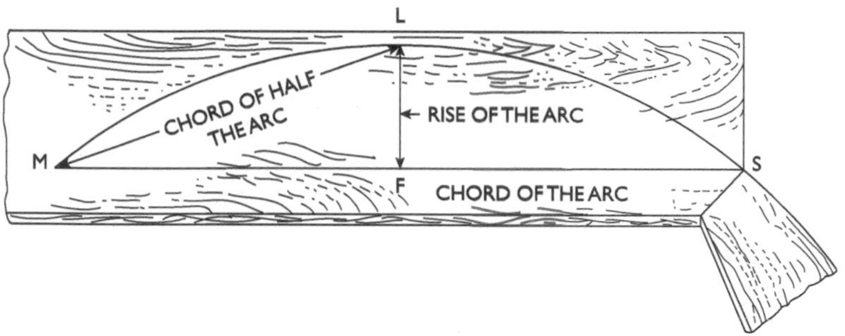

Figure 38-2 Finding the proper size board for Problem 4.

Each plate will subtend an angle of 360 ÷ 6 = 60°.

The "chord of half the arc" (mentioned in rule 1) will subtend 60 ÷ 2 = 30°.

Applying rule 2, "half the angle" = 30° ÷ 2 = 15°.

From table of trigonometric functions, the sine of 15° = .259, which with radius of 5 ft becomes

$$\text{sine } 15°(\text{on 10-ft. circle}) = 5 \times .259 = 1.295$$

Applying rule 2, length of chord MS = 2 ÷ 1.295 = 2.59.

Applying rule 1, rise of arc MS, = $2.59^2 \div 10$ = .671 ft or 8⅟16 in. (approx.). Add to this a 3 in. margin for joints and obtain width of board 8⅟16 + 3 = 11⅟16. Use 12-in. board.

Measurement of Surfaces (Areas)

Problem 5
To find the area of a square.

Rule—Multiply the base by the height. ($A = b \times h$).

Example
What is the area of a square whose side is 5 ft, as shown in Figure 38-3?

$$5 \times 5 = 25 \text{ sq. ft.}$$

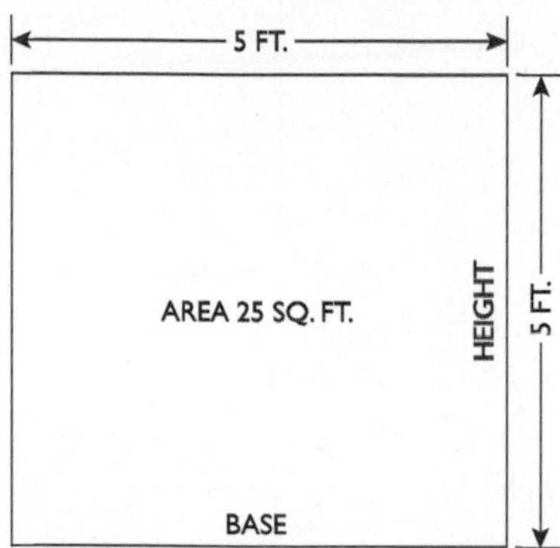

Figure 38-3 Finding the area of a square.

Problem 6
To find the area of a rectangle.

Rule—Multiply the base by the height (i.e., width by length).

Example
What is the area of a rectangle 5 ft wide and 12 ft long?

$$5 \times 12 = 60 \text{ sq. ft.}$$

Problem 7
To find the area of a parallelogram.

Rule—Multiply base by perpendicular height.

Example
What is the area of a parallelogram 2 ft wide and 10 ft long?

$$2 \times 10 = 20 \text{ sq. ft.}$$

Problem 8

To find the area of a triangle.

Rule—Multiply the base by half the altitude.

Example

How many sq. ft of sheet tin are required to cover a church steeple that has four triangular sides, measuring 12 ft (base) × 30 ft (altitude), as shown in Figure 38-4?

½ of altitude = 15 ft.
area of one side = 12 × 15 = 180 sq. ft.
total area (four sides) 4 × 180 = 720 sq. ft.

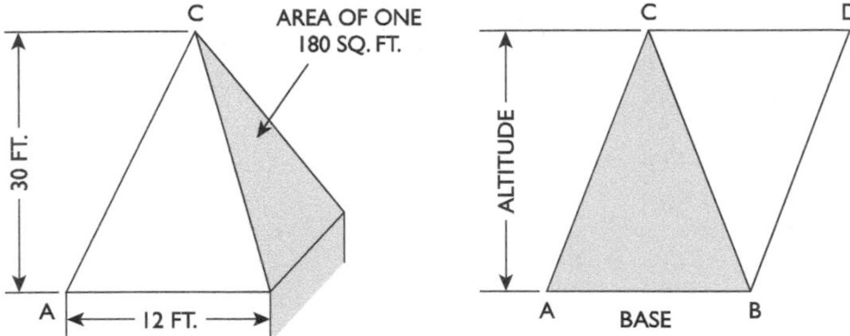

Figure 38-4 Finding the area of a triangle.

Problem 9

To find the area of a trapezoid.

Rule—Multiply one-half the sum of the two parallel sides by the perpendicular distance between them.

Example

What is the area of the trapezoid shown in Figure 38-5?

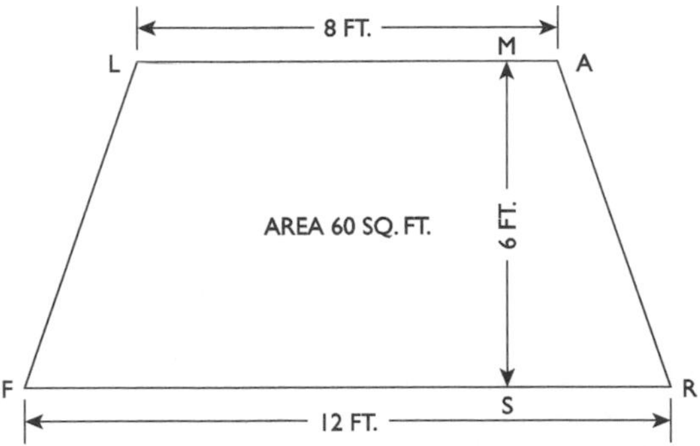

Figure 38-5 Finding the area of a trapezoid.

Here, LA and FR are the parallel sides, and MS is the perpendicular distance between them. Applying rule

$$\begin{aligned} \text{area} &= \tfrac{1}{2}(LA + FR) \times MS \\ &= \tfrac{1}{2}(8 + 12) \times 6 = 60 \text{ sq. ft.} \end{aligned}$$

Problem 10

To find the area of a trapezium.

Rule—Draw a diagonal, dividing figure into triangles; measure diagonal and altitudes and find area of the triangles.

Example

What is the area of the trapezium shown in Figure 38-6, for the dimensions given? Draw diagonal LR and altitudes AM and FS.

$$\begin{aligned} \text{area triangle } ALR &= 12 \times 6/2 = 36 \text{ sq. ft.} \\ \text{area triangle } LRF &= 12 \times 9/2 = 54 \text{ sq. ft.} \\ \text{area trapezium } LARF &= 90 \text{ sq. ft.} \end{aligned}$$

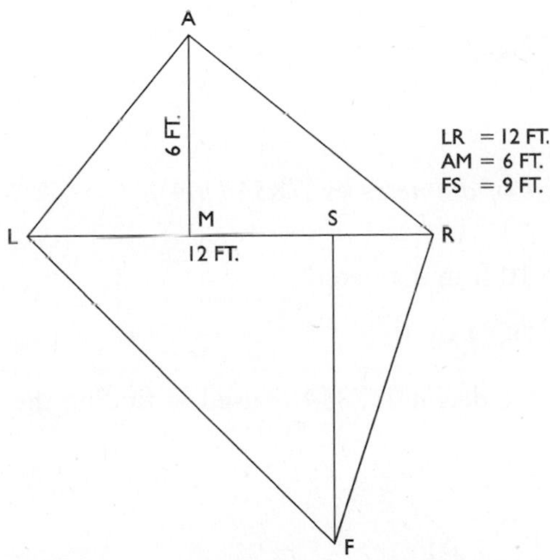

Figure 38-6 Finding the area of a trapezium.

Problem 11

To find the area of any irregular polygon.

Rule—Draw diagonals dividing the figure into triangles and find the sum of the areas of these triangles.

Problem 12

To find the area of any regular polygon when only the length of the side is given.

Rule—Multiply the square of one side by the figure for area when side = 1, shown in Table 38-2.

Table 38-2 Polygon Area

No. of sides	3	4	5	6	7	8	9	10	11	12
Area when side=1	.433	1.	1.721	2.598	3.634	4.828	6.181	7.694	9.366	11.196

Example

What is the area of an octagon (8-sided polygon) whose sides measure 4 ft?

In Table 38-2, under 8, find 4.828. Multiply this by the square of one side.

$$4.828 \times 4^2 = 77.28 \text{ sq. ft.}$$

Problem 13

To find the area of a circle.

Rule—Multiply square of diameter by .7854 ($\pi/4$).

Example

What is the area of a circle 10 ft in diameter?

$$10^2 \times .7854 = 78.54 \text{ sq. ft.}$$

Figure 38-7 shows why the decimal .7854 is used in finding the area of a circle.

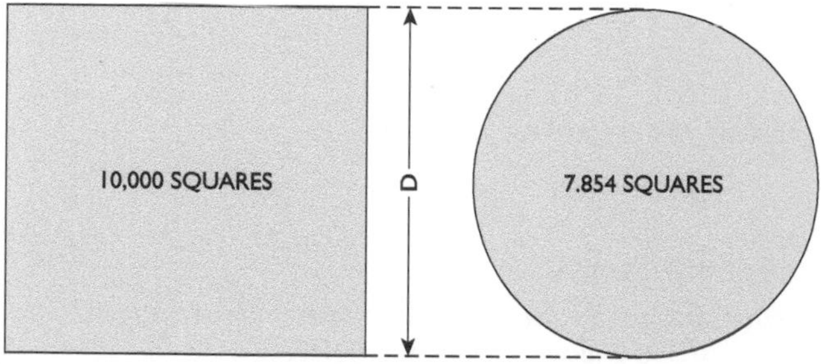

Figure 38-7 Showing why .7854 is used to find the area of a circle.

Problem 14

To find the area of a sector of a circle.

Rule—Multiply the arc of the sector by half the radius.

Example

How much tin is required to cover a 60° sector of a 10-ft circular deck?

$$\text{Length of } 60° \text{ arc} = \frac{60}{360} \times 3.1416 \times 10 = 5.24 \text{ ft.}$$

Applying the rule, tin required for 60° sector = 5.24 × ½ of 5 = 13.1 sq. ft.

Problem 15

To find the area of a segment of a circle.

Rule—Find the area of a sector which has the same arc and also the area of the triangle formed by the radii and chord. Take the sum of these areas if the segment is greater than 180°; *take the difference if less.*

Problem 16

To find the area of a ring.

Rule—Take the difference between the areas of the two circles.

Problem 17

To find the area of an ellipse.

Rule—Multiply the product of the two diameters by .7854.

Example

What is the area of an ellipse when the minor and major axes are 6 and 10 in., respectively?

$$10 \times 6 \times .7854 = 47.12 \text{ sq. in.}$$

Problem 18

To find the area of a cylinder. See Figure 38-8.

Rule—Multiply 3.1416 *by the diameter and by the height.*

Example

How many sq. ft of lumber are required for the sides of a cylindrical tank 8 ft in diameter and 12 ft high? How many pieces 4 in. × 12 ft will be required?

Cylindrical surface $3.1416 \times 8 \times 12 = 302$ sq. ft.
Circumference of tank $= 3.1416 \times 8 = 25.1$ ft.
Number 4″ × 12′ pieces $= 25.1 \div 4/12 = 25.1 \times 3 = 76$

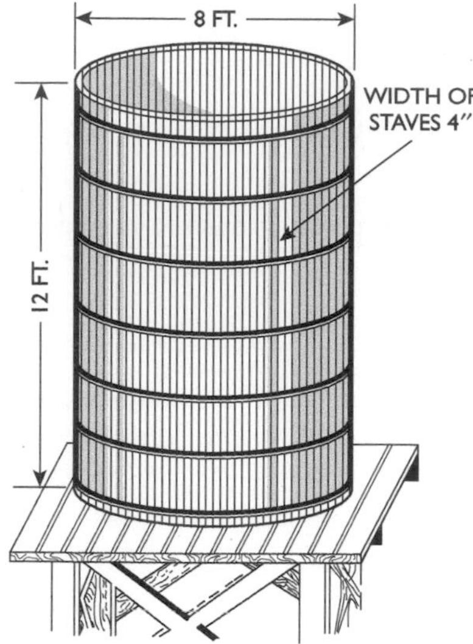

Figure 38-8 Finding the circular area of a cylinder.

Problem 19

To find the slant area of a cone. See Figure 38-9.

Rule—Multiply 3.1416 *by diameter of base and by one-half the slant height.*

Example

A conical spire with a base of 10 ft diameter and altitude of 20 ft is to be covered. Find area of surface to be covered.

In Figure 38-9, first find slant height, thus:

Slant height $= \sqrt{5^2 + 20^2} = \sqrt{425} = 20.62$ ft.
Circumference of base $= 3.1416 \times 10 = 31.42$ ft.
Area of conical surface $= 31.42 \times \frac{1}{2}$ of $20.62 = 324$ sq. ft.

Problem 20

To find the (slant) area of the frustum of a cone. See Figure 38-10.

Rule—Multiply half the slant height by the sum of the circumferences.

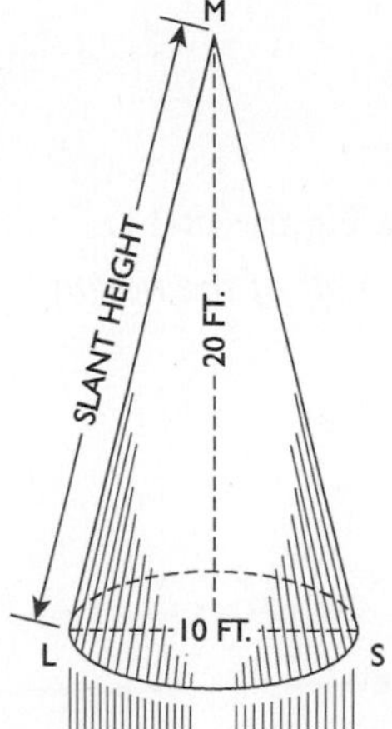

Figure 38-9 Finding the slant area of a cone.

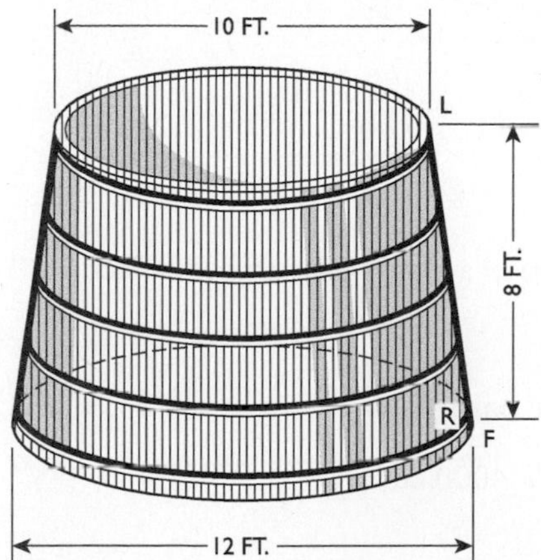

Figure 38-10 Finding the area of the frustum of a cone.

Example

A tank is 12 ft in diameter at the base, 10 ft at the top, and 8 ft high. What is the area of the slant surface?

$$\text{Circumference 10 ft circle} = 3.1416 \times 10 = 31.42 \text{ ft}$$
$$\text{Circumference 12 ft circle} = \underline{3.1416 \times 12 = 37.7 \text{ ft}}$$
$$\text{Sum of circumferences} = 69.1 \text{ ft}$$

$$\text{Slant height} = \sqrt{1^2 + 8^2} = \sqrt{65} = 8.06$$

$$\text{Slant surface} = \text{sum of circumferences} \times \tfrac{1}{2} \text{ slant height}$$
$$= 69.1 \times \tfrac{1}{2} \text{ of } 8.06 = 278.5 \text{ sq. ft}$$

Measurement of Solids

Problem 21
To find the volume of a rectangular wedge. See Figure 38-11.

Rule—Multiply length, breadth, and one-half of the height.

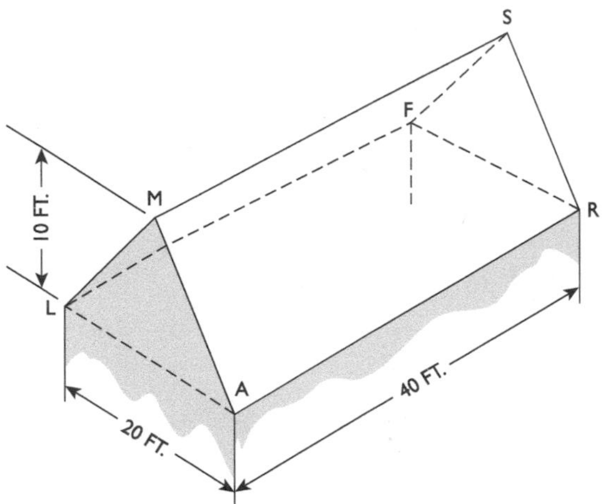

Figure 38-11 Finding the volume of a rectangular-shaped object.

Example
Find the volume *LARFMS* of the barn shown in Figure 38-11.

$$40 \times 20 \times \tfrac{1}{2} \text{ of } 10 = 4000 \text{ cu. ft.}$$

Problem 22
To find the volume of a cylinder.

Rule—Find the area of the base and multiply this by the length.

Example
What is the volume of a cylinder whose diameter is 4 ft and length 7½ ft?

$$\begin{array}{rl} 4^2 = 16 & \\ .7854 & \\ \times\ 16 & \\ \hline 12.5664 & = \text{area of base in sq. ft.} \\ \times\ 7.5 & = \text{length in ft.} \\ \hline 94.2480 & \quad \text{cu. ft.} \end{array}$$

Problem 23

To find the volume of a cone.

Rule—Multiply the area of the base by ⅓ of the altitude, and the product will be the volume.

Example

What is the volume of a cone whose diameter is 12 ft and altitude 10 ft?

Area of a circle = .7854 × sq. of the diameter

Area of base = $.7854 \times 12^2$ = 113.1 sq. ft.

Volume = 113.1 × ⅓ of 10 = 377 cu. ft.

Problem 24

To find the volume of a sphere.

Rule—Multiply the cube of the diameter by .5236 ($\pi/6$).

Example

Find the volume of a sphere whose diameter is 5 ft.

Cube of diameter	Diam.^{3} × .5236
5	.5236
5	× 125
25	65.4500 cu. ft
5	
125 = 5^3	

Table 38-3 Surfaces and Volumes of Regular Solids

No. of Sides	Name	Area, Edge = l	Contents, Edge = l
4	Tetrahedron	1.7320	0.1178
6	Hexahedron	6.0000	1.0000
8	Octahedron	3.4641	0.4714
12	Dodecahedron	20.6458	7.6631
20	Icosahedron	8.6603	2.1817

Mensuration of Surfaces and Volumes

Area of rectangle = length × breadth.
Area of triangle = base × ½ perpendicular height.
Diameter of circle = radius × 2.
Circumference of circle = diameter × 3.1416.
Area of circle = square of diameter × .7854

$$\text{Area of sector of circle} = \frac{\text{area of circle} \times \text{number of degrees in arc.}}{360}$$

Area of surface of cylinder = circumference × length + area of two ends.

To find diameter of circle having given area: Divide the area by .7854, and extract the square root.

To find the volume of a cylinder: Multiply the area of the section in square inches by the length in inches = the volume in cubic inches. Cubic inches divided by 1728 = volume in cubic feet.

Surface of a sphere = square of diameter × 3.1416.
Solidity of a sphere = cube of diameter × .5236.
Side of an inscribed cube = radius of a sphere × 1.1547.

Area of the base of a pyramid or cone, whether round, square or triangular, multiplied by one-third of its height = the solidity.

Diam. × .8862 = side of an equal square.
Diam. × .7071 = side of an inscribed square.
Radius × 6.2832 = circumference.
Circumference = 3.5449 × $\sqrt{\text{Area of circle.}}$

Diameter = 1.1283 × $\sqrt{\text{Area of circle.}}$

Length of arc = No. of degrees × 0.17453 radius.
Degrees in arc whose length equals radius = 57.3°.
Length of an arc of 1° = radius × 0.17453.
Length of an arc of 1 Min. = radius × .0002909.
Length of an arc of 1 Sec. = radius × .0000048.
π = Proportion of circumference to diameter = 3.1415926.

$$\pi^2 = 9.8696044$$
$$\sqrt{\pi} = 1.7724538$$
$$\text{Log } \pi = 0.49715$$
$$1/\pi = 0.31831$$
$$\pi/4 = .7854$$
$$\pi/6 = .5236$$
$$\pi/360 = .008727$$
$$360/\pi = 114.59$$

Lineal feet	×	.00019	= miles.
Lineal yards	×	.0006	= miles.
Square inches	×	.007	= Square feet.
Square feet	×	.111	= Square yards.
Square yards	×	.0002067	= Acres.
Acres	×	4840	= Square yards.
Cubic inches	×	.00058	= Cubic feet.
Cubic feet	×	.03704	= Cubic yards.
Circular inches	×	.00546	= Square feet.
Cyl. inches	×	.0004546	= Cubic feet.
Cyl. feet	×	.02909	= Cubic yards.
Links	×	.22	= Yards.
Links	×	.66	= Feet.
Feet	×	1.5	= Links.
Width in chains	×	8	= Acres per mile.
183,346 circular in			= 1 square foot.
2200 cylindrical in			= 1 cubic foot.
Cubic feet	×	7.48	= U.S. gallons.
Cubic inches	×	.004329	= U.S. gallons.
U.S. gallons	×	.13368	= Cubic feet.
U.S. gallons	×	231	= Cubic inches.
Cubic feet	×	.8036	= U.S. bushel.
Cubic inches	×	.000466	= U.S. bushel.
Cyl. feet of water	×	6.	= U.S. gallons.
Lbs. Avoir	×	.009	= Cwt.(112)
Lbs. Avoir	×	.00045	= Tons (2240)

Cubic feet of water . . . . ×	62.5	= Lbs. Avoir.
Cubic inch of water . . . . ×	.03617	= Lbs. Avoir.
Cyl. feet water . . . . ×	49.1	= Lbs. Avoir.
Cyl. inch water . . . . ×	.02842	= Lbs. Avoir.
13.44 U.S. gallons of water . .		= 1 Cwt.
268.8 U.S. gallons of water . .		= 1 Ton.
1.8 cubic feet of water		= 1 Cwt.
35.88 cubic feet of water		= 1 Ton.
Column of water, 12 inches high, and 1 inch in diameter		= .341 Lbs.
U.S. bushel . . . . ×	.0495	= Cubic yards.
U.S. bushel . . . . ×	1.2446	= Cubic feet.
U.S. bushel . . . . ×	2150.42	= Inches.

Properties of the Circle

Diameter of circle	× .88623	= side of equal square
Circumference of circle	× .28209	= side of equal square
Circumference of circle	× 1.1284	= perimeter of equal square
Diameter of circle	× .7071	= side of inscribed square
Circumference of circle	× .22508	= side of inscribed square
Area of circle × .90031	÷ diameter	= side of inscribed square
Area of circle	× 1.2732	= area of circumscribed square
Area of circle	× .63662	= area of inscribed square
Side of square	× 1.4142	= diam. of circumscribed circle
Side of square	× 4.4428	= circum.
Side of square	× 1.1284	= diam. of equal circle
Side of square	× 3.5449	= circum. of equal circle
Perimeter of square	× .88623	= circum. of equal circle
Square inches	× 1.2732	= circular inches

Problem 25

To find the volume of a rectangular solid.

Rule—Multiply length × breadth × height.

Example

What are the cubical contents of a coal bin whose length is 6 ft, breadth 5 ft, and height 3 ft?

$$\text{Area of base} = 6 \times 5 = 30 \text{ sq. ft.}$$
$$\text{Height} = 3 \text{ ft.}$$
$$\text{Volume} = 30 \times 3 = 90 \text{ cu. ft.}$$

Electrical Horsepower

The unit for measuring the electrical horsepower input to a motor is the watt (W). For direct current motors, watts are the product of the potential in volts (E) and the current in amperes (I) measured at the motor terminals. This product is written $W = E \times I$. Since 746 W is equal to an electrical horsepower (hp), the formula representing the above relation is

$$\text{HP} = \frac{E \times I}{746}$$

Example

If the potential (E) and current (I) measured at the terminal of a direct current motor is 230 and 30, respectively, how many electrical horsepower (hp) is delivered to the motor terminals?

Solution

Inserting numerical values in the formula for horsepower, we obtain hp = (230 × 30)/746 = 9¼ hp (approximately).

Power Factor

In an ac circuit, the true power (W) is equal to the product of volts (E) times amperes (I) times $\cos\phi$, which is written $W = E \times I \times \cos\phi$ When the electrical power is measured by means of a wattmeter, the true power is obtained directly. When, on the other hand, the power is measured by means of an ammeter and voltmeter, their product ($E \times I$) gives the apparent power in watts. Therefore, in order to obtain true watts, this product must be multiplied by the power factor ($\cos\phi$).

Example

In a power station, the voltmeter and ammeter indicate 110 and 50, respectively, whereas the kilowatt-meter reads 3310 W. Find (a) the power factor, and (b) the reactive power of the system.

Solution
With reference to Figure 38-12A,

$$\cos\phi = \frac{3310}{100 \times 50} = 0.6 \text{ or 60 per cent (approximately)}$$

With reference to Figure 38-12B, the reactive power (RVA) may be written

$$RVA = E \times I \times \sqrt{1-\cos^2\phi} = 110 \times 50 \times 0.8 = 4{,}400$$

In other words, if a reactive KVA meter were connected in the circuit, its reading would be 4400 volt amperes.

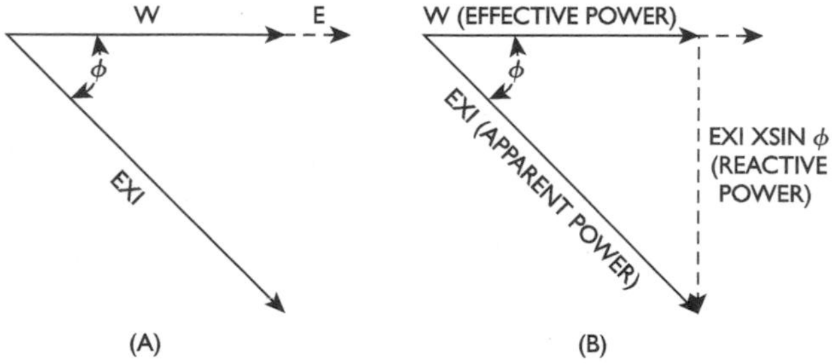

Figure 38-12 Vectorial relations between apparent power and true power.

Alternating Current Motors

If the power is taken from an alternating current system, the following condition will prevail:

If,

W = watts,
E = average volts between terminals,
I = average line current,
pf = power factor expressed as a decimal fraction,

the formula expressing their relations will be

$$HP = \frac{E \times I \times p.f.}{746}$$

$$HP = \frac{2 \times E \times I \times \text{pf}}{746}$$

$$HP = \frac{1.73 \times E \times I \times \text{pf}}{746}$$

Example

What is the power supplied at the motor terminal in a three-phase system when the power factor is 0.8, potential 110 V, and the current 25 amps?

Solution

To obtain the number of horsepower delivered in the example, simply multiply the given units by 1.73 and divide the product by 746, thus:

$$HP = \frac{1.73 \times 110 \times 25 \times 0.8}{746} = 5.1 \text{ horsepower (approximately).}$$

Brake Horsepower

The power output of small engines and motors is best obtained by a brake test, that is, application of the Prony brake, of which a common arrangement is shown in Figure 38-13. From the Prony brake test, the net work of the engine or motor measured in horsepower delivered at the shaft is determined as follows:

Let,

W = Power absorbed per minute;
P = unbalanced pressure or weight in pounds, acting on the lever arms at a distance L;
L = length of lever arm in feet from center of shaft;
N = number of revolutions per minute;
V = velocity of a point in feet per minute at distance L, if arm were allowed to rotate at the speed of the shaft $2\pi LN$

$$\text{Since brake horsepower} = \frac{PV}{33{,}000}$$

Substituting for V,

$$\text{BHP (brake horsepower)} = \frac{P2\pi LN}{33{,}000}$$

Example
A Prony brake is assembled to a motor whose pulley has a diameter of 24 in. The scale indicates 75 lb when the pulley revolves at 600 rpm. What is the output of the motor at this particular load?

Solution
From the previous expression of brake horsepower, a substitution of values gives

$$\text{HP} = \frac{75 \times 2\pi \times 1 \times 600}{33{,}000} = 8.6 \text{ horsepower (approximately).}$$

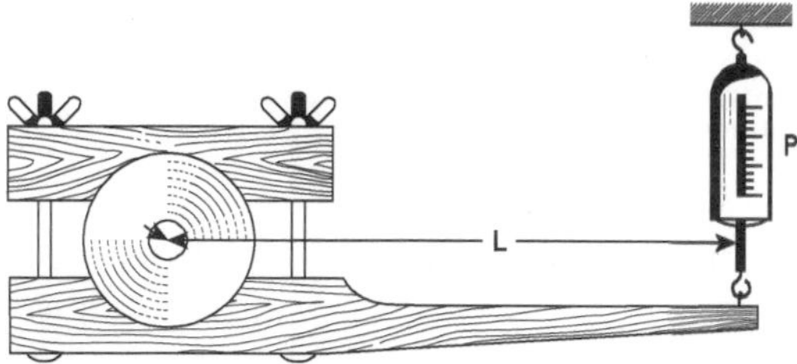

Figure 38-13 Illustrating the Prony brake.

Engine Horsepower

The determination of an engine's horsepower is based on the well-known fact that one mechanical horsepower is equivalent to 33,000 foot-pounds of work in one minute or 550 foot-pounds of work in one second.

To express the indicated horsepower of an engine, the following formula is used:

$$\text{IHP} = \frac{\text{PLAN}}{33{,}000} \times K$$

where P = mean effective pressure in lbs per sq. in. acting on the piston (as shown by indicator diagram)

L = length of stroke in feet
A = area of piston in square inches
N = number of working strokes per min.
K = coefficient equal to ½ times the number of cylinders in gas engines, and in double-acting steam engines, 2 times the number of revolutions.

Example

What is the indicated horsepower of a 4 cylinder 5 × 6 engine running at 750 revolutions per minute and 50 lb per sq. in. mean effective pressure?

Solution

When substituting in the formula for indicated horsepower, the point that should be noted is the treatment of units involved. Thus, for example, the piston area = 0.7854 × its diameter squared and also the stroke L must be reduced to feet by dividing by 12; also, the coefficient $K = 2$.

$$\text{HP} = \frac{50 \times \frac{6}{12} \times (0.7854 \times 5^2) \times 750}{33{,}000} \times 2 = 22.31 \text{ horsepower}$$

Example

What size of piston is required for a 100 hp, 1000-rpm single-acting marine engine when the mean effective pressure (mep) is 26.8 lb per sq. in. and the stroke is 8 in.?

Solution

According to our formula $IHP = (P \times L \times A \times N/33{,}000)$, but since the piston area $A = 0.7854 \times D_2$, a substitution and rearrangement of terms gives

$$D = \sqrt{\frac{100 \times 33{,}000}{26.8 \times \frac{8}{12} \times 0.7854 \times 1000}}$$

$$= \sqrt{235} = 15.3 \text{ in. (approximately).}$$

Work

The purpose of all machinery is to do a certain amount of work. This work may be accomplished in various ways to suit individual requirements. The most common method employed in industrial processes is to use an *electric motor* as the motivating medium through which work is done in various ways, such as that of a punch press stamping holes in metal, a crane lifting commodities from a barge, or a conveyor moving stock from one place to another.

The amount of work done in each instance depends upon the magnitude of the acting force and the distance through which the force acts. This may be written as follows:

$$\text{Work} = \text{Force} \times \text{Distance}$$

The common unit of *work* is known as the foot-pound. A foot-pound of work is done when a force of one pound acts through the distance of one foot.

Example

A man lifts 55 packages each weighing 20 lb from the floor to a truck 3 ft high. How much work is done?

Solution

To place one package on the truck, the work done is 20×3 or 60 foot-pounds. The total work done is therefore 60×55 or 3300 foot-pounds.

In the foregoing example, observe that the amount of time required to do the work is not considered. The same amount of work is done regardless of the time consumed in doing it.

A smaller and weaker man (A), for example, might do the same amount of work as a much stronger man (B) provided he had sufficient time. However, (B) is said to possess greater power because he can do the work in less time.

In a given time, a crane can do much more work than either man, from which it follows that the crane has the greatest *power.* Thus, power differs from work in that time is always involved. *Power means the rate at which work is done.* This may be expressed thus:

$$\text{Power} = \frac{\text{work}}{\text{time}}$$

Common units of power are:

Horsepower (hp)

Watts (W)

Kilowatts (kW)

Horsepower is most widely used to express the mechanical output of machines. It has been experimentally found that for a short time the average horse can work at the rate of 550 foot-pounds per second, or $550 \times 60 = 33{,}000$ foot-pounds per minute. This rate has been standardized throughout the world as the horsepower.

Thus,

$$\text{Horsepower} = \frac{\text{foot-pounds per minute}}{33{,}000}$$

Example

Assume that, in the previous problem, the man did 3300 foot-pounds of work in two minutes. What is the average horsepower expended?

Solution

The work done per minute is 3300/2 = 1650 foot-pounds.

The total power is 1650/33,000 or 0.05 hp.

Torque

The twisting effect produced by a motor, for example, does not depend only upon the magnitude but also upon the radial distance through which the force acts.

The twisting effect is expressed as the product of the force (F) and the radius (R) through which the force acts and is called *torque*. If the torque is represented by (T) and is expressed in pound-feet, it may be written

$$T = F \times R$$

Example

A motor rated at 10 hp runs at 850 rpm; what torque will the motor supply at its rated load?

Solution

$T = HP \times 5252/N$, which after substitution of values becomes $T = (10 \times 5252)/850 = 61.8$ pound-feet.

Shafts

When, for example, two pulleys, two gears, or a pulley and a gear are connected by means of a shaft, the twisting effects of the two are always equal. An illustration of the above conditions is given in the following problem.

Example

Consider Figure 38-14, which shows a 7.5 hp motor running at 1200 rpm. On its shaft is a 4-in. diameter gear running in mesh with a gear that has a diameter of 24 in. The second gear is attached to a pulley, which is 6 in. in diameter. What is the maximum weight that can be raised under the conditions given?

Solution

The force on the smaller gear is

$$F = \frac{HP \times 5252}{NR} = \frac{7.5 \times 5252}{1200 \times \frac{2}{12}} = 196 \text{ lbs. (approximately)}$$

Because the force acting on the larger gear is also 196 lb, the twisting effect of this gear is $\frac{12}{12} \times 196$ pound-feet.

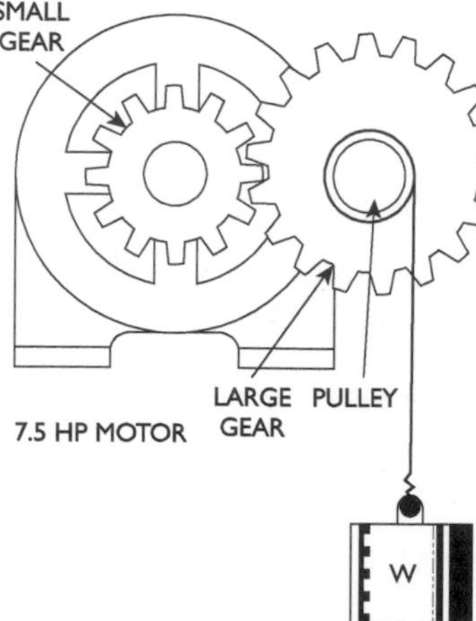

Figure 38-14 Method of hoisting a large load by means of a motor.

The torque rendered by the weight (*W*) is equal and opposite to that of the pulley; hence, the force on the pulley surface will be

$$\frac{196}{3/12} = 784 \text{ pounds}$$

Thus, with this gear combination, the 7.5 hp motor will raise 784 lb at full load.

Appendix

Decimal and Millimeter Equivalents of Fractional Parts of an Inch

Inches		*Inches*	*mm*	*Inches*		*Inches*	*mm*
	1–64	0.01563	0.397		33–64	0.51563	13.097
1–32		0.03125	0.794	17–32		0.53125	13.494
	3–64	0.04688	1.191		35–64	0.54688	13.890
1–16		0.0625	1.587	9–16		0.5625	14.287
	5–64	0.07813	1.984		37–64	0.57813	14.684
3–32		0.09375	2.381	19–32		0.59375	15.081
	7–64	0.10938	2.778		39–64	0.60938	15.478
1–8		0.125	3.175	5–8		0.625	15.875
	9–64	0.14063	3.572		41–64	0.64063	16.272
5–32		0.15625	3.969	21–32		0.65625	16.669
	11–64	0.17188	4.366		43–64	0.67188	17.065
3–16		0.1875	4.762	11–16		0.6875	17.462
	13–64	0.20313	5.159		45–64	0.70313	17.859
7–32		0.21875	5.556	23–32		0.71875	18.256
	15–64	0.23438	5.953		47–64	0.73438	18.653
1–4		0.25	6.350	3–4		0.75	19.050
	17–64	0.26563	6.747		49–64	0.76563	19.447
9–32		0.28125	7.144	25–32		0.78125	19.844
	19–64	0.29688	7.541		51–64	0.79688	20.240
5–16		0.3125	7.937	13–16		0.8125	20.637
	21–64	0.32813	8.334		53–64	0.82813	21.034
11–32		0.34375	8.731	27–32		0.84375	21.431
	23–64	0.35938	9.128		55–64	0.85938	21.828
3–8		0.375	9.525	7–8		0.875	22.225
	25–64	0.39063	9.922		57–64	0.89063	22.622
13–32		0.40625	10.319	29–32		0.90625	23.019
	27–64	0.42188	10.716		59–64	0.92188	23.415
7–16		0.4375	11.113	15–16		0.9375	23.812
	29–64	0.45313	11.509		61–64	0.95313	24.209
15–32		0.46875	11.906	31–32		0.96875	24.606
	31–64	0.48438	12.303		63–64	0.98438	25.003
1–2		0.5	12.700	1		1.00000	25.400

Decimal Inch Equivalents of Millimeters and Fractional Parts of Millimeters

mm	Inches	mm	Inches	mm	Inches	mm	Inches
1–100 =	0.00039	33–110 =	0.01299	64–100 =	0.02520	95–100 =	0.03740
2–100 =	0.00079	34–100 =	0.01339	65–100 =	0.02559	96–100 =	0.03780
3–100 =	0.00118	35–100 =	0.01378	66–100 =	0.02598	97–100 =	0.03819
4–100 =	0.00157	36–100 =	0.01417	67–100 =	0.02638	98–100 =	0.03858
5–100 =	0.00197	37–100 =	0.01457	68–100 =	0.02677	99–100 =	0.03898
6–100 =	0.00236	38–100 =	0.01496	69–100 =	0.02717	1 =	0.03937
7–100 =	0.00276	39–100 =	0.01535	70–100 =	0.02756	2 =	0.07874
8–100 =	0.00315	40–100 =	0.01575	71–100 =	0.02795	3 =	0.11811
9–100 =	0.00354	41–100 =	0.01614	72–100 =	0.02835	4 =	0.15748
10–100 =	0.00394	42–100 =	0.01654	73–100 =	0.02874	5 =	0.19685
11–100 =	0.00433	43–100 =	0.01693	74–100 =	0.02913	6 =	0.23622
12–100 =	0.00472	44–100 =	0.01732	75–100 =	0.02953	7 =	0.27559
13–100 =	0.00512	45–100 =	0.01772	76–100 =	0.02992	8 =	0.31496
14–100 =	0.00551	46–100 =	0.01811	77–100 =	0.03032	9 =	0.35433
15–100 =	0.00591	47–100 =	0.01850	78–100 =	0.03071	10 =	0.39370
16–100 =	0.00630	48–100 =	0.01890	79–100 =	0.03110	11 =	0.43307
17–100 =	0.00669	49–100 =	0.01929	80–100 =	0.03150	12 =	0.47244
18–100 =	0.00709	50–100 =	0.01969	81–100 =	0.03189	13 =	0.51181
19–100 =	0.00748	51–100 =	0.02008	82–100 =	0.03228	14 =	0.55118
20–100 =	0.00787	52–100 =	0.02047	83–100 =	0.03268	15 =	0.59055
21–100 =	0.00827	53–100 =	0.02087	84–100 =	0.03307	16 =	0.62992
22–100 =	0.00866	54–100 =	0.02126	85–100 =	0.03346	17 =	0.66929
23–100 =	0.00906	55–100 =	0.02165	86–100 =	0.03386	18 =	0.70866
24–100 =	0.00945	56–100 =	0.02205	87–100 =	0.03425	19 =	0.74803
25–100 =	0.00984	57–100 =	0.02244	88–100 =	0.03465	20 =	0.78740
26–100 =	0.01024	58–100 =	0.02283	89–100 =	0.03504	21 =	0.82677
27–100 =	0.01063	59–100 =	0.02323	90–100 =	0.03543	22 =	0.86614
28–100 =	0.01102	60–100 =	0.02362	91–100 =	0.03583	23 =	0.90551
29–100 =	0.01142	61–100 =	0.02402	92–100 =	0.03622	24 =	0.94488
30–100 =	0.01181	62–100 =	0.02441	93–100 =	0.03661	25 =	0.98425
31–100 =	0.01220	63–100 =	0.02480	94–100 =	0.03701	26 =	1.02362
32–100 =	0.01260						

Wire Gauge Standards

Decimal Parts of an Inch

Wire Gauge No.	American or Brown & Sharpe	Birmingham or Stubs Wire	Washburn & Moen on Steel Wire Gauge	American S. & W. Co.'s Music Wire	Imperial Wire Gauge	Stubs Steel Wire	U.S. Standard for Plate
0000000	0.651354		0.4000		0.500		0.500
000000	0.580049		0.4615	0.004	0.464		0.46875

00000	0.516549	0.500	0.4305	0.005	4.432		0.43775
0000	0.460	0.454	0.3938	0.006	0.400		0.40625
000	0.40964	0.425	0.3625	0.007	0.372		0.375
00	0.3648	0.380	0.3310	0.008	0.348		0.34375
0	0.32486	0.340	0.3065	0.009	0.324		0.3125
1	0.2893	0.300	0.2830	0.010	0.300	0.227	0.28125
2	0.25763	0.284	0.2625	0.011	0.276	0.219	0.265625
3	0.22942	0.259	0.2437	0.012	0.252	0.212	0.250
4	0.20431	0.238	0.2253	0.013	0.232	0.207	0.234375
5	0.18194	0.220	0.2070	0.014	0.212	0.204	0.21875
6	0.16202	0.203	0.1920	0.016	0.192	0.201	0.203125
7	0.14428	0.180	0.1770	0.018	0.176	0.199	0.1875
8	0.12849	0.165	0.1620	0.020	0.160	0.197	0.171875
9	0.11443	0.148	0.1483	0.022	0.144	0.194	0.15625
10	0.10189	0.134	0.1350	0.024	0.128	0.191	0.140625
11	0.090742	0.120	0.1205	0.026	0.116	0.188	0.125
12	0.080808	0.109	0.1055	0.029	0.104	0.185	0.109375
13	0.071961	0.095	0.0915	0.031	0.092	0.182	0.09375
14	0.064084	0.083	0.0800	0.033	0.080	0.180	0.078125
15	0.057068	0.072	0.0720	0.035	0.072	0.178	0.0703125
16	0.05082	0.065	0.0625	0.037	0.064	0.175	0.0625
17	0.045257	0.058	0.0540	0.039	0.056	0.172	0.05625
18	0.040303	0.049	0.0475	0.041	0.048	0.168	0.050
19	0.03589	0.042	0.0410	0.043	0.040	0.164	0.04375
20	0.031961	0.035	0.0348	0.045	0.036	0.161	0.0375
21	0.028462	0.032	0.0317	0.047	0.032	0.157	0.034375
22	0.025347	0.028	0.0286	0.049	0.028	0.155	0.03125
23	0.022571	0.025	0.0258	0.051	0.024	0.153	0.028125
24	0.0201	0.022	0.0230	0.055	0.022	0.151	0.025
25	0.0179	0.020	0.0204	0.059	0.020	0.148	0.021875
26	0.01594	0.018	0.0181	0.063	0.018	0.146	0.01875
27	0.014195	0.016	0.0173	0.067	0.0164	0.143	0.0171875
28	0.012641	0.014	0.0162	0.071	0.0149	0.139	0.015625
29	0.011257	0.013	0.0150	0.075	0.0136	0.134	0.0140625
30	0.010025	0.012	0.0140	0.080	0.0124	0.127	0.0125
31	0.008928	0.010	0.0132	0.085	0.0116	0.120	0.0109375
32	0.00795	0.009	0.0128	0.090	0.0108	0.115	0.01015625
33	0.00708	0.008	0.0118	0.095	0.0100	0.112	0.009375
34	0.006304	0.007	0.0104		0.0092	0.110	0.00859375
35	0.005614	0.005	0.0095		0.0084	0.108	0.0078125
36	0.005	0.004	0.0090		0.0076	0.106	0.00703125
37	0.004453		0.0085		0.0068	0.103	0.006640625
38	0.003965		0.0080		0.0060	0.101	0.00625
39	0.003531		0.0075		0.0052	0.099	
40	0.003144		0.0070		0.0048	0.097	

Metal Weights

Material	*Chemical Symbol*	*Weight, in lb per Cubic Inch*	*Weight, in lb per Cubic Foot*
Aluminum	Al	0.093	160
Antimony	Sb	0.2422	418
Brass		0.303	524
Bronze		0.320	552
Chromium	Cr	0.2348	406
Copper	Cu	0.323	450
Gold	Au	0.6975	1205
Iron (cast)	Fe	0.260	450
Iron (wrought)	Fe	0.2834	490
Lead	Pb	0.4105	710
Manganese	Mn	0.2679	463
Mercury	Hg	0.491	849
Molybdenum	Mo	0.309	534
Monel		0.318	550
Platinum	Pt	0.818	1413
Steel (mild)	Fe	0.2816	490
Steel (stainless)		0.277	484
Tin	Sn	0.265	459
Titanium	Ti	0.1278	221
Zinc	Zn	0.258	446

Colors and Approximate Temperatures for Carbon Steel

Black red	990°F	532°C
Dark blood red	1050	566
Dark cherry red	1175	635
Medium cherry red	1250	677
Full cherry red	1375	746
Light cherry, scaling	1550	843
Salmon, free scaling	1650	899
Light salmon	1725	946
Yellow	1825	996
Light yellow	1975	1080
White	2220	1216

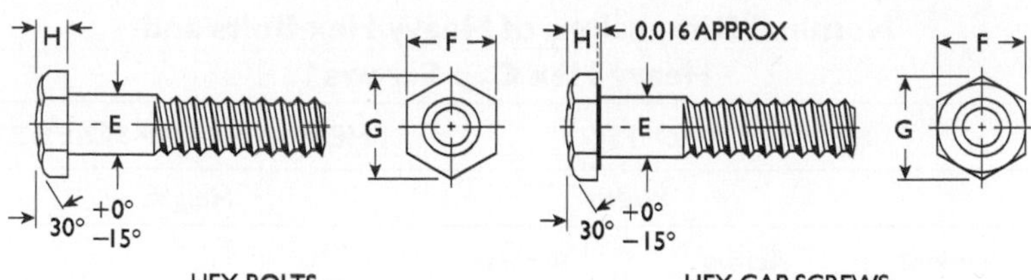

HEX BOLTS HEX CAP SCREWS

Nominal Dimensions of Hex Bolts and Hex Cap Screws

HEX BOLTS		*HEX CAP SCREWS*	
Nominal Size E	***Width Across Flats F***	***Width Across Corners G***	***Head Height H***
1/4	7/16	1/2	11/64
5/16	1/2	9/16	7/32
3/8	9/16	21/32	1/4
7/16	5/8	47/64	19/64
1/2	3/4	55/64	11/32
5/8	15/16	1 3/32	27/64
3/4	1 1/8	1 19/64	1/2
7/8	1 5/16	1 33/64	37/64
1	1 1/2	1 47/64	43/64
1 1/8	1 11/16	1 61/64	3/4
1 1/4	1 7/8	2 11/64	27/32
1 3/8	2 1/16	2 3/8	29/32
1 1/2	2 1/4	2 19/32	1
1 3/4	2 5/8	3 1/32	1 5/32
2	3	3 15/32	1 11/32

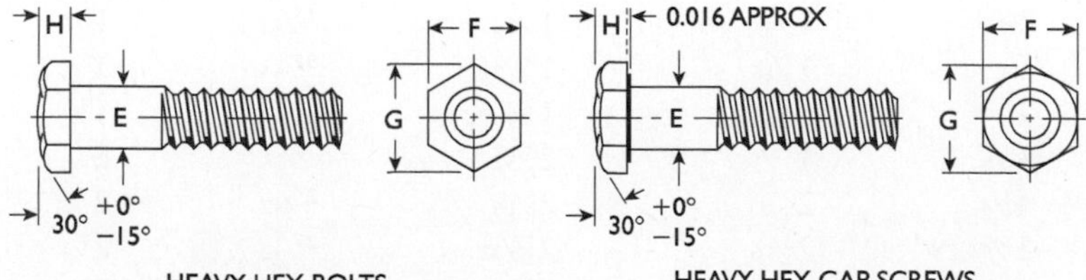

HEAVY HEX BOLTS HEAVY HEX CAP SCREWS

Nominal Dimensions of Heavy Hex Bolts and Heavy Hex Cap Screws

HEAVY HEX BOLTS			*HEAVY HEX CAP SCREWS*	
	Width		*Height*	
Nominal Size	*Across Flats F*	*Across Corner G*	*Bolts H*	*Screws H*
$\frac{1}{2}$	$\frac{7}{8}$	1	$\frac{11}{32}$	$\frac{5}{16}$
$\frac{5}{8}$	$1\frac{1}{16}$	$1\frac{15}{64}$	$\frac{27}{64}$	$\frac{25}{64}$
$\frac{3}{4}$	$1\frac{1}{4}$	$1\frac{7}{16}$	$\frac{1}{2}$	$\frac{15}{32}$
$\frac{7}{8}$	$1\frac{7}{16}$	$1\frac{21}{32}$	$\frac{37}{64}$	$\frac{35}{64}$
1	$1\frac{5}{8}$	$1\frac{7}{8}$	$\frac{43}{64}$	$\frac{39}{64}$
$1\frac{1}{8}$	$1\frac{13}{16}$	$2\frac{3}{32}$	$\frac{3}{4}$	$\frac{11}{16}$
$1\frac{1}{4}$	2	$2\frac{5}{16}$	$\frac{27}{32}$	$\frac{25}{32}$
$1\frac{3}{8}$	$2\frac{3}{16}$	$2\frac{17}{32}$	$\frac{29}{32}$	$\frac{27}{32}$
$1\frac{1}{2}$	$2\frac{3}{8}$	$2\frac{3}{4}$	1	$\frac{15}{16}$
$1\frac{3}{4}$	$2\frac{3}{4}$	$3\frac{11}{64}$	$1\frac{5}{32}$	$1\frac{3}{32}$
2	$3\frac{1}{8}$	$3\frac{39}{64}$	$1\frac{11}{32}$	$1\frac{7}{32}$

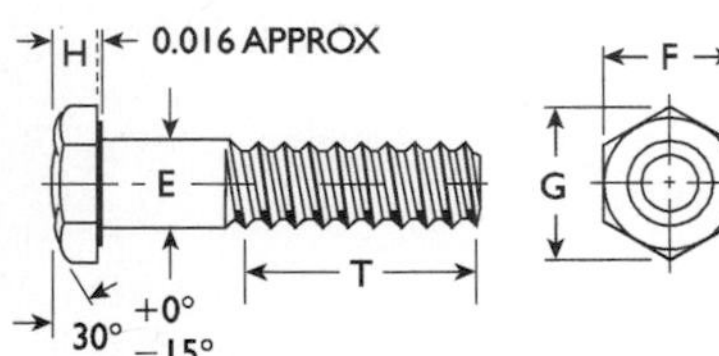

Nominal Dimensions of Heavy Hex Structural Bolts

Nominal Size E	*Width Across Flats F*	*Width Across Corners G*	*Head Height H*	*Thread Length T*
$\frac{1}{2}$	$\frac{7}{8}$	1	$\frac{5}{16}$	1
$\frac{5}{8}$	$1\frac{1}{16}$	$1\frac{15}{64}$	$\frac{25}{64}$	$1\frac{1}{4}$
$\frac{3}{4}$	$1\frac{1}{4}$	$1\frac{7}{16}$	$\frac{15}{32}$	$1\frac{3}{8}$
$\frac{7}{8}$	$1\frac{7}{16}$	$1\frac{21}{32}$	$\frac{35}{64}$	$1\frac{1}{2}$
1	$1\frac{5}{8}$	$1\frac{7}{8}$	$\frac{39}{64}$	$1\frac{3}{4}$
$1\frac{1}{8}$	$1\frac{13}{16}$	$2\frac{3}{32}$	$\frac{11}{16}$	2
$1\frac{1}{4}$	2	$2\frac{5}{16}$	$\frac{25}{32}$	2
$1\frac{3}{8}$	$2\frac{3}{16}$	$2\frac{17}{32}$	$\frac{27}{32}$	$2\frac{1}{4}$
$1\frac{1}{2}$	$2\frac{3}{8}$	$2\frac{3}{4}$	$\frac{5}{16}$	$2\frac{1}{4}$

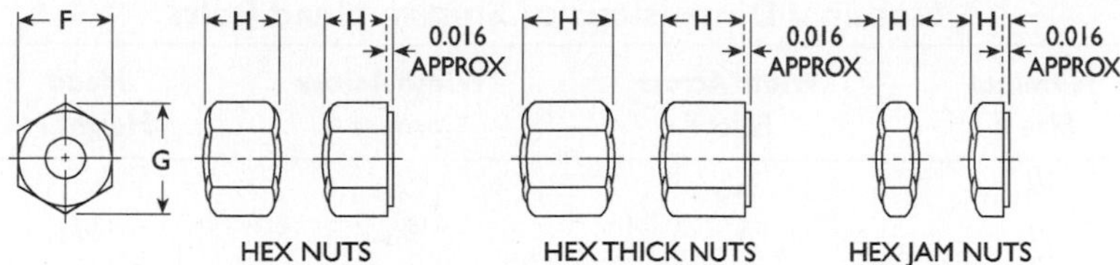

Nominal Dimensions of Hex Nuts, Hex Thick Nuts, and Hex Jam Nuts

Nominal Size	Width Across Flats F	Width Across Corners G	Thickness Hex Nuts H	Thickness Thick Nuts H	Thickness Jam Nuts H
1/4	7/16	1/2	7/32	9/32	5/32
5/16	1/2	9/16	17/64	21/64	3/16
3/8	9/16	21/32	21/64	13/32	7/32
7/16	11/16	51/64	3/8	29/64	1/4
1/2	3/4	55/64	7/16	9/16	5/16
9/16	7/8	1	31/64	39/64	5/16
5/8	15/16	1 3/32	35/64	23/32	3/8
3/4	1 1/8	1 19/64	41/64	13/16	27/64
7/8	1 5/16	1 33/64	3/4	29/32	31/64
1	1 1/2	1 47/64	55/64	1	35/64
1 1/8	1 11/16	1 61/64	31/32	1 5/32	39/64
1 1/4	1 7/8	2 11/64	1 1/16	1 1/4	23/32
1 3/8	2 1/16	2 3/8	1 11/64	1 3/8	25/32
1 1/2	2 1/4	2 19/32	1 9/32	1 1/2	27/32

Nominal Dimensions of Square Head Bolts

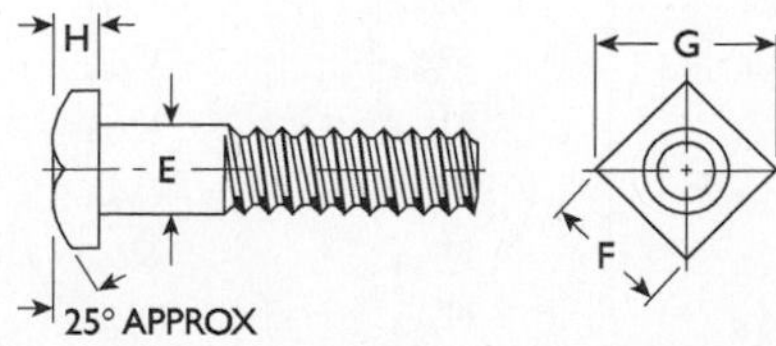

Nominal Dimensions of Square Head Bolts

Nominal Size E	Width Across Flats F	Width Across Corners G	Head Height H
1/4	3/8	17/32	11/64
5/16	1/2	45/64	13/64
3/8	9/16	51/64	1/4
7/16	5/8	57/64	19/64
1/2	3/4	1 1/16	21/64
5/8	15/16	1 21/64	27/64
3/4	1 1/8	1 19/32	1/2
7/8	1 5/16	1 55/64	19/32
1	1 1/2	2 1/8	21/32
1 1/8	1 11/16	2 25/64	3/4
1 1/4	1 7/8	2 21/32	27/32
1 3/8	2 1/16	2 59/64	29/32
1 1/2	2 1/4	3 3/16	1

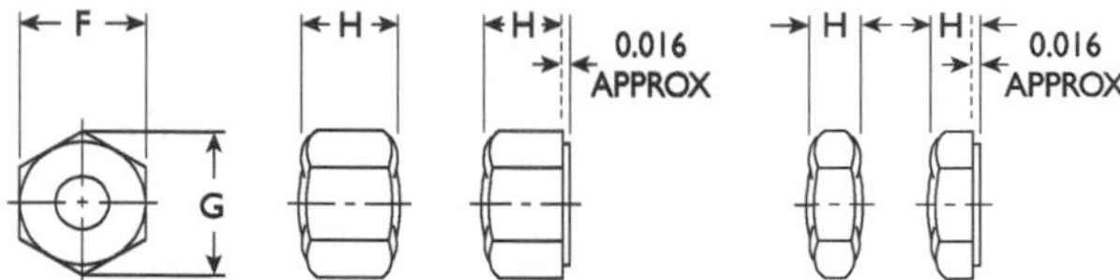

Nominal Dimensions of Heavy Hex Nuts and Heavy Hex Jam Nuts

Nominal Size	Width Across Flats F	Width Across Corners G	Thickness Hex Nuts H	Hex Jam Nuts H
1/4	1/2	37/64	15/64	11/64
5/16	9/16	21/32	19/64	13/64
3/8	11/16	51/64	23/64	15/64
7/16	3/4	55/64	37/64	17/64
1/2	7/8	1 1/64	31/64	19/64
9/16	15/16	1 5/64	35/64	21/64
5/8	1 1/16	1 7/32	39/64	23/64
3/4	1 1/4	1 7/16	47/64	27/64
7/8	1 7/16	1 21/32	55/64	31/64

1	1 5/8	1 7/8	63/64	35/64
1 1/8	1 13/16	2 3/32	1 7/64	39/64
1 1/4	2	2 5/16	1 7/32	23/32
1 3/8	2 3/16	2 17/32	1 11/32	25/32
1 1/2	2 3/8	2 1/2	1 15/32	27/32
1 5/8	2 9/16	2 61/64	1 19/32	29/32
1 3/4	2 3/4	3 11/64	1 23/32	31/32
1 7/8	2 15/16	3 25/64	1 27/32	1 1/32
2	3 1/8	3 39/64	1 31/32	1 3/32

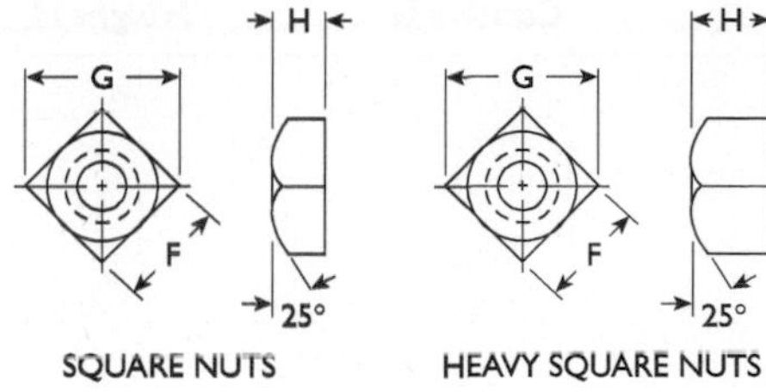

Nominal Dimensions of Square Nuts and Heavy Square Nuts

	Square Nuts		*Heavy Square Nuts*			
	Width Across Flats		*Width Across Corners*		*Thickness*	
Nominal Size	***Regular F***	***Heavy F***	***Regular G***	***Heavy G***	***Regular H***	***Heavy H***
1/4	7/16	1/2	5/8	45/64	7/32	1/4
5/16	9/16	9/16	51/64	51/64	17/64	5/16
3/8	5/8	11/16	57/64	31/32	21/64	3/8
7/16	3/4	3/4	1 1/16	1 1/16	3/8	7/16
1/2	13/16	7/8	1 5/32	1 15/64	7/16	1/2
5/8	1	1 1/16	1 27/64	1 1/2	35/64	5/8
3/4	1 1/8	1 1/4	1 19/32	1 49/64	21/32	3/4
7/8	1 5/16	1 7/16	1 55/64	2 1/32	49/64	7/8
1	1 1/2	1 5/8	2 1/8	2 19/64	7/8	1
1 1/8	1 11/16	1 13/16	2 25/64	2 9/16	1	1 1/8
1 1/4	1 7/8	2	2 21/32	2 53/64	1 3/32	1 1/4
1 3/8	2 1/16	2 3/16	2 59/64	3 3/32	1 13/64	1 3/8
1 1/2	2 1/4	2 3/8	3 3/16	3 23/64	1 5/16	1 1/2

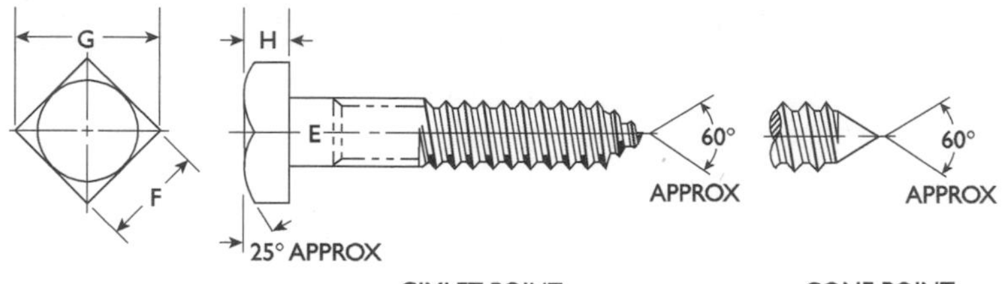

Nominal Dimensions of Lag Screws

Nominal Size E	Width Across Flats F	Width Across Corners G	Head Height H
#10	9/32	19/64	1/8
1/4	3/8	17/32	11/64
5/16	1/2	45/64	13/64
3/8	9/16	11/16	1/4
7/16	5/8	57/64	19/64
1/2	3/4	1 1/16	21/64
5/8	15/16	1 21/64	27/64
3/4	1 1/8	1 19/32	1/2
7/8	1 5/16	1 55/64	19/32
1	1 1/2	2 1/8	21/32
1 1/8	1 11/16	2 25/64	3/4
1 1/4	1 7/8	2 21/32	27/32

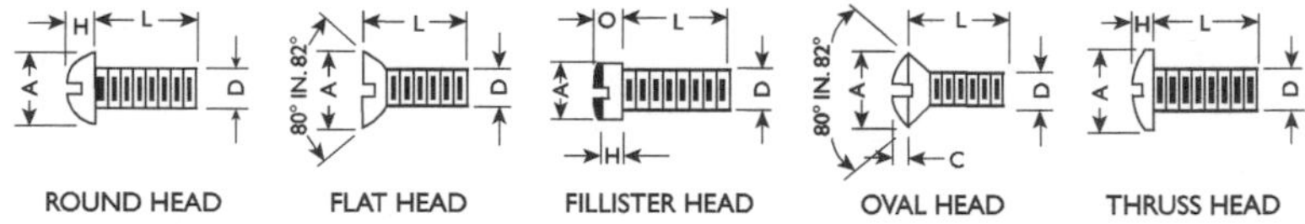

American Standard Machine Screws*—Heads May Be Slotted or Recessed

Nominal Diam.	Round Head A	Round Head H	Flat Head A	Fillister Head A	Fillister Head H	Fillister Head O	Oval Head A	Oval Head C	Truss Head A	Truss Head H
0	0.113	0.053	0.119	0.096	0.045	0.059	0.119	0.021		
1	0.138	0.061	0.146	0.118	0.053	0.071	0.146	0.025	0.194	0.053
2	0.162	0.069	0.172	0.140	0.062	0.083	0.172	0.029	0.226	0.061

3	0.187	0.078	0.199	0.161	0.070	0.095	0.199	0.033	0.257	0.069
4	0.211	0.086	0.225	0.183	0.079	0.107	0.225	0.037	0.289	0.078
5	0.236	0.095	0.252	0.205	0.088	0.120	0.252	0.041	0.321	0.086
6	0.260	0.103	0.279	0.226	0.096	0.132	0.279	0.045	0.352	0.094
8	0.309	0.120	0.332	0.270	0.113	0.156	0.332	0.052	0.384	0.102
10	0.359	0.137	0.385	0.313	0.130	0.180	0.385	0.060	0.448	0.118
12	0.408	0.153	0.438	0.357	0.148	0.205	0.438	0.068	0.511	0.134
¼	0.472	0.175	0.507	0.414	0.170	0.237	0.507	0.079	0.573	0.150
5⁄16	0.590	0.216	0.635	0.518	0.211	0.295	0.635	0.099	0.698	0.183
⅜	0.708	0.256	0.762	0.622	0.253	0.355	0.762	0.117	0.823	0.215
7⁄16	0.750	0.328	0.812	0.625	0.265	0.368	0.812	0.122	0.948	0.248
½	0.813	0.355	0.875	0.750	0.297	0.412	0.875	0.131	1.073	0.280
9⁄16	0.938	0.410	1.000	0.812	0.336	0.466	1.000	0.150	1.198	0.312
⅝	1.000	0.438	1.125	0.875	0.375	0.521	1.125	0.169	1.323	0.345
¾	1.250	0.547	1.375	1.000	0.441	0.612	1.375	0.206	1.573	0.410

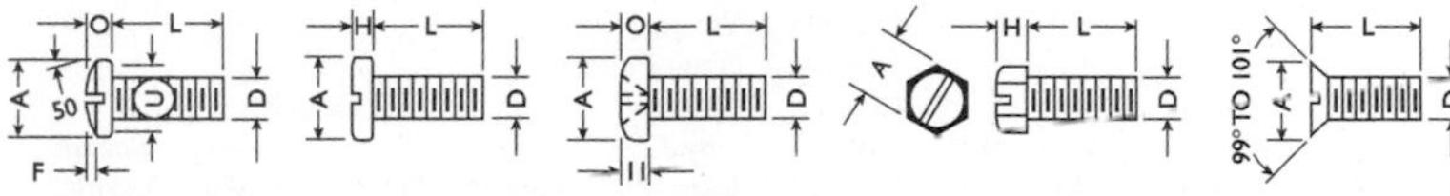

BINDING HEAD PAN HEAD PAN HEAD (RECESSED) HEXAGON HEAD 100° FLAT HEAD

Nominal Diam.	*Binding Head* A	O	F	U	*Pan Head* A	H	O	*Hexagon Head* A	H	*100° Flat Head* A
2	0.181	0.046	0.018	0.141	0.167	0.053	0.062	0.125	0.050	
3	0.208	0.054	0.022	0.162	0.193	0.060	0.071	0.187	0.055	
4	0.235	0.063	0.025	0.184	0.219	0.068	0.080	0.187	0.060	0.225
5	0.263	0.071	0.029	0.205	0.245	0.075	0.089	0.187	0.070	
6	0.290	0.080	0.032	0.226	0.270	0.082	0.097	0.250	0.080	0.279
8	0.344	0.097	0.039	0.269	0.322	0.096	0.115	0.250	0.110	0.332
10	0.399	0.114	0.045	0.312	0.373	0.110	0.133	0.312	0.120	0.385
12	0.454	0.130	0.052	0.354	0.425	0.125	0.151	0.312	0.155	
¼	0.513	0.153	0.061	0.410	0.492	0.144	0.175	0.375	0.190	0.507
5⁄16	0.641	0.193	0.077	0.513	0.615	0.178	0.218	0.500	0.230	0.635
⅜	0.769	0.234	0.094	0.615	0.740	0.212	0.261	0.562	0.295	0.762

**ANSI B18.6—1972. Dimensions given are maximum values, all in inches. Thread length: screws 2 in. long or less, thread entire length; screws over 2 in. long, thread length l = 1¾ in. Threads are coarse or fine series, class 2. Heads may be slotted or recessed as specified, excepting hexagon form, which is plain or may be slotted if so specified. Slot and recess proportions vary with size of fastener; draw to look well.*

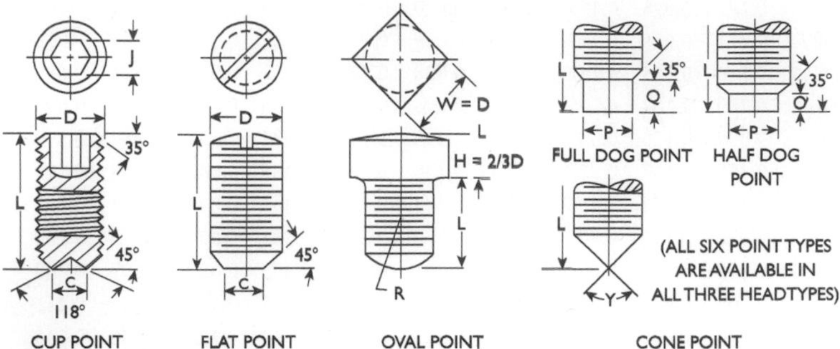

American Standard Hexagon Socket,[a] Slotted Headless,[b] and Square-Head[c] Setscrews

			Cone-Point Angle Y		*Full and Half Dog Points*			
			118° for These Lengths and Shorter	*90° for These Lengths and Longer*		*Length*		
Diam. D	*Cup and Flat-Point Diam. C*	*Oval-Point Radius R*			*Diam. P*	*Full Q*	*Half q*	*Socket Width J*
5	1/16	3/32	1/8	3/16	0.083	0.06	0.03	1/16
6	0.069	7/64	1/8	3/16	0.092	0.07	0.03	1/16
8	5/64	1/8	3/16	1/4	0.109	0.08	0.04	5/64
10	3/32	9/64	3/16	1/4	0.127	0.09	0.04	3/32
12	7/64	5/32	3/16	1/4	0.144	0.11	0.06	3/32
1/4	1/8	3/16	1/4	5/16	5/32	1/8	1/16	1/8
5/16	11/64	15/64	5/16	3/8	13/64	5/32	5/64	5/32
3/8	13/64	9/32	3/8	7/16	1/4	3/16	3/32	3/16
7/16	15/64	21/64	7/16	1/2	19/64	7/32	7/64	7/32
1/2	9/32	3/8	1/2	9/16	11/32	1/4	1/8	1/4
9/16	5/16	27/64	9/16	5/8	25/64	9/32	9/64	1/4
5/8	23/64	15/32	5/8	3/4	15/32	5/16	5/32	5/16
3/4	7/16	9/16	3/4	7/8	9/16	3/8	3/16	3/8
7/8	33/64	21/32	7/8	1	21/32	7/16	7/32	1/2
1	19/32	3/4	1	1 1/8	3/4	1/2	1/4	9/16
1 1/8	43/64	27/32	1 1/8	1 1/4	27/32	9/16	9/32	9/16
1 1/4	3/4	15/16	1 1/4	1 1/2	15/16	5/8	5/16	5/8
1 3/8	53/64	1 1/32	1 3/8	1 5/8	1 1/32	11/16	11/32	5/8
1 1/2	29/32	1 1/8	1 1/2	1 3/4	1 1/8	3/4	3/8	3/4
1 3/4	1 1/16	1 5/16	1 3/4	2	1 5/16	7/8	7/16	1
2	1 7/32	1 1/2	2	2 1/4	1 1/2	1	1/2	1

[a]ANSI B18.3—1976. Dimensions are in inches. Threads coarse or fine, class 3A.

Length increments: 1/4 in. to 5/8 in. by (1/16 in.); 5/8 in. to 1 in. by (1/8 in.); 1 in. to 4 in. by (1/4 in.); 4 in. to 6 in. by (1/2 in.). Fractions in parentheses show length increments; for example, 5/8 in. to 1 in. by (1/8 in.); includes the lengths 5/8 in., 3/4 in., 7/8 in., and 1 in.

[b]*ANSI B18.6.2—1972. Threads coarse or fine, class 2A. Slotted headless screws standardized in sizes No. 5 to 3/4 in. only. Slot proportions vary with diameter. Draw to look well.*

[c]*ANSI B18.6.2—1972. Threads coarse, fine, or 8-pitch, class 2A. Square-head setscrews standardized in sizes No. 10 to 1 1/2 in. only.*

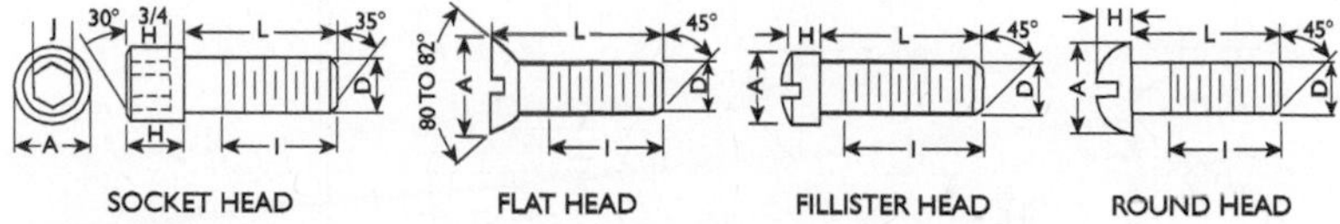

American Standard Cap Screws[a]—Socket[b] and Slotted Heads[c]

Nominal Diam.	Socket Head[d] A	Socket Head[d] H	Socket Head[d] J	Flat Head[e] A	Filister Head[e] A	Filister Head[e] H	Round Head[e] A	Round Head[e] H
0	0.096	0.060	0.050					
1	0.118	0.073	0.050					
2	0.140	0.086	1/16					
3	0.161	0.099	5/64					
4	0.183	0.112	5/64					
5	0.205	0.125	3/32					
6	0.226	0.138	3/32					
8	0.270	0.164	1/8					
10	5/16	0.190	5/32					
12	11/32	0.216	5/32					
1/4	3/8	1/4	3/16	1/2	3/8	11/64	7/16	3/16
5/16	7/16	5/16	7/32	5/8	7/16	13/64	9/16	15/64
3/8	9/16	3/8	5/16	3/4	9/16	1/4	5/8	17/64
7/16	5/8	7/16	5/16	13/16	5/8	19/64	3/4	5/16
1/2	3/4	1/2	3/8	7/8	3/4	21/64	13/16	11/32
9/16	13/16	9/16	3/8	1	13/16	3/8	15/16	13/32
5/8	7/8	5/8	1/2	1 1/8	7/8	27/64	1	7/16
3/4	1	3/4	9/16	1 3/8	1	1/2	1 1/4	17/32
7/8	1 1/8	7/8	9/16	1 5/8	1 1/8	19/32		
1	1 5/16	1	5/8	1 7/8	1 5/16	21/32		
1 1/8	1 1/2	1 1/8	3/4					
1 1/4	1 3/4	1 1/4	3/4					
1 3/8	1 7/8	1 3/8	3/4					
1 1/2	2	1 1/2	1					

[a]*Dimensions in inches.*

[b]*ANSI B18.3—1976.*

[c]*ANSI B18.6.2—1972.*

[d]*Thread coarse or fine, class 3A. Thread length l: coarse thread, 2D + ½ in.; fine thread, 1½D + ½ in.*

[e]*Thread coarse, fine, or 8-pitch, class 2A. Thread length l: 2D + ¼ in.*

Slot proportions vary with size of screw; draw to look well. All body-length increments for screw lengths ¼ in. to 1 in. = ⅛ in., for screw lengths 1 in. to 4 in. = 5 ¼ in., for screw lengths 4 in. to 6 in. = ½ in.

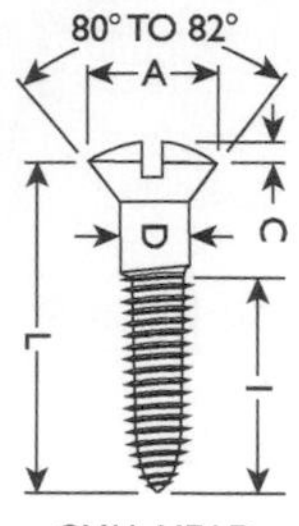

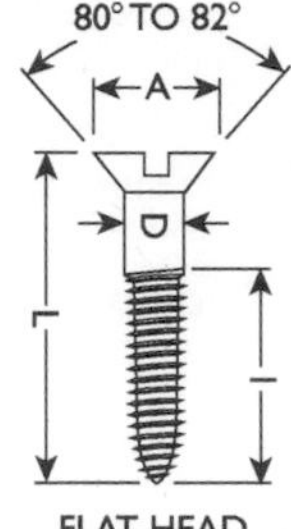

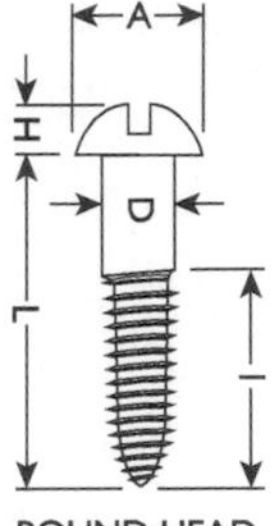

American Standard Wood Screws[a]

Nominal Size	Basic Diam. of Screw D	No. of Threads per Inch[b]	Slot Width[c] J (All Heads)	Round Head		Flat Head	Oval Head	
				A	H	A	A	C
0	0.060	32	0.023	0.113	0.053	0.119	0.119	0.021
1	0.073	28	0.026	0.138	0.061	0.146	0.146	0.025
2	0.086	26	0.031	0.162	0.069	0.172	0.172	0.029
3	0.099	24	0.035	0.187	0.078	0.199	0.199	0.033
4	0.112	22	0.039	0.211	0.086	0.225	0.225	0.037
5	0.125	20	0.043	0.236	0.095	0.252	0.252	0.041
6	0.138	18	0.048	0.260	0.103	0.279	0.279	0.045
7	0.151	16	0.048	0.285	0.111	0.305	0.305	0.049
8	0.164	15	0.054	0.309	0.120	0.332	0.332	0.052
9	0.177	14	0.054	0.334	0.128	0.358	0.358	0.056
10	0.190	13	0.060	0.359	0.137	0.385	0.385	0.060
12	0.216	11	0.067	0.408	0.153	0.438	0.438	0.068
14	0.242	10	0.075	0.457	0.170	0.491	0.491	0.076
16	0.268	9	0.075	0.506	0.187	0.544	0.544	0.084
18	0.294	8	0.084	0.555	0.204	0.597	0.597	0.092
20	0.320	8	0.084	0.604	0.220	0.650	0.650	0.100
24	0.372	7	0.094	0.702	0.254	0.756	0.756	0.116

[a]*ANSI B18.6.1—1972. Dimensions given are maximum values, all in inches. Heads may be slotted or recessed as specified.*

[b]*Thread length l = 2⁄3L.*

[c]*Slot depths and recesses vary with type and size of screw; draw to look well.*

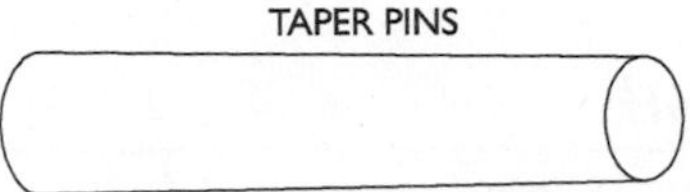

Taper Pins—All Sizes Have a Taper of 0.250 per Foot

Size No. of Pin	*Length of Pin*	*Large End of Pin*	*Small End of Reamer*	*Drill Size for Reamer*
0	1	0.156	0.135	28
1	1¼	0.172	0.146	25
2	1½	0.193	0.162	19
3	1¾	0.219	0.183	12
4	2	0.250	0.208	3
5	2¼	0.289	0.242	¼
6	3¼	0.341	0.279	9⁄32
7	3¾	0.409	0.331	11⁄32
8	4½	0.492	0.398	13⁄32
9	5¼	0.591	0.482	31⁄64
10	6	0.706	0.581	19⁄32
11	7¼	0.857	0.706	23⁄32
12	8¾	1.013	0.842	55⁄64

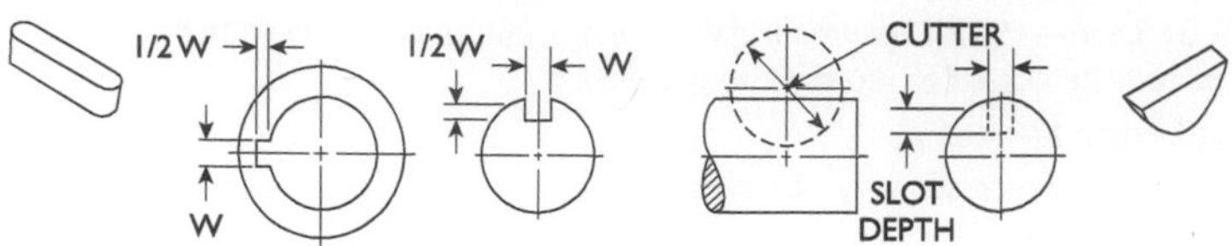

Keyway Data

Shaft Dia.	Square Keyways	Woodruff Keyways* Key No.	Thickness	Cutter Dia.	Slot Depth
0.500	⅛ × 1⁄16	404	0.1250	0.500	0.1405
0.562	⅛ × 1⁄16	404	0.1250	0.500	0.1405
0.625	5⁄32 × 5⁄64	505	0.1562	0.625	0.1669
0.688	3⁄16 × 3⁄32	606	0.1875	0.750	0.2193
0.750	3⁄16 × 3⁄32	606	0.1875	0.750	0.2193
0.812	3⁄16 × 3⁄32	606	0.1875	0.750	0.2193
0.875	7⁄32 × 7⁄64	607	0.1875	0.875	0.2763
0.938	¼ × ⅛	807	0.2500	0.875	0.2500
1.000	¼ × ⅛	808	0.2500	1.000	0.3130
1.125	5⁄16 × 5⁄32	1009	0.3125	1.125	0.3228
1.250	5⁄16 × 5⁄32	1010	0.3125	1.250	0.3858
1.375	⅜ × 3⁄16	1210	0.3750	1.250	0.3595
1.500	⅜ × 3⁄16	1212	0.3750	1.500	0.4535
1.625	⅜ × 3⁄16	1212	0.3750	1.500	0.4535
1.750	7⁄16 × 7⁄32				
1.875	½ × ¼				
2.000	½ × ¼				
2.250	⅝ × 5⁄16				
2.500	⅝ × 5⁄16				
2.750	¾ × ⅜				
3.000	¾ × ⅜				
3.250	¾ × ⅜				
3.500	⅞ × 7⁄16				
4.000	1 × ½				

**The depth of a Woodruff Keyway is measured from the edge of the slot.*

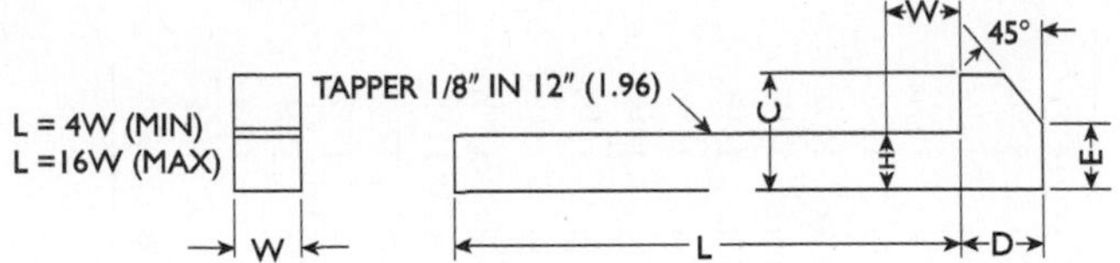

Dimensions of Standard Gib-Head Keys, Square and Flat—Approved by ANSI

	Square Type					*Flat Type*				
	Key		*Gib Head*			*Key*		*Gib Head*		
Diameters of Shafts	***W***	***H***	***C***	***D***	***E***	***W***	***H***	***C***	***D***	***E***
1/2–9/16	1/8	1/8	1/4	7/32	5/32	1/8	3/32	3/16	1/8	1/8
5/8–7/8	3/16	3/16	5/16	9/32	7/32	3/16	1/8	1/4	3/16	5/32
15/16–1 1/4	1/4	1/4	7/16	11/32	11/32	1/4	3/16	5/16	1/4	3/16
1 5/16–1 3/8	5/16	5/16	9/16	13/32	13/32	5/16	1/4	3/8	5/16	1/4
1 7/16–1 3/4	3/8	3/8	11/16	15/32	15/32	3/8	1/4	7/16	3/8	5/16
1 13/16–2 1/4	1/2	1/2	7/8	19/32	5/8	1/2	3/8	5/8	1/2	7/16
2 5/16–2 3/4	5/8	5/8	1 1/16	23/32	3/4	5/8	7/16	3/4	5/8	1/2
2 7/8–3 1/4	3/4	3/4	1 1/4	7/8	7/8	3/4	1/2	7/8	3/4	5/8
3 3/8–3 3/4	7/8	7/8	1 1/2	1	1	7/8	5/8	1 1/16	7/8	3/4
3 7/8–4 1/2	1	1	1 3/4	1 3/16	1 3/16	1	3/4	1 1/4	1	1 3/16
4 3/4–5 1/2	1 1/4	1 1/4	2	1 7/16	1 7/16	1 1/4	7/8	1 1/2	1 1/4	1
5 3/4–6	1 1/2	1 1/2	2 1/2	1 3/4	1 3/4	1 1/2	1	1 3/4	1 1/2	1 1/4

[a]ANSI B17.1—1934. Dimensions in inches.

U Nema Motor Frame Dimensions

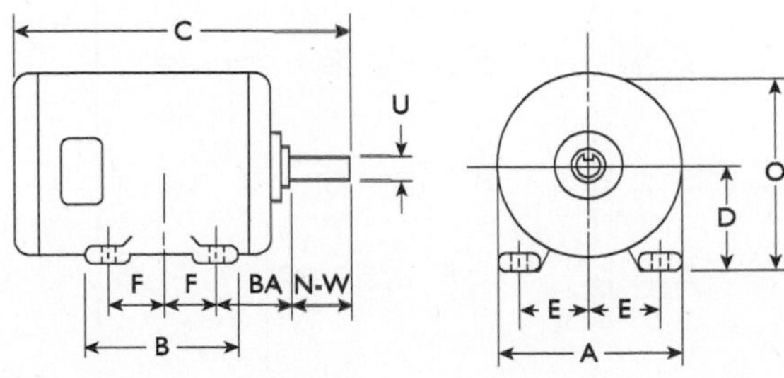

U Nema Motor Frame Dimensions

Horsepower Rating			U Frame		Shaft Keyseat		Key									
3600	1800	1200	No.	U	Width	Depth	Length	N-W	A Max.	B Max.	C	D	E	F	BA	O
1½	1	¾	182	⅞	3/16	3/32	1⅜	2¼	8⅞	6½	12⅛	4½	3¾	2¼	2¾	8 15/16
2 & 3	1½ & 2	1 & 1½	184	⅞	3/16	3/32	1⅜	2¼	8⅞	7½	13⅛	4½	3¾	2¾	2¾	8 15/16
5	3	2	213	1⅛	¼	⅛	2	3	10⅜	7½	15 5/16	5¼	4¼	2¾	3½	10 7/16
7½	5	3	215	1⅛	¼	⅛	2	3	10⅜	9	16 13/16	5¼	4¼	3½	3½	10 7/16
10	7½	5	254U	1⅜	5/16	5/32	2¾	3¾	12 7/16	10¾	20¼	6¼	5	4⅛	4¼	12½
15	10	7½	256U	1⅜	5/16	5/32	2¾	3¾	12 7/16	12½	22	6¼	5	5	4¼	12½
20	15	10	284U	1⅝	⅜	3/16	3¾	4⅞	13⅞	12½	23 11/16	7	5½	4¾	4¾	13 15/16
25	20		286U	1⅝	⅜	3/16	3¾	4⅞	13⅞	12½	25 3/16	7	5½	5½	4¾	13 15/16
	25	15	324U	1⅞	½	¼	4¼	5⅝	15⅞	14	26⅜	8	6¼	5¼	5¼	15 15/16
30			324S	1⅝	⅜	3/16	1⅞	3¼	15⅞	14	24	8	6¼	5¼	5¼	15 15/16
	30	20	326U	1⅞	½	¼	4¼	5⅝	15⅞	14	27⅞	8	6¼	6	5¼	15 15/16
40			326S	1⅝	⅜	3/16	1⅞	3¼	15⅞	14	25½	8	6¼	6	5¼	15 15/16
	40	25	364U	2⅛	½	¼	5	6⅜	17⅝	15¼	29 3/16	9	7	5⅝	5⅞	17 13/16
50			364US	1⅞	½	¼	2	3¾	17⅝	15¼	26 9/16	9	7	5⅝	5⅞	17 13/16
	50	30	365U	2⅛	½	¼	5	6⅜	17⅝	16¼	30 3/16	9	7	6⅛	5⅞	17 13/16
60			365US	1⅞	½	¼	2	3¾	19¾	16¼	27 9/16	9	7	6⅛	5⅞	17 13/16
	60	40	404U	2⅜	⅝	5/16	5½	7⅛	19¾	16¼	32 7/16	10	8	6⅛	6⅝	19⅞
75			404US	2⅛	½	¼	2¾	4¼	19¾	16¼	39 9/16	10	8	6⅛	6⅝	19⅞
	75	50	405U	2⅜	⅝	5/16	5½	7⅛	19¾	17¼	33 15/16	10	8	6⅞	6⅝	19⅞
100			405US	2⅛	½	¼	2¾	4¼	19¾	17¼	31 1/16	10	8	6⅞	6⅝	19⅞

T Nema Motor Frame Dimensions

Horsepower Rating			*T*		*Shaft Keyseat*											
3600	***1800***	***1200***	***Frame No.***	***U***	***Width***	***Depth***	***Key Length***	***N-W***	***A Max.***	***B Max.***	***C***	***D***	***E***	***F***	***BA***	***O***
1½	1	¾	143T	⅞	3⁄16	3⁄32	1⅜	2¼	7	6	12⅝	3½	2¾	2	2¼	7
2 & 3	1½ & 2	1	145T	⅞	3⁄16	3⁄32	1⅜	2¼	7	6	12⅝	3½	2¾	2½	2¼	7
5	3	1½	182T	1⅛	¼	⅛	1¾	2¾	9	6½	12¾	4½	3¾	2¼	2¾	9
7½	5	2	184T	1⅛	¼	⅛	1¾	2¾	9	7½	13¾	4½	3¾	2¾	2¾	9
10	7½	3	213T	1⅜	5⁄16	5⁄32	2⅜	3⅜	10½	7½	15 13⁄16	5¼	4¼	2¾	3½	10½
15	10	5	215T	1⅜	5⁄16	5⁄32	2⅜	3⅜	10½	9	17 5⁄16	5¼	4¼	3½	3½	10½
20	15	7½	254T	1⅝	⅜	3⁄16	2⅞	4	12½	10¾	20½	6¼	5	4⅛	4¼	12½
25	20	10	256T	1⅝	⅜	3⁄16	2⅞	4	12½	12½	22¼	6¼	5	5	4¼	12½
	25	15	284T	1⅞	½	¼	3¼	4⅝	14	12½	23 5⁄16	7	5½	4¾	4¾	14
30	25	15	284TS	1⅝	⅜	3⁄16	1⅞	3¼	14	12½	22	7	5½	4¾	4¾	14
	30	20	286T	1⅞	½	¼	3¼	4⅝	14	14	24⅞	7	5½	5½	4¾	14
40	30	20	286TS	1⅝	⅜	3⁄16	1⅞	3¼	14	14	23½	7	5½	5½	4¾	14
	40	25	324T	2⅛	½	¼	3⅞	5¼	16	14	26½	8	6¼	5¼	5¼	16
50	40	25	324TS	1⅞	½	¼	2	3¾	16	14	24⅝	8	6¼	5¼	5¼	16
	50	30	326T	2⅛	½	¼	3⅞	5¼	16	15½	27¾	8	6¼	6	5¼	16
60	50	30	326TS	1⅞	½	¼	2	3¾	16	15½	26⅛	8	6¼	6	5¼	16
	60	40	364T	2⅜	⅝	5⁄16	4¼	5⅞	18	15¼	28¾	9	7	5⅝	5⅞	18
75	60		364TS	1⅞	½	¼	2	3¾	18	15¼	26 9⁄16	9	7	5⅝	5⅞	18
	75	50	365T	2⅜	⅝	5⁄16	4¼	5⅞	18	16¼	29¾	9	7	6⅛	5⅞	18
100	75		365TS	1⅞	½	¼	2	3¾	18	16¼	27 9⁄16	9	7	6⅛	5⅞	18

Commercial Pipe Sizes and Wall Thicknesses

Nominal Pipe Size	Outside Diam.	Sched. 5[a]	Sched. 10[b]	Sched. 20	Sched. 30	Stan-dard[c]	Sched. 40	Sched. 60	Extra Strong[d]	Sched. 80	Sched. 100	Sched. 120	Sched. 140	Sched. 160	XX Strong
⅛	0.405		0.049			0.068	0.068		0.095	0.095					
¼	0.540		0.065			0.088	0.086		0.119	0.119					
⅜	0.675		0.065			0.091	0.091		0.126	0.126					
½	0.840		0.083			0.109	0.109		0.147	0.147				0.187	0.294
¾	1.050	0.065	0.083			0.113	0.113		0.154	0.154				0.218	0.308
1	1.315	0.065	0.109			0.133	0.133		0.179	0.179				0.250	0.358
1¼	1.660	0.065	0.109			0.140	0.140		0.191	0.191				0.250	0.382
1½	1.900	0.065	0.109			0.145	0.145		0.200	0.200				0.281	0.400
2	2.375	0.065	0.109			0.154	0.154		0.218	0.218				0.343	0.436
2½	2.875	0.083	0.120			0.203	0.203		0.276	0.276				0.375	0.552
3	3.5	0.083	0.120			0.216	0.216		0.300	0.300				0.438	0.600
3½	4.0	0.083	0.120			0.226	0.226		0.318	0.318					
4	4.5	0.083	0.120			0.237	0.237		0.337	0.337		0.438		0.531	0.674
5	5.563	0.109	0.134			0.258	0.258		0.375	0.375		0.500		0.625	0.750
6	6.625	0.109	0.134			0.280	0.280		0.432	0.432		0.562		0.718	0.864
8	8.625	0.109	0.148	0.250	0.277	0.322	0.322	0.406	0.500	0.500	0.593	0.718	0.812	0.906	0.875
10	10.75	0.134	0.165	0.250	0.307	0.365	0.365	0.500	0.500	0.593	0.713	0.843	1.000	1.125	
12	12.75	0.156	0.180	0.250	0.330	0.375	0.406	0.562	0.500	0.687	0.843	1.000	1.125	1.312	
14 O.D.	14.0		0.250	0.312	0.375	0.375	0.438	0.593	0.500	0.750	0.937	1.093	1.250	1.406	
16 O.D.	16.0		0.250	0.312	0.375	0.375	0.500	0.656	0.500	0.843	1.031	1.218	1.438	1.593	
18 O.D.	18.0		0.250	0.312	0.438	0.375	0.562	0.750	0.500	0.937	1.156	1.375	1.562	1.781	

20 O.D.	20.0			0.250	0.375	0.500	0.375	0.593	0.812	0.500	1.031	1.281	1.500	1.750	1.968
22 O.D.	22.0			0.250			0.375			0.500					
24 O.D.	24.0			0.250	0.375	0.562	0.375	0.687	0.968	0.500	1.218	1.531	1.812	2.062	2.343
26 O.D.	26.0						0.375			0.500					
30 O.D.	30.0			0.312	0.500	0.625	0.375			0.500					
34 O.D.	34.0						0.375			0.500					
36 O.D.	36.0						0.375			0.500					
42 O.D.	42.0						0.375			0.500					

The following table lists the pipe sizes and wall thicknesses currently established as standard, or specifically:

1. The traditional standard weight, extra strong, and double extra strong pipe.

2. The pipe wall thickness schedules listed in ANSI B36.10, which are applicable to carbon steel and alloys other than stainless steels.

3. The pipe wall thickness schedules listed in ANSI B36.19, which are applicable only to stainless steels.

All dimensions are given in inches.

The decimal thickness listed for the respective pipe sizes represent their nominal or average wall dimensions. The actual thickness may be as much as 12.5% under the nominal thickness because of mill tolerance. Thickness shown in lightface for Schedule 60 and heavier pipe are not currently supplied by the mills, unless a certain minimum tonnage is ordered.

[a]Thicknesses shown in bold are for Schedule 5S and 10S and are available in stainless steel only.

[b]Thicknesses shown in bold are available also in stainless steel, under the designation Schedule 40S.

[c]Thicknesses shown in bold are available also in stainless steel, under the designation Schedule 80S.

Metric Measures

In accordance with the standard practice approved by the American National Standards Institute, the ratio 25.4 mm=1 inch is used for converting millimeters to inches. This factor varies only two millionths of an inch from the more exact factor 25.40005 mm, a difference so small as to be negligible for industrial length measurements.

The metric unit of length is the meter = 39.37 in.

The metric unit of weight is the gram = 15.432 grains

The following prefixes are used for subdivisions and multiples: milli = 1⁄1000, centi = 1⁄100, deci = 1⁄10, deca = 10, hecto = 100, kilo = 1000, myria = 10,000.

Metric and English Equivalent Measures

Measures of Length

Metric	English
1 meter	= 39.37 inches, or 3.28083 feet, or 1.09361 yards
0.3048 meter	= 1 foot
1 centimeter	= 0.3937 inch
2.54 centimeters	= 1 inch
1 millimeter	= 0.03937 inch, or nearly 1⁄25 inch
25.4 millimeters	= 1 inch
1 kilometer	= 1093.61 yards, or 0.62137 mile

Measures of Weight

Metric	English
1 gram	= 15.432 grains
0.0648 gram	= 1 grain
28.35 grams	= 1 ounce avoirdupois
1 kilogram	= 2.2046 pounds
0.4536 kilogram	= 1 pound 0.9842 ton of 2240 pounds (long tons)
1 metric ton	= 19.68 cwt.
1000 kilograms	= 2204.6 pounds
1.016 metric tons	= 1 ton of 2240 pounds (long tons)
1016 kilograms	

Measures of Capacity	
Metric	***English***
1 liter (= 1 cubic decimeter)	= 61.023 cubic inches .03531 cubic foot .2642 gal. (American) 2.202 lbs of water at 62°F
28.317 liters	= 1 cubic foot
3.785 liters	= 1 gallon (American)
4.543 liters	= 1 gallon (Imperial)

English Conversion Table

Length			
Inches	×	0.0833	= feet
Inches	×	0.02778	= yard
Inches	×	0.00001578	= miles
Feet	×	0.3333	= yards
Feet	×	0.0001894	= miles
Yards	×	36.00	= inches
Yards	×	3.00	= feet
Yards	×	0.0005681	= miles
Miles	×	63360.00	= inches
Miles	×	5280.00	= feet
Miles	×	1760.00	= yards
Circumference of circle	×	0.3188	= diameter
Diameter of circle	×	3.1416	= circumference

Area			
Square inches	×	0.00694	= square feet
Square inches	×	0.0007716	= square yards
Square feet	×	144.00	= square inches
Square feet	×	0.11111	= square yards
Square yards	×	1296.00	= square inches
Square yards	×	9.00	= square feet
Dia. of circle squared	×	0.7854	= area
Dia. of sphere squared	×	3.1416	= surface

Volume

Cubic inches	×	0.0005787	= cubic feet
Cubic inches	×	0.00002143	= cubic yards
Cubic inches	×	0.004329	= U.S. gallons
Cubic feet	×	1728.00	= cubic inches
Cubic feet	×	0.03704	= cubic yards
Cubic feet	×	7.4805	= U.S. gallons
Cubic yards	×	27.00	= cubic feet
Dia. of sphere cubed	×	0.5236	= volume

Weight

Grains (avoirdupois)	×	0.002286	= ounces
Ounces (avoirdupois)	×	0.0625	= pounds
Ounces (avoirdupois)	×	0.00003125	= tons
Pounds (avoirdupois)	×	16.00	= ounces
Pounds (avoirdupois)	×	0.01	= hundredweight
Pounds (avoirdupois)	×	0.0005	= tons
Tons (avoirdupois)	×	32000.00	= ounces
Tons (avoirdupois)	×	2000.00	= pounds

English Conversion Table

Energy

Horsepower	×	33000.	= ft-lb per min.
Btu	×	778.26	= ft-lb
Ton of refrigeration	×	200.	= Btu per min.

Pressure

Lbs per sq. in.	×	2.31	= ft of water (60°F)
Ft of water (60°F)	×	0.433	= lbs per sq. in.
In. of water (60°F)	×	0.0361	= lbs per sq. in.
Lbs per sq. in.	×	27.70	= in. of water (60°F)
Lbs per sq. in.	×	2.041	= in. of Hg (60°F)
In. of Hg (60°F)	×	0.490	= lbs per sq. in.

Power			
Horsepower	×	746.	= watts
Watts	×	0.001341	= horsepower
Horsepower	×	42.4	= Btu per min.
Water Factors (At Point of Greatest Density—39.2°F)			
Miner's inch (of water)	×	8.976	= U.S. gal. per min.
Cubic inches (of water)	×	0.57798	= ounces
Cubic inches (of water)	×	0.036124	= pounds
Cubic inches (of water)	×	0.004329	= U.S. gallons
Cubic inches (of water)	×	0.003607	= English gallons
Cubic feet (of water)	×	62.425	= pounds
Cubic feet (of water)	×	0.03121	= tons
Cubic feet (of water)	×	7.4805	= U.S. gallons
Cubic inches (of water)	×	6.232	= English gallons
Cubic foot of ice	×	57.2	= pounds
Ounces (of water)	×	1.73	= cubic inches
Pounds (of water)	×	26.68	= cubic inches
Pounds (of water)	×	0.01602	= cubic feet
Pounds (of water)	×	0.1198	= U.S. gallons
Pounds (of water)	×	0.0998	= English gallons
Tons (of water)	×	32.04	= cubic feet
Tons (of water)	×	239.6	= U.S. gallons
Tons (of water)	×	199.6	= English gallons
U.S. gallons	×	231.00	= cubic inches
U.S. gallons	×	0.13368	= cubic feet
U.S. gallons	×	8.345	= pounds
U.S. gallons	×	0.8327	= English gallons
U.S. gallons	×	3.785	= liters
English gallons (Imperial)	×	277.41	= cubic inches
English gallons (Imperial)	×	0.1605	= cubic feet
English gallons (Imperial)	×	10.02	= pounds
English gallons (Imperial)	×	1.201	= U.S. gallons
English gallons (Imperial)	×	4.546	= liters

Metric Conversion Table

Length			
Millimeters	×	0.03937	= inches
Millimeters	÷	25.4	= inches
Centimeters	×	0.3937	= inches
Centimeters	÷	2.54	= inches
Meters	×	39.37	= in. (Act. Cong.)
Meters	×	3.281	= feet
Meters	×	1.0936	= yards
Kilometers	×	0.6214	= miles
Kilometers	÷	1.6093	= miles
Kilometers	×	3280.8	= feet
Area			
Sq. millimeters	×	0.00155	= sq. in.
Sq. millimeters	÷	645.2	= sq. in.
Sq. centimeters	×	0.155	= sq. in.
Sq. centimeters	÷	6.452	= sq. in.
Sq. meters	×	10.764	= sq. ft
Sq. kilometers	×	247.1	= acres
Hectares	×	2.471	= acres
Volume			
Cu. centimeters	÷	16.387	= cu. in.
Cu. centimeters	÷	3.69	= fl. drs. (U.S.P.)
Cu. centimeters	÷	29.57	= fl. oz. (U.S.P.)
Cu. meters	×	35.314	= cu. ft
Cu. meters	×	1.308	= cu. yards
Cu. meters	×	264.2	= gal. (231 cu. in.)
Liters	×	61.023	= cu. in. (Act. Cong.)
Liters	×	33.82	= fl. oz. (U.S.J.)
Liters	×	0.2642	= gal. (231 cu. in.)
Liters	÷	3.785	= gal. (231 cu. in.)
Liters	÷	28.317	= cu. ft
Hectoliters	×	3.531	= cu. ft
Hectoliters	×	2.838	= cu. (2150.42 cu. in.)
Hectoliters	×	0.1308	= cu. yds.
Hectoliters	×	26.42	= gals. (231 cu. in.)

Weight			
Grams	×	15.432	= grains (Act. Cong.)
Grams	÷	981.	= dynes
Grams (water)	÷	29.57	= fl. oz.
Grams	÷	28.35	= oz. avoirdupois
Kilograms	×	2.2046	= lbs
Kilograms	×	35.27	= oz. avoirdupois
Kilograms	×	0.0011023	= tons (2000 lbs)
Tonne (metric ton)	×	1.1023	= tons (2000 lbs)
Tonne (metric ton)	×	2204.6	= lbs
Unit Weight			
Grams per cu. cm	÷	27.68	= lbs per cu. in.
Kilogram per meter	×	0.672	= lbs per ft
Kilogram per cu. meter	×	0.06243	= lbs per cu. ft
Kilogram per Cheval	×	2.235	= lbs per hp
Grams per liter	×	0.06243	= lbs per cu. ft
Pressure			
Kilograms per sq. cm	×	14.223	= lbs/in.^{2}
Kilograms per sq. cm	×	32.843	= ft of water (60°F)
Atmospheres (international)	×	14.696	= lbs/in.^{2}
Energy			
Joule	×	0.7376	= ft-lb
Kilogram-meters	×	7.233	= ft-lb
Power			
Cheval vapeur	×	0.9863	= hp
Kilowatts	×	1.341	= hp
Watts	÷	746.	= hp
Watts	×	0.7373	= ft-lb/sec.

Ball Bearing Dimension Tables—100

Basic Bearing Number	Bore		OD		Width	
	mm	in.	mm	in.	mm	in.
100	10	0.3937	26	1.0236	8	0.3150
101	12	0.4724	28	1.1024	8	0.3150
102	15	0.5906	32	1.2598	9	0.3543
103	17	0.6693	35	1.3780	10	0.3937
104	20	0.7874	42	1.6535	12	0.4724
105	25	0.9843	47	1.8504	12	0.4724
106	30	1.1811	55	2.1654	13	0.5118
107	35	1.3780	62	2.4409	14	0.5512
108	40	1.5748	68	2.6772	15	0.5906
109	45	1.7717	75	2.9528	16	0.6299
110	50	1.9685	80	3.1496	16	0.6299
111	55	2.1654	90	3.5433	18	0.7087
112	60	2.3622	95	3.7402	18	0.7087
113	65	2.5591	100	3.9370	18	0.7087
114	70	2.7559	110	4.3307	20	0.7874
115	75	2.9528	115	4.5276	20	0.7874
116	80	3.1496	125	4.9213	22	0.8661
117	85	3.3465	130	5.1181	22	0.8661
118	90	3.5433	140	5.5118	24	0.9449
119	95	3.7402	145	5.7087	24	0.9449
120	100	3.9370	150	5.9055	24	0.9449
121	105	4.1339	160	6.2992	26	1.0236

Ball Bearing Dimension Tables—200

Basic Bearing Number	Bore		OD		Width	
	mm	in.	mm	in.	mm	in.
200	10	0.3937	30	1.1811	9	0.3543
201	12	0.4724	32	1.2598	10	0.3937
202	15	0.5906	35	1.3780	11	0.4331
203	17	0.6693	40	1.5748	12	0.4724
204	20	0.7874	47	1.8504	14	0.5512
205	25	0.9843	52	2.0472	15	0.5906
206	30	1.1811	62	2.4409	16	0.6299
207	35	1.3780	72	2.8346	17	0.6653
208	40	1.5748	80	3.1496	18	0.7087
209	45	1.7717	85	3.3465	19	0.7480
210	50	1.9685	90	3.5433	20	0.7874
211	55	2.1654	100	3.9370	21	0.8268
212	60	2.3633	110	4.3307	22	0.8661
213	65	2.5591	120	4.7244	23	0.9055
214	70	2.7559	125	4.9213	24	0.9449
215	75	2.9528	130	5.1181	25	0.9843
216	80	3.1496	140	5.5118	26	1.0236
217	85	3.3465	150	5.9055	28	1.1024
218	90	3.5433	160	6.2992	30	1.1811
219	95	3.7402	170	6.6929	32	1.2598
220	100	3.9370	180	7.0866	34	1.3386
221	105	4.1339	190	7.4803	36	1.4137
222	110	4.3307	200	7.8740	38	1.4961

Ball Bearing Dimension Tables—300

Basic Bearing Number	Bore		OD		Width	
	mm	in.	mm	in.	mm	in.
300	10	0.0397	35	1.3780	11	0.4331
301	12	0.4724	37	1.4567	12	0.4724
302	15	0.5906	42	1.6535	13	0.5118
303	17	0.6693	47	1.8504	14	0.5512
304	20	0.7874	52	2.0472	15	0.5906
305	25	0.9843	62	2.4409	17	0.6693
306	30	1.1811	72	2.8346	19	0.7480
307	35	1.3780	80	3.1496	21	0.8268
308	40	1.5748	90	3.5433	23	0.9055
309	45	1.7717	100	3.9370	25	0.9843
310	50	1.9685	110	4.3307	27	1.0630
311	55	2.1654	120	4.7244	29	1.1417
312	60	2.3622	130	5.1181	31	1.2205
313	65	2.5591	140	5.5118	33	1.2992
314	70	2.7559	150	5.9055	35	1.3780
315	75	2.9528	160	6.2992	37	1.4567
316	80	3.1469	170	6.6929	39	1.5354
317	85	3.3465	180	7.0866	41	1.6142
318	90	3.5433	190	7.4803	43	1.6929
319	95	3.7402	200	7.8740	45	1.7717
320	100	3.9370	215	8.4646	47	1.8504
321	105	4.1339	225	8.8583	49	1.9291
322	110	4.3307	240	9.4480	50	1.9685
324	120	4.7244	260	10.2362	55	2.1654
326	130	5.1181	280	11.0236	58	2.2835
328	140	5.5118	300	11.8110	62	2.4409
330	150	5.9055	320	12.5984	65	2.5591
332	160	6.2992	340	13.3858	68	2.6772
334	170	6.6929	360	14.1732	72	2.8346
336	180	7.0866	380	14.9606	75	2.9528
338	190	7.4803	400	15.7480	78	3.0709
340	200	7.8740	420	16.5354	80	3.1496
342	210	8.2677	440	17.3228	84	3.3071
344	220	8.6614	460	18.1002	88	3.4646
348	240	9.4488	500	19.6850	95	3.7402
352	260	10.2362	540	21.2598	102	4.0157
356	280	11.0236	580	22.8346	108	4.2520

Ball Bearing Dimension Tables—400

Basic Bearing Number	Bore		OD		Width	
	mm	in.	mm	in.	mm	in.
403	17	0.6693	62	2.4409	17	0.6693
404	20	0.7874	72	2.8345	19	0.7480
405	25	0.9843	80	3.1496	21	0.8268
406	30	1.1811	90	3.5433	23	0.9055
407	35	1.3780	100	3.9370	25	0.9843
408	40	1.5748	110	4.3307	27	1.0630
409	45	1.7717	120	4.7244	29	1.1417
410	50	1.9685	130	5.1181	31	1.2205
411	55	2.1654	140	5.5118	33	1.2992
412	60	2.3622	150	5.9055	35	1.3780
413	65	2.5591	160	6.2992	37	1.4567
414	70	2.7559	180	7.0866	42	1.6535
415	75	2.9528	190	7.4803	45	1.7717
416	80	3.1496	200	7.8740	48	1.8898
417	85	3.3465	210	8.2677	52	2.0472
418	90	3.5433	225	8.8533	54	2.1260
419	95	3.7402	250	9.8425	55	2.1654
420	100	3.9370	265	10.4331	60	2.3622
421	105	4.1339	290	11.4173	65	2.5591
422	110	4.3307	320	12.5984	70	2.5759

List of Vendors

Thanks to the following vendors who provided photos for this book:
Balmac, Inc.
4010 Main Street
Hilliard, OH 43026
(614) 876-1295

Emerson Power Transmission
1248 East 2nd Street
Maysville, KY 41056
(606) 564-2084

Emerson Power Transmission Manufacturing, L.P.
620 South Auroa Street
Ithaca, NY 14850
(607) 274-6078

Emhart Teknologies
50 Shelton Technology Center
P.O. Box 859
Shelton, CT 06484
(203) 924-9341

The Falk Corporation
3001 West Canal Street
Milwaukee, WI 53208-4200
(414) 342-3131

Gardner Denver Inc.
1800 Gardner Expressway
Quincy, IL 62301
(222) 228-5400

Gurley Precision Instruments
514 Fulton Street
Troy, NY 12180-3315
(800) 759-1844

Ingersoll-Rand Company
Air Solutions
800-D Beaty Street
Davidson, NC 28036
(704) 896-4000

John Crane
6400 West Oakton Avenue
Morton Grove, IL 60053
ATTN: Andy Martin
(800) 527-2631

Klozure Oil Seals and Bearing Isolators
A Division of Garlock Sealing Technologies
1666 Division Street
Palmyra, NY 14522
Toll free: 1.866.KLOZURE (1.866.556.9873)
Direct: 315.597.4811

L.S. Starrett Company
121 Crescent Street
Athol, MA 01331-1915
(978) 249-3551

Maintenance Troubleshooting
273 Polly Drummond Road
Newark, DE 19711-4833
(800) 755-7672
www.mtroubleshooting.com

McFeely's Square Drive Screws
3720 Cohen Place
P.O. Box 11169
Lynchburg, VA 24506-1169
(800) 443-7937
www.mcfeelys.com

Milwaukee Electric Tool Corp.
13135 W. Lisbon Road
Brookfield, WI 53005
1-800-SAW DUST (729-3878)

nash_elmo industries
9 Trefoil Drive
Trumbull, CT 06611-1330
(877) 275-6274

Lawton Industries, Inc.
4353 Pacific Street
Rocklin, CA 95677
(800) 692-2600

Poulan Puller-Press, Inc.
120 North Hilton Street
West Monroe, LA 71291
318-396-7771

Rexnord Industries, Inc.
4701 W. Greenfield Avenue
Milwaukee, WI 53214
(414) 643-3000

The Stanley Works
1000 Stanley Drive
New Britain, CT 06053
860-225-5111

TB Wood's Incorporated
440 North Fifth Avenue
Chambersburg, PA 17201
1-888-TBWOODS (829-6637)

TRUARC Company LLC
70 East Willow Street
Millburn, NJ 07041
800-228-4460
www.truarc.com
contact@truarc.com

Index

E

H

M

O

W

Y

Z